Gray • Calvin • Bhatia

Enzymes Biotechnology

- Principles of Industrial Enzyme Production and Utilisation
- Types/Classification of Enzymes
- Application of Enzymes in Industry
- Instrumental Techniques Used in Enzyme Analysis
- Legal and Economic Considerations for the Use of Technical Enzymes
- Patenting Biotechnology Inventions

Enzymes Biotechnology

- Principles of Industrial Enzyme Production and Utilisation
- Types/Classification of Enzymes
- Application of Enzymes in Industry
- Instrumental Techniques Used in Enzyme Analysis
- Legal and Economic Considerations for the Use of Technical Enzymes
- Patenting Biotechnology Inventions

N Gray | M Calvin | SC Bhatia

CBS

CBS Publishers & Distributors Pvt Ltd

New Delhi • Bangalore • Pune • Cochin • Chennai

Enzymes Biotechnology

ISBN: 978-81-239-1829-7 Paperback
978-81-239-1826-6 Hardcover

First Edition: 2010

Published by Satish Kumar Jain and produced by Vinod K. Jain for
CBS Publishers & Distributors Pvt Ltd
4819/XI Prahlad Street, 24 Ansari Road, Daryaganj,
New Delhi 110002, India. Website: www.cbspd.com
Ph: 23289259, 23266861, 23266867 Fax: 011-23243014 e-mail: delhi@cbspd.com

Branches

- Bangalore: Seema House 2975, 17th Cross, K.R. Road, Banasankari 2nd Stage, Bangalore 560070, Karnataka
  Ph: 26771678/79 Fax: 080-26771680 e-mail: cbsbng@gmail.com
- Pune: Shaan Brahmha Complex, 631-632 Basement, Appa Balwant Chowk, Budhwar Peth, next to Ratan Talkies, Pune 411002, Maharashtra
  Ph: 020-24464057/58 Fax: 020-24464059 e-mail: pune@cbspd.com
- Cochin: 36/14 Kalluvilakam, Lissie Hospital Road, Cochin 682018, Kerala
  Ph: 0484-4059061-65 Fax: 0484-4059065 e-mail: cochin@cbspd.com
- Chennai: 20, West Park Road, Shenoy Nagar, Chennai 600030, TN
  Ph: 044-26260666, 26202620 Fax: 044-45530020 email: chennai@cbspd.com

Printed at Somya Printer, Delhi 110053

Preface

Enzymes are biological catalysts, or chemicals that speed up the rate of reaction between substances without themselves being consumed in the reaction. As such, they are vital to such bodily functions as digestion, and they make possible processes that normally could not occur except at temperatures so high they would threaten the well-being of the body. A type of protein, enzymes sometimes work in tandem with non-proteins called co-enzymes. Among the processes in which enzymes play a vital role is in fermentation, which takes place in the production of alcohol or the baking of bread and also plays a part in numerous other natural phenomena, such as the purification of waste-water.

Enzymes are special kinds of proteins that are found in all living matter. Man, animals, plants—all living cells produce and need enzymes to live and grow. However, enzymes themselves are not living organisms. Enzymes are catalysts substances which speed up chemical reactions by being present in very small amounts and without being changed in the reaction.

Modern biotechnology techniques are utilised to improve microbial production strains to increase to enzyme yields and to make minor amino acid changes that improve the functionality of the enzyme. These changes are not known to increase the ability of enzymes to cause allergies.

This reference textbook is divided into five sections. Section I discusses about principles of industrial enzyme production and utilisation. Chapter 1 is devoted to basic concepts of enzymes. Chapter 2 deals with general characteristics of enzymes such as structure, mechanism and kinetics of enzymes. Chemical reactions, including those catalysed by enzymes, depend on environmental conditions. Considering this chapter 3 focuses on environmental effects on environmental activity. Emphasis in this chapter is on micro and macroenvironmental effects. The dominant environmental factors affecting enzyme action are pH, water activity and temperature. Chapter 4 concentrates on modern methods of enzyme expression and design in a simplified and understandable manner—genetically engineered products and implification of genetic engineering for new food related applications. Chapter 5 deals with fermentation process design. The task of the fermentation technologist is to provide and maintain the environment required for carrying out a desired biological process. Thus, a well designed growth environment is needed to obtain the maximum economic yield of the product and this task resolves into environmental design, determining the environmental requirements of the process, and fermentation engineering, providing the means for meeting those requirements of the process, and fermentation engineering. Chapter 6 focuses on extraction and purification of enzymes. Various mediums of extraction and purification are discussed in detail.

Section II concentrates on types of enzymes/classification of enzymes. Chapter 7 concentrates on carbohydrate hydrolysing enzymes. Chapter 8 is devoted to proteases which are enzymes that digest proteins. Various types of proteases such as plant, animal, microbial, serine, etc. are discussed in detail. Chapter 9 deals with lipases which are enzymes that catalyses the hydrolysis of fat or breakdown of lipoproteins. The chapter discusses sources of microbial and industrial sources of lipases along with processes of application of lipases. Chapter 10 concentrates on oxidoreductases which are enzyme catalysing a reaction in which two molecules of a compound interact so that one molecule is oxidised and the other reduced with a molecule of water entering the reaction. Various types of oxidoreductases along with their general characteristics are discussed.

Section III is devoted to application of enzymes in the industry. Enzymes are involved in the formation of starch, gluten, hemicelluloses, and lipids during ripening and are the key to dough fermentation and bread making, since they degrade these polymers. The activities of enzymes in the flour used for bread making depend on the climate, the type of grain, and the harvesting conditions. Keeping this in mind, chapter 11 focuses on milling and baking. Chapter 12 and 13 discusses starches, sugars and syrups and dairy products. Pectic enzymes occur in higher plants and are produced by micro-organisms. Keeping this in mind, chapter 14 concentrates on pectic enzymes in fruit and vegetable juice manufacture. Chapter 15 deals with enzymes associated with savoury flavour enhancement. Chapter 16 and 17 focuses on wine and brewing. Chapter 18 is devoted to fish processing and meat industry. The deliberate use of added enzymes in fish processing is not common as deskinning of fish, membrane removal, and roe purification are examples of such processes. Chapter 19 concentrates on textiles and laundry detergents. Weaving textile fibres exposes the fabric's warp and weft strands to substantial mechanical stress. To prevent breakage of these strands, they are almost always reinforced with sizing. Sizing can consist of various materials. Chapter 20 and 21 deals with pulp and paper and tanning industry. Chapter 22 and 23 concentrates on clinical analysis and organic synthesis. Chapter 24 deals with enzymes and bioremediation. Biological methods or bioremediation promises to provide an ecofriendly solution to this menacing problem. Biodegradation refers to the breakdown of substances or molecules by biological system into smaller units.

Section IV concentrates on instrumental techniques used in enzyme analysis. Chapter 25 is devoted to instrumental techniques in enzymatic analysis. The course of enzyme-catalysed reactions may be monitored by a variety of techniques, of which the most common at the present time are spectrophotometric, spectrofluorimetric and electrochemical procedures. Chapter 26 and 27 deals with enzyme thermistors and enzyme electrodes for food analysis.

Section V focuses on special topics. Chapter 28 focuses on legal considerations. Regardless of whether enzyme preparations are defined as food additives or technical aids, it must be established that they are not health hazards. The purpose of this chapter is on legal aspects of enzyme utilisation is to inform the readers about the measures that the scientific community and legislators have taken to provide maximum safety for both users and consumers of enzyme preparation. It should be noted that regulations for enzyme preparations and their use are by no means uniform throughout the world. Chapter 29 deals with economic considerations for the use of technical enzymes. The economic benefits of using technical enzyme preparations lies in lowered process costs, often increasing product quality, reducing environmental impact, and making use of renewable resources in technology. Chapter 30 is devoted to patenting biotechnology inventions. For biotechnology, patents are the most important form of intellectual property. The chapter discusses number of reasons for issuing patents. For examples, the patent holder obtains exclusive rights for the invention, rights that are a reward for developing procedure, compound or apparatus. For a patent to be granted, an invention must be novel, not obvious and useful.

Glossary and index have been provided at the end for quick references. Diagrams, figures and tables supplement the text. All the topics have been covered in a cogent and lucid style to help the reader grasp the information quickly and easily.

It may not be wrong to hold that the present reference textbook of *Enzymes Biotechnology* is essential reading for all students and teachers of engineering, environment and life sciences. The researchers in analytical and instrumental techniques, chemists and industrialists will also find it highly useful and informative.

It has been prepared with meticulous care, aiming at making the book error-free. Constructive suggestions are always welcome from users of this book.

N. Gray
M. Calvin
S.C. Bhatia

Contents at a Glance

Contents

SECTION I

Principles of Industrial Enzyme Production and Utilisation

Chapter 1

Basic Concepts of Enzymes

INTRODUCTION

Life on earth is based on constant change. Inorganic matter forms the complicated structures that make up the living world, which will again decay into lifeless matter. The lives of plants, animals and human beings are included in this great cycle. The inert substances that form the foundation of life have a very simple structure. Water, carbon dioxide and nitrogen are the basic ingredients that, with the energy supplied by sunlight, permit plants to synthesise the molecular moieties that support life. These compounds are then used to provide more elaborate compounds required by animal life.

The use of enzymes and micro-organisms in processing raw materials from plants and animals has been practiced for a long time. At one time, living micro-organisms were used predominantly. Traditional processes, such as the production of alcoholic beverages and yeast-fermented doughs in baking bread, are displayed in Egyptian wall paintings. Further examples are the processes for preserving food, such as vegetable conservation by fermentation with lactobacilli, or preserving milk by making cheese.

Chemical experiments have shown how specific compounds, namely, the essential amino acids, can be generated from simple mixtures of carbon, nitrogen, oxygen and hydrogen when subjected to extreme pressures and temperatures. Once formed, less extreme conditions can cause the degradation of these complex structures as well. Under the current global conditions, such processes proceed only very slowly. What, then, causes the rapid synthesis of these compounds that are essential for life and the degradation of the high-molecular-weight and highly organised structures that make up life? The catalysts for this assembly and disassembly were for a long time thought to be life itself, or rather a fundamental principle of living cells.

The first observation of an enzymatic degradation reaction was by Spallanzani, a priest and naturalist for Padua, Italy. After placing meat in small porous capsules, he examined the regurgitated pellets of the hawks who had eaten this material and found the capsules to be empty, proving that the meat had been rapidly liquefied by the stomach juices of these birds of prey.

Later, Kirchhoff found that barley contained a substance that was capable of liquefying starch paste into sugar. He assumed that the reaction was caused by the gluten protein of the barley. Subsequently, Payen and Persoz termed the working principle of this saccharification *diastase* (Greek for separation), a term still used for the amylases in the brewing industry. The degradation of starch, which is composed of sugar moieties and the subsequent process of alcoholic fermentation of the sugar into carbon dioxide and alcohol occupied much of this nineteenth-century chemist's time.

Pasteur showed that fermentation is closely associated with live yeast. He distinguished between the actions of 'organised ferments' (cellular) and the 'unorganised ferments' (soluble). These soluble 'ferments', which are not bound to the living cell, were labelled enzymes by Kühne. This term is derived from the Greek enzyme, meaning 'in sour dough'. Concrete evidence for this assumption was provided by E. Buchner, as he showed that the cell-free extract from yeast cells could also produce alcohol from sugars. The enzyme active in fermentation was termed *zymase*. The first book summarising the 'wonderful mechanism' of the enzymes known at that time and outlining the history of their discovery was written by Green and is still read with interest today.

ENZYMES DEFINED AS CATALYSTS

Friederich Wilhelm Ostwald first coined the now-familiar definition of a catalyst, specifically, that a catalyst is a substance that alters the rate of a chemical reaction without being present in the reaction products. Thus, Ostwald recognised enzymes as catalysts.

Today, catalysts and catalysis are familiar terms. The catalysts in internal-combustion engines accelerate the conversion of the pollutants carbon monoxide and nitrogen oxides into the less polluting carbon dioxide and nitrogen, without any change in the catalysts during the process. According to the basic research of Tammann and Van't Hoff, a catalyst must be able to regenerate the initial substrate from its products. Enzymatic reactions also have an equilibrium. An example is the formation and scisson of an ester that forms on reaction between an alcohol with an acid:

$$\text{Acetic acid} + \text{alcohol} \underset{\text{Hydrolysis}}{\overset{\text{Synthesis}}{\rightleftarrows}} \text{acetate ester} + \text{water}$$

If acetic acid and alcohol are mixed in equimolar amounts, some of the molecules will react to form an ester and water (synthesis). Yet, there still will be significant quantities of acetic acid and alcohol in the reaction mixture as well as the products generated, ester and water. A mixture of the same composition can also be obtained when starting with water and an ester (hydrolysis). Even in the absence of a catalyst, the reactions reach equilibrium between ester and synthesis and ester hydrolysis.

When forming and splitting butyl ester, it was found that the equilibrium was rapidly established from either direction on the addition of porcine pancreatic powder (pancreatic lipase). Thus, for life to exist, the biochemical equilibria of essential and exceedingly complex reaction systems must be established rapidly. Therefore, enzymes are the vital catalysts in all life processes.

HISTORY OF THE DEVELOPMENT OF TECHNICAL ENZYMES

Enzyme preparations, from extracts of plants or animal tissue, were used well before much was known about the nature of properties of enzymes. The development of today's industrially important microbial enzymes began with the Japanese Jokichi Takamine. Born in Japan and later a US immigrant, Takamine studied the production of enzymes from moulds. He obtained a process patent for making a diastatic enzyme preparation from fungi. He named the product *Takadiastase*, a mixture of carbohydrases and proteolytic enzymes.

The production occurred either via a surface culture, a semisolid culture, or the *koji* process on moist wheat bran with nutrient salts and buffers added. The sterile culture media, mixed with spores from *Aspergillus oryzae*, was poured onto metal trays to a thickness of a few centimetres and incubated in climate-controlled culture chambers. The culture media, permeated with fungal mycelia, are ground and extracted with water. In a modified form, this process is still used by various manufacturers, although it has largely been replaced by the development of the submerged fermentation process carried out in large fermenters.

Auguste Boidin discovered a new process for the manufacture of alcohol from cereals. Known as the *amyloprocess*, this method involves the extraction of cereal polysaccharides by boiling, then inoculation with a fungus that, in turn, produces saccharifying enzymes and subsequently yeast fermentation of the sugars. Boidin and Professor Jean Effront, who had studied enzymes involved in alcohol production since 1900, founded the Société Rapidase in 1920.

Enzyme preparations from animal organs also continue to play an important industrial role. Otto Röhm discovered the effectiveness of pancreatic proteases in liming and bating of hides in leather manufacture. Until then, slurries of dog feces had been used in bating. Röhm concluded that the bating effect of dog feces was due to excreted pancreatic proteases and subsequently developed the first standardised enzymatic bating agent that, marketed under the name 'Oropon', offered the tanner greater control and reliability in the sensitive leather manufacturing process.

Otto Röhm developed additional uses for the pancreatic enzymes he manufactured in Darmstadt, including the degumming of raw silk. The importance of these enzymes for the silk industry at that time can be shown by the following cost computation: 'One kilogram Degomma S (pancreatic protease) with a value of 4 [Reichsmark] can degum the same amount of silk as 10 kilograms of expensive Marseilles soap' (at that time 2–3 reichsmarks per kilogram). Röhm introduced the first commercially enzymatic detergent. The product, named 'Burnus', rapidly became known as a good prewash and soaking agent. In addition to pancreatic enzymes, it contained sodium carbonate and sodium bicarbonate, which held the pH in the wash water below 9. The pancreatic proteases are sufficiently active and stable at low temperatures in this pH range. However, pancreatic enzymes are inactive in new detergent formulations that contain sequestrants, anionic detergents and oxidising agents.

Jaag in Switzerland developed a detergent that contained, as active components, proteases from *Bacillus subtilis*. The Novo Company [NO] in Copenhagen deserves credit for the subsequent development of this enzyme. This will be discussed in more detail later. Plant enzymes, especially the proteases from papaya fruits (i.e. papain), are industrially important. Wallerstein was the first to use papain for stabilising beer. Beer stabilisation, which prevents protein flocculation or hazing on refrigeration, is particularly important in the United States, where beer is cold-stored.

Boidin and Effront obtained a process patent for manufacturing an enzyme preparation from cultures of *B. subtilis*, grown on the surface of a liquid nutrient medium. The preparation primarily contained amylase, which was able to replace the malt that previously had been used in textile sizing.

Röhm and Haas began producing fungal proteases with semisolid cultures. These enzymes produced gentle bating preparations suited for special applications in leather manufacture. About 1934, the first pectinase preparations for clarifying fruit juices and improving juice extraction from grapes were produced by the same method.

In 1958, Underkofler published a list of commercial enzyme preparations. The most important products were the amylases that were used in bakeries and breweries and in the manufacture of starch hydrolysates. Second were the microbial proteases, which were used in leather manufacture and in removing hazes from cold beer. Other enzymes such as pectinases, lactases, invertases, lipases, and cellulases were, however, produced only in small quantities. The years following World War II saw a rapid development of submerged fermentation technology, triggered by the then-extant antibiotic manufacturing methods. Bacterial and fungal amylases could now be manufactured inexpensively. This had a great influence on the commercial development of enzymatic starch conversion to dextrins and glucose.

In 1960, Novo Nordisk [NO] began production of alkaline bacterial proteases from *Bacillus licheniformis* for use in detergents. A flurry of activity followed that led to a wave of innovations in the entire field of industrial enzymes.

The commercial development of alkaline bacterial proteases was briefly interrupted when cases of lung disease, due to allergic reactions, were reported in the United States. These allergies occurred in workers who had direct contact with dusty enzyme concentrates during detergent manufacture. Granulating the concentrates helped avoid these risks. In 1971, The National Academy of Sciences (USA) confirmed that the use of detergents containing enzymes posed no health hazard to the consumer.

Proteases from *B. lichemformis* are currently the commercially most important microbial enzymes. They represent approximately 60 per cent of all the industrial enzymes, followed by the carbohydrates making up about 30 per cent. The latter include the amylases, glucose isomerases, pectinases and cellulases.

The development of other industrial enzymes will be discussed in subsequent chapters.

APPLICATIONS OF MODERN ENZYME TECHNOLOGY

Major targets of modern enzyme technology continue to be preservation of foods and food components (e.g. vitamins), more efficient use of raw materials and improvement of food quality such as texture and taste. Enzymes are also used in the food industry for process optimisation to reduce process costs, including energy requirements. Other current objectives include the utilisation of new raw materials for feeding humans and animals, the manufacture of dietetic foods and eliminating antinutritive substances from certain nutritional raw materials.

Biotechnological methods have also replaced some of the traditional chemical processes. Enzymatic methods, which include the most recent procedures with immobilised enzymes and immobilised micro-organisms, constitute only one aspect of modern biotechnology. Biotechnology encompasses many processes for producing biologically active compounds, such as amino acids and antibiotics as well as processing waste-water.

Examples of new biotechnological processes include:

1. The use of enzymes in nonaqueous media for the production of chiral compounds and the synthesis of special polymers.
2. The use of enzymes for the synthesis of amino acids, peptides and antibiotics.
3. The use of enzymes for recycling food wastes and in waste-water treatment.
4. The production of sweeteners, such as aspartame, using combined microbiological and enzymatic methods.
5. The production of cyclodextrins from starch with specific enzymes.
6. The use of cellulases and lipases as active components of detergents.
7. The production of rare and expensive mammalian enzymes by micro-organisms. For example, pure calf stomach rennet is currently not available in sufficient quantities.
8. The production of tailored enzymes to serve as specific, process-adapted catalysts. These procedures are part of the so-called protein engineering, which means the design of proteins with specific functions.
9. Large-scale production of enzymes by gene technology is to be expected in future years; recombinant chymosin was a pioneering first. It has since overcome the world shortage of chymosin in the cheese industry. Thus, many limited or costly traditional enzyme sources will be replaceable.

In the past, obtaining the desired metabolites and improved performance depended on the conventional genetic methods of inducing mutations and hybridising sexually propagating micro-organisms. In asexually reproducing micro-organisms, physical and chemical mutagenesis produced the necessary variability. The sought-after characteristic or product was subsequently found by screening large numbers of progeny. Mutations randomly alter the genetic or hereditary information. Success depends on whether one of the random mutations occurs at the correct site within the genome; only then will it be possible to develop better microbial strains. The extensive screening required makes it very expensive to determine whether such random mutations have produced the changes sought in the production of the desired substance. Now, gene technology permits site-specific changes and yields products of great purity under economically feasible conditions.

Chapter 2

General Characteristics of Enzymes

INTRODUCTION

The current state of knowledge on enzymes can be found under captions such as 'enzyme structure', 'enzymatic activity', 'active site', 'mechanism', and 'enzyme technology'.

The enzymes as catalysts are able to rapidly adjust the equilibria of chemical reactions. For example, ester formation between an acid and an alcohol shows that reaction equilibrium is also reached, but extremely slowly, in the absence of an enzyme. Experience has shown, however, that the rate of a chemical reaction depends on the temperature. This also holds true for chemical reactions in food. Butter, for example, becomes rancid more rapidly at room temperature than when refrigerated. At very low temperatures the acid-alcohol system can be regarded as approximately stable in that there will be no reactions between the components.

IMPORTANT ROLE OF ENZYMES

Catalysis: A Lowering of the Energy Barrier

Change in a stable condition can be achieved only by supplying energy. An energy barrier must be surmounted. This can be done in one of two ways:

1. The necessary activation energy can be generated. The energy required is inversely proportional to the temperature; that is, the activation energy decreases with increasing temperature and *vice versa*. This means, however, that very high temperatures are required for certain reactions. Where extremely high temperatures must be avoided, or where the delivery of large amounts of energy is economically infeasible, another possibility remains.
2. The energy barriers may be removed with enzymes that can reduce energy barriers.

Reduction of Energy Barriers by Contact with Active Sites

Enzymes are large, three-dimensional protein molecules with an active site at a defined location on the folded surface. This part of the surface can be envisioned as a pocket that will permit entry only to a specific substrate for a reaction to occur. Emil Fischer stipulated that enzyme and substrate must fit like a lock and key.

The temporary bonds between the enzyme and substrate that form the enzyme substrate complex will loosen the bonds that hold the substrate together. Thus, the energy barrier for cleaving is lowered and the reaction can proceed and reach equilibrium at room temperature. Products A and B are formed from the substrate and the enzyme is liberated again ready to catalyse the next reaction (Fig. 2.1). For the fastest reactions, the contact time between enzyme and substrate is only about 1/85000 of a second.

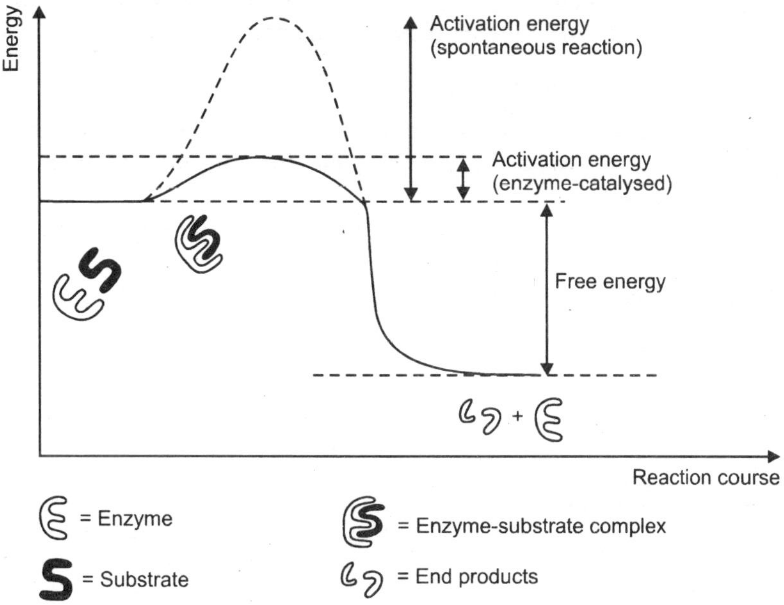

Fig. 2.1. Lowering the activation energy of an enzyme-catalysed decay reaction compared to spontaneous decay.

Enzyme Specificity for Substrate Binding

According to Emil Fischer, the substrate binds to the enzyme only when it matches the active site. In other words, enzymes possess a high degree of specificity. A protease can only recognise and cleave proteins but will not react with starch molecules. However, Fischer's lock and key analogy can be modified in that the substrate, and especially the enzyme, can undergo conformational changes. Zech and Domagk stated that 'the garment can change to fit the figure and the figure to fit the garment'. Thus, the enzyme exhibits not only substrate specificity but also, in some cases, a certain flexibility (Fig. 2.2).

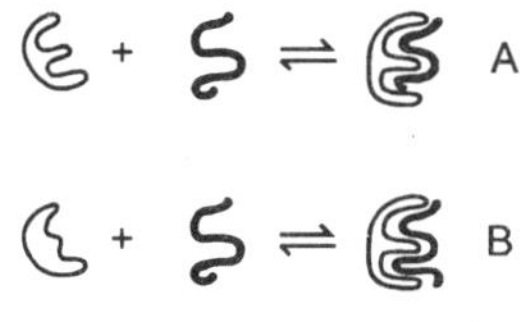

A = Enzyme-substrate complex (after Emil Fisher)

B = Induced-fit theory (after Kosland)

Fig. 2.2. Lock-and-key concept; conformational change of enzymes.

Enzymatic Activity: Conversion Rate per Unit Time

Enzymes are evaluated according to their activities. The following can serve as a simplified analogy. Group A with 10 workers saw 10 cubic metres (m^3) of wood in one hour. Group B, consisting of

20 workers, requires the same time. Thus, work group B is only half as active as group A. Enzymatic activity is determined in a similar way. An enzyme is more active than another when a smaller quantity is required for a specific conversion. A measure of conversion per unit time is the amount of product formed per minute under well-defined, standardised conditions.

Optimal Conditions for Enzymatic Activity

Enzymes need an optimal supply of substrate. The substrate should saturate the enzyme. Again, using our example, this means that the workers can reach their maximum performance only when there is sufficient wood available to be sawed.

Enzyme and substrate must have a constant and unimpaired contact for maximum enzymatic activity. This occurs when the enzyme and substrate are present in dilute aqueous solutions. The performance of the enzyme is reduced or impeded when the substrate is insoluble. Dry solids are enzymatically inert.

Enzymes Work at a Constant Rate

Generally, as long as the reaction conditions do not change, twice the yield of product will be generated in twice the time (Fig. 2.3).

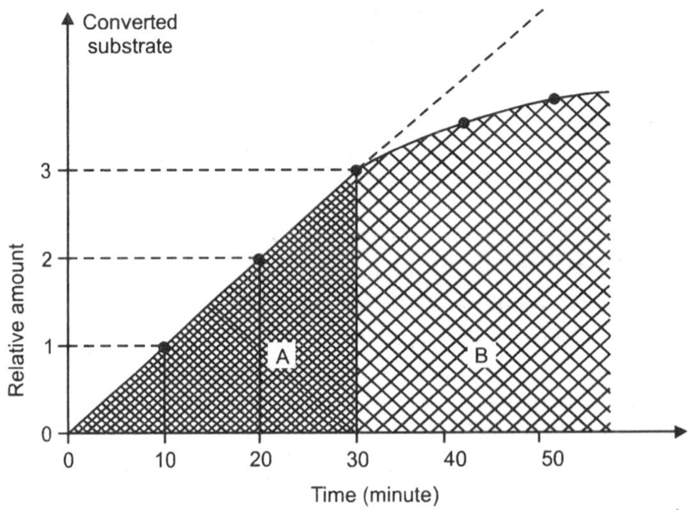

Fig. 2.3. Product formation as a function of time. A = enzyme–substrate complex according to Emil Fischer; B = induced-fit theory according to Kosland.

The curve in Fig. 2.4 shows an idealised curve of pH dependence found for a purified enzyme acting on a defined substrate. The shoulder of this curve suggests that a second enzyme with a different pH dependence is also active (side activity).

The conversion rate is reduced when there is insufficient substrate available to saturate the enzyme or the enzyme is denatured because of an increase in temperature (inactivation).

pH Dependence of Enzymatic Activity

Every enzyme requires a specific pH value or pH range for optimal activity. The pH activity curve shown in Fig. 2.4 is characteristic for an enzyme. This factor is of prime importance when choosing an enzyme for an industrial process. For example, when clarifying an acidic fruit juice with pH <3, an enzyme with optimum activity in the pH range of 4–5 would show only slight activity at pH 3. In order to operate at maximum enzyme efficiency, another enzyme that has optimum activity at pH 3 must be chosen.

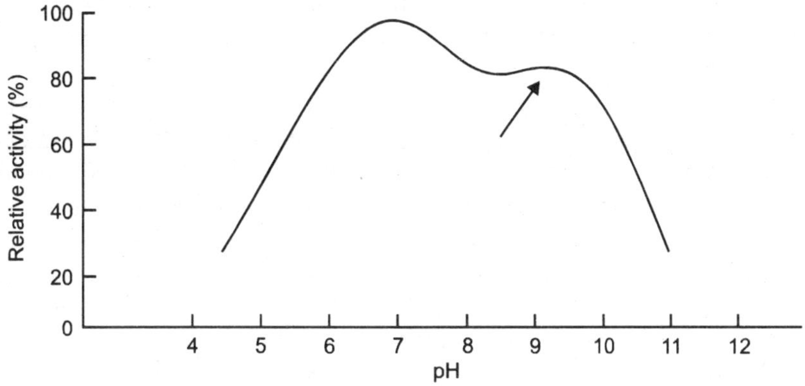

Fig. 2.4. Activity as a function of the pH. Enzyme preparation: neutral bacterial protease containing some alkaline protease (→).

Temperature Dependence of Enzymatic Activity

Enzyme performance usually improves with increasing temperature. Generally, the enzymatic conversion rate doubles on a 10°C temperature increase. This holds true from about 10°–40°C. Enzymes may also be active at relatively low temperatures, as is evident in the development of rancidity in butter on long-term refrigeration. Beyond the optimal temperatures, enzymes may be denatured. The temperature at which inactivation begins is characteristic for every enzyme. In industry the optimum temperature range for a given enzyme reaction is that at which the enzyme is still just sufficiently stable.

Enzyme Stability at High Temperatures

The temperatures at which enzymes are stable or labile is of great significance for many technical processes. In some processes enzymes should be sufficiently labile so as to be completely inactivated at temperatures of 70°–80°C. On the other hand, enzymes that are active above 100°C are needed for the modern production of glucose from starch. Here, high enzyme stability is of significant economic value. Therefore, the search for temperature stable enzymes has a high priority in enzyme research. It should also be mentioned that high substrate concentrations help stabilise enzymes in some industrial processes. For example, starch hydrolysing amylases in the presence of 30–40 per cent starch cannot be inactivated by boiling.

Enzyme Sensitivity and Susceptibility to Inactivation

An example of enzyme sensitivity was mentioned in Chapter 1 during discussion of the use of enzymes in modern detergents. Bacterial protease are insensitive to the action of perborate or anionic detergents, while detergents inactivated the pancreatic proteases employed earlier. In addition to heavy metals, which generally destroy enzymes, there are a number of chemicals and many natural substances that inhibit or completely inactivate enzymes.

ENZYME STRUCTURE AND ACTION MECHANISMS

The isolation of crystalline enzymes was a deciding factor in resolving the long-ongoing debate on the nature of enzymes. Sumner was the first person to produce an enzyme in crystallised form. Because this enzyme catalysed the hydrolysis of urea to carbon dioxide and ammonia, he called it urease. In subsequent years, additional crystalline enzymes were isolated and hydrolysed into their basic amino acid building blocks. Many enzymes, such as the hydrolases, consist entirely of protein. Most (industrial) enzymes also belong to this category. Others contain, besides the protein component, a non-protein prosthetic group that is essential for their activity.

Proteins are amino acid polymers, joined by peptide linkages (polypeptide chains). In contrast to starch or cellulose, where the polymer chains are composed of linked building blocks of the same type, proteins can contain up to 21 different amino acids. In proteins, there are usually hundreds of amino acids bound together. Their sequence is random and a great variety of different proteins exist. In addition, the polypeptide chains (primary structure) adopt a more or less rigid, helical configuration (secondary structure). A three-dimensional shape develops on the folding of these helixes; this is designated as the tertiary structure.

The specific way in which the protein is folded is dictated by the primary sequence of the amino acids. Most proteins fold spontaneously into their correct form. Heat or other environmental changes can lead to the loss of this conformation and cause the proteins to become denatured.

All proteins in living cells have specific functions. These functions are governed by their interactions with other molecules or by the given amino acid sequences of the polypeptide chain. These are, for example, the regions that bind to other molecules. The location of catalytic activity in enzymes is called the active site. Other functions of proteins in living cells include the transport of metabolites and the formation of cell structures. All enzymes are proteins, but not all proteins are enzymes.

Most technical enzymes have molecular weights between 20,000 and 70,000 daltons. The active centre, however, is composed of only a few amino acids. At least three amino acid residues of trypsin (Ser, His and Asp) participate in the proteolytic activity of this enzyme [Fig. 2.5(a)]. Initially the protein substrate becomes associated with the enzyme to form an enzyme-substrate complex [Fig. 2.5(b)]. Interaction between the serine residue and the histidine residue within the active centre leads to activation of the serine residue. The activated serine residue then reacts with a peptide linkage in the substrate protein. A serine ester between the peptide chain of the substrate and the enzyme is formed [Fig. 2.5(c)], whereby, in one step, the peptide-linkage in the peptide chain is broken. Peptide 2 is released [Fig. 2.5(d)]. Finally, the ester linkage in the active serine is hydrolysed by the addition of water and peptide 1 is released. The enzyme has thus returned to its original state [Fig. 2.5(e)].

ENZYME ANALYSIS AND ENZYME UNITS

Several methods are available for determining the hydrolytic activity of enzymes. The substrates are primarily naturally occurring polymers. The same polymers that are used as industrial substrates are employed. These polymers are usually diverse with respect to their chemical composition and molecular weight. Defining enzyme activity with these substrates is more difficult than when low molecular weight substrates are used. In industry the latter are seldom used; therefore, allowances are generally made for problems associated in defining activity. Thus, there are very few internationally recognised methods for determining hydrolase activities. Most methods are specific for each manufacturer, and it is difficult to directly compare various activity specifications from different sources.

Automated Enzyme Analysis

Enzyme analyses are time-consuming manual procedures that have demanding personnel requirements. The individual steps must be precisely timed.

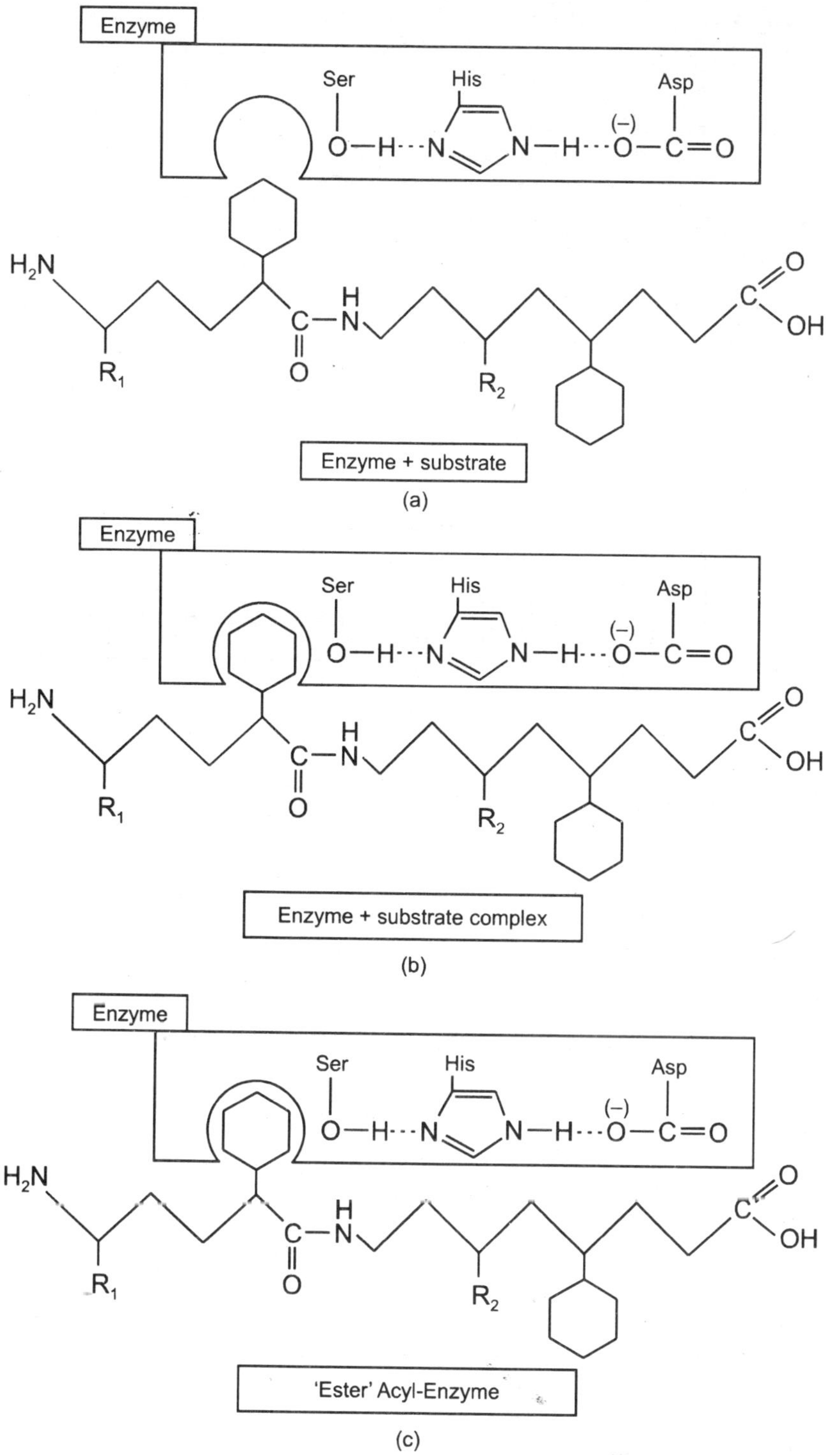

Fig. 2.5. Proteolysis catalysed by trypsin. **(contd...)**

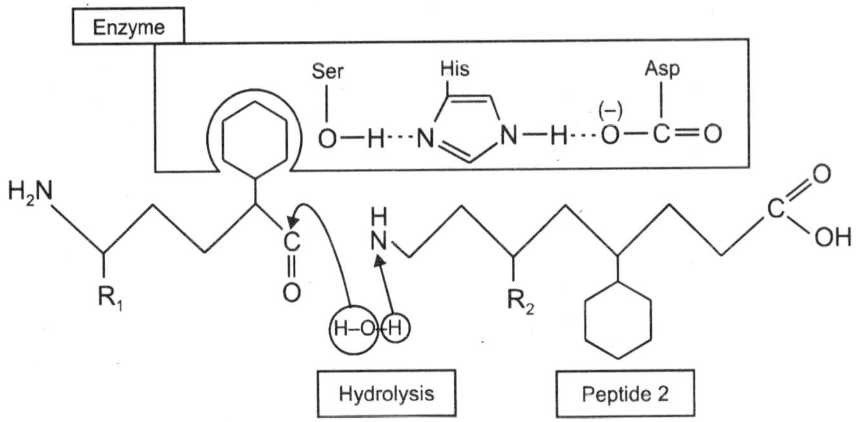

(d)

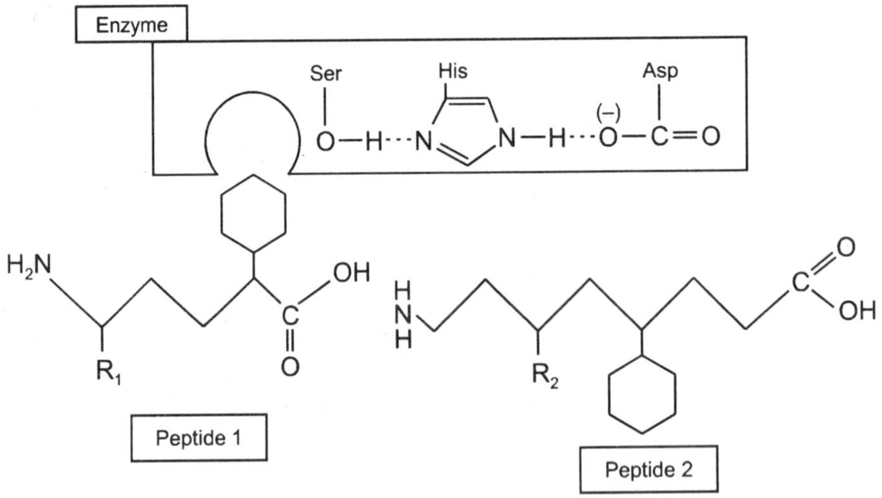

(e)

Fig. 2.5. (contd...) Proteolysis catalysed by trypsin.

For processing several samples simultaneously, the various steps must be coordinated. If numerous samples need to be analysed, it is worthwhile to employ autoanalysers. Two different systems are commonly used:

1. Fixed-volume analysers work with fixed controllable volumes of each reaction compound, similar to manual procedures. Automation allows the processing of more samples per unit time. In a thermostated reaction chamber, the sample, substrate, and reagents are added from dosing stations for the wet chemical analysis of the reaction products. The timer-regulated sample carousel rotates, passing each sample by the various dosing stations. A detection system at the end of the reaction sequence measures the reaction and transfers the signals to a central control unit, which records the data and computes the results.
2. Alternative systems are continuously operating analysers known as autoanalysers, or the flow-injection processors such as those used in clinical analysis.

The importance of analytic methodology for enzyme research demands continuous development in microbial selection and enzyme production and application. Future innovations will be primarily technical. Large numbers of equal samples make programmable robots and pipetting stations economical. Another important consideration is to put existing methods on line to continuously monitor and control individual process steps. This is of particular interest in the production of enzymes from live micro-organisms.

Enzyme Units

Some of the methods used internationally for industrial enzymes are listed below. Because of the variation in reaction conditions, different results are often obtained; therefore, it is necessary to be cautious when making comparisons.

Proteases

1. AU: The Anson unit (AU) defines protease activity. The test uses haemoglobin at pH 4.7 (original method). There are modifications of this method at pH 5, 7.5, and so on. Crystalline subtilisin has 25–30 AU/g. Technical enzyme preparations contain 2.5–3 AU/g.
2. BAPA: One BAPA Na-benzoyl-L-arginine-*p*-nitranilide) unit corresponds to the amount of protease required to convert 1 μmol of substrate in 1 minute under standard reaction conditions.
3. NU: One Northrop unit (NU) is the amount of protease required to hydrolyse 40 per cent of one litre of casein substrate in 60 minutes under standard reaction conditions.
4. PU: One protease unit (PU) is the amount of enzyme that liberates trichloroacetic acid-soluble fragments from casein, equivalent to 1 μg of tyrosine in 1 minute at 30°C under standard reaction conditions.
5. HbU: One haemoglobin unit (HbU) is the amount of protease that releases 0.0447 mg of nitrogen from amino acids and peptides (determined in a spectrometer at 275 nm) in 30 minutes under standard reaction conditions. The original method calls for pH 4.7. This method is of particular interest for a protease used in beer stabilisation. The Anson method is more appropriate for the neutral pH range. Purified acid fungal protease has an activity of ~1 mil HbU/g protein.
6. LVU: One Löhlein-Volhard unit (LVU) corresponds to the amount of protease that increases casein fragments in 20 ml of filtrate of a 4 per cent casein solution, equivalent to the action of 5.75×10^{-3} ml 0.1 N NaOH.

Carbohydrases

1. DP°: The unit is defined as 'degrees of diastatic power' (DP°) and is the amount of enzyme present in 0.1 ml of a 5 per cent enzyme solution yielding the amount of reducing sugar (in 100 ml of a 1 per cent starch solution at 20°C for 1 hour), which reduces 5 ml of Fehling's solution.
2. DU: The dextrinogen unit (DU) is used to measure α-amylase activity, particularly bacterial α amylase. One DU is the amount of enzyme that catalyses the conversion of 1 mg of starch to dextrins under defined reaction conditions. Conditions (Roehm): 1.2 per cent soluble starch (amylum soluble, Merck AG); acetate buffer pH 5.7; reaction time 10 minutes, 30°C.

According to this method, crystalline bacterial α-amylase has an activity of ~30000 DU/mg protein. Technical enzyme powders show activities between 100 and 1000 DU/mg; liquid preparations, about 100–300 DU/ml.

1. GAU: One glucoamylase [amyloglucosidase (AMG)] unit (GAU) is the amount of enzyme that liberates 1 gram of glucose from soluble starch in 1 hour.

2. SKB-U: One SKB unit (SKB) is the amount of amylase that hydrolyses 1 gram of β-limit dextrin to the iodine-negative point in 1 hour under defined reaction conditions. In more recent modifications of the method, the β-limit dextrin is replaced by soluble starch. Crystalline fungal α-amylase has an activity of ~2,50,000 SKB/g; the activity of technical preparations ranges from 50,000 to 1,20,000 SKB/g.
3. NF-U: According to NF (The National Formulary, 1960, USP XIV), one unit (NF) is defined as the time required for I gram of enzyme preparation to hydrolyse 100 grams of starch in a 3.75 per cent starch solution at 40°C. The end-point is reached when the iodine colour disappears.
4. Liquefon-U: The liquefon method by Sandstedt determines the time required to reduce the relative viscosity of a defined amount of starch paste by 50 per cent.
5. MWU: One modified Wohlgemuth unit (MWU) is the amount of enzyme needed to hydrolyse 1 mg of soluble starch to specific dextrins under standard reaction conditions in 30 minutes.

Pectinases

Numerous methods are based on either the decrease in viscosity of a pectin solution or the complete depectinisation of apple juice.

1. PG-U: One polygalacturonase unit (PG) is defined as the enzymatic activity required to lower the viscosity of a standard pectin solution by $1/\eta_{specific} = 0.000015$ under standard reaction conditions.
2. AJD-U: One apple juice depectinisation (AJD) unit is the amount of enzyme that completely depectinises 1 litre of a standard apple juice under standard reaction conditions.
3. PA-U: Mixtures of pectinases have a maximum of ~1500 PA/g protein. This means that 1 gram of such a preparation can depectinise 1500 litres of apple juice under standard reaction conditions.
4. PE-U: One pectinesterase unit (PE) is the amount of enzyme needed to liberate 1 μmol of titratable carboxyl groups per minute under standard reaction conditions.
5. PTE-U: Pectin transeliminase (PTE) units are determined by hydrolysis of pectin and measurement of the fragments at 235 nm.

Cellulases

1. CU: One cellulase unit (CU) is the amount of enzyme that releases 1 μmol of glucose from a 1.5 per cent solution of carboxymethylcellulose (CMC) at 30°C and pH 4.5. Determination of the reducing sugars can be measured calorimetrically using, for example, *p*-hydroxybenzoic acid.
2. Xyl-U: The activity of a xylanase unit (Xyl-U) corresponds to the amount of enzyme that can release reducing sugar equivalent to 1 mmol of xylose per minute from a xylan solution at 30°C and standard reaction conditions.

Oxidoreductases

Oxidases

The glucose oxidase unit definition varies with the enzyme source and supplier.

Glucose oxidase

1. GO-U: One glucose oxidase (GO) unit will liberate 1 μmol of H_2O_2 per minute at 25°C and pH 7.
2. GO-U: One glucose oxidase unit (GO) is defined as the amount of enzyme that will oxidise 1 μmol of β-D-glucose per minute at 25°C and pH 4.1.

3. GO-U: One glucose oxidase unit (GO) will oxidise 1 μmol of β-D-glucose to D-gluconic acid and H_2O_2 per minute at 35°C and pH 5.1.

Catalase

1. BU: One Baker unit (BU), is the amount of catalase (fungal catalase) that will decompose 264 mg of hydrogen peroxide under the reaction conditions defined by Scott and Hammer.
2. CaU: One catalase unit (CaU; Japanese method [UA], [AM]) is the amount of catalase that degrades 1 mmol of hydrogen peroxide in 1 minute at 30°C (iodine thiosulphate titration).
3. KU: One Keil unit (KU) is the amount of catalase (liver catalase) necessary to decompose 1 gram of 100 per cent hydrogen peroxide in 10 minutes at 25°C and pH 7 in an inert atmosphere of CO_2 or N_2.

Lipases

References to other methods of determining enzymatic activity can be found in the chapters on individual enzymes.

Generally, in commercial usage, the unit of lipase activity (LU) is measured as the amount of enzyme needed to produce a certain amount of acid that is determined by a pH-stat. Commercial manufacturers employ two kinds of assay: one using soluble esters and another one with a substrate such as olive oil in a standardised emulsion.

A more sensitive assay has been developed for clinical applications in which fluorescence is used to follow the hydrolysis of substances such as umbelliferyl heptanoate. Apparently, this assay is also useful in revealing lipases on gel electrophoresis.

ENZYME KINETICS

The kinetic behaviour of enzymes has been studied in detail for a century, beginning with the classic work of Henri and Michaelis and Menten. The objectives have been three-fold: to gain an understanding of the mechanisms of enzyme action; to illuminate the physiological roles of enzyme-catalysed reactions; and, mainly in recent years, to manipulate enzyme properties for biotechnological ends. Experimental practice has been overwhelmingly dominated by the first of these aims; most experiments have been designed as if shedding light on the mechanism was the principal, or even the only, objective. However, although much valuable information has been obtained in this way, there are some important aspects of enzyme function that are obscured when working mainly with isolated enzymes in conditions far removed from those that exist in the cell. For this reason the latter part of this chapter will be devoted to a discussion of enzymes as they behave in complex mixtures.

Michaelis-Menten Kinetics

The starting point for any discussion of enzyme kinetics is the Michaelis-Menten equation, which expresses the initial rate v of a reaction at a concentration a of the substrate transformed in a reaction catalysed by an enzyme at total concentration e_0:

$$v = \frac{k_0 e_0 a}{K_m + a} = \frac{Va}{K_m + a} \qquad \text{... (2.1)}$$

The parameters are k_0, the catalytic constant, and K_m, the Michaelis constant. The form shown in the middle is more fundamental than that on the right, but the second form, in which k_0e_0 is written as the limiting rate V, is often used because the enzyme concentration in meaningful units is often not known.

V is the limit that v approaches as the enzyme becomes saturated (that is, when a becomes very large) and K_m is the value of a at which $v = 0.5V$ (that is, at which the rate is half-maximal). The ratio k_0/K_m is called the specificity constant and given the symbol k_A: It is a more fundamental constant than K_m in the analysis of enzyme mechanisms (i.e. it has a simpler mechanistic meaning), but the equation is usually written in terms of V (or k_0) and K_m nonetheless.

The principal assumption implied by equation 2.1 is that the rate of the reverse reaction is negligible: This may be because the reaction is irreversible for practical purposes, but even with reversible reactions the reverse reaction can be made negligible by measuring the rate in the absence of products and by extrapolating the rate back to zero time, that is, by estimating the initial rate at a time when no products have accumulated. Even if there is no significant reverse reaction, products can still affect the rate of the forward reaction because product inhibition (discussed later) is a common phenomenon.

Another assumption is that the reaction is allowed sufficient time to reach a steady state because all reactions pass through an initial acceleration phase known as the transient state. This phase is normally very brief (a few milliseconds) and in practice enzymes are usually studied under steady-state conditions. Note, however, that this is made experimentally possible by working with extremely small enzyme concentrations compared with those that may exist in the cell. The use of very low enzyme concentrations has two important consequences: First, it normally means that the enzyme concentration can be neglected in comparison with the substrate concentration, and second it makes the steady-state rate sufficiently slow to be easily measured and the steady-state phase long enough to be meaningful. If high enzyme concentrations were used, the transient state would be as brief as before, but the steady state would also be very brief, so that there would be no period in which one could adequately treat the rate as constant.

Equation 2.1 may be derived from the following model:

$$A + E \underset{k_{-1}}{\overset{k_1}{\rightleftharpoons}} EA \underset{k_{-2}}{\overset{k_2}{\rightleftharpoons}} EP \underset{k_{-3}}{\overset{k_3}{\rightleftharpoons}} E + P \quad \ldots (2.2)$$

which assumes that the reaction passes through an enzyme–substrate complex EA, which undergoes catalytic transformation to an enzyme–product complex EP, which then breaks down to form products. Although real enzyme mechanisms may be more complicated than this, every reaction passes through steps of substrate binding, chemical transformation, and product release. In simple introductory treatments the second and third steps are often treated as a single step, but conceptually they are clearly distinct. In reactions of more than one substrate, the steps do not necessarily occur in the order one might guess; that is, some products may be released before all substrates have bound, but the general principle that any reaction involves the same three kinds of step remains valid. The Michaelis–Menten parameters can be defined as follows in terms of the rate constants shown in equation 2.2:

$$k_0 = \frac{k_2 k_3}{k_{-2} + k_2 + k_3}; \; k_A = \frac{k_1 k_2 k_3}{k_{-1}k_{-2} + k_{-1}k_3 + k_2 k_3}; \; K_m = \frac{k_{-1}k_{-2} + k_{-1}k_3 + k_2 k_3}{k_1(k_{-2} + k_2 + k_3} \quad \ldots (2.3)$$

Note that none of the three parameters has a simple transparent meaning. The interpretations commonly attributed to them depend on additional simplifying assumptions that are not always correct. For example, K_m is often said to be equal to the equilibrium constant k_{-1}/k_1 for dissociation of A from EA, from the expression in equation 2.3 does not take this form unless k_2 is very small. As there is no good reason for k_2 to be small, and indeed ideas of evolutionary optimisation of enzyme function lead one to expect the opposite when the enzyme is acting on its natural physiological substrate, *m* follows that K_m should not, in general, be regarded as a measure of the equilibrium dissociation constant.

Despite these difficulties in providing a detailed mechanistic meaning to K_m, it does provide a measure of the tightness of substrates binding in the steady state, as it is quite correct to take K_m as equal to the ratio of the sum of concentrations of all enzyme complexes (i.e. both EA and EP) over the concentration of free enzyme. Similarly k_0 provides a valid measure of the capacity of the enzyme–substrate complex to mean to give products, even if it cannot be interpreted as the rate constant for a unique step in the mechanism. The reason for the term specificity constant for k_A, that is, the relationship to enzyme specificity, will become clear once inhibition has been considered.

Graphical Analysis

Equation 2.1 defines a metabolic dependence of rate on substrate concentration, as illustrated in Fig. 2.6. The initial steep rise in v as a measures from zero is rapidly transformed into the phenomenon of saturation, whereby further increases in a procure smaller and smaller increases in v, which approaches but does not reach or exceed, the limiting rate V. The rectangular hyperbola makes this type of plot inconvenient for estimating the values of the kinetic parameters (because the line does not approach the saturation limit closely enough at reasonable values of a to allow direct measurement of V). For this purpose, therefore, it is usual to transform equation 1 into one of the following three forms, which underlie the three straight-line plots illustrated in Figs 2.7 through 2.9:

$$\frac{1}{v} = \frac{1}{V} + \frac{K_m}{V}\frac{1}{a} \qquad \text{... (2.4)}$$

$$\frac{a}{v} = \frac{K_m}{V} + \frac{1}{V}a \qquad \text{... (2.5)}$$

$$v = V \quad K_m \frac{v}{a} \qquad \text{... (2.6)}$$

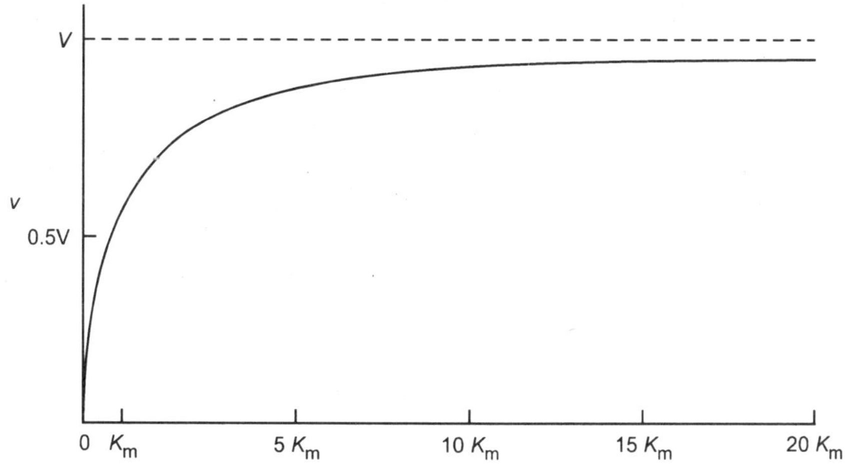

Fig. 2.6. Michaelis–Menten dependence of rate on substrate concentration. The curve is a rectangular hyperbola through the origin, approaching a limit of $v = V$ at saturation. The rate is 0.5V at a substrate concentration equal to the Michaelis constant, K_m, but note that the approach to the limit is slow, so that, for example, even at $a = 10\ K_m$ the rate is still nearly 10 per cent less than V.

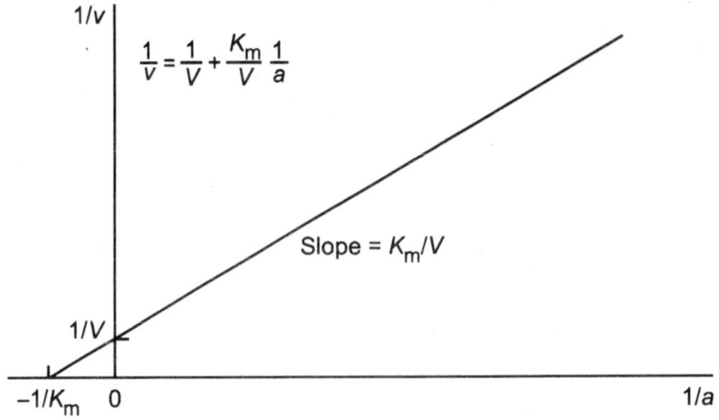

Fig. 2.7. The double-reciprocal plot. This is the most widely used method of plotting the Michaelis-Menten equation as a straight-line. However, the severe distortion of any experimental errors in the original data causes it to give a misleading impression.

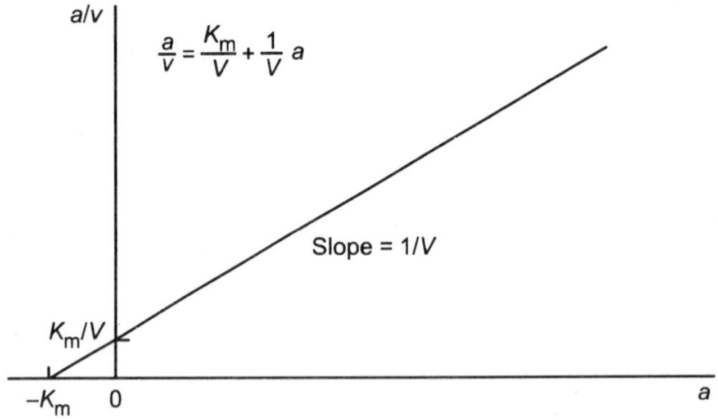

Fig. 2.8. The plot of *a*/*v* against *a*. This alternative to the plot shown in Fig. 2.7 produces much less distortion of the experimental error.

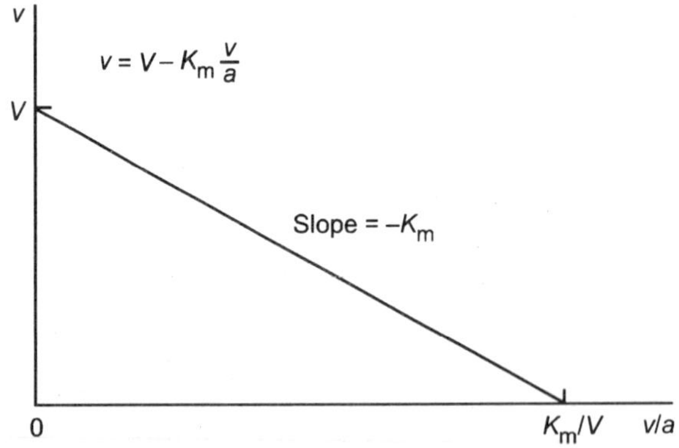

Fig. 2.9. The plot of '*v*' against *v*/*a*. The third way of plotting the Michaelis-Menten equation as a straight-line also avoids the error-distorting property of the plot shown in Fig. 2.7, and maps the entire range of observable rates (from 0 to *V*) onto a finite range of paper. This is a desirable property because it makes it impossible to disguise deficiencies in the experimental design.

The double-reciprocal plot, illustrated in Fig. 2.7 and based on equation (2.4), is the mostly widely used, but it is also the least satisfactory because it distorts the effect of experimental error to such an extent that it is difficult to form any visual judgement of where the best line should be drawn. The other two plots are better, and the plot of v against v/a (Fig. 2.9, eq. 2.6) has the particular advantage that the entire observable range of v values, from 0 to V, is mapped onto a finite range of graph; this makes it easy to judge by eye if an experiment has been well designed. On the other hand, it has the disadvantage that v, normally the less reliable measurement, contributes to both coordinates, and errors in v cause deviations along lines through the origin, rather than parallel with one or the other axis.

In modern practice it is usually best to regard these plots as for illustration purposes only, and to use suitable computer programmes for the actual parameter estimation. For this purpose, it is not sufficient just to apply unweighted linear regression to the straight-line plots, as this suffers from the same statistical distortions as the plots themselves. Full treatment would require more space than is available here. The following two equations for calculating best-fit values of K_m and V give satisfactory results if the v values have uniform coefficient of variation (uniform standard deviation expressed as a percentage), as is usually at least approximately correct:

$$K_m = \frac{\sum v^2 \sum(v/a) - \sum(v^2/a)\sum v}{\sum(v^2/a^2)\sum v - \sum(v^2/a)\sum(v/a)} \qquad \ldots (2.7)$$

$$V = \frac{\sum(v^2/a^2)\sum v^2 - [\sum(v^2/a)]^2}{\sum(v^2/a^2)\sum v - \sum(v^2/a)\sum(v/a)} \qquad \ldots (2.8)$$

Each summation is made over all observations.

Two-Substrate Reactions

Enzymes that catalyse reactions of a single substrate are only a small minority of all the enzymes known, but the Michaelis-Menten equation remains useful for examining the kinetics of the more common case of a reaction with two substrates and (often but not necessarily) two products, because such a reaction normally obeys Michaelis-Menten kinetics when only one substrate concentration is varied at a time. This is illustrated by the following typical equation for such a reaction:

$$v = \frac{k_0 e_0 ab}{K_{iA}K_{mB} + K_{mB}a + K_{mA}b + ab} \qquad \ldots (2.9)$$

Although at first sight this appears quite different from equation (2.1), it can be arranged in the same form if one of the two substrate concentrations, for example b, is treated as a constant:

$$v = \frac{\left(\dfrac{k_0 b}{K_{mB} + b}\right) e_0 a}{\dfrac{K_{iA}K_{mB} + K_{mA}b}{K_{mB} + b} + a} \qquad \ldots (2.10)$$

The two fractions in this equation can be regarded as 'apparent' values of the Michaelis-Menten parameters for A, that is, the equation can be written as:

$$v = \frac{k_0^{app} e_0 a}{k_{mA}^{app} + a} \qquad \ldots (2.11)$$

with

$$k_0^{app} = \frac{k_0 b}{k_{mB} + b};\ k_A^{app} = \frac{(k_0/K_{mA})b}{(K_{iA}K_{mB}/K_{mA}) + b};\ k_{mA}^{app} = \frac{K_{iA}K_{mB} + K_{mA}b}{K_{mB} + b} \quad \text{... (2.12)}$$

Notice that the expressions for the apparent values of k_0 and k_A are both individually of Michaelis-Menten form with respect to b, whereas that for the apparent value of K_m is more complicated: This behaviour is quite typical and is one of the reasons why k_0 and k_A should regarded as more fundamental parameters than K_m. More generally, the concept of apparent parameters pervades the analysis of simple cases in steady-state enzyme kinetics, being important for the study of reactions with multiple substrates and inhibition.

INHIBITION AND ACTIVATION

Inhibition

For most enzyme-catalysed reactions, molecules exist that resemble the substrate closely enough to bind to the enzyme, but not closely enough to undergo a chemical reaction. Such a molecule is known as a competitive inhibitor and causes competitive inhibition, characterised by a rate equation of the following form:

$$v = \frac{k_0 e_0 a}{k_m(1 + i/K_{ic}) + a} \quad \text{... (2.13)}$$

in which i is the concentration of the inhibitor and K_{ic} is the competitive inhibition constant. (The qualification 'competitive' and the second subscript c are usually omitted if only this simplest kind of inhibition is being considered.)

Inhibitors can interfere with catalysis as well as with substrate binding. In the simplest case, an inhibitory term affects the variable term in the denominator of the Michaelis-Menten equation, instead of the constant term:

$$v = \frac{k_0 e_0 a}{k_m + a(1 + i/K_{iu})} \quad \text{... (2.14)}$$

This is called uncompetitive inhibition, and the inhibition constant K_{iu} is the uncompetitive inhibition constant. This is important as a limiting case of inhibition, but in its pure form it is not at all common. Much more often one has mixed inhibition, when both competitive and uncompetitive effects occur simultaneously:

$$v = \frac{k_0 e_0 a}{k_m(1 + i/K_{ic}) + a(1 + i/K_{iu})} \quad \text{... (2.15)}$$

There is no particular reason for the two inhibition constants K_{ic} and K_{iu} to be equal, and most of the mechanisms one might propose to account for mixed inhibition lead one to expect them to be different, yet the case where $K_{ic} = K_{iu}$ is often given an undeserved prominence in discussions of inhibition, largely because experiments done many years ago suggested that it was a more common phenomenon than it is. This is called non-competitive inhibition and its rate equation is the same as equation 2.15, but with both K_{ic} and K_{iu} written simply as K_i.

All of these kinds of inhibition are conveniently discussed in terms of apparent Michaelis-Menten parameters. In the general case (equation 2.15), these are as follows:

$$k_0^{\text{app}} = \frac{k_0}{1 + i/K_{\text{iu}}};\; k_{\text{A}}^{\text{app}} = \frac{k_{\text{A}}}{1 + i/K_{\text{ic}}};\; k_{\text{m}}^{\text{app}} = \frac{K_{\text{m}}(1 + i/K_{\text{ic}})}{1 + i/K_{\text{iu}}} \quad \text{... (2.16)}$$

Note that the first two expressions have the same form, and both simplify to independence of i in the event that one or other inhibition term is negligible. The expression for the apparent value of K_m is more complicated, especially when one considers how it varies with the different types of inhibition: It increases with the concentration of a competitive inhibitor, it decreases as the concentration of an uncompetitive inhibitor increases, it may change in either direction as the concentration of a mixed inhibitor increases, or it is independent of inhibitor concentration if the inhibition is non-competitive. In general, it is simplest to regard k_A as the parameter affected by competitive inhibition, negligibly so when the competitive component is negligible, k_0 as the parameter affected by uncompetitive inhibition, negligibly so when the uncompetitive component is negligible, and K_m just as the ratio of the two, so $K_m = k_0/k_A$.

The effects of the different kinds of inhibition on the common plots as illustrated in Figs 2.7 through 2.9 follows naturally from equation (2.16). Any competitive effect affects the apparent value of k_A, hence, it increases the slope of the plot of $1/v$ against $1/a$ (Fig. 2.7), it increases the ordinate intercept of the plot of a/v against a (Fig. 2.8), and it decreases the abscissa intercept of the plot of v against v/a (Fig. 2.9). Conversely, any uncompetitive effect increases the ordinate intercept of the plot of $1/v$ against $1/a$, increases the slope of the plot of a/v against a, and decreases the ordinate intercept of the plot of v against v/a. When both components of the inhibition are present, both kinds of effects occur. As an illustration we may consider just one example, the effect of competitive inhibition on the plot of $1/v$ against $1/a$: Plots made at various different inhibitor concentrations produce a family of straight-lines intersecting on the ordinate axis, as shown in Fig. 2.10, the lack of effect on the ordinate intercept being a direct consequence of the lack of effect on the apparent value of V.

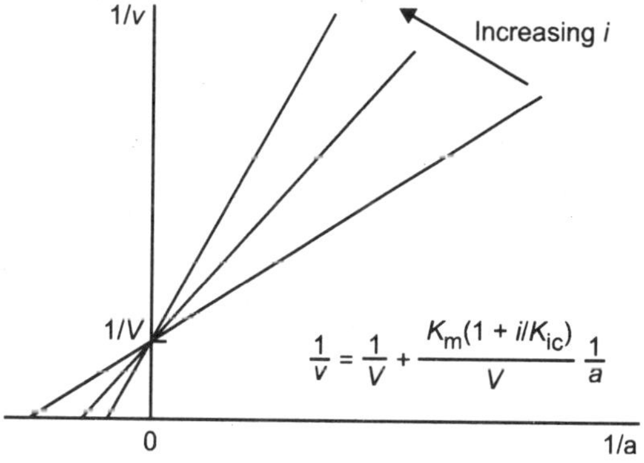

Fig. 2.10. Effect of competitive inhibition on the double-reciprocal plot.

Specificity

Specificity is the most fundamentally important property of enzymes. Although one is often impressed by the catalytic effectiveness of enzymes, accelerating a reaction is not, in reality, difficult: Heating the reaction mixture in a sealed tube is an efficient way of accelerating virtually any reaction, essentially without limit. What is difficult is to accelerate one selected reaction without at the same time accelerating a great mass of unwanted reactions. What is important about an enzyme, therefore, is not that it is an excellent catalyst for a small set of reactions, but that it is an extremely bad catalyst—virtually without any activity—for all other reactions. In other words, the essential properties of enzymes are that they act under very mild conditions and are highly specific.

In the past, specificity was often assessed by comparing the kinetic parameters for different reactions measured in isolation from one another, which led to sterile arguments as to whether specificity was best measured in terms of k_0, K_m, or some combination of the two. This type of argument was resolved once it was realised that the only meaningful way of defining specificity is as a property of an enzyme that allows it to discriminate between substrates that are mixed together. The simplest way to consider this is with a model similar to that for competitive inhibition, except that one assumes that both molecules are capable of reacting. The equation for reaction of one substrate A in the presence of a competing substrate A' follows an equation similar to that for competitive inhibition (eq. 2.13):

$$v = \frac{k_0 e_0 a}{K_m(1 + a'/K'_m) + a} \qquad \text{... (2.17)}$$

with the inhibitor concentration replaced by a', the concentration of A', and the inhibition constant by K'_m, the Michaelis constant for the reaction of A' considered in isolation. The rate of reaction of A' is given by the same equation with an obvious transposition of symbols:

$$v' = \frac{k'_0 e_0 a'}{K'_m(1 + a/K_m) + a'} \qquad \text{... (2.18)}$$

It can then be seen that the ratio of rates is the ratio of substrate concentrations multiplied by the ratio of specificity constants:

$$\frac{v}{v'} = \frac{k_0/K_m}{k'_0/K'_m)}\frac{a}{a'} = \frac{k_A a}{k'_A a'} \qquad \text{... (2.19)}$$

This result, which is still less well known than its importance merits, is the reason for the term specificity constant. Note that although inspection of equation 2.1 suggests that k_A is no more than the parameter that defines the rate at very low substrate concentrations, no assumption about the magnitudes of the concentrations was made in arriving at equation 2.19. It is thus valid at all concentrations, and the specificity constant measures specificity at all concentrations, not just low ones.

Activation

Activation is the opposite from inhibition, in which a reaction proceeds more rapidly in the presence of a particular molecule than in its absence. It is less common than inhibition, and discussion is complicated by the fact that a variety of quite different phenomena have been termed 'activation'. The most important of these is a confusion between true activation and the case where the activator is really a component of the substrate. Numerous ATP-handling enzymes are said to be activated by magnesium ions, when in reality the complex MgATP is the true substrate, that is, the species that reacts with the enzyme. Other metal ions, such as the zinc in a number of enzymes, may be true activators as they bind to the enzyme itself and confer catalytic properties on it.

Another misuse of the term activation relates to coenzymes such as NAD in many dehydrogenases: Alcohol dehydrogenase, for example, may be said to be activated by NAD, but although consideration of the metabolic pathway in which the reaction occurs may make it convenient to regard ethanol as the substrate and NAD as the coenzyme, this is meaningless when the reaction is considered in isolation. So far as alcohol dehydrogenase is concerned, it catalyses a reaction that requires two substrates, ethanol and oxidised NAD; the reaction will not proceed unless both are present, and neither has any more reason to be called the substrate than the other.

When such improper uses of the term are excluded, there remain a number of enzymes for which the true inverse of inhibition occurs. In the simplest cases the equations are just the inverse of inhibition equations, with terms of the form i/K_i replaced by ones of the form K_x/x (for an activator X with activation constant K_x). However, the simplest cases constitute a smaller proportion of the whole than they do for inhibition. This is because whereas most inhibitors inhibit completely, in the sense that enzyme species with inhibitor bound retain no activity as long as the inhibitor remains bound, many enzymes subject to activation retain some activity in the absence of the activator. As a result, full analysis of activation is often more complicated than it is for inhibition, but this will not be discussed further here.

Irreversible Inhibition

The types of inhibition considered so far are examples of reversible inhibition; the inhibitor binds reversibly and catalytic activity returns when the inhibitor is released. Irreversible inhibition also occurs, in which the inhibitor either binds so tightly that for practical purposes it cannot be removed, or reacts with the enzyme and converts it irreversibly to a form that has no catalytic activity. These two cases are conceptually different, and the former is more correctly called tight-binding inhibition, rather than irreversible inhibition. However, they are not easy to distinguish in practice, and have similar practical effects and, hence, similar practical uses.

Although irreversible inhibition has played a smaller part than reversible inhibition in the academic study of enzyme mechanisms, it has far greater industrial and pharmacological importance. This is because competitive inhibitors, the most common kind of reversible inhibitors, are almost completely ineffective in complete physiological systems, for reasons to be considered shortly. By contrast, whenever irreversible (or tight-binding) inhibition occurs in a physiological system, it can be expected to have profound effects. Many toxic and pharmacologically active substances owe their effects to irreversible inhibition.

Both tight-binding and irreversible inhibition manifest themselves in ways that allow them to be confused with non-competitive inhibition, as in equation 2.15 with the two inhibition constants equal. This is because the practical effect of irreversible inhibition is not on any of the kinetic parameters in the Michaelis-Menten equation, but on e_0, the total enzyme concentration. However, a decrease in e_0 can be confused with a decrease in the apparent value of k_0, as they occur as a product in equation 2.1. Although uncompetitive inhibition affects k_0, it does so without affecting k_A, and thus also changes K_m. Decreasing k_0 without affecting K_m, similar to what one would observe if e_0 decreases, is a definition of non-competitive inhibition.

Inhibitory Effects in Metabolic Systems

Competitive and uncompetitive inhibition are sufficiently similar in their effects in artificial experiments on isolated enzymes that they are often not distinguished, and an uncompetitive component in mixed inhibition often passes unnoticed.

Many inhibitors are described in the literature as competitive inhibitors in the absence of any real evidence of the type of inhibition. This sort of confusion can easily lead to the entirely false idea that they are similar in their effects in systems where the inhibited enzyme is mixed with other enzymes and catalyses a step in the middle of a pathway.

In a typical experiment *in vitro*, one decides the concentrations of the various components in advance and measures the rate that results; however, this is very different from what happens in the cell. To a first approximation, an enzyme catalysing a step in the middle of a pathway must transform its substrate at the rate at which it arrives, that is, within certain limits it has little or no effect on the rate of its reaction, but instead determines the concentrations of the metabolites around it. (This is an oversimplification: but is useful for discussion.)

It is useful therefore to transform equations (2.13) and (2.14) into expressions for a in terms of i:

$$a = \frac{vK_m(1 + i/K_{ic})}{k_0e_0 - v} \quad \text{... (2.20)}$$

$$a = \frac{vK_m}{k_0e_0 - v(1 + i/K_{iu})} \quad \text{... (2.21)}$$

However, similar equations 2.13 and 2.14 may seem, their transformed versions are drastically different. Equation 2.20 shows a linear dependence of a on i, which means that increasing i can never result in uncontrolled increases in a. This is illustrated in Fig. 2.11(a). Even at an inhibition concentration equal to the inhibition constant, the substrate concentration is only doubled. By contrast, the curve defined by equation 2.21 is a rectangular hyperbola [Fig. 2.11(b)] that produces a steep and uncontrolled rise in substrate concentration at quite moderate inhibitor concentrations. The point is that in competitive inhibition, rises in substrate and inhibitor concentrations oppose one another— not only does the inhibitor compete with the substrate, but equally, the substrate competes with the inhibitor. In uncompetitive inhibition, however, these effects potentiate one another.

It follows that although it is relatively easy to find molecules that will act as competitive inhibitors, it is also largely useless as a strategy for designing pesticides or drugs because it is correspondingly easy for the organism to counteract the effect of the inhibition. To produce major metabolic effects one needs uncompetitive inhibitors, irreversible inhibitors, or tight-binding inhibitors: None of these are as easy to produce as weakly binding competitive inhibitors, but they are far more effective.

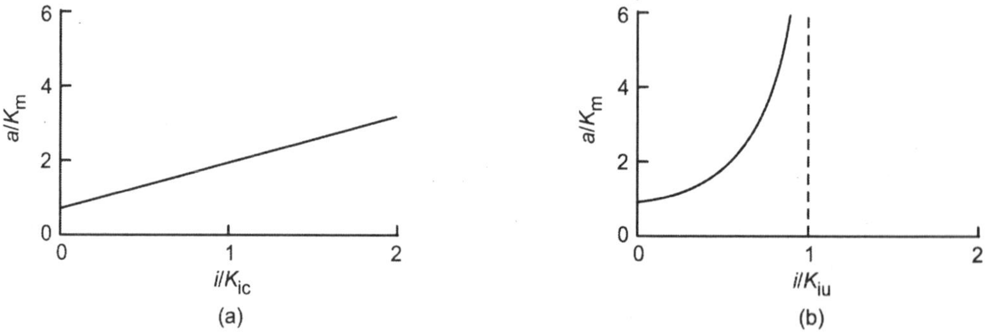

Fig. 2.11. Effects of (a) competitive and (b) uncompetitive inhibition on the concentration of substrate in a constant-rate system. Both curves are drawn for the case of $a = K_m$ in the absence of inhibitor. Note that both kinds of inhibitor have quantitatively equal effects at very low concentrations, but the initial slope is maintained indefinitely if the inhibition is competitive, whereas it rapidly becomes infinite if the inhibition is uncompetitive.

NON-MICHAELIS-MENTEN BEHAVIOUR

All of the cases considered so far can be regarded as generalisations of the Michaelis-Menten equation (equation 2.1). However, although many enzymes do behave in this way, at least as a first approximation, there are some important exceptions. It is simple to calculate from equation 2.1 that if $a = K_m/9$ then $v = 0.1V$ and if $a = 9\ K_m$ then $v = 0.9V$; in other words, spanning the 10–90 per cent range of available rates requires an 81-fold range of substrate concentrations, almost two orders of magnitude. Similar calculations may be done with any of the equations of the Michaelis-Menten type for additional substrates, inhibitors, or activators. Their implication is that as long as enzymes follow Michaelis-Menten kinetics, their rates cannot be adequately varied by manipulating concentrations of substrates, for example, because effective regulation will often require sensitivity to small changes in signals—certainly changes much smaller than two orders of magnitude. A second difficulty arises from the fact that inhibition of the types considered commonly derives from structural similarities between inhibitors and substrates or products, whereas there is no reason to expect the molecules needed for metabolic signals to resemble the substrates or products of the enzymes that need to respond to the signals. In reality, the concentration of the end product of a pathway often serves as such a signal: too low, and the pathway needs to be activated; too high, and it needs to be inhibited. It is often found, therefore, that the enzyme that catalyses the first committed step of a pathway, that is, the first step after a branch point, in the branch that leads to the end product in question, is inhibited by that end product. For inhibition of this kind to be possible, the enzyme must have a specific binding site for the end product, independent of the binding sites for substrates and products. Such a site is called an allosteric site, and the phenomenon is called allosteric inhibition. Because the need for it often coincides with the need for higher sensitivity than is provided by Michaelis-Menten kinetics and the common kinds of inhibition, allosteric inhibition is often cooperative. This means that the equations that define it are more complicated than those considered above, allowing, for example, a change from 10 per cent to 90 per cent inhibited over a concentration range much smaller than 81-fold, and typically less than 10-fold (though rarely, if ever, less than 3-fold) (Fig. 2.12).

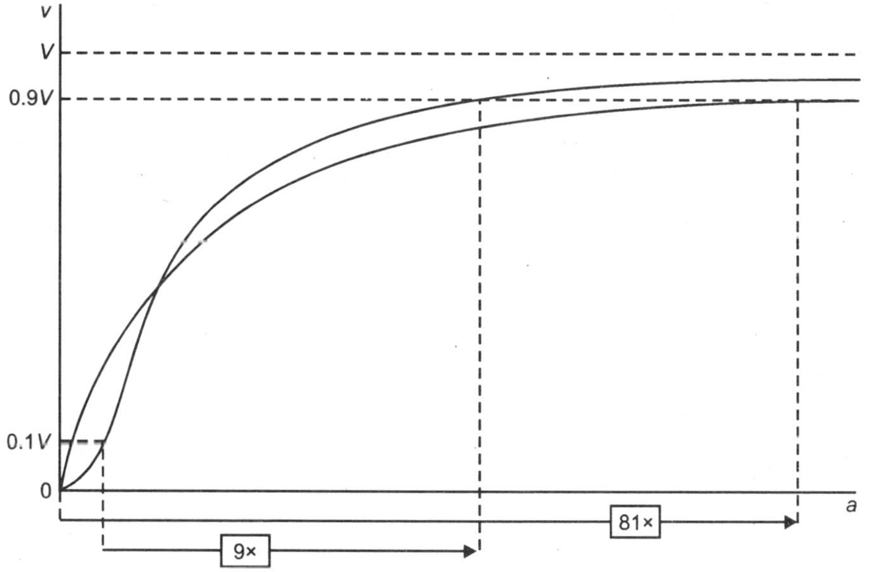

Fig. 2.12. Non-Michaelis-Menten kinetics. For an enzyme obeying the Michaelis-Menten equation (Fig. 2.6), an 81-fold increase in substrate concentration is needed to bring the rate from 10 per cent to 90 per cent of *V*. If the enzyme shows positive cooperativity the curve typically becomes sigmoid (S-shaped), and this range of substrate concentrations is decreased (to nine-fold in the example, but in strongly cooperative cases it can be as small as three-fold).

The fact that allosteric and cooperative behaviour are often found together has led many authors to treat the two terms as synonymous, a tendency encouraged by the fact that in one of the most widely accepted models of non-Michaelis-Menten behaviour, that of Monod, Wyman, and Changeux, both result from the same structural properties attributed to the enzyme. Nonetheless, most careful authors consider the two properties to be conceptually distinct and not necessarily occurring together, so the two terms should be considered distinct as well.

KINETICS OF MULTIENZYME SYSTEMS

As noted in the introduction, nearly all kinetic studies of enzymes have been carried out using isolated enzymes, and although this has been very valuable for arriving at a good understanding of the nature of enzyme catalysis, it is quite inadequate as a guide to how systems of enzymes will behave. One cannot assume that the flux through a metabolic pathway is a property of a unique enzyme catalysing the rate-limiting step, and that the properties of the pathway as a whole can be deduced from studies of the kinetics of this one enzyme in isolation.

Space does not permit a full analysis of this subject, but it should suffice to examine an example of a pathway in which biotechnologists have attempted to increase the flux by identifying the rate-limiting enzyme and using genetic manipulation to increase its activity. Tryptophan biosynthesis in yeast provides such an example, tryptophan being a commercially valuable metabolite for which increased production would be very desirable. The tryptophan pathway consists of five enzymes, and in the classical model anyone of these could be the 'key' enzyme catalysing the rate-limiting step. However, when the activity of each of these enzymes was increased in turn, either singly or at the same time as others in the pathway, the results were trivial: Increases of enzyme activity of 20-fold or greater produced flux increases of perhaps 30 per cent. Only when all five enzyme activities were increased (or all but one, apparently unimportant, activity) was there a substantial increase in flux, which even then was much smaller than the increase in enzyme activity.

The object here is not to analyse in detail why manipulation of tryptophan biosynthesis did not produce the desired results, but to use it to illustrate the point that the whole approach is misconceived. One cannot treat the kinetic behaviour of systems of enzymes as if it were determined by the properties of a single component. Moreover, abundant evidence exists to show that all organisms have evolved regulatory mechanisms to control metabolic fluxes and concentrations to satisfy their own requirements. Artificially trying to force more activity in a pathway by increasing the activities of certain enzymes simply stimulates the regulatory mechanisms to resist. Tryptophan biosynthesis is just one example of a general result, and similar efforts in other pathways—for example, increased alcohol production in yeast and increased starch production in potatoes—have produced similar results.

The essential point is that flux control is not a property of a single enzyme in a pathway, but is shared among all the enzymes in the pathway. Strictly speaking, it is shared among all the enzymes in the cell, but it is usually safe to assume that enzymes catalysing reactions remote from the pathway of interest have very little effect on the flux through it, so to a first approximation one can consider flux control to be shared among the enzymes of the pathway.

To express this idea in quantitative terms, one can define a flux control coefficient for any enzyme by the following equation:

$$C_i^J = \frac{\partial \ln J}{\partial p} \bigg/ \frac{\partial \ln v_i}{\partial p} \quad \text{... (2.22)}$$

This definition compares the effect on the flux J through a pathway of some perturbation of the activity of the ith enzyme, represented by a change in the parameter p, with the effect the same perturbation would have on the rate v_i of the same reaction if it were considered in isolation. The identity of the parameter p does not have to be specified, because as long as it affects only one enzyme, the control coefficient defined by equation 2.22 is independent of the manner in which the flux and isolated rate are perturbed. However, to make the definition more concrete, consider the case where p is the logarithm of the enzyme concentration, e_i. As most reactions are considered under conditions where the rate is proportional to the enzyme concentration, it will often be true to write $d\ln v_i = d\ln e_i$, so that the denominator in equation 2.22 has a value of unity and the whole equation simplifies to the following:

$$C_i^J = \frac{\partial \ln J}{\partial \ln e_i} \quad \text{... (2.23)}$$

This equation is less abstract and simpler to understand than equation 2.22, and in the past was often used as a primary definition of a control coefficient. However, this is not recommended, first because it is not always true that the isolated rate is proportional to the enzyme concentration, and second because equation 2.23 can give the false impression that control coefficients are concerned only with effects brought about by changes in enzyme concentration.

Flux control coefficients are measures of how much the flux through a pathway is dependent on the activities of the individual enzymes. Mathematical analysis shows that they satisfy a property called the summation relationship:

$$\sum C_i^J = 1 \quad \text{... (2.24)}$$

The interpretation of this relationship would be straightforward if all flux control coefficients were always positive, but in reality negative flux control coefficients also occur; for example, if a pathway contains a branch-point, the enzymes in one branch normally have negative flux control coefficients for flux through the other. Nonetheless, in practice, negative flux control coefficients are nearly always small in magnitude, so although the sum of positive flux control coefficients may be greater than 1, it is rarely much greater than 1, so that the idea of sharing flux control among all the enzymes of a system is reasonably accurate. It follows that we should not expect any enzyme to have complete control over flux, and the closer to complete control any enzyme approaches, the less control all of the others have, taken together.

Moreover, a flux control coefficient is not a constant, but tends to decrease when the activity of an enzyme is increased. In other words, even if one can identify one enzyme that has a large proportion of the total flux control, increasing its activity will tend to decrease that proportion, so that the amount the flux can be increased by increasing the activity of one enzyme will always be small. Once this is understood, the failure to increase the flux to tryptophan (in the example mentioned) or to achieve significant flux increases (in other similar examples), ceases to be a mysterious result, but is simply what was to be expected.

ENZYME UTILISATION INDUSTRY

Enzymes offer the potential for many exciting applications in industry. Some important industrial enzymes and their sources are listed in Table 2.1. In addition to the industrial enzymes listed below, a number of enzyme products have been approved for therapeutic use. Examples include tissue plasminogen activator and streptokinase for cardiovascular disease; adenosine deaminase for the rare severe combined immunodcliciency disease; b-glucocerebrosidase for Type I Gaucherdisease; L-asparaginase for the treatment of acute lymphoblastic laeukemia: DNAse for the treatment of cystic fibrosis and neuraminidase, which is being targeted for the treatment of influenza.

Table 2.1. Some important industrial enzymes and their sources.

Enzyme	*EC number*	*Source*	*Industrial use*
Chymosin	3.4.23.4	Abomasum	Cheese
α-amylase	3.2.1.1	Malted barley, *Bacillus*, *Aspergillus*	Brewing, baking
β-amylase	3.2.1.2	Malted barley, *Bacillus*	Brewing
Bromelain	3.4.22.4	Pineapple latex	Brewing
Catalase	1.11.1.6	Liver, *Aspergillus*	Food
Penicillin amidase	3.5.1.11	*Bacillus*	Pharmaceutical
Lipoxygenase	1.13.11.12	Soyabeans	Food
Ficin	3.4.22.3	Fig latex	Food
Pectinase	3.2.1.15	*Aspergillus*	Drinks
Invertase	3.2.1.26	*Saccharomyces*	Confectionery
Pectin lyase	4.2.2.10	*Aspergillus*	Drinks
Cellulase	3.2.1.4	*Trichoderma*	Waste
Chymotrypsin	3.4.21.1	Pancreas	Leather
Lipase	3.1.1.3	Pancreas, *Rhizopus*, *Candida*	Food
Trypsin	3.4.21.4	Pancreas	Leather
β-glucanase	3.2.1.6	Malted barley	Brewing
Papain	3.4.22.2	Pawpaw latex	Meat
Asparaginase	3.5.1.1	*E. chrisanthemy*, *E. carotovora*, *Escherichia coli*	Human health
Xylose isomerase	5.3.1.5	*Bacillus*	Fructose syrup
Protease	3.4.21.14	*Bacillus*	Detergent
Aminoacylase	3.5.1.14	*Aspergillus*	Pharmaceutical
Raffinase	3.2.1.22	*Saccharomyces*	Food
Glucose oxidase	1.1.3.4	*Aspergillus*	Food
Dextranase	3.2.1.11	*Penicillium*	Food
Lactase	3.2.1.23	*Aspergillus*	Dairy
Glucoamylase	3.2.1.3	*Aspergillus*	Starch
Pullulanase	3.2.1.41	*Klebsiella*	Starch
Raffinase	3.2.1.22	*Mortierella*	Food
Lactase	3.2.1.23	*Kluyveromyces*	Dairy

There are also thousands of enzyme products used in small amounts for research and development in routine laboratory practice and others that are used in clinical laboratory assays. This group also includes a number of DNA and RNA modifying enzymes (DNA and RNA polymerase, DNA ligase, restriction endonucleases, reverse transcriptase, etc.), which led to the development of molecular biology methods and were a foundation for the biotechnology industry. The clever application of one thermostable DNA polymerase led to the polymerase chain reaction (PCR) and this has since blossomed into numerous clinical, forensic and academic embodiments. Along with the commercial success of these enzyme products, other enzyme products are currently in commercial development.

Another important field of application of enzymes is in metabolic engineering. Metabolic engineering is a new approach involving the targeted and purposeful manipulation of the metabolic pathways of an organism, aiming at improving the quality and yields of commercially important compounds. It typically involves alteration of cellular activities by manipulation of the enzymatic functions of the cell using recombinant DNA and other genetic techniques. For example, the combination of rational pathway engineering and directed evolution has been successfully applied to optimise the pathways for the production of isoprenoids such as carotenoids.

The new era of the enzyme technology industry is growing at a constant rate. The potential economic, social and health benefits that may be derived from this industry are unforeseen and therefore future development of enzyme products will be unlimited.

Chapter 3

Environmental Effects on Enzyme Activity

INTRODUCTION

Chemical reactions, including those catalysed by enzymes, depend on environmental conditions. This characteristic of reactions is particularly important in food systems and processing applications, in which a broad spectrum of conditions is encountered. Knowledge of how a specific enzyme reaction is affected by its environment, whether the conditions are imposed or can be manipulated, is required to allow optimisation or control of any enzyme-mediated process.

Enzyme activity is affected by long-range and short-range, or macro- and microenvironmental, conditions. Emphasis in this chapter will be on macroenvironmental effects. However, important microenvironmental effects will be addressed also.

The dominant environmental factors affecting enzyme action are pH, water activity, and temperature. The individual effects of these factors are often interdependent. Therefore, an attempt will be made to identify the primary influences of each environmental factor. Effects of environmental conditions other than those just mentioned will be discussed briefly.

EFFECT OF pH

General Considerations

The effects of pH on enzymatic reactions are caused largely by the reversible ionisation of substrate or amino acid residues of the enzyme. These effects are manifested as changes in enzyme activity (catalysis), stability, or interaction (binding) with ligands, or a change in reaction equilibrium. The effects of pH on enzymes also depend on the presence or absence of specific ions or ligands, temperature, dielectric constant, and ionic strength.

The pH of foods can range from very acidic (pH 3 in fruit juices or purees) to near neutral pH (milk, vegetables, grains, and some muscle foods) to mildly alkaline (hominy and ripe olives). The pH optima of enzymes can vary similarly, from pH 2 (pepsin) to pH 10 (alkaline phosphatase), most enzymes exhibiting optima near neutral pH.

The same enzyme isolated from various sources can exhibit different pH optima, probably reflecting the physiological requirements of the organism from which the enzyme was isolated. Fungal enzymes often have slightly acidic pH optima, whereas those from yeasts often display pH optima near neutral pH. In fruit tissues, several forms of invertase with various pH optima are known to exist.

Mechanistic Considerations

Effect of pH on enzyme activity

The effect of pH on enzyme catalysis is caused by the ionisation of substrate or enzyme, which can affect substrate binding or transformation to product directly or affect enzyme stability. Many of these ionisable groups exhibit a range of pK values because of microenvironmental effects that influence the ionisation potential of the amino acid side chains. Verification of the role of these amino acids in activity can be achieved by evaluating the pH dependency of enzyme photoinactivation, enzyme activity following specific chemical modifications, and temperature dependency of the pH dependency of enzyme activity. This last feature is based on the enthalpies of ionisation of amino acid functional groups. Studies on point mutations in the polypeptide chains of enzymes also can lead to an improved understanding of the requirement far specific amino acid residues in enzyme catalysis. The identification of the vital amino acid residues can be useful because modification of these residues, by chemical or biotechnological means, potentially yields an added dimension of control over enzyme-catalysed reactions.

Effect of pH on ligand ionisation

For some enzymes, the ionisation of ligands (substrate, product, cofactor, or inhibitor) must be considered to explain the pH dependency of activity.

The pH dependency of enzyme-inhibitor interaction is characterised by pK values of 3.0 and 5.6. Whether the inhibitory action depends on the ionisation state of the inhibitor, the enzyme, or both is not clear. Since this inhibitor is of the non-competitive type, the difference in pK values simply may reflect the ionisation properties of the different binding sites for substrate and inhibitor. However, since the inhibitor is also a protein, its ionisation state is likely to be related to its inhibitory action.

Effect of pH on enzyme stability

In some cases, the effect of pH on enzyme activity can be explained in terms of enzyme conformational stability. This effect is assessed most easily by preincubating an enzyme solution at a range of pH values, then assaying enzyme activity at the optimum pH. In the case of peroxidase extracted from red beet root, the loss of activity in the acidic pH region can be attributed largely to a lack of enzyme stability (Fig. 3.1).

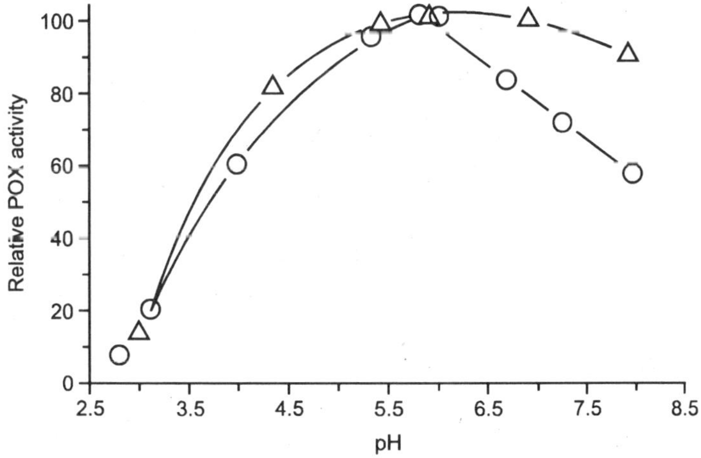

Fig. 3.1. Effect of pH on stability (Δ) and activity (O) of peroxidase isolated from red beet roots. Incubation period was 5 minutes; activity was measured within 2 minutes at 25°C.

The pH dependency of beet peroxidase above the optimum pH appears to be caused by a direct effect of pH on enzyme activity, rather than on enzyme stability. The composition of the medium also can affect the pH stability of enzymes. For β-amylase, the presence of calcium stabilises the enzyme at pH 5–10.

Effect of pH on enzyme reaction equilibria

An effect of pH on enzyme activity can be observed for many dehydrogenases that use NAD to catalyse the reversible oxidation of a substrate (AH_2):

$$AH_2 + NAD^+ \qquad A + NADH + H^+$$

and

$$K_{eq} = \frac{[A]\,[NADH]\,[H^+]}{[AH_2]\,[NAD^+]}$$

If the K_{eq} of the reaction remains fairly constant over a broad range of pH, the equilibrium is shifted to the left and in the direction of reduction of substrate *A* as pH is reduced. Therefore, at fixed levels of initial substrate and cofactor concentrations, the extent of the reaction (the equilibrium state) at any pH can be predicted by simple calculation. In doing so, the effect of temperature on K_{eq} also must be considered.

EFFECT OF THE AQUEOUS ENVIRONMENT

General Considerations

Water is usually the continuous phase in most foods and food systems. It functions as a diffusion medium for solutes, controls the concentration of dissolved solutes, allows protein and enzyme function and serves as a co-substrate in hydrolytic reactions. Dehydration and freezing are used to preserve foods by the removal or 'immobilisation' of water, thereby restricting microbial growth. Usually, enzyme activity is attenuated greatly by such processes. However, some enzymes remain active and often display altered patterns of activity in low moisture environments.

Water Activity

Effect on enzyme stability

Generally, enzymes in solutions or in food systems are stable to dehydration, provided the temperature is controlled to prevent thermal inactivation. Indeed, lyophilisation is used routinely to prepare active enzyme powders from solutions. During the controlled drying of grains, many endogenous enzymes remain active (or dormant). Removal of water from biomaterials with organic solvents (acetone, alcohols, ether) also can lead to retention of enzyme activity, provided the temperature is maintained at 0°C or less. The composition of the medium in which the enzyme is suspended can influence retention of enzyme activity; methylamines, carbohydrates, and amino acids can stabilise proteins in dry or high osmotic environments.

Effect on enzyme activity

Hydrolytic enzymes

Hydrolytic reactions are reversible, although in foods in which water is 35–50 M, hydrolysis is favoured heavily. Several enzymes of importance in food processing—including carbohydrases, proteases, and lipases—have been shown to catalyse reverse reactions in water-restricted environments. The rate and extent of many enzyme reactions also depend on the availability of water. An application of the control of invertase activity by controlling humidity (a_w) in foods is illustrated in the study of the occurrence of the defect known as 'sugar wall' in dates. The 'sugar wall' defect arises from the crystallisation of sucrose during the storage and handling of dates. Invertase, added as a spray or by vacuum infiltration, can delay or eliminate the occurrence of this defect. Hydrolysis by invertase of sucrose into its constituent sugars, glucose and fructose, interferes with the crystallisation of sucrose. A water spray devoid of invertase also was found to delay the onset of the 'sugar wall' defect. The water spray probably elevated the a_w in the date tissue and potentiated the action of endogenous invertase. Glucoamylase, one of the enzymes used in the hydrolysis of starch, can catalyse synthetic reactions under the conditions used during industrial saccharification (30 per cent starch, *w/v*). In the presence of 2.07 M glucose, glucoamylase forms a limited amount of maltose and appreciable quantities of isomaltose (Fig. 3.2). This behaviour may account for the 6–8 per cent oligosaccharide content that remains in corn syrup.

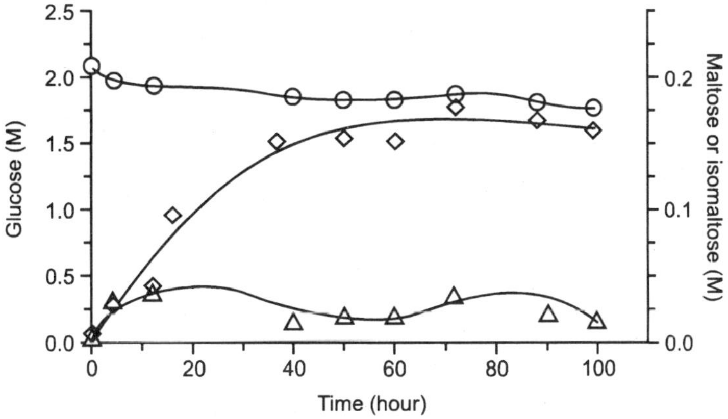

Fig. 3.2. Time course of the conversion of glucose (O) to isomaltose (◇) and maltose (Δ) by glucoamylase (from *Aspergillus niger*) at pH 5.0 and 40°C.

Proteases have been reported to synthesise peptide bonds in the presence of high concentrations (30–50 per cent *w/v*) of protein hydrolysates. This behaviour is the basis of the plastein reaction, in which the rearrangement of protein structures is proposed to improve the functionality of relatively non-functional proteins, to reduce bitterness in protein hydrolysates, or to improve the nutritional quality of proteins through the incorporation of essential amino acids. The reversal of such proteolytic reactions may not be due to a true a_w effect, but may be driven more by the high concentration of reactants.

One commercial application of the control of reaction equilibria by a_w involves the action of lipase. Under conditions of restricted water content, lipase catalyses the esterification of fatty acids to glycerol, monoacylglycerols, and diacylglycerols. Water-restricted environments also can be used to catalyse the interesterification and transesterification of a mixture of acylglycerides.

This process has been used to modify oils to improve their physical characteristics and to derive a cocoa butter substitute from a fraction of palm oil. One method of refining food grade oils is treating them with lipase under restricted water conditions to esterify any free fatty acids present in the oil. Because of their specificity, enzymes often are preferred to nonenzymatic catalysts in obtaining directed modifications of lipids by transesterification.

Non-hydrolytic enzymes

The activities of lipoxygenase and phenol oxidase in water-restricted environments have been the subject of much study. In food-related model systems, soya lipoxygenase activity is linearly dependent on increasing a_w until maximal activity is observed at a_w 0.55–0.65. At low a_w, capillary water was concluded to facilitate enzyme action on both free linoleate and linoleate esterified to phospholipids. Phenol oxidase activity exhibits a hyperbolic dependency on a_w; activity is observed initially at a_w as low as 0.5. A decrease in a_w from 1.0 to 0.35 also shifts the pH optimum of phenol oxidase almost 1 pH unit to the alkaline side. Progress curves of activity indicate that the extent of phenol oxidase action also is controlled by a_w (Fig. 3.3). Thus, a_w or the system appears to regulate activity in this case by controlling the diffusibility of reactants. Adjustment of the a_w from a lesser value to a greater one increases the extent of the reaction accordingly. The practical relevance of these phenomena is that, in the storage of intermediate moisture foods, fluctuation in a_w may potentiate undesirable chemical changes representative of storage at an intolerably high a_w.

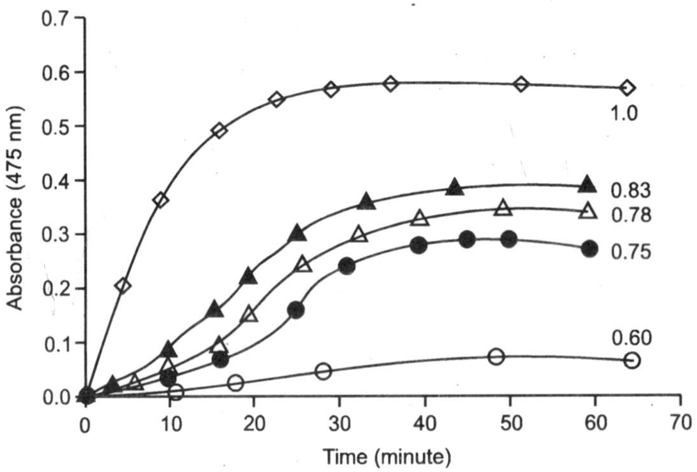

Fig. 3.3. Phenoloxidase activity as a function of water activity measured as A_{475}. Water activity was adjusted with glycerol and enzyme activity was measured on tyrosine at pH 7.0 and 30°C.

The nature of the products formed during enzyme reactions can be influenced by a_w. For soya lipoxygenase activity on free linoleate at a_w 0.65, the initial product formed is the linoleate hydroperoxide. As the reaction progresses, the hydroperoxide appears to be converted, to a significant degree, into polar oligomeric secondary products. This behaviour may be important in achieving the desirable effect on texture that lipoxygenase has in dough mixing and bread baking, where a reduced a_w would prevail.

Altered reaction pathways at low a_w also have been noted for hydrolytic enzymes. At a_w 0.95, α-amylase (an endoamylase) forms a mixture of oligosaccharides (dextrins) from starch, whereas at a_w 0.75, glucose formation is highly favoured. The lowered a_w appears to restrict the access (diffusibility) of the enzyme to the internal sites of the starch polymer, reducing dextrin formation. In dry beans, the thermal potentiation of endogenous phytase activity converts phytate to inositol and inorganic phosphate, with little accumulation of the intermediate inositol phosphates. Again, the restriction of reactant diffusibility in the low a_w environment is likely to favour a more exhaustive hydrolysis of a single phytate molecule than a high a_w system.

Ionic Strength

Ionic strength of a food system can be increased through the addition of electrolytes or as a result of freezing or dehydration. The effects of ionic strength on enzyme activity can be caused by specific or nonspecific effects of electrolytes. Ionic strength is well known to affect protein and, therefore, enzyme solubility by salting-in and salting-out phenomena. In most cases, enzyme solubility is a requisite for activity, since reactions require the diffusion of substrate to enzyme. Specific effects of electrolytes depend on the relative concentrations of agents that stabilise, activate, inhibit, or inactivate a particular enzyme. Metal ions are particularly important, since many of them (e.g. magnesium, calcium, zinc) are cofactors or co-substrates for enzyme reactions. Other metal ions such as the heavy metals (e.g. lead, mercury, cadmium) can inactivate enzymes by virtue of their ability to react with free sulphydryl groups (cysteine residues).

Ion-specific effects may result from the presence of ions that are lyotrophs or chaotrophs. Lyotrophs (principally cations) preserve water-solute interactions and water polarity and enhance the stability of soluble proteins. Chaotrophs (principally anions) disrupt water-solute interactions and reduce water polarity, often compromising the stability of soluble proteins. The relative chaotropic power of anions generally is related to charge density and follows the Hofmeister series (in decreasing order of strength):

$$SCN^- > I^- > ClO_4^- > NO_3^- > Br^- > Cl^- > CH_3COO^- > F^- > SO^-_4$$

Organisms that thrive in, or can adapt to, high osmotic environments have mechanisms to allow for normal enzyme functioning. Some enzymes have evolved to become halo tolerant and are unimpaired by high osmotic environments. The evolution of osmolyte systems in organisms indigenous to high osmotic environments is a second mechanism of allowing normal enzyme functioning during osmotic stress. The mechanism(s) by which enzymes are stabilised by these osmolyte systems (amino acids, methylamines, and polyols) is unknown. However, the effect of polyhydroxy compounds apparently is caused by an increase in hydration of soluble proteins.

Exploiting enzyme adaptation to high osmotic conditions in food processing situations in which high concentrations of electrolytes exist may be possible. The addition of protein stabilising osmolytes is known to preserve enzyme function in systems that contain high levels of chaotropic ions or protein denaturants (urea) or have high osmolalities. Alternatively, enzymes adapted to high osmotic environments can be used in selected food processing situations. This adaptation is illustrated by the study of a neutral protease, with halophilic properties, isolated from *Bacillus subtilis*. This enzyme is activated by the addition of NaCl and shows maximal activity on casein substrates at 4 per cent NaCl (Fig. 3.4). At least 80 per cent activity is retained on casein substrates at 10 per cent NaCl. This behaviour may prove valuable in cheesemaking, because high NaCl concentrations are known to inhibit chymosin proteolysis of β-casein.

Freezing

General considerations

Although the activities of many enzymes are attenuated by freezing, most enzymes survive freezing and some can be very active at high frozen storage temperatures. The net effect of freezing on enzyme stability and activity is a function of the combined effects of 'dehydration', concentration of solutes, and reduced temperature. The 'dehydration' caused by freezing is not the same as that caused by dehydration of foods at temperatures above freezing (or in freeze-drying), since high a_w conditions prevail (> 0.82) in the former at or above recommended commercial frozen storage temperatures.

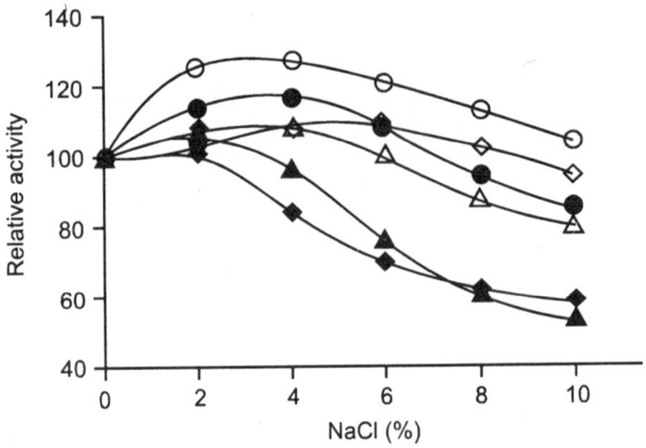

Fig. 3.4. Proteolytic activity on milk proteins as a function of NaCl concentration. Substrates were β-(O), κ-(◇), αs-(Δ), and whole casein (•), whey protein(♦), and β-lactoglobulin (▲). Activity was measured by the production of TCA-soluble nitrogen at 35°C and pH 6.5 (caseins) or 8.5 (whey proteins).

The physical changes brought about by freezing include the concentration of solutes (including enzymes, substrates, inhibitors, and cofactors), an increase in viscosity of the unfrozen phase, and a disruption of (sub)cellular structures. The extent to which each of these changes takes place depends on the freezing temperature and freezing rate. Fast freezing tends to keep pools of unfrozen water segregated and minimises the extent of cellular damage. The importance of cellular damage to the control of enzyme action lies in the decompartmentalisation of previously separated enzymes and substrates. Although this effect may be of greatest importance during or after thawing, when the temperature becomes elevated, cellular systems (tissue foods) often exhibit an acceleration of certain, sometimes undesirable, enzymatic activities on freezing.

Effect on enzyme stability

Much of our understanding of the behaviour of enzymes in frozen foods comes from the study of model non-cellular systems. For catalase, retention of activity after freezing is favoured by high initial enzyme concentrations, quick thawing rates, and slow freezing rates. Invertase also retains greater activity at lower freezing rates, although myosin B ATPase activity is more stable at faster freezing rates. This difference may be caused by the macromolecular organisation of the enzymes and by microenvironmental effects; invertase and catalase were studied as solutions whereas ATPase was associated with a complex muscle protein fraction. Retention of enzyme activity during freezing also depends on the composition of the medium. Retention of invertase and lactate dehydrogenase activities depend on pH. The concentration of solutes on freezing and their eutectic behaviour can lead to changes in pH of the frozen system of as much as 1 pH unit in foods and in simple solutions. Polyhydroxy compounds are very effective in stabilising phosphofructokinase to freeze-induced inactivation.

Effect on enzyme activity

The need to blanch vegetables prior to freezing is a constant reminder that food-deteriorating enzymes remain active in frozen environments. In some cases, freezing can accelerate enzyme action in simple solutions and in foods.

Simple solutions

Freezing of dilute solutions of trypsin at –23°C and invertase at –5°C accelerates enzyme activity with respect to that occurring at non-freezing or corresponding supercooled temperatures. This effect has been attributed to the several-fold increase in reactant concentrations brought about by freezing, and may offset any detrimental effects of freezing on enzyme stability.

Lipoxygenase and lipase exhibit activity at –12° to –15°C in frozen solutions. In both cases, elevation of the temperature of the frozen system results in an increased accumulation of reaction products (Fig. 3.5). These observations indicate that the extent of enzyme activity may be limited by the amount of unfrozen water available as a diffusion medium. Fast freezing only marginally reduced (by 10–20 per cent) the extent of lipoxygenase activity in model systems at –5° to –15°C, compared with slow freezing.

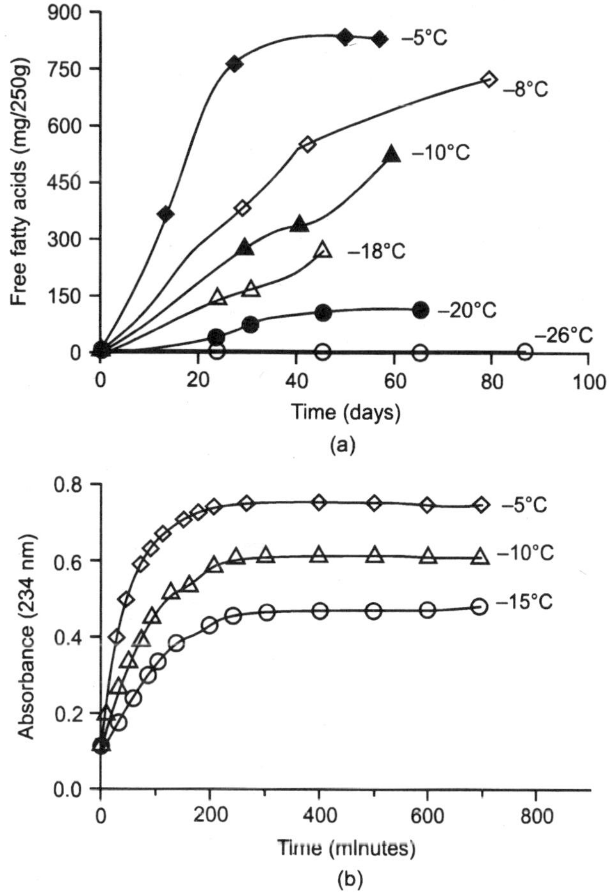

Fig. 3.5. (a) Lipolysis in unblanched frozen peas as a function of temperature; and (b) lipid oxidation (conjugated diene formation) in frozen solutions of lipoxygenase and linolenic acid as a function of temperature.

Tissue systems

Several examples of enzyme action that is accelerated in tissue systems at high freezing temperatures relative to low non-freezing temperatures are presented in Table 3.1. In muscle foods, textural and organoleptic qualities can be compromised during improper frozen storage. Significant extents of phospholipase action can be detected in fish stored at temperatures between –10° and –14°C. This activity can produce free fatty acids and potentiate oxidation or possibly cause the insolubilisation (denaturation) of muscle proteins. Freezing of muscle of fishes of the Gadoid family is known to cause a rapid and unacceptable loss in textural quality if storage temperature is not maintained at –18°C or below. This effect is proposed to be caused by the 'activation' of the endogenous enzyme trimethylamine-*N*-oxide demethylase via decompartmentalisation. Behnke also observed an increase in the rate of ATP depletion and lactate accumulation in red muscle frozen at –3°C, compared with unfrozen muscle. These investigators attempted to account for the relative contributions of temperature and solute concentration on reaction rates during the freezing of foods (Table 3.2). The two forces controlling enzyme activity in frozen systems are temperature and solute concentration. For each of the four situations presented, the relative influence of these two forces and the overall effect on reaction rate can be predicted.

Table 3.1. Examples of potentiation of enzyme activity on freezing of food tissue systems.

Reaction catalysed	*Tissue*	*Temperature range of enzyme potentiation (°C)*
Glycogen loss and lactate accumulation	Muscle	–2.5 to –6
Loss of high energy phosphates	Muscle	–2 to –8
Phospholipid hydrolysis	Muscle	–4 to –14
Peroxide decomposition	Vegetable	–0.8 to –5
L-Ascorbic acid oxidation	Vegetable and fruit	–2.5 to –10
Trimethylamine-*N*-oxide demethylation	Muscle	–5 to –18

Enzyme reactions also can be accelerated on freezing in plant foods (Table 3.1). However, because of the long shelf-life expectancy of frozen vegetables, the mere persistence of enzyme action at low temperatures can be devastating to quality. Lipase activity in unblanched frozen peas is significant at temperatures as low as –20°C [Fig. 3.5(a)]. After 7 months at –15°C, up to 32 per cent of the total esterified fatty acid is hydrolysed. Traditionally, peroxidase has been used as a blanching indicator.

However, some research has indicated that lipid-degradative enzymes (palmitoyl-CoA hydrolase), may be more suitable as blanching indicators. To optimise blanching, the endogenous enzyme that limits the retention of quality must be identified first. Because of differences in enzyme composition and thermal resistance, the target enzyme is likely to vary between specific products.

EFFECT OF TEMPERATURE

General Considerations

All chemical reactions depend on temperature. The relationship between reaction rates and temperature is described by the Arrhenius equation:

$$k = Ze^{\mathrm{E_a}/2.3\mathrm{RT}} \qquad \text{... (3.1)}$$

where, k is the rate constant, T is the temperature (K), R is the gas constant, and Z is the frequency of favourable collisions between reactants. For most non-enzymatic reactions, reaction rates continually increase with increasing temperature. For enzymatic reactions, a temperature optimum is observed. This optimum results from the opposing effects of increasing temperature on the reaction rate and on the rate of enzyme thermal inactivation (both as predicted by Eq. 3.1). Other effects of temperature on the rate of enzymatic reactions involve changes in environmental conditions (e.g. pH), the nature or availability of substrate(s), the ionisation state of prototropic groups, kinetic constants of the reaction (e.g. K_m), the association of the subunits of oligomeric enzymes, and the thermal sensitivity of substrate(s). The effect of temperature on enzyme reactions is related intimately to the composition of the medium. Ionic strength, a_w, and pH can exert profound effects on the temperature response of enzyme activity. Therefore, although the general effects of temperature on enzyme reactions will be presented here, the reader should bear in mind that the effect of temperature on, a specific enzyme depends on the environment. For example, an enhanced thermal stability and deviation from first-order kinetics destruction is observed for endogenous pectinesterase activity in orange juice as solids are increased from 10°–30° to 35°–50°Brix. In addition, alkaline phosphatase becomes less heat resistant when lactose is removed from milk by ultrafiltration.

Table 3.2. Rates of reactions as influenced by temperature and concentration of solutes during freezing.

	Change in rate of reaction caused by			
Situation	*Lowering of temperature (effect T)*	*Concentration of solutes (effect S)*	*Relative influence of the two effects*	*Total effect of freezing on reaction rate*
1	Decrease	Decrease	Cooperative	Decrease
2	Decrease	Slight increase	T > S	Slight decrease
3	Decrease	Moderate increase	T = S	None
4	Decrease	Great increase	T < S	Increase

Effect on Enzyme Activity and Stability

Increasing temperature

Although many enzymes are described as having an optimum temperature for activity, this optimum is defined within certain constraints. These constraints can be appreciated most easily by referring to Fig. 3.6, which shows the progress of a hypothetical enzyme reaction as a function of temperature. In this example, enzyme 'reactivity' can be measured in at least three ways: initial velocity and the extent of reaction at times (t) 1 and 2.

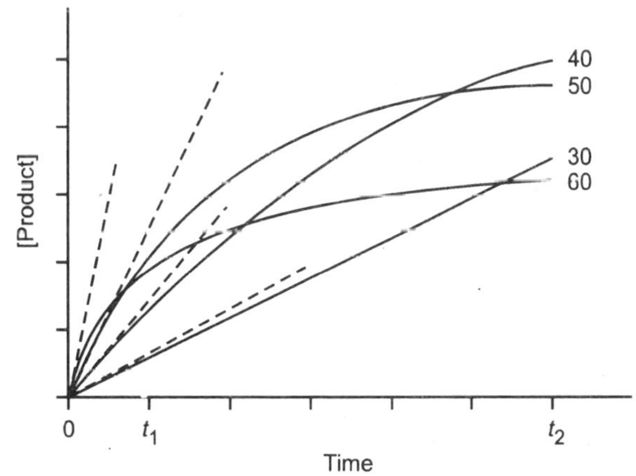

Fig. 3.6. Time course of the reaction of a hypothetical enzyme as a function of temperature (°C). Initial velocities are represented as dashed lines.

If one uses initial velocity as a criterion, the temperature optimum is identified as 60°C. However, if t_1 or t_2 is used, the temperature optima for the reaction are identified as 50° and 40°C, respectively. The decline in reaction progress with time (*dp*/*dt*) at the higher temperatures can be explained by the progressive inactivation of enzyme during incubation, provided the substrate does not become rate limiting. Therefore, in a food processing setting, optimum temperature for a reaction often should be identified in the context of the extent of substrate transformation to product desired.

The opposing effects of increasing temperature on reaction rate acceleration and enzyme inactivation are illustrated in Fig. 3.7 for the caseinolytic activity of a protease isolated from melon fruit. The enzyme is stable (after a 10 minutes incubation) at temperatures up to 50°C, whereas the optimum temperature for activity (amount of product formed after 10 minutes) is defined as 70°C.

Often, the Arrhenius plot is used to analyse the temperature dependency of enzyme activation and inactivation. When the rate constants of an enzyme reaction are being compared, the logarithm of the Arrhenius equation (Eq. 3.1) is taken and integrated between the limits of k_1 and k_2 and a specific temperature range [T (K)] to yield the following relationship:

$$\log\frac{k_2}{k_1} = \frac{E_a(T_2 - T_1)}{2.3\,RT_2T_1} \qquad \text{... (3.2)}$$

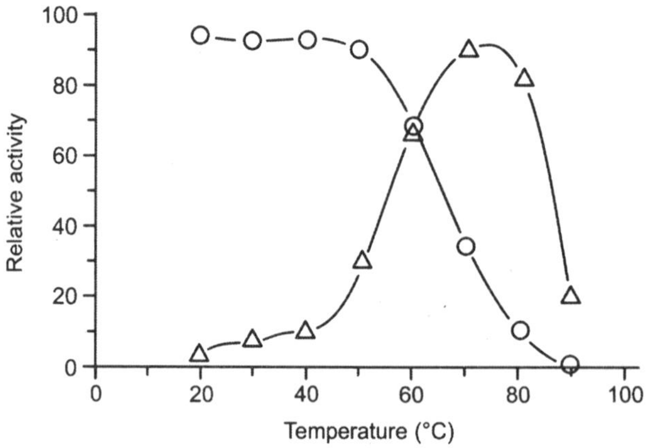

Fig. 3.7. Caseinolytic activity (Δ) and stability (O) of a 54-kDa protease isolated from melon fruit as a function of temperature.

The slope of a plot of log k vs $1/T$ has a value of $\pm E_a/2.3R$. (This example is for illustrative purposes only and does not advocate the use of two points to determine a straight-line for calculation of an E_a value for inactivation.) Using the data on enzyme activity in Fig. 3.7 and transforming to an Arrhenius plot yields Fig. 3.8. Taking the slopes of the plot on either side of the temperature 'optimum' allows the calculation of $E_{a\,act}$ and $E_{a\,act}$ as 13 and 37 kcal/mol, respectively.

This pattern is generally representative of all enzymes; the temperature coefficient for thermal inactivation is greater than that for the acceleration of reaction rates. The range of values for $E_{a\,act}$ and $E_{a\,act}$ for most enzymes is 6–15 and 50–150 kcal/mol, respectively. This feature accounts for the general observation that, as temperature increases, reaction rate increases exponentially to a point, after which it declines more rapidly (Figs. 3.7 and 3.8). Although the Arrhenius plot defines the ordinate as log k, any reaction rate parameter that is directly proportional to k (e.g. V_{max}) can be used, as long as the relationship between the substituted parameter and k does not change over the temperature range evaluated.

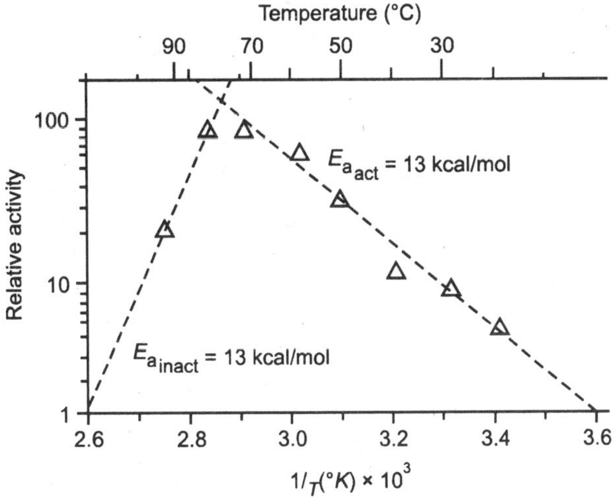

Fig. 3.8. Arrhenius plot of caseinolytic activity of a melon protease as a function of temperature. Data are from the activity curve of Fig. 3.7.

A better alternative to using the Arrhenius plot of enzyme activity described in Fig. 3.8, exists to evaluate the temperature effect on enzyme inactivation. The approach described in Fig. 3.8 does not distinguish between the effects of temperature on reaction activation and enzyme inactivation. A method that evaluates only the effect of temperature on enzyme deactivation involves preincubation of the enzyme solution at preselected temperatures and times followed by assaying residual enzyme activity under a standard (and often optimum) condition under which no further enzyme deactivation is expected. When residual enzyme activity is compared with incubation time on a semi-log plot, often straight-line relationships exist over a range of temperatures, indicating that enzyme thermal inactivation can be modelled as a first-order process. The slope of the plots at each temperature is proportional to the rate constant k for inactivation. The k values then can be presented on an Arrhenius plot and the E_a for enzyme inactivation determined.

Temperature sensitivity for enzyme inactivation is related directly to the water content of the system. Enzymes in dry or semi-moist food systems tend to be more heat stable. As water is removed near the monolayer of an enzyme, the enzyme can become remarkably stable at 100°C, as illustrated for lipase. Water is required to facilitate the unfolding of proteins during thermal denaturation. The enhanced thermal stability of enzymes in dry systems has potential food applications. This tendency is illustrated in the comparison of the temperature dependencies of hydrolytic and transesterification activities of lipase in aqueous and non-aqueous environments, respectively (Fig. 3.9). Transesterification reactions can be achieved at high temperatures in non-aqueous media, taking full advantage of the temperature coefficient of the reaction, whereas hydrolytic reactions possess a lower temperature optimum.

Although increased temperatures often are used to denature enzymes, several cases show it to be used to potentiate enzyme activity thermally and yield beneficial consequences. Probably the most common example used in the food industry is the preincubation of green beans at moderate temperatures prior to canning or blanching. This step potentiates endogenous pectin methylesterase activity and the attendant calcium reactivity with deesterified pectin polymers to induce a firming of the tissue and prevent sloughing in the final product. The same effect has been reported in other vegetable tissues, including potatoes and carrots. Similarly, the control of exogenously added α-amylase activity at 70°C has been reported to improve the consistency of a sweet potato puree product.

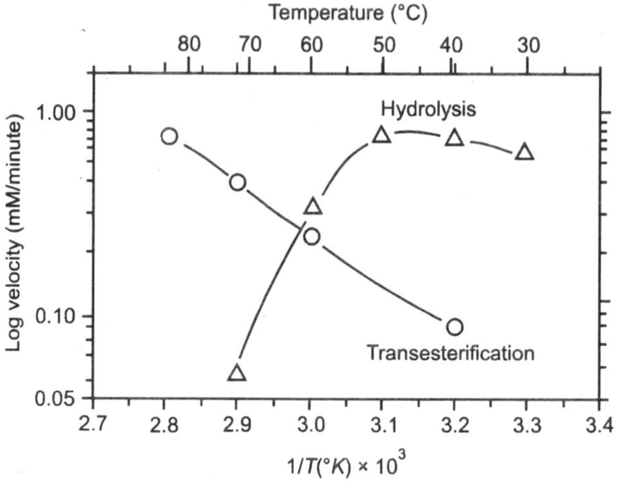

Fig. 3.9. Arrhenius plots for the hydrolysis (Δ) and transesterification (O) reactions catalysed by pancreatic lipase. Hydrolysis was measured on a 2 per cent tributyrin suspension at pH 8.0 and transesterification was measured with myristic acid and 5 per cent butter oil in hexane without the addition of water.

The thermal sensitivities of enzymes also have been exploited as indicators of adequate thermal processing of food materials. Peroxidase and alkaline phosphatase are indicator enzymes commonly used to insure the adequate blanching of vegetables and the pasteurisation of dairy products, respectively. Proper and timely estimations of residual peroxidase and alkaline phosphatase activities are important in these situations, since both are known to renature or become reactivated after thermal processing.

In general, enzymes with low molecular weights, a single polypeptide chain, and the most extensive disulphide cross-links (cystine residues) are more thermally stable than others. The mechanisms of thermal inactivation have begun to be investigated. Protein unfolding appears to be responsible for the initial loss of enzyme activity during exposure to high temperatures.

However, irreversible inactivation appears to depend on the extent to which chemical reactions are allowed to proceed during extended high temperature exposure. Some of the chemical reactions involved in irreversible thermal inactivation are deamidation of asparagine residues, hydrolysis of Asp–X peptide bonds, and the destruction of cystine residues (β elimination). On cooling, the refolding of 'thermally scrambled' enzymes to yield inactive structures can also contribute to irreversible inactivation. The relative contributions of these various mechanisms to thermal inactivation are pH dependent. This type of fundamental information is critical to the design of enzymes with enhanced thermal stability. The replacement of asparagine and aspartate residues by genetic engineering with residues that are less chemically reactive or have greater nonpolar character should enhance enzyme stability. Polyhydroxylation and glycosylation are proposed chemical modifications that can enhance enzyme thermal stability.

Decreasing temperature

Low and non-freezing temperatures also can have profound effects on enzyme activity. Generally, the prediction of reaction rates at low temperatures follows the relationship described by the Arrhenius equation (Eq. 3.1). However, several cases of deviation from this predicted behaviour are known.

Phosphofructokinase (PFK) from potato tubers has been shown to undergo a reversible dissociation of sub-units at 0°–5°C. The low temperature is likely to weaken the hydrophobic interactions involved in the assembly of the PFK sub-units. The resultant impairment of enzyme activity and, thus, glycolysis may be partially responsible for the tendency of potato tubers to accumulate sugars during low-temperature storage. This low-temperature sweetening is well known to be associated with the propensity of cold-stored tuber stocks to darken extensively during thermal processing.

'Breaks' in the Arrhenius plots at low temperatures have been observed also. Such breaks imply a change in the physical state of an enzyme or its microenvironment, so the reaction becomes impaired (has a higher E_a) at low temperatures. The relevance of this observation to food science is in the proposed roles that membrane rigidification and the resultant impairment of membrane-associated enzyme activities (e.g. mitochondrial oxidase activity) have in the development of chilling injury in sensitive fruits and vegetables. Whether or not these 'breaks' truly represent a 'change in state' of a membrane enzyme remains to be determined. However, a possible explanation for such 'breaks' at low temperatures may be the temperature dependence of K_m for some enzymes. K_m for ATP of a membrane-associated ATPase increases with increasing temperature. Thus, when reaction velocity at a fixed level of substrate is plotted on an Arrhenius plot, a 'break' occurs at 20°–25°C, showing a higher E_a at the lower temperatures. In this case, the 'break' is more likely to be a function of the dependency of reaction velocity on a change in $[S_0]/K_m$ than of a temperature-dependent change in the intrinsic activity of the ATPase. This example reemphasises the need to use the appropriate values for the ordinate on an Arrhenius plot.

Other Effects

As already mentioned, temperature influences the ionisation potential of prototropic groups of the enzyme, which may influence activity.

Temperature also can affect the disposition of the substrates for a reaction. In multienzyme systems, competing reactions may influence the overall pattern of reactivity as controlled by the relative temperature dependencies of the constituent enzymes. Substrates or cofactors that are temperature sensitive may also become rate limiting to reactions at high temperatures. Temperature also affects the solubility of substrates. Although substrate solubilities generally increase with increasing temperature, an exception to this trend is exhibited by gases. Therefore, the reactivity of oxygen-requiring enzymes may be attenuated at high temperatures not only by a direct effect on intrinsic enzyme activity or stability, but also by a decrease in the availability of substrate.

One last example of the effect of temperature on substrates can be found for enzymes acting at interfaces. Lipases are generally more active on acylglycerides in the fluid state. This feature is central to the concept of directed transesterification reactions in which reaction equilibria can be 'pulled' toward the formation of acylglycerides that solidify below a specific reaction temperature, thereby effectively removing them from the reaction mixture. Phospholipases exhibit a somewhat different dependency on the physical state of phospholipid micellar substrates. Reactivity appears to be most efficient when a mixture of liquid-phase and solid-phase phospholipids exists. This tendency is demonstrated in Fig. 3.10, which shows the dependency of K_m and V_{max} on the phase transition behaviour of ternary dispersions of dipalmitoylphosphatidylcholine with a 0.18-mol fraction of 1-palmitoyllysophosphatidylcholine plus palmitic acid.

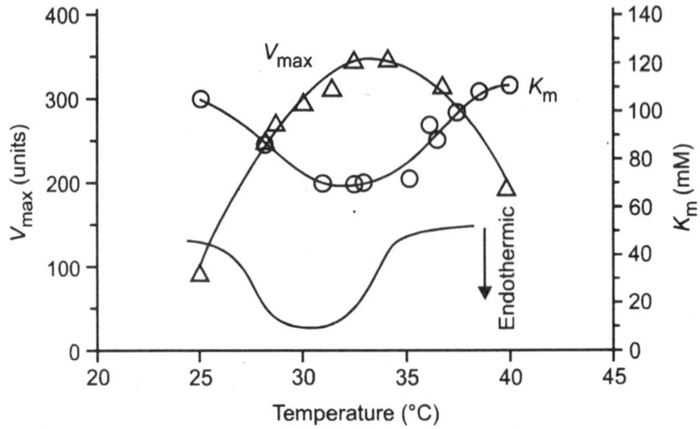

Fig. 3.10. Effect of the physical state of a phospholipid-fatty acid ternary dispersion on the kinetics of phospholipase A_2 activity. Bottom curve represents the endothermic profile of the lipid dispersion.

EFFECT OF OTHER ENVIRONMENTAL CONDITIONS

Viscosity, sonic energy, pressure, and irradiation are other environmental parametres that can influence enzyme reactivity. Viscosity probably plays a role in the reactivity of enzymes in frozen foods since more than 90 per cent of the 'free' water is frozen at commercial storage temperatures. Although the direct evaluation of the effect of viscosity in frozen systems is obviously problematic, the observations of enzyme systems in non-frozen media provide some insight. Invertase activity is known to deviate from the behaviour predicted by the Michaelis-Menton relationship (substrate saturation and attainment of V_{max}) at sucrose levels above 6 per cent. Bowski concluded that this phenomenon could be explained by the increased viscosity of the system and by a direct inhibitory effect of sucrose on the enzyme.

Pressure also can have an influence on enzyme stability. Very high pressures (up to 6100 bar) often are required to inactivate enzymes. This requirement is likely to preclude any widespread commercial application of pressure to control enzyme activity in foods.

Sonic energy can be used to inactivate enzymes. Sonic energy-induced enzyme inactivation is most likely caused by cavitation and the resultant interfacial denaturation of proteins. This effect has been shown for the protease papain. The extent of destruction of papain activity by sonic irradiation is inversely proportional to protein concentration and cannot be characterised by a first-order process.

Because of the resurgent interest in irradiation, some general comments on its effect on enzyme activity are in order. The damaging effect of ionizing radiation is directed randomly and is based on probability. Larger molecular weight structures are more likely to be damaged by ionizing radiation than smaller molecular weight structures. For enzymes, first-order rate constants for inactivation by ionizing radiation increase with increasing molecular weight. Thus, the eradication of micro-organisms in spices and the prevention of sprouting in vegetables can take place at doses of irradiation lower than those required to inactivate enzymes fully. DNA presents a larger 'target' for γ-rays than most enzymes.

Several enzymes have been shown to be sensitive to γ-irradiation in model systems. Dry enzymes are more stable than those in solution. For example, 7 Mrad are required to inactivate 63 per cent peroxidase activity in a 'dry' system, whereas only 0.24 Mrad are required to inactivate 85 per cent peroxidase activity in a moist system. This result is caused by the added effect of the production of radiolytic products from water, resulting in the indirect destruction of enzymes by ionising radiation.

Enzyme damage is enhanced at high oxygen tensions also, because of the production of oxy radical species. The secondary radiolytic products formed from water and oxygen also contribute to the continuing loss of enzyme activity often observed during postirradiation incubation.

In foods, the effect of ionizing radiation is less predictable. Although endogenous enzymes often are destroyed, their activity in foods depends on several other factors. The destruction of inhibitors, activators, substrates, or cofactors can affect residual enzyme activity. In addition, the radiation-induced damage to endogenous proteases may impair the natural turnover process for endogenous enzymes, allowing some to accumulate to levels higher than those observed in non-irradiated tissues. Finally, ionising radiation is certain to result in some degree of cellular damage and the possible potentiation of enzyme activities on previously segregated substrates. One study illustrates the complexity of the effects that γ-irradiation has on food systems. Sweet potato roots irradiated at 0.1–0.2 Mrad showed a transient increase in β-amylase, sucrose synthase, sucrose-phosphate synthase, phosphoglucomutase, phosphoglucose isomerase, and phosphorylase activities, and an increase in sucrose content during the post-irradiation period. Higher levels of irradiation (0.3–0.5 Mrad) led to a decrease in these enzyme activities, although the sucrose content still increased during the postirradiation period.

Thus, to exercise maximum control over enzyme-mediated food modification, a thorough understanding of the environmental parameters that can be manipulated and how they affect enzyme action is required. In tissue foods, the manipulation of the environment is limited to rudimentary parameters such as humidity (a_w), temperature, and the 'immobilisation' of water by freezing. Even these simple controls can have a great impact on maintaining the quality of foods. The proper conditioning of muscle into edible meat relies on the fine balance of control of temperature, rate of postmortem glycolysis, and progressive decline in pH. The proper control of temperature and humidity during the melting of barley in preparation for mashing results in the accumulation of endogenous amylolytic activities and the partial hydrolysis of the starch granule. The application of water (increasing a_w) to dates is a simple yet partially effective means of delaying the 'sugar wall' defect.

In non-tissue food systems, added dimensions of reaction control are available. Exogenous enzymes of varying thermal stability can be chosen on the basis of the extent of activity that is desired, an example being the use of α-amylase from various sources in dough conditioning and bread baking. Water content can be changed to control or alter the progress of an enzyme reaction; acidification has long been used as a mean of attenuating enzyme action in food systems. However, with the ability to manipulate the environment comes the added complexity of the interaction between environmental factors. The effects of pH, temperature, a_w, and other environmental conditions on enzyme action are interdependent. Thus, the optimisation of any enzyme-mediated food modification requires the integration of all the environmental effects on enzyme action for any process to be achieved consistently and uniformly.

Chapter 4

Modern Methods of Enzyme Expression and Design

INTRODUCTION

Until the advent of genetic engineering, enzyme producers were limited in their ability to produce innovative products for the marketplace. They were constrained to isolate enzymes from organisms approved for the food industry. The desired characteristics could be enhanced only using classical mutagenesis techniques. When these methods failed, no alternatives were available. Commercialisation depended on incremental yield improvements gained by continuous programmes of strain development. The ability to use recombinant DNA (rDNA) techniques has removed many of these barriers.

Enzyme producers recognised early the potential for commercialisation of new products using genetic manipulation. They worked with a wide variety of single-celled organisms that were simpler and, thus, easier to understand than the higher orders of plants and animals. The organisms already were well characterised for growth and expression rates. Short life cycles allowed rapid testing. These systems were ideal for genetic manipulation using rDNA techniques. Genetic engineering, combined with an understanding of biocatalysis to predict alterations for enzyme improvements, is revolutionising the production and use of enzymes in the marketplace. Offering a recombinantly produced product represents the culmination of a long and complex effort on the part of a multitude of disciplines: molecular and microbiology, X-ray crystallography, enzymology, protein and organic chemistry, biochemistry, fermentation and formulation engineering, assay chemistry, and technical service/applications, marketing, and sales. Because of the variety of disciplines required, a 'critical mass' is needed to innovate products successfully and bring them to market. The continued proliferation of novel enzyme products requires development of core technologies so complex and expensive that they can be justified only if they can be used in multiple applications. In addition, companies pursuing rDNA technology must consider regulatory issues, ownership protection, and consumer acceptance.

The purpose of this chapter is to review the current art and science of enzyme production in a simplified and understandable manner; to describe genetically engineered (GE) products available in the marketplace; and to present the power and implications of genetic engineering for new food-related applications.

BACKGROUND

In the late 1800s, discoveries in biology, chemistry, and enzymology culminated in what is now recognised as the beginning of the biotechnology age. On one hand were researchers who concentrated on understanding the mechanics of life. On the other were chemists who sought to elucidate the structures of organic products. At that time, these disciplines were considered vastly different.

The first theories concerning life functions were drawn from observations of natural systems. Two of the earliest contributors were the German microscopists Matthias Schleiden and Theodor Schwann, who proposed that all plants and animals are composed of fundamental units called cells. Gregor Mendel proposed the basic rules of heredity, that traits were controlled by elements we call genes. As early as 1908, the gene-enzyme relationship was postulated by A.E. Garrod, who suggested that genes work by controlling the synthesis of specific enzymes.

As analytical procedures improved, progress was made in defining the actual structure of proteins. Early in the 20th century, German chemist Emil Fischer established that proteins were polymers of nitrogen-containing organic molecules called amino acids. Proteins could contain up to 20 different amino acids in a highly varied and irregular order, but any given type of protein maintained a unique amino acid sequence. If the sequence determined the functionality of the protein, then a mechanism for choosing and ordering the sub-units was needed in the cell.

Since all enzymes were shown to be proteins, the key question became how genes participated in the synthesis of enzymes. Although it was known that genes possessed a unique molecular constituent, deoxyribonucleic acid (DNA), there was no way to show that DNA carried genetic information. James D. Watson and Francis Crick sorted out the molecular structure of DNA. More importantly, they recognised and proposed its function: the storage and replication of genetic data in the sequence of the base pairs. The tools were close at hand for manipulating genes in organisms, producing proteins of interest, and altering them for different characteristics. Methodologies were soon developed to implement the postulated ideas.

Our understanding of enzyme catalysis also advanced. In the early 1950s, Frederick Sanger established the first definitive order of amino acids in a protein by sequencing insulin. In 1967, D.C. Phillips and colleagues, knowing the primary structure and the three-dimensional conformation, proposed a precise chemical mechanism for the hydrolysis of a substrate by an enzyme (lysozyme). With succeeding work on other enzymes, this effort indicated that the chemical mechanisms of enzyme-catalysed reactions could be understood.

Simplified Model

Crick, hypothesised a central dogma: that chromosomal DNA functions as the template for synthesis of ribonucleic acid (RNA) molecules (transcription), which move to the cytoplasm where they determine the arrangement of amino acids in proteins (translation):

The arrows indicate the direction for transfer of genetic information. The arrow encircling DNA shows that it is the template for its own replication. The arrow between DNA and RNA shows that RNA molecules are made on (transcribed off) DNA templates. Correspondingly, all protein sequences are dictated by (translated on) RNA templates. Thus, the genetic code, the relationship between nucleotide sequence and amino acid sequence, has been deciphered and appears to be structurally and functionally identical in all living organisms: microbes, fungi, plants, and animals.

The DNA molecule consists of two molecular chains coiled in a helix and held together by hydrogen bonds (a weak non-covalent chemical bond). The subunits of each chain are nucleotides, which contain phosphoric acid and the sugar deoxyribose coupled to one of four nitrogenous bases: adenine (A), guanine (G), cytosine (C), and thymine (T). In the double helical DNA molecule, A always pairs with T and G always pairs with C in the opposite chain. Watson and Crick first postulated the significance of this base pairing: the sequence of one strand exactly defines that of its partner and allows self-replication.

A gene is a DNA segment that carries information for synthesis of a specific RNA molecule and, therefore, usually of a specific protein (Fig. 4.1). The gene defines the amino acid sequence in the form of triplets of bases, called codons. The gene also contains various signal or control regions, called promotors and terminators, that are involved in the regulation and expression of the gene product (when and how often the RNA corresponding to the protein will be synthesised). Start and stop codons define the beginning and end of the amino acid sequence. The terminator region defines the end of the gene in the DNA molecule.

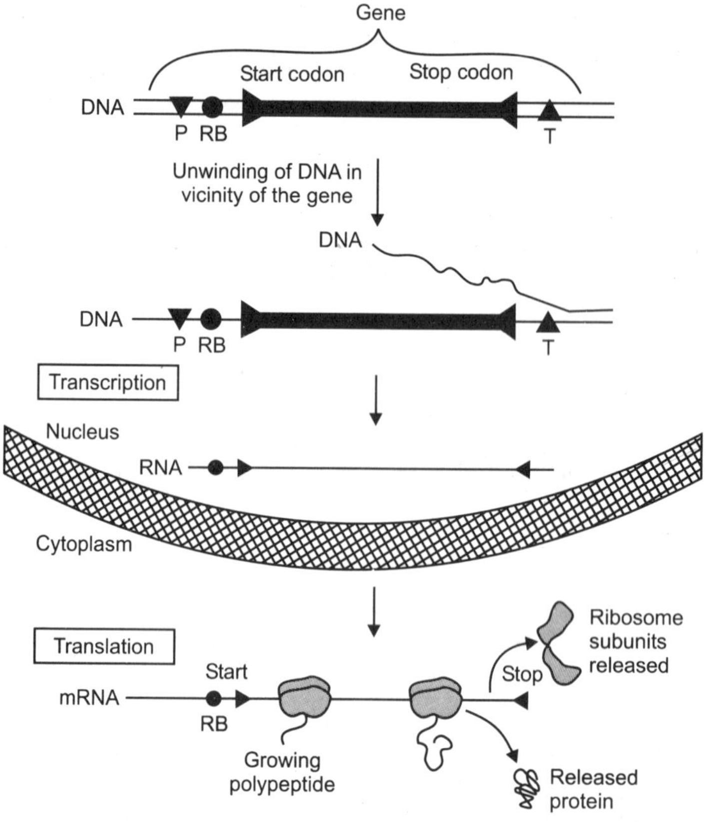

Fig. 4.1. Expression of a gene to produce protein. P, promoter; RB, ribosome binding site; and T, terminator.

Protein production requires several steps. First, the double-stranded DNA unwinds and separates in the vicinity of the gene. Specific enzymes move along a single strand, 'reading' the bases. Using base pairing, they construct another nucleic acid molecule, RNA, as a complementary copy of the information in the original sequence (a process called transcription). RNA is chemically similar to DNA, but contains the sugar ribose rather than deoxyribose and the base uracil (U) rather than thymine. Once RNA containing genetic information from DNA is constructed, it moves from the nucleus to the ribosomes located in the cell cytoplasm and functions as a template for protein synthesis (translation). When the RNA reaches the cytoplasm, it is called messenger RNA (mRNA). The ribosomes attach themselves at the ribosome binding sequence and synthesise the protein from individual amino acids available in the cytoplasm.

Classical Mutagenesis

Since DNA contains all the information required for proper growth and maintenance of the cell, copying DNA must be essentially foolproof to insure correct reproduction and replication of the organism. Mutation is a change in DNA, a variation in the nucleotide sequence, that is passed to succeeding generations. This process happens continuously in nature at very low levels. In the laboratory, mutations can be induced by:

1. Errors in base pairing at transcription or translation.
2. Interference with the enzymes that synthesise or repair DNA.
3. Mutagens, chemicals that interfere with DNA function (an analog incorporated into DNA instead of a base).
4. Exposure to ultraviolet radiation that is absorbed by the DNA and damages it.
5. Direct chemical attack on the DNA.

A programme of directed mutation plays an important role in enzyme production improvement. Such a programme is very effective and used extensively. Most production organisms were developed this way. Mutagenesis programmes in the past have given a steady increase in productivity over time by selecting higher yielding mutants or preferential producers of a selected activity.

The mutagenesis process involves stressing an organism under near-fatal conditions and growing the survivors, which often show a relatively high rate of mutation. A successful programme requires a predictive assay for identification and selection of promising mutants as well as rapid laboratory methods. Modern manufacturers often use automated methods with robots performing the tedious procedures.

The main disadvantage of classical mutagenesis is that it is random and time-consuming. Organisms can be manipulated and the selection process strictly adhered to, and still the objective is not achieved. Also, a characteristic cannot be manipulated if it is not expressed in the organism at all. Classical methods combined with genetic engineering offer a more complete approach to enzyme improvements.

GENETIC ENGINEERING

Recombinant DNA Techniques

In the late 1970s, a set of methodologies emerged that permitted isolation of discrete DNA segments that could then be inserted into living cells. The key to this method was the discovery of restriction and ligase enzymes to cut and join DNA. The ability to manipulate and recombine DNA allowed a more focused approach to mutagenesis, and created the potential for the manipulation of industrial micro-organisms to synthesise valuable new enzymes or food ingredients.

Gene cloning

Figure 4.2 summarises the steps in cloning (producing an identical copy of) a gene in a new host organism. The remainder of this section provides a more detailed description of those steps.

Step 1. *Obtain an enzyme of interest with desired characteristics*: Often enzymes with characteristics needed for a specific application are produced at low levels, in obscure organisms unsuitable for modern fermentation processes, or in hosts unacceptable for food applications. Such enzymes are likely candidates for cloning into production organisms.

Step 2. *Determine the amino acid sequence of the enzyme*: Automatic analysers are now available that perform the chemical analysis steps once done manually. Phenylisothiocyanate is used to react with the N terminal of the amide bond. Trifluoroacetic acid is added to cleave one amino acid residue at a time, releasing a cyclic compound. This isothiocyanate derivative is analysed, usually by high performance liquid chromatography (HPLC). The protein sequence then can be read by comparing the results against standards. Because of degradation of the protein in the system over time, less than 30 amino acids in a sequence can be analysed in a single run. Overlapping sections of the enzyme protein must be analysed so the sequence can be assembled. The process often takes weeks of effort; a degree of luck also helps accomplish the complete sequencing of an enzyme.

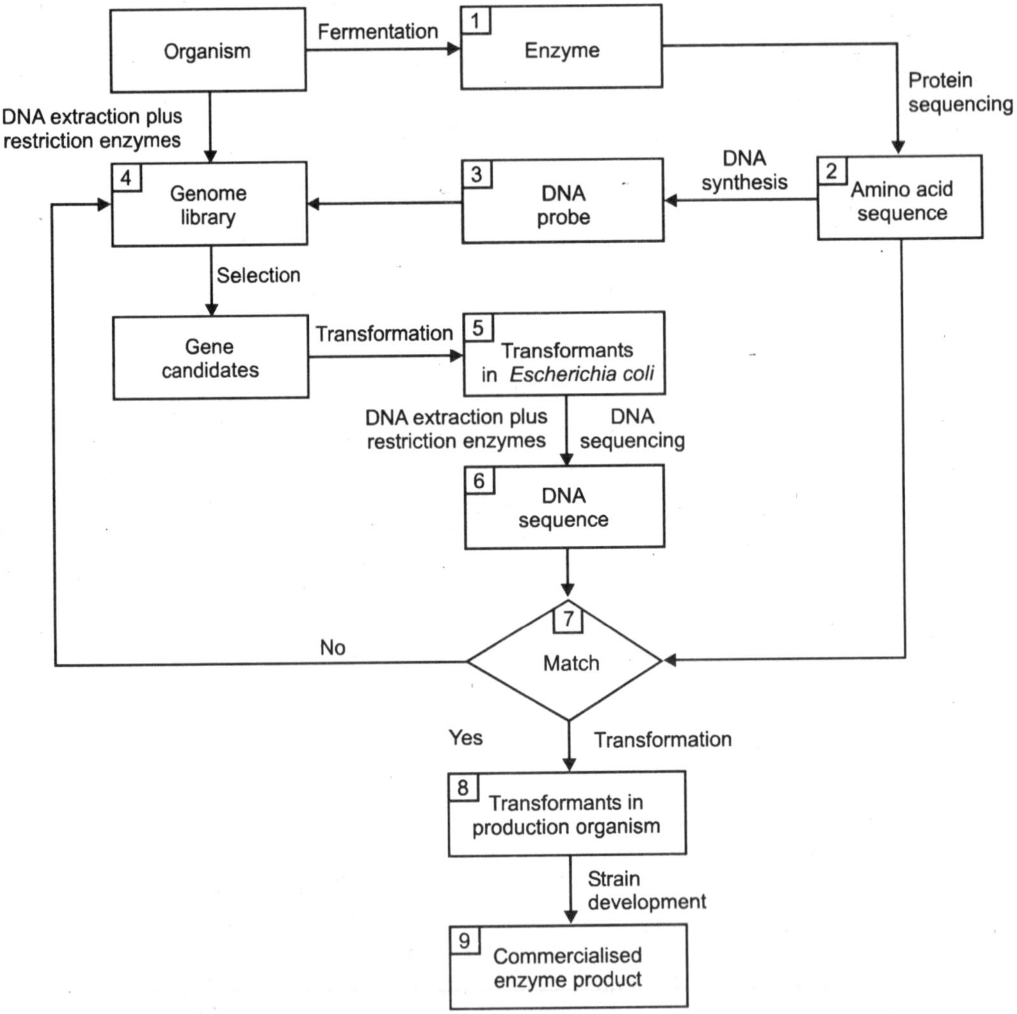

Fig. 4.2. Summary of steps to clone a gene into a new host. (See text for explanation.)

In the DNA molecule, three bases, called a codon, encode a particular amino acid. Using Table 4.1, a DNA sequence can be suggested to match an enzyme's protein sequence. The start codon is always

AUG, which also encodes methionine. (All polypeptides start with methionine.) Three separate codons (UAA, UAG, and UGA) serve as stop signals. Unfortunately, many amino acids are specified by more than one codon. This characteristic of the code is called degeneracy and makes tracing backward to find the correct DNA sequence of a gene difficult. The number of possible sequences can become astronomical.

Table 4.1. The genetic Code[a].

First position	Second position: U		C		A		G		Third position
U	UUU	Phe	UCU	Ser	UAU	Tyr	UGU	Cys	U
	UUC		UCC		UAC		UGC		C
	UUA	Leu	UCA		UAA	Stop	UGA	Stop	A
	UUG		UCG		UAG	Stop	UGG	Trp	G
C	CUU	Leu	CCU	Pro	CAU	His	CGU	Arg	U
	CUC		CCC		CAC		CGC		C
	CUA		CCA		CAA	Gln	CGA		A
	CUG		CCG		CAG		CGG		G
A	AUU	Ile	ACU	Thr	AAU	Asn	AGU	Ser	U
	AUC		ACC		AAC		AGC		C
	AUA		ACA		AAA	Lys	AGA	Arg	A
	AUG	Met	ACG		AAG		AGG		G
G	GUU	Val	GCU	Ala	GAU	Asp	GGU	Gly	U
	GUC		GCC		GAC		GGC		C
	GUA		GCA		GAA	Glu	GGA		A
	GUG		GCG		GAG		GGG		G

[a]Abbreviations for amino acids: glycine, Gly; alanine, Ala; valine, Val; leucine, Leu; tryptophan, Trp; methionine, Met; phenylalanine, Phe; isoleucine, Ile; serine, Ser; threonine, Thr; tyrosine, Tyr; asparagine, Asn; glutamine, Gln; aspartic acid, Asp; glutamic acid, Glu; lysine, Lys; arginine, Arg; histidine, His; cysteine, Cys; proline, Pro.

Some mutations (in nature or induced in a laboratory) may result in codon changes so a different amino acid is inserted into a protein. One base change, from UUU to UUG, changes the resultant amino acid from phenylalanine to leucine. A change from UCU to UCG, however, still specifies serine. Once in a great while, mutagenesis may improve catalytic activity and benefit the organism. Other changes give rise to stop codons that terminate proteins prematurely and confer no benefit at all.

Step 3. *Construct a DNA probe*: A probe is a single strand of DNA that encodes a section of the desired gene. A probe is made by synthesising a short segment of DNA using radio-labelled nucleotides. Since the probe will be used to find a specific gene in a pool of DNA, a sequence is chosen to match a unique section of the gene of interest. Sequences containing codons for methionine or tryptophan are often used because only one codon specifies each of these amino acids. Sometimes sequences can be guessed because of a species tendency to use some codons preferentially.

Step 4. *Screen a genome library*: A genome library is a pool of DNA segments that is likely to contain the gene of interest. Using standard procedures, DNA is extracted from the organism that makes the desired enzyme. The DNA is cut using restriction enzymes. Segments too small to contain a full gene are discarded. The mix is separated by size using gel electrophoresis. A quantity of the DNA probe is added and the material is warmed. Under controlled heating, the double-stranded DNA will 'melt' and disassociate. After cooling, it will come back together; however, the single-stranded probes will bind to the sections they match. If bands can be detected autoradio-graphically, they contain sequences similar to those of the gene being sought. The band is cut out and the DNA it contains is replicated.

Probing becomes more difficult when dealing with DNA from eukaryotic cells because of introns (noncoding sections of DNA in a gene). The presence of introns creates more complexity when working with higher species; this is one reason bacteria are preferred.

Step 5. *Transformation*: The process of genetic transfer of DNA is called transformation. DNA fragments are inserted into plasmids or phages, which are used because they are easy vectors with which to transfer foreign DNA into intact cells. The DNA of interest, inserted into the plasmid or phage, is mixed with organisms whose cell walls have been made receptive by chemical treatment. These organisms are usually special strains of *Escherichia coli* that have been characterised and modified carefully so that they can be transformed easily. The DNA enters some of the organisms and is replicated with the host DNA. The transformation process is random and is not successful for every organism in the experiment. Therefore, an important step is screening the population for the organism that has taken up the new DNA from among the many who have not.

Transformed organisms are selected by adding markers to the gene before transformation. Markers are sections of DNA inserted with the desired gene to make selection and detection of successfully transformed organisms a straightforward process. Often the marker confers an ability to survive under specific growth conditions and must be used with specially modified hosts that lack the specific trait. The markers will be removed before transformation of the gene into a production organism. Table 4.2 shows some common markers. The transformed organisms, which contain the DNA of interest, are grown to provide a source of recombinant DNA for further testing and manipulation.

Table 4.2. Selection of transformed mutants.

Marker	*Description*
Antibiotic resistance	Addition of a gene to produce a protein that hydrolyses an antibiotic and allows the organism to grow; candidates are cultured in the presence of the antibiotic
Supplementation of an auxotroph	Addition of a gene to supply a needed growth factor (a vitamin or specific amino acid) that compensates for a deficiency in the host whose gene has been deleted; candidates are cultured in the absence of the growth factor
Hydrolysis of a specific sugar	Addition of a gene to hydrolyse a specific carbon source (e.g. lactose); candidates are cultured on the single carbon source

Step 6. *Sequencing the chosen* DNA: The ability to sequence DNA was one of the more important breakthroughs that made genetic engineering practical. Two methods for accurately establishing the order of the four bases A, G, C, and T along a given DNA segment have been developed. DNA strands are separated (since only one strand needs to be sequenced) and labelled at one end by enzymatically removing the phosphorus atom and replacing it with a radioisotope of phosphorus. The solution of single-stranded DNA is then divided into four pools.

In Maxam and Gilbert's method, each pool is treated with chemical reagents, for example, dimethysulphate or hydrazine, using conditions that selectively destroy one or two of the four nucleotides in each pool. The conditions are adjusted so the reaction does not proceed to completion; at most one nucleotide per fragment is destroyed. These reactions result in four pools of radioactively tagged strands of all possible lengths, ending in the nucleotide destroyed. In practice, nucleotides destroyed are G, G + A, T + C, and C, each in a separate pool.

Sanger's method makes use of the enzyme DNA polymerase to produce fragments ending in specific bases. The pieces of single-stranded DNA serve as templates for synthesis of complementary strands of DNA. The nucleotides necessary to form the complementary strand are added to each of the four pools along with the polymerase enzyme. In addition, a small amount of a nucleotide analog, a radioactive 2′,3′-dideoxynucleotide, containing one of the four base types is added to each pool. The concentration is controlled so only one base will be incorporated per strand. The polymerase builds the complementary strand along the DNA fragments until one of the base analogs is inserted in the chain. Further strand elongation is blocked, since the analog interferes with placement of the next nucleotide. The result is four pools of radioactive complementary strands of DNA of all possible lengths, ending in a specific base. Gel electroporesis is used in both methods to separate each pool by length, with a size resolution of a single base. The gels are slabs of poly acrylamide with sample applied at one end. An electric potential causes the fragments to move through the gel in a track with the smallest fragment moving most quickly. The gel is then used to expose X-ray photographic film. The resulting autoradiograph shows four tracks of bands. Since the gel has a resolution of one base, and all possible fragment lengths exist in the four pools, the DNA sequence can be read directly from the gel (Fig. 4.3).

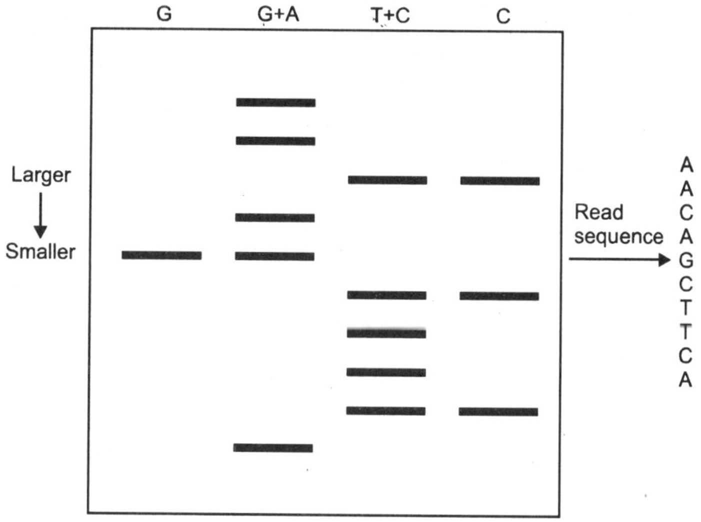

Fig. 4.3. Reading the sequence of a DNA fragment.

Because of the ease of the enzymatic method and the ready availability of dideoxynucleotide reagents, the Sanger method is becoming the method of choice for sequencing DNA.

Step 7. *Verify the DNA has the right sequence to make the enzyme*: The DNA sequence is compared with the known protein sequence of the desired enzyme. If the sequence does not match, a new genome library must be screened. The DNA fragment is examined to determine if the entire gene is present. Start and stop codons allow recognition of whether the gene is intact. With some additional effort, the gene can be isolated using restriction enzymes to cut and trim it until only the coding sequence remains.

Step 8. *Transform the gene into an appropriate host*: A toxin-free organism with a demonstrated high production capacity for foreign gene products is chosen. The highly produced native genes are deleted, while the transcriptional, translational, and secretory control elements are retained. The gene of interest is transferred into the chosen production organism. Transformed organisms are selected and cultured. The expressed enzyme is purified and analysed using general protein chemistry methods to insure that it is identical to the native enzyme.

Step 9. *Improve the yield to levels acceptable for commercialisation*: Living cells possess intricate control systems to insure efficient use of material and energy resources. Often, cell regulatory elements limit the amount of enzyme produced, making commercialisation economically unfeasible. However, changes in the control systems can be used to boost yield. This area is of great importance to enzyme producers and will be discussed in greater detail in a later section.

Instability of recombinant production strains is always a possibility. Monitoring of stock cultures is necessary to avoid costly production failures.

Complementary DNA

Another option for obtaining DNA suitable for enzyme production is the construction of complementary or copy DNA (cDNA). Messenger RNA is a complementary copy of a DNA sequence. When an organism is actively making a protein, reasonable amounts of mRNA are available in the cytoplasm and can be recovered. The mRNA can be used as a template to make a single strand of DNA using reverse transcriptase enzymes. Polymerase enzymes add the complementary strand, yielding a DNA fragment that encodes the desired gene. This DNA is not complete because it does not have introns or transcription signals (promotors), but often cDNA is the only alternative available for generating a gene.

Site-directed mutagenesis

The commercial advantages of transferring a gene from one organism to a more productive one are obvious. The first products made available to consumers are of this type. The techniques can be used to manipulate a segment of DNA of any length (even a codon, to change one amino acid in a protein sequence), although the benefit of small changes was not immediately obvious. Bott discuss the results of substituting a single amino acid for the glycine at the 166 position (in a peptide chain of 275 amino acids) in subtilisin from *Bacillus amyloliquefaciens*. The 166 position is not part of the catalytic site, but is implicated in substrate binding. Figure 4.4 shows the activity of the protease mutants containing various amino acids as well as that of the protease containing glycine. All 19 mutants retained measurable enzyme activity, but differed in activity from the naturally occurring enzyme. These results suggest the potential for the deliberate design of enzymes for specific applications.

Enzyme Design

Organisms evolve functionally optimal enzyme systems for their environment because doing so confers a survival advantage. However, the optimal enzyme system for chemical or food processes may require characteristics not generally found in nature. The greatest opportunity for improvement, therefore, lies in engineering those properties that have not been selected naturally.

Structure-function relationships

Subtilisins are among the most highly studied and well-understood enzymes. They are secreted in large amounts by a variety of Bacillus species. The protein sequences from several commercially important species and their three-dimensional structures have been determined. The enzyme mechanism is known, and the catalytic and substrate binding residues have been identified. Extensive kinetic and chemical modification and crystallographic studies have been performed. Subtilisin is an ideal model for the application of protein engineering to better understand the theoretical basis for catalytic systems.

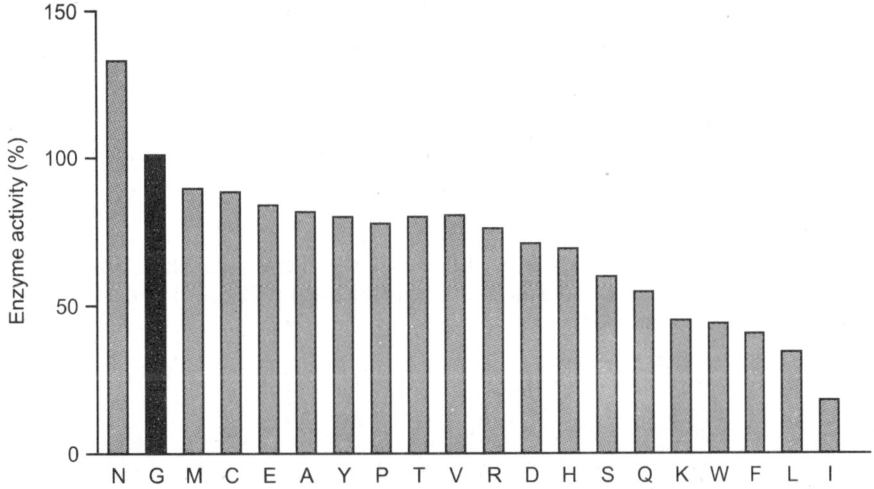

Fig. 4.4. Substitution of various amino acids in subtilisin. The glycine at the 166 position in subtilisin from *Bacillus amyloliquefaciens* was substituted with various amino acids using recombinant DNA techniques. The resulting protease mutants were assayed for their ability to hydrolyse casein at specific conditions. The activity of the native enzyme (glycine-166) is set to 100 per cent. Abbreviations: G, glycine; A, alanine; V. valine; L, leucine; W, tryptophan; M, methionine; F, Phenylalanine; I, isoleucine; S, serine; T, threonine; Y, tyrosine; N, asparagine; Q, glutamine; D, aspartic acid; E, glutamic acid; K, lysine; R, arginine; H, histidine; C, cysteine; P, proline.

X-ray crystallography provides a valuable tool for improving our understanding of structure-function relationships. The crystal structure can be incorporated into a three-dimensional modelling system that enables enzymologists to study the active site, the binding regions, how substrates fit, the possible conformations of the enzyme, and stearic hindrances of suggested changes. The active sites of similar enzymes from different organisms can be overlapped to see structural similarities that may correspond to similarities in function. By understanding the relationship between catalytic properties and structure, improvements in function eventually may be predicted for suggested changes in the structure.

Over 400 site-specific structural changes have been made in subtilisin to verify computer-modelled predictions of improvements to enzyme properties. Some results were predictable; others still need explanation. Almost every property of the enzyme has been altered by protein engineering: catalytic site, substrate specificity, pH/rate profile, and stability to oxidative, thermal, and alkaline inactivation. Examples from this work are presented in Figs. 4.5, 4.6, and 4.7.

Random mutagenesis

Enzymes are large and complicated catalysts. The three-dimensional structures of fewer than 100 (of ~10,000) have been determined to date. Painstaking study, prediction, and expression as described for subtilisin is not always feasible. Researchers with limited resources perform random mutagenesis on a selected gene followed by expression and testing. When little is known about the enzyme structure, numerous mutagenised enzymes are generated rapidly and screened for indication of change in the characteristics.

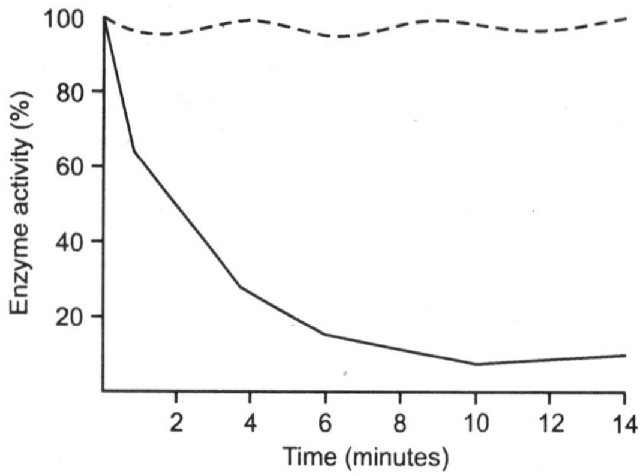

Fig. 4.5. Engineering resistance to oxidising agents. Effect of 1 M hydrogen peroxide (bleach) on the activity of the parent subtilisin, methionine-222 (——) and engineered mutant alanine-222 (----). The mutant has dramatically improved resistance to the oxidising agent.

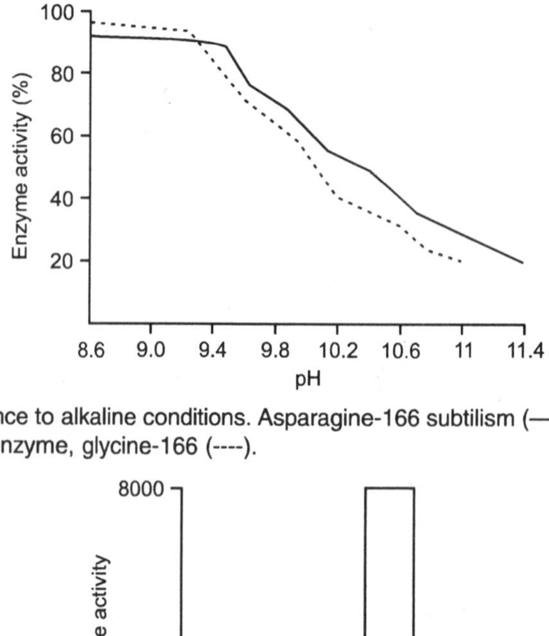

Fig. 4.6. Engineering tolerance to alkaline conditions. Asparagine-166 subtilism (——) works more efficiently at higher pH than the parent enzyme, glycine-166 (----).

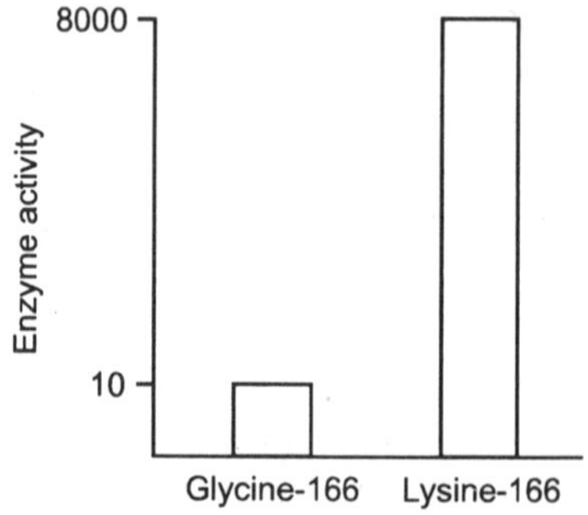

Fig. 4.7. Engineering higher efficiency, lysine-166 subtilisin hydrolyses a specific substrate 800 times faster than the parent enzyme (glycine-166).

The interesting mutants are sequenced to suggest the region in the enzyme that influences the characteristic. Although still a random process, this method is an improvement on classical mutagenesis, since it focuses on the gene of interest.

Because this approach is more precise, fewer resources and less time are required to yield an acceptable result. Successful enzyme engineering depends on the ability to screen for mutants. Development of a screen for rapid and accurate selection is central to the process of mutagenesis. Temperature stable mutants are among the first mutants commercialised from wild-type enzymes, because testing the enzymes at higher temperatures was an easy screen. Temperature stability demonstrated in the laboratory also scaled to the plant level. Some properties pose more difficulty in screening. One example is alteration of pH optima, because enzyme activities depend on the assay conditions (buffers, their ionic strength, and specific substrate conditions). Assays may not represent the variety of end-user processing conditions and substrates. Therefore, screening for commercialisable candidates requires extensive testing, which is time and resource consuming.

Table 4.3 shows some common methods for detection and selection of mutants that allow the researcher to find the transformed organism. If selection methods can be employed during the growth phase, the number of assays for detection of the desired enzyme can be reduced significantly.

Table 4.3. Some methods of detection and selection of mutants.

Characteristic	*Selection and detection*
Higher enzyme yield	Organisms are plated on agar with dye-linked substrate; higher producers give larger clearing zones; enzymes are tested for desired activity
Higher temperature stability	Enzymes are assayed for activity at desired temperature
Different pH optima	Enzymes are assayed for activity at the desired pH using a variety of buffers, substrates, concentrations, etc.
Ability to use a cheaper carbon source	Organisms are cultured only on the new carbon source; enzymes are assayed for desired activity
Resistance or tolerance to inhibitory chemicals (e.g. sulphite tolerance)	Organisms are cultured in the presence of the chemicals; enzymes are assayed for desired activity

Metabolic Pathway Modification

Pathway engineering is the process of modifying metabolic pathways in organisms by changing or adding enzymes to an existing pathway or by creating new ones. Food scientists can manipulate intermediate steps in the production of certain foods to alter final product characteristics. Many examples of pathway engineering encompassing numerous industries are recognised (Table 4.4).

Table 4.4. Examples of potentials for pathway engineering.

Introduction of nitrogen-fixing capabilities in certain plants to reduce/eliminate requirements for nitrogen-based fertilisers
Design of metabolic systems capable of degrading certain hazardous waste compounds
Control of methane-producing metabolism in bacteria to make fuel from organic material/waste
Introduction of gene encoding the enzyme necessary for ascorbic acid precursor synthesis in micro-organisms
Manipulation of pathways for the creation of phage-resistant microbial strains used by the food industry
Alteration of sugar utilisation pathways in microbes to allow fermentation of novel products

A specific example is the metabolic pathway engineering of *Lactobacillus bulgaricus* as a means of preventing postfermentation acidification of yogurt. Like many perishable foods, yogurt develops off-flavours during refrigerated storage. In this case, the phenomenon responsible is the continued production of lactic acid by the organism *L. bulgaricus* used in the production of yogurt. Specifically, lactose in the yogurt is converted to glucose and galactose by the enzyme β-galactosidase. Glucose, used by the organism to generate energy, results in the production of lactic acid, the metabolic end product, which is secreted into the yogurt and changes the flavour.

If lactic acid production were limited, yogurt would be stable under refrigeration. Yoast describe the *in vitro* random mutagenesis of the *L. bulgaricus* β-galactosidase gene using chemical mutagens. After the plasmid was cloned into an *Escherichia coli* expression system, enzymes were produced that had randomly changed amino acid sequences. They were screened to select for those that showed normal activity at yogurt production temperature (37°C), but little or no activity at refrigeration temperatures (<10°C). A number of mutant cold-sensitive enzymes were identified to replace the wild-type *L. bulgaricus* β-galactosidase. The genetically engineered host organism will exhibit little or no ability to acidify yogurt during refrigerated storage, resulting in longer shelf life, because no off-flavour develops.

Engineering for Production Improvement

Current enzyme production encompasses fermentation, recovery, purification, and formulation technologies. An in-depth understanding of the fermentation parameters for each organism is required to maximise growth and enzyme production. Submerged-culture batch fermentation predominates, but fed-batch and continuous fermentation are taking hold rapidly. A different set of conditions generally is required for efficient recovery and purification of each enzyme. As a result, recovery processes frequently are designed as dedicated trains, reducing plant utilisation and increasing capital investment.

Use of recombinant techniques has begun to influence how enzymes are manufactured. Innovations in rDNA techniques and protein engineering allow producers to design a strategy for future production strains: the development of well-defined and well-understood bacterial and fungal hosts for overproduction of foreign gene products. The ability to express new proteins in 'workhorse' organisms using common induction systems allows use of fermentation processes similar to existing schemes and rapid scale-up of new products. Such consistencies will expedite transition from the laboratory to the marketplace.

Knowledge of repression and induction systems permits generation of targeted alterations within the organism to develop more desirable fermentation processes. Organisms can be modified to contain expression systems that are triggered by temperature shifts, certain growth rates, low oxygen concentrations, or inexpensive inducing agents, as well as being insensitive to feedback inhibition from a build-up of media components over long fermentation cycles. Protein will be expressed during earlier growth phases (necessary for continuous fermentation processes in fungi). Use of rDNA techniques allows dramatic changes in fermentation process design and economics.

Several other molecular biology options exist to boost productivity. Yields can be increased by inserting multiple genes copies with a combination of promotors to achieve optimal secretion. Non-active sections of an enzyme can be deleted to reduce the biological burden to produce it in the cell. Sporulation controls can be modified to make an organism produce longer than it would in its natural cycle.

Synthesis of extra protein represents a waste of valuable energy and intermediary metabolites for the host. In hyperproducing organisms, the new gene product may constitute a large portion of the total cellular protein. Since this product serves no useful purpose for the organism, diversion of so much biosynthetic activity seriously limits the capacity of the cell for growth and even survival. New fermentation schemes and higher yields make organisms more sensitive to processing excursions. These organisms will demand a higher level of process control than previously required.

Alterations can be made in the protein products themselves to improve production. Yields of a protease that autolyses during fermentation and recovery can be improved by shifting the pH optimum of the enzyme away from the pH used during production. Changes in enzyme stability to temperature or certain compounds (oxidants) could allow wider selection of fermenter growth conditions. Amino acid substitutions that alter the surface charge may allow more efficient extraction, for example, by enhancing ultra filtration membrane efficiency. Production strategies that incorporate improved enzyme characteristics for end-user requirements and increased yield for improved economics, and that better utilise recovery and purification equipment, will be a major new focus for rDNA technology application.

COMMERCIALISATION OF GENETICALLY ENGINEERED PRODUCTS

In the biotechnology age, the ability to offer a new product to the marketplace successfully will require the combination of enzyme design and product information capabilities of enzyme producers, and application information, requirements, and acceptance of enzyme users and consumers. This section cites examples of products available in the marketplace and, for one specific case, describes some of the complexities in development and production.

Example

Genetically engineered (GE) products are now available in the marketplace and will continue to proliferate. Table 4.5 shows a listing of GE enzymes in commerce. This list excludes the large number of restriction enzymes and other small-scale enzymes currently in use. Products with applications for GRAS (generally recognised as safe) status with the United States Food and Drug Administration (FDA) have been noted.

Table 4.5. Genetically engineered enzymes in commerce.

Enzyme	*Source*	*Host*	*Use*
Subtilisin	*Bacillus* sp.	*Bacillus subtilis*	Detergents
Subtilisin	*Bacillus alcalophilus*	*Bacillus alcalophilus*	Detergents
Hydrolase	*Bacillus* sp.	*Bacillus* sp.	–
Lipase	*Humicola*	*Aspergillus* sp.	Detergents
α-Amylase	*Bacillus megaterium*	*Bacillus subtilis*	Starch modification
α-Amylase	*Bacillus stearothermophilus*	*Bacillus subtilis*	Starch modification
Chymosin	Calf stomach	*Escherichia coli*	Cheese production
Chymosin	Calf stomach	*Kluyveromyces marxianus* var. *lactis*	Cheese production
Chymosin	Calf stomach	*Aspergillus niger* var. *awamori*	Cheese production

An example of the application of genetic engineering to solve an enzyme supply problem is illustrated by developments in the production of milk-clotting enzymes for the manufacture of cheese. Historically, the prevalent milk coagulant has been calf rennet, derived from the fourth stomach of suckling calves. The principle active proteases in calf rennet are chymosin (EC 3.4.23.4) and pepsin (EC 3.4.23.1). These two enzymes are closely related—both initiate the clotting process in milk by cleavage of casein. Chymosin is produced predominantly only in the first weeks after birth; thereafter, it is replaced increasingly by pepsin. Of the two enzymes, chymosin is preferred by cheesemakers because of its limited proteolysis, leading to minimal bitterness in aged cheese.

Sometimes high-quality calf rennet is in short supply because of a decline in the market demand for veal and an increase in cheese production. These conditions have led to unstable rennet prices. Alternative sources of milk-clotting enzymes have been offered commercially to meet market demands, including microbial coagulants derived from *Mucor miehei*, *Mucor pusillus*, and *Endothia parasitica*. However, unweaned calf rennet, specifically chymosin with very low levels of pepsin, is considered the preferred coagulant by the cheese industry worldwide.

Cloning and producing GE chymosin by controlled fermentation can provide a consistent supply of enzyme with properties analogous to those of the native chymosin. The proteins are identical, despite the fact that the enzyme is produced in three different micro-organisms: the bacterium *Escherichia coli*, the yeast *Kluyveromyces marxianus* var. *lactis*, and the fungus *Aspergillus niger* var. *awamori*. Table 4.6 summarises the advantages and disadvantages of each expression system.

Table 4.6. Heterologous expression systems for calf chymosin.

Advantages	*Disadvantages*
Bacteria, especially *Escherichia coli*	
Well characterised production strain	Chymosin is deposited as insoluble refractile bodies and must be solubilised and renatured to be active
	Products from *E. coli* must undergo extensive purification prior to use in food
Yeast, such as *Saccharomyces cerevisiae*	
Chymosin is secreted as an active enzyme	Current published production levels are uneconomical
Safe production organism	
Filamentous fungi, such as *Aspergillus niger* var. *awamori*	
Chymosin is secreted as an active enzyme	Low frequency transformation system
Safe production organism	

The rest of this section describes the heterologous (cross-species) expression in a food-grade filamentous fungus of the gene encoding chymosin from the fourth stomach of a calf, and the subsequent processing of the gene product to a commercially useful product.

Fungi produce prodigious amounts of protein under proper fermentation conditions. For example, glucoamylase in *A. niger* is induced by the presence of maltose or isomaltose and can be secreted in excess of 50 g/litre in batch fermentations regulated by starch induction. Until recently, however, very little was known about how to transform fungi and use them as hosts. New expression systems had to be identified and developed. Since the bovine promotion system was not needed, the gene could be constructed from RNA (cDNA). Chymosin was secreted using expression units that contain prochymosin (the inactive chymosin precursor) cDNA coupled to the transcriptional, translational, and secretory control elements of the *A. niger* glucoamylase gene. Protein expression was regulated by starch induction and production processing was similar to that of glucoamylase.

Transformation

Expression and secretion of chymosin involved these steps.

1. Development of a host strain with markers for transformation experiments: *Aspergillus niger* var. *awamori* was chosen as a host because it is toxin free and a hyperproducer of glucoamylase. The organism was subjected to UV mutagenesis, which produced a mutation in the *pryG* gene, resulting in a requirement for uridine. Chemical mutagenesis of the same precursor produced a

strain with arginine auxotrophy (inability to synthesise a needed growth factor) caused by a mutation in the *argB* gene. A double auxotrophic strain was selected from the progeny of a parasexual cross of the two single auxotrophic strains.

2. Deletion of the host's protease, aspergillopepsin A, which causes degradation of chymosin or produces bitter peptids in cheese: To generate strains that were deficient in the production of aspergillopepsin A, a gene replacement strategy similar to that reported by Miller was employed. Briefly, the aspergillopepsin DNA coding sequence was replaced by the *Aspergillus nidulans argB* gene and used for transformation of the double auxotrophic strain. Of the transformants selected, four lacked the aspergillopepsin gene. Absence of aspergillopepsin activity was verified by several tests.
3. Construction of a plasmid vector containing a selectable marker and an expression unit for prochymosin: A vector called pGAMpR was constructed (Fig. 4.8) in which prochymosin cDNA sequences encoding the prochymosin B isozyme were fused to the last codon of the *A. niger* var. *awamori* glucoamylase gene. The selectable marker used in the vector was the *Neurospora crassa pyr4* gene, which corrects for uridine auxotrophy when its gene product (the enzyme orotidine-5′-monophosphate decarboxylase) is expressed. The gene product of this construction is a fusion protein consisting of the *awamori* glucoamylase enzyme covalently coupled to prochymosin.

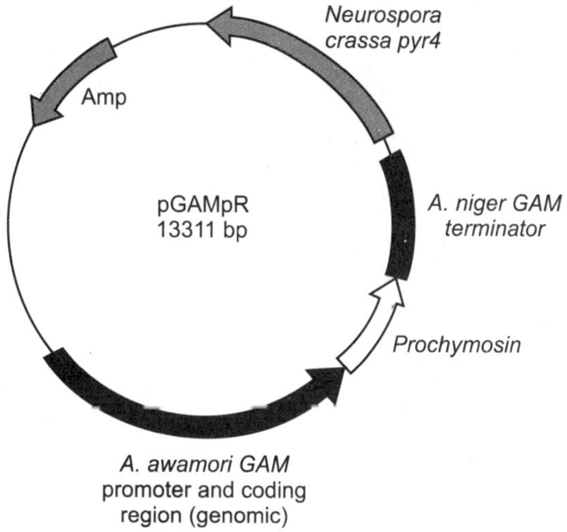

Fig. 4.8. Map of the glucoamylase-prochymosin fusion vector.

4. Transformation of the host strain with the expression vector and subsequent molecular analysis of the transformed strains: Aspergillopepsin-deficient strains were transformed and a number of chymosin-producing strains identified and selected. The highest producing strains were analysed and shown to contain the glucoamylase prochymosin expression unit integrated into the genome, possibly in multiple copies.
5. Mutagenesis and screening of improved strains: The strains were subjected to further rounds of classical mutagenesis, screening, and selection for overproduction of chymosin. In addition, all possible production strains were tested carefully to insure that they produced no mycotoxins or antibiotics.

Characterisation

The enzyme product of the transformed strain is chymosin B, an isozyme containing aspartic acid at position 290 instead of glycine as found in chymosin A. Chymosin B has a lower specific activity than chymosin A but is reported to be more stable. Both isozymes are found in natural products.

Figure 4.9 illustrates the production of mature chymosin by the recombinant organism. Only mature chymosin and mature glucoamylase are observed in the fermentation medium. This suggests that the fusion protein undergoes an *in vivo* processing step, resulting in the cleavage of glucoamylase from prochymosin B and the subsequent self-activation of prochymosin to chymosin. The cleavage of the two enzymes can be effected either by prochymosin itself or by other endogenous *awamori* proteases.

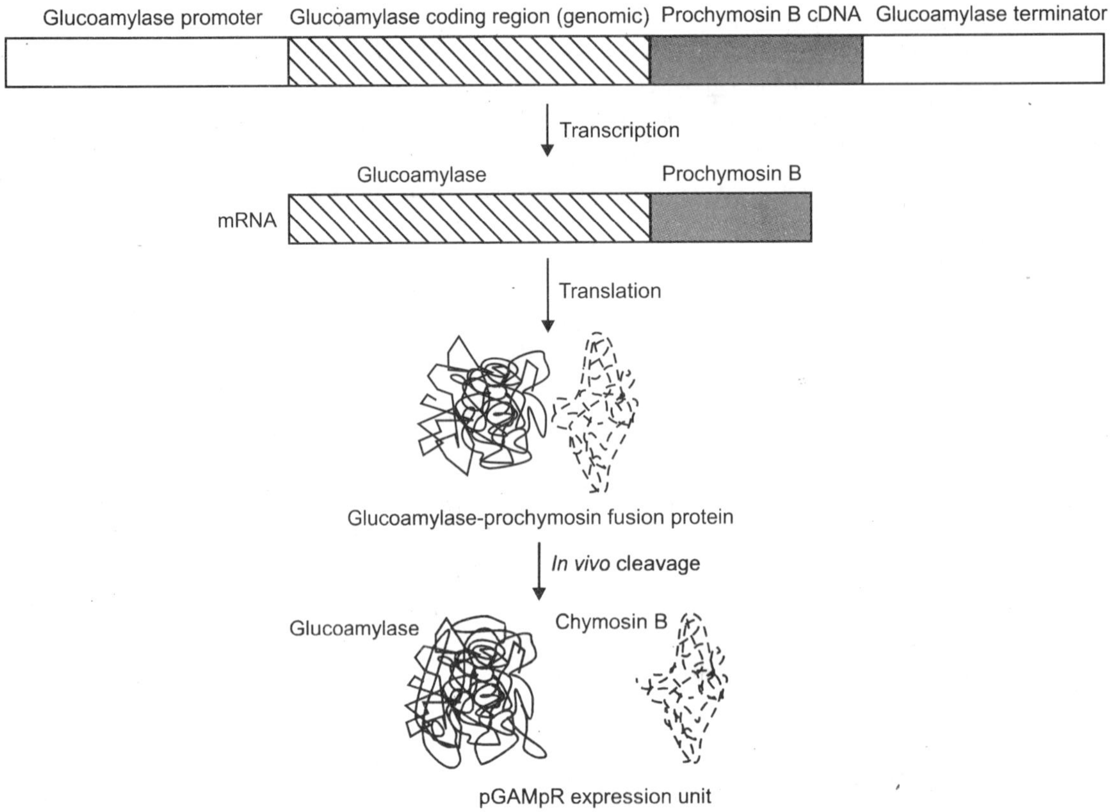

Fig. 4.9. Schematic model for the production of chymosin from the glucoamylase-prochymosin expression unit.

Chymosin B was purified using standard biochemical methods. The enzyme was compared exhaustively to animal-derived chymosin using physical, immunological, and functional tests. All results confirmed there was no significant difference between native and recombinant chymosin expressed in *A. niger* var. *awamori*.

Fermentation and recovery

Production of chymosin is accomplished by submerged fermentation. On completion, an *in situ* cell inactivation protocol is used in which the cells are killed before the fermenter broth leaves the contained fermenter. The pH of the broth is lowered by acid addition while maintaining normal agitation and aeration. After 1 hour, viable cell counts are reduced to acceptable levels.

First, the broth is filtered to remove cells and other debris. Further processing is required because the harvested broth contains residual nutrients as well as endogenous enzymes produced by the organism. Two processing schemes were developed to recover high quality chymosin from the broth. The first consists of addition of sodium chloride followed by separation on a hydrophobic chromatography column. Chymosin binds to the resin while impurities pass through the column. The resin is washed and the chymosin eluted with a low ionic strength buffer. The product is then formulated, sterile filtered, and packaged.

The second process involves chemical extraction using a two-phase liquid-extraction system. Selected conditions allow efficient extraction of the chymosin into one phase, leaving the impurities in the other. The phases are separated by centrifugation. The chymosin-containing phase is passed over an ion exchange resin to bind the chymosin, which allows complete removal of the extraction phase chemicals. The resin is washed and the chymosin eluted. The product is then formulated, sterile filtered, and packaged. The finished product meets an extensive list of specifications chosen to insure that a consistent, safe, high-quality product is produced.

Regulatory Considerations

The anticipated appearance in the marketplace of the first products made using rDNA techniques raised questions concerning whether or not new regulations were required for their sale. If new regulations were needed, the substantial time required might delay benefits that could be realised by the food industry. Also, the risk of investment for those companies actively pursuing the technology might be increased. Food and food ingredients are regulated in the United States by the FDA under the Federal Food, Drug, and Cosmetic Act (1982).

Ownership Protection

The use of biotechnology can be a justified expense only if the developer can prevent the invention from being used by the competition or, at least, cause the competition to pay the same costs to compete. For enzymes—as is the case to date for pharmaceuticals the battle for market supremacy will be fought in the patent courts, not in the marketplace. Those companies with the best technology and the appropriate patent portfolio will be the leaders of tomorrow.

Industry has two alternatives for protecting valuable 'intellectual property': Trade secrets and patents. A 'trade secret' is information, which, if kept secret, provides a competitive advantage to its owner. The advantage exists only as long as the owner actually keeps it secret and no other individual independently discovers and uses it. The decision thus depends on the ease with which the secret can be kept, meaning not only an assessment of a competitor's ability to invent the product on its own or to discover it through 'reverse engineering', but also of the ability of employees and the company to maintain secrecy. Theoretically, secrecy can be maintained indefinitely if enough care is taken. Patents, on the other hand, are exclusive rights granted to an inventor to exclude others from making, using, or selling an invention and are enforceable against all other individuals, including independent discoverers, but are limited in the time of their enforceability. Industrial enzyme inventions, subject to either trade secrets or patents, can be of two types: compositions of matter (an enzyme, intermediate, or the like) and processes for making and using the composition of matter.

Trade secrets become the primary method of protection for enzyme production when the likelihood of another individual discovering the method or of the inventor policing others is low, or when the ability to engineer around the process cheaply and efficiently is high. Trade secrets are also sometimes useful for the protection of individual microbial strains that produce high amounts of protein product or have some unique characteristic, because of the need to deposit such strains publicly for patent purposes, which enables competitors to obtain and improve on these strains to avoid patent infringement. Trade secrets are also important when processes are fairly obvious but optimum production conditions are not. Since the patent must contain the best mode known of practicing the invention, the inventor may feel that a patent is not enough benefit to warrant disclosure.

Patents are useful for protecting novel enzymes and their intermediates as compositions of matter, as well as protecting certain processes of making and using the enzymes. Composition-of-matter patents have always been viewed as a major goal in research and development. They cover a product (regardless of the process used to make the product), are easy to police, and are very difficult (if not impossible) to engineer around without producing an entirely new compound.

Although process patents have been seen as relatively weak and often worthless, new patent legislation in the United States strengthens courts to enforce process patents. Process patents are useful as a roadblock when engineering around the process may be difficult, when the process is costly or time consuming, or when a core technology can be protected by the process patent. A key example is the basic patent that covers recombinant technology issued to Stanford University. The patent applies to all companies that want to produce products recombinantly in the United States. Stanford University licenses the company for an initial fee and receives royalties that currently amount to tens of thousands of dollars yearly, an amount that is expected to grow tremendously as more and more biotechnology products are introduced. All patent holders may not be so generous. There is no longer a requirement in the United States that a patent holder license a patent: the holder may exclude others from ever practicing the invention.

Until recently, enzymes were not subject to much active patenting. Most enzymes and their uses were known and process patents were viewed as weak. These factors combined to minimise the amount of research and development done on such systems. The advent of biotechnology changed the face of enzyme biochemistry. Now it is possible to produce mutant enzymes with new DNA sequences and processes for production that meet the United States Patent and Trademark Office requirements for novelty, non-obviousness, and utility.

The first patent issued for a novel mutant enzyme produced by genetic changes at certain positions in the DNA was U.S. Patent no. 4,760,025 issued to Genencor. The patent claims novel subtilisin enzymes with changes front a naturally occurring to a non-naturally occurring amino acid residue at specific positions in the enzyme. These changes result in new enzymes with properties different from those of the original enzyme. Figure 4.5 shows data on the oxidative stability of one of those novel subtilisin enzymes.

Process patents also will become extremely important. Although a recombinant process may not be the only method by which to obtain a commercially desirable enzyme, it may be, by far, the cheapest method of producing a product, making it the only 'commercially viable' method. The new aspects of process patent law also will discourage infringing uses of process patent technology, both in the United States and abroad.

Consumer Attitudes

Success of food industry ventures depends on the ability to interpret consumer attitudes accurately and produce products that meet their needs and desires. Some individuals suggest that processing the most historically accurate and, therefore, 'natural' way is what customers want. Laws such as Germany's Reinheitsgebot, which legislates that only pure water, hops, malt, and yeast can be used to make beer, is a case in point. In addition, Standards of Identity describe in detail what a product must contain and, sometimes, how it can be made to meet labelling requirements. Although these well-intentioned regulations are meant to protect consumers from unscrupulous manufacturers, they tie products to historical definitions of processing conditions and prevent improvements based on advancing technology.

Innovation provides consumers with speciality products that have functionalities that mean differentiation and high margins, for example, low-calorie, low-fat, high-fibre, low-alcohol, and improved flavour. Enzymes—especially the new enzymes made available by rDNA technology—can generate such innovation. A degree of comfort with and acceptance of the science by the food industry is needed. That comfort and acceptance also must be translated to the marketplace by educating general consumers about the benefits of biotechnology. Public concerns must be recognised and answered effectively. Harlander cites the food irradiation issue and warns of the dangers if public concerns over food biotechnology are ignored or trivialised.

IMPLICATIONS FOR THE FUTURE

Like many breakthrough technologies, the basic discoveries of the biotechnology age did not, at the time, seem to be of great industrial importance. In 1953, DNA structure and its implications were proposed. In the 1960s, the basic mechanics of translation, transcription, and gene expression were elucidated. The 1970s brought development of the techniques to manipulate, splice, and sequence DNA. However, these abilities have not led automatically to a large impact on the food processing sector. Only in the late 1980s were new products cautiously introduced into the food industry to test the regulatory climate and the acceptance by consumers, and to establish the benefits of the technology.

Less frequently discussed are the developments in technology necessary to exploit the new science, or the developments in consumer attitudes to accept the benefits of the science readily. The mechanics of life are fascinating, but do not necessarily allow production of a cheaper fuel, plastic, or foodstuff. Enzymatic processes proceed under milder conditions, give purer products, and impact the environment to a lesser extent, yet still must compete against established chemical methods.

A debt is owed to individuals who pioneered the discoveries leading to the commercial use of genetic engineering. Although the methods described in this chapter seem relatively simple, they have been developed painstakingly over a number of years. Once defined, however, they are straightforward and easily accomplished, requiring only care and attention to achieve results. Indeed, high school biology classes perform cloning experiments as part of their laboratory requirements. However, the directed use of these techniques to introduce commercially viable products has been a long struggle; practice is a combination of extensive training and understanding, careful persistent work, a large amount of both skill and art, and some luck as well.

The successful biotechnology company must be proficient in all phases of enzyme engineering and design, process development, and day-to-day production, marketing, distribution, and administration. The technology required to exploit microbes also includes development of more efficient and creatively engineered fermentation facilities and product separation processes. These areas all must be handled in-house, adding costs to already strained research and development budgets. The industry is still young and cannot support engineering, design, and construction firms similar to those available to well-established industries (for example, the petrochemical or food processing industries). For these reasons, the application of biotechnology has not yet reached its full potential.

Returns from research and development programmes cannot be realised quickly. Therefore, target markets must be large to justify the expense involved. Non-food applications will be favoured to avoid the risk associated with regulatory and consumer acceptance hurdles. High value-added industries will be the primary, markets for enzyme manufacturers, for example, the petrochemical, energy, or environmental industries, in which historically favoured processing requirements do not exist. Already the food industry has been neglected by biotechnical companies that have directed their resources to human health-care applications, in which the high value pays handsome returns and the life-saving benefits reduce the reluctance of the general consumer. The food industry may be among the last industries to embrace the benefits offered by rDNA technology.

The use of enzymes for alternative processes or new products has been discussed at great length. However, enzymes currently have limited application in the food industry. They are characterised by relatively undifferentiated non-proprietary products, forced to compete on a price basis. The majority of useful enzymes are hydrolytic in nature and are used as processing aids. Enzyme-modified products must compete with commodity processing of foodstuffs. Limited proprietary protection has reduced willingness to invest in research and development. Enzyme production and use often requires extensive capital expenditure.

The advent of rDNA technology changes this outlook. Genes can be transferred into high-yielding organisms for more economical applications. Enzyme engineering can alter substrate specificities and pH or temperature optima for improved efficiency. Enzymes for synthesis reactions are on the brink of commercialisation.

However, biotechnology is a highly skilled and specialised science. The food industry cannot rely on micro-biologists to propose new applications. Insight is needed to identify opportunities in which novel enzymes might excel. Such insight requires knowledge of the benefits and willingness by food scientists to work jointly with enzyme suppliers to pioneer new applications. The opportunities are unlimited.

Chapter 5

Fermentation Process Design

INTRODUCTION

Production of any material on a large scale involves extracting it from a source relatively rich in the desired material, from metals to enzymes. Enzymes have the immense advantage that their sources can be created almost at will in large quantities, usually in industrial fermentation processes. This account therefore, deals with fermentation systems in enzyme production. It should be borne in mind that macro-organisms, plants and animal products are still useful and sometimes unique sources of particular enzymes, often utilising an otherwise waste product, such as animal offal. Agriculture is notoriously susceptible to political interference, and on occasions even terrorist attack, but industrial microbiology offers a stable source of enzymes and other biochemical substances whose range and productivity increases daily. Biotechnology may have been held back by unfounded fears of genetically modified organisms.

PROCESS DESIGN

The task of the fermentation technologist is to provide and maintain the environment required for carrying out a desired biological process. After the organism development phase of a project, a well-designed growth environment is needed to obtain the maximum economic yield of product to be obtained from the organism synthesising it. This task then resolves into environmental design, determining the environmental requirements of the process, and fermentation engineering, providing the means for meeting those requirements. An example of organism development has recently been reported. A recombinant *Escherichia coli* was created to improve production of penicillin G amidase (PGA), used in the industrial production of penicillin derivatives from penicillin G. PGA catalyses the hydrolysis of penicillin G into 6-aminopenicillanic acid (6-APA) and phenylacetic acid (PAA) and is commonly obtained for industrial use from *Escherichia coli* ATCC11105, *Bacillus megaterium* ATCC14945 and their mutants. Production of PGA by *E. coli* ATCC11105 is induced by phenylacetic acid, repressed by glucose and suppressed at temperatures above 30°C.

A mutant strain of *E. coli* produced by treating *E. coli* ATCC11105 with *N*-methyl-*N*-nitro-*N*-nitrosoguanidine was found to be less susceptible to glucose repression. This was transformed with plasmid pXL20, a derivative of pBR322 containing the PGA gene, and PGA production by the resulting transformed *E. coli* was stimulated by glucose rather than repressed, and gave very high PGA production in the presence of both PGA and PAA. This indicated that the performance of the transformed organism depends on both the properties of the host cell and the plasmid structure.

Environmental Design

Environmental design is concerned with determining the reaction conditions needed for formation for the required product in useful quantities. Reaction conditions include pH, temperature, dissolved oxygen, water activity, nutrient availability and supply of precursors and substrates for product formation, and these in turn imply that homogeneity is an environmental requirement, so that all parts of the reactor system are provided with the required environment. All these are likely to vary with time, and the controlled variation of the reaction environment is now developing as an important means of process enhancement. Microbial products involve two phases: microbial growth and product formation. The two phases may coincide, when the product is growth-associated, or may be in sequence, where the product is non-growth-associated.

Many enzymes are not growth-associated products, so that maximum biomass production does not necessarily occur at the same time as maximum enzyme yield. Of course all products are growth-associated in the sense that the organism must grow in order to create the mechanisms for product formation, but organism growth and product formation are likely to have different environmental requirements. In practice, the phases overlap to a greater or lesser extent, but separation of the organism growth phase from the product formation phase enables the most appropriate environment to be implemented for each phase. In the production of intracellular cytochrome P-450 enzymes in brewer's yeast, specific enzyme concentration in yeast biomass follows the same pattern in time as the biomass in shake-flask culture and was considered to be a growth-associated product. In stirred-tank culture, yeast growth is faster, and the enzyme peak appears after the biomass peak.

The aim of the overall process is to obtain maximum economic yield of product, which includes product separation as well as product formation. The growth conditions used can affect product recovery, notably with intracellular enzymes that are not secreted into the growth medium. Disruption of *Candida utilis* becomes easier as the organism growth rate increases, probably because the average cell age and/or cell size decreases. *Candida utilis* grown on ethanol in continuous culture became more difficult to disrupt in conditions of nitrogen limitation, and yeasts grown on alkanes are in general difficult to disrupt. The purpose of cell disruption is to release the intracellular product, which may be damaged by the disruption process. In recovering microsomal cytochrome P-450 from brewer's yeast using a homogeniser, the enzyme activity released appeared to be unaffected by the homogenisation pressure used. This was attributed to a compensating effect between enzyme release and enzyme denaturation, such that high homogenisation pressures released more enzyme, but damaged more, while low pressures released less but damaged less. Assessing disruption efficiency is in any case problematical, not least in differentiating between disrupted and undisrupted cells. While protein release seems the obvious criterion to use, it may include protein and peptide material release from the cell wall, and partially degraded macromolecules can effect protein solubility.

The environment must be free of undesirable organisms which may be unproductive or produce substances which contaminate the product or inhibit the desired organism. Failures due to contamination are not generally widely reported, but a failure rate of one batch in every 20 to 50 has been estimated. The cost and risk of contamination need to be balanced against the cost of preventing contamination, which can be very expensive.

Traditional brewing, for example, manages with hygienic conditions, rather than sterile, by using heavy inocula which can outgrow competition and secrete an inhibitory product, although these proved insufficient for continuous brewing.

Fermentation Engineering

Fermentation engineering involves the design, development, construction and operation of the process plant maintaining the required environment, including the fermenter vessel—the container for the organisms and their required environment—the means of supplying the nutrients and substrates and the systems for controlling the environment.

An industrial fermentation process may appear in principle to be a laboratory procedure carried out with larger amounts of materials in large steel tanks instead of shake-flasks. In practice, the eventual industrial operation differs from the original laboratory-scale operation because account has to be taken of the ways in which a process changes as the scale of operation changes. Most obviously, large quantities of material have to be handled differently from small quantities. A large volume of liquid cannot just be popped into an incubator or a water-bath like a shake-flask to change its temperature: the liquid has to be pumped through heat exchangers. A process is also likely to behave differently in different scales of operation. It is more difficult to maintain uniform conditions in a large body of liquid. Large tanks of liquid have large differences in hydrostatic pressure between top and bottom, for example, and organisms carried about the tank by movement of the liquid will be subjected to fluctuations of pressure. A recently reported example of differences in behaviour between different scales of operation is in the induction of gene expression of the restriction endonuclease EcoRI in a recombinant *Escherichia coli*. Chemical induction produced higher enzyme activity than temperature-shift thermal induction in a 60-litre fermenter, while in shake-flasks, thermal induction was more productive than chemical induction. This is presumably because it is more difficult to produce a rapid change of temperature in a large volume of liquid than in a small volume.

Ideally, the three development phases should be coordinated throughout the development of a fermentation project, although in practice they are often sequential. Hybridoma cells are reputed to be sensitive to shear, which restricts the means of mixing that can be used in a large-scale project. It may be possible to select shear-resistant strains at an early stage of organism development and widen the choice of equipment used. Similarly, high-permeability organisms with 'leaky' cells to facilitate product excretion were too fragile for large-scale cultivation. Another example of this has been recently reported, where the choice of fermenter system for scaling up the production of lignin peroxidase from white-rot fungus cells was determined by the shear regime obtainable. There has been considerable recent interest in enzymes produced by tree-rot fungi for applications in degrading environmentally hazardous aromatic substances as well as lignin-related compounds. *Phanerochaete chrysosporium* is a white-rot basidiomycete which degrades lignin in conditions of nitrogen, carbon or sulphur deprivation, when it produces two groups of peroxidases, lignin peroxidases (LiPs) and manganese peroxidases (MnPs). The lignin-degrading system then consists mostly of these two groups of peroxidases plus a system for generating hydrogen peroxide. A mutant *P. chrysosporium* has been characterised which overproduces lignin-degrading enzymes under nutrient-rich conditions, although this has not so far been grown on an industrial scale. Growth under high oxygen tension increases both lignin breakdown to carbon dioxide (called *mineralisation*) and synthesis of lignin peroxidases, but agitation has been shown to inhibit both ligninolytic activity and secretion of lignin peroxidases. A major scale-up problem is thus to develop a bioreactor with good oxygen transfer in a low-shear environment, and growth of the fungus in stirred tanks has not so far produced useful amounts of lignin peroxidases. Some success has been achieved by immobilising the fungus on polyurethane foam in submerged agitated culture. In this work, air-lift fermenters produced much higher yields of lignin and manganese peroxidases, and other extracellular protein, than mechanically agitated systems, which produced biomass rather than enzymes.

MODE OF OPERATION

The choice of mode of operation is basically between batch processing and continuous operation. In contrast to the chemical process industries, batch operation is generally preferred in the biological industries.

Closed-batch Fermentation

In a closed-batch fermentation, most of the materials required are charged to the fermenter, decontaminated before the process starts and then removed at the end. The only material added and removed during the course of a batch operation is fermenter exhaust gas, pH control fluids and air if the fermentation requires oxygen. A continuous system has in addition a continuous input of fresh growth medium and output of culture, which constitute a continuous risk of contamination.

In batch fermentation, conditions are continually changing with time, and the fermentation is an unsteady-state system, although in a well-mixed reactor conditions will be uniform throughout the reactor at any instant of time.

Growth Kinetics

Numerical methods are becoming increasingly important in fermentation environment design, but it must be borne in mind that even the most sophisticated mathematical analyses and computer programs are only as good as the assumptions on which they are based. It is, therefore, opportune to review some of the basic relationships on which most mathematical models are based.

The definition of specific growth rate, denoted by μ, follows from Malthus' law:

$$\frac{dw}{dt} = \mu w \qquad \text{... (5.1)}$$

so that

$$\mu = \frac{dw/dt}{w} = \frac{d(\ln w)}{dt} \qquad \text{... (5.2)}$$

where, w is the organism or biomass concentration and t is time. It is worth noting that the specific growth rate is often mistakenly called the growth rate, which is actually dw/dt and a very different quantity. This is very confusing if no further definition is given.

A growth equation can be derived by considering microbial growth as a second-order reaction between the nutrients, of concentration S, and the organism, so that:

$$\frac{dw}{dt} = k_2 w S \qquad \text{... (5.3)}$$

where, k_2 is the second-order reaction rate coefficient. This equation can be integrated if S can be replaced by an expression in w, such as the growth yield coefficient, Y, describing the conversion rate of nutrient into biomass:

$$Y = \frac{dw}{dS} = \frac{dw/dt}{dS/dt} \qquad \text{... (5.4)}$$

The minus sign allows for the fact that the growth in biomass produces a decrease in nutrients. This differential definition of Y is always true, but is useful only if Y can be assumed constant over a particular period of time, so that:

$$Y = \frac{w - w_0}{S_0 - S} \quad \text{... (5.5)}$$

where, w_0 is the organism or biomass concentration and S_0 the nutrient concentration at $t = 0$, at the start of the time period, such as the start of the fermentation. This expression can be used for replacing S in the second-order reaction rate equation, so that:

$$\frac{dw}{dt} = k_2 w \left\{ S_0 - \frac{(w - w_0)}{Y} \right\} = \frac{k_2 w (YS_0 + w_0 - w)}{Y} \quad \text{... (5.6)}$$

Now the group $(YS_0 + w_0)$ can be considered, to be the maximum possible concentration of biomass, w_M that could be obtained in the system, when all the nutrients are converted to biomass and added to the starting concentration, so that:

$$w_M = w_0 + YS_0 \quad \text{... (5.7)}$$

and equation (5.6) becomes:

$$\frac{dw}{dt} = \frac{k_2 w (w_M - w)}{Y} \quad \text{... (5.8)}$$

which is known as the logistic equation. It has the advantage of involving only biomass and time as variables, and gives a good description of biomass growth in complex media where the growth-controlling substrate may be unknown. The logistic equation has been used as the growth model in optimising the penicillin fermentation. The growth pattern is one of nearly exponential growth at the start, slowing down as the biomass concentration approaches maximum. The logistic equation can now be integrated to give:

$$M.t = \ln \left[\frac{w(w_M - w_0)}{w_0 (w_M - w)} \right] \quad \text{... (5.9)}$$

or

$$w = \frac{w_0 w_M (\exp M.t)}{w_M + w_0 (\exp M.t - 1)} \quad \text{... (5.10)}$$

where,

$$M = \frac{k_2 w_M}{Y} \quad \text{... (5.11)}$$

This approach has the disadvantage that the specific growth rate decreases steadily as the biomass concentration increases:

$$\frac{dw/dt}{w} = \mu = \frac{k_2 (w_M - w)}{Y} \quad \text{... (5.12)}$$

and, from equation (5.3) the specific growth rate increases in proportion to nutrient concentration without limit:

$$\frac{dw/dt}{w} = \mu = k_2 S \quad \text{... (5.13)}$$

In practice, there must be some upper limit to the specific growth rate, not least because high nutrient concentrations become inhibitory to growth through the consequent decrease in water activity.

If the upper limit to the specific growth rate is μ_M the logistic equation can then be modified by the condition:

$$\mu = \frac{k_2(w_M - w)}{Y} \text{ for } \mu < \mu_M; \text{ otherwise, } \mu = \mu_M \quad \text{... (5.14)}$$

This corresponds to a growth model known as Blackman kinetics, originally derived to described growth resulting from photosynthesis. This model gives a good description of many microbial growth systems, but the mathematical discontinuity makes it difficult to deal with. The problem of growth in complex media was avoided by Monod, who specified a system in which all nutrients were in ample supply except one, which was then the growth-controlling nutrient, with concentration denoted by S on which was growing a single organism. He then found that the specific growth rate could be described by a saturation relation, a rectangular hyperbola similar in form to two other well-established expressions, the Langmuir isotherm describing adsorption kinetics and the Michaelis-Menten for enzyme kinetics:

$$\mu = \frac{\mu_M S}{K_S + S} \quad \text{... (5.15)}$$

where, K_S is saturation coefficient, usually called the 'Monod constant', and is equal to the substrate concentration when the specific growth rate is half the maximum value. Monod's original data have been subjected to much scrutiny, and have been found to fit other growth models as well as or better than the Monod relation, including Blackman kinetics.

At extreme values of substrate concentration, the Monod model gives a similar pattern to the Blackman model. At high substrate availability, when S is very much larger than K_S, the Monod relation approximates to $\mu = \mu_M$. At very low substrate availability, when S is small compared to K_S, then the Monod relation approximates to:

$$\mu = \frac{\mu_M S}{K_S} \quad \text{... (5.16)}$$

which is the same as the logistic equation if (from eq. 5.13).

$$\frac{\mu_M}{K_S} = k_2 \quad \text{... (5.17)}$$

The Monod relation has been much abused, with attempts to apply it in situations not fulfilling the basic precepts, by engineers with insufficient knowledge of biology and by biologists with insufficient knowledge of mathematics. Following the same procedure as for the logistic equation, the Monod relation can be integrated to give an expression for the time t taken to achieve a biomass concentration W:

$$\mu_M t = \ln\left(\frac{w}{w_0}\right) + \left(\frac{Y.K_S}{w_M}\right) \ln\left\{\frac{w(w_M - w_0)}{w_0(w_M - w)}\right\} \quad \text{... (5.18)}$$

This expression describes biomass growth as being virtually exponential until the biomass concentration approaches the maximum level, or more accurately as the residual substrate concentration falls to a level close to the saturation coefficient K_S where it sharply becomes asymptotic to the maximum biomass concentration. The general form of the integrated Monod relation is similar to that of the integrated logistic equation which it approaches as the saturation coefficient K_S increases. The Monod equation has provided the basis for a large number of variations, with extra terms to account for substrate inhibition:

$$\mu = \frac{\mu_M S}{K_S + S + S^2/K_I} \quad \text{... (5.19)}$$

where, K_I is an inhibition coefficient, or low substrate availability with dense biomass:

$$\mu = \frac{\mu_M S}{w.K_W + S} \quad \text{... (5.20)}$$

where, K_W is a biomass inhibition coefficient, or inhibition by the product formed:

$$\mu = \frac{\mu_M S}{K_S + S + K_P p} \quad \text{... (5.21)}$$

where, p is product concentration and K_P is a product inhibition coefficient. In considering the validity of these expressions, it should be borne in mind that adding additional terms to an expression makes it easier to fit to a set of data.

Batch critique

The principal disadvantage of batch processing is the high proportion of unproductive time, called down-time, between batches, when expensive plant is unused while the fermenter is charged and discharged, cleaned sterilised and restarted. Continuous operation gives long production runs with relatively little down-time, and thus high process productivity. Long production runs require the organism grown to be genetically stable, because any faster-growing mutant will replace the original strain. In batch processing, each batch is started with a fresh inoculum, so that the correct strain is closely controlled and the genetic stability of the organism needs to be sufficient to last the length of the batch. One version of batch processing retains part of the culture from one batch to inoculate the following batch, sometimes called 'fill-and-draw' operation. This is useful where the organism is genetically stable, and can reduce the inter-batch down-time, but it does subject the organism to repeated cycles of starvation conditions at the end of a batch, to plentiful conditions at the start of the next batch. This environmental cycling can induce synchronous growth and consequent loss of product-forming capability. It has, however, been used successfully in large-scale production of monoclonal antibodies from mouse-mouse hybridoma cells at concentrations of 4 million cells/ml in 1.3 m^3 vessels.

Batch processing does exert an uneven demand on utilities such as refrigeration and cooling water, steam for sterilization, power and compressed air, but these fluctuations can be evened out by suitable phasing of multiple fermenters. Ancillary processes such as growth medium make-up and sterilisation and product separation are usually batch operations which can be coordinated with batch fermentations. In continuous operation, an interface between batch and continuous stages, such as a buffer tank, has to be provided, which is not only an additional equipment cost, but is also a risk point for contamination and deterioration.

An advantage of batch processing that makes it valuable in the health care industry is the ability to identify the sources of all materials involved in the manufacture of any specified sample of the product.

Product Formation

The biomass is produced so that it forms the required product. The rate of product formation is usually assumed to be proportional to the amount of biomass present, and their rate of product degradation proportional to the product concentration, so that the nett rate of product accumulation, *dp/dt*, is given by:

$$\frac{dp}{dt} = k_P w - k_D p \qquad \text{... (5.22)}$$

where, k_P and k_D are the specific product formation rate and specific product decay rate coefficients respectively. The values of both the coefficients are affected by the environment, especially temperature. In modelling penicillin production, for example, the production rate coefficient was taken to decrease parabolically with temperature from a maximum at 20°C, while the decay rate coefficient was assumed to follow the Arrhenius law and double with every 10°C increase in temperature.

The production rate coefficient is likely to be a function of the specific growth rate of the organism as well. A group of scientists used equation 5.22 in combination with the logistic equation to describe the production of intracellular cytochrome P-450 enzymes in brewer's yeast using a closed-batch stirred-tank fermenter. The model was used to develop optimal time profiles for temperature, pH and stirrer speed, which controlled the dissolved oxygen level, using the Pontryagin continuous maximum principle. Optimal constant settings for temperature, pH and stirrer speed had previously been determined, and the optimal profiles predicted a further doubling of enzyme yield. This predicted improvement was not obtained, although yields were still good.

Continuous Culture

Continuous culture can take two forms, completely mixed or plug flow, the difference being in the events within the fermenter vessel.

Continuous plug-flow reactors

In a plug-flow reactor, the reagents enter one end of a reactor and react while passing together through the system and can be considered as a series of elemental batch reactors moving along the length of the plug-flow system. Indeed, in kinetic studies, a plug-flow system can conveniently be simulated by a set of batch reactors in series. A plug-flow continuous system can be both steady state and unsteady state simultaneously. At equilibrium, the input and output conditions remain steady, as do the conditions at any particular point in the reactor. Each element of liquid passing through a plug-flow reactor experiences a continually changing environment, but each element undergoes the same experience. Plug-flow enzyme reactors are used industrially in saccharification of starch to glucose and continuous mashing of malt to maltose, and some biological waste-water treatment systems are virtually plug flow in operation. In biological growth reactors an inoculum must be provided at the inlet end, since there is, in principle, no mixing with the culture further along the reactor, in contrast to a completely mixed continuous system.

Simple continuous culture

The term 'continuous' generally implies a completely mixed system if not otherwise specified, and implies that the culture conditions within the fermenter are nominally uniform throughout the culture volume. The output from the fermenter then has the same composition as the main bulk of the culture in the fermenter. A completely mixed continuous fermentation can generally be controlled to give an apparent steady state, where the biomass growing is balanced by the biomass dying and the biomass being removed in the output. The environmental conditions are then fairly constant with time, which makes process control considerably simpler. The conditions at the state of balance depend on the average residence time spent by elements of fluid passing through the vessel, the vessel volume divided by the liquid flow rate. In practice, there is a distribution of residence times round the average value, as some elements pass straight from the input to the output and have a short residence time, while others are carried round for long periods before leaving in the output stream. The composition of the output is then an average value made up of all the different elements.

The Monod relation works well in describing simple continuous culture in a continuous single-stage straight-through reactor (CSSSTR). With a cell-free (sterile) feed at a flow rate f containing a concentration S_0 of growth-controlling nutrient into a well-mixed working volume of culture V containing an organism with specific growth rate μ and a biomass concentration w, no cell death, and the same flow rate f of culture medium out of the reactor, the rate of increase of biomass concentration is given by:

$$\text{Rate of increase} = \text{Input rate} + \text{Growth rate} - \text{Removal rate} \quad \text{... (5.23)}$$

$$\frac{dw}{dt} = f \times 0 + V\mu w - fw \quad \text{... (5.24)}$$

At steady state, when the growth just balances the biomass removed in the output,

$$fw = V\mu w \quad \text{... (5.25)}$$

and

$$\mu = \frac{f}{V} = D \quad \text{... (5.26)}$$

where, D is called the dilution rate and in this system is the reciprocal of the mean hydraulic residence time. This simple relation is also much abused, and the specific growth rate is assumed to equal the dilution rate in situations which do not fulfil the prerequisite conditions. To obtain a description of the biomass concentration obtained at different dilution rates, some kinetic model is required. In practice, the biomass concentration is fairly constant and roughly independent of dilution rate except at extreme values of dilution rate. The biomass falls to zero at zero dilution rate as this implies an infinite residence time, during which the biomass dies off. At high dilution rates, the biomass falls to zero as it is removed faster than it can grow—a condition called 'washout'. The logistic model predicts a linear fall of biomass concentration with increasing dilution rate to washout, which is far from realistic. The Monod relation is closer to reality, although the assumption of no cell death means that the maximum biomass concentration occurs at zero dilution rate—really an infinitely long batch:

$$w = Y[S_0 - DK_S(\mu_M - D)] \quad \text{... (5.27)}$$

and the washout point, the critical dilution rate at which the biomass concentration falls to zero, occurs when the dilution rate equals or exceeds D_C, given by:

$$D_C = \frac{\mu_M S_0}{K_S + S_0} \quad \text{... (5.28)}$$

This simple analysis demonstrates a couple of important points about straight-through continuous culture, insofar as the Monod model is correct. The specific growth rate of the organism cannot reach the maximum value for those conditions, because the washout dilution rate D_C must be less than the maximum specific growth rate μ_M, and therefore the biomass concentration must always be less than its maximum possible value of YS_0. This is disadvantageous if formation of a desired product is associated with a high specific growth rate.

The response of a continuous culture at nominal steady state to perturbations in the form of pulsed injections of key nutrients can provide information for growth-medium development. Pulse injection of different nutrients into a continuous culture was used to identify responses to nutrients of an *Escherichia coli* strain carrying the plasmid pZ_3. After identifying essential growth nutrients, their individual yield coefficients were determined in a series of nutrient-limited runs in continuous culture. The values of the yield coefficients were used for developing an optimal glucose-limited minimal growth medium for supporting balanced growth in continuous culture.

Biomass and substrate concentrations were well described by the Monod relation, although heterologous protein production was suppressed and acetate was produced at specific growth rates approaching the maximum of 0.87 hr^{-1}.

Continuous critique

Monitoring and controlling an unsteady-state system is obviously more difficult than in a steady-state system, although the development of computerised control systems has solved some of the problems. In addition, a continuous system disturbed from its steady state can take a long time to restabilise, during which time the product formed may well be substandard and contaminate previously collected normal output.

The balance between organism rate of growth and rate of removal in completely mixed continuous systems has other implications which are disadvantageous compared to batch systems. It has already been pointed out that the biomass concentration obtained is always lower than that obtainable in batch fermentation, and the maximum specific growth rate of the organism can never be achieved. Most importantly, in fermentations where the product formation occurs when the organism is in a declining growth phase, it is difficult to stabilise in a continuous system. Batch processing allows successive growth phases to be run in the same vessel, with environmental conditions changed as appropriate. This makes batch operation more flexible. A continuous system would need an additional reactor in which the new environment is maintained. As already discussed, mathematical modelling of β-lactamase overproduction in an *Escherichia coli* host-plasmid system showed that a single-stage continuous reactor would be unsuitable and that a two-stage system is necessary for stable protein production. The expression of β-galactosidase in recombinant *Saccharomyces cerevisiae* was successfully investigated in continuous culture. Specific enzyme activity and biomass concentration increased as dilution rate was decreased, and overall enzyme productivity was unexpectedly higher at low dilution rates. A strain of *Saccharomyces cerevisiae* unable to synthesise invertase due to deletion of the chromosomal SUC gene was complemented by the pRB58 plasmid containing the entire yeast SUC2 gene and its growth studied in continuous culture. A strategy for maximum invertase production was developed comprising high cell growth rate in aerobic growth with high glucose levels followed by anaerobic growth with low glucose levels. This suggests that batch fermentation would be preferable and work with this organism in fed-batch fermentation has also been reported.

Apart from high productivity, the main advantage of completely mixed continuous fermentation is the control it gives over the specific growth rate of the organism. The specific growth rate is affected by all the environmental parameters such as temperature, pH and dissolved oxygen concentration, but if they are maintained at their required values, then the nutrient availability is the determining factor. In a continuous fermenter this is determined by the rate of growth-medium input, whereas in a closed-batch reactor, only the starting nutrient concentration can be varied—after the batch starts it is uncontrolled.

In simple straight-through continuous fermentation, the average biomass residence time is inevitably the same as the mean hydraulic residence time. To separate the two parameters, biomass must be recycled, but it must be more concentrated than in the bulk of the culture otherwise recirculation merely provides additional mixing as well as an additional contamination risk. Recirculation in continuous culture has recently been reported which involved growing the fission yeast *Schizosaccharomyces pombe* in a continuous fermenter system where the yeast was separated from the fermenter output by crossflow filtration and recycled to the 2-litre fermenter vessel. In this way, the specific growth rate can be manipulated independently of the dilution rate, in contrast to conventional simple straight-through systems. Growth yield coefficients were determined under glucose-limited and glucose-excess conditions. In glucose-limited cultures, no maintenance requirement for carbon source or energy were found, but in glucose-excess with respirofermentative conditions, the maintenance glucose requirement was growth linked.

Fed-batch Operation

The nutrient concentration in batch processing can be controlled to a limited extent by feeding in supplementary nutrients during the course of the reaction. This system is known as a fed-batch process and is very commonly used in industrial fermentations for producing baker's yeast, penicillin, some enzymes and amino acids and other organic acids. Batch feeding is used to control the growth rate of an organism so as to keep it in a productive phase and to supply nutrients and substrates which are only sparingly soluble or are too toxic to add the whole requirement for the batch at the start.

It is also used as a means of controlling the viscosity of the culture as the viscosity of a mycelial culture can be considerably decreased by the addition of a relatively small quantity of water. A low viscosity improves the transfer of oxygen into the medium, the transfer of heat out of the medium to the cooling system for temperature control and the mixing needed to maintain uniform conditions throughout the culture.

The penicillin fermentation utilises all these features of batch feeding. The fermentation requires oxygen, which is the most common example of a sparingly soluble nutrient which must be supplied throughout a batch operation. The precursor for penicillin G, sodium phenylacetate, is dosed in continually during the fermentation to avoid the toxic effects of putting the whole batch requirement in at the start. As the biomass grows, both the oxygen demand and viscosity of the culture increase. The increased demand and decreased oxygen transfer efficiency result in a decrease in dissolved oxygen concentration. When the dissolved oxygen concentration falls to a level approaching that for oxygen limitation, a quantity of the culture is removed from the vessel and replaced by fresh nutrient medium. This reduces the oxygen demand and the viscosity of the culture and the dissolved oxygen level rises again until the biomass has grown back to oxygen-limiting conditions. This type of fed-batch operation is sometimes called a repeated fed-batch process. Penicillin formation is associated with the stationary phase of batch culture, after biomass growth, when there are insufficient nutrients remaining for further biomass growth. Nutrients are added at a rate sufficient to maintain the culture at this level of activity giving maximum penicillin yield, but not to add so much that the biomass resumes growth.

The alternative type of fed-batch process is where supplementary growth medium is added during the fermentation, but no culture is removed until the end of the batch. This system is sometimes called a single fed-batch process, and its principal disadvantage is that much of the fermenter volume is not utilised until the end of the batch, implying low productivity, and the corollary of this, that the duration of the batch is limited by the fermenter volume. The single fed-batch system is used in baker's yeast production, where the medium is fed in continually to maintain yeast growth at a rate which just maintains aerobic conditions.

If the conditions become anaerobic, then nutrients are converted to ethanol instead of biomass, which in this case represents lost biomass. Conditions that are too aerobic decrease substrate conversion and also waste air compression and agitation power.

Modelling fed-batch processes

The increasing use of fed-batch processes has made them the subject of considerable attention in the form of mathematical analysis to determine the best feeding patterns. The feed rate of growth medium to the culture is the key process variable. The addition of liquid increases the nutrient concentration, but decreases the concentration of biomass and product (and any other species not present in the feed). The rate of change in concentration of a component is given by the production rate less the rate of decrease due to dilution by the feed:

$$\frac{dw}{dt} = \mu w - \frac{fw}{V} \qquad \text{... (5.29)}$$

$$\frac{ds}{dt} = -\frac{\mu w}{Y} + \frac{f(S_0 - S)}{V} \qquad \text{... (5.30)}$$

$$\frac{dp}{dt} = k_P w - k_D p - \frac{fp}{V} \qquad \text{... (5.31)}$$

These can be combined with the definitions of μ and Y in equations 5.2, 5.4 and 5.5 as the basis for a numerical model of biomass and product formation. As mentioned earlier, the specific product formation and decay rate coefficients are dependent on temperature and pH, and the specific product formation rate coefficient is also likely to be a function of the specific growth rate. These relationships can easily be built into the model if they are known or can be assumed.

Fed-batch feeding time profiles

In computing substrate feeding profiles from equations 5.29 to 5.31 the growth yield coefficient Y and functions for the specific growth rate in terms of substrate concentration $\mu(S)$ and the specific product formation rate in terms of the specific growth rate $k_P(\mu)$ need to be determined experimentally, and if possible the value of the specific decay rate coefficient k_D, for the conditions of temperature and pH appropriate to the fermentation being considered. A substrate feed profile was computed in this way for maximising the production of lysine from an auxotrophic bacterial mutant in single fed-batch fermentation, and product decay was assumed to be zero. The specific growth rate μ was found to be proportional to the substrate concentration, which indicates logistic kinetics, and the specific lysine formation rate k_P correlated parabolically with the specific growth rate μ. These two correlations effectively described the production process, and the optimal feed pattern was computed by applying Green's theorem. The productivity predicted from the optimised feeding profile was only slightly larger than that obtained in practice by using a constant feed rate. A similar approach to finding the feed pattern for maximum biomass production in a repeated fed-batch fill-and-draw fermentation assumed Monod kinetics and zero product degradation.

The volume of culture remaining from one batch to act as the starting volume for the next batch has the strongest effect on productivity. Maximum productivity was obtained by filling the fermenter as quickly as possible to its maximum capacity, then operating as a closed-batch fermentation, removing the culture as quickly as possible down to the residual volume, then starting again. The optimal residual volume was surprisingly large, at about three-quarters of the total volume, giving a 'little-and-often' feed pattern which maintains a high average biomass concentration and thus high biomass growth rate. A 'little-and-often' feed pattern does of course approach the conditions of continuous culture which gives the maximum obtainable productivity.

Glutathione production by *Saccharomyces cerevisiae* KY5711 was optimised in fed-batch fermentation by controlling the specific growth rate according to an optimal profile. The profile was generated by assuming a relation between the specific production rate of glutathione and the specific growth rate of the yeast, and carrying out mass balances round the process. Fed-batch culture offers the opportunity of separating the cell growth phase from the cloned gene expression phase in the culture of recombinant organisms. This separation may be affected by different feeding patterns, and this was investigated with a recombinant strain of *Saccharomyces cerevisiae*.

The host yeast strain could not produce invertase due to a deletion of the chromosomal copy of the SUC2 gene, but the recombinant strain contained the pRB58 plasmid covering the entire SUC2 gene, including promoter, signal sequence and structural gene. Fed-batch operation is advantageous because a suitable glucose-feeding strategy enables the system to operate with low glucose concentrations for prolonged periods, thus avoiding the suppression of the SUC2 promoter by high glucose concentrations. In fed-batch fermentation, the SUC2 promoter was derepressed and responded rapidly to low glucose levels and specific invertase activity increased as the glucose level decreased, whereas in closed-batch aerobic culture, the yeast exhibited diauxic growth. Determinations of the yeast biomass yield coefficient appeared to be higher at low glucose concentrations, which suggests that there is a shift to a more energy-efficient pathway.

Fed-batch feeding time profiles with inhibition

Fed-batch operation is particularly useful when the fermentation substrate inhibits growth at high concentrations, but is metabolised at low concentrations. There will then be some substrate concentration at which the specific growth rate is a maximum, and fed-batch fermentation can be used to maintain the substrate at that maximum level. Where the product is growth-inhibiting, such as in ethanol production, several approaches were reviewed by Winkler. Penicillin fermentations suffer oxygen transfer problems at high biomass concentrations, substrate inhibition and side reactions at high substrate concentrations. Feeding patterns, taking maximum possible biomass and substrate concentrations and substrate-inhibited kinetics into account, have been developed.

Applications of the ability of cytochrome P-450 enzymes to activate hydrophobic substances being investigated include hazardous waste destruction, synthesis of organic chemicals and electrochemical analysis. Cytochrome P-450 in immobilised liver microsomes is now being used on an industrial scale to produce drug metabolites in Japan. A potentially important application is in the detoxification of trace polynuclear aromatics and low molecular weight halocarbons, such as 1,2-dibromo-3-chloropropane (DBCP) and naphthalene, in waste waters or the environment. Although there are differences in substrate specificity, soluble microbial cytochrome P-450 enzymes offer the advantages of bulk production and ease of separation over membrane-bound mammalian enzymes. Growth conditions, such as growth-medium composition, dissolved oxygen tension and growth phase in batch culture, strongly influence the type and quantity of cytochromes present in cells. Cytochrome P-450, designated 'cytochrome P-450_{cam}', induced in *Pseudomonas putida* PpG 786 by D(+)-camphor was produced in shake flasks and fed-batch and continuous culture in a 2-litre fermenter. In response to D(+)-camphor. *Pseudomonas putida* expresses intracellular cytochrome P-450_{cam} at high levels. In fed-batch culture, the extended linear growth phase characteristic of hydrocarbon fermentations was found and the iron content of the growth medium was shown as a key factor in production of the terminal protein of the cytochrome P-450_{cam} enzyme system. Enzyme production rates in continuous culture were improved by reducing the gas input but enriching it with oxygen.

ADAPTIVE CONTROL

Adaptive control is the name given to a control system in which the controller learns about the process as it controls it. Data taken from the process itself is used for continual updating of the control model. A parameter estimator monitors the process inputs and outputs at preset time intervals and computes estimates of the process dynamics in terms of the parameters of a previously defined mathematical model of the process from the change in the values over the sample period.

The control design algorithm is then used to generate controller coefficients from these estimates, and the controller computes the required control signals to the devices controlling the process. An extremely important feature of an adaptive controller is the structure of the model used by the parameter estimator to compute estimates of process dynamics. The process can be represented by a set of mass balance equations, in terms of quantities which can be measured directly, such as pH and dissolved oxygen concentration, or quantities which are inferred from other measurements, such as biomass concentration. The production of baker's yeast in fed-batch fermentation has been a favourite system for experimenting with adaptive control and several reports have been summarised by Winkler. The required product is the biomass, which is inevitably growth-associated, and the control objective can be formulated simply as maintaining the respiration quotient at 1. The aim is to avoid producing ethanol, which represents lost nutrients and inhibits growth, while avoiding excessive aeration. One or more of the sugar feed, the air sparge rate and the agitator speed are therefore controlled to hold the dissolved oxygen level as low as possible while still avoiding anaerobic conditions.

Successful operation and scale-up of fed-batch culture using adaptive control of feeding an inhibitory inducer of cytochrome P-450 has been reported. Pravastatin is a specific inhibitor of the rate-limiting enzyme in cholesterol biosynthesis, hydroxymethylglutaryl-coenzyme-A reductase (HMG-CoA reductase) used in cholesterol lowering therapy, and is produced in a two-stage fermentation. *Penicillium citrinum* produces ML-236B which is treated with sodium hydroxide to produce a sodium salt, (ML-236B Na) followed by microbial hydroxylation to pravastatin by *Streptomyces carbophilus*.

The conversion is catalysed by a cytochrome P-450 enzyme, whose hydroxylation activity in *S. carbophilus* is induced by ML-236B Na, but inhibited by high concentrations, so that continuous feeding throughout the fermentation is necessary to maintain the optimum level of ML-236B Na for maximum conversion. This was achieved by using an adaptive control system for feeding the substrate to the fermentation, with both substrate (ML-236B Na) and product (pravastatin) measured on-line using HPLC. The substrate feed rate was computed by predicting the conversion rate over a time interval from regression analysis of five previous determinations of conversion rates. Measurements of cytochrome P-450 levels were found to be directly proportional to the concentration of ML-236B up to a level where the ML-236B became inhibitory. The feed was then controlled to maintain the ML-236B at the level just giving maximum conversion rate. Following successful operation of the feed control system with a 30-litre fermenter, the process was scaled up, using a 6000-litre fermenter, to give a three-fold increase in pravastatin productivity compared with manual feed control, and is now working well on a full industrial scale.

KNOWLEDGE-BASED CONTROL

While computers have obvious advantages in process control in being able to handle vast quantities of process information almost instantaneously, there is also a valuable store of human experience associated with an established process that can be harnessed. Much experience is of a qualitative nature, and the concept of the fuzzy set has been used in developing methods of utilising it. The technique has been reviewed by Hitzmann. The approach has proved successful in the cement industry and a state estimator based on fuzzy logic has been applied to a fed-batch fermentation.

Knowledge-based control has the advantage of flexibility in dealing with the uncertainties and non-quantitative information which characterises biological systems. A component of a knowledge-based control system is a fuzzy logic controller (FLC), which is particularly attractive for controlling fermentations, such as yeast cultures, where the dynamics are not well understood but a large amount of process information has been accumulated. A fuzzy logic controller can be developed to recognise emergency states, such as glucose accumulation or deficiency, and take appropriate control action. A conventional fuzzy control uses the control error and its rate of change, which works well with small control errors but tends to overreact to large errors. The fuzzy controller can be improved by changing part of the set of rules to diagnose states of glucose emergency, accumulation or deficiency, and take control action.

In recently reported work, a fuzzy logic controller was developed and applied to implementing an optimal control strategy for the production of glutathione (GSH) in fed-batch yeast fermentation. The key control problems resolve into prevention of the Crabtree effect so as to maximize yeast biomass production, and thus biomass-related products, which is of obvious commercial importance in baker's yeast production as well. One strategy for preventing the Crabtree effect is to control ethanol concentration so as to maintain zero nett production. A conventional control system, such as proportional plus integral (PI) control, has a limited capacity to cope with the nonlinear behaviour of fermentations, so that human intervention may be necessary for strict implementation of the control strategy.

In glutathione production, the maximum specific production rate occurs at a critical value of the specific growth rate (μ) at which no ethanol production occurs, so that minimising ethanol production maximises glutathione production. The control task was thus to formulate and implement a control strategy to set and maintain the culture at this critical specific growth rate to coincide with the injection of the cysteine substrate. After sufficient biomass has grown and the specific growth rate is established at its maximum level, the specific growth rate is decreased to the value critical for glutathione production and the cysteine injected. An approach to fermentation control, called 'physiological state control' has been developed using fuzzy set theory and pattern recognition theory, whereby a process is represented in terms of the prevailing physiological state of the culture. No conventional mathematical model is needed, although expert knowledge of the physiology of the organism and the behaviour of the process is required, and newly acquired process knowledge can be utilised with a fuzzy recognition procedure.

STATISTICAL OPTIMISATION

The mechanisms involved in the formation of secondary metabolites such as enzymes can be too complex and/or insufficiently understood for mathematical modelling and computer-based optimisation. Empirical methods such as statistical analysis can provide process improvement without disruption of the production scheme if the experimental programme is well designed. The essence of the experimental programme is to make a carefully planned sequence of small changes in the process conditions, measure the response of the process to them, and use statistical analysis to interpret the meaning of the responses.

Factor Analysis

The first step is to decide which process conditions can be altered and which are likely to affect the process noticeably. This may be on the basis of expert knowledge of the process or by means of a preliminary statistical procedure known as factor analysis, which can be thought of as 'optimising the optimisation'. A problem with empirical methods is that process variables are unlikely to be independent from each other. Changing growth-medium composition, for example, will change not just its nutrient properties but also its viscosity, and thus its oxygenation, mixing and heat-transfer properties. Factor analysis gets round this by assuming that there are a number of hypothetical reference factors which are independent of each other, and each has a clearly identifiable effect on the process. From the results of a series of experiments, the contribution to the real-life process variables by the hypothetical reference factors can be worked out, and thence their significance to the process. Insignificant variables can then be left out of the optimisation, saving a great deal of experimentation. Factor analysis was used in optimising the production of intracellular cytochrome P-450 from brewer's yeast, and the concentration of one apparently minor growth-medium constituent proved to be unexpectedly significant.

Single Factor Optimisation

The simplest form of optimisation is single-factor optimisation, where all the variables in a process are held constant except one. The remaining variable is adjusted until the best value of the process response is obtained. This is repeated for each of the variables in turn and the result is a set of 'optimum' values.

This method can give a rapid answer when there are insufficient time and resources for a better planned experiment, but is misleading if there are interactions among the variables, as the results will then depend on the order in which the variables are tested. Fairly straightforward optimisation principles were used successfully in the scale-up of the production of human proinsulin in *Saccharomyces cerevisiae* in fed-batch fermentation. The optimal temperature for secretion of non-glycosylated human prourokinase by *Saccharomyces cerevisiae* was measured in a 'perturbed batch' fermentation from which it was concluded that the conditions for optimal biomass productivity and for optimal secretion of heterologous protein are not necessarily the same.

Factorial Experimental Design

A statistical procedure called factorial experimental design can be used to assess the relative effect on a set of process variables and their interactions. This is also a series of carefully planned experiments, with different values of each variable in different combinations. The method can also deal with 'yes-no' or 'either-or' choices, such as different strains of organism or with or without antifoam. The investigation of only a small number of variables can require an enormous number of experiments. For example, if four variables are tested, each with two alternative choices, then $2 \times 2 \times 2 \times 2 = 2^4 = 16$ experiments are needed, which may need to be replicated several times to allow for random variations. This illustrates the importance of selecting the variables to be investigated very carefully.

Fractional designs

A way of reducing the number of experiments in a factorial design is to use a fractional design, where some of the less important interactions may be left untested. These must be chosen with great care, so as to leave sufficient degrees of freedom for the estimation of experimental error and because, in fractional designs, an unwanted effect called an *alias* can take its value from the same data as a required effect. Fractional designs were used in the optimisation of cytochrome P-450 production by *Saccharomyces cerevisiae*, and eventually produced increases of 173 per cent of total enzyme yield and 255 per cent in specific enzyme yield compared with the original shake-flask conditions. Part of this improvement must be ascribed to the improved control and homogeneity in a stirred-tank fermenter. Factorial experimental design was used in investigating growth environment control of nuclease P1 synthesis by

Penicillium citrinum in shake flask culture. Enzyme synthesis was shown to be affected by both carbon and nitrogen sources as well as by their interaction. The phosphorus source was significant but its interactions with other components were not. Nuclease P1 synthesis was repressed by the nucleotide 5′-uridine. Factorial experimental design was also used for optimising the stereoselective reduction by *Saccharomyces cerevisiae* in batch culture of ethyl acetoacetate to (+)-(S)-ethyl-3-hydroxybutyrate, which is a precursor for the antibiotic griseoviridin and for (S)-(+)-sulcatol, an insect pheromone.

VARIABILITY

Nominally identical procedures in biological systems can produce variable results, usually accepted as 'normal' batch-to-batch variation. This wastes resources, because equipment has to be overdesigned to allow for this, and it is an expensive nuisance in process analysis because high levels of replication are needed to separate genuine changes from fortuitous ones. The standard technological approach to variability is to identify the cause of the variation and correct it. In many cases, the fault may be impossible or too expensive to correct. In microbiological systems there may be variations in the composition of complex raw materials and strain degeneration in the organism. The approach pioneered by the Japanese statistician Genichi Taguchi is then to find a means of suppressing the effects of variations if their cause cannot be corrected. This has achieved considerable success in the manufacturing industry, notably the manufacture of ceramic tiles. The Taguchi approach has been applied to the variability in yield of cytochrome P-450 enzymes from brewer's yeast with some success. The approach was to consider the variability of the process as a feature of the process in the same way as yield or productivity. The effect of process conditions on variability was then tested using one of Taguchi's factorial designs. The main difficulty is that large numbers of experimental results are needed to assess variability.

Nomenclatures used in the Chapter

D	dilution rate (f/V)	time^{-1}
D_c	critical dilution rate, washout point	time^{-1}
f	flowrate	m^3/time
K_I	substrate inhibition coefficient	concentration, kg/m^3
K_P	product inhibition coefficient	product concentration, kg/m^3
K_S	saturation coefficient	concentration, kg/m^3
K_w	biomass inhibition coefficient	kg (substrate)/kg (biomass)
k_2	second-order reaction-rate coefficient	time/kg (substrate)
M	composite coefficient ($= k_2 w_M/Y$)	time^{-1}
p	product concentration	kg/m^3
S	substrate concentration	kg/m^3
S_0	starting substrate concentration	kg/m^3
t	time	
V	working culture volume	m^3
w	biomass concentration	kg/m^3
w_M	maximum biomass concentration	kg/m^3
w_0	starting biomass concentration	kg/m^3
Y	growth-yield coefficient	kg (biomass)/kg (substrate)
μ	specific growth rate	time^{-1}
μ_M	maximum or asymptotic specific growth rate	time^{-1}

Chapter 6

Extraction and Purification of Enzymes

INTRODUCTION

Before attempting to extract an enzyme from an organism, some preliminary considerations should be made. For a commercial venture the most important requirement would be to use a source which enabled large amounts of a suitable enzyme to be extracted by some convenient procedure: any source which fulfilled this requirement could be chosen. On the other hand, a programme of scientific investigation might require the extraction of a given isoenzyme from a specified source, no alternatives being permissible. In the present chapter, therefore, we will discuss principles of extraction which are of general application.

First of all, it is essential to know the location of the enzyme within the cell, and whether the enzyme is present in free solution or bound to a membrane.

EXTRACTION OF ENZYMES

Extraction of Soluble Enzymes

Soluble cytoplasmic enzymes are the simplest to extract, for any disruption of the plasma membrane will enable the enzyme to pass into the surrounding medium. Soluble enzymes present in the organelles of eukaryotic cells are also easy to liberate, but the disruption of intracellular membranes often requires harsher conditions than disruption of the plasma membrane. To minimise the extraction of unwanted enzymes, cell fractionation may be carried out prior to the disruption of organelles.

The extraction procedures used depend on the type of organism acting as source.

Animal tissue

Animal tissue is removed as soon after death as possible and kept cold to minimise autolysis. Obvious non-enzyme matter (e.g. fat deposits and connective tissue) is removed and the remaining tissue treated to disrupt membranes using a technique appropriate to the location of the enzyme being extracted: homogenisation in an isotonic medium disrupts the plasma membrane but leaves most organelles intact; the techniques of freezing and thawing, sonication and drying *in vacuo* disrupt all plasma and intracellular membranes.

The extraction of soluble enzymes from higher plants varies considerably according to the type of source material. Enzymes may be extracted from cereal flours simply by placing in an aqueous medium and stirring, and from soft storage organs (e.g. the potato) by mincing and pressing out the juice through muslin or cheesecloth. Extracts from leaves, dried seeds, etc. can be obtained by grinding in a mortar and then treating with a suitable medium in a pestle homogeniser. Leaves and other plant tissue may also be homogenised directly in medium by the use of a Waring blender. Alternatively, leaves may be disrupted in a Waring blender in the presence of several volumes of cold acetone (to displace water) followed by careful drying to produce an acetone powder: such powders may subsequently be treated with suitable medium, and the extracts will contain less chlorophyll and resinous material than would otherwise be the case. Tougher tissues, such as fibrous roots, are disrupted by grinding in a mortar with sand or ground glass. This procedure, like that of homogenisation in a Waring blender, tends to disrupt organelles.

Micro-organisms

Micro-organisms, in common with plants, have a cell wall and so tend to be more resistant to disruption than are animal cells. In general, methods of extraction fall into two main categories: drying and mechanical disruption. Drying may cause the cell to become permeable, enabling soluble enzymes to be extracted.

Air-drying

Air-drying (particularly of yeast) where the cells are allowed to dry in air for several days at around 30°C, and drying *in vacuo*, where a cell suspension is placed in a vacuum desiccator for one or more days, are both accompanied by some degree of autolysis.

Lyophilisation (freeze-drying)

Lyophilisation (freeze-drying) procedures, where a frozen cell suspension is dried by sublimation *in vacuo*, is not accompanied by autolysis, and soluble enzymes are less easily extracted from lyophilised cells than from air-dried or vacuum-dried preparations. Micro-organisms may also be dehydrated with acetone to produce powders containing easily extractable enzymes. Mechanical disruption of micro-organisms may involve grinding with carborundum beads or other abrasives, applying pressure (e.g. with Hughes or French presses), sonication, or a combination of these methods. Another procedure, which is becoming increasingly important, is to weaken the bacterial cell wall with enzymes such as lysozyme and glucuronidase, particularly where these are available in an insolubilised form and can be recovered: this treatment makes the cell more susceptible to disruption by relatively mild techniques, e.g. osmotic shock.

In general, the method of choice depends not only on the type of cell being disrupted and the subcellular location of the enzyme being extracted but also on the characteristics of this enzyme (e.g. its stability under the conditions of the extraction procedure): it must be admitted that the final choice is often made empirically, but a consideration of the principles discussed above can help to limit the number of possibilities.

Extraction of Membrane-bound Enzymes

Enzymes which are bound to plasma or intracellular membranes cannot normally be extracted into the surrounding medium simply by the disruption procedures discussed already in this chapter. However, some enzymes are loosely bound and some tightly bound (many actually forming an integral part of the membrane), so there can be no general rule about their extraction.

Membranes, besides containing enzymes and other proteins, are made up largely of amphipathic lipids, i.e. lipids which contain hydrophilic and hydrophobic regions. The main classes of membrane lipids are phospholipids (fatty acid/phosphoric acid esters of glycerol and sphingosine) and glycolipids (fatty acid esters, mainly of sphingosine, attached to a sugar). Some membranes in eukaryotic cells also contain sterols (Fig. 6.1). In each case the hydrophilic portion of the molecule is small compared with the hydrophobic part (note that in Fig. 6.1 the R-groups of the fatty acid residues are long hydrocarbon chains).

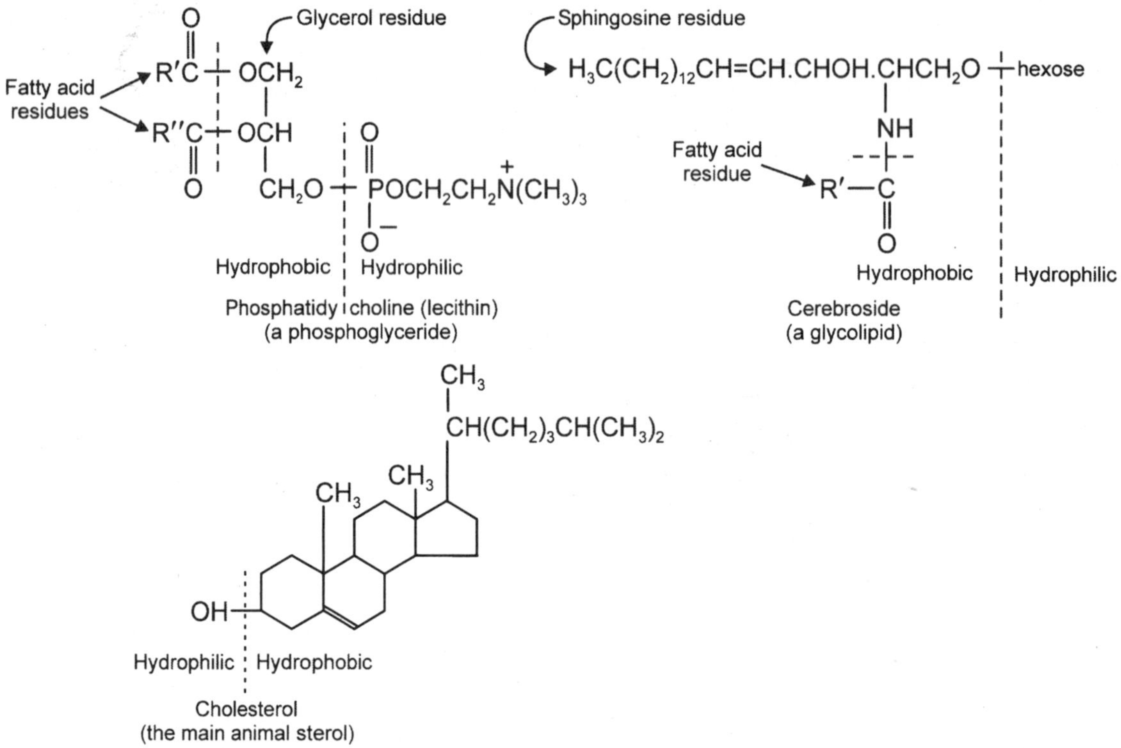

Fig. 6.1. Some important examples of lipids found in membranes.

An amphipathic lipid may, therefore, be represented as a molecule with a hydrophilic head and one or more hydrophobic tails, for example:

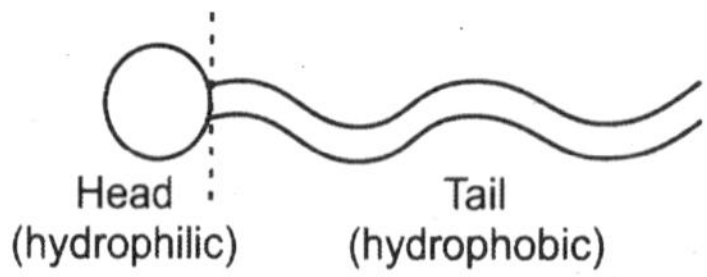

Although the detailed structures of membranes have yet to be established beyond doubt, the model which best fits the available evidence (including that obtained by electron microscopy and NMR spectroscopy is that of Singer and Nicolson: this is termed the fluid-mosaic model (Fig. 6.2).

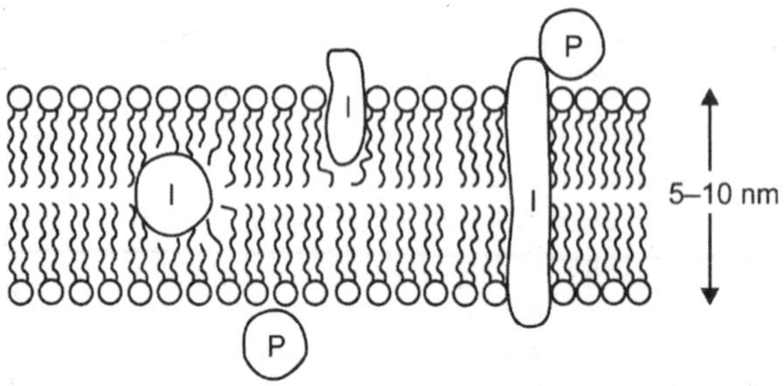

Fig. 6.2. Representation of a typical membrane, according to the fluid-mosaic model of Singer and Nicolson (P = peripheral protein and I = integral protein).

Proteins may be peripheral (or extrinsic), when they are bound to the surface of the membrane, or they may be integral (or intrinsic) when they are wholly or partly embedded in the lipid layers. Some evidence has been produced to show that the parts of polypeptide chains embedded in membranes are rich in non-polar amino acid side chains, which may form hydrophobic bonds with the lipids, whereas those parts which stick out, thus coming into contact with water, contain more polar side chains and, in some instances, are attached to hydrophilic carbohydrate units. It seems that at least some of the pores through the membrane may be lined with protein.

According to this model, lipids and proteins can move laterally about the membrane, to some extent, but find much more difficulty in moving in a transverse direction, from one membrane surface to the other.

Peripheral proteins are apparently linked by electrostatic or hydrogen bonds to the polar heads of lipids or to integral proteins and can easily be dissociated from the membrane (e.g. by treating with a solution of high ionic strength, such as 1 M NaCl, by freezing and thawing, or by sonication). Integral proteins, on the other hand, can only be extracted by breaking the hydrophobic interactions between lipid and protein (i.e. by dissociating a lipoprotein complex). Furthermore, for the extraction of an integral protein which is an enzyme, this must be done in such a way that enzyme activity is not lost during the process. A major problem is that an integral protein in its natural environment has some parts of its surface in contact with the hydrophobic membrane and other parts in contact with water, so structural changes are likely to take place no matter whether it is extracted into an organic or an aqueous medium. Again, the best method for each enzyme must be found largely by trial-and-error.

In some cases, controlled autolysis has been used to successfully free enzymes from membranes; this relies on the membrane structure being broken down and the enzyme released before the activity of the enzyme is lost, so it cannot be held to be a reliable procedure.

Some lipid-protein interactions are broken during the formation of acetone powders, so it may be possible to extract the required enzyme from these. Alternatively, the lipid-protein interactions may be broken by the action of specific enzymes: for example, phospholipase and triacylglycerol lipase have been used to extract alkaline phosphatase.

Of more general significance, detergents have been used to disrupt membranes and extract lipoprotein fragments into aqueous media as components of micelles (aggregates of amphipathic molecules arranged with their hydrophobic parts in the interior of the micelle and their hydrophilic parts on the surface). These detergents include natural bile salts such as cholate; non-ionic synthetic detergents like Triton X-100; and ionic synthetic detergents which may be zwitterionic, as in the case of CHAPS, acidic like sodium dodecyl sulphate (SDS) or basic, e.g. CTAB. In addition to the protein-detergent micelles, molecules of a detergent will themselves form micelles if present at a concentration above the critical micelle concentration (CMC). In general, the bile salts form the smallest micelles (molecular weight below 2000) and have the highest CMC (1–10 mM) while the non-ionic detergents form the largest micelles (molecular weight up to 1,00,000) and have the lowest CMC (in the order of 0.1 mM). A knowledge of these characteristics may be helpful in the further purification of the extracted enzyme. The choice of detergent will be influenced by whether or not the enzyme is required in an active form, and what further purification procedures are intended. A detergent of particular importance in the investigation of membranes is SDS, which will sever most protein-lipid (and protein-protein) interactions and enable the molecular weights of the liberated proteins to be determined. However, for extraction of an active enzyme, non-ionic detergents are usually preferred.

Lipoprotein fragments may also be extracted by the use of organic solvents, and some free proteins may be liberated on subsequent dispersion in aqueous media: for example, xanthine oxidase has been obtained from lyophilised milk protein by a technique involving ether extraction. Such procedures might require evaporation of the organic medium and dissolving the enzyme in a detergent solution. The solvent most widely used for the extraction of enzymes from membranes has been *n*-butanol. This has both lipophilic and hydrophilic character, so can act like a detergent in removing lipoproteins from membranes and, further, tends to dissociate the protein and lipid components. Most membrane proteins are found in the aqueous phase of a butanol-water two-phase system, with the lipids in the butanol phase. Different groups of enzymes may be separated from membrane lipids by treatment with *n*-butanol under different conditions: for example, alkaline phosphatase has been extracted at pH 5–6, γ-glutamyltransferase at pH 7 and urate oxidase at pH 10. In general, however, organic solvents tend to cause some degree of protein denaturation, so are not ideal for the extraction of active enzymes.

It will be realised that enzymes can exist in cells complexed to nucleic acids or carbohydrates, but, unless they are also linked to lipids in membranes, their extraction will be relatively straightforward.

Nature of the Extraction Medium

Once cells have been disrupted, by whatever method, the contents are extracted with several volumes of aqueous medium. In the case of enzymes which are an integral part of membranes the original extraction may be in organic solvent but the liberated enzyme must then be further extracted into aqueous medium.

In general, a considerable amount of debris (mainly membrane fragments) is likely to be present, so it is essential that the enzyme of interest can be easily separated from this by being soluble in the extraction medium. Four main factors govern the solubility of proteins: salt concentration, pH, the organic solvent content and temperature. The temperature of the medium is usually kept below 4°C, despite the reductions in solubility that this entails, in order to minimise loss of activity of the enzyme.

Proteins are least soluble at their isoelectric point, so the extraction of an enzyme must be carried out at a pH value far from its isoelectric point, but nevertheless at a value where the enzyme is stable. The salt and organic contents of the medium are chosen, largely on an empirical basis, to ensure that the enzyme is soluble but many other proteins are not, thus achieving a preliminary purification. For example, Cohn and co-workers showed that glutamate dehydrogenase, arginase, alkaline phosphatase and acid phosphatase from a bovine liver homogenate were soluble in 19 per cent ethanol at pH 5.8 and ionic strength 0.02, but catalase, peroxidase and D-amino-acid oxidase were not; however, these three enzymes were soluble at pH 5.8 in the absence of ethanol and at ionic strength 0.15; also, various proteinases which were insoluble in both of these media were soluble if the pH of the second medium was increased to 7.4.

The solubility of an enzyme can sometimes be increased by the addition of its substrate: for example, alkaline phosphatase is more soluble in the presence of β-glycerophosphate than otherwise. Hence the extraction of an enzyme can be facilitated by adding its substrate to the medium. Stabilisers, such as dithiothreitol (Cleland's reagent) or mercaptoethanol are sometimes added to extraction media to prevent oxidation of sulphydryl groups and thus loss of enzyme activity. Inhibitors of proteolytic enzymes (e.g. DFP) may also be added to prevent these attacking the enzyme of interest. Polyols (e.g. glycerol) or chelating agents (e.g. EDTA) also have a use in stabilising solubilised enzymes.

PURIFICATION OF ENZYMES

Preliminary Purification Procedures

As mentioned already in this chapter, a preliminary purification step is usually included in the procedure used to extract an enzyme. For example, some degree of cell fractionation is often carried out before extraction, so that only enzymes in the same sub-cellular compartment as the relevant enzyme are extracted. Also, the composition of the extraction medium is chosen so that the enzyme of interest is soluble but many others are not.

Extracted nucleic acids may be precipitated by treatment with basic substances such as streptomycin or protamine, or with $MnCl_2$ or $MgCl_2$.

All precipitates and cell debris are then removed by centrifugation and discarded. Polysaccharides may also be removed by high-speed centrifugation, although the results are not always satisfactory. An alternative way to dispose of polysaccharides is to break them down into small units with enzymes such as α-amylase; nucleic acids may similarly be broken down by the use of nuclease enzymes.

The next stage of purification is usually to precipitate the enzyme of interest from solution, thus separating it from mono- and oligo-saccharides, nucleotides, free amino acids, etc. and from many other proteins which remain in solution. This may be achieved by altering the pH and the organic or salt concentrations of the medium. For example, the pH may be adjusted to the isoelectric point of the enzyme. Alternatively a high salt concentration can be introduced: ammonium sulphate is often used for this purpose, because it is extremely soluble in water. The salt concentration is usually increased in stages, the aim being to precipitate other proteins (which are then discarded) before precipitating the fraction containing the relevant enzyme. This is then redissolved in water, and residual ammonium sulphate removed by dialysis.

Although a considerable degree of purification can be achieved by these procedures, many other proteins will still be present because of overlap of solubility ranges. One possible way to remove some of these is to raise the temperature of the medium for a few minutes to a value where the enzymes being purified is known to be stable, but others might be denatured and precipitate from solution. Enzymes are often particularly stable in the presence of their substrates, so the relevant substrate can be added to increase the effectiveness of this procedure.

Further Purification Procedures

The crude extract, partially purified as described already, may be treated in a variety of ways to increase the purification of the relevant enzyme.

Adsorbent gels

Adsorbent gels (e.g. zinc hydroxide) have been used to remove pigments from enzyme preparations. A mixture of enzymes may also be adsorbed by a suitable gel (e.g. aluminium hydroxide) and then fractionated by elution with buffers of increasing ionic strength. Partition chromatography has similarly been used to separate mixtures of enzymes: for example, ribonuclease was purified by Martin and Porter on columns packed with kieselguhr, using a two-phase system consisting of ammonium sulphate, ethyl cellosolve and water; the sample was applied to a column which had been pre-equilibrated with the organic phase, and then eluted by the aqueous phase, separation of components being on the basis of relative solubility in the two phases. Such techniques played an important part in the development of enzymology, but have now been superseded by others.

Electrophoresis

Electrophoresis is mainly an analytical procedure, since it is ideally suited to the separation of small amounts of material, but it has also been used for the purification of proteins. The rate and direction of migration of a protein in an electric field depends on its net-charge at the pH used and also on the size of the molecule, since this imposes some resistance to movement. In zone electrophoresis, a mixture of proteins is introduced at a common point (the origin) and allowed to move in an electric field, usually in a horizontal direction, each molecule travelling in the same zone as others of the same charge and size. To minimise diffusion the whole process is usually carried out in a solid but porous support medium (e.g. starch gel) through which the buffer and proteins permeate. When separation has been achieved, as indicated by cutting a narrow longitudinal section from the support medium and staining for proteins (e.g. with Amido Black) then the rest of the support medium may be cut into strips across the direction of travel and the contents of each extracted by elution. Alternatively, zone electrophoresis may be carried out in a vertical direction in a closed column with, for example, powdered cellulose as the support medium; after electrophoresis has been completed, a tap at the bottom of the column is opened to enable the liquid contents to be run out and collected in fractions.

Some forms of electrophoresis capable of giving extremely high resolution, e.g. isotachophoresis and isoelectric focusing, which are employed at present mainly for analytical purposes, are likely to find increasing use in preparative work.

Cation exchange chromatography

Cation exchange chromatography on columns packed with carboxylated polystyrene has been used in the successful purification of stable, low molecular weight basic proteins such as lysozyme and ribonuclease. However, cellulosic ion exchange chromatography has proved of more general application to the separation of proteins. Such resins, in contrast to most other ion exchange resins, are relatively hydrophilic and have an open structure readily penetrated by large molecules such as proteins. The most frequently used are diethylaminoethyl (DEAE)-cellulose which is an anion exchanger, and carboxymethyl (CM) cellulose, a cation exchanger. The DEAE group, $-OC_2H_5NH(C_2H_5)_2$, is highly positively charged at pH 6–8, so DEAE-cellulose is most useful for the chromatography of proteins which are negatively charged in this range; similarly, CM-cellulose, cellulose–$OCH_2CO_2^-$, is most applicable for the separation of proteins which are positively charged at around pH 4.5.

Elution of proteins from the columns may be brought about by changes in either salt concentration or pH: as the concentration of salt (e.g. NaCl) increases, protein is displaced from DEAE-cellulose by the anion (Cl^-) and from CM-cellulose by the cation (Na^+); if the pH is altered over the relatively narrow working range, proteins are eluted as their isoelectric point is reached, since they then have no net charge with which to bind to the resin. Ion exchange chromatography has been used in the purification of membrane proteins in non-ionic or zwitterionic detergents as well as for the purification of soluble enzymes. The buffering capacity of ion exchangers may be exploited to create a linear pH gradient in the column and separate proteins by chromatofocusing. This technique has features in common with isoelectric focusing.

Gel filtration

Gel filtration (size exclusion or molecular-sieve chromatography), another important technique in enzyme purification, is carried out with columns packed with swollen gels which separate components of a sample on the basis of molecular size: molecules too large to enter the pores of the gels will pass quickly through the columns; smaller molecules will pass through the column more slowly, the actual speed for each component being dependent on the ease with which its molecules can pass into the gels and be thus retarded. Gels commonly used include cross-linked dextrans (Sephadex), cross-linked agarose (Sepharose) and cross-linked polyacrylamide (Biogel). These are graded according to pore size, and thus to the size (and hence molecular weight) of protein molecule which can enter. For a successful separation, the enzyme of interest must not be too large to penetrate the gel, because it would then pass straight through the column together with all other proteins of similar and greater molecular size.

Eluting buffers should be of high ionic strength (e.g. 20 mM) to counteract the few charges which may be present on the gel: apart from that, the only criterion for the buffer is that the proteins are stable in it: no pH or salt gradient is employed, since the separation is by size alone. However, it should be noted that some of these gels are available with attached ion exchange groups (an example is DEAE-Sephadex) in which case the principles of elutions are as for cellulosic ion exchange chromatography.

In the case of membrane proteins solubilised in detergents, gel filtration is widely used to separate protein-detergent micelles from excess detergent: this is particularly easy to achieve if detergent molecules are kept in the monomeric form, e.g. by ensuring that the detergent concentration is less than the CMC, or by adding a small amount of a bile salt.

Affinity chromatography

Affinity chromatography is a bio-specific process and thus ideally suited for the separation of one protein from all others, including those which resemble it so closely in physical characteristics that separation by any other procedure is extremely difficult. The column is packed with an inert matrix (e.g. agarose) to which ligands for the required enzyme have been attached. These immobilised ligands may be, for example, substrate analogues, and must be covalently linked to the matrix in such a way that they can still bind to their enzyme, if this is present: this usually means that a spacer arm (e.g. a hydrocarbon chain) must be included between matrix and ligand, joining to the ligand at a point which does not interact with the enzyme, if steric hindrance to the formation of an enzyme-ligand complex is to be avoided. The attachment of the spacer arms to –OH groups of the matrix is often achieved by the use of CNBr reagent.

The protein mixture is applied to the column, and the relevant enzyme is trapped by the immobilised ligands while all other proteins pass through and are discarded. The enzyme is then liberated from the column either by eluting with a deforming buffer at a pH which changes the characteristics of the enzyme and no longer allows it to bind to the immobilised ligand, or by the use of a competitive counter-ligand, which displaces the immobilised ligand on the enzyme. In both cases, the enzyme passes through the column and can be collected, now free of other proteins (Fig. 6.3). An example of an enzyme extracted by the use of such a technique is β-galactosidase with the substrate-analogue *p*-aminophenyl β-D-galactoside as ligand, the deforming buffer being 0.1 M borate at pH 10.

Immunoaffinity chromatography

A closely related procedure is immunoaffinity chromatography, where the immobilised ligand is an antibody specific for the enzyme which is being purified. The use of an immobilised monoclonal antibody as ligand gives even greater specificity than use of a more conventional antibody.

The bio-specific nature of these techniques means that a high degree of purification can be achieved; on the other hand, it also means that a polymer-ligand complex prepared for the purification of one enzyme is limited to that particular application.

Another rapidly developing form of affinity chromatography involves the use of reactive triazine-based dyes, e.g. Cibacron Blue F3G-A and dyes of the Procion series, as ligand. Under alkaline conditions the chlorotriazine dye is linked, usually directly, to a matrix such as agarose by a triazine bond; the reaction involves a hydroxyl group on the matrix and a chloride on the dye. Such immobilised dyes bind a wide range of NAD- and NADP-dependent dehydrogenases and other enzymes. They are, of course, less specific than biological group-specific ligands, and their specificity is difficult to predict, but they have many advantages: they have a greater protein-binding capacity, they are extremely resistant to chemical and enzymic degradation, they can be used and re-used in a variety of applications, and the triazine bond is more stable than the isouronium linkage introduced by the CNBr activation of agarose. Elution of enzyme from the ligand is usually achieved by changing pH or increasing salt concentration.

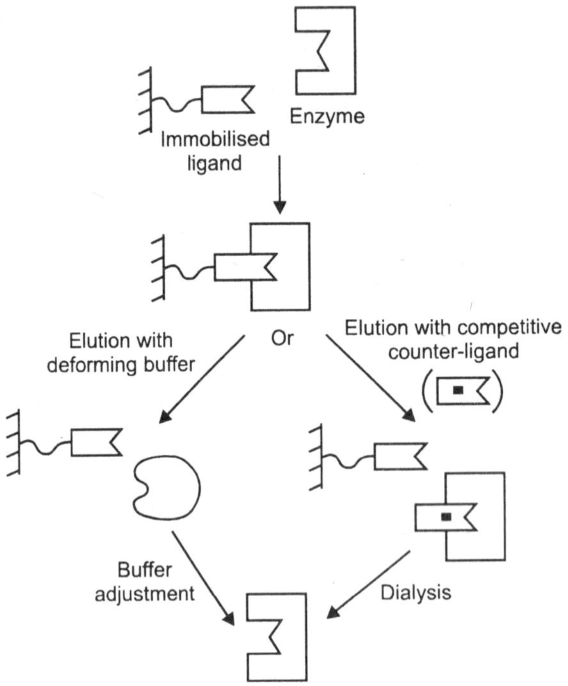

Fig. 6.3. Diagrammatic representation of the processes involved in affinity chromatography; the treatment with deforming buffer and with competitive counter-ligand are alternatives.

Even less specific, though still useful, is hydrophobic interaction chromatography. This technique depends on the fact that although amino acids with polar side chains predominate at the surface of enzymes, some non-polar ones will be present which can form hydrophobic interactions with column packings such as phenyl-Sepharose, particularly when the ionic strength is high. Elution can be brought about by decreasing the ionic strength or introducing an organic solvent.

A major advance over the past few years has been the application of HPLC technology to protein purification. The use of stainless steel columns and high-quality robust packing material of particle size 10 μm or less has allowed separations to be obtained similar to those for conventional low-pressure chromatography but in minutes rather than hours. HPLC columns for preparative work are wider than those for analytical use (at least 20 mm internal diameter), which allows milligram or even gram amounts of protein to be applied; a fraction collector is usually included in the system (e.g. the Waters Delta Prep 3000 system) after the detector. High-performance size-exclusion chromatography (HPSEC) is based on the same principle as conventional gel filtration, but Sephadex, Sepharose and other non-rigid or semi-rigid gels can only withstand low pressures. Instead, HPSEC often utilises rigid spherical beads of porous silica with bonded hydrophilic polar groups. High-performance ion exchange chromatography

(HPIEC) utilises amines as anion exchangers and sulphonic or carboxylic acids as cation exchangers, each bonded to some rigid support such as silica. High-performance liquid affinity chromatography utilises ligands bound to supports such as epoxy-silica:

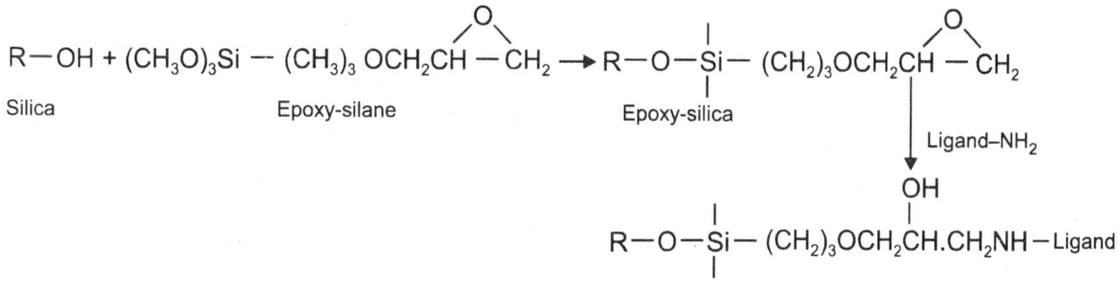

Proteins may also be separated by reverse-phase HPLC on alkylsilica columns, the eluting solvents being buffered aqueous and organic mixtures.

When after the use of one or more of these procedures it is considered that the relevant enzyme has been separated completely from other proteins mineral salts and any small molecules present may be removed by dialysis, if this is desired.

The purified enzyme preparation is likely to be quite dilute, so it might require concentrating before being used for the purpose for which it was prepared. One convenient method for doing this is lyophilisation, followed by the redissolving of the enzyme in a small volume of liquid. Another suitable procedure is ultrafiltration: this involves forcing the solvent molecules through a membrane of chosen porosity (e.g. by the application of nitrogen gas pressure) and thus separating them from the protein molecules, these being too large to pass through the pores.

The ideal way to complete a purification procedure is to crystallise the enzyme, if this is possible. Sometimes it may be achieved by adding not quite enough ammonium sulphate to cause precipitation of the enzyme and leaving this solution in a coldroom for several days. In general, successful conditions have to be determined empirically by attempting crystallisation at a variety of salt and organic solvent concentrations and pH values, and by showing much patience.

Membrane enzymes may be inactive after purification unless re-introduced into a phospholipid environment. For example, phospholipid may be added to purified protein-detergent micelles and concentrations adjusted so that the phospholipid replaces detergent in the micelles, released detergent, in the monomeric form, may then be removed by gel filtration, dialysis or ultrafiltration. An example of a membrane enzyme that has been characterised is Na^+–K^+ ATPase.

Criteria of Purity

At each stage of a purification procedure, an assay for the enzyme being purified should be performed on all fractions and its specific activity in each fraction determined. This will show, irrespective of preconceived ideas, precisely which fractions contain the important enzyme and will enable the degree of purification to be calculated. The total activity of this enzyme will be unchanged as a result of the purification step, unless some has been lost during purification; under no circumstances can it be increased. However, some of the contaminating proteins which were originally present in the same preparation as the enzyme should now be in different fractions. Hence, if the fractions containing the relevant enzyme are combined, the total activity of the enzyme in the combined fraction should be the same as that started with, whereas the total amount of protein in this fraction will be less than that originally present. The specific activity of the enzyme in the combined fraction should, therefore, be greater than that in the preparation before purification, and the increase in specific activity will be a measure of the purification achieved. With each successive purification step, the specific activity of the fraction (or fractions) containing the enzyme should be greater than before until complete purification is achieved and the specific activity reaches a limiting value.

However, the finding of the same specific activity value before and after a purification step does not necessarily mean that the enzyme preparation is completely pure: it could simply mean that contaminating proteins have passed through the procedure in the same fraction as the enzyme. Similarly, crystallisation cannot be taken as proof that only one protein is present, for many mixed protein crystals have been found. Hence other criteria of purity have to be considered.

If a particular enzyme is known to be a possible contaminant, then a logical step is to carry out an assay for that enzyme and demonstrate its absence in that way. However, that would not exclude the possibility that other contaminants were present.

Analytical ultracentrifugation

Analytical ultracentrifugation is widely used to investigate the purity of enzyme preparations. With high centrifugal fields (e.g. 5,00,000 grams), diffusion effects are largely overcome; hence, as a randomly distributed group of identical protein molecules sediment in an analytical cell under the effect of such a field, a sharp boundary is formed between the portion of the medium which has been cleared of protein molecules and that which still contains them. This boundary refracts light, and its position can be determined through a quartz window in the cell: for example, it shows up as a peak if the refractive index gradient at each point in the cell is measured using the Schlieren optical system. The movement of this boundary as sedimentation proceeds can be constantly monitored, and automatically recorded, and the data used to calculate the characteristics of the molecule in question. If the protein is pure, only one boundary should be observed: detection of extra boundaries suggests the presence of contaminants.

Electrophoresis

Electrophoresis in various forms is very effective in separating mixtures of proteins so it is used extensively to investigate the purity of enzyme preparations. Of particular importance in this context is discontinuous (disc) electrophoresis, where the electrodes are placed in separate buffer reservoirs linked vertically by gel (usually polyacrylamide) supported by glass plates or cylinders. The discontinuity is in the gels and buffers used, pH differences being used to concentrate the sample into a very narrow band as it passes through a large-pore spacer gel on its way to the running gel where the actual separation of the sample components takes place (Fig. 6.4).

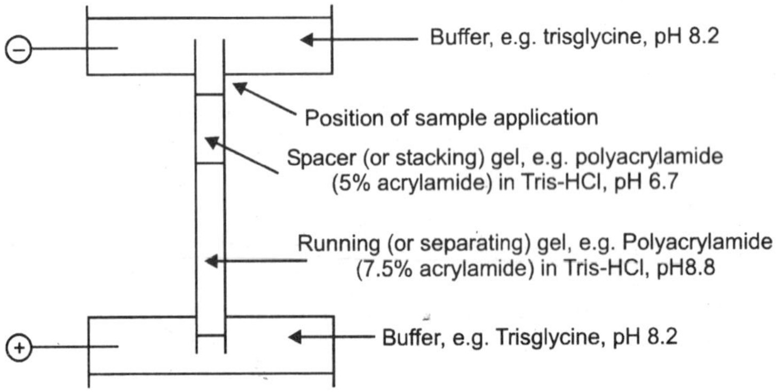

Fig. 6.4. Diagrammatic representation of an apparatus set up for discontinuous (disc) electrophoresis of proteins.

This running gel has smaller pores than the spacer gel, and separation is achieved by a combination of electrophoresis and gel filtration. On completion of the electrophoresis and after staining for proteins, it is easy to see if more than one band is present, for each is extremely narrow and well-defined.

Capillary electrophoresis

Capillary electrophoresis, in which charged components are separated in media inside small capillaries, is becoming increasingly important. This provides a rapid and efficient analytical technique.

Another relevant technique is isoelectric focusing, which is an example of moving boundary electrophoresis rather than zone electrophoresis. A pH gradient is set up between the electrodes by allowing an acid (e.g. phosphoric acid) to diffuse in from the anode end and a base (e.g. ethanolamine) from the cathode end. This can be stabilised by introducing a mixture of ampholytes (e.g. synthetic and natural amino acids) selected so that their individual isoelectric points cover the required pH range. Each ampholyte will move in the electric field to the position in the pH gradient corresponding to its isoelectric point and then remain there, since it will no longer have a net charge.

The protein sample is then introduced and electrophoresis continued, possibly for several days, until each component reaches a stationary position at its isoelectric pH. The zones containing each protein are very sharp as a result of this focusing and proteins whose isoelectric points differ by as little as 0.02 pH units can be distinguished by this method.

Isotachophoresis

Isotachophoresis is another example of moving boundary electrophoresis, and it also has some similarities to discontinuous electrophoresis. For the separation of proteins at mildly alkaline pH, when most would be anions, a leading anion (e.g. phosphate) is added which has a faster mobility towards the anode than any of the sample anions. A trailing or terminating anion (e.g. glycine), with a slower mobility than any in the sample, is also added. The system is buffered by a common-counterion, e.g. Tris. The leading and terminating anions are applied at different sides of the sample, the leading anion nearest the anode. A high voltage is applied and, when steady-state is achieved, all components will migrate towards the anode in discrete zones at the same velocity ('iso tacho' being Greek for 'same speed'), the sample anions arranged in order of their mobilities.

Spacers similar to the ampholytes used in isoelectric focusing may be added to separate the zones of sample anions. Isotachophoresis may be carried out in a purely aqueous environment in a capillary tube or, particularly for preparative work, in columns packed with gel (e.g. polyacrylamide) of pore size large enough to minimise molecular sieving.

Analytical HPLC

Analytical HPLC procedures may also be used to investigate the purity of a protein preparation.

No single technique, however sensitive, can establish enzyme purity because of the possibility that a contaminating protein might behave in an identical fashion to the important protein under the conditions used. However, if several characteristics are investigated, and the conditions of investigation varied, the chances of the contaminant avoiding detection are reduced. Therefore, evidence of purity must be obtained by the use of several of the above procedures before it can be concluded that the sample is indeed pure.

DETERMINATION OF MOLECULAR WEIGHTS OF ENZYMES

When an enzyme has been purified, its molecular weight may be determined.

Gel Filtration

Gel filtration, which separates molecules on the basis of size, provides one way of doing this. A packed column is calibrated by applying proteins of known molecular weight to the top of the column and determining the volume of buffer required for the elution of each protein: as each protein leaves the column there should be an absorbance peak at 280 nm in the column eluate, thus providing a simple way of monitoring the elution. From the data obtained, a graph may be drawn of elution volume against molecular weight. The protein of unknown molecular weight is then passed through the column and its elution volume determined exactly as for the marker proteins. Hence, by reference to the calibration graph, the molecular weight of the protein may be estimated. One source of error in this procedure is that it takes no account of molecular shape: in fact, a protein molecule which is somewhat elongated will pass down the column more slowly than a spherical one of the same molecular weight.

Ultracentrifugation

Ultracentrifugation is also widely used for the determination of molecular weights of proteins. For a spherical protein of molecular weight M_r sedimenting through a medium in a centrifugal field, the centrifugal force per gram-mole is $M_r(1 - \bar{v}r)\omega^2 r$ where, $\bar{v}$ is the partial specific volume of the molecule (i.e. the increase in volume in cm^3 when 1 gram solute is added to a large volume of solvent; the value for a protein is usually about 0.74). The centrifugal force is opposed by an equal frictional force per gram-mole of:

$$\frac{RT}{D} \times \text{sedimentation rate}$$

where, R is the gas constant (in ergs per mole per degree), T is the absolute temperature and D is the diffusion constant of the molecule.

$$\therefore\ M_r(1 - \bar{v}\rho)\omega^2 r = \frac{RT}{D} \times \text{sedimenation rate}$$

But sedimentation rate = $s.\omega^2 r$

$$\therefore\ M_r(1 - \bar{v}\rho)\omega^2 r = \frac{RT}{D}\, s\omega^2 r$$

$$\therefore\ M_r = \frac{RTs}{D(1 - \bar{v}\rho)}$$

This is known as the Svedberg equation, after the pioneer of ultracentrifugation.

As discussed already the rate of movement of the boundary separating the protein-free zone from the rest of the medium can be determined, and this is a measure of the sedimentation rate. Hence the sedimentation coefficient (s) can be calculated, since ω^2 and r are easily determined: for most proteins, s lies in the range 1–200 S. The diffusion constant (D) can similarly be calculated from moving boundary determinations, this time in the absence of a centrifugal field.

Once these and the other relevant factors have been determined, the molecular weight of the protein can be calculated by the use of the Svedberg equation. However, as with gel filtration, there is a possible source of error in the assumption that all protein molecules are spherical.

A technique which is commonly used to determine molecular weights of proteins, and which does not assume that the molecules are spherical is SDS-gel electrophoresis. SDS (sodium dodecyl sulphate,

$CH_3(CH_2)_{11}$ $SO_3^-Na^+$) is an anionic detergent whose hydrocarbon chain can become linked to hydrophobic regions in the interior of globular proteins, leaving the ionised sulphonate (—SO_3^-) group jutting out into the surrounding medium. Many SDS molecules can bind in this way to a single protein molecule, disrupting its natural shape and giving it a large negative charge, regardless of the net charge originally possessed by the protein. Large protein molecules will complex with more SDS than small ones, so all will have roughly the same charge/mass ratio, and all will have much the same shape. Therefore, if molecules of different proteins are complexed with SDS and subjected to zone electrophoresis in an unrestrictive medium, all should travel together, regardless of their original characteristics. However, if electrophoresis is performed in a support medium with a restrictive pore size (e.g. in polyacrylamide gel), the complexes from each of the different proteins will be separated, entirely on the basis of size. Since the size of each protein-SDS complex must be dependent on the original size of the protein molecule, a system can be calibrated by the use of marker proteins of known molecular weight; hence the molecular weight of a test protein may be determined. One point that should be borne in mind, however, is that SDS tends to break all weak polypeptide-polypeptide interactions, so, for an oligomeric protein, this technique may well give the molecular weight of the sub-unit rather than of the complete protein.

To sum up, soluble enzymes may be extracted from cells simply by the disruption of membranes the actual technique employed depending on the nature of the cell, the intracellular location of the enzyme and the stability of the enzyme. In each case, the pH, salt and organic solvent concentration of the medium must be carefully chosen to ensure that the enzyme being extracted remains in solution while the cell debris is removed by centrifugation.

Special procedures have to be employed to extract enzymes which form an integral part of membranes (e.g. extraction with a detergent): otherwise they would be discarded with the cell debris.

Techniques for the purification of the extracted enzyme include electrophoresis, ion exchange chromatography, gel filtration and affinity chromatography. Enzyme assay should be performed after each purification step, and the specific activity of each fraction determined. As purification proceeds, the specific activity of the enzyme preparation should rise to a limiting value. When this has been achieved, the purity of the preparation should be checked by a variety of techniques, including ultracentrifugation, electrophoresis, isoelectric focusing and by high performance liquid chromatography.

After purification, the molecular weight of the enzyme may be determined by gel filtration, ultracentrifugation or SDS-gel electrophoresis.

SECTION II

Types of Enzymes/Classification of Enzymes

Chapter 7

Carbohydrate Hydrolysing Enzymes

INTRODUCTION

Carbohydrates are widely distributed in nature. Fifty to eighty per cent of the dry weight of plants consists of carbohydrates, which include the important structural components cellulose, pectin, and starch. The term carbohydrate is taken from the French hydrate de carbone. Carbohydrates are composed principally of carbon and water, as shown in the chemical formula for glucose: $C_6(H_2O)_6$. Polymeric carbohydrates are built from simple carbohydrate monomers such as glucose. Oligosaccharides, such as sucrose (cane sugar) or maltose, are composed of only a few of the monomers linked together. Additional linkage of oligosaccharides leads to long-chain polysaccharides, which play an essential role in human and animal nutrition and are the basis of many natural fermentation products, such as beer, wine and vinegar. They also provide the raw materials for the production of many industrial products, such as antibiotics and enzymes. Carbohydrates give cotton cloth, writing paper and a valuable source of fuel. Enzymes can efficiently degrade carbohydrates without combustion for many practical applications.

ACTION OF ENZYMES

The action of enzymes can be experienced at breakfast time. The physician tells clients 'chew your bread well', and when they do, they experience a sweet taste. The enzymes in human saliva, the salivic amylases, convert tasteless starch in bread into sweet-tasting sugar. Cooked starch forms rigid gels, which are called pastes. A trace amount of salivic amylase converts such a paste into a liquid in seconds. The high-molecular-weight, hydrophilic starch is hydrolysed into low-molecular-weight sugars and dextrins. This hydrolysis is essential for our nutrition because humans, animals, and even micro-organisms absorb carbohydrates only in the hydrolysed form as sugar.

Scientifically, the carbohydrate cleaving enzymes are called glycosidases or more commonly, carbohydrases. They hydrolyse polysaccharides such as starch, cellulose, and pectin. Their action is apparent in lowering the viscosity of watery solutions of carbohydrates such as pectin and carboxymethyl-cellulose solutions. The activity of the enzyme is determined by the amount of free reducing sugars produced by its action.

Glycosidic Linkage

The monomeric building blocks of carbohydrates and their polymerisation to oligomers and polymers is well known. The linkage between monomeric sugars, such as glucose, to disaccharides, is termed a glycosidic linkage (Fig. 7.1).

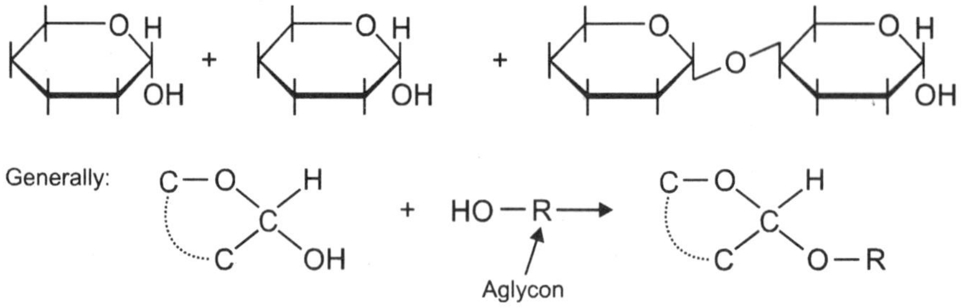

Fig. 7.1. The glycosidic linkage.

A glycoside is scientifically defined as a full acetal of a sugar, where the free acetal hydroxy group of the sugar is replaced by an 'R' residue. If 'R' is another sugar residue, then the molecule is a disaccharide. If both the sugar and 'R' are glucose, then maltose is formed. 'R' can also be a different molecule containing a hydroxyl group, such as methanol or phenol. These are, in general, termed aglycons. Such glycosides occur widely in nature. They are found in flavour substances and in natural dyes.

Enzyme Specificity

Carbohydrases hydrolyse the specific glycosidic linkages of certain monosaccharide residues. They are able to cleave short-chain oligosaccharides as well as polysaccharides with various structures. Table 7.1 shows some of the products of enzymatic hydrolysis of some oligosaccharides. Table 7.2 depicts some naturally occurring polysaccharides, their monomeric building blocks, their linkages and the enzymes used for their scission.

Table 7.1. Oligosaccharides.

Substrate	*Hydrolysis product*	*Linkage*	*Enzyme*
Maltose	2 Glucose molecules	α-1,4	α-Glycosidase
Sucrose	Glucose, fructose	α-1,2	–
α-Methylglycoside	Glucose, methanol	–	–
Cellobiose	2 Glucose molecules	β-1,4	β-Glycosidase
Raffinose	Galactose, sucrose	α-1,6	α-Galactosidase

Table 7.2. Polysaccharides.

Polysaccharide	*Monomer unit*	*Linkage*	*Enzyme*
Starch	Glucose	α-1,4	α-Amylases
–	–	–	β-Amylases
–	–	α-1,6	Isoamylases
Cellulose	Glucose	β-1,4	Cellulases
Dextran	Glucose	α-1,6	–
Xylan	Xylose	β-1,3; β-1,4	Xylanases
Pectin	Galacturonic acid	α-1,4	Pectinases
Araban	Arabinose	α-1,5; α-1,3	Arabanases
Inulin	Fructose	β-1,2	Inulinase

The mechanism of scission is consistent with a transfer of a substrate glycosyl residue to water (Fig. 7.2).

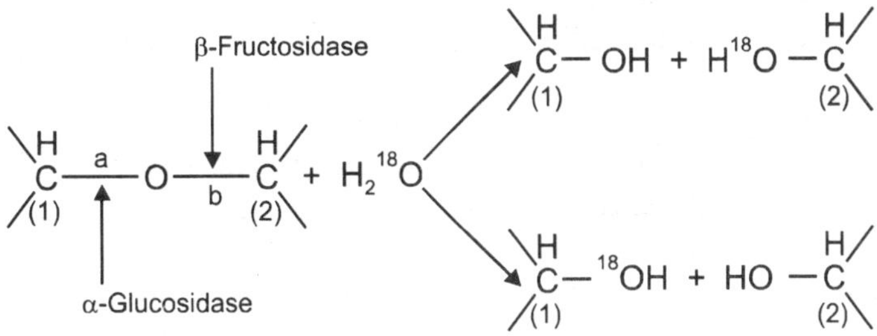

Fig. 7.2. Sucrose hydrolysis with enzymes of different specificities.

The position of transfer can be located using ^{18}O-labelled water. An α-glycosidase cleaves at the α-position, whereas a β-fructofuranoside cleaves at the β-position. Free reducing sugars are formed by hydrolysis.

The specificity of carbohydrases depends on:

1. The configuration of the glycosidic linkage. For example, maltose can be hydrolysed only by α-glucosidase and not by β-glucosidase.
2. The chemical structure of the monomeric sugars joined by the linkage.
3. The molecular weight of the substrate, α-amylases are able to hydrolyse high-molecular-weight amylose very rapidly, but maltose or maltotriose are hydrolysed only very slowly, if at all. Enzymatic hydrolysis is, by definition, reversible. The aqueous reaction equilibrium favours the hydrolysis products.

In non-aqueous systems, glycosyl residues can be transferred to other sugars instead of water. This reaction, termed transglycosylation, yields new oligoglycosides.

By volume and value, carbohydrases are the most important of the industrially produced enzymes. Their major application is in starch processing, which encompasses the production of dextrins, glucose, maltose and different types of fructose syrup. These products are needed in all areas of the food industry. Carbohydrases are used in distilleries, breweries, and the production of baked goods. The production of fruit juices or fruit juice concentrates without these enzymes is unimaginable.

AMYLASES

Starch

As a reminder, a few facts about starch are given here. Starch is the major component of most foods and the most important energy source in our food supply. Starch products are obtained from grains, roots, and tubers. Glycogen is stored as a reserve carbohydrate in the liver of animals and humans. Starch granules contain amylose and amylopectin, which are bound to each other by hydrogen bonding. The types of cross-linking and the degree of polymerisation determine the various characteristics of the different types of starch. These factors dictate, among other properties, the different swelling power and gelatinisation temperature of various starches, properties that are of great importance for the food processing industry. Amylose consists of a linear chain of 500 or more α-1,4-bound glycosidic linkages. By contrast, amylopectin is highly branched. The branches occur in intervals of 20 glucose residues and are formed via α-1,6 linkages. So-called waxy starch in corn consists of almost 100 per cent amylopectin.

If amylopectin is treated with pure β-amylase, the hydrolysis stops at the 1,6 linkages, giving so-called limit dextrins. When treated with iodine solution, starch forms a deep-blue-coloured iodine complex, while amylopectin gives a red colourisation. The point at which the characteristic blue colour disappears during starch degradation is called the iodine-negative value or the achroic point.

Gelatinisation of Starch

On heating a suspension of starch in water, the starch swells when the specific gelatinisation temperature is reached. The sudden water uptake leads to a strong increase in viscosity and to a change in the chemical and enzymatic vulnerability of the starch. Gelatinisation plays an important role in the process of bread baking, in the production of thickening agents based on starch, and in the production of textiles where starch and dextrins are used as glue or sizing. As the gelatinised starch cools, the amylose fraction crystallises rapidly by forming intermolecular hydrogen bonds. This process, known as retrogradation, is one of the reasons why daily bread stales so rapidly.

Stepwise Starch Degradation

The amylases split starch into dextrins and sugars by hydrolysis of the α-1,4 glycosidic linkages. They are common in nature, occurring in saliva, pancreas, and many plants such as cereals. In industry, amylases are produced from cultures of bacteria and fungi. There are α- and β-amylases. α-Amylases split the glycosidic linkage in the interior of the starch chain. On the other hand, β-amylases split maltose from the non-reducing end of the chains. In addition, some amyloglucosidases liberate glucose residues stepwise from the end of the chain (Fig. 7.3).

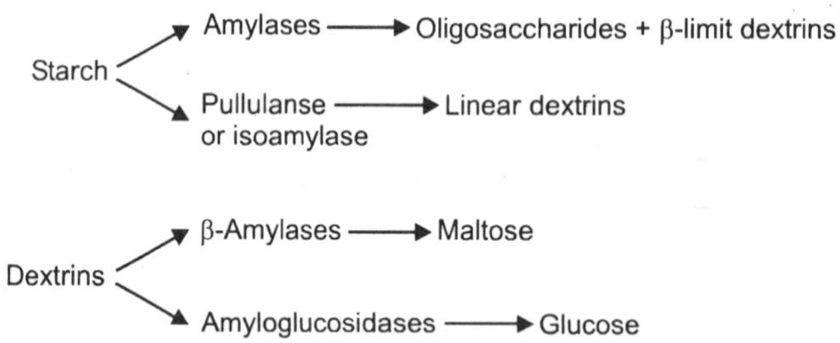

Fig. 7.3. Enzymatic starch hydrolysis.

Endoamylases (EC 3.2.11)

The endoamylases hydrolyse the α-1,4 glycosidic linkages in starch, glycogen and derivatised speech. Depending on the source, the properties of different amylases, vary widely. Many α-amylases can be produced in pure crystalline form, such as those from malt, pancreas, *Aspergillus oryzae* and *Bacillus subtilis*. Their properties have been well investigated. They differ in their molecular weights, temperature stability, optimum-activity pH, and hydrolytic specificity. Because of this specificity, different α-amylases produce oligosaccharides of various chain lengths and yields. The fast and complete degradation of starch requires pre-gelatinisation of the substrate. Bacterial α-amylase degrades gelatinised starch 300 times faster and fungal α-amylase 1,00,000 times faster than native starch. The rate of hydrolysis depends primarily on the degree of polymerisation of the starch, dropping markedly with a lower degree of polymerisation. Amylose is hydrolysed faster than amylopectin. Amylase-induced starch hydrolysis is shown in Fig. 7.4.

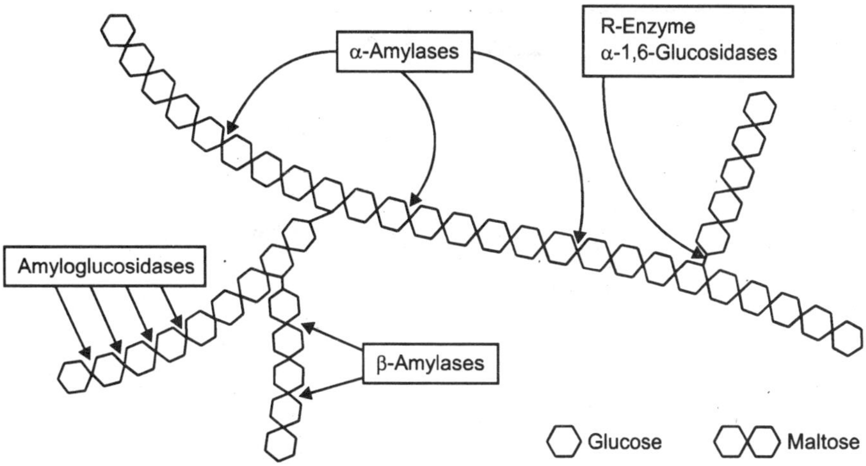

Fig. 7.4. Starch hydrolysis by amylases.

Bacterial α-amylase (EC 3.2.1.1. CAS 9000-90-2)

In industry, bacterial α-amylases are produced mainly from cultures of *B. subtilis* var. *amyloliquefaciens*. In addition, industrial preparations from *B. subtilis* or *B. subtilis* var. *amylosaccharíticus* are available. The latter produces more glucose from starch than the liquid α-amylase from *B. amyloliquefaciens*.

Properties

1. Molecular weight: 49000 daltons (Fischer and Stein); 60000 daltons (Granum) from *B. amyloliquefaciens*; 48000 daltons from *B. subtilis* (Kakiochi, 1965). In the presence of zinc, a dimer is formed with a molecular weight of 1,00,000 daltons.
2. Co-factors: α-Amylase contains no co-enzymes, but as a calcium metalloenzyme, it requires at least one gram-atom of calcium per mole of enzyme in order to maintain its activity. This calcium also stabilises the enzyme against heat denaturation and attack from proteases. Complexing calcium with EDTA leads to denaturation of the free amylase by heat, urea, or acid and rapid degradation by proteases.
3. Activity: The K_m (starch) is between 1 and 3.4×10^{-3} for amylases of different *B. subtilis* strains. The specific activity is about 4×10^5 SKB units per gram of enzyme protein (Figs. 7.5 and 7.6).
4. Temperature stability: Ninety per cent of the activity is maintained on incubation with 150 ppm Ca^{2+}, at pH 6.5–7.5 for one hour at 65–70°C (Fig. 7.7). High substrate concentration, such as 25–40 per cent starch, also stabilises the enzyme, so that total inactivation, even at boiling temperatures, requires a long treatment time.
5. Inactivation: The enzyme, depending on temperature, is rapidly inactivated at pH values lower than 4.5–4.7. In the presence of high substrate concentrations, the pH must be adjusted to below 4.0. Treatment for 10–15 minutes at 120°C is required for heat denaturation of pH 6–8.
6. Inhibitors: Chelating agents such as polyphosphate, EDTA, oxalate, free chlorine, and oxidising agents.

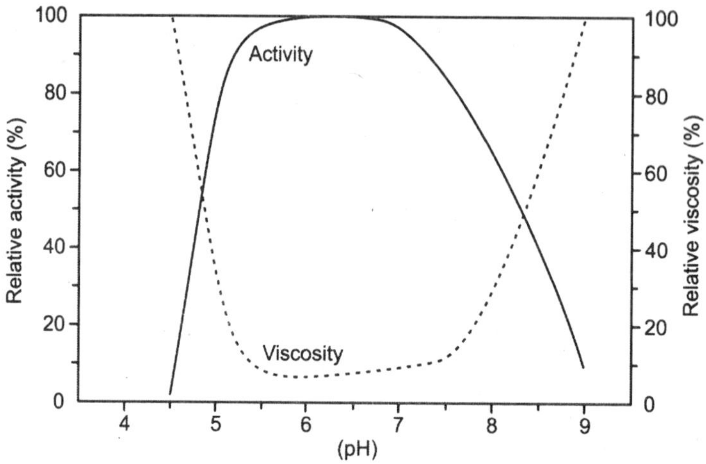

Fig. 7.5. pH dependence of bacterial amylase from *B. amyloliquefaciens*; measured by loss of viscosity and increase in reducing sugars.

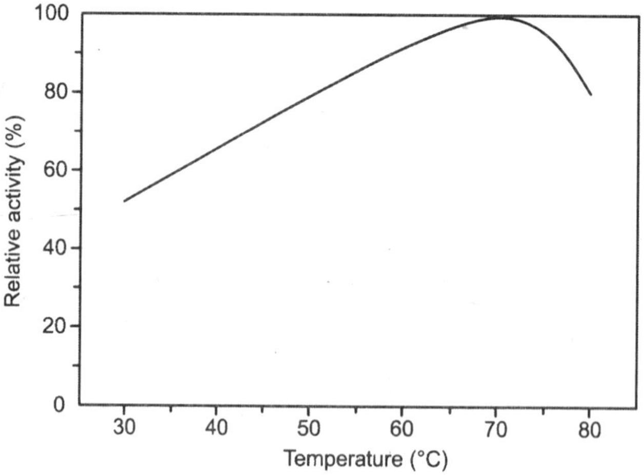

Fig. 7.6. Temperatures dependence of bacterial amylase (2 per cent soluble starch, pH 6.0).

Technical preparations

1. Activity: Industrial α-amylase preparations are offered in liquid and powder forms. The liquid preparations are mainly just centrifuged, filtered culture liquids, stabilised with sodium chloride. Their activity is about 500–3000 SKB units per gram. Powdered products, prepared by precipitation or spray drying, are generally standardised with sodium chloride, calcium sulphate, starch, dextrin, or sugar. Their activity is about 5000–50,000 SKB/g.
2. Stability: Activity measurements of stored samples show that the enzymes are highly stable under dry storage. Loss in activity is about 5–10 per cent per year at 20°C. Liquid products lose about 10 per cent of their activity in 6 months at 15°–20°C.
3. Secondary activities: Technical enzyme preparations contain, in addition to α-amylases, neutral and alkaline bacterial proteases, β-glucanases, and other hemicellulases.

4. Hydrolysis pattern: The reaction rate of bacterial α-amylase increases with soluble starch and decreases with chain length. The closer the cleavage site is to the α-1,6 branching, the slower the reaction. Maltosyl-α-1,6-maltotriose cannot be degraded to lower-molecular-weight compounds. The initial products of amylose are G_2, G_3, G_6, and G_7 oligomers (Fig. 7.11).
5. Products and manufacturers: Canalpha [BN], Nervanase [AM], Bactamyl [GB], Optamyl [SO], Aquazym [NO], Rohalase A [RM].

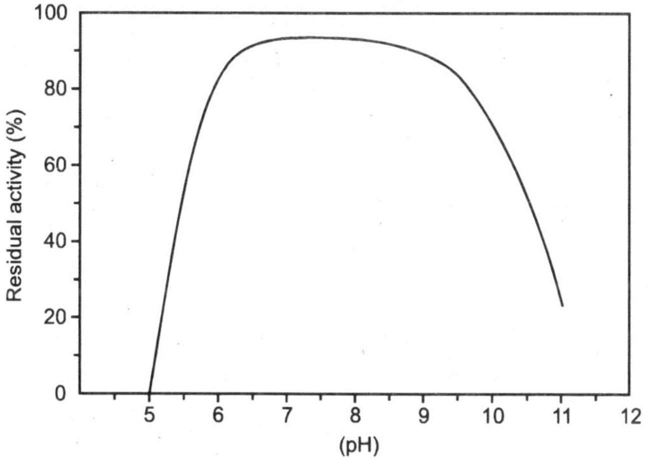

Fig. 7.7. Effect of pH on stability of bacterial α-amylase (temperature 70°C; reaction time 1 hour, 0.5 gram $CaCl_2$ and 6.0 grams NaCl per litre).

Intermediate stable bacterial amylase

A bacterial amylase with a stability intermediate between that of the fungal α-amylase and that of the bacterial amylase produced by *B. megaterium* was described by Hebeda. The optimum pH is 6.0, with 80 per cent activity at pH 5.5–6.5 and an optimum temperature of ~70°C. The enzyme is inactivated between 85° and 90°C and serves as an antistaling enzyme in baking.

Thermostable bacterial α-amylases (EC 3.2.1.1; CAS 9000-91-2)

The properties of thermostable α-amylases, and foremost the degree of heat stability, depend on the source of the enzyme. The data on heat stability in the literature can be misleading if exact details of the experimental conditions are not given. It is particularly important to know whether stability was determined in aqueous solution with calcium ions and sodium chloride or in the presence of substrate. Currently, the α-amylase used most frequently as a technical enzyme is the heat-stable α-amylase from cultures of *B. licheniformis*.

The enzyme from *B. stearothermophilus* shows a temperature optimum at about 80°C in the presence of substrate. The optimum pH is about 3.0. Tamura, Kanno, and Ishii compared different α-amylases with that from *B. stearothermophilus*, under identical reaction conditions in the presence of 5 mM Ca^{2+} ions (Table 7.3).

Srivastava isolated a strain from *B. stearothermophilus*, which produces both α- and β-amylases. Molecular weights were 48000 daltons for α-amylase, containing 13.5 per cent carbohydrate; 57,000 daltons for the β-amylase, which contained 0.8 per cent carbohydrate.

Table 7.3. Comparison of various bacterial α-amylases.

Enzyme source	*pH Range*	*Molecular weight (Daltons)*	*Temperature optimum (°C)*
B. stearothermophilus	4.0–5.2	96000	80
B. subtilis	4.5–6.5	49000	60
B. licheniformis	5.0–9.0	22500	78
B. stearothermophilus (Ogasakawa)	5.0–6.0	48000	70

Optimal activity for both enzymes is at pH 7.0. Calcium improves the heat stability, so that for incubation at 70°C in the presence of 15 mM Ca^{2+}, the preparation still showed 80 per cent activity after 40 minutes. This stability value is higher than that for the amylase from *B. amyloliquefaciens* and lower than that for the enzyme from *B. licheniformis*.

α-Amylase from B. licheniformis (EC 3.2.1.1)

This enzyme, which is presently used in starch liquefication, was first used industrially in 1973.

Properties

1. Molecular weight: 28000 daltons (SDS-gel electrophoresis; Krishnan and Chandra).
2. Activation energy: 5.1×10^5 J/mol.
3. Co-factors: Na^+, Ca^{2+}, Mg^{2+} activate.
4. Activity: Eighty per cent of maximum activity is shown between pH 4.2 and 8.0 (4.3 mM Ca^{2+}, 60°C, 10 minutes, 46 per cent soluble starch) (Fig. 7.8). In excess of 90°C, the optimum activity under these conditions is at pH 5.7 for a reaction time of 20 minutes. In a 30–40 per cent starch slurry, typical for industrial processes, short reaction times at temperatures of up to 110°C can be used (Fig. 7.9).
5. Stability: Figure 7.10 illustrates the effect of calcium ion concentration on enzyme stability. Addition of 3.4 ppm Ca^{2+} is sufficient to stabilise the enzyme. In contrast, for the α-amylase from *B. amyloliquefaciens*, addition of 150 ppm Ca^{2+} is necessary. The enzyme is stable in the presence of sodium tripolyphosphate.
6. Inhibitors: NO_3^-, F^-, SO_3^{2-}, $S_2O_3^{2-}$, MoO_4^-, *p*-chloromercuribenzoate, sodium iodoacetate.

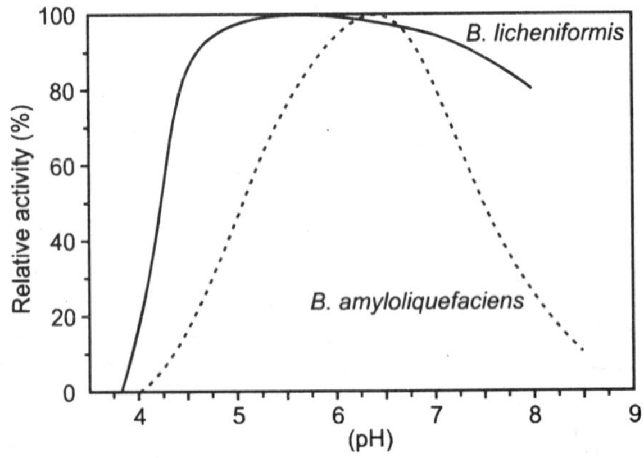

Fig. 7.8. pH dependence of bacterial α-amylases from *B. amyloliquefaciens* and *B. licheniformis* at 60°C. Substrate: 0.4 per cent soluble starch in 0.05 M acetate buffer, pH 4–5.6; 0.05 M tris-maleate buffer, pH 6–8.5; and 4.3 mM Ca^{2+} [NO].

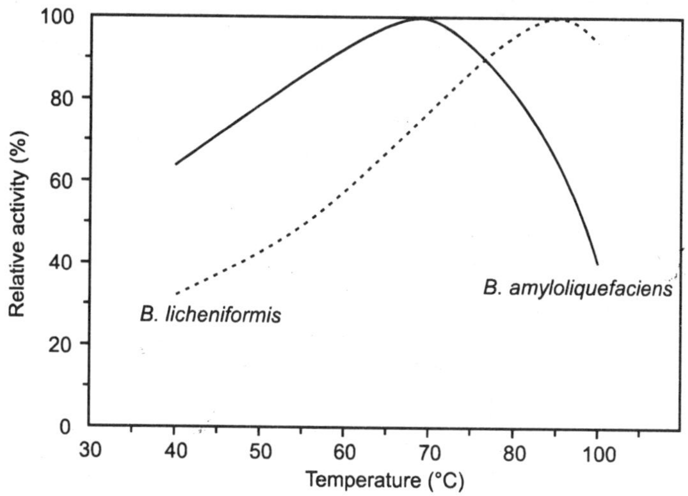

Fig. 7.9. Temperature dependence of bacterial α-amylases. Substrate: 0.46 per cent soluble starch, pH 5.7 and 4.3 mM Ca^{2+}; reaction time 7–20 minutes [RM].

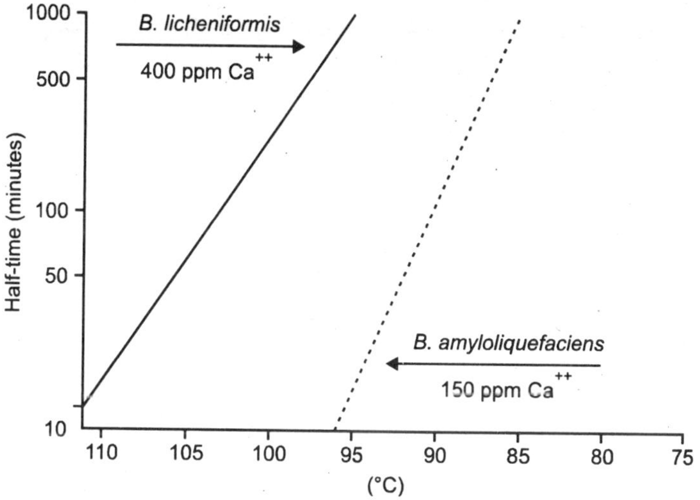

Fig. 7.10. Effect of temperature on half-life of bacterial α-amylases. Substrate: 33 per cent starch.

7. Stabilisers: Cysteine, thiourea, mercaptoethanol, sodium glycerophosphate.
8. Inactivation: The enzyme is totally inactivated by adjusting the pH to 3.5–4.0 and incubating for 5–30 minutes at 90°C.
9. Hydrolysis pattern: Although it was earlier assumed that starch degradation followed strict statistical laws, it was found later that during the first hydrolysis phase, G_3 (maltotriose) was preferentially released from the non-reducing end. *B. amyloliquefaciens* amylase preferentially yields G_6-oligosaccharides, which subsequently are further hydrolysed extremely slowly. *B. licheniformis* amylase produces mainly G_5 oligosaccharides (Fig. 7.11); these are also hydrolysed further.

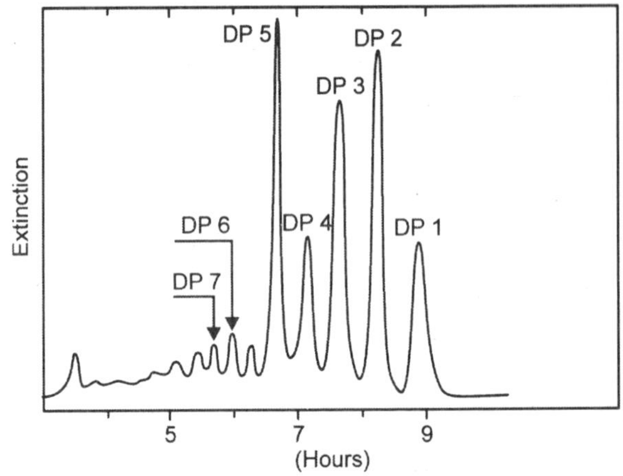

Fig. 7.11. Oligosaccharides produced from corn starch on hydrolysis by bacterial α-amylases. (DP = degree of polymerisation.)

Technical preparations

1. Activity: Most preparations are liquids with activities ranging from 2000 to 6000 SKB/g.
2. Manufacturers: [NO], [GB], [RM], [SO], [DA], [BN], [SE], and others.

Other α-amylases: In recent years, innumerable α-amylases have been described, some of which have found technical applications. These include special oligosaccharide-producing α-amylases. Certain specific oligosaccharides are of interest for the manufacture of glycosides with pharmaceutical applications and for the preparation of dietetic products.

Maltotetraose producing α-amylase: An amylase isolated from *Pseudomonas stutzeri* preferentially produces maltotetraose from high-molecular-weight dextrins.

Properties

1. Molecular weight: 62,000 daltons.
2. Isoelectric point: pH 4.7.
3. Activity: Optimum pH 6.7–7.0; maximum temperature 45°C. Potato starch was liquefied with bacterial amylase and reacted with the *P. stutzeri* enzyme for 17 hours at pH 7.0 and 45°C. The potato starch used yielded 70 per cent maltotetraose.
4. Maltopentaose-producing α-amylase: Fermentation of *B. cereus* yields an amylase that produces maltopentaose from soluble starch with a 65 per cent yield.

Properties

1. Molecular weight: ~90,000 daltons.
2. pH optimum: pH 6–7.
3. Temperature optimum: 50°–55°C.
4. Hydrolysis pattern: Amylose, amylopectin, and maltohexaose are cleaved. β-Cyclodextrin, pullulan, dextran, and maltopentaose are not cleaved.

Maltohexaose producing β-amylases

Aerobacter aerogenes is known as a producer of pullulanase. Kainuma obtained an amylase from cultures of this organism, which produced maltohexaose from amylose and amylopectin. The yield, based on the amylopectin, was about 36 per cent. During the first phase, high-molecular-weight limit dextrins were formed; thereafter, only maltohexaose was released from the non-reducing end of G_7–G_9 oligosaccharides. This mechanism of action categorises this enzyme as a β-amylase.

Maltogenic amylases (See also 'β-amylases')

Koaze found an amylase from *Streptomyces* sp. that cleaved starch to maltose (71 per cent), glucose (4 per cent), and oligosaccharides by an endo mechanism. The mixture can be hydrogenated to sugar alcohols, predominantly maltitol. Pure maltose, which can be obtained using ion exchangers, is used to replace glucose in infusion solutions, where the same osmotic pressure can be obtained by using twice the weight of glucose. Another, mainly maltose-forming, α-amylase is obtained industrially from *Streptomyces hygroscopicus*. Optimum activity is at pH 5–6. Above 60°C the enzyme is inactivated.

Manufacturer: [MJ]

Fungal α-amylases (EC 3.2.1.1; CAS 9000-90-2)

Fungal cultures were cultivated on rice in China and Japan during the nineteenth century. Those cultures, known as *koji*, were extracted with water and were used as Takadiastase for the production of fermentable sugars, which, in turn, were used to make alcohol. Dried preparations were used as digestive aids in medicine.

Fungal α-amylases were prepared from various strains of *Aspergillus*. *A. oryzae* is the most commonly used organism, but *A. niger*, *A. awamori*, and *A. usamii* are also described as being good producers. Technically, the enzyme is mainly prepared using *A. oryzae*. In Europe and in the United States, the enzyme is manufactured by submersion fermentation. In East Asia, some processes still use fixed-bed cultures (koji) with wheat or rice bran as substrate.

Fungal α-amylase from A. oryzae

Properties

1. Molecular weight: 51,000 daltons.
2. Molecular structure: The fungal amylase is a glycoprotein containing mannose residues.
3. Co-factors: The *A. oryzae* amylase binds 10 grams-atoms of calcium per mole of enzyme; 9 grams-atoms of these are bound only loosely and can be removed without loss of catalytic activity. Activity is lost with the removal of the remaining calcium with EDTA.
4. Specific activity: 50,000 SKB/mg of enzyme protein.
5. pH optimum: Fungal amylases from different sources yield different pH-activity curves. The optimum pH for the enzyme from *A. oryzae* is 5.0, that from *A. usamii* is 4 (Fig. 7.12).
6. Temperature optimum: 50°–60°C at pH 5.0 (*A. oryzae*) (Fig. 7.13).
7. Stability: Fungal α-amylases are less heat-stable than bacterial amylases and are inactivated before the gelatinisation temperature of starch is reached (68°–70°C). The enzyme from *A. oryzae* is stable for about 30 minutes at pH 5.7 and 50°C. Below pH 3.5 and at higher temperatures (30°–40°C), the enzyme is rapidly inactivated. This must be taken into consideration when the enzyme is used in fruit juices. Ca^{2+} ions stabilise the enzyme. The enzyme from *A. usamii* shows somewhat better stability in the acidic pH range.
8. Hydrolysis pattern: During hydrolysis of starch with fungal amylases, no glucose is formed from dextrin polymers larger than G_8 (Fig. 7.14).

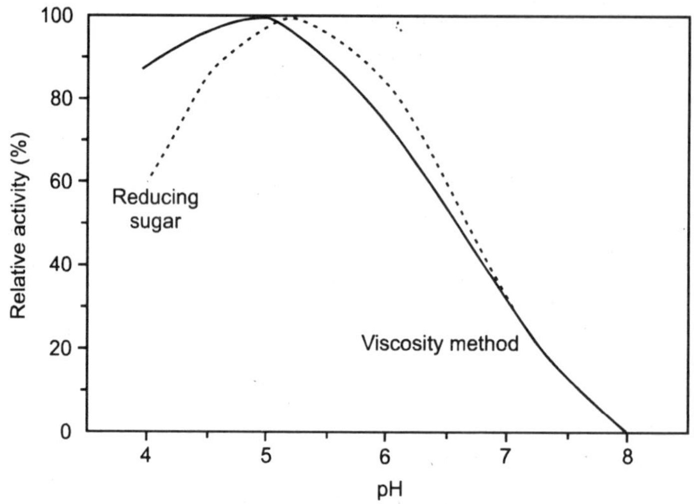

Fig. 7.12. pH dependence of fungal α-amylase (*A. oryzae*); measured by: (i) loss of viscosity; and (ii) increase in reducing sugars.

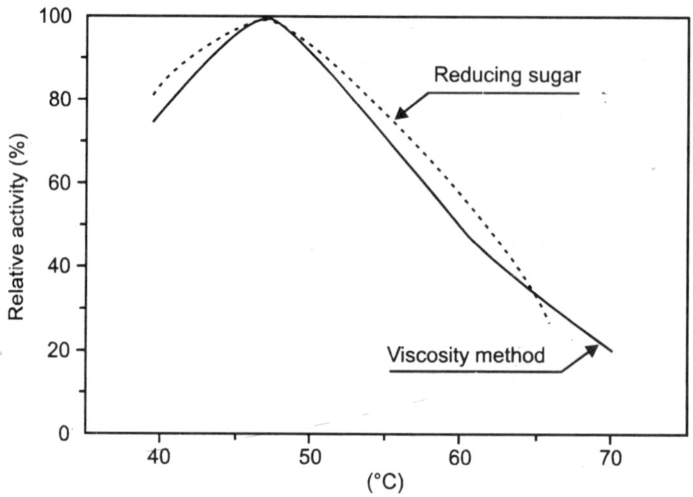

Fig. 7.13. Temperature dependence of fungal α-amylase; measured by: (i) loss of viscosity; and (ii) increase in reducing sugars. Substrate: 4 per cent soluble starch, pH 5.0; reaction time 10 minutes.

Technical preparations

1. Activity: Powdered preparations, obtained by spray drying or precipitation with alcohol, are standardised according to the type of application with salts, starch, maltodextrin, lactose, or glucose. The activity of the preparations used in flour milling is about 500–5000 SKB/g. Concentrates for the starch or baking industry are set to 50,000–1,20,000 SKB/g. Liquid concentrates for the starch industry are offered with activities from 6000 to 40,000 SKB/g.
2. Stability: The drop in activity using dry storage at about 20°C is approximately 5–10 per cent per year.
3. Secondary activities: Of particular practical relevance are companion activities due to acidic proteases and cellulases. In certain applications, many of these additional activities are desirable,

whereas for others they may be less desirable or even harmful. Protease activities between 100 and 30000 HUT units can be present, and this must be taken into account for applications in flour processing. In addition, preparations from fixed-bed cultures, such as the well-known Takadiastase, are a valuable source of many enzymes. Secondary enzymes found included α- and β-glycosidases, nucleases, hemicellulases, various proteases and peptidases.

4. Manufacturers: [RM], [NO], [GB], [AM], [UA], [SN], [SA], [SO], [SOE], and others.

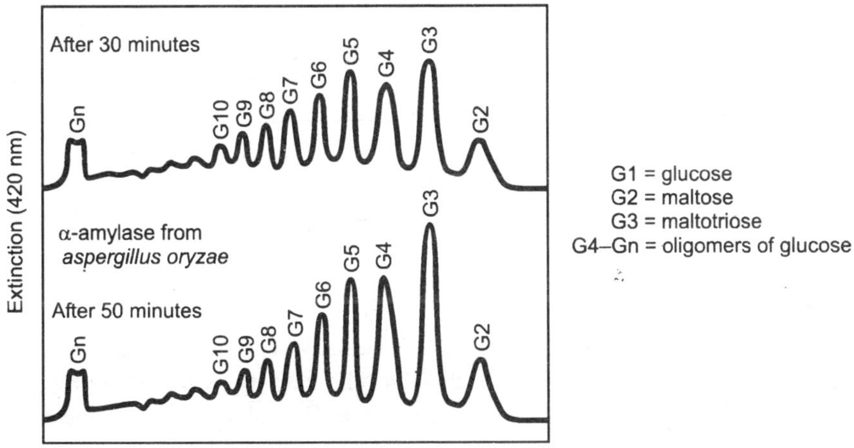

Fig. 7.14. Products formed during amylose hydrolysis by fungal α-amylase from *A. oryzae.*

Fungal β-amylase from A. fumigatus K 27

Smith described an *Aspergillus* α-amylase that is particularly effective in degrading raw starch. The fungal strain, which presumably belongs to the *A. fumigatus* group, produces a mixture of about 70 per cent amyloglucosidase and 30 per cent α-amylase. The pH optimum for raw-starch degradation is between 4.6 and 5.5 at 55°C. Cornstarch can be completely degraded at 55°C in 7–24 hours, depending on the amount of enzyme used. The amount of enzyme required, however, is relatively high. About 1 ml of culture solution of a 5–6-day-old culture growth is required to degrade 250 mg of cornstarch in 24 hours. By comparison, 1 ml of amyloglucosidase culture medium from 5-day-old culture of *A. niger* (a 5-day culture) can cleave about 1 kg of gelatinised starch in 24 hours at 60°C.

α-Amylase from Chalara paradoxa

An enzyme that is particularly effective in the degradation of raw starch is obtained from cultures of *Chalara paradoxa*. The optimum activity range lies between pH 4.5 and 5.5 at 45°–55°C.

Properties and manufacturer

1. Activity: Five milligrams of the only commercial preparation produces 2000 mM of reducing sugar from raw cornstarch at pH 5.0 and 40°C. One enzyme unit is defined as the amount of enzyme required to produce 1 mM reducing sugar (in form of glucose) from raw cornstarch at pH 5.0 and 40°C (Fig. 7.15).
2. Manufacturer: [MJ].

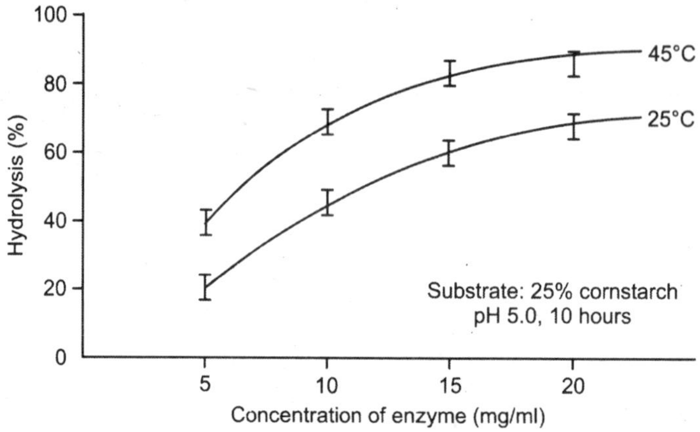

Fig. 7.15. Hydrolysis of raw starch by amylase from *Chalara paradoxa* [MJ]. Substrate: 25 per cent cornstarch, pH 5.0, 10 hours.

Malt α-amylase

Those α-amylases important in breweries and distilleries coexist with β-amylases in malted grains. The terms used for the grain amylases are diastase and diastatic activity, which describe the cooperative action of both enzymes. The presence of the two enzymes was first unequivocally shown by Sumner and Somers. While β-amylase is already present in large amounts in unmalted grains, α-amylase is generated only during the malting process.

During malting, α-amylase activity increases several-hundred-fold in 4–5 days. 'Green' grain α-amylase differs from that of unmalted grain in its properties in that is able to attack intact wheat starch.

Properties

1. Molecular weight: 52,000 daltons (malted barley). Different types of grain contain multiple forms of the enzyme, with a molecular-weight distribution of 42,000–54,000 daltons.
2. pH optimum: pH 5.5 (malted barley).
3. Temperature: Optimum is 70°C; inactivation is rapid at higher temperatures. At high substrate concentrations, this α-amylase is effective beyond the gelatinisation point of starch.
4. Stability: The enzyme has a good stability between pH 4.5 and 9.5 at 25°C. Inactivation is rapid below pH 4.5.
5. Inhibitors: Potassium periodate, oxidising agents.
6. Hydrolysis pattern: β-Limit dextrins are formed during the first phase; on more extensive hydrolysis, oligosaccharides from G_2 to G_6.

Pancreatic α-amylase

Pancreatic α-amylase was formerly prepared by fractional precipitation of aqueous extracts from hog or bovine pancreas. The technical enzyme preparation was used principally for textile sizing. Because of their low temperature stability, the preparations are not suitable for today's rapid high-temperature sizing processes for which heat-stable bacterial α-amylases are now used.

Pancreatic amylase is used as a digestive substitute in a pharmaceutical preparation of pancreatic enzymes called pancreatin (Fig. 7.16).

Properties, products and manufacturers

1. Activity: Activity optimum is between pH 6.0 and 7.0. The enzyme is activated by NaCl and stabilised by Ca^{2+}.
2. Inactivation: The enzyme is inactivated in aqueous solutions if accompanied by pancreatic protease. It is also inhibited by chelating agents and by SH-group blocking substances, such as iodoacetamide or heavy-metal ions.
3. Products and manufacturers: [RM], soluble preparations; [HÖ], [KC], [BF], and others. Product: pancreatin.

Amylases cleaving α-1,6 linkages

The important raw starches contain about 80 per cent amylopectin. As was mentioned previously, α-amylases cleave the α-1,4 linkages, whereas the α-1,6 linkages resist attack. Intensive hydrolysis of amylopectin with α-amylases leads to the 'α-limit dextrins'. In similar fashion, treatment with β-amylases leads to the 'β-limit dextrins', as cleavage here also terminates at the α-1,6 branches.

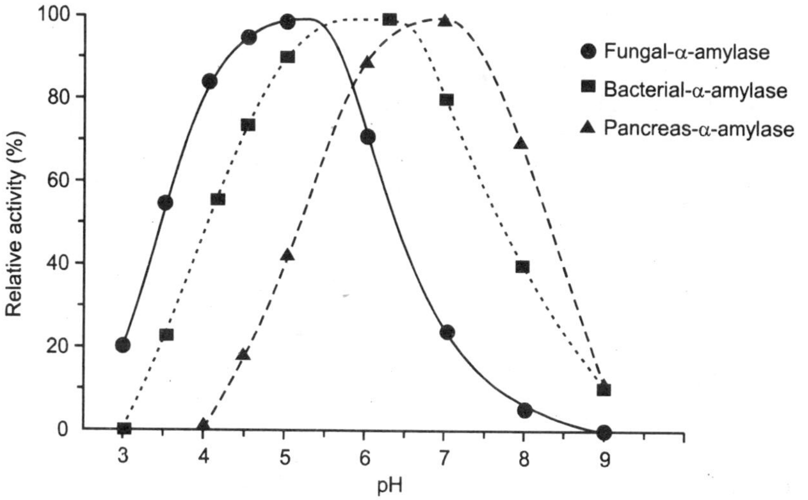

Fig. 7.16. Comparison of pH dependence of α-amylases from different sources.

The amylases that cleave α-1,6 linkages can be divided into three groups:

1. Amylo-1,6-glycosidases (in higher organisms).
2. Pullulanases, also R-enzyme or debranching amylases.
3. Isoamylases.

Pullulanase (EC 3.2.1.41; CAS 9075-68-7)

Pullulan (Fig. 7.17), which is produced by the fungus *Aureobasidium pullulans*, consists of maltosyl-(G_2) and maltotriosyl-(G_3) residues, which are joined by α-1,6 linkages. Tetraosyl residues are also present.

Pullulan degrading enzymes are produced by plants and micro-organisms. Even purified enzymes cleave not only pullulan but also amylopectin, β-dextrins, and glycogen.

The first isolation of pullanase was from *Aerobacter aerogenes* by Bender and Wallenfels. The enzymes from cultures of *Klebsiella pneumoniae* (L) and *B. cereus* are of technical importance.

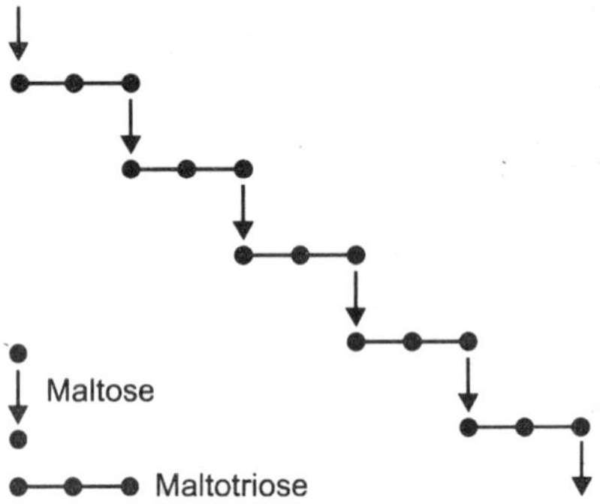

Fig. 7.17. Schematic of pullulan structure.

Pullulanase from K. pneumoniae

Properties

1. Activity: pH optimum at 5.5–6.0 (50°C).
2. Temperature: Maximum 50°–55°C at pH 5.0.

See Figs. 7.18 and 7.19. A technical pullulanase preparation from *Bacillus* sp. shows 80 per cent of maximum activity at pH 5–7 (Fig. 7.20).

Low-molecular-weight α-1,6-oligosaccharides such as maltotriosyl-α-1,6-maltotriose are cleaved 10 times faster than amylopectin.

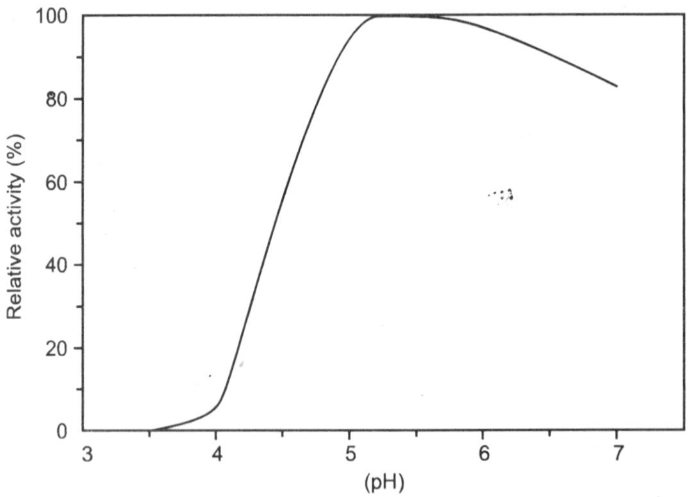

Fig. 7.18. Effect of pH on pullulanase activity. Substrate: 2 per cent pullulan, 50°C, 30 minutes.

Pullulanase from Aerobacter aerogenes

Properties

1. Molecular weight: 1,45,000 daltons.
2. Molecular structure: Single-stranded polypeptide chain.
3. Isoelectric point: pH 3.88; 4.46; 7.6. The crystalline enzyme is separated by electrofocusing into three protein bands.

4. Optimal activity: At pH 6.0 and 50°C.
5. Activators: Ca^{2+}, ~5 × 10^{-4} M.
6. Inhibitors: Ca^{2+}, > 10^{-2} M; Co^{2+}, Ni^{2+}, Cd^{2+}, Zn^{2+}, Hg^{2+}, EDTA.
7. Stability: pH 4.0–8.0 at ≤ 50°C.

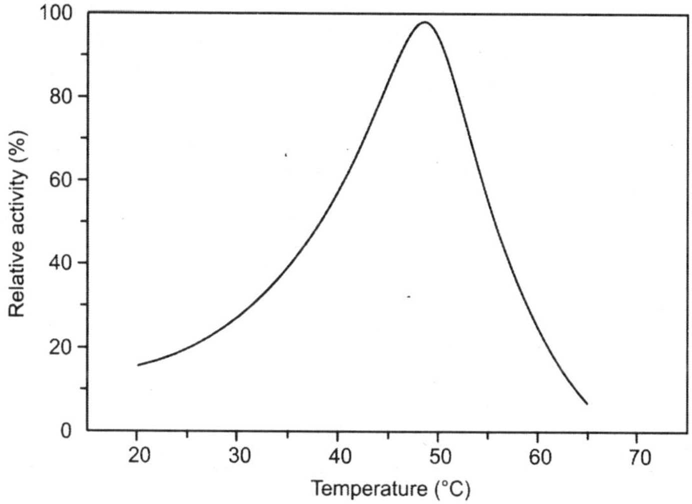

Fig. 7.19. Effect of temperature on pullulanase activity. Substrate: 2 per cent pullulan, pH 5.0, 30 minutes.

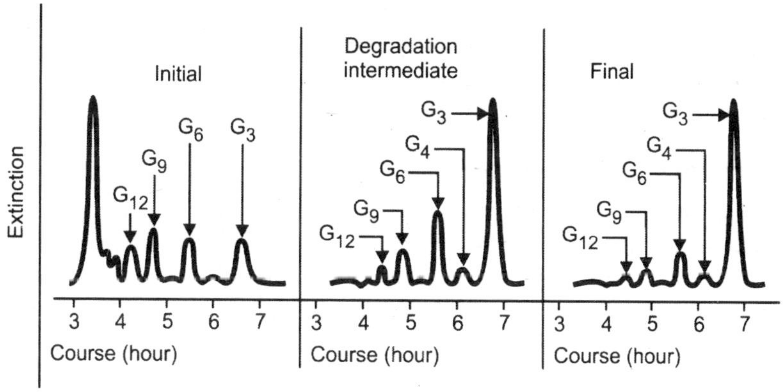

Fig. 7.20. Products formed during pullulan hydrolysis by an enzyme from *Bacillus* sp. (50°C, pH 4.5).

Pullulanase from Bacillus deramificans

According to Amory and Konieczyny-Janda, this enzyme has a better thermostability in the acidic pH range than do other pullulanases from *Bacillus*. The gene coding for the pullulanase was isolated from *B. deramificans* and expressed in *B. licheniformis*.

Properties

1. Molecular weight: 1,05,000 daltons.
2. Molecular structure: Single polypeptide chain.
3. Isoelectric point: pH 4.5.

4. Optimal activity: pH 4.0–4.5 at 55°–60°C.

In the saccharification process the formation of disaccharides can be minimised by reduction of the amyloglucosidase dosage, which is used together with the pullulanase.

Product: Optimax (SO).

Isoamylases (EC 3.2.1.68; CAS 9067-73-6)

These debranching enzymes, of which there are many different types, cleave pullulan slowly and amylopectin rapidly, as shown in Table 7.4.

Table 7.4. Effect of isoamylase and pullulanase on various substrates.

	Relative rate of hydrolysis	
Substrate	*Isoamylase*	*Pullulanase*
Pullulan	-	100
Amylopectin	100	15
Glycogen	124	1
6^3-0-α-Maltosylmaltotriose	2.8	22
6^3-0-α-Maltotriosylmaltotriose	9.7	162
6^3-0-α-Maltotriosylmaltotetraose	33	146

Isoamylase from Pseudomonas sp.

Properties

1. pH activity: 80 per cent maximum activity in the pH range of 3.5–5.5.
2. Temperature: Maximum 50°C at pH 4–4.5.
3. Inactivation: In 10 minutes at pH 2.0 and 50°C.

Isoamylase from B. cereus var mycoides

This organism, which produces isoamylase as well as β-amylase and pullulanase, was first described by Y. Takasaki. The characteristics of the enzyme mixture are very similar regarding optimum pH, optimum temperature and stability. This mixture is thus ideally suited for the production of maltose by the simultaneous action of different enzymes, and 88–90 per cent maltose can be obtained from liquefied starch (DE = 1) with only 0.1 per cent glucose as a by-product.

Manufacturers: [NO], [GB], and others.

Exoamylases

The β-amylases (EC 3.2.1.2; CAS 9000-91-3), also known as saccharogenic or maltogenic amylases, are found in cereals, malted cereals, sweet potatoes, and other plants. β-Amylases hydrolyse maltose residues at the α-1,4 linkage from the non-reducing end of the starch chain. The designation does not relate to the configuration of the glycosidic linkage that is hydrolysed, but rather to the free hydroxyl of the cleavage product: β-maltose. β-Amylase does not split α-1,6 linkages. The reaction products from amylopectin are maltose and β-limit dextrin (40–45 per cent). The β-amylases occur in different plants as isozymes with various characteristics. Molecular weights vary from 57,000 to 64,000 daltons in cereals and 1,52,000 daltons for the enzyme in sweet potatoes. The heat stability of β-amylases depends on their origin.

When malt, which contains a mixture of α- and β-amylases, is heated to 70°C in the presence of Ca^{2+}, the β-amylase is inactivated. The β-amylase from soyabeans is inactivated in 30 minutes at 70°C.

β-Amylase from malted wheat flour

Properties

1. Structure: Plant β-amylases are sulphydryl enzymes that do not need metal ions for activation.
2. Molecular weight: 57,000 daltons.
3. Activity: One mole of the enzyme cleaves 2,52,000 glycosidic linkages per minute at 30°C and pH 4.8.
4. pH activity: Optimum pH 5.0; 90 per cent activity in the pH range of 4.0–6.0.
5. pH stability: Optimum pH 4.5–9.0.
6. Temperature: Maximum 55°–57°C, inactivation above 60°C.
7. Inhibitors: Heavy metals and SH-group blocking agents.

Technical preparations

Industrial preparations are isolated from barley; in East Asia, from soyabeans. They differ from typical malt extracts. Barley β-amylase is used in the saccharification of liquid starch to maltose syrup. Depending on whether the starting material is obtained by acid liquefaction or enzymatic degradation, a sugar mixture with the following composition is obtained: 2–8 per cent glucose, 45–60 per cent maltose, and 10–25 per cent maltotriose. The mixture has a dextrose equivalent (DE) of 35–50.

1. Activity: The 'diastatic power' is defined according to FFC (Food Chem. Codex). The unit is called 'degrees of diastatic power' of DP°. This method of determination cannot be used for malt extracts because 20–40 per cent of the malt extract activity can be attributed to α-amylase. Traditionally, Windisch-Kohlbach units are still being used in the brewing industry. Liquid products with 1500 DP°/ml are commercially available.
2. pH optimum: 5.3 for saccharification (Fig. 7.21).
3. Optimum stability: At pH 6–7.
4. Storage stability: ~6 months at 4°C.

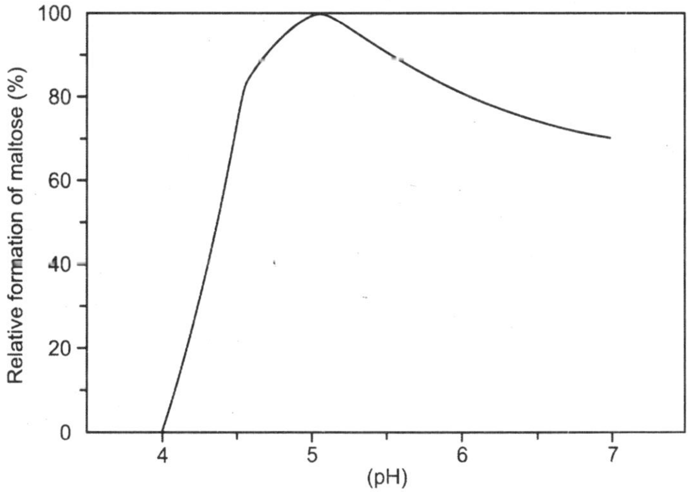

Fig. 7.21. pH dependence of maltose formation [FR]. Substrate: 34 per cent enzymatically hydrolysed starch (14 DE); temperature 60°C. Enzyme: 0.2 litre of Spezyme BBA [FR] per tonne of dry solids.

β-Amylases from micro-organisms

It is known that β-amylases are produced by a number of micro-organisms. Examples are *B. cereus*, *B. megaterium*, *Pseudomonas* sp., and *Streptomyces* sp.

The amylase from *Bacillus polymyxa* can hydrolyse amylopectin to maltose at a 94 per cent yield. Originally, it was assumed that the enzyme possessed α-amylase as well as β-amylase properties. Fogarty and Griffin were able to show that the high conversion yield of starch to maltose was due to two types of activity, one from a α-1,6-linkage debranching enzyme and the other from a β-amylase, with properties similar to those of plant β-amylases.

β-Amylase from B. megaterium

Okada and Higashibara produced β-amylases from various strains of *B. megaterium*. These differed slightly in their pH optima and stability. The pH optima were in the pH range of 6.0–6.5. The enzymes were inhibited by SH-group blocking agents such as *p*-chloromercuribenzoate and were re-activated by cysteine. The molecular weight of the pure β-amylase is about 50000 daltons; the isoelectric point is at pH 9.14.

β-Amylase from Streptomyces sp.

Koaze investigated various *Streptomyces* strains, which produced maltose from starch. The glucose: maltose ratio averaged 0.055:1, the pH optimum was 4.5–5.0, and the optimum temperature was at 50°–60°C.

Maltogenic amylase from B. stearothermophilus

Outtrup and Norman described a thermostable maltogenic amylase from a gen of *B. stearothermophilus* expressed in *B. subtilis*. The enzyme is produced by Novo Nordisk. The optimum pH is 5.0 (80 per cent of activity between pH 4.5 and 6.0). The optimum of temperature is 75°C. Qi Si investigated the functional mechanism of the antistaling effect of the enzyme in baking.

Product: Novamyl [NO].

Amyloglucosidases (EC 3.2.1.3; CAS 9032-08-0)

Amyloglucosidase (AMG) has for some time been one of the most important technical enzymes for the manufacture of glucose and for conversion of carbohydrates to fermentable sugars. It is used in distilleries, the manufacture of baked goods, and the production of fruit juice. Ueda, in Japan, discovered different active forms of amyloglucosidase in culture filtrates from *Aspergillus awamori*.

Beginning with the manufacture of large quantities of glucose syrup in the 1970s in the United States, the importance of this enzyme grew significantly. Presently, several thousand tonnes of enzyme are produced annually.

The enzyme, also called glucoamylase, α-1,4-glucan-glucohydrolase, or α-amylase, cleaves β-D-glucose via an exo mechanism for non-reducing chain ends of amylopectin, amylose, and glycogen.

For the technical production, two types of micro-organisms are used. These micro-organisms produce different enzymes. The majority of glucoamylases are produced today from *Aspergillus* sp. by a submersion process. A smaller quantity is produced in fixed-bed processes with *Rhizopus*, mainly in Japan. A commercial product from cultures of *Endomycopsis capsularis* is also in use.

Amyloglucosidase from *A. niger* is the most important enzyme of this type for technical manufacture and applications. Besides α-1,4 linkages, this enzyme also cleaves α-1,6 and α-1,3 linkages (Table 7.5).

Table 7.5. Relative efficiency of amyloglucosidase in disaccharides.

Disaccharide	*Linkage*	*Relative rate of hydrolysis*
Maltose	α-1,4	100
Isomaltose	α-1,6	1.0
Nigerose	α-1,3	0.2

The oligosaccharides maltose and maltotriose are hydrolysed much more slowly than amylose. This is the case for the enzyme from *Rhizopus* as well as from *Aspergillus*, as shown by Fukumoto (Table 7.6).

Table 7.6. Relative efficiency of fungal amyloglucosidase in oligosaccharides.

	Relative rate of hydrolysis	
Substrate	*R. delemar*	*A. niger*
Amylose	312	300
Amylopectin	1222	1260
β-Limit dextrin	937	800
Maltotriose	182	142
Maltose	100	100

Most strains of *Aspergillus niger* produce at least two, and sometimes three, isozymes. Isozymes AMG 1 and AMG 2 were described by Hayashida and Yoshino. According to Svensson, all commercially available products contain at least two isozymes, the distribution of which changes during fermentation and from batch to batch. Both enzymes are glycoproteins. The carbohydrates are *O*-glycosidically bonded to the peptide chain. The major component is mannose, which is linked at about 44 sites to seryl, threonyl, or asparagyl residues in the form of di-, tri-, or tetrasaccharides. In the literature, there is occasional confusion about the classification of the two isozymes. In Table 7.7, the AMG with the higher molecular weight is labelled as AMG-HM; the one with the lower molecular weight, as AMG-LM. According to Svensson, the isozymes can be described as shown in Table 7.7.

Table 7.7. Characteristics of two amyloglucosidases.

	Isozyme	
Characteristics	*AMG[a]-HM*	*AMG[a]-LM*
Molecular weight (daltons), from AA[b] composition	82000	70000
Molecular weight (daltons), from sedimentation	1,10,000	81000
Molecular weight (daltons), per Pazur and Ando	1,12,000 (AMG 11)	99000
pI	3.25	3.58
pI per Pazur and Ando	4.0	3.4
Monosaccharide equivalents per mole of enzyme mannose, glucose, galactose	99–100	99–102

[a]Amyloglucosidase; HM, high molecular weight; LM, low molecular weight.
[b]Amino acid.

AMG-HM has strong debranching activity, and is absorbed on raw starch, which it efficiently degrades. AMG 1 from *Rhizopus* sp. shows a similar action. Highly branched waxy cornstarch is more readily hydrolysed than amylose.

According to Hayashida and Yoshino, a gradual degradation by protease and glycosidases (fungal α-mannosidases) occurs during cultivation of the micro-organisms, which results in the formation of multiple forms of AMG.

Partial hydrolysis of AMG-HM with subtilisin results in AMG-LM and an enzymatically inactive glycopeptide with a molecular weight of 13200 daltons.

AMG-LM shows only weak debranching activity and degrades raw starch only slightly. Similar properties were found for AMG from *Rhizopus*. According to Konieczny-Janda [SO], the molecular structures of the two enzymes can be described as shown in Fig. 7.22.

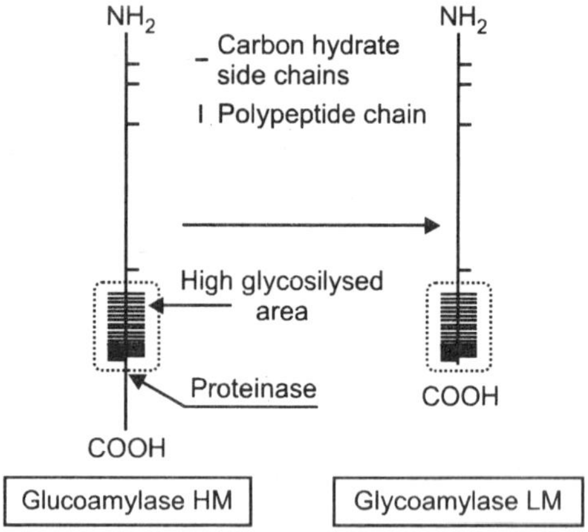

Fig. 7.22. Schematic representation of AMG-HM being converted to AMG-LM by protease.

Both enzymes show a slightly different temperature dependence. At temperatures in excess of 65°C, the loss of activity for AMG-HM is somewhat less than that for AMG-LM. Also, under acidic conditions (i.e. pH < 3.5), the larger enzyme molecule loses its activity more slowly. An AMG that is particularly alkali and heat-stable is produced by the thermophilic fungus *Humicola lanuginosa*. The pH optimum is around 6, the enzyme is stable up to pH 11, and the temperature optimum is 65–70°C.

Degradation of raw starch

While gelatinised starch is rapidly degraded by endo- and exoamylases, non-swollen starch grains are difficult to degrade. According to Walker and Hope, crystallised α-amylases from human saliva, hog pancreas, *B. subtilis*, and *A. oryzae* all, to a greater or lesser extent, degrade various raw starches. For example, hog pancreatic enzyme, at 30°C and pH 6.5, hydrolyses 5 per cent of potato starch and 55 per cent of cassava starch in 24 hours. The size of the starch granules is important because small granules are more rapidly attacked. It is important for the manufacture of glucose or the production of alcohol to save the energy that is used to gelatinise starch. These costs represent 50 per cent of the total energy costs in alcohol production. The degradation of raw starch with amyloglucosidase from *Rhizopus* is the subject of many publications and patents. Takaya found that amyloglucosidases from *R. niveus*, *R. amagasakiense*, and *Endomycopsis* hydrolyse cornstarch in similar fashion. Enzymatic activity takes place over the surface of the starch grain, thereby producing holes that run radially from the surface inward.

A fungal amylase that, besides other activities, also contains amyloglucosidase (isoelectric point 8.0), seems to be especially suitable for the degradation of raw starch: it is described by Dwiggins. In the manufacture of bread, enzymatic hydrolysis of raw starch also plays a role in dough preparation. During milling, about 6–8 per cent of the starch granules are damaged. They subsequently swell in cold water and are hydrolysed by amylases.

Amyloglucosidase from A. niger

Properties

1. Molecular weight: See Table 7.7.
2. Structure: Various glycosylated isozymes; *A. saitoi* has a content of 18 per cent neutral sugars.
3. Isoelectric point: See Table 7.7.
4. pH activity: Depending on the strains used and culture conditions, slightly shifted pH optima are found. Substrate concentration and temperature influence the optimum reaction conditions. At 33 per cent substrate concentration (dextrin 2 DE), 90 per cent of maximum activity is obtained between pH 3.3 and 4.7 at 60°C.
5. Stability: Optimum stability is at pH 4.0 and 60°C. Most preparations are stable in solution for several hours with substrate at pH 2.0 and 30°C. AMG-LM is more stable than the isozyme AMG-HM (Figs 7.23–7.25).
6. Inactivation: In solutions with less than 10 per cent starch, total inactivation occurs in 10 minutes at pH 3–5 and 80°C, with higher starch concentrations, 90°–95°C is necessary.

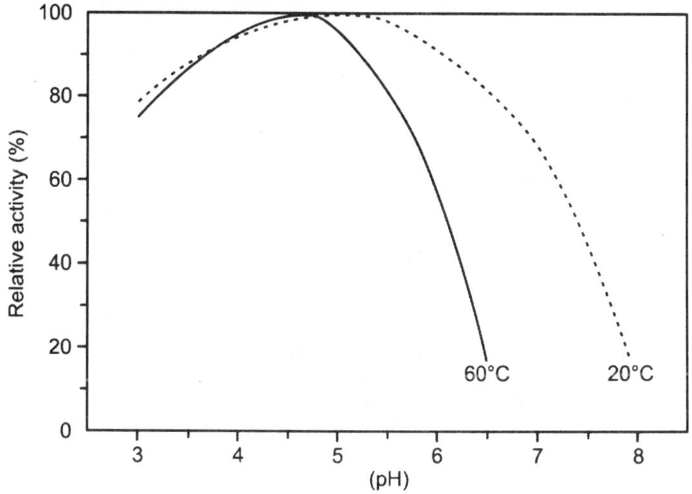

Fig. 7.23. pH dependence of amyloglucosidase at two different temperatures [RM].

Technical preparations

1. Activity: The technical preparations used for the production of glucose from starch are mostly stabilised liquid products with an activity between 200 and 400 AMG units/g. Salt and benzoate are used as preservatives, and glucose, maltodextrins, or sorbitol in concentrations of ≤30 per cent is used to stabilise activity. Powdered products with 500–2000 AMG units/g are produced by spray drying.
2. Side activities: Technical enzymes contain, in addition to α-amylase, a transglycosidase activity that lowers the yield of glucose in starch saccharification.
3. Manufacturers: [MC], [GB], [RM], [NO], [AM], [SO], [SOEG], and others.

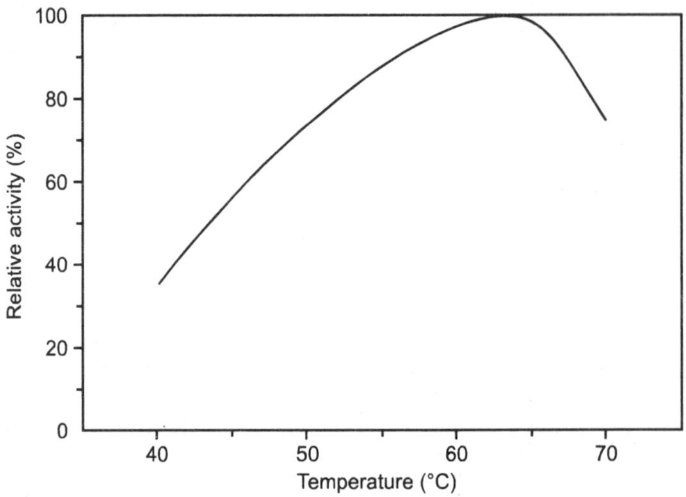

Fig. 7.24. Temperature dependence of amyloglucosidase [RM].

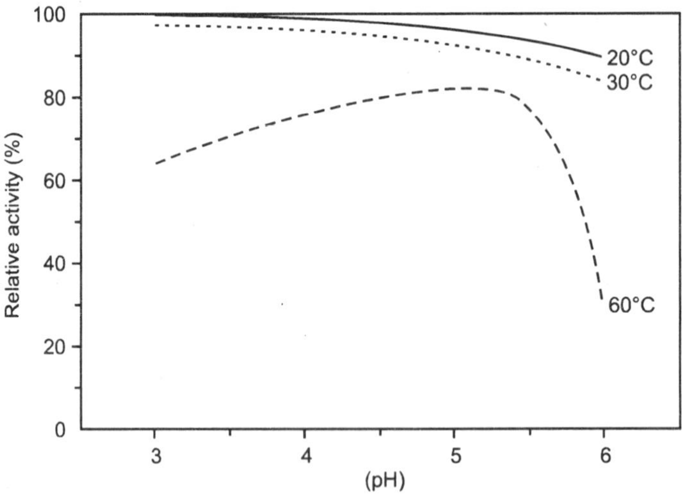

Fig. 7.25. pH-dependent stability of amyloglucosidase at various temperatures (in aqueous solution) [RM].

Amyloglucosidase from Rhizopus sp.

Properties

1. Molecular weight: 1,00,000 daltons (*R. delemar*).
2. Structure: Glycoprotein, innumerable isozymes.
3. Isoelectric point: pH 7.2 (*R. delemar*).
4. Activity: Optimum pH 4.5; optimum temperature 40°C (*R. delemar*).
5. Stability: Optimum stability at 55°C is pH 3.5–4.5.

Technical preparations

Rhizopus AMG preparations are generally produced in powder form by precipitation with alcohol. Liquid preparations are unstable, presumably because of their protease content. The protease is largely denatured by heating to 60°C in the presence of glycerol, whereas the AMG remains intact. In this way, Plainer and others were able to impart good stability to liquid products.

Powdered products with about 1000 AMG units are stable when kept dry.

1. Manufacturers: [SN], [AM].
2. Side activities: Amylases, transglycosidases, proteases, esterases, lactase, and cellulases.
3. Enzyme affinity: *Rhizopus delemar*; K. maltose, 1.16 mM; K. maltotriose, 0.2 mM; K. maltopentaose, 0.12 mM; K. amylose, 0.00384 mM.

Transglucosidase (EC 2.4.1.24; CAS 9030-12-0) α-1,4-D-glucan-6-α-glucotransferase

Transglucosylation reactions have been known for some time. The reaction is essentially similar to enzymatic hydrolysis. The enzymatic hydrolysis of a glycoside can be conceived of as a transfer of a glucose residue (donor) to water (acceptor). Other molecules such as sugar can be acceptors instead of water.

The reverse reaction of amyloglucosidase at higher substrate concentrations and longer incubation time has already been mentioned. The reversion products are maltose and isomaltose. Disaccharides can also be synthesised through the mechanism of glucose transfer. The enzyme that catalyses this reaction is a transglucosidase, formed as a by-product in the production of amyloglucosidase. The extent of transglycosylation activity can be estimated by adding the enzyme to a 30–50 per cent glucose solution and determining the amount of isomaltose formed.

The transfer of a glucose residue from a α-1,4 position on maltose or maltotriose to a α-1,6 position leads to panose and isomaltose.

Transglucosidase from A. niger

Properties

1. Molecular weight: 1,50,000 daltons.
2. pH optimum: 4.5–5.0.
3. Temperature optimum: 60°C.

Because transglucosidase reduces the yield of glucose during the saccharification of starch, many researchers have worked on the removal or inactivation of this enzyme. Bentonite is an effective absorbent over a range of pH values and is commonly employed. According to US Patent 3,108,928 (1963), the pH is raised to 9.9 with magnesium oxide, and then after incubation at 30°C for 40 minutes, the pH is re-adjusted to 5.0. Recently developed genetic methods have led to the production of transglucosidase-free *Aspergillus* strains.

Fructosyltransferase (EC 2.4.1.9; CAS 9030-16-4)

This enzyme, generated by *A. niger*, is used by Adachi and Hidaka at Meiji (Japan) for the production of fructooligosaccharides under the trade name Neosugar. The product is a mixture of α-1,2-linked fructose residues (F) with a terminal glucose (G), as, for example, F-F-G (1-kestose) or F-F-F-G (nystose). The product has special dietetic properties and is used as a low-calorie sweetener. These fructooligosaccharides are not hydrolysed by the human digestive system, but are absorbed by the intestinal *Bifidobacterium* sp., the *Bacterium fragilis* group, and *Peptostreptococcus*.

The growth of the bacteria is strongly stimulated, thereby eliminating digestive problems caused by lack of fermentation.

Cyclodextrin glucosyltransferase (CGT) (EC 2.4.1.19; CAS 9030-09-5), cyclodextrinase

Cyclodextrins are circular oligomers with 6-, 7-, or 8-glucose molecules, known as α-, β-, and γ-cyclodextrin, respectively. They are formed by the action of the enzyme (CGT) on starch or dextrins. They are named *Schardinger dextrins* after the scientist who discovered them.

Cyclodextrins have gained importance in recent years because of their ability to form inclusion compounds with other molecules. In this way, volatile concentrated aroma molecules that are sensitive to the action of oxygen, light, or heat can be stabilised by complexing with cyclodextrins. In pharmaceuticals, this encapsulation has many advantages, such as improving solubility or dispersability. New applications for cyclodextrins are continually being developed in the area of cosmetics and perfumes.

Cyclodextrins have been known for some time. The nature of their formation by an enzyme from *B. macerans* was first closely investigated by Cramer and Steinle. The reaction rate was monitored by the disappearance of the iodine colour during the enzyme's reaction with starch. It was believed, at one time, that the formation of cyclodextrin depended on the complex action of three enzymes: a dextrinising amylase, a cyclodextrin synthetase, and a glucosyltransferase. However, after completely extracting and purifying the hydrolytic activity from a crude enzyme preparation of *B. macerans*, Kobayashi found that the pure crystalline enzyme contained all three activities (Table 7.8).

Table 7.8. Properties of cyclodextrin glucosyltransferase from *B. macerans*.

Property	*Depinto and Campbell*	*Kobayashi*	*Stavn and Granum*
Molecular weight (daltons)	1,39,300[a]	1,45,000[a]	67000[b]
pI	–	–	5.4
pH optimum	6.1–6.2	6.0	5.4–5.8
Temperature optimum	–	60°C	60°C
K_m (starch)	3.33 mg/ml	–	5.7 mg/ml

[a]Two sub-units with identical amino acid compositions.

[b]Single chain.

A cyclodextrinase (CGT) from *B. megaterium* with high transglucosidase activity was described by Kitahata and Okada. Alkalophilic enzymes are also sources for the CGT. Nakamura and Horikoshi found that unpurified enzymes from such organisms showed two pH optima, one at pH 4.5–5.0 and the other at pH 7.5–8.5. A purified enzyme isolated from the organism had a molecular weight between 85,000 and 88,000 daltons and showed optimum activity at pH 7 and 50°C.

On binding to a carrier, the enzyme with the two pH optima lost its alkaline optimum, and the temperature stability was not improved by immobilisation.

CARBOHYDRASES OTHER THAN AMYLASES

Inulinase (EC 3.2.1.7; CAS 9025-67-6)

2.1-(1,3:1,4)-β-D-Glucan-3(4)-glucanohydrolase

Inulin (see structure in Fig. 7.26) is a linear, β-2,1-linked polymer of fructose. A sucrose residue is bound to the end of the chain. The molecular weight is about 6000 daltons. Inulin serves as an energy reserve in many plants, especially the composites, such as chicory, Jerusalem artichoke, and dahlia. Pure inulin is almost insoluble in cold water, but can be extracted from plant material with hot water. Inulin can be hydrolysed using inorganic acids at pH 1–2 in 1–2 hours at 80°–100°C. Because fructose is unstable at low pH values, such treatment readily leads to by-products that are deeply coloured. Therefore, it is desirable to completely hydrolyse inulin to fructose under gentle conditions. Fructose, with its high degree of sweetness, is important in the dietetic and beverage industries. The enzymatic hydrolysis of inulin offers an alternative method of fructose production beside the standard procedure with starch as the source material.

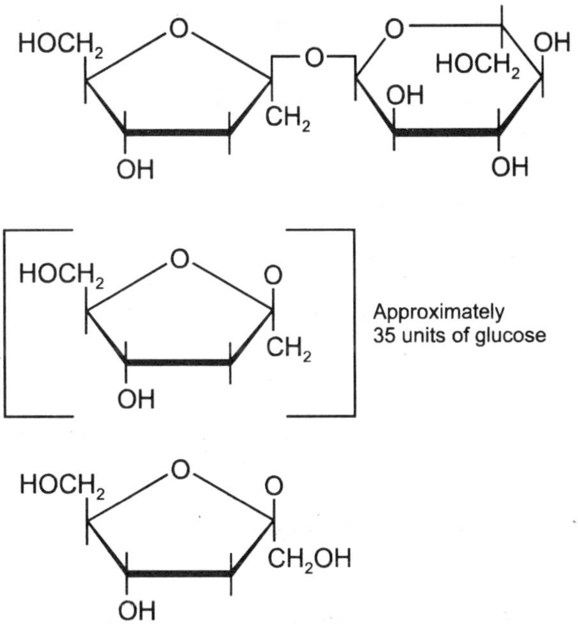

Fig. 7.26. Inulin structure.

Inulinase is produced by several micro-organisms; those from yeasts (*Candida* sp. and *Kluyveromyces fragilis*) and *Aspergillus* are of particular interest for the manufacture of technical enzyme products. The yeast inulinase from *Candida* sp. has an optimum activity at pH 4.5 and exhibits 80 per cent of maximum activity in the pH range of 3.5–6.0. Optimum temperature is 45°C at pH 4.5, the enzyme is inactivated above 55°C. Inulinase from *Aspergillus* was isolated, crystallised, and characterised by Nakamura. According to Zittan, there are two enzymes in *A. ficuum*—an endoinulase (EC 3.2.1.7) and an exoenzyme (EC 3.2.1.80)—a fact known for other carbohydrases as well.

The enzyme described by Zittan has the following properties: Optimal activity is at about pH 4.5 and 60°C. Figure 7.27 shows the dependence of *Aspergillus* and yeast inulinase activity on pH.

Products and manufacturers: Inulinase from *Aspergillus* [NO], [MJ], and others.

Dextranase (EC 3.2.1.11; CAS 9025-70-1)

In the manufacture of sugar (sucrose) from sugarcane or sugarbeet, dextran can be formed by *Leuconostoc mesenteroides*. This causes not only a loss of sugar content but also significant problems in filtration and crystallisation during the production process. The enzyme dextran sucrase produced by *Leuconostoc* synthesises α-1,6 dextran from sucrose (nF-G) according to:

$$\text{Sucrose (nF-G)} \rightarrow \text{dextran (Gn)} + \text{fructose (nF)}$$

This enzyme was isolated by Tilburg from *Penicillium funiculosum*, who co-cultivated *Penicillium* with the dextran producer *Leuconostoc* in a sugar medium.

Dextran forms chains of approximately 50,000 glucose residues, has extremely high viscosity, and interferes with filtration and clarification of sugar juices. Dextranase cleaves the α-1,6 linkages of dextran in a stepwise fashion. The reaction products are isomaltose and isomaltotriose.

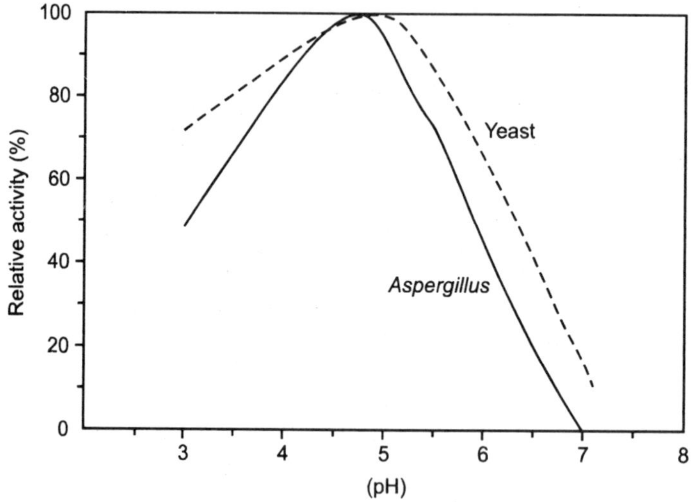

Fig. 7.27. pH dependence of *Aspergillus* inulinase.

Properties

1. Activity: One unit of enzyme is the amount that produces the reducing sugar equivalent of 1 gram of maltose in one hour from dextran at 40°C [NO].
2. Temperature optimum: 60°C.
3. Stability optimum: In buffer, pH 5.4, the enzyme is stable in sugar solution for about one hour at 50°C. At 70°C, 50 per cent of activity is lost in about 15 minutes (see Fig. 7.28).

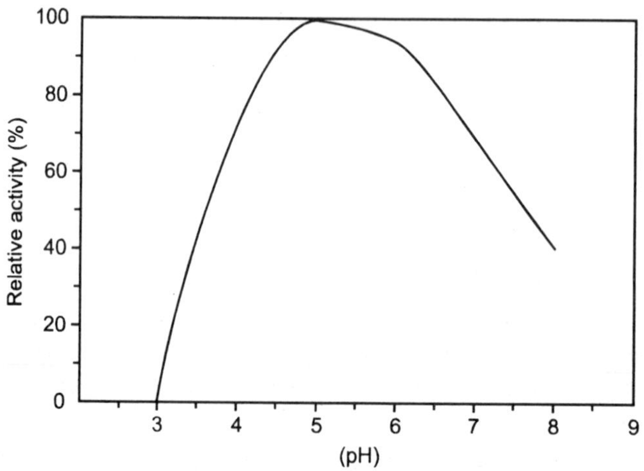

Fig. 7.28. pH dependence of dextranase from *Penicillium liliacinum* (McIlvaine buffer, 40°C [NO]).

Penicillium funiculosum forms two isozymes (pI 3.98 and pI 4.19). The molecular weight is about 44000 daltons. The enzymes are activated by CO^{2+} and Mn^{2+} and inactivated by Hg^{2+} and CU^{2+}.

Dextranase from Arthrobacter sp.

This enzyme cleaves insoluble glucans more rapidly than the enzyme from *Penicillium* sp. It is active between pH 5.5 and 7.5, with an optimum at pH 6.5–7.0. The optimum temperature is between 37 and 45°C. Molecular-weight determinations: 40000 (gel filtration) and 62000 daltons [sodium dodecylsulphate (SDS)-electrophoresis].

Manufacturer: [NO], [GO], and others.

Glycosidases

The scission of the glycosidic linkage for polysaccharide-cleaving carbohydrases and the endo- and exoamylases was discussed earlier. In this section, the enzymes cleaving oligosaccharides will be described. The enzymes in this group show definite specificities:

1. Either α- and β-specificity, according to which the enzymes are actually classified.
2. Sugar or structure specificity, which means that the enzymes recognise the configuration of the glycosidic hydroxyl group.
3. Specificity can depend on the degree of polymerisation of the oligomers.
4. Specificity relative to the aglycon that is the non-sugar moiety of a glycoside.

A large number of important natural products that can be cleaved by specific glycosidases belong to this class of substances. An example is amygdalin, the hydrolysis of which has been studied for a number of years (Fig. 7.29). See also Table 7.9.

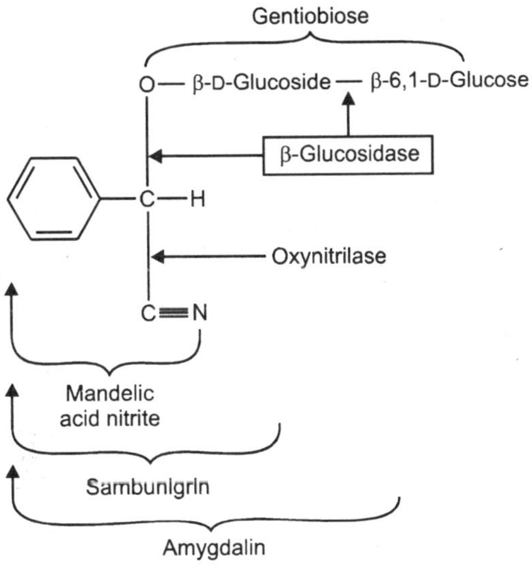

Fig. 7.29. Enzymatic hydrolysis of amygdalin.

α-Glucosidases (EC 3.2.1.20; CAS 9001-42-7) α-D-glucoside glucohydrolase

The best studied enzyme of this group is maltase. This enzyme splits one molecule of maltose into two molecules of glucose and always accompanies the amylases in the animal digestive tract. α-1,4-linked glucose residues are cleaved from the nonreducing end of the oligosaccharides and α-D-glucosides.

Polysaccharides are hydrolysed only very slowly. α-Glucosidases are very specific with respect to the glucose residues of the glucoside, but less specific with respect to the aglycons. Maltase from animal organs is able to cleave sucrose; however, raffinose is not cleaved by the intestinal mucous membrane.

Table 7.9. Glycosidases, substrates and products.

Enzyme	*Substrate*	*Hydrolysis product*
α-Glucosidases	Maltose	2 Glucose molecules
	Sucrose	D-Glucose + D-fructose
	Melezitose	2 Glucose molecules + D-fructose
β-Glucosidases	Cellobiose	2 Glucose molecules
	Prunin	Glucose + naringenin
	Gentobiose	2 Glucose molecules
α-Galactosidases	Raffinose	D-Galactose + saccharose
	Stachyose	D-Galactose + raffinose, galactose branches in galactomannan chains
	–	D-Galactose and polymers
β-Galactosidases	Lactose	D-Galactose and D-glucose
β-Rhamnosidase	Naringin	D-Rhamnose + prunin
β-Fructosidase	Sucrose	D-Fructose + D-glucose
β-Xylosidase	Xylobiose	D-Xylose
α-L-Arabinofurnanosidase	D-Nitrophenyl α-L-arabinofuranoside	D-Nitrophenyl and L-arabinose

Yeast is a good source of α-glucosidase. This enzyme hydrolyses maltose and maltose derivatives, aryl- and alkyl-α-D-glucoside and sucrose.

The pH optimum for the maltases from *A. niger* and malted barley is 4.5.

Invertase (EC 3.2.1.26; CAS 9001-57-4) β-fructosidase, β-D-fructofuranoside fructohydrolase

The disaccharide sucrose (cane sugar) consists of glucose and fructose, coupled by a glycosidic linkage between the C-1 atom of glucose and the C-2 atom of fructose. Hydrolysis of this bond leads to a mixture of reducing sugars that is sweeter than cane sugar because of the high sweetening power of the liberated fructose (Fig. 7.30).

The term for this reaction is inversion, and the resultant monosaccharide mixture is called invert sugar. The term inversion relates to the observation that an aqueous sucrose solution rotates the polarised plane of light to the right whereas invert sugar rotates the plane to the left because of the fructose, which shows a high degree of rotation to the left. During hydrolysis, an inversion in the direction of rotation of the polarised light is observed.

In industry, hydrolysis is usually accomplished with inorganic acids. The drawback of this traditional method is that hydrolytic by-products are produced that have to be removed by purification. As invertase is easily immobilised and highly efficient with regard to product formation, the enzymatic process can be considered superior to the chemical method. The candy industry in particular has shown interest in this enzyme. It is used in the production of marzipan and soft chocolate fillings, as well as for artificial honey and sugar invert syrup. The latter is used in many branches of the food industry. One special application for invert syrup is as winter food for bees.

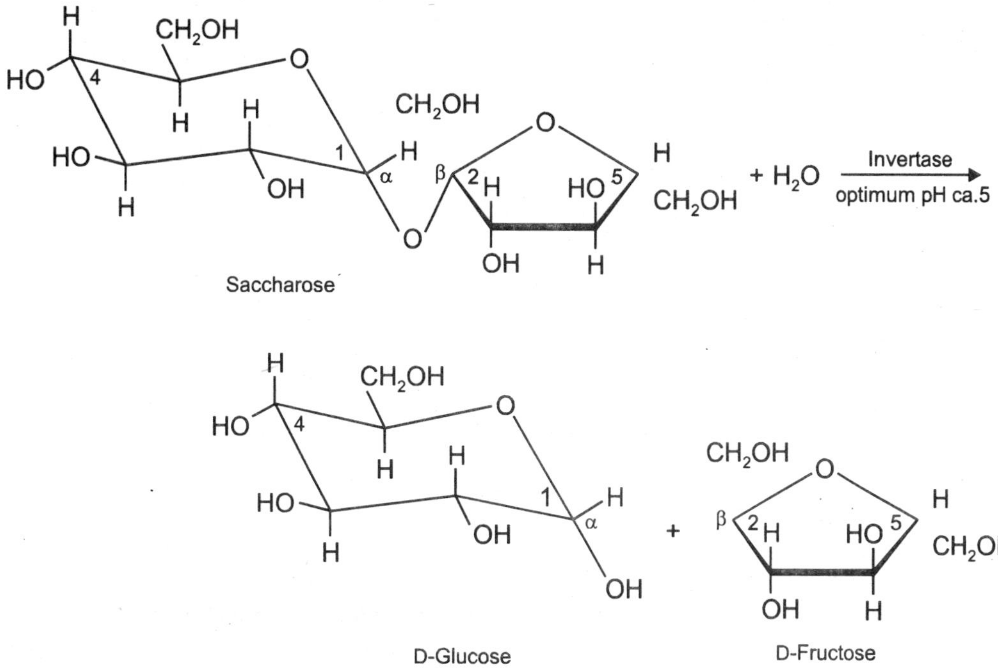

Fig. 7.30. The action of invertase.

Berthelot was the first to isolate the enzyme from yeast and it was subsequently used in extensive studies on enzyme kinetics. Invertase is an intracellular enzyme. This means that the enzyme remains within the cell during the yeast cultivation. The free enzyme is obtained by lysis of the cell wall.

Today, yeast is used almost exclusively for the production of commercial invertase.

Properties, products and manufacturers

1. Molecular weight: For the pure intracellular invertase, 1,35,000 daltons; for the extracellular enzyme, 2,70,000 daltons, of which ~50 per cent consists of bound mannan from yeast.
2. pH activity: The optimum activity ranges between pH 4 and 5.5.
3. Temperature: Because various commercial preparations show different stabilities, exact data for optimum reaction temperatures are not available. In most cases 55°C is the optimum temperature for dilute, and 65–70°C for concentrated sugar solutions (Fig. 7.31).
4. Stability: There are significant differences between the intracellular and the extracellular enzymes. Technical preparations are stable at 30°C between pH 3 and 7.5.
5. Immobilised invertase.
6. Products and manufacturers: Invertin [MK], [SE]; Maxinvert [GB]; Rohalase I 10 [RM].

α-D-Galactosidase (EC 3.2.1.22; CAS 9025-35-8), α-D-galactoside galactohydrolase

This enzyme is present in plants and animals and is also produced by various micro-organisms. The enzyme has been isolated from alfalfa, guar beans, and green coffee beans. Several strains of *A. niger* produce the enzyme together with several hemicellulases.

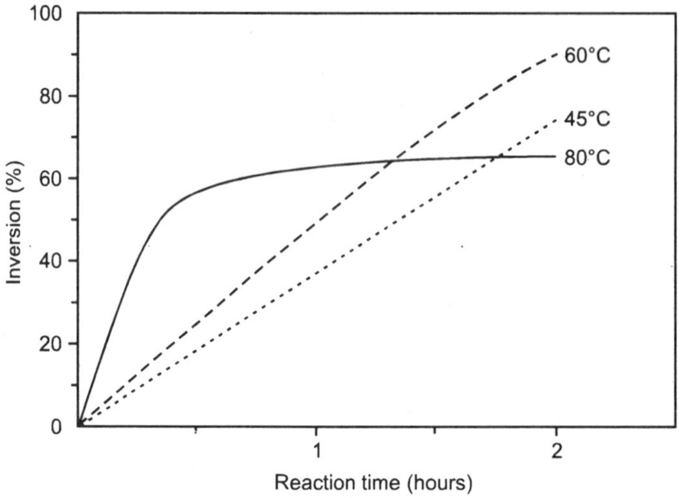

Fig. 7.31. Temperature-dependent inversion of a 70 per cent sucrose syrup [GB]. K_m values: 26 mM for sucrose, 150 mM for raffinose.

β-Glucosidases (EC 3.2.1.21; CAS 9001-22-3): β-D-glucoside glucohydrolase

Naringinase (α-L-rhamnosidase)

Citrus fruits contain two different types of bitter substances known as flavonoids (naringin) and the limonoids (limonin). Naringin, the 7-(2-rhamnosido-β-glucoside) of 4′,5,7′-trihydroxy-flavonone, occurs in large amounts in grapefruit and bitter oranges. For the average taste, a small content of naringin is desirable. However, if grapefruit is reduced to juice, an undesirable level of bitterness may be generated during extraction.

In Japan, Kishi discovered an enzyme from *A. niger* that degraded naringin. Today's commercial products contain two different glycosidases: a rhamnosidase (EC 3.2.1.43) that hydrolyses naringin to prunin by releasing rhamnose, and a β-glucosidase, which cleaves prunin to naringenin and glucose:

$$\text{Naringin} \xrightarrow{\text{rhamnosidase}} \text{prunin} + \text{rhamnose} \xrightarrow{\beta\text{-glucosidase}} \text{naringenin} + \text{glucose}$$

Naringinase preparation must be free of pectin-depolymerising enzymes; otherwise an undesirable clarification of the citrus juice will result. In a study on the hydrolysis of naringin by Pilnik, it was shown that as naringin disappears, the intermediate products, prunin and the end-product naringenin, were formed.

Enzyme characteristics

The optimum activity of commercial preparations is around pH 3.5; 80 per cent of the maximum activity lies between pH 3 and 4.7. The effect of pH and temperature on activity can be seen in Fig. 7.32. The loss in stability of naringinase with decreasing pH and increasing temperature can be seen in Fig. 7.33.

Technical preparations

1. Fungal sources: *A. niger, Penicillium* sp. and others.
2. Products and manufacturers: Naringinase [AM]; Kumitanase [SA], and others.

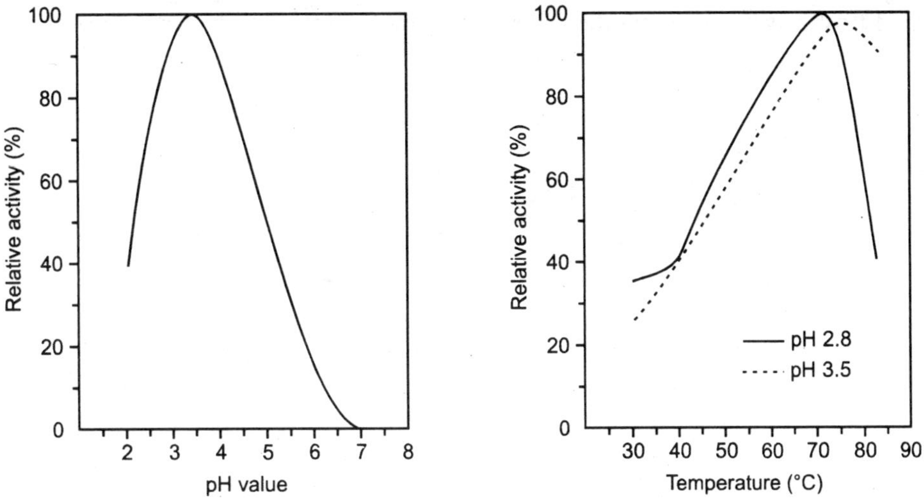

Fig. 7.32. pH and temperature dependence of naringinase activity [AM].

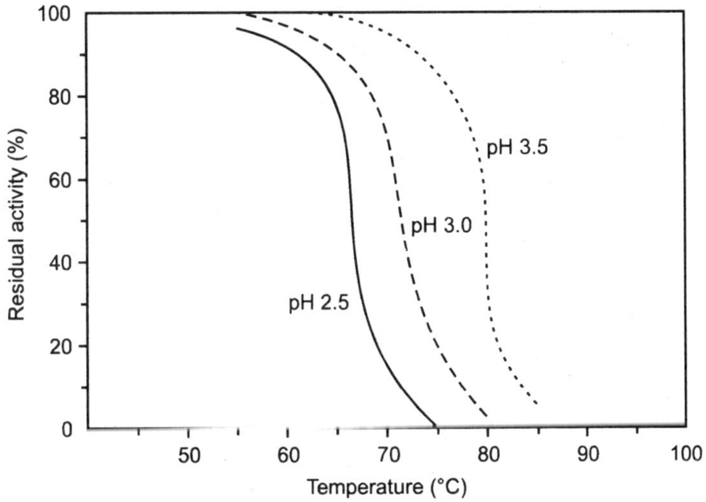

Fig. 7.33. Temperature stability of a naringinase preparation [AM] at various pH in 0.1 M buffer, 10 minutes.

Hesperidinase

Hesperidin is another glycoside that occurs in large quantities in citrus fruits. It is the 7-(2-rhamnosido-β-glucoside of 4′-methoxy-3′,5,7′-trihydroxyflavanone (hesperidin). It sometimes crystallises in frozen fruits or during the manufacture of orange juice concentrations and is almost tasteless. Hesperidin crystals can interfere with the stability of turbidity in orange juice.

Cleavage of limonin

When discussing the bitter substances in citrus fruits, limonin (Fig. 7.34) must also be mentioned. Limonin is an exceptionally bitter triterpenoid dilactone that is present in navel oranges as a nonbitter precursor molecule, the limonic acid A-ring lactone. Limonin is released from the precursor molecule by acid. The precursor, which is bound to the albedo layer of the fruit, can be released into the juice on cell rupture during a hard frost or when pressing the fruits. It is then converted by the acidic juice into the bitter limonin.

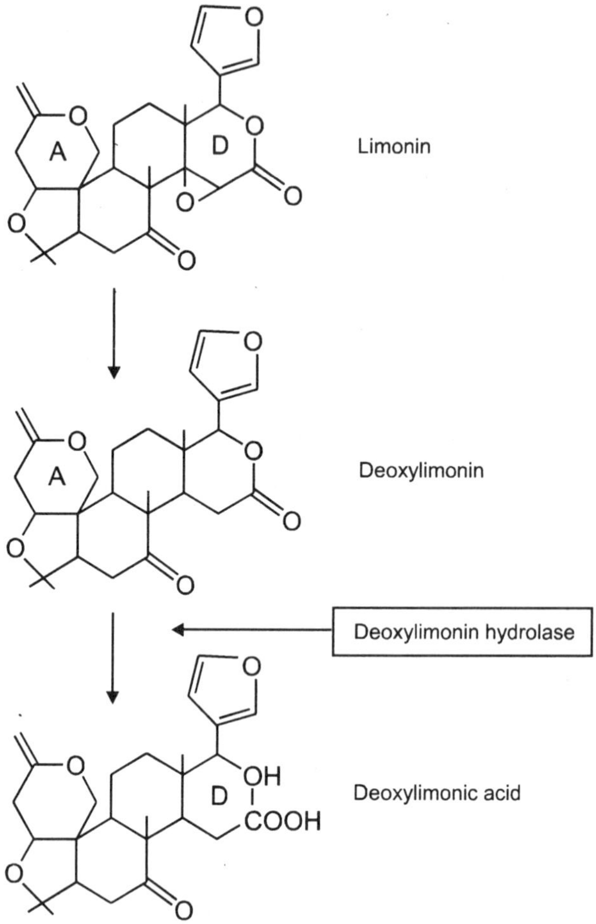

Fig. 7.34. Limonin hydrolysis.

A number of methods have been sought to eliminate these types of bitterness. Limonin dehydrogenase (or limonoate dehydrogenase, CAS 37325-58-9), found in *Arthrobacter globiformis* and *Pseudomonas*, convert limonin and limonin-A-ring-lactone to a 17-dehydro derivative. In this way, formation of the bitter D-ring lactone cannot occur. This interesting possibility of solving a very important problem for the citrus industry still fails because the dehydrogenases described depend on expensive co-enzyme systems (NAD) and further, the pH optimum is not in the range of acidic citrus juices.

Lysozyme (EC 3.2.1.17; CAS 9001-63-2), muramidase

Lysozyme is widely distributed in animal tissue, plants, and micro-organisms. It was first isolated by Fleming. It is also known as an antibiotic because it is able to kill certain bacteria by hydrolysing their cell walls. Lysozyme is present in milk (especially human milk), tears, and egg albumin. Human milk contains about 3000 times more lysozyme than cow milk. This antibacterial activity is one advantage of breastfeeding over using cow milk. Lysozyme is currently produced on a large scale from egg albumin. Mainland China is a major producer. According to Scott, it is estimated that 100 tonnes are produced annually worldwide. Lysozyme is isolated by isoelectric precipitation from albumin and adsorption on ion exchangers, then concentrated by ultrafiltration with repeated recrystallisations.

Unpurified papain is also rich in lysozyme. Twenty per cent of the soluble protein in the latex consists of lysozyme. There are various applications for the enzyme in food and medicinal technologies. Among these is the use of lysozyme in the production of certain cheeses, where it is important to kill harmful micro-organisms such as *Clostridium tyrobutyricum.*

J. Jolles and others have described the structure of lysozyme in human milk. Weidel found that part of the cell wall of certain gram-positive bacteria consists of a single sacklike macromolecule. This murein sack is very specifically cleaved by lysozyme. It hydrolyses the β-1,4 linkage between the *N*-acetyl muramic acid (Fig. 7.35), and the 2-acetamido-2-deoxyglucose residues in mucopolysaccharides or mucopeptides.

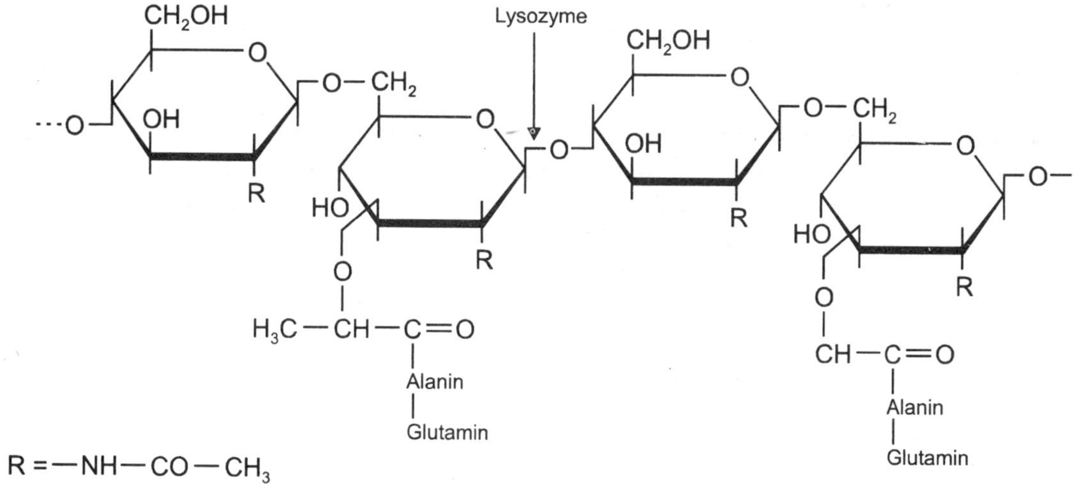

Fig. 7.35. Muramic acid hydrolysis with lysozyme

Properties and manufacturers

Lysozyme is marketed as a white crystalline powder. It is readily water-soluble but is insoluble in organic solvents. It is a polypeptide with 129 amino acid residues with a molecular weight of 14388 daltons. Lysozyme is a strongly basic protein with an isoelectric point between pH 10.5 and 11.7.

1. K_m: 4.0 × 10^{-3} with *p*-nitrophenyl-β-D-chitobioside as substrate.
2. Activity: The pH optimum is 7–8 with *Micrococcus* cells as substrate; the temperature optimum is 35°C.
3. Stability: Optimum is pH 4–5. Aqueous solutions are stable under aseptic conditions at 2°–4°C.
4. Inhibitors: Surfactants such as dodecylsulphate, oxidising agents, and heavy metals.
5. Analysis: Determination of lysozyme by the Federation International Pharmaceutique (FIP), or by the Calbiochem lysozyme assay.
6. Substrate: *Micrococcus luteus* cells, determined by the decreased absorption at 450 nm. Crystalline lysozyme shows about 25000 FIP units per milligram.
7. Manufacturers: [GB], [NG], [SOE], [WB].

Lysozyme from micro-organisms

A number of micro-organisms have been cited as producers of lysozyme, or lytic enzymes with the specificity of lysozyme; examples are *B. subtilis*, *B. cereus*, and *Lactobacillus casei*.

According to Woehner, a bacteria lysing enzyme showing lysozyme-type activity can be produced by fermentation of *Streptomyces coelicolor*.

Lactases, β-Galactosidases (EC 3.2.1.23; CAS 9001-22-3)

Lactase catalyses the hydrolysis of lactose (milk sugar), a disaccharide present in milk, to glucose and galactose.

The enzyme often occurs together with β-glucosidases in yeast, fungi and bacterial cultures. It is formed in the pancreatic and lactal glands of mammals. Lactase is immobilised on the brush-border of the intestinal wall (epithelium of the jejunum). Lactase deficiency, due to biochemical or genetic aberrations leads to flatulence, diarrhea, and bleeding. Various populations, particularly those of East Asia and Africa, suffer from lactase deficiency. Only adults are affected. Because of the minor nutritional importance of milk products in these regions, lactose intolerance presents no significant problem.

Milk contains about 4.5–5 per cent lactose. This sugar is relatively insoluble, having only about half the sweetness of sucrose. Increasing interest in the enzymatic scission of lactose into a more valuable mixture of monosaccharides is due to the fact that large amounts of lactose containing whey are produced annually. Until recently, whey was used either as animal feed or simply discharged, where it becomes a source of high-cost waste-water treatment. Even today, where much of the whey is ultrafiltered for the extraction of milk protein, there is still the problem of how to utilise the lactose in the permeate. Lactose is used, albeit to a limited extent, by the pharmaceutical industry. The annual global yield of whey is estimated at about 50 million tonnes, which corresponds to 2.5 million metric tonnes of lactose.

Enzymatic cleavage of lactose can be achieved with free or carrier-bound lactases. Yeast lactase is used to make milk tolerable for people with lactase deficiency. The enzyme can also be used for the manufacture of yogurt and for the hydrolysis of lactose in whey. The advantages of whey hydrolysates to that of lactose are:

1. The hydrolysate is sweeter (~70 per cent of sucrose sweetness).
2. The glucose-galactose mixture is significantly more soluble than lactose, a property important in food processing.

The lactose hydrolysate can be used as a syrup for the manufacture of baked goods, in the production of ice cream and desserts, and in dressings and lemonades. Lactase hydrolyses lactose according to the scheme in Fig. 7.36.

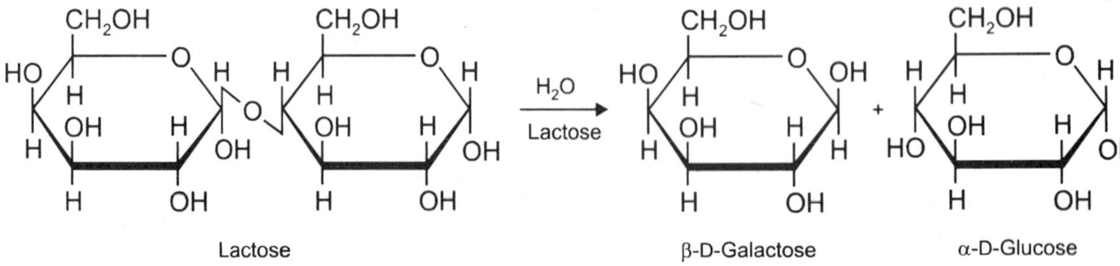

Fig. 7.36. Lactose hydrolysis.

Innumerable micro-organisms produce lactases of different properties. The lactase produced by *Escherichia coli* has been used by many researchers involved in research on the genetics of *E. coli.*

E. coli lactase is produced in industrial quantities for diagnostic purposes [BM]. For applications in food processing, the enzyme is extracted from yeast and *Aspergillus* sp. These two enzymes differ in their activity and stability optima as a function of pH (Table 7.10).

Table 7.10. Enzyme characteristics.

Characteristics	*Yeast lactase*	*Aspergillus lactase*
Organism	*K. fragilis*	*A. oryzae*
Activity	5000 EU/g (pH 6.2/30°C)	8600 EU/g (pH 4.5/30°C)
pH optimum	6.0–6.2	4.0–5.0
pH stability	5.5–8.5	4.0–8.5
Temperature optimum	43°C	50–60°C
Temperature stability	43°C (max)	60°C (max)
pI	4.0/6.2	3.5/4.4
Molecular weight (daltons)	1,30,000	1,30,000
K_m	20 mM	60 mM

Because the reaction parameters are strongly dependent on the enzyme, comprehensive pH and temperature dependences are presented in detail.

Optimal activity of the *A. oryzae* enzyme is at pH 4.5, which corresponds to the pH value of acid whey (pH 4–5). Enzymatic activity is only about 40 per cent of that of the optimum in the hydrolysis of lactose in milk or sweet whey (pH 6.2). The enzyme from *S. lactis* shows optimal activity in the pH range of milk (Fig. 7.37).

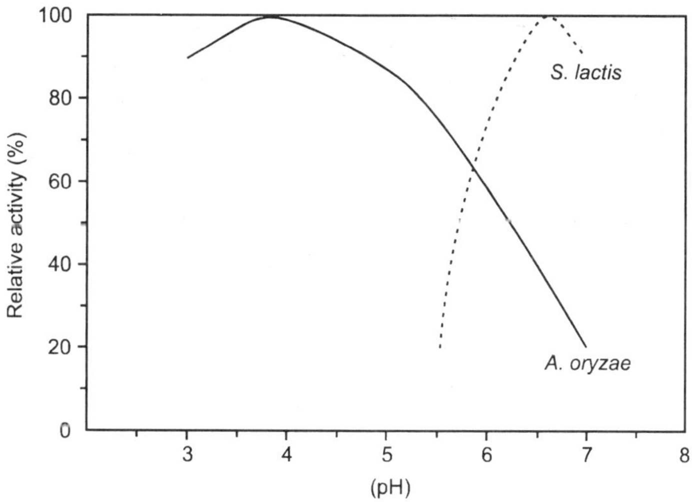

Fig. 7.37. pH dependence of lactase from *A. oryzae* and *S. lactis* (5 per cent lactose solution, 30°C).

Yeast lactase

The enzymes from *K. lactis* (Fig. 7.38) or *K. fragilis*, together with the fungal lactase from *A. oryzae* (Fig. 7.39), cover the whole spectrum of lactase applications. The use of yeast lactase is difficult in some cases because of its sensitivity to pH, temperature, and ions. Calcium ions inhibit—whereas manganese, magnesium, and potassium ions activate—yeast lactase. Fungal lactase is more stable but has the disadvantage in that the galactose produced by the hydrolysis inhibits product formation. It is recommended that hydrolysis be undertaken in dilute ~10 per cent substrate solutions.

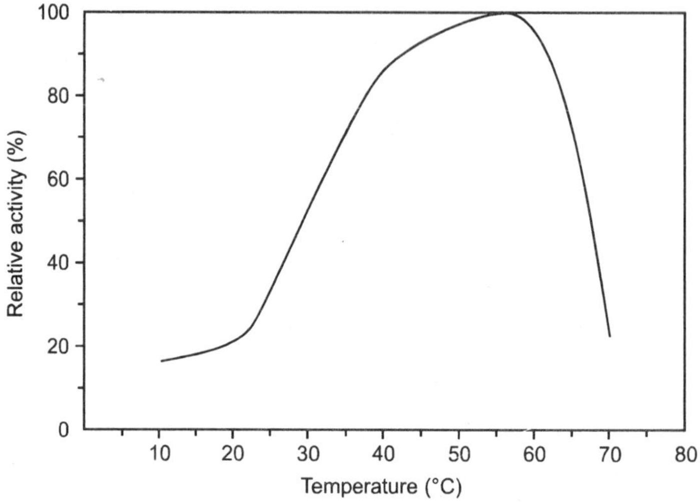

Fig. 7.38. Temperature dependence of lactase from *A. oryzae* [5 per cent solution, reaction time 15 minutes; Plainer, (RM)].

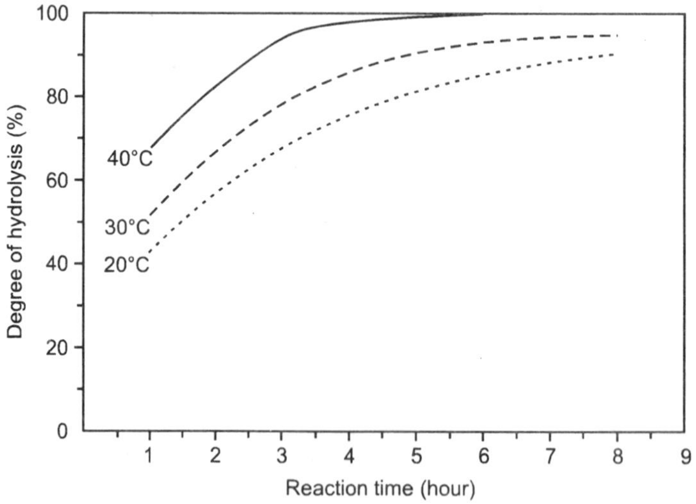

Fig. 7.39. Temperature dependence of lactase from *K. lactis* [AM].

Lactase from A. niger

Technical lactase from *A. niger* shows optimum activity at pH 3.5–4.5 and can tolerate temperatures up to 50°C. Product inhibition with this enzyme is also high.

Lactase from E. coli

1. Molecular weight: 5,40,000 daltons.

2. Optimum activity: pH 7.0; K_m (lactose), 3.8×10^{-3}.
3. Stability: After incubation at 30°C for 30 minutes, at pH 6.6–7.6, a residual activity of 80 per cent remains.

Lactase from B. stearothermophilus

The enzyme described by Goodman and Pederson has a molecular weight of 2,15,000 daltons and an optimum activity at pH 6–6.4. Activity drops off rapidly at acidic pH. The optimum temperature is 65°C; the K_m value for lactose (at 55°C) is 0.11 mM, and that for *O*-nitrophenyl-β-D-galactopyranoside is 9.8 mM.

Lactase from B. circulans

The enzyme is temperature stabile, but is unsuitable for practical use because of extremely strong product inhibition. Because lactose is currently a waste product, efforts are being made to make lactose hydrolysis more cost-effective; a lactose hydrolysate product needs to be competitive with sugar or glucose-fructose syrup. One possibility would be to use an immobilised enzyme for the hydrolysis.

Immobilised lactase

Some typical data from an immobilised lactase, Plexazym LA 1 [RM], is given here. Plexazym LA 1 consists of macroporous methacrylate beads to which lactase from *A. oryzae* is bound. The hydrolysis of lactose, either in the form of whey or from filtrates, occurs in a fixed bed at pH 4–5 and 30°–35°C. The flow rate amounts to 35–50 times the fixed-bed volume per hour. The capacity of the immobilised lactase is in the order of 200 tonnes sour whey per kilogram of Plexazym LA 1.

Analysis of the lactases

1. Determination with *o*-nitrophenyl galactopyranoside (ONPG) as a substrate: With the method according to Morisi, the release of *o*-nitrophenyl is measured photometrically at 420 nm. The reaction proceeds with the *A. oryzae* enzyme at pH 4.5, with the yeast lactase at pH 6.5 and 30°C.
2. Determination with lactose as substrate: The enzymatic activity is assayed measuring the glucose formed during hydrolysis. It is determined either by enzymatic glucose test (offered by [BM], [MK]), or more rapidly with an enzyme electrode. The conditions are 5 per cent lactose solution, 30°C at pH 4.5 or 6.5, depending on the type of enzyme.

Products and manufacturers: Mostly stabilised liquid products with 1000–4000 U/ml (lactose) or 3000–12,000 ONPG units/ml are available. Products: Lactase 7041 C [RM]: Plexazym LA 1 [RM]; Maxilact [GB]; Hydrolact [SE]; Lactase Y-L [AM]: Sumizyme [SN]; Lactozym [NO].

CELLULOSE DEGRADATION

Substrate

The polymers cellulose and hemicellulose constitute the major portion of the organic matter on our planet. They form the cell walls of plants and are consequently the principal components of wood, cotton and grass. Cellulose, as a renewable raw material, in an ideal resource for the industrial production of paper, textiles, and synthetic products. If economical processes could be developed for its enzymatic hydrolysis, cellulose would also be an ideal source of fermentable carbohydrates from which food and fuel could be produced. Along with cellulose, wood contains large quantities of hemicellulose, of which xylan is a major component. This polysaccharide and its monomeric building block, xylose, are raw materials for industrial fermentation, and xylose has been used as a dietetic sugar for a number of years.

The enzymatic degradation of plant tissue and plant constituents has been investigated for some time by different food processors. To understand the enzymes that make the degradation of high-molecular-weight carbohydrates possible, one should be familiar with their structures and compositions.

Plant cell structure

The plant cell consists of cytoplasm, which contains soluble and insoluble substances and is surrounded by a cell membrane. The cell wall lies tightly against the outer surface of this cytoplasmic membrane. The thickness of the rigid cell wall varies between 0.1 μm and several micrometers. Neighbouring cells have a common wall, similar to the inside walls of a house. The cell walls not only join the single cells to form a tissue but also contain channels for circulating liquids within the tissue. The cell wall is porous and penetratable for practically every type of molecule, while the plasma membrane has selective permeability. Water, which diffuses through the cell membrane by osmotic pressure, increases the intracellular pressure and thereby forces the membrane against the cell wall.

After cell division in the growing region (meristematic area), the cells expand in one dimension. The cell wall in young cells that are still growing is thinner and more pliable than walls in mature cells. This is the primary cell wall (Fig. 7.40). On completion of growth, the cell wall thickens and develops the secondary cell wall. The composition of cell walls differs according to the cell's function. The walls obtain their rigidity through a network of cellulose microfibrils (CMF) that are intrically interwoven with a matrix of other polysaccharides and proteins (Table 7.11).

Table 7.11. Cell wall polysaccharides of higher plants.

Polysaccharide	*Structure*	*Monomer unit*
Cellulose	β-1,4-D-Glucan	Glucose
Pectins	Polygalacturonic acids	Galacturonic acid and its methyl ester
	Rhamnogalacturonic acids	Rhamnose
	Galactans, arabinogalactans	Arabinose
Hemicelluloses	Xylans	Xylose
	Xyloglucans	Xylose, glucose
	β-1,3-; β-1,4-D-Glucans	Glucose
	Galactomannans	Galactose, mannose
Other polysaccharides	Arabinogalactans	Galactose, arabinose
	Glucuronomannans	Glucuronic acid, mannose

Cellulose

Cellulose consists of linear chains of several thousand glucose residues, linked by β-1,4 glycosidic bonds. The chains are stabilised by internal hydrogen bonds. Bundles of about 40–60 chains are aligned in parallel and linked via hydrogen bonds. Such lengthy, crystalline aggregates, known as microfibrils, are, in turn, sub-units of macrofibrils that have a cross-sectional diameter of about 0.5 nm (Fig. 7.41). Twenty to thirty per cent of the primary plant cell wall consists of cellulose.

The cellulose in native cotton has a degree of polymerisation (DP) of about 10,000; this means that about 10,000 glucose residues are glycosidically linked in each molecule. In contrast, wood pulp has a

DP of only 600–1000 and regenerated cellulose, about 200–600. Cotton consists of approximately 90 per cent cellulose; wood, approximately 60 per cent.

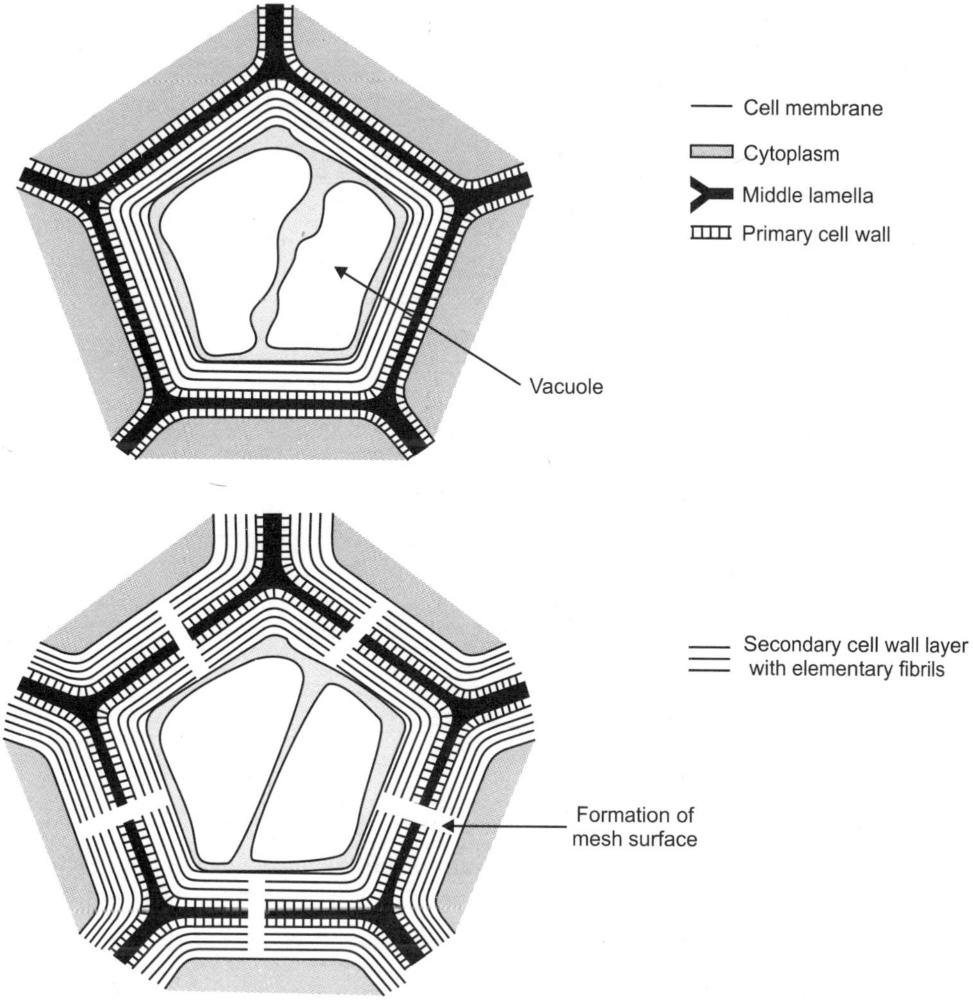

Fig. 7.40. Changes during cell wall development.

All glucose chains run in the same direction; that is, the reducing ends lie at the same end of the microfibrils. These are, in turn, surrounded by an ordered network of cellulose and polysaccharide chains that are tightly, but not covalently, bound to the microfibrils.

Hemicelluloses

Good literature reviews on hemicelluloses and the so-called plant mucins can be found in Pomeranz, Feldman, and Wucherpfennig and Dietrich. The last review refers largely to colloids that occur in fruit juices and wine.

In addition to cellulose, the plant cell wall contains a number of other polysaccharides, some of which are soluble in alkali. The insoluble fraction is also known as α-cellulose; the alkali soluble and acid precipitable fraction, as β-cellulose. The fraction that remains soluble in the acidic solution contains predominantly hemicellulose.

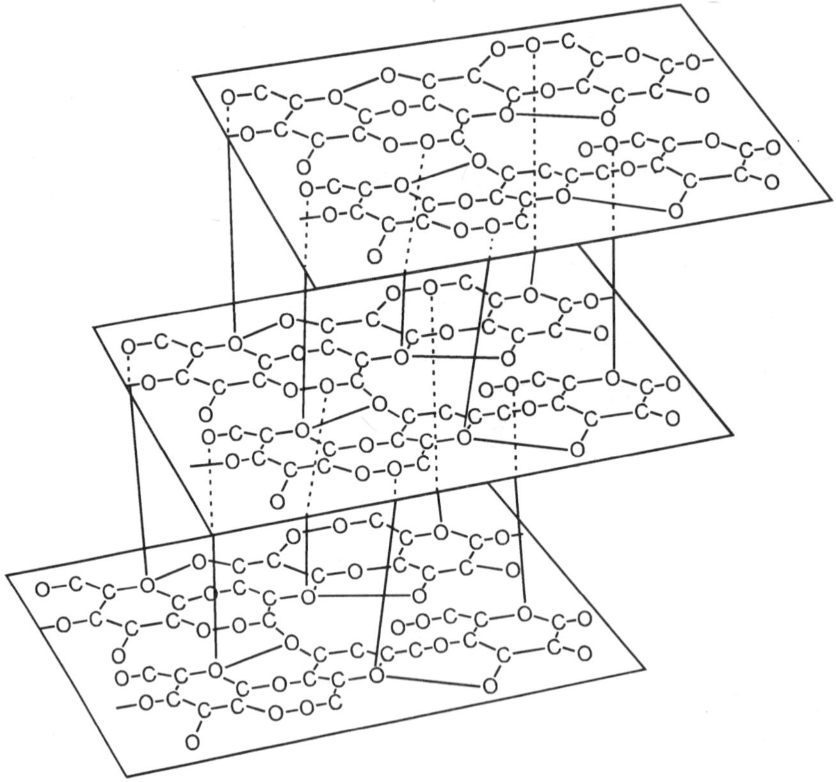

Fig. 7.41. Structure of cellulose microfibrils.

Hemicelluloses consist of a heterogeneous group of polymers of mannose, galactose, certain β-glucans, the heteropolymers from uronic acid, and methyl- or acetyl-substituted hexoses. Also included are the pentosans that are polymers of xylose and arabinose (Fig. 7.42). The cell wall is made up of about 20 per cent hemicellulose.

Xyloglucans

Xyloglucans are bound directly to microfibrils by hydrogen bonds. The main chain consists of β-1,4-linked glucose residues, branched primarily with xylose residues (β-1,6-linked) as well as some arabinose residues. Additional side chains contain galactose and occasional fucose residues (Fig. 7.43).

Arabinogalactans

The main chain of this polymer consists of alternating sequences of galactose and arabinose oligomers, to which both monomers are bound as side chains. They are bound to the glycoproteins of the cell wall. Serine and hydroxyproline residues form the bridges between the protein and the carbohydrates. There are two types of arabinogalactan, which differ in their linkage patterns:

1. Arabinogalactan I has β-1,4-linked D-galactose residues in the main chain and side chains with α-1,5-L-arabinose residues linked to the main chain with α-1,3 linkages.

2. Arabinogalactan II has β-1,3-linked D-galactose residues with β-1,6-D-galactose and α-1,3 and β-1,3-L-arabinose residues in the side chains (Fig. 7.44).

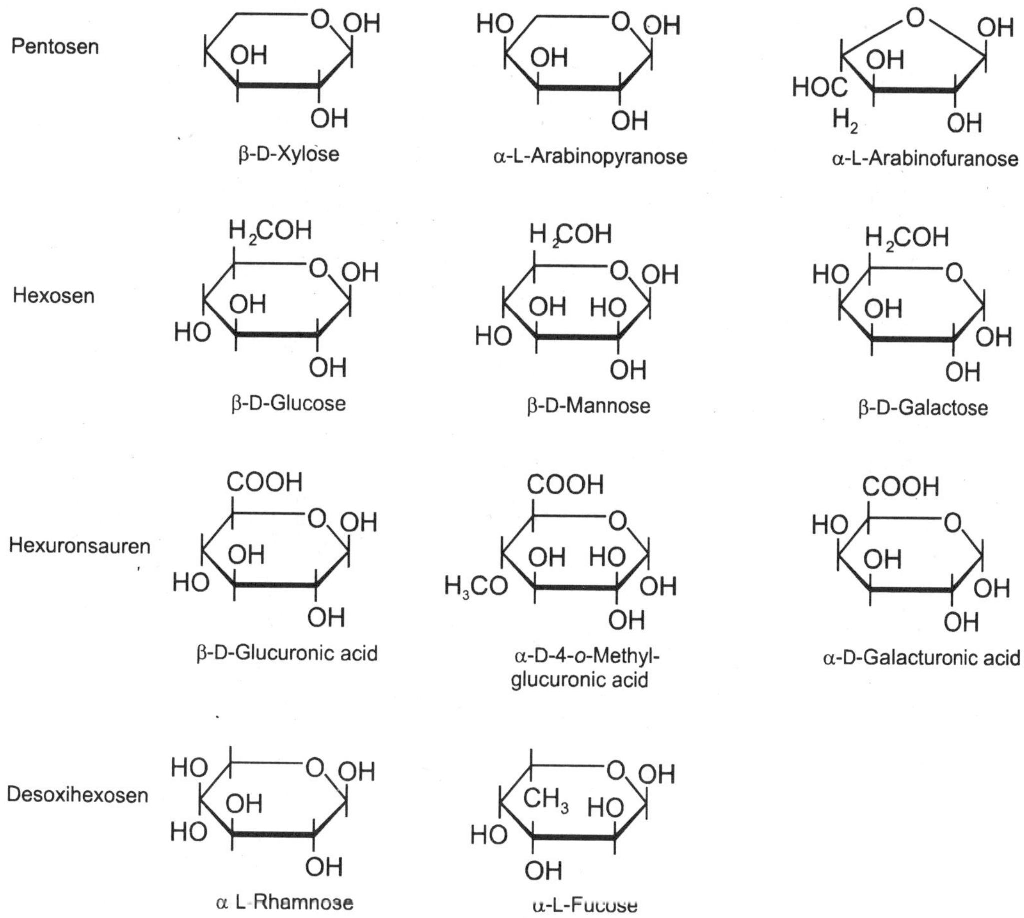

Fig. 7.42. Plant cell wall monosaccharides.

Rhamnogalacturonans

The rhamnogalacturonans can be categorised as pectic substances (Fig. 7.45).

Pectins

Pectins are a group of acidic polysaccharides that occur in the middle lamella of the primary plant cell wall. In the middle lamella, the pectins are bound to calcium and are responsible for the cohesion and texture of the plant tissue. The degradation of the middle lamella by enzymes leads to disintegration of the plant tissue into single cells. This type of hydrolysis is called maceration.

The pectins are not uniform chemical substances, but are very heterogeneous mixtures of polysaccharides with different molecular weights and degrees of esterification. The major building blocks are nonesterified or (anhydro)galacturonic acid units esterified with methanol. Galacturonic acid polymers are formed by α-1,4 linkages.

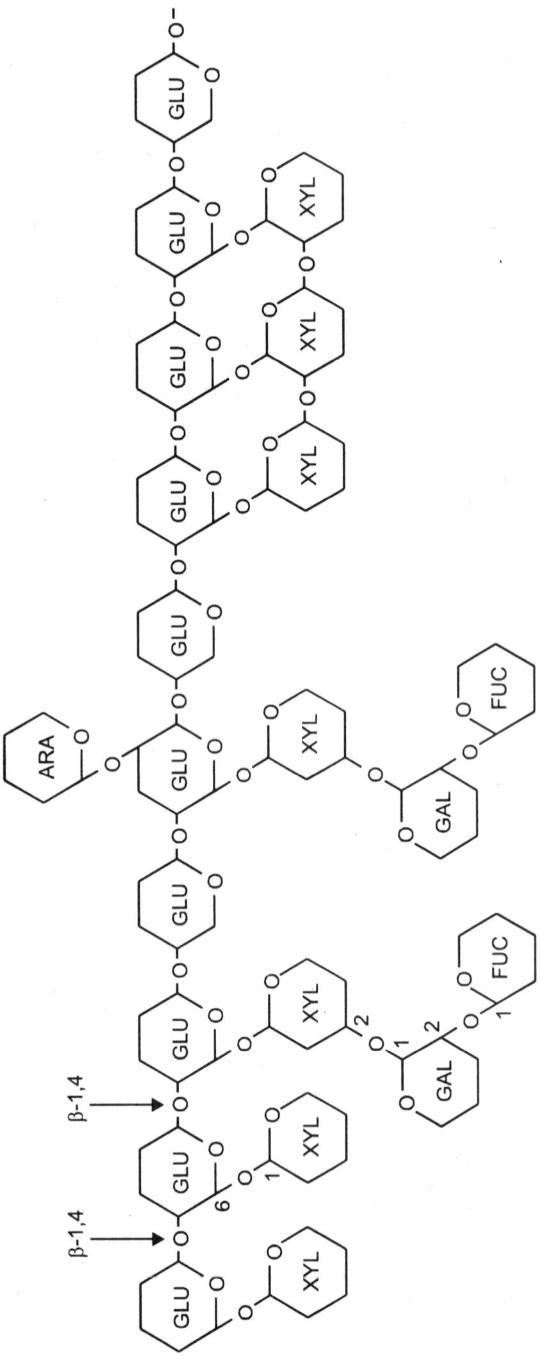

Fig. 7.43. Structure of xyloglucan.

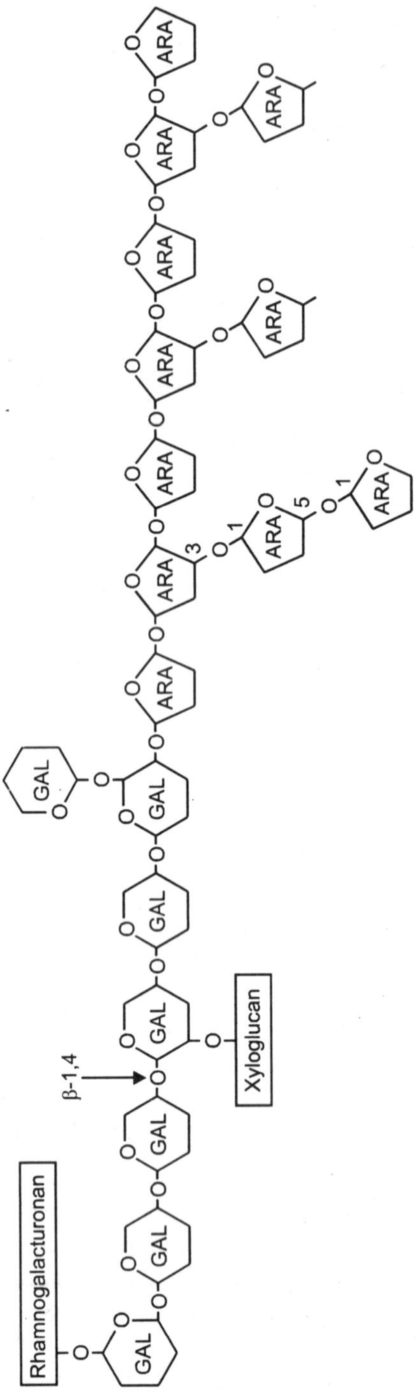

Fig. 7.44. Structure of arabinogalactan.

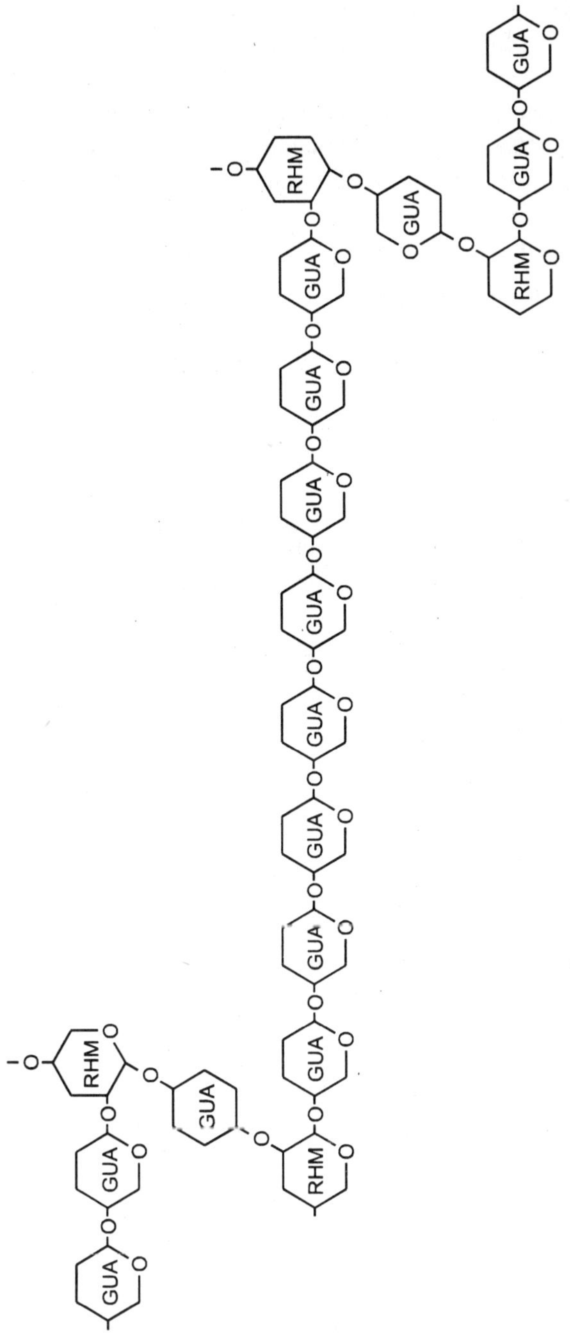

Fig. 7.45. Structure of rhamnogalacturonan.

Bound to these chains are additional rhamnose-rich polymers with side chains containing arabinose, galactose, and xylose residues. They are accompanied by hemicelluloses such as arabinans and galactans. Extracted pectins show molecular weights from 3000 to 3,00,000 daltons. The solubility of the pectins increases with increasing degree of esterification and lower molecular weight. The negatively charged polymers are cross-linked by calcium ions forming rigid gels. The insoluble pectin that constitutes the middle lamella is, because of its function, known as 'cell cement'. It plays a special role in fruit ripening, which is related to structural changes in the tissue. In the manufacture of cloud-stable fruit and vegetable juices, attempts are made to specifically dissolve the middle lamellar pectin, also called protopectin. The water-soluble pectins, formed mainly during fruit ripening, are highly esterified with methanol.

Pectins play a vital role in processing raw plant materials into foods (Table 7.12).

Table 7.12. Pectin content of plant material.

Source	*Pectin, dry weight (%)*
Apple pomace	15–18
Lemons	30–35
Oranges	30-40
Beets	25–30
Carrots	~7

The plant cell wall components discussed above and their structural composition are illustrated in Fig. 7.46 (see also Table 7.13).

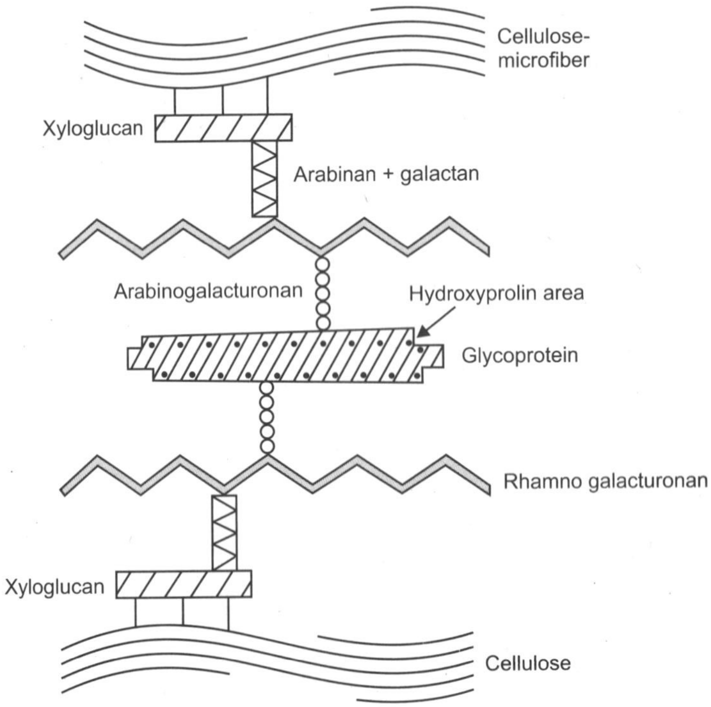

Fig. 7.46. Schematic illustration of the primary plant cell wall.

Table 7.13. Composition of some plant cell walls.

Apples	Galacturonic acid, 8.2%; glucose, 75.9%; galactose, 8.95%; arabinose, 6.0%; xylose, 1.2%
Tomatoes	Cellulose, 17%; pectin, 22%; protein, 17%; arabinogalactan, 13–23%; xylose and glucose, 13–23%
Pears	Glucan, 21.4%; galactan, 3.5%; mannan, 1.1%; xylan, 21%; arabinan, 10%; polygalacturonic acid, 11.5%; lignin, 16.1%
Wheat endosperm	Arabinoxylan, 2–3%
Rye	Pentosans, ~8%

Fungal Cellulases

In nature, cellulose is degraded by micro-organisms that rot wood and by soil bacteria. Bacterial flora present in the intestinal tracts of plant eaters rapidly degrade cellulose. Fungi and bacteria produce cellulases; they are also present in the seeds of higher plants and malt. Further, they are found in worms, caterpillars, and snails. The enzyme complex that has been isolated from snails (*Helix pomatia*) contains a number of very efficient cellulases in addition to β-glucosidases, hemicellulases, and α-glycosidases. While native cellulose in its natural environment is rapidly degraded by certain enzymes, degradation by enzymes isolated from organisms occurs very slowly. Enzymatic activity is significantly improved if cellulose is chemically or physically pretreated. Cellulose that has been preswollen in phosphoric acid or alkali is attacked very rapidly. This applies even more so to chemically modified cellulose derivatives such as carboxymethylcellulose (CMC) or hydroxyethylcellulose (HEC), which are often used to determine cellulase activity. Liquefaction of wallpaper glue that contains CMC can be observed after fungal growth on the wallpaper. It is beyond the scope of this book to relate the history of cellulase degradation; however, a brief summary of the literature of the last 100 years reveals some interesting studies on the degradation of cellulose during plant seed germination. The plant-cell-degrading enzymes, called cytases, are less thermostable than diastases (amylases).

De Bary described the action of a fungus that he had grown on beets, 'the mycelium of which poisoned the cells, and softened the beet tissue as if it had been cooked'. Karrer and Schubert and their school intensively studied the action of snail enzymes on regenerated cellulose (artificial silk). Selby and Maitland reported on the cellulases of the soil fungus *Myrothecium verrucaria*—a group of cellulases that has been studied especially extensively.

The work of Reese helped elucidate an understanding of the complex action of cellulases. The two-step degradation process suggested requires an activation reaction, followed by hydrolytic cleavage. The activating but not hydrolysing enzyme was named 'C_1 activity'. According to this concept, micro-organisms that are capable of degrading crystalline cellulose have C_1 activity. This activity is not present in enzymes that attack only substituted cellulose (CMC). These have only CMCase or C_x activity as shown below illustrating the enzymatic hydrolysis of cellulase:

$$\text{Cellulose} \xrightarrow{C_1} \text{reactive cellulose} \xrightarrow{C_x} \text{cellobiose} \xrightarrow{\beta\text{-glucosidase}} \text{glucose}$$

According to current knowledge, the degradation of native cellulose is believed to require the synergistic action of three different hydrolytic enzymes. This means that the attack occurs simultaneously by at least the first two enzymes:

1. Exocellulase, also called exobiohydrolase.
2. Endocellulase, also known as endoglucanase.
3. β-Glucosidase, also called cellobiase.

According to this scheme, the initial reaction at the amorphous area on the surface of cellulose microfibrils is conducted by an endoglucanase (with a special C_x activity?). This leads to new terminal groups, thus providing substrate for the next step (Fig. 7.47).

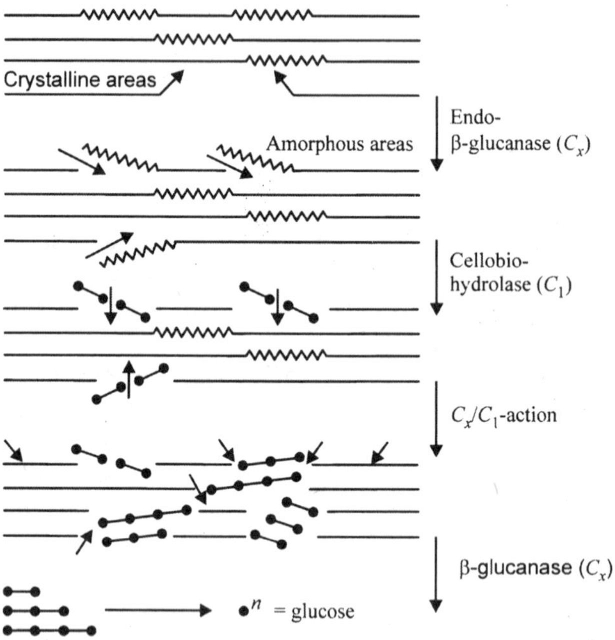

Fig. 7.47. Schematic illustration of the stepwise enzymatic hydrolysis of cellulose.

The terminal groups are formed on cleaving the cellobiose from the non-reducing end of glucan chains by cellobiohydrolase. This enzyme shows the highest affinity for cellulose. According to Wood and McCrae, the isolated and purified cellobiohydrolase is capable of cleaving up to 80 per cent of microcrystalline cellulose. As the reaction with cellobiohydrolase progresses, substrate is produced for the endoglucanase (C_x activity).

As this reaction sequence proceeds, the released cellobiose molecules are steadily hydrolysed by the third enzyme, cellobiase. This step is very important because the accumulation of cellobiose strongly inhibits the cellulose degradation reaction (step 2).

Lactones such as gluconolactone or cellobionolactone are strong inhibitors of cellobiase. According to Bruchmann, the lactone ring hydrolysing lactonases that occur in *A. niger* or in *T. reesei* play an important role in cellulose degradation.

A number of other influencing factors as well as other reaction mechanisms have been suggested for the initial attack described above. Eriksen claims that an oxidative degradation is possible in the first step. A cellobiose oxidase could thus be involved in the reaction.

Further information on the degradation of cellulose can be found in excellent reviews by Enari and Gosköyr and Eriksen.

A description of the three enzymes involved in cellulose degradation follows. The sources of microbial cellulase that are important for technical production are also described. The principal strains are *Trichoderma reesei* and *Aspergillus* sp.

Cellobiohydrolase (EC 3.2.1.91; CAS 37329-65-0), 1,4-β-glucan cellobiohydrolase

This exocellulase cleaves cellobiose from the non-reducing end of the glucan chain and from cellodextrins (cellulose fragments with three to six glucose residues). A number of cellobiohydrolases can be isolated by using modern fractionation methods (Table 7.14).

Table 7.14. Cellobiohydrolases from *Trichoderma reesei.*

Enzyme source	*Molecular weight (daltons)*	*pI*	*Carbohydrate (%)*
Cellulase			
Onozuka	42000	3.79	9.2
(5 isozymes)	53000	–	–
Meicellase	48000	–	0.1
T. reesei	65000	4.2–3.6	–
'CBH I'	60000	3.9	–
'CBH II'	60000	3.6	–

The cellobiohydrolases CBH I and CBH II differ in their immunologic reactivity. CBH I is more active, and there are five isozymes, all of which have similar molecular weights and amino acid profiles. They contain 0.1–10 per cent carbohydrate. The amino acid sequence was determined by Fagerstam.

Wood and McCrae have shown that the purified C_1 activity is a cellobiohydrolase that degrades phosphoric acid-treated cellulose well but acts on CMC very slowly. The reaction product is almost totally cellobiose (95 per cent).

Endoglucanase (EC 3.2.1.4; CAS 9012-54-8), 1,4-β-D-Glucan-4-glucanohydrolase

This enzyme splits the β-1,4 glycosidic linkages randomly in the internal regions of cellulose chains of cellodextrins, phosphoric acid-treated cellulose, and cellulose derivatives (CMC and HEC).

Table 7.15. Fungal endoglucanases.

Enzyme no./source	*Molecular weight (daltons)*	*pI*	*Carbohydrate (%)*
I/*T. reesei*	20000	7.5	–
II/*T. reesei*	40000	4.6	–
	12500	4.6	21
	50000	3.9	12
	47000	5.74	–
T. reesei	43000	4.0	–

The amorphous regions of crystalline cellulose are apparently attacked by a specific endoglucanase.

According to Beldman, six specific and non-specific endoglucanases can be distinguished in *T. viride.* The non-specific types also cleave xylan.

Aspergillus sp. strains, especially *A. niger* (see Table 7.15), form endoglucanases but only very small, amounts of exoglucanases (C_1 activity).

The degradation of cellulose by bacteria differs from that by fungi. Bacteria produce extracellular endoglucanases but no exoglucanases and only intracellular β-glucosidase.

β-Glucosidase (EC 3.2.1.21; CAS 9001-22-3), cellobiase, *β*-D-glucoside-glucohydrolase

Cellobiose and cellodextrins, including cellohexanose, are hydrolysed to glucose by this enzyme. Cellulose and higher cellodextrins are not attacked. The enzyme is produced by several different *Trichoderma* species, especially by *Aspergillus* species, primarily *A. niger* (Table 7.16).

Table 7.16. Fungal β-glucosidases.

Enzyme source	*Molecular weight (daltons)*	*pI*
T. reesei	47000	5.74
T. reesei	35000	–
	1,30,000	–
A. oryzae	2,18,000	4.3
A. niger	1,50,000	–

Cellulases from Humicola sp.

During the last few years several different cellulases, some of which are thermostable, have been found in *Humicola* species. They include *H. insolens*, *H. grisea*, and *H. grisea thermoidea*.

The enzyme from *H. lanuginosa* is a CMCase and has a molecular weight of 70,000 daltons and optimum activity at pH 5 and 45°C.

A cellulase from *H. insolens* (Celluzyme [NO]) that is used in the textile industry has an optimum activity at pH 7–8.5.

Bacterial cellulases

Very little is known about the cellulases from bacteria. Hagget reported on cellulases from a *Cellulomonas* strain. Cellulases from bacteria active in the alkaline range were described by Horikoshi and Akiba and Kawai and in Japanese Patent 007511.

Properties

1. Molecular weight: About 35,000 daltons (by gel filtration).
2. Activity: Endoglucanase activity ranges within 80–100 per cent at pH 5–10.5 with CMC as substrate. The enzyme shows no activity toward cellulose swollen in phosphoric acid. After incubation at pH 9 for 1 hour and 30°C, the activity is reduced by only 10 per cent. The maximum temperature tolerated at pH 9 is 55°C. *p*-Nitrophenyl-cellobioside is cleaved.
3. Activators: NaCl, KCl, $CaCl_2$, activation at ~10 per cent.
4. Stability: Complexing agents do not hinder activity, nor do soaps, Softanol 70 H, or Hostapur SAS. The cellulase is completely resistant to attack by proteases from *B. subtilis* or *B. licheniformis* (Alkalase [NO]; Maxatase [GB]).

After it was discovered that bacterial cellulases can be utilised as active components in detergents, industrial research and production groups focused on this new area.

An alkaline-stable cellulase from *Bacillus* with a molecular weight of ~16,000 daltons and a temperature optimum at 60°C was described by Oshino. Cellulase from *Acidothermus cellulyticus*: Poutanen discovered a thermophilic bacterial strain that was isolated from the hot springs in Yellowstone National Park. The cellulase from this bacterium has a temperature optimum at 80°C and could be of great interest to industry. Here again, three types of cellulose-degrading enzymes were found. The endoglucanase activity seems to be the most stable (83°C) (Fig. 7.48).

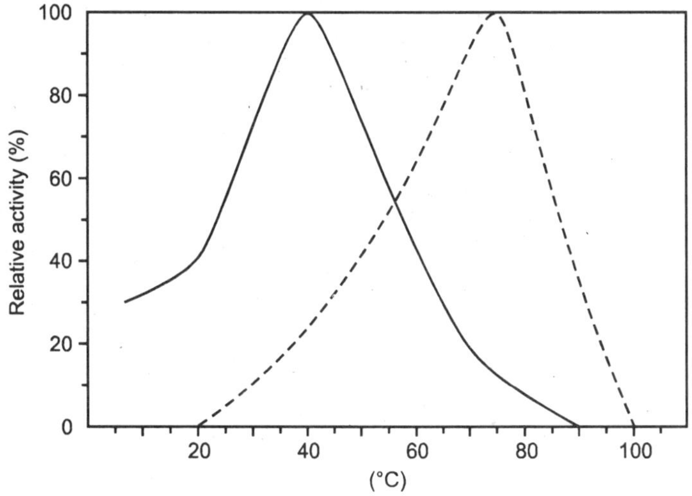

Fig. 7.48. Temperature dependence of cellulase from (i) *Acidothermus* and *Trichoderma reesei* (wild strain), and (ii) on filter paper.

Analysis of Cellulases

Because of the complex nature of both substrate and enzyme, analysis of cellulases is not simple. Cellulase preparations from different micro-organisms show very different enzyme profiles. The composition of an enzyme preparation from the same organism depends strongly on the conditions under which the organism was cultivated. The type and composition of the culture medium are an important factor.

The activity of the individual enzymes is of primary interest to researchers. Very specific analytic methods are necessary to describe their properties. For industry, the complex action of an enzyme preparation and its specific activity in a particular area of application are more relevant. Consequently, different methods have been developed for either biochemical investigations or for industrial applications.

The substrates for analysing the complex, synergistic action of endoglucanases and exoglucanases, the so-called C_1 activity, are insoluble crystalline celluloses such as Avicel, Solkafloc, and especially filter paper (see Table 7.17). The standardisation of these substrates is difficult because of their macromolecular structure. Degree of polymerisation, degree of crystallinity, purity, and substrate source all influence the analytic results.

Small molecule substrates such as *p*-nitrophenyl-β-glucoside, cellobiose, or oligocellulodextrins would be ideal for standardisation. However, such substances are not readily cleaved by all cellulase preparations and would also be sensitive to β-glucosidase activity.

Methods for analysing C_1 activity (Filter paper activity)

Filter paper activity according to Mandels is defined as the amount of sugar that is formed in one hour when Whatman no.1 filter paper (50 mg) is incubated with 0.5 ml of enzyme solution at pH 4.8 (citrate buffer) in a total volume of 1.5 ml.

Because the reaction does not run linearly, it is recommended that the enzyme be so diluted, that ~2.0 mg of reducing sugar is formed per hour. This method yields between 0.2 and 2 FPU per millilitre in the culture liquids.

Table 7.17. Methods for determination of cellulase activity.

Measured activity	*Substrate*	*Determination method*
Total cellulase activity (C_1)	Filter paper	Glucose
	Avicel	
	Solkafloc	
	Linters	
Solubilising activity	Filter paper	Weight reduction
	Avicel	Optical density
	Cellulose azur	Optical density
	Coloured cellulose	
Endoglucanase	CMC	Reducing sugar
	HEC	
	Phosphorous acid	
	Soaked cellulose	Decrease of viscosity
Exoglucanase	Linters	Reducing sugar
	Avicel	
	Filter paper	
β-Glucosidase	Cellobiose	Glucose
	p-Nitrophenyl-β-D-Glucosid	

To circumvent the problems that arise in the analysis of insoluble substrates, water-soluble derivatives such as carboxymethylcellulose (CMC) or hydroxyethylcellulose (HEC) are used extensively. It is important to remember, however, that these methods determine not the desired 'real' cellulase activity but mainly endoglucanase activity. Although neutral HEC is better suited for the analytic determination, CMC with a defined degree of substitution is most commonly employed.

Methods for determining endoglucanase activity

1. Viscometric determination according to FIP. HEC and CMC form highly viscous solutions in water. The viscosity drops rapidly on the action of cellulases. It was found that several different endoglucanases show changes in specific viscosity proportional to enzyme concentration and reaction time. This is valid only for a certain range of viscosity decrease, namely, around 50 per cent relative viscosity. The FIP standard 'endocellulase' is used as reference. Moreover, it is difficult to determine the molecular reaction rate of a cellulase from the change in viscosity. However, Almin and Eriksen have described a method in which absolute enzyme units can be calculated from the viscosity change of a precisely standardised HEC substrate.
2. Analysis of endoglucanase activity by measuring the reducing sugars. In general, the amount of reducing sugar produced during hydrolysis is measured either with dinitrosalicylic acid or by Nelson-Somogy procedures. For the analysis of cellulase, the ferric cyanide method of Wood and McCrae is preferred.

Technical Cellulase Preparations

Technical cellulase preparations are usually obtained from *Trichoderma reesei*, *Aspergillus niger*, and more recently, bacterial strains. In Japan, preparations were initially obtained from *Trichoderma viride* using surface cultures in the well-known Koji process.

Steamed wheat bran was used as culture medium, which, after inoculation with the spores or mycelia of the given micro-organism, was incubated in trays. Optimal culture conditions were determined by performing an extensive series of tests in which the pH, temperature, moisture, and other parameters were varied. On harvest of the culture (after 3–5 days), the growth medium was extracted with water, the extract concentrated, and the enzyme precipitated with alcohol. Even today, some companies still produce *Trichoderma* cellulases using the Koji process. Despite some inconvenience, the process offers advantages such as in the manufacture of certain products with a low production volume. The same can be said for cellulase prepared from *A. niger*. According to Yamada, about 45 tonnes of cellulases were produced from *T. viride* and *A. niger*.

Trichoderma cellulases

In the laboratories of the US Army in Natik and at Rutgers University, the wild strain of *Trichoderma viride* QM 6 was the first source for the production of cellulases. Later, mutations and selections were able to increase the productivity of this strain. Mutant QM 9414 was chosen as the basis for further high-performance mutants.

Use of cellulase preparations from *T. reesei* was limited to the manufacture of digestive aid products only a few years ago. In research, the cellulases are needed in combination with pectinases for the production of protoplasts (cellulase Onuzuka, Rohament CT/Rohament P).

Cellulase prices have been considerably lowered by improving cellulase production yields. Consequently, new areas of application for cellulases in food production and agriculture have become economically feasible. These applications include producing silage from grasses, addition to animal feed, the degradation of hemicelluloses during starch production from corn and wheat, and processing fruit and vegetables.

Commercial products, in both liquid and solid state, differ with respect to composition of cellulase activities and accompanying enzymes. The latter are of importance for applications in specific areas. The following side activities were found in enzyme preparations from *Trichoderma*: various xylanases, proteases, and a whole array of pectinases. In addition, the presence of β-1,3-glucanases, mannases and galactomannases, amylases, and glycosidases has been described.

Especially, loaded with side activities are the preparations from *Trichoderma harzianum*, which contain a series of cell wall lysing enzymes. These enzymes can be used for the lysis of yeast cells and for the production of protoplasts from *Aspergillus* and *Penicillium*.

There has been much literature on the relationship between the enzymatic composition of cellulases in technical preparations and the practical applications of these cellulases. Van Belle's group investigated the cellulase complexes in order to find a relationship between analytic composition and practical activity in the grass silages. A systematic analysis revealed the following enzymatic activities: cellulases, mannase, arabinogalactanase, arabanase, and pectinase. These studies were intended to help develop a specific working enzyme preparation for silage treatment.

Liquid cellulase preparations have an activity of 5–10 FPU/ml, the solid preparations about 10–50 FPU/g. The CMCase (C_x activity) differs from the filter paper units (C_1 activity). This activity ranges within 1000–1500 CU/ml in liquid preparations and 200–4000 CU/g in solid ones.

Properties

The relation of temperature and pH to the activity of some commercial preparations is shown in Figs. 7.49–7.54.

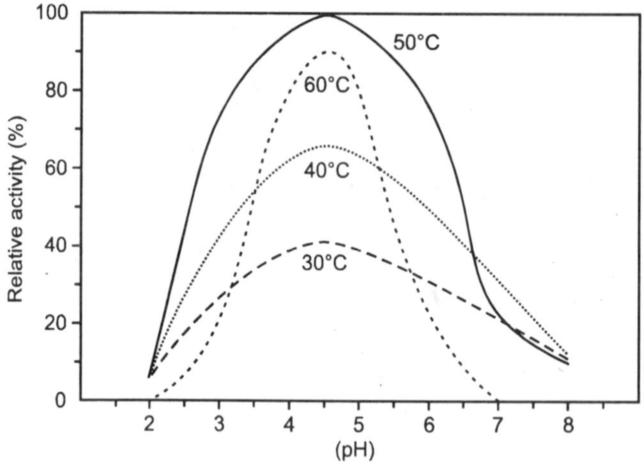

Fig. 7.49. pH and temperature dependence of cellulase during Avicel hydrolysis.

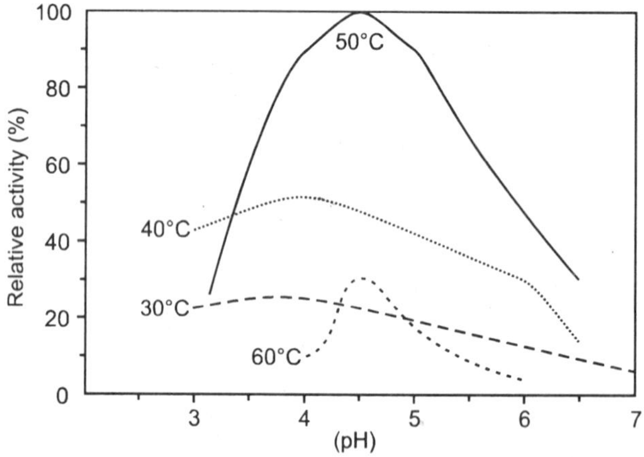

Fig. 7.50. pH and temperature dependence of cellulase during filter paper hydrolysis.

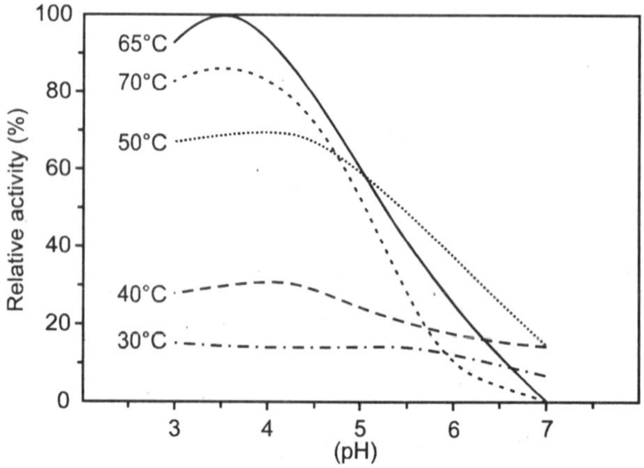

Fig. 7.51. pH and temperature dependence of cellulase (CMCase) during carboxymethylcellulose hydrolysis (measured as increase in reducing sugars).

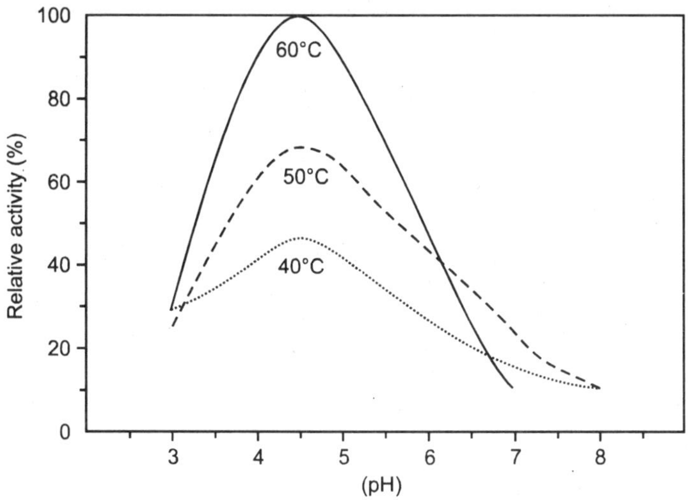

Fig. 7.52. pH and temperature dependence of β-glucosidase from *T. viride*.

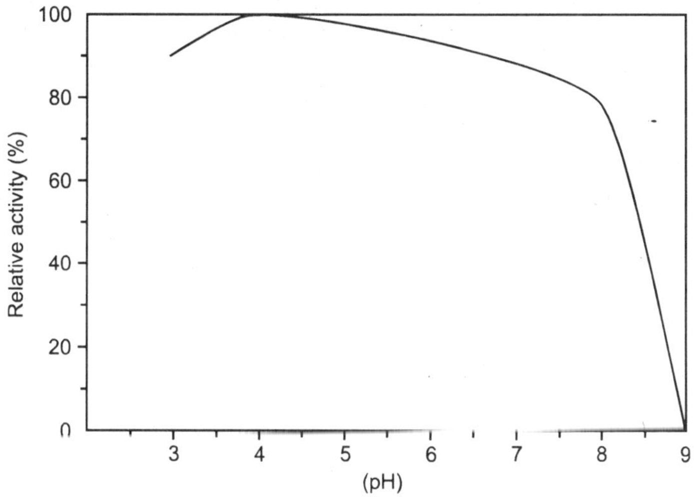

Fig. 7.53. pH dependence of cellulase stability; reaction time 12 hours; pH 5.0; 25°C.

Products and manufacturers

Celluclast and Novozyme 234 [NO], Econase C15 [AO], Rohament CT [RM], Avizyme [FR], Sumizyme C [SN], Meicelase MC [MJ], Cellulase TAP [AM], Cytolase 123 (GR).

Aspergillus cellulases

In the scientific literature, only those enzymes that are capable of degrading native cellulose are called cellulases. However, in industry, other carbohydrases capable of splitting the β-1,4-glycosidic linkage are also often described as cellulases or CMCases. Such cellulase complexes contain numerous types of hemicellulases, amylases, and proteases in addition to standard cellulase activity. These products are very manufacturer-specific, and their composition depends not only on the producing *Aspergillus* strain but also on the culture conditions used by the manufacturer.

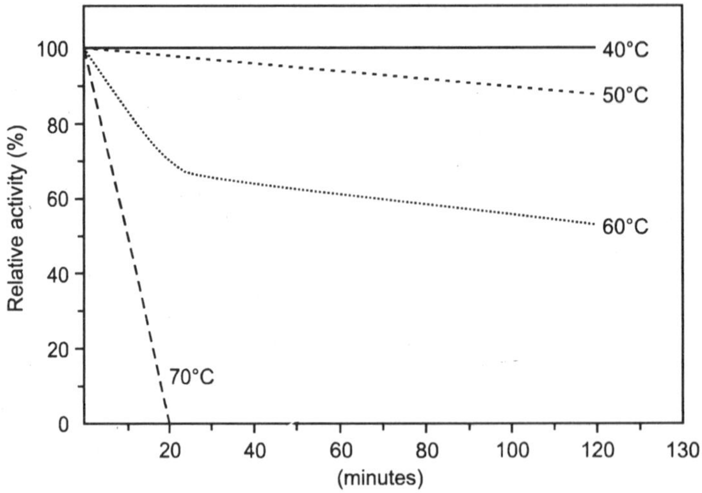

Fig. 7.54. Effect of temperature and time on stability of cellulase from *T. reesei* in aqueous solution [RM].

Because of the diverse activities of the various hydrolases, these preparations are used to improve the filtration rates of hydrolysates from barley and corn (breweries and distilleries). In addition, the efficiency of pectinases in fruit and vegetable processing can be improved or modified by *Aspergillus*-derived cellulase preparations.

Technical preparations are most often produced from cultures of *A. niger* (Figs. 7.55 and 7.56). Activity determinations for these enzymes are standardised using CMC as substrate. The activity of the dry products ranges from 1000 to 20000 CU/mg.

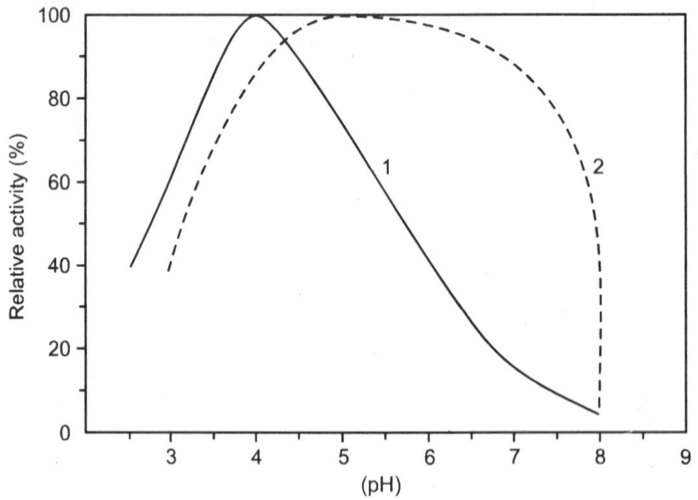

Fig. 7. 55. pH dependence of a technical-grade cellulase preparation from *A. niger* [RM]. The activity was determined viscosimetrically (CU/mg, curve 1); and by increase in reducing sugars (curve 2).

Those preparations are accompanied by enzymatic activity from pectinases, acidic fungal proteases, α- and β-glycosidases, β-glucanases, and pentosanases. Thus, these enzymes can be designated as multifunctional hydrolases.

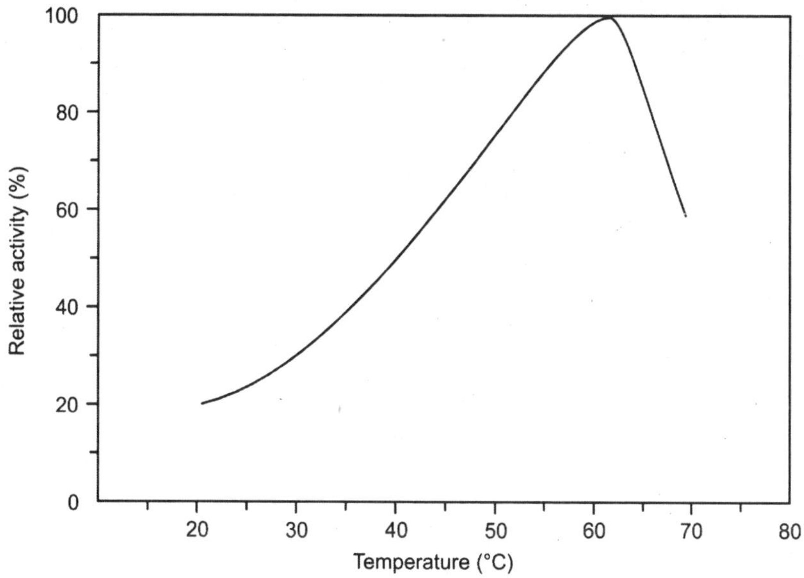

Fig. 7.56. Temperature dependence of CMCase from *A. niger* (pH 4.0; reaction time 15 minutes [RM].

The cellulase preparation, Rohament CW [RM] from *A. wentii*, for example, has yielded several interesting enzymes, including a β-glycosidase (Table 7.18).

Table 7.18. Enzymes in ROHAMENT CW [RM].

Cellulase (C_x)	900 CU/mg
Pectin glycosidase (PG)	560 PGU/mg
Pectin esterase (PE)	95 PEmU/mg
α-Glucosidase	10 mU/mg
β-Glucosidase	200 mU/mg
β-Galactosidase	170 mU/mg
Galactomannanase	3 mU/mg
Xylanase	160 mU/mg
Xylobiase	2 mU/mg
Protease	30 HbmU/mg

The pH dependence of CMCase stability is illustrated in Fig. 7.57.

Products and manufacturers: Rohament CA and CW [RM], Cellulase AP[AM], Cellulase P 4000 [SO], Cellusin [UA], Pancellase [YT].

Cellobiase preparations

Cellobiase preparations are used in combination with cellulase preparations from *Trichoderma* to hydrolyse cellulose in order to obtain optimum conversion of cellulose into glucose. In silage, for example, the application of *Trichoderma* cellulases leads to the formation of cellobiose, which cannot be utilised by most micro-organisms (Figs 7.58 and 7.59).

Manufacturer: [NO].

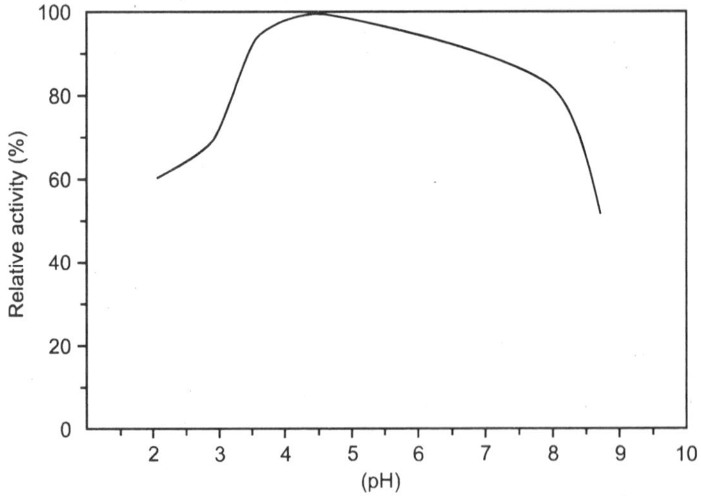

Fig. 7.57. pH dependence of CMCase stability in aqueous solution (40°C, reaction time: 30 minutes [RM]).

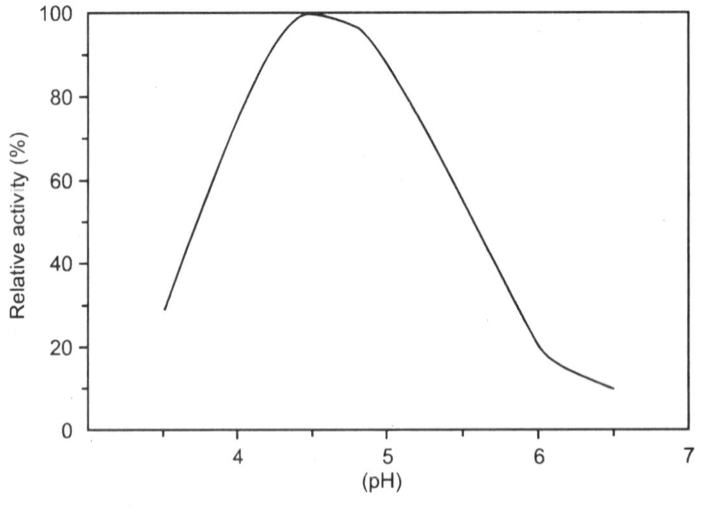

Fig. 7.58. pH dependence of cellobiase [NO].

Cellulase from Humicola insolens

The enzyme from this thermophilic fungal strain differs from other cellulases in its pH-activity profile, which extends into the alkaline range. Stability is surprisingly good at pH 9, even at a temperature of 60°–65°C. The enzyme is used as an additive in detergents and is commercially available in granulated form (Figs 7.60 and 7.61)[RM].

HEMICELLULASES

The heteropolymer carbohydrates that are associated with cellulose in plants are, with the exception of pectic substances, hemicelluloses. Included are polymers of hexoses, such as glucose, rhamnose, or mannose, as well as the polymers of pentoses: xylose and arabinose. Furthermore, plant mucins and other polymers obtained from plants are included in this group and are utilised in the food industry as thickening agents. From this large area only those substances, the enzymatic degradation of which is of commercial importance, will be discussed. There are the β-glucans and pentosans from cereals, hemicelluloses from fruits, seeds, and legumes, and also the β-glucans produced by yeasts and fungi.

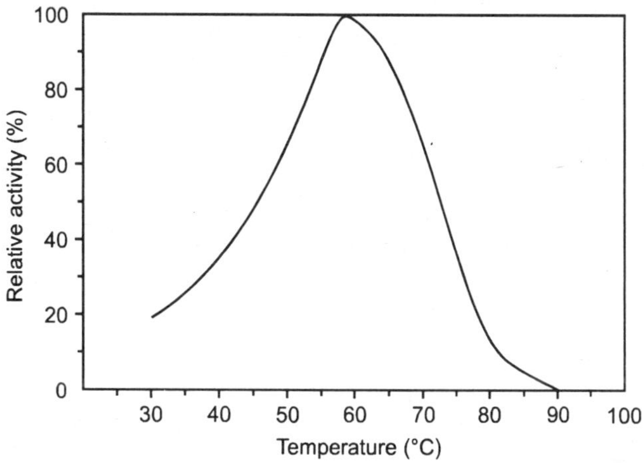

Fig. 7.59. Temperature-dependent cellobiase activity [NO].

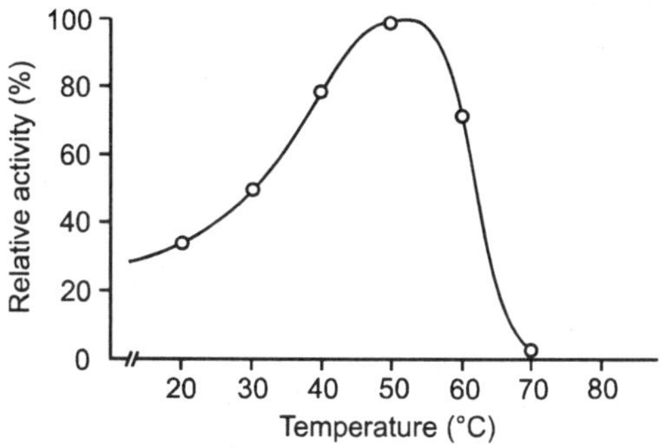

Fig. 7.60. Temperature-dependent cellulase activity (substrate: 7.5 grams CMC/litre; phosphate buffer; 40°C; reaction time 20 minutes [NO]).

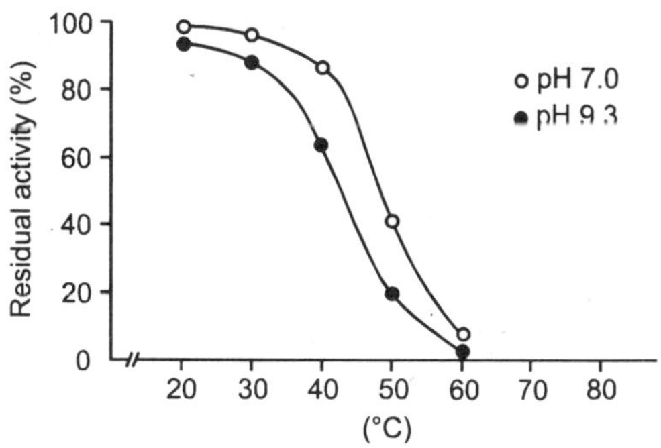

Fig. 7.61. Temperature-dependent stability of *Humicola* cellulase at pH 7 and 9.3 [NO].

Wheat endosperm contains arabinoxylan and a 1,3- and 1,4-β-D-glucan, which constitute 2–3 per cent of the weight of wheat grain. About one-third of this fraction is water-soluble. The other two-thirds can be extracted with alkali. Rye flour contains 8 per cent pentosan, which contributes to the high water-binding capacity and viscosity of rye dough. The β-1,3;1,4-glucans are found in barley and are responsible for filtration problems encountered during brewing beer. In corn, there are mucins that reduce starch yield in the separation of starch from gluten. Undesirable turbidity, attributable to insoluble arabinose polymers, is encountered in the production of modern pear and apple juice concentrates. For processing of berry fruits it is necessary to degrade not only pectin but also the various polymers of unknown structure. During extraction of coffee, the degradation of mannan polymers by enzymes can improve soluble yield. Turbidity that occurs in coffee concentrates can also be prevented by enzyme treatment. Similar problems are encountered during the extraction of tea for instant-tea or liquid-tea production.

This area of application is generally an interesting one for enzymes; however, their targeted application is made difficult by the fact that very complex heteropolymers with various monomeric compositions and linkages are present. During the past few decades, basic research has made such progress in this area mainly by determining the structures, of the natural polymers, as well as through the isolation, purification, and characterisation of the necessary enzymes. Numerous problems still exist, and these can be solved by the cooperative efforts of natural product chemists, enzymologists, and last, but not least, the technical people working in the field.

In the following section, the enzymes will be emphasised and the natural products and their associated problems will be dealt with only briefly.

β-Glucanases

β-Glucans constitute, on a weight basis, 2.8–5.5 per cent of the total weight of the cell walls of barley endosperm. They are heterogeneous, linear polymers of glucose that are linked by β-1,4 (75 per cent) and β-1,3 glycosidic bonds. Depending on the cereal source, molecular weights are as follows: barley glucan ~55000, rice glucan ~64000, and wheat glucan ~34000 daltons. Pentosan preparations from barley have molecular weights from 50000 to 2,00,000 daltons. β-Glucans are of particular interest in the brewing industry because they can cause difficulties during clarification of wort and the filteration of beer. During the malting barley, the β-glucans and pentosans are degraded by the enzymes endogenous to barley. If the activity of the endogenous β-1,4; 1,3-glucanase is not high enough, then high-molecular-weight glucans remain in the mash and cause the difficulties described above. In addition, the degradation of the β-glucans is dependent on the temperature of the various mash processes.

The endogenous barley β-glucanases are inactivated above 45°C so that in the typical infusion processes in brewing at ~65°C, only a part of the β-glucans is hydrolysed. An improvement in solubility can be achieved by addition of temperature-stable microbial β-glucanase preparations. Problems with β-glucans are also encountered in wine production. These β-1,3; 1,6-glucans have structures different from those of the cereal β-glucans. They are produced by fungi (on wine production). Commercial preparations involve a number of different types of β-glucanases, which are described below.

Cellulases (EC 3.2.14)

The cellulase enzymes that attack amorphous cellulose and hydrocellulose can also attack barley β-1,3;1,4-glucans. They are capable of hydrolysing highly branched β-1,4-glucans, such as the tamarind amyloid, a highly branched xyloglucan. Commercial bacterial β-glucanases do not have this enzyme, which appears in *Trichoderma* and *Penicillium* cellulases.

Endo-1,4-β-glucanase (EC 3.2.1.73; CAS 37288-51.0)

This enzyme acts on 1,3;1,4-β-glucan but shows no activity with 1,4-β-glucan. It cleaves the β-1,4 linkage adjacent to the reducing end of a β-1,3 linkage, which is absolutely necessary for cleavage to occur. This cleavage pattern is characteristic of bacterial b-glucanases (lichenase) and the malt β-glucanases. These enzymes also show no action on linear β-1,3-glucans, as can be seen in Fig. 7.62.

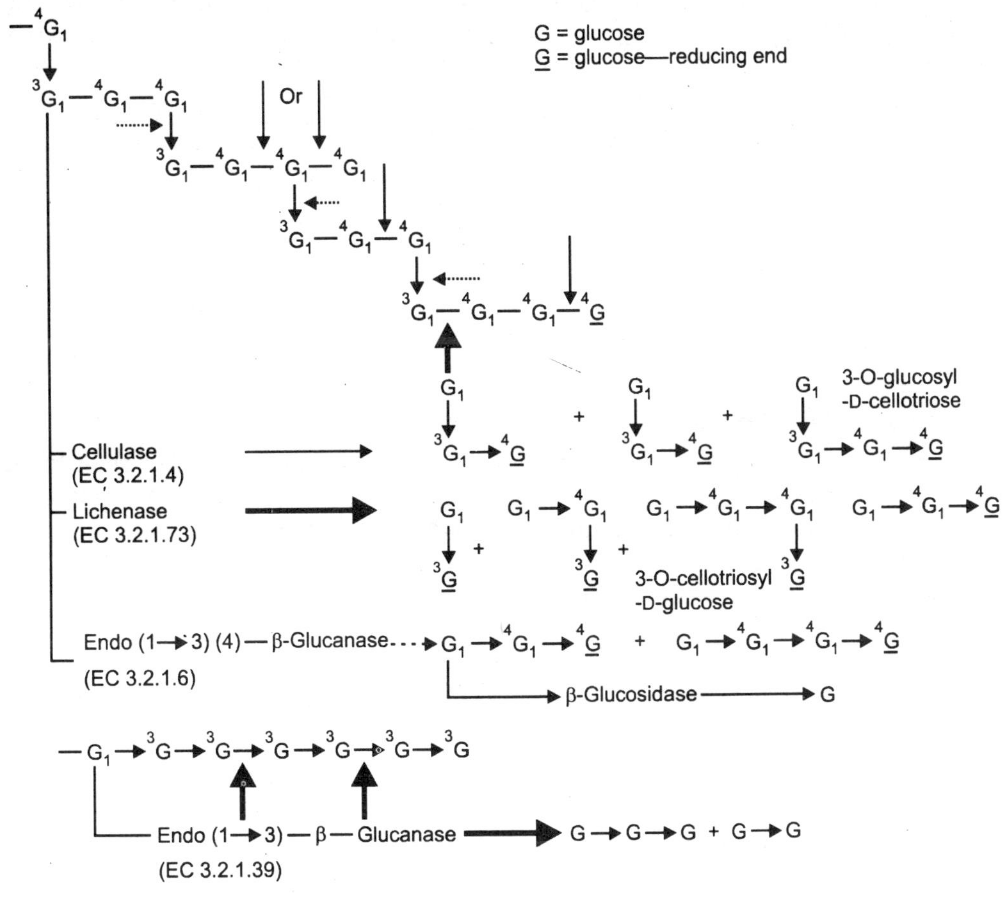

Fig. 7.62. Action of various enzymes on β-glucan.

Endo-1,3-β-glucanase (EC 3.2.1.39; CAS 9025-37-0), also called laminarinase and endo-1,3(4)-β-glucanase (EC 3.2.1.6; CAS 62213-14-3)

The first enzyme cleaves the β-1,3 linkage at the reducing end of β-1,3-bound glucose chains. The 1,3(4)-β-glucanase, in contrast, is also capable of cleaving a 1-4-β-linkage at the reducing end of a β-1,3-bound glucose chains so that it can also act on those β-glucans with both linkages similar to the action of lichenase.

Endo-1,6-β-D-glucanases

These enzymes attach the β-1,6-bound glucose residues that appear, for example, in yeast glucan. Lutean, a β-1,6-glucan, is degraded by this enzyme and can be used to test for it.

Technical Preparations

Technical β-glucanase preparations are prepared from different *Bacillus* and fungal species, such as *Penicillium, Aspergillus, Rhizopus,* and *Trichoderma.*

Sauter and Sprössler isolated pure β-glucanases from industrial preparations and characterised them according to their action on various substrates. The enzyme isolated from *B. subtilis* had a molecular weight of 29000 daltons, a pH optimum at pH 7.5, and an optimum temperature of 40°–50°C. The enzyme from *A. niger* also has a molecular weight of 29000 daltons and optimum activity at pH 5.0 and 50°–60°C (Table 7.19).

Table 7.19. Comparison of various β-glucanases by sugar oligomers generated with different substrates.

		Substrate	
Enzyme source	*CMC*[a]	*Laminarin*	*Barley β-glucan*
B. subtilis	–	–	G_3, G_4
A. niger	G_2, G_3	G_3, G_4, G_5	G_3, G_4
T. reesei	G_2, G_3	G_2, G_3, G_4	G_2, G_4, G_5

[a]Carboxymethylcellulose

β-Glucanases from B. subtilis

These enzymes have specificities similar to those of the glucanases from malt. They are active against lichenin, have an affinity for the trimeric glucose unit of a β-glucan chain, and cleave the linkage adjacent to the next β-1,4 linkage. Neither acts on laminarin or carboxymethylcellulose.

1. Molecular weight: 26,000 daltons; 29000 daltons.
2. Activity: pH optimum is 6.0–7.5. Other types show an optimum between 7.0 and 9. A third variation has a narrow optimum at pH 7.5.
3. K_m values: Lichenin, 1.43 mg/ml; barley glucan, 1.15 mg/ml.
4. Temperature optimum: 58°–60°C. Temperature-stable types are active up to 65°C. Inactivation at temperatures above 75°C. The temperature stability is improved by Ca^{2+} ions.

Another *B. subtilis* β-glucanase was described by Rafalovskaya:

1. Molecular weight: 35,000 daltons, pI at pH 5.2.
2. Activity: Optimum pH 6 and 55°–60°C.
3. Stability: pH range 6–9.
4. Inhibition: SH reagents inhibit, as well as Ca^{2+}, Cu^{2+}, Al^{3+}, Pb^{2+}.

β-Glucanase from B. licheniformis

Lloberas isolated and characterised an endo-β-1,3; 1,4-D-glucanase from the organism described above. The molecular weight was between 27000 and 28000 daltons; the isoelectric point was at pH 4.7. The activity optimum was very broad, ranging from pH 4.0 to 10.5. The optimum temperature was 55°C.

β-Glucanase from Penicillium emersonii

There are also at least two types, which differ from the bacterial enzymes with respect to both specificity and activity-stability range.

The enzyme complex is particularly heat-stable; at least 75 per cent of the activity is retained after heating the enzyme solution to 95°C for 10 minutes (Figs 7.63 and 7.64).

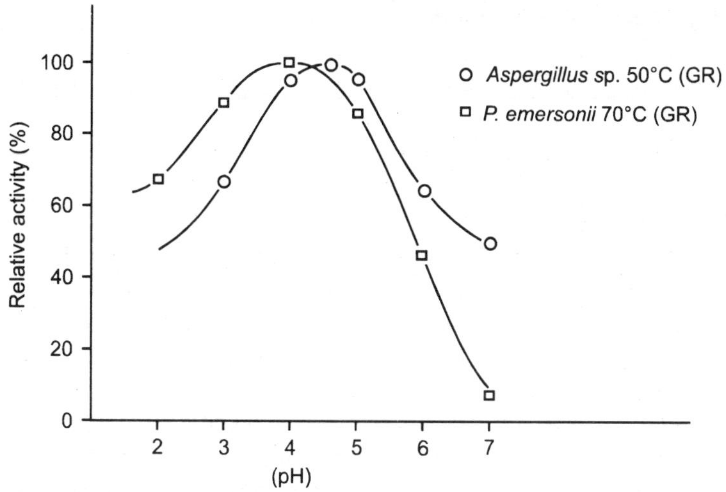

Fig. 7.63. pH dependence of fungal β-glucanase from *P. emersonii* and *Aspergillus* sp.

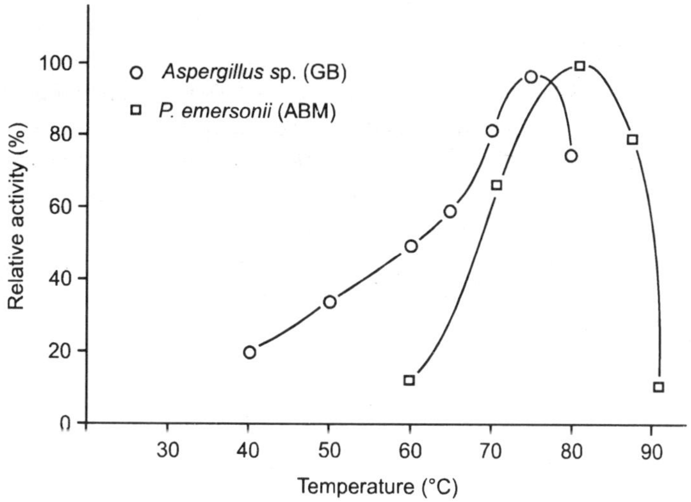

Fig. 7.64. Temperature dependence of fungal β-glucanase from *P. emersonii* and *Aspergillus* sp.

β-Glucanase from A. niger

This enzyme has an optimum activity at pH 5, with 80 per cent of the maximum activity between pH 4 and 6.2. Optimum temperature is 65°C; maximum temperature is 75°C (Glucanase GV [GT]).

β-Glucanase from Trichoderma reesei

A β-1,3-glucanase was isolated by Ogawa from a fixed-bed culture:

1. Molecular weight: ~40,000 daltons.
2. Activity: The maximum activity is at 55°C and pH ~5.
3. Stability: The enzyme is stable between pH 4 and 6 at ~50°C.
4. Specificity: Laminarintetraose is cleaved to glucose, laminarinbiose, and laminarintriose.

β-Glucanases from Trichoderma harzianum

Enzymes from this organism contain a series of very active β-glucanases. After fractionation by Dubourdieu, two exo-1,3-glucanases and one endo-1,6-D-glucanase were found. This enzyme can degrade β-1.3; 1,6-D-glucans, such as cinerean.

Molecular weight: ~40000 daltons; pI at pH 7.8.

β-Glucanase from Arthrobacter sp.

So-called yeast cell lytic enzymes are produced by some enzyme manufacturers from various organisms such as *Trichoderma* sp. These were not particularly effective when used for the lysis of yeast cell walls; thus enzymes that were more specific were sought. Various *Arthrobacter* sp. were found that formed a yeast-lysing enzyme system for yeast cells or yeast cell walls. β-1,3-Glucanase, protease, and mannanase activities were identified in the enzyme complex. 'Tunicase' [DA] is a commercial enzyme preparation prepared from *Arthrobacter* cultures. It is a cellulase that acts specifically on the β-1,3-glucans of yeast cell walls. Cell walls form *Candida*, *Saccharomyces cerevisiae*, *Torula*, and *Kluyveromyces fragilis* are lysed. The enzyme has an optimum activity between pH 7 and 8 and is stable only below 35°C. The enzyme is stable in the pH range 4.0–9.0. A similar enzyme from *Archromobacter* is active at pH 7 up to 55°C.

Products and manufacturers

Technical preparations are standardised by using β-glucan as substrate. The amount of reducing sugar produced or the reduction in viscosity is determined. Another method is described by McCleary in which the activity of lichenase can be determined by using a soluble chromogenic substrate.

Most commercial preparations are in liquid form. They contain amylases and, depending on source of origin, protease side activities:

1. Bacterial β-glucanases from *B. subtilis* [AM], [NO], [RM], [SOE], [GB].
2. Fungal β-glucanase from *P. emersonii* [GX], [BN], [AM].
3. *Trichoderma reesei* β-glucanase [GR].
4. Fungal β-glucanase from *A. niger* [RM], [GT], [NO].
5. Glucanase from *Arthrobacter* sp. [DA].
6. Glucanase from *Achromobacter* sp. [AM].

Mannanases and Galactomannases

Mannans, linear α-1,2 or α-1,6 polymers of mannose, occur in yeasts and other micro-organisms and also in higher plants. Mannose is, however, more frequently found as a component of heteropolymers such as glucomannan and galactomannan. Galactomannans occur as reserve carbohydrates in guar beans and in locust bean seeds. They consist of a main chain of β-1,4-linked mannose residues that carry side chains of α-1,6-bound-D-galactose residues. In addition to monomeric galactose residues, digalactosyl residues or gal (3,1)-ara-(3,1)-ara side chains can also occur.

Depending on their origin, the galactomannans differ in their galactose: mannose ratios. The galactose content varies between 10 and 50 per cent. Galactomannan containing more than 25 per cent galactose (e.g. guar bean galactomannan contains 35 per cent galactose) are readily soluble in cold water and give highly viscous solutions. Those galactomannans with a galactose content of 18–24 per cent are soluble in hot water (e.g. locust bean meal, 23 per cent galactose). When these polymers are mixed with other polysaccharides such as agarose or xanthan, gels with a very low polymer concentration (solid content) are formed.

Mannanase (EC 3.2.1.78; CAS 37288-54-3) endo-β-1,4-D-mannanase

This enzyme cleaves the linear mannan chain of locust bean galactomannan to oligomers with galactose branches, which serve as a basis for the production of food additives. Guar galactomannan is only slightly attacked by this enzyme. The enzyme is produced by *B. subtilis*, *A. awamori*, *A. niger*, and *T. reesei* and can be isolated from commercial galactomannanase preparations.

Galactomannanase

Technical galactomannanase preparations that can be obtained from *A. niger* contain, in addition to the mannanase, a series of additional enzymes such as α-galactosidase (EC 3.2.1.22) and β-D-mannosidase (EC 3.2.1.25). These preparations are used to reduce the high viscosities of galactomannan solutions such as in the process used to extract secondary yields from crude oil. Galactomannan modification is also desirable for some applications in the food industry. For the extraction of coffee, the mannan-containing matrix of the coffee beans can be ruptured. This not only leads to an increase in extraction yield but also prevents the accumulation of mannan-containing sediments in the extracts that are difficult to filter.

Standardisation is accomplished by determining the reducing sugar produced calculated as mannose, from locust bean gum galactomannan (Fig. 7.65).

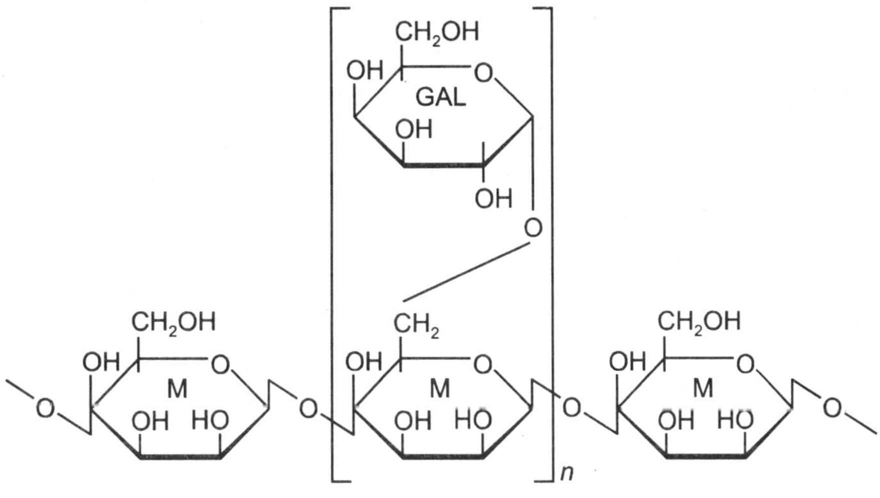

Fig. 7.65. Structure of locust bean gum galactomannan.

The hemicellulose preparation Rohament AC [RM] is stable in the absence of substrate over a wide pH range: 95–98 per cent of the enzyme remains active after incubation at 20°–25°C for 24 hours between pH 3 and 8.5. The enzyme is stable up to 60°C at pH 4–5; in the presence of 2 per cent substrate, up to 70°C. See Figs 7.66 and 7.67.

Chitosanase

Chitin and chitosan occur in crustaceans, where they are bound to calcium and covalently linked to proteins. The proteins serve as binding links to other carbohydrates. Chitin and chitosan are copolymers of *N*-acetylglucosamine and glucosamine, which are linked in the β-1,4 position. The copolymer is referred to as chitin if it contains less than 7 per cent nitrogen. The chitin deacylase, for example, from

Mucor rouxii, catalyses the hydrolysis of chitin to chitosan and acetic acid. The soluble chitosan has a molecular weight of several hundred thousand to two million daltons. Chitinases are formed by higher plants and micro-organisms. The enzymes have been isolated from plants such as tomatoes, soyabeans, and wheat bran. *Streptomyces*, especially *S. griseus*, *Serratia* and *Aeromonas* species are potential producers.

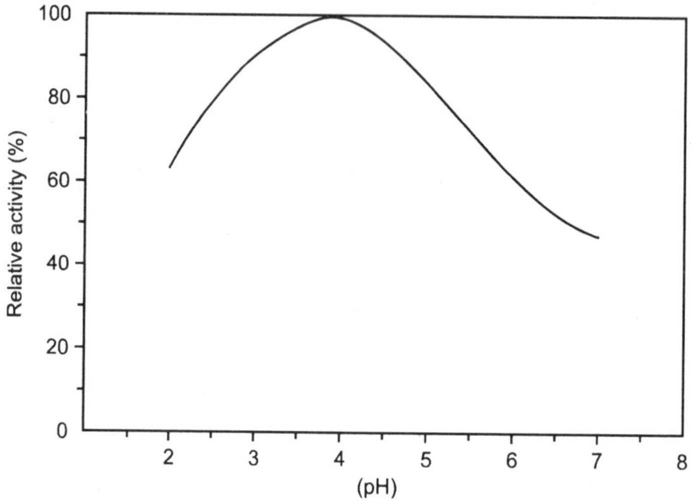

Fig. 7.66. pH dependence of galactomannase activity (40°C; reaction time 10 minutes).

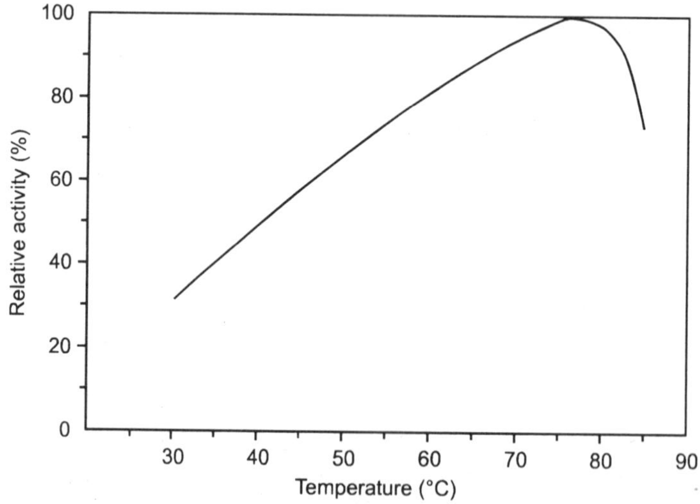

Fig. 7.67. Temperature dependence of galactomannase activity (pH 4.0; reaction time 10 minutes).

The enzyme from *S. griseus* has a molecular weight of 35000 daltons, and optimum pH is 8.0. The enzyme is stable at 37°C between pH 6.0 and 8.0. Inactivation occurs at higher temperatures.

Chitinase, similar to other carbohydrases, consists of a complex of three enzymes: endochitinase, exochitinase, and chitobiase. A commercial chitosanase is obtained by fermentation of *B. pumilus* (Chitosanase BP [MJ]). The activity of this chitosanase is determined by lowering the viscosity of a chitosan solution or determining of the reducing sugar groups according to the Somogyi-Nelson method.

Chitooligosaccharides (DP = 2–10) and, eventually, after longer reaction times, small amounts of the glucosamine monomer are produced. The enzyme is also capable of degrading the cell walls of micro-organisms that contain glucosamine polymers. The activation optimum ranges within pH 5–6.5 and 40°–60°C. Another very active preparation is produced by *Aeromonas hydrophilia* sub-species *anaerogenes* cultures [GO].

PECTIN DEGRADATION

The following pectin substances can be distinguished:

1. Pectic substances: Pectic substances are those materials that contain principally polygalacturonic acids, partially or completely esterified. A completely 100 per cent esterified polygalacturonic acid has a theoretical methoxyl content of 16.3 per cent.
2. Protopectin: According to Joslyn protopectins are water-insoluble pectins that are chemically and physically bound to other cell wall components such as cellulose, xylan or glucomannan.
3. Pectic acids: Very lightly esterified or nonesterified polygalacturonic acids are called pectic acids. They form rigid gels with sugars and acids or divalent ions. The English literature differentiates between pectinic acid, for low degrees of esterification (up to 7 per cent); and pectic acid, for totally deesterified pectin.

As already mentioned, the texture of fruit and vegetables is regulated by the content and the properties of the pectic substances (Table 7.20). In immature fruits they appear principally as insoluble protopectin. This converts to soluble pectin during ripening and is associated with the softening of the fruit. The loosening of the rigid cell wall can be counteracted by using calcium salts.

Table 7.20. Pectin content of some fruits and juices.

Source	*Pectin (%)*	*Esterification (%)*
Grapes	0.2–1.0	10–65
Grape juice	0.01–0.09	–
Apples	0.5–1.6	70–90
Apple juice (non-clarified)	0.2	–
Black currant mash	1.6	–
Grapefruit	1.6–4.5	–
Lemon (rind)	32.0	50–65 (albedo)
Lemon (pulp)	25.0	–
Orange (rind)	20.0	–
Orange (membrane)	29.0	–
Orange (juice sacs)	16.0	–
Turnips	10.0	–
Sugarbeet pulp	30.0	–

Pectinases

Pectinases are important in processing fruit and vegetables and in wine production. Specifically, these enzymes are used for clarification of fruit juice, enzymatic fruit mash treatment, improvement of juice yield and colour in wine production, and the maceration and liquefaction of plant tissue. The pectinases are also used to improve the extraction yield of oils from citrus fruit and olives.

Pectinases are usually mixtures of several enzymes, which are capable of cleaving high-molecular-weight pectin within the polygalacturonic chain (depolymerases) and deesterifying pectins (esterases). Depolymerisation can progress in one of two ways: either by a hydrolytic reaction, as occurs generally in the hydrolysis of glycosidic linkages, or by means of a very specific β-transelimination reaction. This class of depolymerases are called lyases. These different cleavage reactions can be used to classify the pectinases (Fig. 7.68 and Table 7.21).

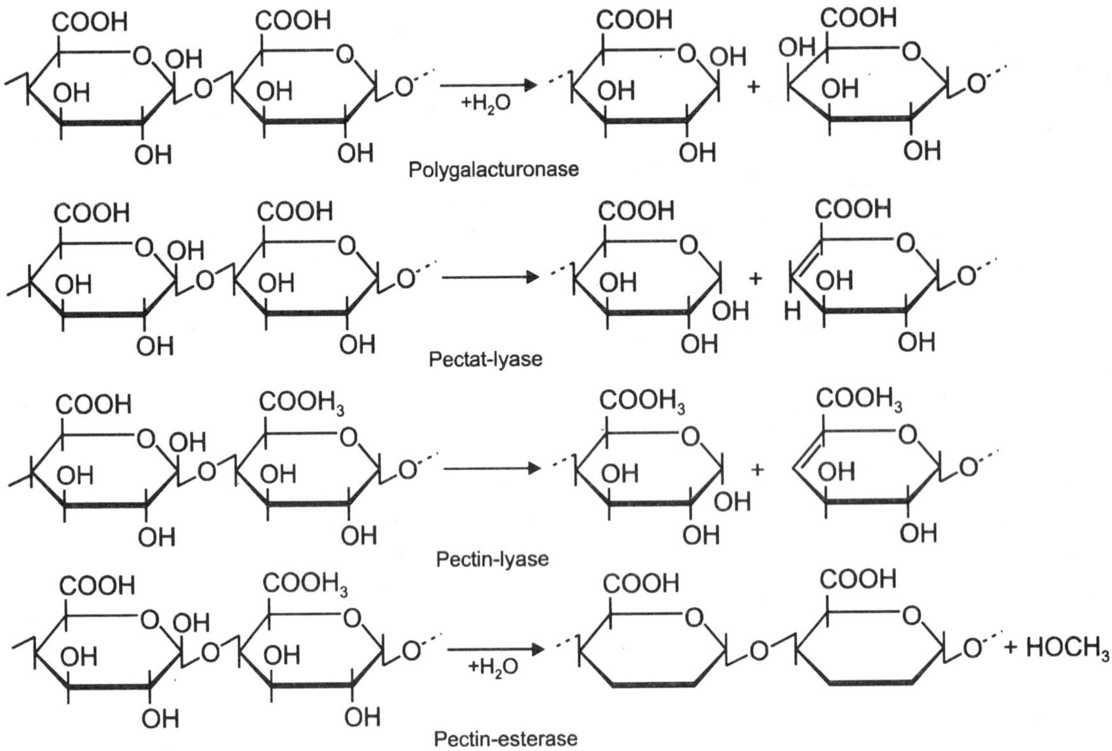

Fig. 7.68. Ester hydrolysis and depolymerisation of pectin.

Table 7.21. Pectinase classification.

Name	*EC number*	*Preferred substrate*	*Cleavage site*
Hydrolases			
Endopolygalacturonase	3.2.1.15	Pectate	Randomly within the chain
Exopolygalacturonase	3.2.1.67	Pectate	Terminal
Lyases			
Endopectate lyase	4.2.2.2	Pectate	Randomly within the chain
Exopectate lyase	4.2.2.9	Pectate	Terminal
Endopectin lyase	4.2.2.10	Pectin	Randomly within the chain
Esterase			
Pectinesterase	3.1.1.11	Pectin	Deesterification

Endopolygalacturonases (Endo-PG; EC 3.2.1.15; CAS 9032-75-1)

Endopolygalacturonases (PG) (Table 7.22) are formed by a large number of micro-organisms such as fungi, bacteria, and yeasts. They are found in many plants. Tomatoes are particularly rich in PG, which is responsible for the steady decrease in the viscosity of cold-pressed tomato juice.

Table 7.22. Properties of some industrially important endopolygalacturonases.

Enzyme source	*Molecular weight (daltons)*	*pH optimum*	*pI*
A. niger	35000	4.1	–
	85000	3.8	–
A. niger	46000	5.0	–
A. japonicus	35500	4.5	–
Trichoderma koningii	32000	5.0	6.41

The best substrate for PG is pectate from which random cleavage leads to oligogalacturonides, such as mono-, di-, and trigalacturonides. Rapid reduction in the viscosity of pectate solutions is an indication of PG activity. The cleavage of only a few linkages causes a viscosity reduction of 50 per cent. PG activity drops with increasing degree of esterification of the substrate, as free carboxyl groups are necessary for catalytic action. In addition, PG activity also decreases with reduction of the degree of polymerisation. Digalacturonides are rarely cleaved, and trigalacturonides are cleaved only slowly.

Aspergillus, *Penicillium*, and *Rhizopus* form different PG isozymes. In addition, there are two types of PG: a group of liquefying PGs and macerating PGs. The latter reduce the viscosity of pectate solutions and are capable of disintegrating plant tissues. An industrial process for the separation of macerating PG from *A. niger* from a liquefying PG on an ion exchanger has been described by Löffler. The macerating PG from *A. saitoi* showed optimum activity at pH 4.8–5.0 and 45°C and cleaved 60 per cent of the linkages in polygalacturonic acid and 17 per cent of the linkages in pectin. The enzyme was inactivated in 10 minutes at 50°C. The macerating activity of a PG from *A. alleacus* (Rohament P [RM]) is, according to Bock, distinguishable from other PGs (see Figs. 7.69 and 7.70) by its inhibition by various plant extracts.

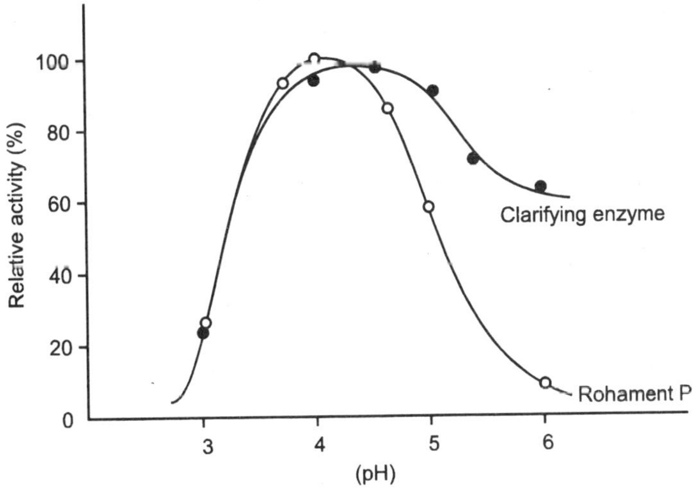

Fig. 7.69. pH dependence of a PG enzyme activity (Rohament P [RM]) and a commercial clarifying enzyme; reaction temperature 30°C.

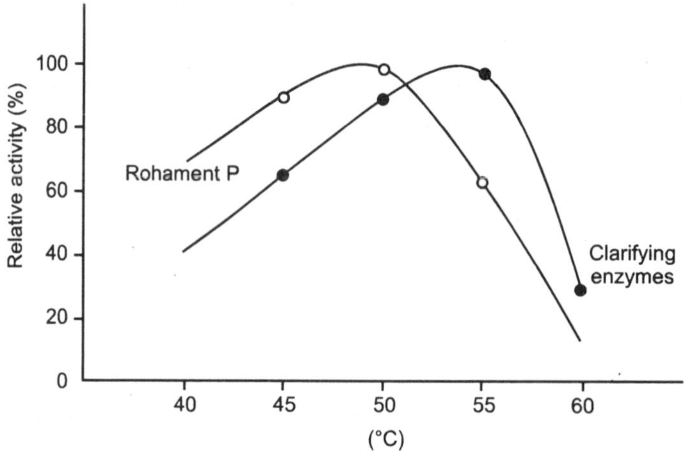

Fig. 7.70. Temperature-dependence of a PG activity (Rohament P [RM]) and commercial clarifying enzyme; reaction at pH 4.2.

It is, in contrast to PG from tomatoes, almost totally inhibited by compounds found in cucumbers and pears. The macerating PG from *A. alleacus* (Rohament P) cleaves protopectin into pectin fragments with molecular weights of about 1,00,000 daltons. These are then only slowly further degraded. A similar observation was made by Kaiji for the PG isolated from *C. felsineum*. Under strong macerating conditions, the reduction in viscosity of pectin at pH 4.7, as well as the production of reducing sugars, was small. Various macerating PGs were also found in *Rhizopus* sp.

Analysis of PG

Polygalacturonases can be analytically determined by the reduction of the viscosity of pectate solutions. The pH used in the measurements should lie near the PG pH optimum. This optimum is dependent on temperature.

The activity of PG is determined by the amount of reducing sugars formed. This amount can be measured with the dinitrosalicylic acid method of Miller.

Exopolygalacturonases (Exo-PG, EC 3.2.1.82)

Monomeric galacturonic acids from pectate, oligogalacturonates, and digalacturonates are cleaved in stepwise fashion from the nonreducing end of the polymer chains by exopolygalacturonases (exo-PG). These enzymes occur in higher plants, some bacteria, and fungi. An extracellular exopolygalacturonase produced by *A. niger* was described by Heinrichovà and Rexová-Benkovà. The formation of monomeric galacturonic acids by the action of commercial enzymes from *A. niger* or *A. wentii* on pectate can be attributed to the action of exo-enzyme. Omran was able to obtain and subsequently characterise an exo-PG from the commercial preparation Pectinol C [RM] by inactivating the endopolygalacturonase at pH 8.5 and purifying by chromatography. After cleaving 75 per cent of the glycosidic linkages in polygalacturonic acid (PGA) by the isolated enzyme, Omran observed that the viscosity of the PGA solution fell to only two-thirds that of the starting material.

Nevertheless, the molecules remaining after the degradation still had a high degree of polymerisation. During hydrolysis of oligogalacturonides, the trimer is cleaved at the highest maximum reaction rate. The optimum pH of this exo-PG is about 4.32. The stability (80 per cent residual activity) extends from pH 3.6 to neutral. The maximum temperature for activity and stability is 45°C.

Endopectate lyases (PATE; EC 4.2.2.2; CAS 9015-75-2)

These lyases, which are produced by some bacteria and phytopathologic fungi, show a pH optimum above pH 8.0 and require CA^{2+} ions to be active.

The enzyme from *B. subtilis* isolated by Cesson and Codner has a molecular weight of 33,000 daltons, an isoelectric point at pH 9.2, and an optimum pH of 8.5. The endopectate lyase from an alkalophilic *B. subtilis* has an optimum pH of 10.0 and is completely inhibited by EDTA.

Analysis of PATE

Endopectate lyase is best determined by method of Macmillan and Pfaff using 1 per cent polygalacturonate solution in buffer in the presence of Ca^{2+} ions. The increase in extinction at 235 nm at 40°C is measured using a spectrophotometer.

Endopectin lyases (PTE; EC 4.2.2.10; CAS 9033-35-6), endopolymethylgalacturonate lyases

PTE was first found in the technical preparation Pectinol R 10 [GR] by Albersheim and was later isolated and characterised. These types of depolymerases are important components of commercial enzymes and are especially important in practical applications. They are formed mainly by fungi and depolymerise highly esterified pectin in a random manner. They cause rapid decrease in the viscosity of pectin solutions. They specifically cleave at the glycosidic linkages adjacent to a methylated galacturonic acid residue. Highly esterified pectins are the best substrate, while pectate, pectin acid amide, and the glycyl ester of polygalacturonic acid are not degraded. Activity decreases at lower chain lengths and lower degrees of esterification. The smallest substrates still being degraded are the completely esterified tetra- and trigalacturonates. Most pectin lyases are activated by calcium. This activation is dependent on pH and the degree of esterification of the pectin (Fig. 7.71).

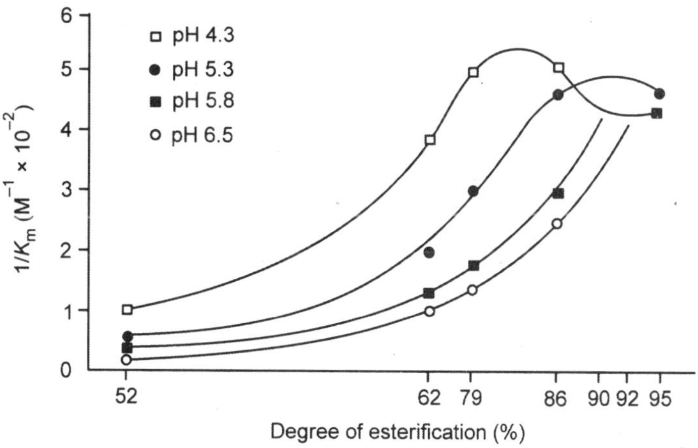

Fig. 7.71. Endopectin lyase affinity ($1/K_m$ as a function of the amount of esterified pectin substrate at various pH).

Analysis of PTE

Pectin lyase (Table 7.23) can be determined with a spectrophotometric test by using pectin that is at least 70 per cent esterified as substrate and measuring the increase in absorption at 235 nm. The presence of pectinesterase strongly influences the analytic results. A less specific but more commonly used method is a viscometric determination with solutions of highly esterified pectin as described by Bush and Codner.

Table 7.23. Properties of some pectinlyases.

Enzyme source	*Molecular weight (daltons)*	*pI*	*pH optimum*
A. niger	35000	3.65	6.0
A. niger	33100	3.75	6.0
A. japonicus	32000	7.7	6.0
A. sojae	32000	–	5.5
A. niger	–	3.9	6.0[a]
			8.5[b]

[a]Containing 71 per cent esterified pectin.

[b]Containing 94 per cent esterified pectin.

In PTE analysis, the dependence of the pH optimum on temperature and the degree of esterification of the substrate must be considered. The optimum activity (viscometric) is at pH 6.5 for 90 per cent esterified pectin and a reaction temperature of 30°C. The optimum is at pH 4.3 for 80 per cent esterified pectin. The pH optimum for citrus pectin is at pH 5.1–5.3 (30°C).

The PTE isolated by Zhou from the commercial preparation Pectinex Ultra [NO] showed an optimum activity with apple pectin (71 per cent esterified) at pH 6.0. Optimum activity with a synthetic highly esterified (94 per cent) pectin is at pH 8.5. The optimum temperature is 60°C; higher temperatures cause rapid inactivation.

Pectinesterase (PE; EC 3.1.1.11; CAS 9025-98-3)

Pectins are enzymatically deesterified by PE, and converted to low esterified pectin or pectic acid with release of methanol. PE occurs in many higher plants, in various fungi, and in some yeasts and bacteria. PE preparations contain a large number of isozymes. These have similar action, but their isoelectric points and molecular weights differ. The preferred point of attack of PE in the pectin chain is at an esterified galacturonic acid residue adjacent to a free carboxyl group. Plants, especially tomatoes and citrus fruits, show very high PE activity (Table 7.24). This can have a major effect during the production of canned foods and juices. For example, citrus pectin is rapidly deesterified in the manufacture of orange juice by endogeneous PE. This causes coagulation of particulates and subsequent undesirable clarification of the juice. This can be prevented only by inactivation of the enzyme.

Table 7.24. Properties of some pectinesterases.

Enzyme source	*Molecular weight (daltons)*	*pI*	*pH optimum*	*K_m pectin (mg/ml)*
Tomatoes	27500	–	6–9	0.74
Oranges	36200	10.0	7.6	0.083
Fusarium sp.	35000	–	7.0	–
A. niger	–	–	4.32	0.215 with pectin 68.4% VE

Pilnik and Rombouts have reported on the different pectinesterases that occur in oranges and their mutual synergistic interaction with depolymerases. Thus, orange PE shows optimum activity at pH 7, but is still effective at pH 2–2.5 and is very heat-stable. The affinity of this PE for pectin increases with reduction in degree of esterification of the pectic substrate. Ishii and others have described a PE that they isolated from *A. japonicus*. Commercial PE preparations have been prepared from cultures of mould fungi, particularly *A. niger*, by inactivation of the accompanying depolymerase and chromatographic purification. As many of the previous characterisation and kinetic studies had been undertaken on crude or only partly purified preparations. Döerreich prepared a highly purified PE from *Aspergillus* using ion exchangers in an attempt to characterise the enzyme (Fig. 7.72).

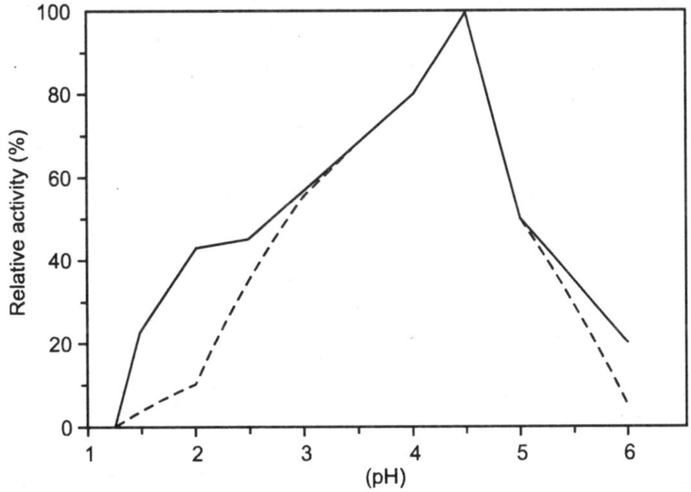

Fig. 7.72. Effect of pH on the activity of crude (1) and purified (2) pectin esterase in Rohapect C [RM].

The crude enzyme shows optimum activity at pH 4.4, while that of the pure enzyme is pH 4.3. Under acidic conditions, the enzyme is active to pH 1.5–2 and shows relatively good stability.

The crude enzyme is fully inactivated in 5 minutes at 65°C; the purified enzyme, at 70°C. The phenolic compounds tannin and gallic acid inhibit activity only slightly. The affinity of this PE for pectin with a degree of esterification between 50 and 80 per cent is considerably less than that for pectin with 33 per cent esterification.

Analysis of PE

A rapid method for the determination of PE involves the monitoring of the colour change of a pH indicator such as bromocresol green. Agar plates can be used for the test. For quantitative assays, the free carboxyl groups released can be measured titrimetrically or, alternatively, the methanol produced can be estimated by gas chromatography.

Technical Preparations

Technical preparations are obtained principally from fungi, including *A. niger*, *A. alleacus*, *A. wentii*, *A. aculeatus*, *A. usamii*, and *Rhizopus* sp. The preparations have been obtained since the 1930s by the cultivation of fungi in fixed-bed cultures. The major portion of the culture medium consists of wheat bran with a moisture content between 45 and 60 per cent.

Following extraction, ultrafiltration is used to remove metabolites and culture medium fragments and at the same time to concentrate the extract. After sterile filtration, liquid preparations are standardised with water and stabilisers such as glycerol or sorbitol. The addition of approved food preservatives is allowed in most countries. Solid, powdered, or granulated preparations are produced by spray or agglomeration drying with the addition of glucose, sucrose, or maltodextrin.

Fixed-bed or surface processes are currently used by various international enzyme manufacturers for the preparation of pectinases. At one time it was assumed that a multienzyme complex of pectinases could be prepared only by the fixed-bed process. The work of Zetelaki and Nyiri showed that the enzyme complex necessary for industrial applications could also be prepared using submerged fermentation. The composition of the enzyme complexes with respect to the activities of the individual enzymes—PG, PTE, and PE—is dependent on the fungal strain as well as the culture conditions. Specific conditions are required in preparations for various applications of pectinases in the production of fruit and vegetable juices. This is relevant for the characteristics of the pectinase complex, as well as for the side activities. Because of the different amounts of PG, PTE, and PE, the characteristics of the preparations vary with respect to stability at various pH values and to dependence of activity on pH and temperature.

In addition, enzymatic activities other than those of the pectinases are necessary. For maceration, the presence of special PG enzymes and cellulases is required. In order to remove araban precipitation in apple juice concentrates, it is important to have standardised arabanase activity in commercial preparations.

Special proteases work synergistically with the pectinases and hemicellulases in clarification of red wine. Because of these requirements, the first generation of commercial preparations that were initially used in fruit processing have been developed for more specific applications. Standardisation of the side activities is difficult because various manufacturers use their own analytic methods; therefore, it is not easy to compare commercial preparations. In practice, preparations are selected on the basis of one or more analytically controlled production tests.

In addition to the specific practical applications, an analytic comparison of technical products is of interest to establish a correlation between analysis and practice. Gierschner's group investigated a series of technical preparations and attempted to define the relationship between the analytically determined product composition and the results of practical trials. Some interesting qualitative data but no quantitative relationships resulted. Some of the analytic results are reproduced in Fig. 7.73.

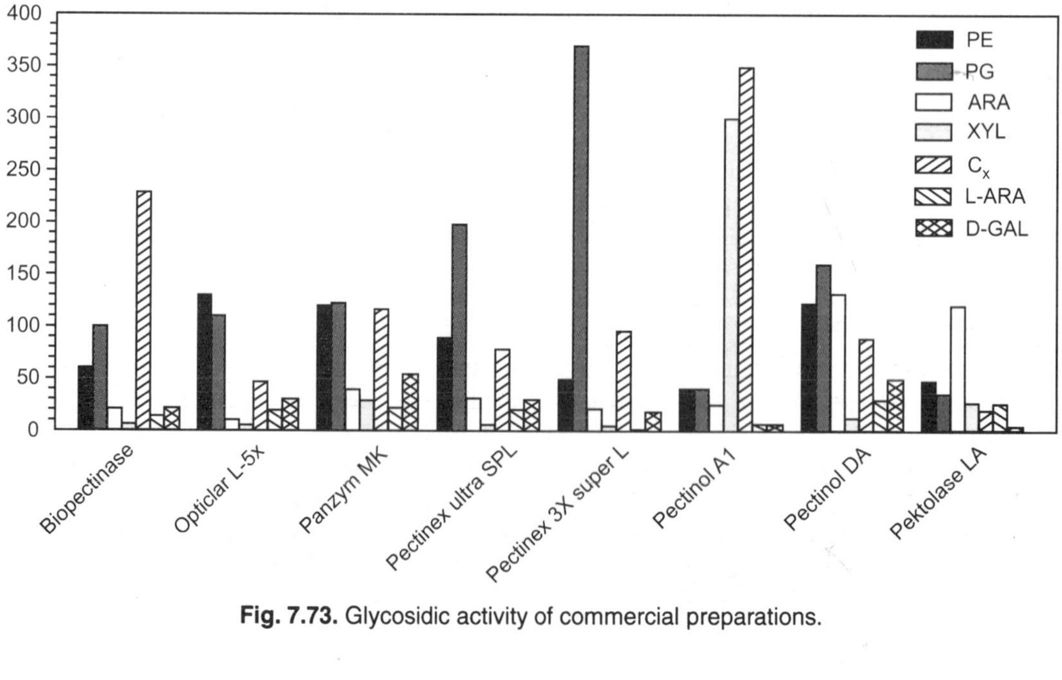

Fig. 7.73. Glycosidic activity of commercial preparations.

Practical information on the properties of technical products include the relationships of activity and enzyme stability to various reaction parameters such as pH, temperature, and reaction time. Descriptions of these properties can differ according to the particular enzyme manufacturer. For example, some of the factors on which reaction rates depend are given in Figs. 7.74–7.76.

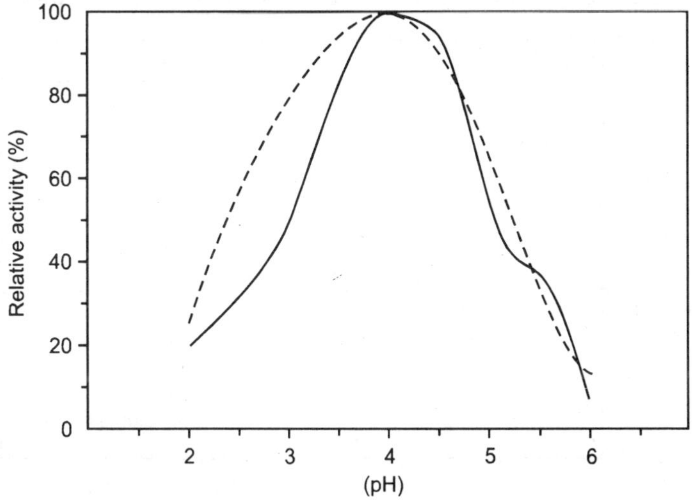

Fig. 7.74. Effect of pH on the activity of two technical preparations.

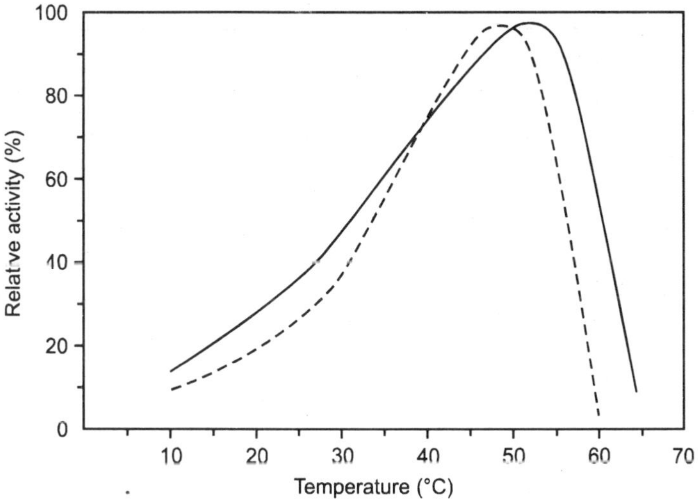

Fig. 7.75. Effect of temperature on the activity of two technical preparations [RM].

Preparations with specific activity

Macerase (Rohament P [RM])

This preparation (Figs. 7.77 and 7.78), in which the primary activity is a polygalacturonase with special macerating properties, was prepared from *A. alleacus* prior to 1988. Today it is prepared from *A. niger*. Various authors, including Grammp and Radola, have described the properties of this preparation, which is used for the production of macerates from vegetables, fruit concentrates, and pulpy fruit juices. The macerating properties depend principally on the activity of a particular PG and less on the accompanying enzymes such as cellulases and hemicellulases. Because of the lack of pectinesterase, the preparation is not capable of clarifying apple juice or depectinising fruit juice. The temperature stability maximum is between 45° and 50°C and can be increased to 65°C in the presence of substrate.

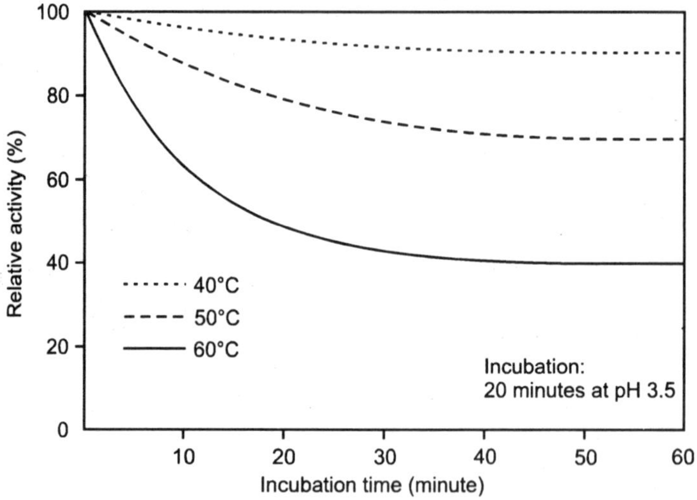

Fig. 7.76. Effect of temperature on enzyme stability [RM].

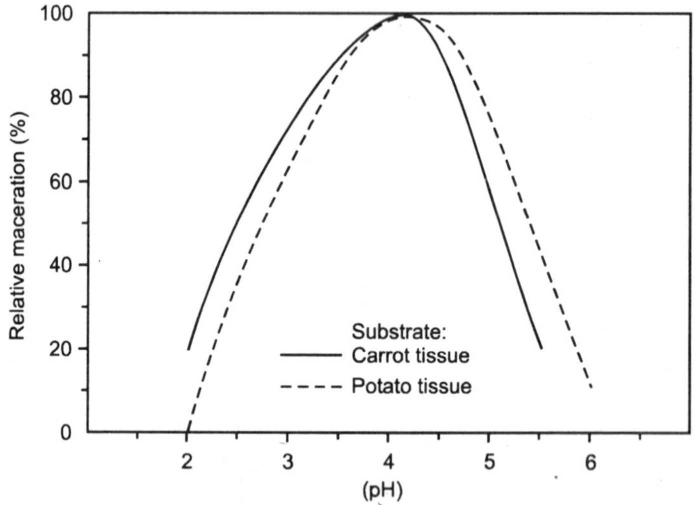

Fig. 7.77. Effect of pH on Rohament P activity [RM].

Pectinesterase preparations

It is difficult to compare information on the activity of technical preparations as given by the manufacturers because their standardisation is specific for the company's own methodologies. Their specifications depend on the particular reaction parameters used and the nature of the substrates. Thus, for a particular application, only on use can it be determined whether the material is an enzyme concentrate or a dilute preparation (Fig. 7.79).

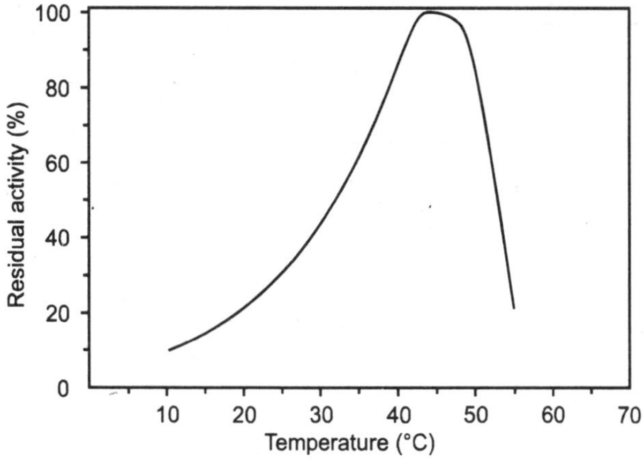

Fig. 7.78. Effect of temperature on Rohament P stability [RM].

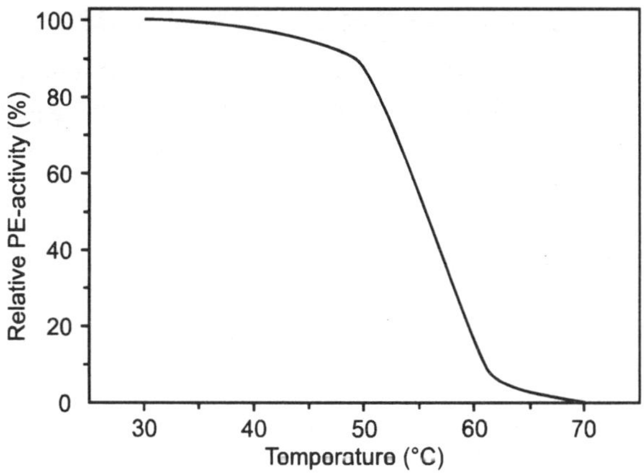

Fig. 7.79. Effect of temperature on pectin esterase activity [RM].

Products and manufacturers: Rohapect [RM], Rohament [RM], Pectinex [NO], Ultrazym [NO], Biopectinase [BN], Panzym [BÖI], Pectolase [GT], Klerzyme [GB], Spark [SOE], Sumizyme [SN], Pectinol [GR], [AM].

PENTOSANASES

Arabanases

In addition to cellulose and pectin, the arabinogalactans and the arabans belong to the major components of the plant cell wall. Here, similar to other carbohydrases, are endoenzymes, which cleave the linkages within the polymer chain, and exoenzymes, which release monomeric units from the end of the chains. The endoarabinases cleave α-1,5 linkages within the araban chain to give di- and trisaccharide products. In addition, there are numerable exoenzymes, α-L-arabinofuranosidases, some of which attack only the α-1,3 linkages in the side chains and others, that cleave α-1,3 as well as α-1,5 linkages.

Following developments in fruit juice extraction technology, the arabanases have become important in recent years, especially in the production of pear and apple juice concentrates. By treating the apple mash with complex pectinase preparations, low-solubility arabanases are released. These cause turbidity in the juice concentrates. According to Pilnik and Voragen, the native araban in apples is a branched water-soluble α-araban. Exoarabanase, which occurs in most commercial pectinase preparations, cleaves off the α-1,3 side chains. The residual α-1,5-araban then aggregates into an insoluble precipitate, which is analogous to the retrogradation of starch (Fig. 7.80).

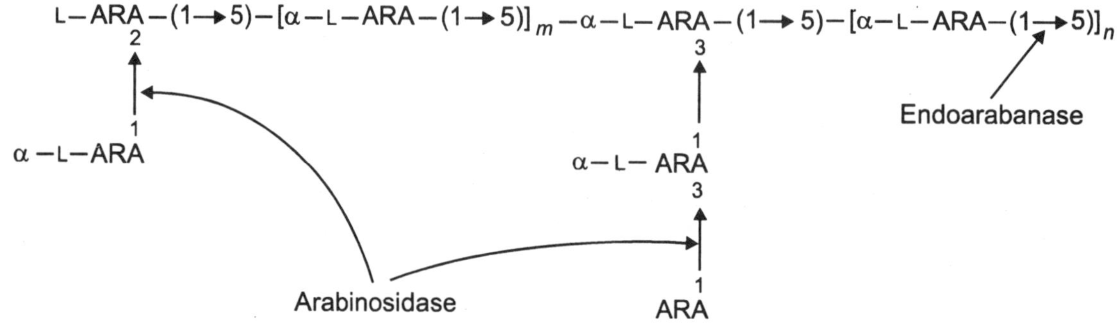

Fig. 7.80. Sites of enzymatic attack in arabinan.

The araban in apple juice was isolated by Churms. The degradation of araban from pear concentrate was described by Babsky and Scobinger. Rombouts and others investigated the sugar composition as well as the distribution of linkages in several arabanases. They showed that 90 per cent of linear α-1,5-L-araban consisted of arabinose.

Eighty-eight per cent of the arabinose residues are α-1,5-glycosidically linked. The araban in beets (beet araban) has about the same arabinose content, but only 28 per cent of the arabinose residues are α-1,5-glycosidically linked. According to Ducroo, the molecular weight of the insoluble araban in apple juice concentrates is about 10000 daltons, but molecular weights of 3000 daltons were also found.

Endoarabanases (EC 3.2.1.99)

Some endoarabanases from *B. subtilis* with preferred specificity for α-1,5 linkages were described by Weinstein and Albersheim. The molecular weights of these enzymes lie between 32000 and 35000 daltons; optimum activity is between pH 5 and 6. Because the stability of these enzymes is low at acidic pH, they are not suitable for the manufacture of fruit juice.

The endoarabanase isolated from *A. niger* by Rombouts has the following characteristics:

1. Molecular weight: (SDS electrophoresis) 35000 daltons.
2. Isoelectric point: pH 4.5–5.5.
3. pH optimum: 5.0.
4. Temperature optimum: 50°C.

The linear α-1,5-arabans, beet araban, potato araban, and the arabinose-pentasaccharides are also hydrolysed. Disaccharides are less readily attacked; *p*-nitrophenylarabinoside remains intact.

Exoarabanases, α-L-arabinofuranosidase (EC 3.2.1.55)

This enzyme is produced by fungi and bacteria. Currently, the enzymes from *A. niger* are of commercial importance. One of these was isolated by Kaji and Tagawa.

It resembles one of the enzymes isolated by Rombouts from the commercial preparation Pectinase 29 [GB] (enzymes A and B, Table 7.25).

Table 7.25. Properties of arabinofuranosidases A and B.

	Enzyme	
Property	*A*	*B*
Molecular weight (daltons)	1,28,000	60000
pI	6–6.5	5.5–6
pH optimum	4.1	3.7
Temperature optimum	50°C	60°C
Cleaves		
p-Nitrophenyl arabinoside	+++	+++
Arabinose disaccharide	++	++
α-1,5-Araban	–	+
Araban from beets	+	+++

Analysis of arabanases

Technical preparations: Although a pure arabanase is not available at present, different industrial enzyme preparations contain various amounts of arabinase activities such as, for example, Rohapect DA 3L [RM] and Pectinex AR [NO], [GB].

Xylans and Xylanases

Xylan is one of the most widely occurring polysaccharides. In terms of abundance, xylan occupies third place after cellulose and starch. It is present in wood and grasses and is, as was described in the beginning of this chapter, a component of the plant cell wall. Xylans from different plant sources are commercially available, although some of these products contain significant quantities of starch and other carbohydrates.

Xylans have chains of 1,4-linked β-D-xylopyranose residues. Xylans from hardwoods, softwoods, and grasses are substituted differently. In grasses, especially in straw, single arabinofuranosyl residues are linked at the C-3 positions in the xylan chain. Xylans are also substituted in the C-2 position with D-glucopyranosyl-uronic acid residues or their 4-methyl esters. There are also a few arabinofuranosyl residues bound to the chains of softwood xylan. Hardwood xylan is acetylated.

As already discussed, substituted xylans are components of the plant cell wall. D'Appolonia and Kim have suggested a strcutural scheme for the water-soluble pentosans in wheat (Fig. 7.81).

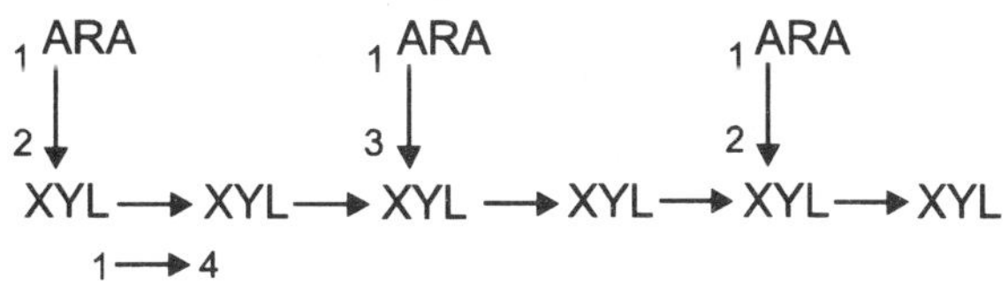

Fig. 7.81. Structure of the branched arabinoxylan.

Anhydro-L-arabinofuranose residues are bound at the 2- or 3-position of the β-1,4-xylopyranose chain. The coupling of such arabinoxylans with the protein in the cell wall leads to glycoprotein complexes that are dispersible in water to give highly viscous solutions (Fig. 7.82).

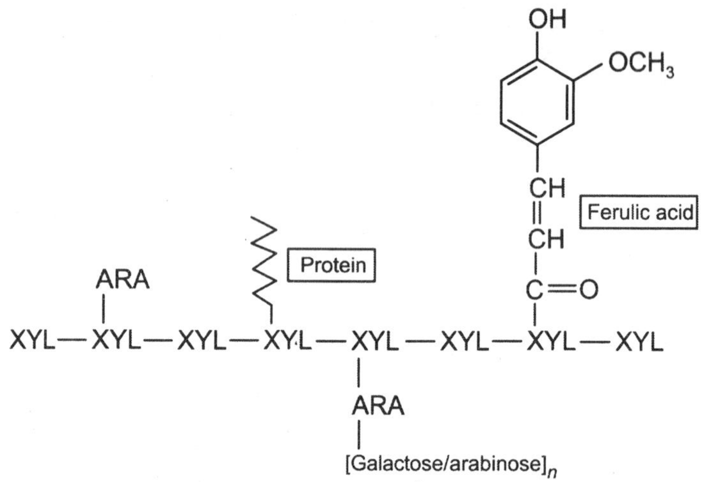

Fig. 7.82. Hypothetical structure of wheat flour glucoprotein.

The phenolic groups of ferulic acid can be oxidatively cross-linked to yield diferulic acid residues. Gels are formed (Fig. 7.83). In this manner, horseradish peroxidase and H_2O_2 can effect the cross-linking of proteins in wheat doughs, thereby raising dough viscosity. According to Kiefer, this improves the baking properties of flour.

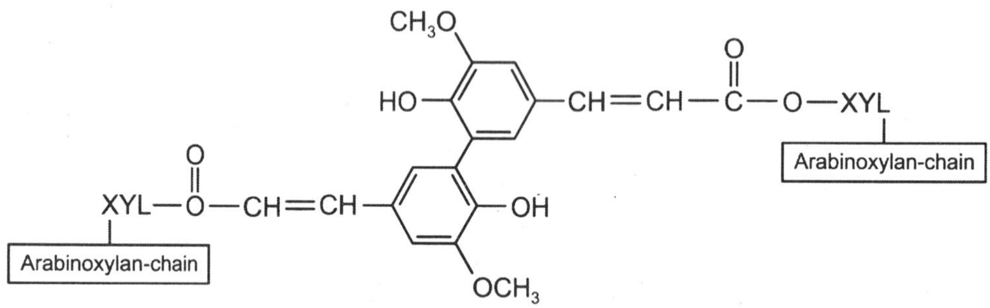

Fig. 7.83. Oxidative phenolic cross-linking of arabinoxylan chains.

There are various types of xylanases with different specificities:

1. Endo-1,4-β-xylanase (EC 3.2.1.8; CAS 9025-57-4); (1,4-β-D-xylan-xylanohydrolase).
2. Endo-1,3-β-xylanase (EC 3.2.1.32; CAS 9025-55-2); (1,3-β-D-xylan-xylanohydrolase).
3. Exo-1,4-β-xylosidase (EC 3.1.2.37; CAS 9025-53-0); (1,4-β-D-xylan-xylohydrolase).

Xylanases are formed by many *Aspergillus*, *Bacillus*, and *Trichoderma* species. In technical enzyme products, they are accompanied by many different cellulases and hemicellulases. Only a few of these organisms are used for the technical manufacture of such complex preparations. See also Tables 7.26 and 7.27 and Figs 7.84–7.86.

Xylanases from Bacillus sp.

Numerous bacteria produce xylanases; these have been studied in connection with the microbial degradation of wood. Technically, these enzymes are currently of little importance. A few characteristics of bacterial xylanases are given in Table 7.28.

Table 7.26. Xylanases from *Trichoderma* sp.

Enzyme no./ source	*Molecular weight (daltons)*	*Carbohydrate (%)*	*pI*	*pH optimum*	K_m *(%)*
I/*T. reesei*	32000	14	4.1–4.2	4–5	–
II[a]	23000	8	6.4–6.5	4–5	–
I/*T. harzianum*	20000	–	–	–	0.16
I/*T. harzianum*	29000	–	–	–	0.066
I/*T. harzianum*[b]	22000	–	8.5	4.5–5 (50°C)	–
III/*T. viride*	1,02,000	4.5	4.45	4.5 (55°C)	–

[a]This enzyme produces xylobiose and large xylose oligomers from xylan. Laminarin is also hydrolysed.

[b]Three isoenzymes with molecular weights of 20000–29000 daltons were detected. It has been reported that all three enzymes are exclusively xylan-specific. They do not digest xylobiose and do not convert arabinoxylan to arabinose.

Table 7.27. Xylanases from *Aspergillus* sp.

Enzyme no./ source	*Molecular weight (daltons)*	*Carbohydrate (%)*	*pI*	*pH optimum*	K_m *(%)*
I/*A. niger*	14000	–	–	4–4.5	2.4×10^{-2}
II/*A. terreus*[a]	11000	–	–	4–5.5 (40°–55°C)	–
	14500	–	–	–	–
	20000	–	–	–	–
III/*A. niger*[b]	1,22,000	0	4.92	3.8–4 (70°C)	–

[a]Six enzymes of varying molecular weight but with strict β-1,3 specificity were isolated. Pure β-1,3-xylan from seaweed is hydrolysed. β-1,4-Xylan, laminarin, cellulose powder, and β-1,3-xylobiose are not digested.

[b]The enzyme is stable between pH 3–8 and shows no loss in activity during heating at 50°C for 1 hour. D-Xylose is a competitive inhibitor.

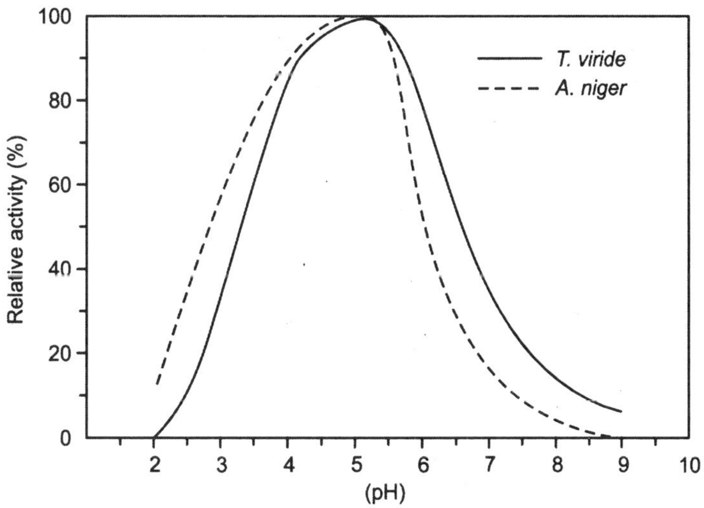

Fig. 7.84. Effect of pH on activity of xylanase from *A. niger* [RM] and *Trichoderma viride* [UA].

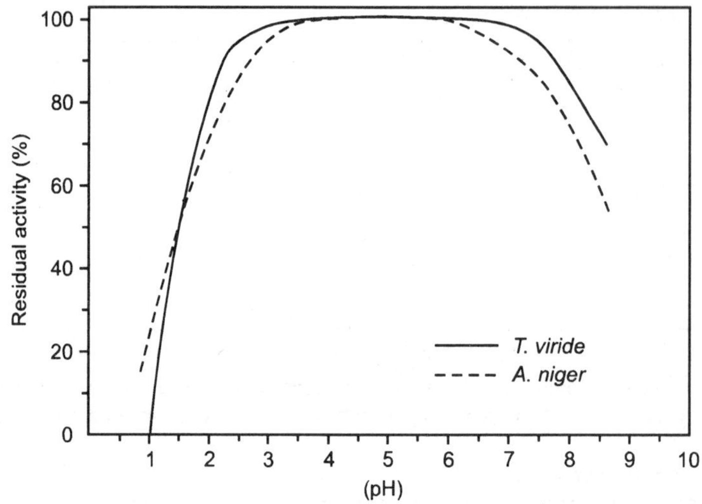

Fig. 7.85. Effect of pH on stability of xylanases from *A. niger* and *T. viride* (1 per cent buffered enzyme solution, 4 hours, 30°C).

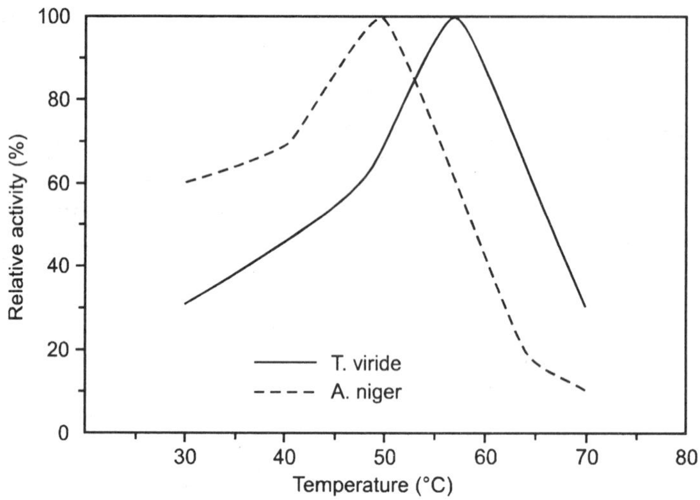

Fig. 7.86. Effect of temperature on activity of xylanases from *A. niger* [RM] and *T. viride* [RM].

Table 7.28. Xylanases from *Bacillus* sp.

Enzyme no./Source	*Molecular weight (daltons)*	*pI*	*pH optimum (Temperature)*
I/*B. pumilus*	20000	–	6.5 (40°C)
I/*B. coagulans*	22000	10	6.0 (37°C)
III/*Bacillus* sp.	1,90,000	4.9	6.5 (30°C)
I/*B. subtilis*	20000	–	6–7

In addition, wood-destroying fungi belonging to the *Basidiomycetes* produce xylanases that are manufactured technically in small amounts and used as components in digestive aid preparations. The xylanases from alkalophilic *Bacillus* sp. are active in the pH range 5.5–9 and stable for a duration of 10 minutes at 60°C. More xylanases of this type are described by Akiba and Horikoshi. A new thermostable xylanase (type I) from *B. stearotheremophilus* was produced by Grueninger and Fichter. This enzyme has a broad activation optimum from pH 5.5 to 8.5 and a half-life of 15 hours at 75°C and is stable at 68°C for 5 days.

A new xylanase produced by a recombinant strain of *B. subtilis* (molecular weight 20,000 daltons) was reported in a patent of Röhm. The enzyme is produced for a baking application, especially to obtain a high volume in bread.

Analysis of xylanases

1. Determination of reducing sugars: Xylan from wood or straw is swelled by incubation for approximately 12 hours in 0.5 N NaOH and adjusted with buffer to the desired pH. About 2 ml of a solution containing approximately 1 gram of xylan in 100 ml of buffer is required for hydrolysis. Following hydrolysis (30°C, 15 minutes), the reducing sugars produced are analysed according to the method of Somogyi-Nelson.
   Enzyme unit: A xylanase unit is determined as the amount of enzyme needed to liberate from xylan the amount of reducing sugars equivalent to 1 μM xylose per minute at 30°C under standard conditions.
2. Xylanase determination with a nephelometric method. Nummi established a method that allows the determination of endoxylanase as well as exoxylanase using freshly precipitated insoluble birch wood xylan.
3. A valuable analytical method for the xylanase has been commercialised by Megazyme Ltd., Sydney, Australia.

Manufacturers: [RM], [SN], [GT], [NO] and others.

Chapter 8

Proteases

INTRODUCTION

The proteins from plants and animals are important for human and animal nutrition as well as for clothing (leather, wool, and silk). They are used as organic fertiliser in the cultivation of plants and are a necessary component of culture media for industrial fermentations. Some proteins serve as raw materials for industrial products as a paint base or for the manufacture of synthetics.

In nutrition, proteins are usually 'naturally modified', that is, processed using physical methods such as heating or homogenisation. The changes induced in the native proteins serve to improve digestibility or storage stability.

It is also desirable in different branches of food technology to change the physical and chemical properties of proteins. This can be done by the physical methods mentioned above, as well as by chemical and biological fermentation methods, such as in the production of milk and cheese. Many of these traditional processes are based on the activity of naturally occurring enzymes, including those produced by the organisms in the original product. The enzymatic activities in raw foods, such as in fruit, vegetable or cereal products, depend on the variety and the climate and are thereby subject to natural variation. To achieve targeted and specific changes in proteins, commercial protease preparations with different specificities are used. Such preparations are currently employed to produce many foods in which enzymes can replace potentially carcinogenic or otherwise harmful chemicals. Table 8.1 shows some processes used in the manufacture of foods in which enzymes have specific applications.

Table 8.1. Application of proteases in food and industrial technology.

Product or process	*Use*
Beer	To solubilise grain proteins; to stabilise beer
Cheese	To coagulate milk proteins and to ripen cheese
Meat tenderising	To partially separate connective tissues
Bread	To increase gluten elasticity
Cookies and crackers	To improve crispness
Leather	To remove wool, hair and pigments; to soften skins
Laundry detergents	To remove protein stains

In addition, there are numerous processes used in Asia for the modification of fish or soya protein with living micro-organisms or with microbially produced proteases. For example, there is the production of soya sauce using *Aspergillus sojae* or *Aspergillus oryzae* and the production of tofu or tempeh with *Mucor* or *Rhizopus*, which form proteases during their growth.

Protease preparations are, on a commercial basis, the most important of the currently produced enzymes. They are obtained from plants and animal organs and micro-organisms, with the majority obtained from bacteria.

PLANT PROTEASES

Plant proteases occur in numerous, mainly tropical, plants. Proteases are found in papaya (*Carica papaya*), pineapple (*Anana sativa*), figs (*Ficus carica*, *Ficus glabrata*), artichokes (*Cynera cardunculus*) and soyabeans (*Soya hispidus*). All of these proteins belong to the same group of enzymes. The juice of some agave species is also strongly proteolytic.

Plant proteases are characterised by sulphydryl groups in the active site. These groups are responsible for the catalytic activity.

Proteolytic activity can be impressively demonstrated by the action of fresh pineapple on meat. During the preparation of Indonesian dishes of meat with fresh pineapple, pieces of chicken are amazingly liquefied after warming for only 1–2 hours. The main source of industrial enzymes is green papaya, from which papain is obtained. Bromelain is produced by alcohol precipitation from extracts of pineapple or pineapple stalk. The ficin preparations, in the form of ficus latex, are of limited commercial importance. The plant proteases are used as meat tenderisers and in other areas of food production. Examples including brewing, cookie baking and the production of protein hydrolysates. Other applications are in tanning and pharmaceuticals (digestive-aid preparations).

Papaya Proteases

Brief note on papain manufacture

Papain is obtained from papaya tree plantations in India, Sri Lanka, Zaire, and Uganda. The still-green young fruit is scratched with a sharp knife along the surface. A colourless latex immediately appears and turns milky white after a few minutes before coagulating. The dripping latex is caught in pieces of cloth or tubs and is air-dried. One kilogram of this latex yields approximately 200 grams of crude papain. On average, 0.45 kg of latex per tree is obtained. Newer production methods utilise purification of latex at the point of origin.

This is accomplished by adding sodium hydrogen sulphite to raw juice to prevent oxidation, followed by centrifugation or filtration. Hemicellulose addition can facilitate filtration. After concentrating the juice *in vacuo* and then spray drying, a powderlike product is obtained.

Crude papain contains, in addition to papain (10 per cent of the protein in the latex), a number of other enzymes—namely, 45 per cent chymopapain A and B, which are similar to papain but differ in their specificity. Besides endo-β-1,4-glucanase, a chitin-splitting hydrolase and a carboxypeptidase, a very active lysozyme (~10 per cent) is also present. The milk-curdling activity of crude papain seems to be a characteristic of the protease. This activity is not destroyed by heating to 70°C for 90 minutes. Hydrogen peroxide inactivates this activity; cystein reactivates it. Lipase is also found in papain. The activity optimum with olive oil is 40°C at pH 5.8–6.2. Amylase activity can be measured in fresh papaya latex.

The plant lysozyme has an activation optimum between pH 3.2 and 4.2, while the optimum for egg lysozyme lies at pH 5.3.

Papain (EC 3.4.22.2; CAS 9001-73-4)

1. Molecular weight: 21000 daltons.
2. Structure: A chain of 159 amino acids including three cystein bridges. The presence of cystein at the active site is required for activity. The SH group of cystein easily reacts with oxygen or heavy metals. Consequently, to obtain full activity, the presence of reducing agents or heavy-metal-complexing agents is necessary.
3. Isoelectric point: pH 8.5–9 (different values are found; these are presumably a result of autolytic processes).
4. Activity: Ninety per cent of the maximum enzymatic activity is obtained in a pH range of 6.5–7.8 with haemoglobin denatured by urea. With native haemoglobin, activity is pH 3–4. If the substrate is casein, maximum activity lies between pH 7.5 and 8; and with gelatine, between pH 4.8 and 5.2. Optimum activity with ovalbumin is pH 7–7.5; with serum albumin, pH 5.7–5.9; with peptone, pH 5.0. At pH 6–7, the maximum temperature tolerated is 60°–70°C (casein). The proteolytic activity of papain is given in Table 8.2.
5. Inhibitors: Oxidising substances, H_2O_2, O_2 and heavy metals; for example, Hg^{2+}, Fe^{3+} and Cu^{2+}.
6. Re-activation: Cystein, $NaHSO_3$, glutathione, EDTA.
7. Specificity: Papain is a relatively non-specific protease. Proteins, peptides, amides and esters are cleaved by the enzyme. Benzoyl-L-arginine-*p*-nitroanilide (BAPA) is a good substrate that is used for analytic purposes.

Table 8.2. Determination of proteolytic activity of papain with various methods.

Method	*Crude papain*	*Purified papain*
BAPA[a]U/mg	–	420–470
BAPA U/mg	2000–3000	5000–6000
Anson-haemoglobin units	1–2	5–6
Tyrosine units	100–200	800–900
Haemoglobin units (HbmU/mg)	2000–3500	≤11000
USP[b]/mg	10000	56000

[a]Na-benzoyl-L-arginine-*p*-nitranilide.

[b]US Pharmacopoeia units.

Chymopapain (EC 3.4.27.6; CAS 9001-09-6)

Chymopapain has characteristics similar to those of papain but differs in its activity toward proteins. It is less heat-sensitive and is more stable than papain at pH 2.

1. Molecular weight: 36,000 daltons (three isozymes are known).
2. Isoelectric point: pH 10.1.
3. Specificity: Very similar to papain. On the basis of equal BAPA activity, the milk-curdling activity of chymopapain is twice as high as that of papain.

Technical papain preparations

Products in powder form are not desirable today because of the danger of allergic reactions. The production of liquid preparations that contain sorbitol or other sugars and glycerol as stabilisers is described in numerous publications.

In order to stabilise enzymatic activity, reducing agents such as $NaHSO_3$, cystein, or thiosulphate are employed. The effect of pH on the enzymatic activity of papain is shown in Fig. 8.1.

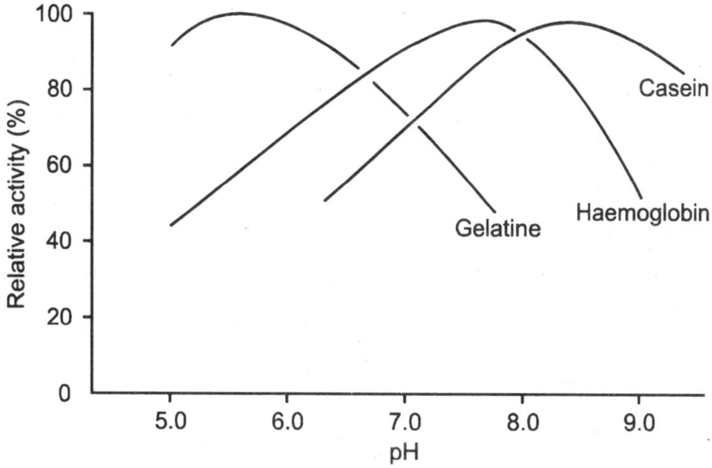

Fig. 8.1. Effect of pH on the activity of papain with various substrates (Reisner, technical information [RM]).

1. Activity: 500–1000 BAPA units.
2. Analysis: Papain is standardised with casein or haemoglobin or more commonly, as in brewing, for example, with BAPA. In addition, a very sensitive immunologic determination that uses papain antibodies is available.
3. Stability: Liquid products are more stable than powder. At a storage temperature of 4°–10°C, activity is reduced by approximately 10 per cent per year. Binary combinations of papain with microbial proteases are not compatible in various proportions. When papain is combined with fungal proteases in aqueous solution, an inactivation of fungal protease of ≤40 per cent was observed in 1.5 hours.
4. Products and manufacturers: Colourase L 10 [RM], Auxillase [MK], Collupin [GB] and others.

Other Plant Proteases

Bromelain (EC 3.4.22.4; CAS 9001-00-1)

There are two enzymes in pineapples that show very similar specificities. Most industrial preparations are obtained from the pulverised stems of the whole plant (stem-bromelain). The fruits also have a high protease activity (juice-bromelain, EC 3.4.22.5). A commercial enzyme concentrate can contain up to 0.1 per cent of this enzyme. Technical preparations are used for the production of digestive aids, for healing preparations, in diet foods and for the stabilisation of beer.

1. Molecular weight: 33,000 daltons.
2. Structure: Glycoprotein, SH protease; five isoenzymes have been found.
3. Activity: See Table 8.3.
4. Optimal pH: The optimal pH for haemoglobin and casein is 6–8 with gelatine; with papain, pH 5. The optimal temperature is 50°–55°C for casein and 60°C for gelatine.
5. Products and manufacturers: [TMKB]: Preparations with 1200 and 1400 CDU/mg; [MK], [GB].

Table 8.3. Comparison of relative activity of various sulphydryl proteases.

Substrate	*Stem Bromelain*	*Fruit Bromelain*	*Papain*	*α-Chymotrypsin*
BAA[a]	0.14	6.3	–	–
BAPA[b]	0.9	14.7	–	–
Casein	9.2	23.7	24.2	23.8

[a]Benzoyl-L-arginine amide.

[b]Defined in Table 8.2.

Ficin (EC 3.4.22.3; CAS 9001-33-6)

Figs have a very high protease activity. A green fig weighing 10–15 grams contains about 100–150 mg ficin, based on a commercial preparation. On drying, most of the activity is lost. Ficin is currently produced in only a few South American countries. At Röhm, calabashes were received from Peru that contained a very active ficin that stabilised with a high concentration of SO_2. The hides in the beamhouse of the tannery were strongly attacked. Glycoproteins were hydrolysed and the collagen also degraded. Applications are similar as those for papain and bromelain. These include the hydrolysis of fish waste and for plant proteins.

1. Molecular weight: 26,000 daltons.
2. Structure: The active site is similar to that in papain.
3. Activity: The pH optimum against casein is pH 6.7. Stability is high in the pH range 3.5–9.
4. Inactivation: At 80°C.
5. Inhibitors: Oxidants, heavy metals, 10^{-4} M sorbic acid.
6. Manufacturers: [MK], [GB].

ANIMAL PROTEASES

The protein-hydrolysing enzymes of the mammalian digestive tract have been known for a long time and have been researched intensively since the beginning of enzymology. Even before the chemical nature of proteins and specifically enzymes were determined, the protein-cleaving action of pepsin and pancreas proteins was known.

According to the current classification, which relates to the structure of the active site of proteases, the animal proteases belong to various classes. These enzymes are considered together here because of historical developments. Typical for these enzymes is the fact that they occur as inactive precursors, the proenzymes, in the organs in which they are produced. The precursor is not activated to the active enzyme until the enzyme is secreted into the digestive tract. In addition to the endoenzymes, proteases that cleave the proteins inside the peptide chain, a series of exoenzymes that are important in certain industrial applications also exist.

Uses for these enzymes are found in food production, in the manufacture of protein hydrolysates, and for processing meat and fish residues. Today large amounts of pancreatic proteases are needed for the leather manufacturing industry. In medicine, these enzymes are a part of digestive medical aid preparations. The animal proteases and peptidases that are still used in the industry are summarised in Table 8.4. The carboxypeptidases and aminopeptidases include a number of very specific single enzymes, which are active at different pH ranges. Some proteases and peptidases are called cathepsins. Today, this terminology is used only for the protein-cleaving enzymes that occur in tissue (liver, kidneys and muscle).

Table 8.4. Animal proteases.

Protease	*Proenzyme*	*pH range*	*Activator*
Pepsin	Pepsinogen	1.5–4.0	H^+, proteases
Chymosin (gastric)	Prochymosin	5–6	Ca^{2+}
Trypsin	Trypsinogen	6.5–9	Enderokinase, Ca^{2+}
Chymotrypsin	Chymotrypsinogen	7–8.5	Ca^{2+}
Carboxypeptidases	Procarboxypeptidases	7.4 (6–8.5)	Trypsin
Aminopeptidases	–	7.5–9.0 (6.5–8)	–

Endoproteases

Pepsin (EC 3.4.23.1-3; CAS 9001-75-6)

Pepsin, is secreted from the gastric mucous membrane as the non-active proenzyme, pepsinogen. This is converted into active protease by the action of stomach acid.

This autocatalytic action occurs below pH 6.0, with the maximum rate of conversion at pH 2.0. Activity can be induced by unrelated pepsins or microbial acidic proteases. At alkaline pH levels, the pepsinogen is stable, whereas pepsin is rapidly inactivated.

In a procine stomach, different proteins with similar action but different molecular weights are found. Pepsins of various animal species are immunologically different. In addition, an enzyme that resembles pepsin is found in the stomach mucous membrane. This is gastricin, which has a pH optimum at pH 3.2.

Properties of pepsin

1. Molecular weight: 35,000 daltons; pepsinogen, 41,000 daltons (hog).
2. Isoelectric point: pH ~1.0. Values of pH 3 and 1.7 are also mentioned in the literature. A pI at pH 1.7 was determined following removal of a covalently bound phosphoryl group.
3. Structure: A single-chain polypeptide with 321 amino acids. The amino acid sequence and the tertiary structure are well known.
4. Activity: Many proteins, including keratin, are cleaved at pH 1.0–2.5. Mucoids such as those of the stomach mucous membrane or ovomucoid are not attacked and are strong inhibitors. Silk fibroin and protamine are not hydrolysed. The activity optimum of short peptides and synthetic peptide derivatives lies between pH 2 and 4.
5. Stability: Pepsin is relatively stable in aqueous solution from pH 5–5.5. It is more stable in glycerol, in which it dissolves. It is irreversibly inactivated at alkaline pH.
6. Specificity: Peptide linkages with aromatic amino acids as well as glutamic acid, cysteine and cystine peptides are preferentially cleaved. Polyglutaminic acid is a good substrate, whereas oligomers such as Glu_2 to Glu_4 are not attacked. Relatively specific synthetic substrates include Ac-Phe-Tyr, with optimum at pH 2.0. Z-His-Phe-Phe, optimum at pH 3.0.
7. Technical preparations: Pepsin 1:10000 is equivalent to NF 11, or ~10000 mHbU/mg (pH 3).
8. Applications: In digestive-aid medicine, for the preparation of protein hydrolysates, in mixtures with chymosin in rennet preparations and occasionally for stabilisation of beer. (See also Fig. 8.2.)

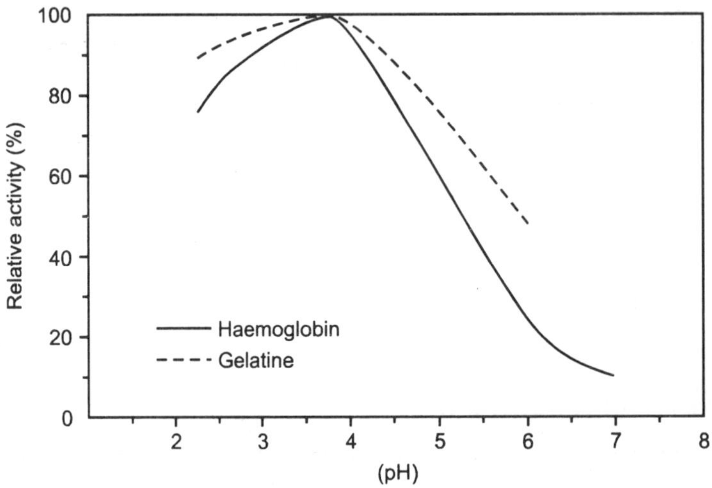

Fig. 8.2. Effect of pH on the activity of pepsin with haemoglobin and gelatine substrates (25°C; 10 minutes [RM]).

Chymosin (EC 3.4.23.4; CAS 9001-98-3), chymase, renin, rennet

Properties and manufacturers

Chymosin, like pepsin, is an acidic protease that occurs in the stomach and is required for the manufacture of cheese. Chymosin is produced in stomach mucous membranes as its inactive precursor, prorennin (molecular weight 36000 daltons). Activation occurs at hydrogen ion concentration below pH 5.

1. Extraction: The enzyme is obtained by extraction from stomachs of young calves, goats or lambs. The extraction from older animals contains significant quantities of pepsin.
2. Molecular weight: 31000 daltons.
3. Activity: One part of a highly purified chymosin can coagulate 72 million parts of milk in 10 minutes. Optimum proteolytic activity is at pH 3.8. The optimum of the preparation from rennet is above pH 6.5, at an optimum temperature of about 40°C. Comparison of X-ray structural studies of homologous enzymes have shown that chymosin belongs to the carboxyl proteases. Fractionation of chymosin has revealed two to three isozymes with similar properties.
4. Specificity: The peptide linkage between phenylalanine and methionine is preferentially cleaved in soluble calcium caseinate in milk. This is the first stage of hydrolysis; in the second stage, calcium-induced coagulation occurs. Apart from the peptide linkage between phenylalanine and methionine, 22 other linkages in α-s_1-casein are cleaved, although the rate of hydrolysis is much slower.
5. Analysis: The rennet strength or the Soxhlet unit per milliliter is defined as the number of milliliters of milk that can be coagulated in 40 minutes at 35°C. The standard rennet in Great Britain has 15,000 rennet units per milliliter and contains ~75–80 per cent pure rennet. The rennet preparations are mostly liquid and contain salt, glycol and benzoate as stabilisers. Because milk is a difficult substrate to standardise, Berridge suggested that milk powder as substrate in 0.01 M calcium chloride solution should be used. One rennet unit is defined as follows:

$$\text{Rennet activity} = \frac{10 \times \text{volume of milk}}{\text{Coagulation time} \times \text{rennet volume}}$$

Trypsin (EC 3.4.21.4; CAS 9002-07-7)

Trypsin is a serine protease. The precursor of this enzyme is produced in the pancreas. This precursor, trypsinogen, is transformed into active trypsin by a protease in the intestinal mucous membrane known as enterokinase. This transformation of trypsinogen, with a molecular weight of 24,000 daltons, occurs on the hydrolysis of a single peptide linkage between lysine and isoleucine with the release of a hexapeptide (Val-Asp_4-Lys). Activation by trypsin also occurs autocatalytically. The reaction is strongly accelerated by calcium.

Properties and application

1. Molecular weight: 24,000 daltons.
2. Structure: Trypsin consists of a polypeptide chain with 233 amino acids.
3. Activity: The activation optimum for most proteins and synthetic substrates lies between pH 7 and 9. In terms of cleavage reactions, trypsin is very specific. Only peptide linkages in which the carboxyl group of lysine or arginine is linked with the amino group of other amino acids are attacked. This strict specificity was very helpful in determining the amino acid sequences of several proteins. In the case of the insulin B-chain, only two peptides are formed. These can be easily separated and further sequenced.
4. Stability: Optimal stability of trypsin is at pH 3. Incubation of trypsin preparations at pH 1.5 for several hours destroys other proteolytic activities; only about 20 per cent of the trypsin activity is lost. The stability of aqueous solutions in alkali is low because of self-digestion.
5. Inactivation and inhibitors: Alkylating reagents, such as DFP (diisopropyl fluorophosphonate) or PMSF (phenylmethylsulphonyl fluoride), react stoichiometrically with a seryl residue at the active site of trypsin. This leads to rapid inactivation. Natural inhibitors are found in soyabeans (two soyabean inhibitors), egg white (ovomucoid), potatoes, and cereals. The soyabean inhibitor is destroyed during toasting. Kunitz crystallised a soyabean inhibitor with a molecular weight of 20100 daltons.
6. Application: Used medically in pure or crystalline form in wound treatment.

Chymotrypsin (EC 3.4.21.1; CAS 9004-07-3)

Several forms of chymotrypsin are known. Chymotrypsin is a serine protease and, like trypsin, is found in the pancreas as an inactive precursor. Trypsin converts chymotrypsin into the first active form: the unstable α-chymotrypsin. Further proteolytic processes subsequently lead to the α-, β-, and γ-chymotrypsins (Fig. 8.3).

Properties

1. Molecular weight: α-Chymotrypsin, 25000 daltons.
2. Activity: Activation optimum is at pH 8–9, with, at pH 7, ~50 per cent maximum activity at pH 7. Maximum temperature is 40–45°C; at high substrate concentrations, up to 55°C.
3. Stability: Chymotrypsin is less stable in acid than trypsin. Aqueous solutions are stable in the pH range 2.5–6 at 10°C for 20 hours; at 40°C, for 1 hour.
4. Inhibitors: Heavy metals, natural trypsin inhibitors, polyphosphates, strong oxidation agents, DFP and PMSF.
5. Application: Pure forms are used in medicine. In some countries, α-chymotrypsin, because of its antiinflammatory action, is prescribed to be taken orally against infections, broken bones,

and so on. Technical applications include mixtures with other pancreas enzymes to be used, for example, in the leather industry.

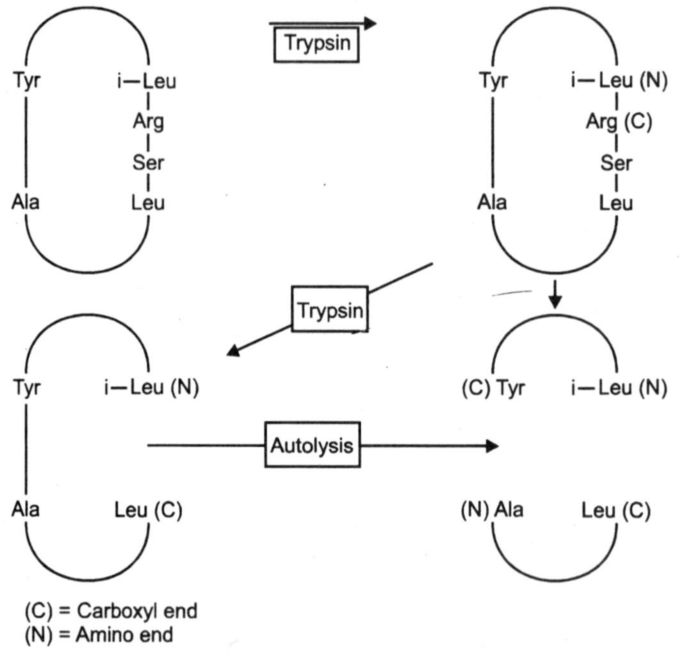

Fig. 8.3. Conversion of α-chymotrypsinogen into α-chymotrypsin.

Pancreatic elastase (EC 3.4.21.11; CAS 9004-06-2), pancreatopeptidase E

Elastin is a component of animal connective tissue. It differs from other proteins in that it has a very high content of amino acids with non-polar side chains, such as Leu, Val and Ala. The cervical ligament of steers (ligamentum nuchae) has a high elastin content, from which preparations can be extracted, which may be of value in enzyme research.

The elastase, a serine protease, is formed in pancreas as an inactive proelastase. Pure elastase consists of a polypeptide chain with 240 amino acids and resembles the active centre of a trypsin or chymotrypsin.

Properties

1. Activity: Optimum pH 6–8.
2. Specificity: Peptide linkages between aliphatic amino acids, such as Ala-Leu, Leu-Val, Val-Glu, and Val-Cys (SO_3H), are preferentially cleaved.
3. Synthetic substrates: *N*-Benzoyl-L-Ala-methylester and *N*-acetyl-L-Ala-L-Ala-L-Ala-methylester.
4. Stability: Elastase is relatively stable below pH 6 and is degraded by autolysis above pH 8.

Peptidases

Peptidase activity is found in technical pancreatic enzyme preparations, accompanied by proteases. The carboxypeptidases cleave amino acids from the carboxyl end and the amino peptidases amino acids from the amino end of the peptide chain. The reduction of the chain length of peptides by the action of endoproteases is of interest commercially. In other applications, protein hydrolysis is used to produce the maximum number of amino acids or di- or tripeptides. A high degree of hydrolysis (DH) can be achieved only by the combined action of numerous proteases and peptidases. The enzymatic release of amino acids plays an important role in fermentation processes such as in cheese ripening. In this case, the peptidases are produced by *Lactobacillus* or fungi. These peptidases are described in the next section of this chapter.

Highly purified peptidases are used in research for peptide sequencing. Because of their importance in providing side activities in technical preparations, some information on the most important peptidases is given here.

Carboxypeptidase (EC 3.4.2.1; CAS 11075-17-5)

Numerous carboxypeptidases occur in pancreatic glands. Carboxypeptidase A is formed by the action of trypsin on the inactive precursor of the procarboxypeptidase A. The enzyme was first crystallised by Anson and contained several other proteases.

Properties

1. Molecular weight: 34,000 daltons.
2. Structure: The single chain with 307 amino acids contains one tightly bound zinc atom per mole of enzyme. Following elucidation of the primary structure, the tertiary structure was determined by Lipscomb.
3. Activity: Optimum pH 7–8.
4. Specificity: Amino acids other than Pro, Arg, Hyp, and Lys are cleaved from the C-terminal of peptides. In the case of the amino acids mentioned, sequential cleavage is interrupted. Peptide linkages that include proline (Pro) or hydroxyproline (Hyp) are cleaved by prolidase and prolinase.

Aminopeptidases (EC 3.4.1.2; CAS 9054-63-1)

Aminopeptidases can be obtained from animal organs, especially pig kidneys. These animal aminopeptidases are of little commercial importance, although those produced microbiologically could become important for food production.

Technical Pancreatic Protease Preparations

The commercial product 'pancreatin' contains all the enzymatic activities of the pancreatic glands. It is produced by grinding, defatting and drying the glands of calves and pigs. The insoluble pancreatin, where the enzymes are still bound to tissue fibres, is standardised against different activities, such as protease or lipase. For pharmaceutical applications, standardisation is done according to the analytic guidelines of the Federation International Pharmaceutique (FIP) or the US Pharmacopoeia (USP).

Other products were prepared from the glands of pigs or also from calves and sheep by pulverising, squeezing, and subsequent salt precipitation of the expressed fluids. In these soluble preparation, the proteins were fully activated, whereas a part of the lipase activity destroyed.

A further means of production involves the by-products of insulin manufacture. Here the pulp of ground glands is extracted at acidic pH, yielding dried preparations that are relatively rich in trypsin, whereas the chymotrypsin is partly denatured. Such preparations represent good starting material for the manufacture of pure trypsin. See also Figs. 8.4–8.6.

1. Application: A large amount of pancreatin is used in the pharmaceutical industry in digestive-aid preparations. The accompanying lipase activity is particularly important here. Soluble

products are preferred in leather manufacturing for the tanning of hides. They are also used in the manufacture of protein hydrolysates and for the liquefaction of fish protein and meat residues.

2. Products and manufacturers: Soluble preparations; Corolase PP [RM]; Novotrypsin [NO]; Pancreatin [KC].

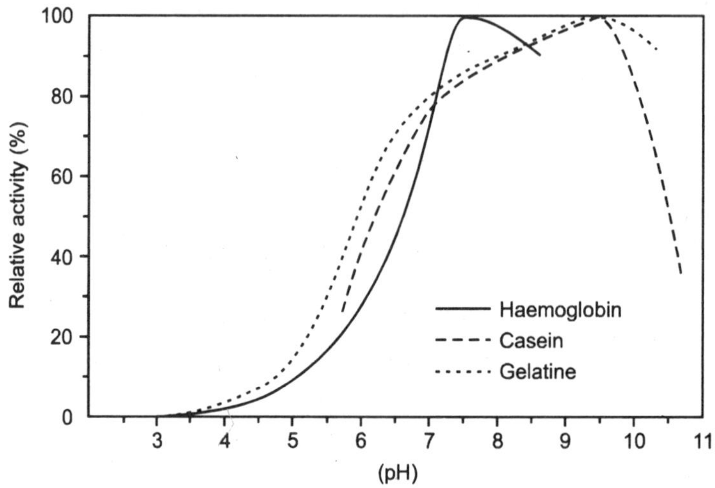

Fig. 8.4. Effect of pH on activity of pancreatic protease (Corolase PP [RM]) with various substrates. Activities determined with a haemoglobin substrate by a modified Anson method; with gelatine by measuring the drop in viscosity and with casein by the PU method.

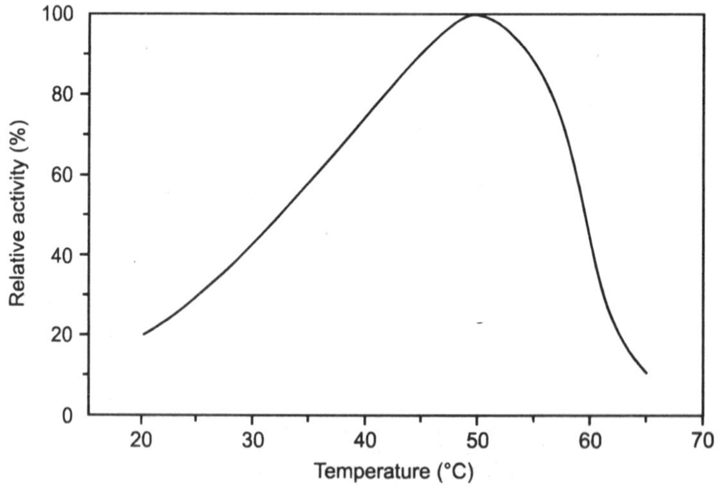

Fig. 8.5. Effect of temperature on the activity of pancreatic protease [RM] (pH 8.0; reaction time 10 minutes).

MICROBIAL PROTEASES: BRIEF OVERVIEW

Of all the proteases, those produced microbiologically have the largest commercial importance. Seventy five per cent of these proteases were used in the manufacture of detergents, and about 10 per cent were used in cheesemaking. Other applications are in tanneries and numerous food processes. Proteases are formed by a large number of different micro-organisms, although only a few are useful for the production of commercial preparations. Only those organisms that produce enzymes in very high yield and release the proteases into the culture medium are suitable.

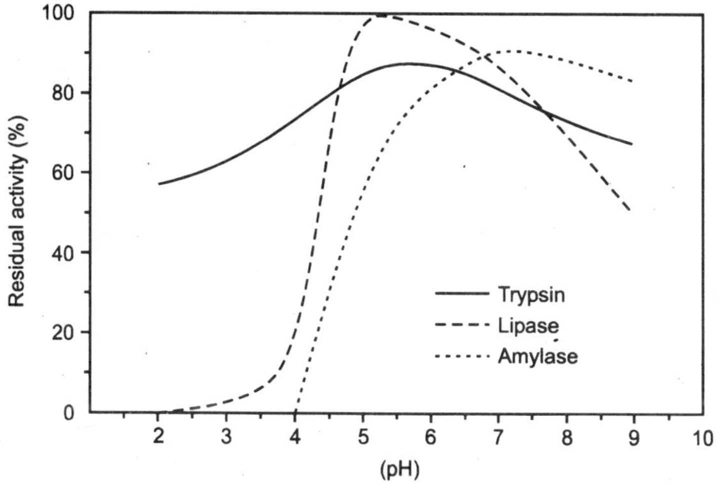

Fig. 8.6. Effect of pH on the enzyme stability in a pancreatin preparation [NO] (reaction time 4 hours; 25°C).

As was mentioned in the introduction, proteases are classified according to their structure or the properties of the active site. The most important microbial proteases belong to the serine-, metallo- and carboxyl proteases. In industrial products, the classification relates to the pH optimum of the respective enzyme. There are acidic, neutral and alkaline proteases, with some overlapping in definition. The serine proteases generally have optimum activity at alkaline pH levels, although 'neutral' serine proteases have also been described.

In addition, there are classifications in the literature derived from various applications. There are references to 'gelatinases', 'keratinases' and 'caseinases'. For example, the protease from *Streptomyces fradiae* was named *keratinase*.

It is indeed very active against keratin, as are other proteases, but it also cleaves casein, gelatine and haemoglobin.

This case clearly demonstrates the difference between analytic determination and practical application. It was originally hoped that this enzyme could be used for dewooling, but it was later shown that other bacterial proteases, when used in combination with reducing chemicals, are much better suited for this purpose.

The collagenase from *Clostridium histolyticum* owes its name to the fact that it can only cleave collagen. The collagenase-specific substrate (Z–Gly–Pro–Gly–Gly–Pro–Ala) is also cleaved by so-called unspecific collagenases. Compared to trypsin, the microbial enzymes have, in general, a significantly broader activity; specifically, they cleave proteins much less specifically (Fig. 8.7 and Table 8.5).

The specific activities of the crystallised enzymes were compared using the Kunitz test. Casein was hydrolysed at the pH optimum for 10 minutes at 30°C. Measurements were made at 280 nm. The technical proteases are produced by bacteria and fungi. The different types of enzymes, the producing micro-organisms and examples of the respective commercial products are shown in Table 8.6.

Table 8.5. Properties of bacterial proteases compared to trypsin.

Enzyme	*Molecular weight (daltons)*	*Inhibitor (activator)*	*Effective pH Range (80%)*	*Activity/ mg protein*
Subtilisin Carlsberg	26000	DFP[a]	8.5–11	2200
B. subtilis, alkaline protease	25000	DFP	8.5–11	2300
Neutral bacterial protease (Tsuru I)	44700	EDTA[b] (Zn, Ca)	6.5–7.5	13600
Neutral bacterial protease (Tsuru II)	34000	EDTA	7.5–8.0	12500
A. oryzae, alkaline protease	35000	DFP	7.5–9.5	3500
Trypsin	24000	DFP	7.0–9.0	1700

[a]DFP, isofluorophate.

[b]Ethylenediaminetetraacetic acid.

Table 8.6. Technical enzymes, microbial sources and manufacturers.

Enzyme source	*Commercial product/[Producer]*
Acid carboxyl proteases	
Rhizopus sp.	Sumizyme RP [SN]; Newlase [AM]
A. oryzae	Corolase PS [RM]; Veron PS [RM]
	Sumizyme LP [SN]; Sanzyme [SA]
A. niger	Sumizyme FP [SN]; Proctase [MJ]
A. saitoi	Seishin Pharm. Co., Noda, Japan [SP]
Mucor miehei	Rennilase [NO]; Fromase [GB]; Marzyme [SOE]; Morcurd Plus [NO]
Mucor pusillus	Meito Rennet [MSJ]; Noury Lab [MSJ]
Metalloproteases	
B. amyloliquefaciens (*B. subtilis*)	Neutrase [NO]; Corolase N [RM]; Veron P [RM]; Bioprase [NG]; Rhozyme [GR]; Orientase [UA]
B. thermoproteolyticus	Thermoase [DA]
Serine proteases	
A. sojae	Corolase PN [RM]
A. melleus	Sumizyme MP [SN], [AM]
B. licheniformis	Alkalase [NO]; Optimase [SO]; Maxatase P [GB]
Bacillus sp.	Esperase [NO]; Savinase [NO]
B. griseus	Pronase [KK]

Table 8.6 is not comprehensive, in that other enzyme preparations are commercially available that contain more than one enzyme. In addition, all the source organisms for the commercial products listed have not been cited.

Serine Proteases from Bacteria

The extraction of a serine protease from *B. subtilis* and its purification and crystallisation are described by Güntelberg and Ottesen. The organism was later classified as *B. lichenformis*. The enzyme became known under the name 'Subtilisin Carlsberg' and has been produced since 1960. The enzyme gained great commercial significance following its introduction as an active component in detergents.

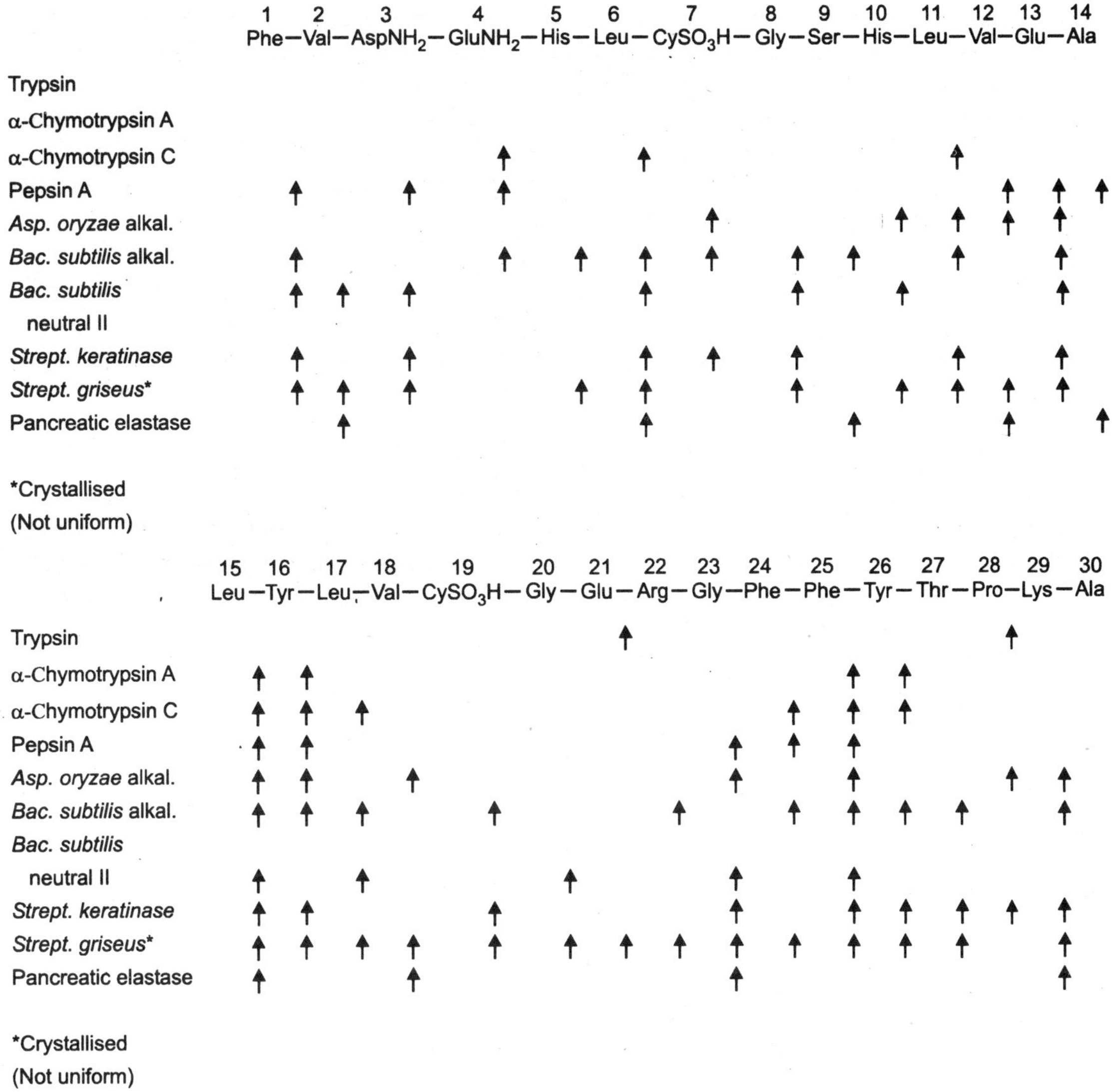

Fig. 8.7. Proteolysis of the oxidised insulin B-chain by various proteases.

In 1959, a detergent containing alkaline bacterial protease (Bio 40) was developed by Jaag at the Snyder Company in Switzerland. Prior to this, attempts had been made to use pancreatic proteases for this application by Otto Röhm. However, the pancreatic proteases were not sufficiently stable at high pH levels in the presence of tripolyphosphate and anionic detergents. After the introduction of the protease from *B. licheniformis*, which is stable in alkali even in the presence of complexing agents, a period of rapid development of technical enzymes followed.

Today, the protease-containing detergents are used in many countries. Other enzymes, such as cellulases and lipases, are also being added in increasing measure.

The serine proteases include a number of very specific proteases from different organisms that, until now, were used only for structural determinations such as amino acid sequencing of proteins. One of these is the serine protease from *Staphylococcus aureus*. This enzyme cleaves peptide linkages C-terminal from Glu and Asp (Glu-X, Asp-X) (endoprotease Glu-C, [BöM]). The endoprotease Lys-C [BöM] from *Lysobacter enzymogenes* specifically cleaves peptides with lysine linkages.

The molecular weights of serine proteases range from 15000 to ~30000 daltons and the isoelectric points are at approximately pH 9.0.

The following gives the properties and the most significant constants of the three most commercially important subtilisin enzymes.

Subtilisin Carlsberg (EC 3.4.21.14; CAS 9014-01-1) (Organism: B. licheniformis)

Properties

The molecular weight is 27277 daltons, calculated from the 274 amino acid residues of the single-chain molecule. It contains no cysteine and thus no disulphide bridges. The isoelectric point is at pH 9.4.

1. Activity: Optimum pH 8–9. (See also Figs. 8.8 and 8.9.)
2. Stability: Inactivated under pH 5.0 and above pH 11. Calcium is not necessary for activation or for stability. The enzyme remains stable in the presence of complexing agents.
3. Specificity: Largely non-specific; cleaves most peptide linkages; prefers those with aromatic amino acids and many esters. Up to 40 per cent of the peptide linkages in casein are cleaved.
4. Inhibitors: Diisopropyl fluorophosphate (DPF), phenylmethylsulphonyl fluoride (PMSF).

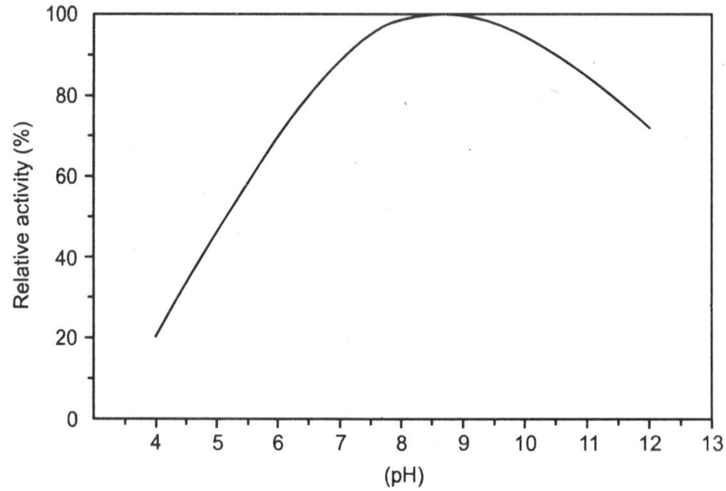

Fig. 8.8. Effect of pH on the activity of subtilisin Carlsberg (haemoglobin, denatured with urea, phosphate buffer, 25°C).

Subtilisin BPN

Haghihara first isolated and crystallised this enzyme from a commercial preparation of the bacterial protease 'Nagarse' (NG). In 1960, Ottesen and Spector isolated an identical enzyme (Subtilisin Novo) from the bacterial protease Novo. The producing strain was identified as *B. amyloliquefaciens*. Subtilisin BPN has 275 amino acid residues in a chain that does not contain cysteine. The enzyme has an isoelectric point at pH 9.1 and has a high homology (identical amino acid sequence) to Subtilisin Carlsberg. Only

58 of the 274 peptide linkages are different. The stability of Subtilisin BPN is slightly more dependent on calcium ions at high temperatures and high pH values than is Subtilisin Carlsberg. In addition, it shows different specificity against esters.

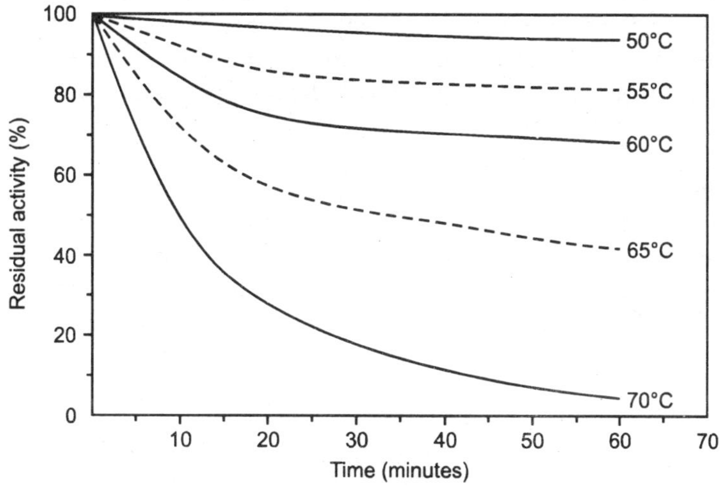

Fig. 8.9. Effect of temperature on the stability of subtilisin Carlsberg (pH 8.5, Trismaleate buffer).

Proteases from alkalophilic Bacillus sp.

These enzymes are also serine proteases with a molecular weight of 20,000–30,000 daltons and an isoelectric point at approximately pH 11. The characteristics are similar to the above-mentioned subtilisins with the exception of a higher activity in the alkaline pH range.

1. Activity: 80 per cent of the maximum activity between pH 8 and 12.
2. Stability: pH 6–12.

Endoprotease from Streptomyces griseus

This pure enzyme, isolated from the commercial product 'pronase' [KK], is used in research for total hydrolysis of proteins.

Properties

1. Molecular weight: 20000 daltons (gel filtration), 22000 daltons [sodium dodecylsulphate-polyacrylamide gel electrophoresis (SDS-PAGE)].
2. Isoelectric point: At pH 8.4.
3. Activity: Activation optimum at pH 8.4.
4. Specificity: Preferably cleaves peptide C-terminal of Glu and Asp.
5. Inhibitors: DPF, no inhibition by EDTA, *p*-chloromercuribenzoate, or SH blockers.

Jurasek found that in many respects, the enzyme is similar to trypsin. It has a similar amino acid sequence, a similar active site and is sensitive to trypsin inhibitors. He found the enzyme cleaves preferentially at Lys and Arg in the oxidised B-chain of insulin.

Technical preparations

The most important application for serine proteases is in the detergent industry and for this area most preparations are produced as granulates. The activity of the granulates lies between 0.1 and 2 Anson units. In the new liquid detergents, the enzymes are delivered in a stabilised liquid form.

Serine Proteases from Fungi (EC 3.4.21.15)

Technical fungal proteases contain a mixture of serine proteases and proteases active in neutral and acidic pH ranges. The enzyme known as *Aspergillus* protease B is produced by the *Aspergillus* strains: *A. flavus, A. sydowi, A. candidus, A. oryzae, A. sojae* and *A. melleus*. The enzyme is usually obtained with the solid-bed culture process and has a specificity similar to that of the subtilisins. This statement has to be treated with caution because the enzymes from various *Aspergillus* strains differ strongly in their characteristics. Incomplete purification and the tendency of proteases to autolyse can be attributed to inconsistencies in some of the literature.

Protease from *A. candidus*

This protease has an isoelectric point at pH 4.9. The pH optimum with casein lies at pH 11–11.5 (30°C). The maximum activity at 47°C is reached at pH 7.0. Enzyme stability is at pH 5.9–9. Ions have no influence on activity, but calcium stabilises against heat denaturation. The inhibition through DPF and potato inhibitor points to a serine protease. There is no inhibition by SH regents.

Protease from *A. flavus*

This enzyme resembles in many of its properties the enzyme from *A. candidus*. It can be incubated without loss of activity in 8 M urea at 20°C but is rapidly inactivated by 4 M guanidine + HCl. It is stable between pH 5 and 10.

Protease from *A. melleus*

This enzyme mixture is virtually identical to that from *A. oryzae* or *A. sojae*. Differences, however, are apparent in practical applications; for example, in the cleavage rates of different proteins.

Protease from *A. oryzae*

One of the most fundamental studies on the protease complex from *A. oryzae* was done by Bergkvist, who isolated the enzyme from a submersion culture by precipitating it with tannin. After purifying on DEAE cellulose, 28 per cent of the protease activity was found in three fractions. These were separated according to their pH optima into three sub-fractions: alkaline (1), neutral (2) and acidic (3).

1. Protease 1: ~29 per cent of the activity with casein at pH 6; 66 per cent activity with casein at pH 7.4.
2. Protease 2: ~24 per cent with casein at pH 6; 35 per cent with casein at pH 7.4.
3. Protease 3: ~47 per cent yield with the test at pH 6 and 35 per cent at pH 7.4. The pH curves were recorded using gelatine as substrate. The drop in viscosity of a 5 per cent gelatine solution in 5 minutes at 37°C was monitored (Fig. 8.10).

In 1950, Crewther and Lennox were able to crystallise a protease from Takadiastase (Takamine). The enzyme was named *Aspergillopeptidase A* (EC 3.4.4.17). Because of its differences from protease 1 of Bergkvist, the classification of this protease is still uncertain. Table 8.7, from Bergkvist shows some characteristics of the three enzymes.

Protease from *Tritirachium album Limber*

This enzyme, which is also extensively used for protein hydrolysis in research, is known as 'Protease K' [MK]. The 'K' refers to its ability to hydrolyse keratin. The primary and tertiary structures of this enzyme are known. The active site is similar to that in subtilisin.

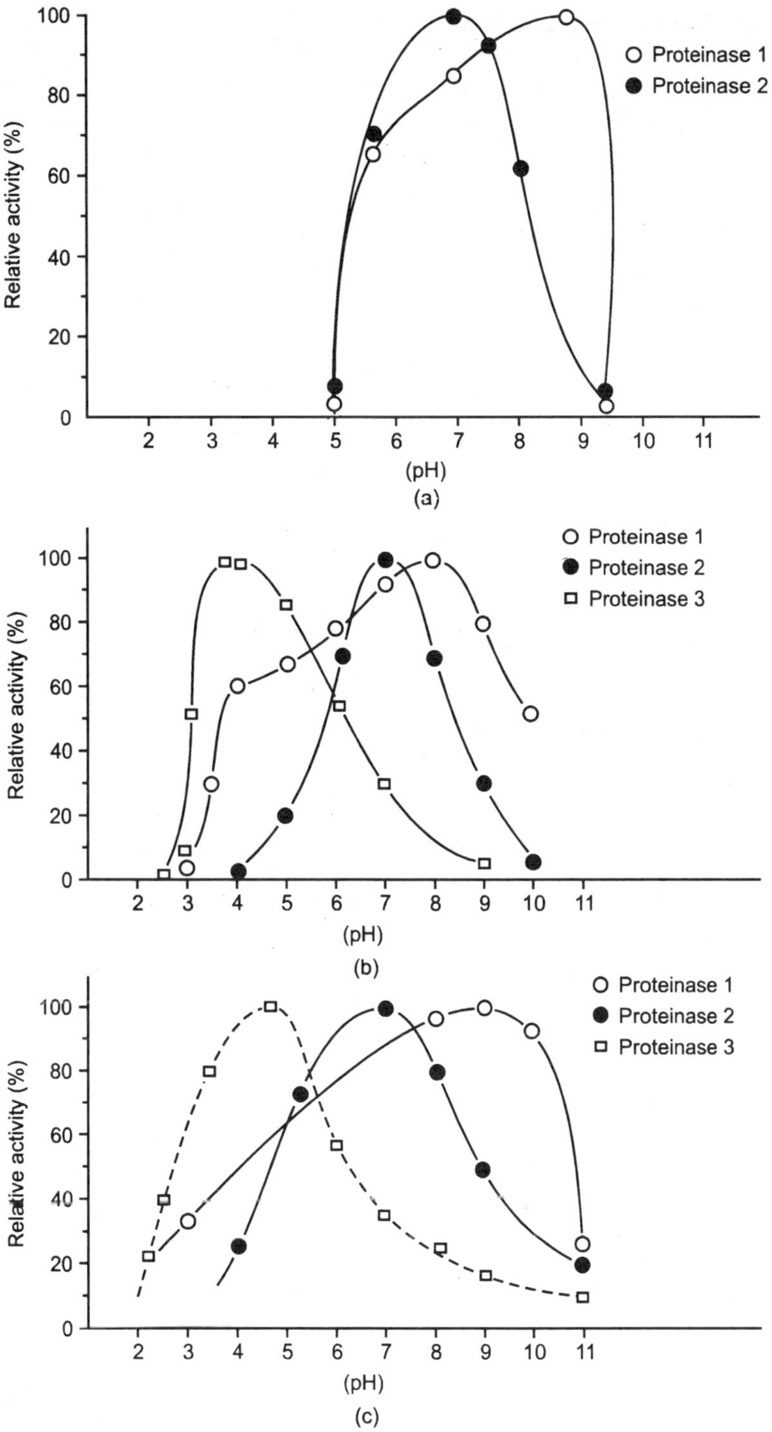

Fig. 8.10. Effect of pH on the activities of three proteases from *A. oryzae*; (a) caseinase activity; (b) haemoglobinase activity; and (c) gelatinase activity (viscosimeter).

Table 8.7. Properties of purified proteases from *A. oryzae*.

Property	*Protease* 1	2	3
pH optima for			
Casein	8.2	6.8	–
Denatured haemoglobin	7.6	6.8	4.3
Gelatine	9.0–9.5	6.3	4.5
TAME[a]	8.5	–	–
Temperature optima (°C)	50	50	45
pH-stability	5–8.5	4.5–10.5	3.0–6.3
Inhibitors			
Na-laurylsulphonate	+	–	+
Laurylamine	+	+	+
L-Cysteine	–	+	–
EDTA[b]	–	+	–
6-Aminocaproic acid	–	–	–
Soyabean inhibitor	–	–	–
Ascorbic acid	–	+	+

[a]N-α-*p*-tosyl-L-arginine methylester hydrochloride.

[b]Ethylenediaminetetraacetic acid.

Properties

1. Molecular weight: 28,500 daltons, polypeptide chain with 277 amino acids. Contains two disulphide bridges and one free cysteine group.
2. Specificity: The enzyme cleaves peptide linkages C-terminal to aliphatic, hydrophobic and aromatic amino acids. Native proteins are rapidly hydrolysed and similarly, enzymes are also rapidly inactivated by unspecific degradation or limited proteolysis.
3. Specific substrates: Glutaryl–Ala–Gly–Phe–pNA or glutaryl–Ala–Ala–pNA.

Technical preparations

Technical fungal proteases that contain primarily alkaline proteases (Fig. 8.11) are used to a limited extent in different branches of the food industry. A special application is in leather manufacturing where a constant enzyme complex composition is of great importance. When properly applied, the enzyme works well in a number of processes without damaging the leather.

Products and manufacturers: Corolase PN [RM], Sumizyme MP [SN], Rhozyme P 11 [RH].

Metalloproteases: Neutral Proteases (EC 3.4.24.4; CAS 9068-59-1)

The neutral proteases from bacteria and fungi that are used in many applications belong to this group. These enzymes distinguish themselves in that their activity is dependent on one or more metal atoms at the active site. The activation optimum lies at pH 7.0. These enzymes generally have zinc as the metal ion. They cleave peptide linkages relatively unspecifically but prefer amino acids with hydrophobic side chains.

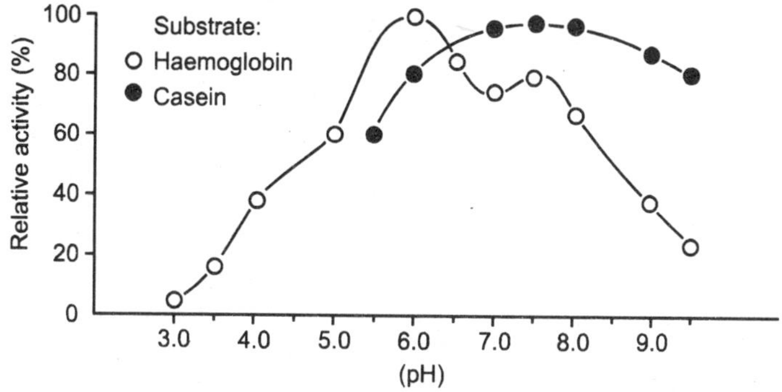

Fig. 8.11. Effect of pH on a predominantly alkaline fungal protease (Corolase PN [RM]).

Amides and esters are not cleaved, and this characteristic can serve as a means of distinguishing them from the serine proteases: Model substrates eluded discovery for a long time. Ward mentioned a substrate, furylacrolylglycyl leucinamide (FAGLA), which is cleaved only by neutral but not serine proteases. The most important technical neutral proteases are produced from the following organisms: *B. amyloliquefaciens*, *B. subtilis* (it can be assumed that, in many technical preparations, the enzyme was not obtained from *B. subtilis* but from *B. amyloliquefaciens*), *B. stearothermophilus*, *A. oryzae*, *A. niger*.

Neutral protease from B. amyloliquefaciens

Properties

1. Molecular weight: 37,000 daltons.
2. Structure: One zinc atom and two calcium atoms per mole of enzyme.
3. Stability: Optimal stability lies at pH 4.8–6.6. Calcium and protein stabilise the enzyme. Strong complexing agents (EDTA) inhibit it completely. If calcium ions are removed during the purification process, for example, during precipitation with sulphates, the stability is seriously reduced. Therefore, it is recommended that the enzyme is used in the presence of Ca^{2+} ions. Dancer and Mandelstam showed that neutral protease was already degraded during fermentation by alkaline serine protease. By inhibiting the latter with PMSF in the medium, larger fractions of neutral protease are protected.
4. Inhibitors: Heavy-metal salts such as Hg^{2+}, Ag^{2+}, Fe^{3+}, EDTA, and phosphate.
5. Specificity: Cleavage at amino end of Phe, Leu and Val. See also Figs. 8.12 and 8.13.

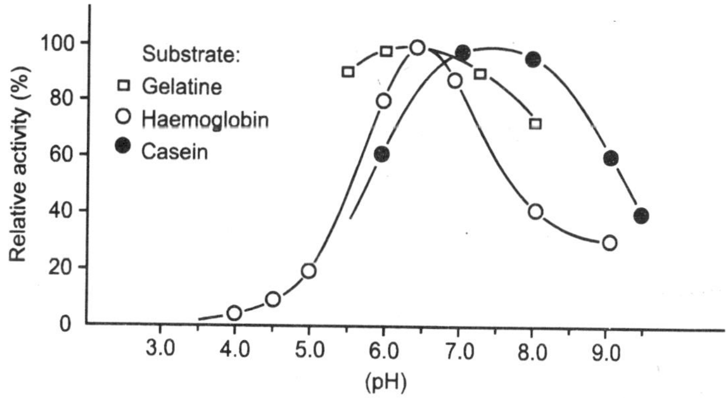

Fig. 8.12. Effect of pH on proteolytic activity with casein, haemoglobin and gelatine substrates (Corolase N [RM]).

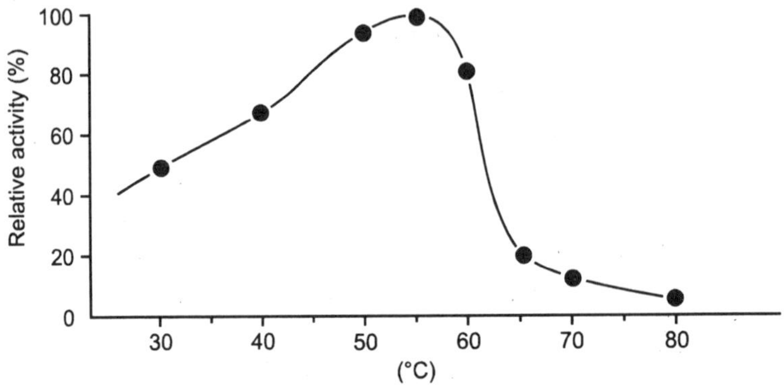

Fig. 8.13. Effect of temperature on the proteolytic activity of neutral bacterial protease.

Neutral protease from B. thermoproteolyticus

This protease is known as 'Thermolysin' or as 'Thermoase' [DA], which are the pure and technical enzyme preparations, respectively.

Properties

1. Molecular weight: 34000 daltons.
2. Structure: One zinc atom and four calcium atoms per mole. Zinc is necessary for activity and calcium for stability.
3. Activity: 90 per cent of optimum activity between pH 7 and 9. At pH 7.5 the optimum temperature is 80°C.
4. Stability: Of the optimum stability, 90 per cent is between pH 5.5 and 7.5 at 40°C. When an aqueous solution at pH 7.2 is heated to 84°C for 15 minutes, 50 per cent residual activity remains. For the enzyme from *B. liquefaciens*, this point is reached at 59°C under the same conditions.

Neutral protease from A. oryzae

The enzymes isolated by Bergkvist contain a neutral protease that is inhibited by EDTA and cysteine. Two neutral proteases from *A. oryzae* with different heat stability were found by Nakadai.

1. Protease 1: pH optimum pH 7.0. Stability: pH 5.5–12. Inactivation above 50°C.
2. Protease 2: pH optimum pH 5–6. Stability: 70 per cent residual activity remains on heating for 10 minutes at 90°C.

In addition, two types of neutral proteases are produced from *A. sojae* that do not cleave esters or amides. These enzymes act specifically on peptide linkages containing amino acids with hydrophobic side chains and on various synthetic peptides containing Phe, Leu, Tyr and other amino acids.

Technical preparations

Neutral bacterial protease

Neutral bacterial proteases in powder form show activities between 0.5 and 3 Anson units. The following side activities have been observed: α-amylase, alkaline protease and β-glucanases. Preparations formulated with sugar or salts are offered for special applications in food and leather manufacturing.

The preparations are very stable during dry storage conditions. At a storage temperature of 20°C, activity decreases by less than 10 per cent per year. Liquid preparations with 0.5–1.0 Anson units are stabilised with glycerol or for commercial purposes, with ethylene glycol.

Products and manufacturers: Corolase N [RM], Neutrase [NO], GODO-BNP [GO], HT Proteolytic [SOE].

Neutral fungal protease

In technical proteases, the neutral protease is only one component among various proteases, and thus activity ranges from pH 5 to 8.5 depending on the source of the preparation.

Products and manufacturers: Panazyme [AM], Veron PS [RM], Fungal Protease P [SO], Maxatase [GB], [SA], [SN], [UA].

Acid Proteases: Carboxyl Proteases

Acid proteases in industry are those in which the activity depends on the presence of a carboxyl group, such as in aspartic acid, at the active site. Enzymes are active in the pH range 2–6. In the systematic classification, acidic proteases are those with an activity equal or similar to pepsin.

These enzymes are produced by several fungi. Those from the *Aspergillus* and *Rhizopus* strains are important in industrial processing.

Acidic proteases are not inhibited by alkylating agents, such as DPF, chelating agents (EDTA) and SH-group blocking reagents.

This group of proteases also includes the microbiologically produced milk coagulating enzymes that have, to some extent, replaced rennet in cheese production. These proteases are obtained from *Mucor pusillus*, *Mucor miehei*, and *Endothia parasitica*. In terms of value, these enzymes are the most important of the acid proteases. Other applications are in the manufacture of baked goods, the hydrolysis of proteins and the clarification of wine and beer.

Acid protease from A. saitoi (EC 3.4.23.6; CAS 9025-49-4)

This enzyme is known as *Aspergillopeptidase A*. It was first isolated and characterised by Yoshida. According to Ichishima, the enzyme can be produced either in surface cultures or by submerse fermentation.

Properties

1. Molecular weight: 34000–35000 daltons.
2. Structure: Single-stranded peptide chain with 283–289 amino acids, with two cysteine residues that form a disulphide bridge.
3. Activity: The pH optimum lies between pH 2.5 and 3.0 with casein and soyabeans. The temperature optimum lies at approximately 30°C.
4. Stability range: pH 2–5.
5. Specificity: Preferably cleaved are Glu, Asp and Leu peptides, as well as Leu–Val, Ala–Leu and Glu–Arg linkages.

Protease from A. niger var. macrosporus

Two acidic proteases are produced from this fungal strain. Acid stability and pH optimum are different for each enzyme (Table 8.8).

Table 8.8. Properties of two acid proteases from *A. niger* var. *macrosporus.*

Property	*Enzyme A*	*Enzyme B*
Molecular weight (daltons)	20,000–21,000	40,000
pI	1.5–2	~4
Activity optimum at pH	2	2.5–3
Temperature optimum (with casein substrate)		
At pH 1.5	60°C	<30°C
At pH 2.0	65°C	40°C
At pH 2,6	70°C	55°C
Gelatine liquefaction	Strong at pH 2	Weak at pH 2
Collagen digestion	Very strong	Strong

Acid protease from A. oryzae

Takadiastase, mentioned earlier, contains at least one acid protease. The characteristics (pH-activity profile and temperature stability) of this enzyme are similar to the acid protease component described by Bergkvist.

Although several isoenzymes can be isolated during the fractionation of these enzymes, only two different acid proteases from *A. oryzae* seem to exist. The isozymes are presumably formed by autolytic processes. One of the acidic proteases has a pH optimum at pH 4.0–4.5 with haemoglobin as substrate. It activates trypsinogen and has a limited esterase activity. The second protease has an activity optimum from pH 4.8–5.2 and is less heat-stable than the first one. It is the main component of the technical preparation Rhozyme A4. Both enzyme types can be isolated from preparations produced in surface cultures. Products and manufacturers: Proctase [MJ], Sumyzyme FP [SN], Molsine [Tanabe], Corolase PS [RM].

Acid protease from Rhizopus sp. (EC 3.4.23.4)

This enzyme is produced by different *Rhizopus* strains. *R. oligosporus* forms two proteases with optimum activity at pH 3.0 and 5.5. The *R. chinensis* protease has a molecular weight of 35000 daltons and consists of a single peptide chain with 324 amino acid residues. No histidine residues are present. The pH optimum lies between 2.9 and 3.3 and the optimum stability, between pH 3 and 6. The protease has a broader specificity than pepsin or rennin.

Microbiological Rennet Preparations

Microbiologically produced rennet has been available for the last 25 years. Whether natural rennet should be replaced was, at one time, an issue of much discussion. The introduction of a replacement was supported by the fact that pure rennet was becoming rare and expensive and the production of cheese after World War II had increased significantly worldwide. Over the last 15 years, the new products have been improved.

Now they are accepted by the dairy industry and have found their place next to natural rennet. In addition, there are ethical and religious reasons for renouncing animal rennet. Consequently, various plant proteases were also used for milk coagulation. Because these proteases do not have the specificity of natural rennet, problems occurred in the production of cheese.

These days, microbiological rennet preparations are prepared from cultures of *Mucor pusillus*, *Mucor miehei*, and *Endothia parasitica* (Table 8.9).

Table 8.9. Producers and sources of some commercial microbial rennet preparations.

Enzyme (Tradename)	*Producer*	*Source*
CHY-MAX	Pfizer Inc. [PF]	*E. coli* K-12
EMPORASE	Sanofi Bio-Industries Inc. [SBI]	*M. pusillus lindt*
FROMASE	Gist Brocades N.V. [GB]	*M. miehei*
HANNILASE	Chr. Hansen's Laboratories A/S [CH]	*M. miehei*
MARZYME	Solvay Enzymes Inc. [SOE]	*M. miehei*
Meito-RENNET	Meito Sangyo Co., Ltd. [MSJ]	*M. pusillus lindt*
MORCURD PLUS	Novo Nordisk A/S [NO]	*M. miehei*
RENNILASE	Novo Nordisk A/S [NO]	*M. miehei*

Rennet from Mucor pusillus

The *Mucor* strain for the production of the first satisfactory rennet preparation was found in 1964 by Arima, an important Japanese researcher in applied biotechnology. Because of the developmental work of Arima, the enzyme is still obtained today in very high yields from surface cultures. After extraction of the solid, mycelia-penetrated culture medium, the enzyme is purified and a solid product is obtained with ethanol precipitation or freeze drying.

Properties

1. Molecular weight: 30000 daltons.
2. Structure: Contains two cysteines without formation of a disulphide bridge.
3. Activity: pH optimum with haemoglobin pH 4; with casein, pH 4.5.
4. Stability: Optimum from pH 3 to 6.

Rennet from Mucor miehei

The strain isolated by Aunstrup is quite suitable for rennet production in submersed cultures.

Properties

1. Molecular weight: 38000 daltons.
2. Structure: Single-stranded peptide chain, contains ~6 per cent carbohydrate.

The enzyme is very similar to the enzyme in *M. pusillus*, although it differs significantly with respect to the effect of calcium on coagulation. At the same calcium concentration, the relative coagulation capacity of the *M. pusillus* rennet is 50 per cent greater.

Rennet from Endothia parasitica

The enzyme found by Sardinas gives a less specific protease action than the enzyme from *Mucor* sp., as shown by the cleavage spectrum of the oxidised insulin B-chain. Its application is limited as it readily produces bitter peptides.

Properties

1. Molecular weight: 34000–39000 daltons. Single-stranded peptide chain, without a carbohydrate fraction.

2. Isoelectric point: At pH 5.5.
3. Optimum activity: At pH 2.0–4.0. Denatured haemoglobin, pH 2.0.
4. Stability optimum: pH 4.0–4.5.

Exopeptidases from *Aspergillus* sp.

Fungi and yeasts, apart from producing a complex spectrum of endoproteases, also produce a series of exopeptidases. These include some very specific carboxypeptidases and aminopeptidases.

An aminopeptidase from *A. niger* was isolated from the mycelia of a surface culture by Uhlig, purified and characterised. The enzyme, which was stabilised by CO^{2+} ions, was produced free of protease activity. Enzymatic activity was inhibited by complexing agents, but not by iodoacetamide or PMSF. The molecular weight was 60000 daltons as determined by gel chromatography. Noteworthy of these peptidases is their ability to degrade native proteins, such as egg albumen-lysozyme and native ribonuclease, from the amino end.

The preparation (aminopeptidase-*O*) was used to sequence the amino-terminal sequence of lysozyme. An industrial application was not apparent at that time.

Today, there is interest in the use of peptidase preparations for the debittering of protein hydrolysates, in cheese ripening, and in the preparation of amino acid mixtures from proteins.

Plainer and Sprössler reported on the peptidase activity of various industrial protease preparations during their work on the debittering of protein hydrolysates. As tne peptidase activity varied with each production batch, as can be seen in Table 8.10, listed broad activity ranges.

Table 8.10. Peptidase activities of technical protease preparations from various sources.

		Relative activity							
		Amino peptidase							*Carboxy-*
Enzyme source	*Tradename*	*Ala**	*Gly**	*Leu**	*Phe**	*Pro**	*Tyr**	*Arg**	*peptidase B*
Papaya	Corolase L10	–	–	–	–	–	–	2	–
Pancreas	Corolase PP	3	2	2	3	4	4	–	3
B. subtilis	Corolase N	–	–	1	–	–	–	–	–
A. sojae	Corolase PN	2	1	5	2	2	1	3	1
A. oryzae	Corolase 7092	4	3	6	2	2	2	2	–
A. oryzae	Röhm 40–88	3	2	5	1	2	2	1	–
A. oryzae	Röhm 51–86	2	1	5	1	1	1	–	–

*As the L-amino acid *p*-nitroanilides:

Level	*Activity range (units/mg)*
1	$0–10^2$
2	$10^2–10^3$
3	$10^3–10^4$
4	$10^4–10^5$
5	$10^5–10^6$
6	$10^6–10^7$

ESTER CLEAVAGE: FAT HYDROLYSIS

Brief Overview

Fats and oils

The fats and oils that occur widely in the plant and animal kingdoms are compounds (esters) of free fatty acids with glycerol. They are known as lipids or lipoids. They play important roles in human nutrition and affect the taste of foods. They are also important as carriers of aroma compounds and influence the richness of taste and texture of foods, such as meat or baked goods. Fats and oils are used in a number of applications (Table 8.11) as commercial raw materials without any modifications. On hydrolysis, they are converted into glycerol and fatty acids, which are basic chemicals for the chemical industry. Soap is obtained by saponifying the fats with alkali.

The conversion of natural fats and oils to those with special functional properties was previously achieved only by chemical or physical methods. As enzymes are able to catalyse not only the degradation of fat but also, under specific conditions, the synthesis of fats and esters, many new possibilities have been opened up for lipid chemistry. Keywords here are enzymatic esterification, transesterification and the stereospecific cleavage of racemates by esterases and lipases.

The lipolytic enzymes have a physiological importance in the degradation and formation of lipids in organisms. Lipases are formed mainly by the pancreatic glands, while various esterases are produced in the liver. These enzymes are found in blood, muscle tissue and milk. The milk lipases are especially significant in the formation of aroma substances in milk products and cheese.

Table 8.11. Industrial use of lipids and oils.

Raw materials for laundry detergents, softeners for synthetics	~40%
Rubber and textiles	~22%
Paints and coatings	~13%
Food additives	~11%
Lubricants, flotation, metal protectants, etc.	~11%

Lipases occur in many plant cells and seeds, such as *Rhicinus*, grains and beans; they are also generated by fungi, yeasts, and bacteria. Until the late 1960s, lipase preparations in medicine were limited to use as digestive aids. Now, with the new technologies mentioned, these applications have been considerably extended. The development of new enzyme technologies, which is driven by new lipase preparations from micro-organisms, is fully under way. Also worth mentioning is the current introduction of lipases as active ingredients in detergents.

Esterases and lipases

Esterases contain a number of enzymes. These are classified according to the chemical nature of the respective acid components. Generally, esterases (EC 3.1.1.1) cleave carboxylic acids from esters according to the formula:

$$R—CO—G—R^1 + H_2O \longrightarrow R—COOH + R^1—OH$$

For this large group of enzymes, the term aliesterases is still frequently used. Simple carboxylic esters, especially those of acetic acid, are also cleaved by proteases. Reference is often made to the esterase activity of protease. Trypsin, chymotrypsin and papain all show relatively non-specific esterase activity. Besides these general esterases, there are a number of specific esterases, which are, in part, already used commercially. For example, the phospholipases are used for the specific cleavage of lecithin. If the alcohol component of the ester is a glycerol residue, the acid component is a fatty acid, and the triglyceride is water-insoluble, then the effective enzyme is a lipase (EC 3.1.1.3). There are still some problems of definition, but lipases can be described as enzymes that hydrolyse esters at the oil-water interface in an insoluble or heterogeneous system.

Lipases

Nomenclature

Glycerides are triester of glycerol with two primary groups and one secondary ester group. The two primary ester groups can be distinguished sterically. The carbon atoms of glycerol are labelled 1, 2 and 3 from top to bottom. In the Fischer projection, the secondary alcohol group is placed on the left. The triglycerides are represented in a simplified manner in Fig. 8.14. The fat acids are labelled by the first letter of their names or by the number of their carbon atoms. The number of double bonds present in a given fatty acid provides added classification.

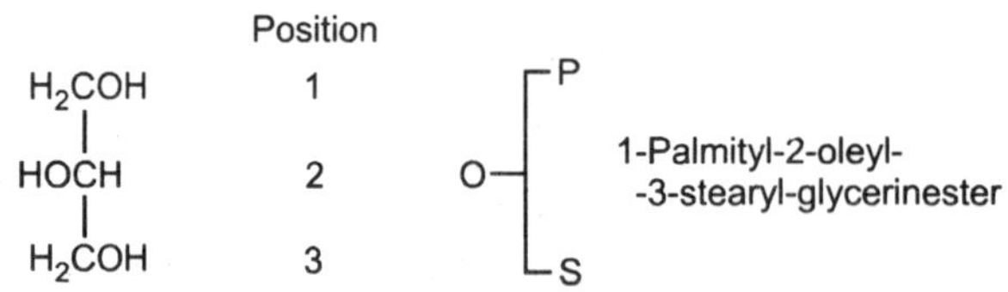

16:0 = P = Palmitic acid
16:0 = S = Stearic acid
18:1 = O = Oleic acid
18:2 = L = Linoleic acid
18:3 = Ln = Linolenic acid

Fig. 8.14. Acylglycerol nomenclature.

Substrates

Plant and animal fats and oils are substrates for the lipases. For the analysis of lipases, purified or synthetic triglycerides, such as triolein or tributyrin, are used. The latter is easier to use in routine analysis as it forms relatively stable emulsions with water or buffer solution without the use of stabilisers. It should be noted that tributyrin is also cleaved by esterases that are not real lipases according to the preceding definition.

Factors influencing lipase activity

Lipid hydrolysis depends on different parameters such as pH, temperature; water content and the phase boundary area. It is, therefore, difficult to compare the results of different authors on this subject. The pH optimum of most lipase lies between pH 7.5 and 9. Different microbial lipases are active between pH 5.5 and 8.5 and milk lipases, between pH 4.1 and 6.5. The lipase activity depends on the temperature. Lipase activity has been detected in frozen foods at –30°C.

Salts in different concentrations influence lipase activity. NaCl is needed to activate pig pancreas lipase. The presence of calcium ions is absolutely necessary for the hydrolysis of long-chain triglycerides so that the free fatty acids produced can be removed in the ionic form as their insoluble calcium salts.

Sodium taurocholate stimulates the lipolysis of triolein in a weakly alkaline medium but inhibits it in the weakly acidic range. In general, bile acids inhibit lipolysis but can, in low concentrations, stabilise the pancreatic lipases.

The colipase, a protein with a molecular weight of 10,000 daltons seems to be essential for lipolysis in the intestines of vertebrates. One possible function of this enzyme is to neutralise the bile acid inhibition of lipid cleavage. Characteristics of the pancreatic lipases are their affinity for the oil-water interface and the high rate of hydrolysis. The lipase activity depends strongly on the inside surface of an oil-water emulsion. Extreme care in the production of emulsions has to be exercised to obtain reproducible hydrolysis rates. Stable emulsions from olive oil can be produced by vigorous stirring or by homogenising the oil with an aqueous solution of gum arabic, carboxymethylcellulose or polyvinyl alcohol.

Specificity

Many lipases are unspecific and cleave various natural lipids and oils, such as olive oil, soyabean oil, coconut oil, butterfat and pork and beef fat as well as synthetic substrates, with very similar conversion rates. Nevertheless, two kinds of specificities can be distinguished:

1. Position specificity or regiospecificity.
2. Fatty acid specificity, depending on the structure of the fatty acid residues.

Regiospecificity

Two groups of lipases can be distinguished by referring to their regiospecificity. This specificity can be determined through lipase-catalysed *trans*-esterification of triolein (OOO) with palmitic acid (P).

The formation of tripalmitin (PPP) shows that the position specificity is not exhibited while the absence of PPP points to 1,3-regiospecificity. A method for determining the 1,3-regiospecificity of lipases is shown in Fig. 8.15:

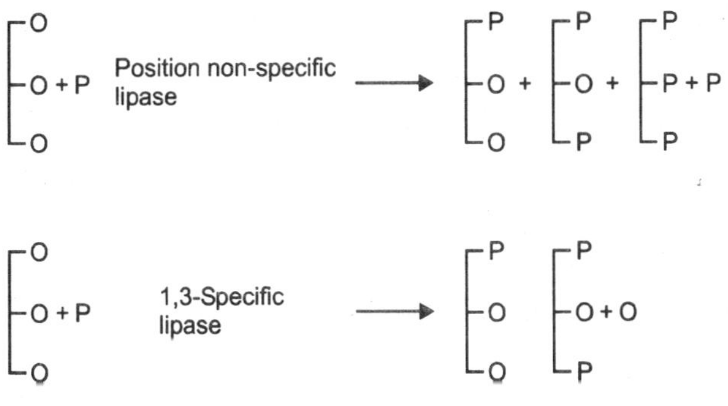

Fig. 8.15. Method for determining the 1,3 regiospecificity of lipases.

1. Non-specific lipases that cleave triglycerides into free fatty acids produce intermediate 1,2- or 2,3-diglycerides. The first reaction (I) is very fast, while the next step (II), namely, the hydrolysis into 2-monoglycerides, proceeds very slowly. The last step (III), the hydrolysis to free glycerol and fatty acid, is also a slow reaction (Fig. 8.16). The 1,2- or 2.3-diglycerides and the 2-monoglycerides are chemically unstable; these glycerides are transformed into 1,3-diglycerides or 1-monoglycerides, respectively, through an intramolecular acyl migration. For technical use, the most important representatives of this group are the lipases from *Candida cylindracea* and *Humicola lanuginosa*. Bühler and Wandrey obtained 95 per cent cleavage of soyabean and olive oil with a lipase preparation from *Candida cylindracea* under technically feasible conditions.
2. The 1,3-specific lipases, including the pancreatic lipase, can completely hydrolyse certain triglycerides. This 1,3 specificity occurs relatively frequently in the microbial lipases, such as

in the lipases from *A. niger*, from *R. delemar*, from different *Mucor* species and from *Pseudomonas* sp. According to the investigations of Slotboom, microbial lipases, particularly those from *R. arrhizus*, show stereospecificity with respect to positions 1 and 3 of the glycerol residue.

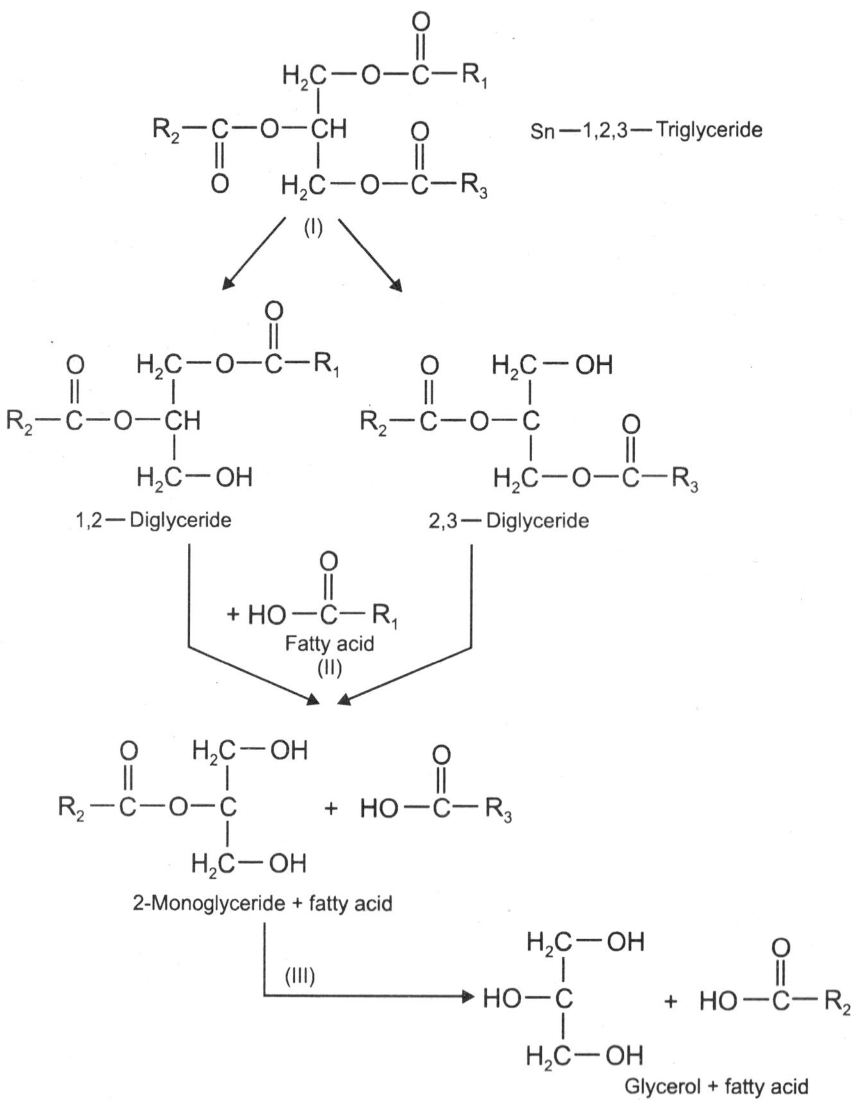

Fig. 8.16. Reaction steps in triglyceride hydrolysis.

Fatty acid specificity

It is common to investigate fatty acid specificity using the alkyl esters of the fatty acids as substrate. However, the speed of the reaction—which is very slow if triglycerides are used as substrate—must be considered. The problem with triglycerides is that some are liquid at normal reaction temperature, whereas those with long-chain fatty acid residues are solids. The lipase from *Geotrichum candidum* shows a remarkable fatty acid specificity. This lipase preferably liberates long-chain fatty acids with a *cis* double bond in the 9 position from triglycerides. Thus, the methylesters from oleic acid, linoleic acid, and linolenic acid are hydrolysed relatively fast, while the esters of the saturated fatty acids such as palmitate and stearate are attacked very slowly. The triglycerides-triolein (*cis*-d_9-18:1), trilinolin (*cis*-$d_{9,12}$-18:2) and trilinolen (*cis*-$d_{9,12,15}$-18:3) are cleaved relatively rapidly. Tributyrin, trimyristin (14:0), and tristearin (18:0) are poor substrates.

Many microbial lipases show only a low specificity in the hydrolysis of natural oils and fats. Fish oils and butterfat are the exception. Differences in the specificity of different *Aspergillus lipases* can be observed; some prefer to cleave short-chain fatty acids whereas others prefer the long-chain fatty acids.

Animal lipases

Pancreatic lipase

Willstätter and Memmen isolated a lipase from pig pancreas, although complete purification of the enzyme was not accomplished until many years later by Verger. Problems were encountered as a result of the instability of the enzyme during the purification steps. Stability is improved if the activity of the pancreas protease can be inhibited during the purification procedure. The specific activity with respect to defatted pancreas powder was improved by a factor of 30.

Properties

1. Molecular weight: Hog pancreatic lipase, 45000 daltons, 50000 daltons (UC). Two species exist: LA, a more acidic protein; and LB, a more alkaline protein. Both exhibit the same specific activity. Beef pancreas lipase, 48000 daltons.
2. Isoelectric point: At pH 4.9–5.0.
3. Structure: The enzyme molecule contains six disulphide bridges and two free SH groups. One of the SH groups appears in the vicinity of the active site, but does not seem to be necessary for the catalytic activity. SH reagents in low concentrations do not influence the activity. Because of the inhibitory action of organophosphates, it was concluded that the enzyme was of a serine type. As the activity is also inhibited or destroyed by photo-oxidation, it was concluded that a histidine participates in the catalytic activity.

Pancreatic lipase requires NaCl in order to be active (7 mM for optimal activity). Even though many substances influence the lipase activity, no co-factors are necessary. The role of calcium, the influence of bile acids and the co-lipase have already been mentioned.

Lipase action

According to the model by Brockerhoff, the pancreatic lipase, on approaching a 'supersubstrate', such as the surface of an oil droplet, is bound in the correct orientation for the reaction. The hydrophobic area of the enzyme is linked to the hydrophobic substrate. The active site lies close to this area but is also accessible to hydrophilic regions because water is necessary for the hydrolysis of the intermediate acylenzyme (Fig. 8.17). Specificity: Analysis of the lipolysates from glycerol-1-oleate-2,3-dipalmitate and from 1,2-dipalmitate-3-oleate showed that the two primary positions, 1 and 3, are cleaved equally. Pancreatic lipase does not cleave stereospecifically. The enzyme is regiospecific for the primary ester position (1,3). If the glycerides are present in the emulsified form, the short-chain fatty acids (*tributryn*) are hydrolysed faster than the long-chain ones. A pronounced structure specificity does not exist.

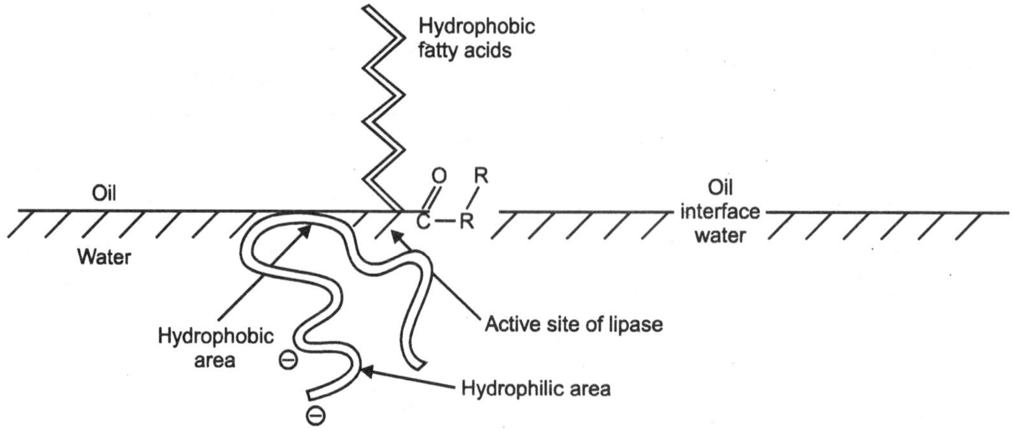

Fig. 8.17. Orientation of pancreatic lipase at an oil-water interface (model according to Brockerhoff).

Technical preparation

Pancreatin preparations are used in technical as well as medical applications. The lipase activity is standardised according to the analytic procedures of the FIP or the NF. There are pancreatins with a six- to eight-fold lipase NF activity, or 30000–35000 FIP units.

One preparation with an industrial application, 7023 C [RM], contains ~8000 FIP units. In all preparations, the lipase is available only in a partially soluble state. A significant part is bound to the fibres of the pancreatic tissue and is insoluble until contact is made with the substrate; thus, the enzyme acts as if it were immobilised. Such preparations are ideally suited for conversions in minimally aqueous media.

Pre-gastric lipases or lipase esterases

These enzymes are obtained from the parotid glands of calves, goats and lambs by extraction with glycerol or salt solutions. Molecular weights lie between 1,50,000 and 1,72,000 daltons. The optimal activity range stretches from pH 4.8 to 5.5 at 32–42°C. The optimal activator is 0.5 M NaCl. The regiospecificity targets preferentially the Sn-3-position of the glycerol. In the case of butterfat, the short fatty acids C_4–C_8 are primarily cleaned off.

The lipase from pre-gastric calf glands exhibits the following fatty acid specificity, with the activity determined with a tributyrin standard at 100: triacetin = 2; tricaproin = 113; tricapryl = 58; tricaprin = 49; trilaurin = 13; trimyristin = 1; tripalmitin = 9; tristearin = 0. *Aspergillus* and *Mucor* lipases show similar specificities. Comparison with the pancreas and fungal lipases reveals pronounced differences in the specific hydrolysis of mono- and dibutyrins, relative to activity with tributyrin at 100 (Table 8.12):

1. Activity: Tributyrin, tricaproin, β-naphthylbutyrate, or milk fat, emulsified in 10 per cent gum arabic, are used as substrates. The pH optimum of these esterases lies within pH 4.5–5, at 37°C to a maximum of 40°C. Optimal stability is pH 5.0–7.0. In the United States, these preparations are used to enhance aroma and flavour in the production of Italian cheeses, such as provolone or romano. The activity of gastric lipases obtained from calf stomachs differ from pre-gastric lipases. A combination of the two enzymes is used in the manufacture of provolone and cheddar cheese.
2. Products and manufacturers: From animal organs: Italase [SOE], Kapalase [SOE], Picantase [GB], Lipase PGE [AM]. From micro-organisms: Palatase [NO], Lipolact [RM].

Table 8.12. Relative fatty acid specificity of lipases from different sources.

Enzyme source	*Tributyrin*	*Dibutyrin*	*Monobutyrin*
Pre-gastric esterase	1.00	0.04	0
Pancreatic lipase	1.00	0.74	0.95
Fungal lipase	1.00	0.26	0.15

Microbial lipase preparations

Lipases are produced by numerous bacteria and fungi. Until about 15 years ago these enzymes were known only as preparations for scientific research. Later, Japanese producers started to offer lipases from *Rhizopus* or *Aspergillus* to the pharmaceutical industry as substitutes for pancreatin. With the use of this new technology, several new lipase preparations from different micro-organisms have been developed. The characteristics of some of these enzymes, now produced on an industrial scale, are discussed in the following section.

Lipases from *Aspergillus* sp.

Several lipases from *A. niger* with different specificities are known. An enzyme isolated by Fukumoto following eight-fold enrichment and recrystallisation has been well described. The enzyme has an optimal activity at pH 5.6 at ~25°C and hydrolyses triglycerides without regiospecificity.

Two isozymes from *A. niger* were isolated and characterised by Hofelmann. Some typical properties are given in Table 8.13 (see Figs 8.18 and 8.19).

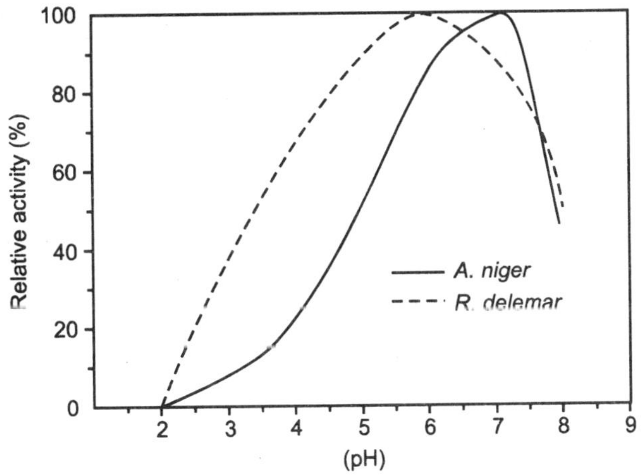

Fig. 8.18. Effect of pH on the activities of lipases from *A. niger* and *R. delemar* [AM].

During butterfat hydrolysis, scientists discovered differences in aroma development and aroma spectra between the two enzymes. Other *A. niger* lipases with molecular weights between 22000 and 25000 daltons have been described by various authors. The classification of technical preparations is, however, difficult. Products and manufacturers: Lipolact [RM], Lipase AP [AM], Lipozyme [NO].

Lipases from *Mucor* sp.

Mucor miehei produces a lipase with 1,3 regiospecificity, which cleaves off short- and long-chain fatty acids; 70 per cent of the activity is reached between pH 7 and 8.2. The optimal temperature is 40°C. Up to 93.7 per cent of olive oil can be hydrolysed.

Table 8.13. Characteristics of two lipase isozymes from *Aspergillus niger.*

Characteristic	*Lipase I*	*Lipase II*
Molecular weight (daltons)	31000	19000
pI	4.0	3.5
pH optimum	5–6	5–6
Substrate	Long-chain fatty acids	Short-chain fatty acids
Hg^{2+}	No influence	Inactivates
Iodoacetamide	Inhibits	No influence

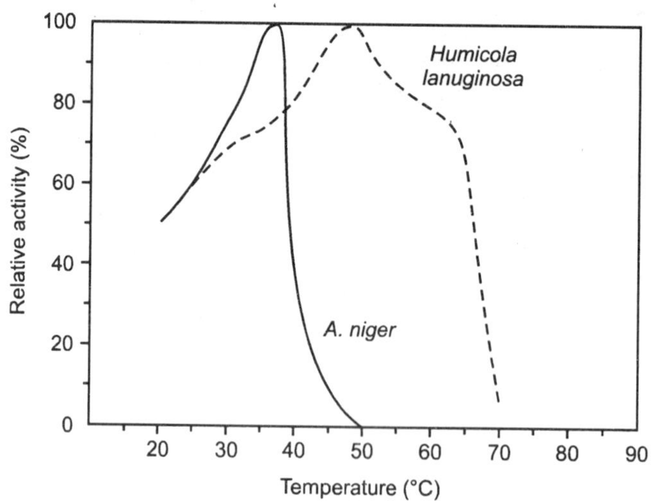

Fig. 8.19. Effect of temperature on the activities of lipases from *A. niger* (lipase AP [AM]) and *Humicola lanuginosa* [NO].

This lipase is a good catalyst for transesterification in the 1,3 position. *Mucor javanicus* (molecular weight 21000 daltons, 2.6 per cent carbohydrate) is another strain used for the production of lipase products.

Manufacturers: [AM], [NO].

Lipase from Candida cylindracea

1. Molecular weight: 1,20,000 daltons, isoelectric point at pH 4.2, 4.2 per cent carbohydrate.
2. Structure: The amino acid composition was determined by Tomizuka following crystallisation of the enzyme.
3. Activity: Range pH 5–8.5. Optimum (90 per cent) pH 7–8, optimum temperature ~45°C.
4. Specificity: Regiospecificity does not exist; short-chain to long-chain fatty acids are hydrolysed at comparable speeds.
5. Product and manufacturers: Lipase AY: [AM], [SA].

Lipases from Rhizopus sp.

A series of lipases from *Rhizopus* has been described. These include *R. arrhuzus*, *R. javanicus*, *R. niveus*, *R. delemar* and *R. rhizopodiformis*. Of these, the lipase from *R. arrhizus* has been characterised in most detail.

The other enzymes show a 1,3 specificity, the activity optimal lies within pH 5.0–7.0 and the optimum temperature range is 30°–45°C.

Lipase from R. arrhizus

1. Molecular weight: 43000 daltons, isoelectric point at pH 6.3.
2. Structure: Autolysis of the purified lipase gives a glycopeptide and a carbohydrate-free protein with molecular weights of 8500 and 32000 daltons, respectively. The latter shows full lipase activity. The glycopeptide contains 13–14 per cent mannose and 2 mol of hexosamine per mole.
3. Activity: The pH activity range and the specificity are similar to that of the pancreatic lipases. Fatty acids are cleaved off the primary ester groups. There is no stereospecificity with respect to the sn-1 or sn-2 positions.

Lipase from R. delemar

The lipase preparation purified by Mohsen had a molecular weight of 48720 daltons. For activity, see Fig. 8.18.

Product and manufacturer: Lipase D [AM].

Lipases from Penicillium sp.

The most important lipase source is *P. roquefortii*, which forms at least two different exocellular enzymes, both of which are 1,3-specific. Both lipases cleave, in order of decreasing rate, tributyrin, tricapryl, tricaprin, trimyristin and triolein. The two lipases differ with respect to optimum activity. The acidic lipase has optimum activity at pH 5.5; the alkaline lipase, at pH 8.0.

A. *Penicillium* lipase has been described in which synthesis reactions, such as esterification, shows a strong specificity for the sn-2 position. However, with respect to hydrolysis, the enzyme is non-specific, which is presumably related to the different reaction conditions.

Manufacturer: [AM].

Lipase from Pseudomonas sp.

Pseudomonas aeruginosa, the enzyme purified by Stuer, has a molecular weight of ~29000 daltons and isoelectric point at pH 5.8. The *Pseudomonas* lipases are recommended for application in detergents because of their activity and stability in alkaline pH levels.

Manufacturer: [AM].

Lipase from a transformed fungal strain

According to the manufacturer's information, the donor strain was an alkalophilic fungal strain. The lipase gene was inserted into the *Aspergillus oryzae* genome. A particular property of this enzyme is its high activity at pH 10–pH 11, as well as an adequate stability at 50°–55°C in this pH range. This lipase is manufactured for use in detergents.

Manufacturer: [NO].

Lipase from Humicola lanuginosa

Omar described an enzyme from *Humicola lanuginosa* with enhanced temperature stability that was suitable for the hydrolytic as well as the esterification reaction. The need for a higher temperature stability is due to the fact that hard fats such as tallow are solid at 25–30°C and are difficult to hydrolyse, and that syntheses involving palmitic or stearic acid should be performed above their melting-point temperatures.

The optimum conditions for hydrolysis are at 45°–50°C, at pH 7.0 (in buffer) and a water content of at least 60 per cent. For esterification, the optimum is at 45°C, pH 7.0, and 8 per cent water. The enzyme is relatively specific for C_{12} to C_{18} fatty acids.

No synthesis is observed for C_2 to C_8 fatty acids. The best solvents for synthesis reactions are *n*-octanol, *n*-decanol, and isoamyl alcohol.

Phospholipases

Lecithin

The phospholipid, lecithin, is indispensable in food processing technology. It is found in many plants and animal organs and can be obtained from soyabean, sunflower, rapeseed, and corn oils. Gobley, in the nineteenth century, isolated lecithin from egg yolk. The name is derived from the Greek word for egg yolk, *lekitos*.

The lecithin used today in industry and food processing is obtained from raw soyabean oil. Lecithin swells if water is added at high temperature and can be separated from the oil phase by centrifugation. The raw lecithin contains, besides 60–70 per cent polar lipids, approximately 27–37 per cent soyabean oil and 0.5–1.5 per cent water.

Phospholipids are insoluble in acetone; therefore, oil-free powder products can be produced from raw lecithin by means of acetone precipitation.

Structure of lecithins and enzymatic attack

Phospholipids are cleaved in different ways by two groups of enzymes. These enzymes are lipases (phospholipase A_1 and A_2) and lecithin-phosphodiesterases (phospholipase C and D) (Fig. 8.20).

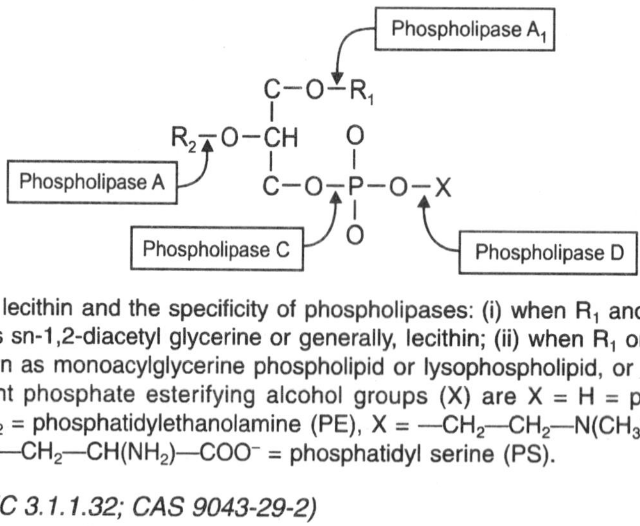

Fig. 8.20. Structure of lecithin and the specificity of phospholipases: (i) when R_1 and R_2 are fatty acids, the compound is known as sn-1,2-diacetyl glycerine or generally, lecithin; (ii) when R_1 or R_2 is an H atom, then the compound is known as monoacylglycerine phospholipid or lysophospholipid, or simply lysolecithin; and (iii) the most important phosphate esterifying alcohol groups (X) are X = H = phosphatidic acid (PA), X = $-CH_2-CH_2-NH_2$ = phosphatidylethanolamine (PE), X = $-CH_2-CH_2-N(CH_3)_2-CH_3$ = phosphatidyl choline (PC), and X = $-CH_2-CH(NH_2)-COO^-$ = phosphatidyl serine (PS).

Phospholipase A_1 (EC 3.1.1.32; CAS 9043-29-2)

This enzyme is C-1-(α-)-specific and is found in the venoms of snakes, scorpions, and bees, as well as in the pancreas and in micro-organisms. It requires Ca^{2+} for activity. At one time it was thought that the action of these venoms was traceable to the phospholipase. However, in the case of rattlesnake venom, the enzyme-free toxin, crotoxin, was later separated from the enzyme. Alpha- or sn-2-lysolecithins that are formed by the action of phospholipase A_1 are capable of lysing erythrocytes.

Phospholipase A_2 (EC 3.1.1.4; CAS 9001-84-7)

This enzyme occurs in pancreatic glands as its inactive precursor, pro-a-phospholipase A_2. It is activated by trypsin, hydrolysis of an arginine-alanine linkage, and the release of a heptapeptide.

It is isolated as a technical enzyme for the production of lysolecithin from porcine pancreatic glands.

Properties, product and manufacturer

The enzyme has a molecular weight of, 14000 daltons. It consists of a peptide chain of 123 amino acids, which are linked by six disulphide bridges. The isoelectric point is at pH 6.3.

1. Activity: Calcium is the cofactor for this metalloenzyme. The pH-optimum at 40°C lies between pH 8 and 9, the optimal temperature at pH 8.0 is about 50°C.
2. Stability: See Fig. 8.21.
3. Product and manufacturer: Lecitase [NO].

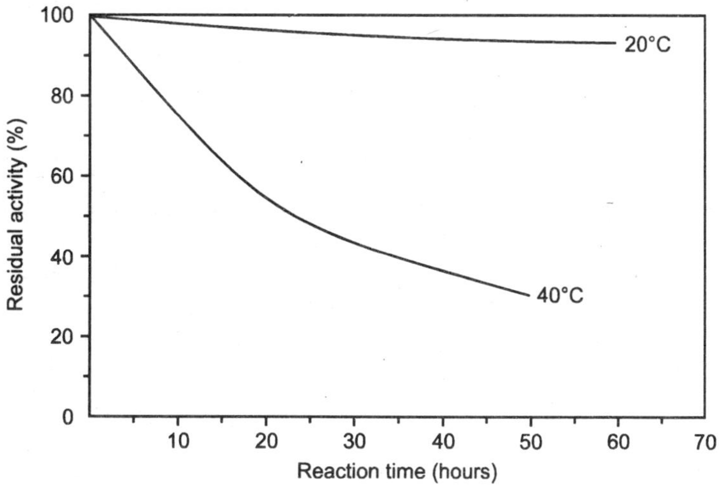

Fig. 8.21. Effect of temperature on the stability of phospholipase A-2 [NO] (enzyme concentration: 1 mg/ml; 0.2 M Tris HCl, pH 8.0; 0.01 M $CaCl_2 \cdot 2H_2O$).

Phospholipase C

This enzyme occurs in *Clostridium* sp., especially *Clostridium perfringens* (*C. welchii*). The molecular weight is between 20000 and 24000 daltons, and the pH optimum is around neutral. Other enzymes with equal specificities are isolated from *A. oryzae*, *B. cereus* and *Pseudomonas* sp., as well as from animal tissue.

Phospholipase D

This enzyme occurs in cabbage and carrot leaves and is also produced by *Corynebacteria.* It requires calcium for its activity. The plant enzyme has a pH optimum of pH 5.6–5.8, the microbial enzyme from pH 7–8. The latter has a molecular weight of about 90000 daltons.

Phytase (EC 3.1.3.16; CAS 9001-89-2), phytate-6-phosphatase

The hydrolytic release of phytic acid from myoinositol hexaphosphate ester was the first phosphatase-catalysed reaction known. Because of the nutritional significance of phytase, especially in dietary fibre-rich materials, some aspects of the enzyme will be detailed here.

Phytate occurs in plant materials, the endosperms of different cereals (particularly wheat bran), and also in immature potatoes, where it is bound to starch. It serves as a phosphate reserve for the seeds and grains. Corn contains about 1 per cent; sesame, about 5 per cent phosphorous. The calcium-magnesium salt is called phytin. The highly ionised orthophosphate groups form salts with cations and complexes with proteins. The poor availability of mineral salts, especially calcium and magnesium, of certain protein hydrolysates could be attributable to the formation of protein-phytate-mineral salt complexes. Other substances, such as fibrous bulk fillers, oxalates and phenolic substances, may also be responsible for the poor availability. Nevertheless, the presence of phytate seems to be a factor in the antinutritive properties of beans and bran products. In World War II, there was a calcium deficiency illness in children, which was associated with phytate intake. This deficiency was compensated for with bread that was produced from debranned wheat that is low in phytate, and by addition of calcium.

Erdmann and Frobes showed that in rats, piglets, and chickens, the availability of phosphate, calcium, iron, and zinc was limited. Also in human nutrition, diets containing large amounts of unfermented cereals with high phytin contents can be a problem.

Phytase occurs in cereals, in the blood of amphibians and birds, in the intestinal membranes of rats, and in innumerable micro-organisms. The phytase from wheat bran is inhibited by an excess substrate and by phosphate ions. Phosphatides do not affect the enzyme. The phytase from wheat bran shows optimal activity from pH 7 to 7.5. This enzyme can be activated by autolytic treatment of wheat bran (6 hours, 50°C, pH 4.5). In breads made from flour to which such pre-treated bran has been added, calcium, magnesium, and phosphate are more available than without such pre-treatment. In bread production without such phytate-rich materials—that is, without an excess of bran—the small amount of phytate disappears as a result of the action of the native wheat enzymes and enzymes in the yeast.

The claim that phytate is an antinutritive substance is questionable because the intestines of most monogastrics, including that of humans, contain phytase. This activity is usually weak, but is induced by phytate. Calcium inhibits phytase activity. Processes have been developed, for example, with beans, which reduce phytate by autolytic processes. Incubation for ~24 hours and a temperature of ~50°C is required. The phytate content of beans is also lowered by germination. A phytase, which was obtained by Rojas and Scott from *Aspergillus ficuum*, was used for the hydrolysis of phytate in linseeds.

Product and manufacturer: A technical phytase is now available [AO], [GB].

Chapter 9

Lipases

INTRODUCTION

As were most enzymes applied in food processing, lipases originally were used for preparation of foods in an entirely empirical fashion. The early users of lipases were unaware of the involvement of lipases in traditional, well-established processes for food preparation and processing. This statistic applies to the use of micro-organisms for development and enhancement of flavours in cheese manufacturing, in which lipases generated by the applied micro-organisms play a key role by liberating flavour precursors or actual flavour compounds from esters present in milk fat.

The early uses of enzyme preparations recognised as lipases were developed mainly to mimic such established uses of lipases while improving their reproducibility and, therefore, the consistency of products made by processes involving these lipases. Preparations of lipases applied in this fashion initially were derived from animal sources and later from industrial fermentation of micro-organisms. Advances achieved using such lipase preparations, in addition to the reproducibility of food processing, included the possibility of using alternative raw materials for cheese production or of modulating the characteristics of the finished product to meet market requirements.

These early developments in the use of lipases in food processing and manufacturing are quite important from the perspective of the consumer, the manufacturer and the enzyme producer. More recent developments in lipase technology, however, must be expected to exert an even stronger influence on the food industry. Currently, lipase technologies are so advanced that production of entirely new food ingredients must be considered industrially feasible, as must the production of a wide variety of compounds, such as flavours and emulsifiers that are of importance to the food sector.

SOURCES OF MICROBIAL LIPASES

The number of known microbial species that produce lipases is rather large and is, in fact, still increasing. The most important of these lipase-producing micro-organisms are listed in Table 9.1. In Table 9.2, important industrial sources of lipases are given, with the actual commercial producers of the enzymes. Many of these industrial lipases have been developed for uses outside the food industry. Some enzymes were selected and industrialised because of their possible application in the detergent industry whereas others were developed because of their potential as catalysts in organic chemical processing. Table 9.3 lists major application areas of lipases as well as the most important uses of the enzymes in the food sector. Several discussions of the mode of action of lipases, their various uses in and out of the food area, and their properties have been published, as excellent reviews that discuss most aspects of these enzymes.

Table 9.1. Sources of microbial lipases.

Micro-organism
Absidia corymbifera
Absidia hyalospora
Achromobacter sp.
Achromobacter lipolyticum
Acinetobacter calcoaceticus
Acinetobacter pseudoalcaligenes
Alcaligenes sp.
Alcaligenes denitrificans
Amylomyces rouxii
Arthrobacter sp.
Aspergillus awamori
Aspergillus flavus
Aspergillus fumigatus
Aspergillus japonicus
Aspergillus niger
Aspergillus oryzae
Bacillus cereus
Bacillus megaterium
Bacillus stearothermophilus
Candida antarctica
Candida auricularia
Candida curvata
Candida cylindraceae
Candida deformans
Candida foliorum
Candida humicula
Candida lipolytica
Candida rugosa
Candida tsukubaensis
Chaetomium thermophile
Chromobacterium chocolatum
Chromobacterium viscosum
Corynebacterium acnes
Flavobacterium arborescens
Fusarium oxysporum
Fusarium solari
Geotrichum candidum
Humicula grisea

(Contd...)

Micro-organism
Humicula insulens
Humicula lanuginosa
Lactobacillus sp.
Leishmania donovani
Malbrancheae pulcella
Micrococcus freudenreichii
Mucor sp.
Mucor javanicus
Mucor lipolyticus
Mucor miehei
Mucor pusillus
Myxococcus xanthus
Penicillium crustosum
Penicillium cyclopium
Penicillium roquefortii
Phycomyces nitens
Pichia miso
Propionibacterium acnes
Propionibacterium granulasum
Proteus sp.
Pseudomonas sp.
Pseudomonas aeruginosa
Pseudomonas fluorescens
Pseudomonas fragi
Pseudomonas pseudoalcaligenes
Pseudomonas stutzeri
Rhizopus sp.
Rhizopus arrhizus
Rhizopus chinensis
Rhizopus delemar
Rhizopus japonicus
Rhizopus microsporus
Rhizopus niveus
Rhizopus nodosus
Rhizopus oligosporus
Rhizopus oryzae
Rhodotorula minuta
Rhodotorula pilimonae
Saccharomyces fragilis

(Contd...)

Micro-organism
Saccharomyces lipolytica
Schizosaccharomyces pombe
Sporotrichum thermophile
Staphylococcus aureus
Staphylococcus carnosus
Streptococcus lactis
Streptomyces coelicolor
Streptomyces fradiae
Streptomyces panayensis
Talaromyces thermophilus
Thielavia minor
Torula thermophila

Table 9.2. Industrial sources of lipases.

Lipase	*Source*
Achromobacter sp.	Meito Sangyo
Alcaligenes sp.	Meito
Arthrobacter sp.	Sumitomi
Aspergillus niger	Amano
	Novo Nordisk
	Röhm
Candida cylindraceae	Amano
	Enzyme Development
	Meito
Chromobacterium viscosum	US Biochemicals
	Toyo Jozo
Humicula languinosa	Amano
	Novo Nordisk
Mucor miehei	Amano
	Gist Brocades
	Röhm
	Novo Nordisk
Phycomyces nitens	Takeda
Pseudomonas sp.	Amano
Pseudomonas aeruginosa	Amano
Pseudomonas fluorescens	Amano
Rhizopus sp.	Serva
	Nagase
Rhizopus arrhizus	Precibio (France)

(Contd...)

Lipase	*Source*
	Boehringer-Mannheim
	Rapidase
	Gist Brocades
Rhizopus delemar	Chemical Dynamics
	Tanabe
Rhizopus japonicus	Amano
	Nagase
	Osaka Saiken
Rhizopus niveus	Amano
Rhizopus oryzae	Amano

Several lipases have been cloned and expressed in industrial yields in hosts that are particularly amenable to industrial fermentations. Lipases derived from such fermentations are, in some cases, of very high purity, for example, Lipolase, which was marketed in 1988 as a detergent lipase. The lipases of *Mucor miehei* and *Candida antarctica* also have been cloned and expressed in industry-friendly hosts. The experience gained from lipase manufacturing using genetically engineered strains increases the likelihood that industrially important lipases of the future will be produced in this fashion. The purity and price of lipases manufactured using genetically engineered strains are unlikely to be surpassed by lipases produced using conventional techniques.

LIPASES IN FOOD MANUFACTURING PROCESSING

For some important applications of lipases in food manufacturing and processing, immobilised rather than free lipases are required. The lipases listed in Table 9.2 may be immobilised for such processes, if necessary. Some lipases are manufactured and marketed in an immobilised state that is well suited to application in food-related processes such as modification of fats and oils. The immobilised lipase of *Mucor miehei*, marketed under the trade name Lipozyme, is an example of an immobilised lipase available on an industrial scale.

Table 9.3. Industrial application areas for microbial lipases.

Industry	*Effect*	*Product*
Dairy	Hydrolysis of milk fat	Flavour agents
	Cheese ripening	Cheese
	Modification of butter fat	Butter
Bakery	Flavour improvement and shelf life prolongation	Bakery products
Beverage	Improved aroma	Beverages
Food dressing	Quality improvement	Mayonnaise, dressings, and whipped toppings
Health food	Transesterification	Health foods
Meat and fish	Flavour development and fat removal	Meat and fish products
Fat and oil	Transesterification	Cocoa butter, margarine
	Hydrolysis	Fatty acids, glycerol, mono- and diglycerides
Chemical	Enantioselectivity	Chiral building blocks and chemicals
	Synthesis	Chemicals

(Contd...)

Industry	*Effect*	*Product*
Pharmaceutical	Transesterification	Speciality lipids
	Hydrolysis	Digestive aids
Cosmetic	Synthesis	Emulsifiers, moisturing agents
Leather	Hydrolysis	Leather products
Paper	Hydrolysis	Paper products
Cleaning	Hydrolysis	Removal of cleaning agents, e.g. surfactants

Some of the most fascinating developments in the application of lipases in the food industry concern the use of lipases for the modification of triglycerides. A key factor in this context is the specificity of the enzymes for the position and the nature of the acyl residues in triglyceride molecules. Some lipases preferentially attack the 1 and 3 positions in triglyceride molecules, whereas other lipases exhibit no preference for any of the three acyl residues. In addition, most lipases show some specificity based on the nature of the acyl residue in triglycerides by preferentially attacking unsaturated fatty acids or short chain fatty acids. Some examples of 1,3-specific lipases are the *Mucor miehei* lipase and Lipolase, which cleave the 1- and 3-ester bonds in triglycerides almost exclusively. Human pancreatic lipase and human lipoprotein lipase likewise exhibit 1,3-specificities.

The possible industrial exploitation of these specificities has been recognised since the late 1960s in the fat and oil industry. A major target has been the production of new triglyceride types with, for example, desirable melting properties by interesterification of readily available triglycerides. For comparison, see the example shown in Fig. 9.1. This technological potential of lipases first was envisioned by Macrae and the Unilever research team, as was the possibility of applying lipases in non-aqueous environments for industrial purposes—a requirement for the processes depicted in Fig. 9.1 since lipases, in the presence of water, will hydrolyse triglycerides rapidly.

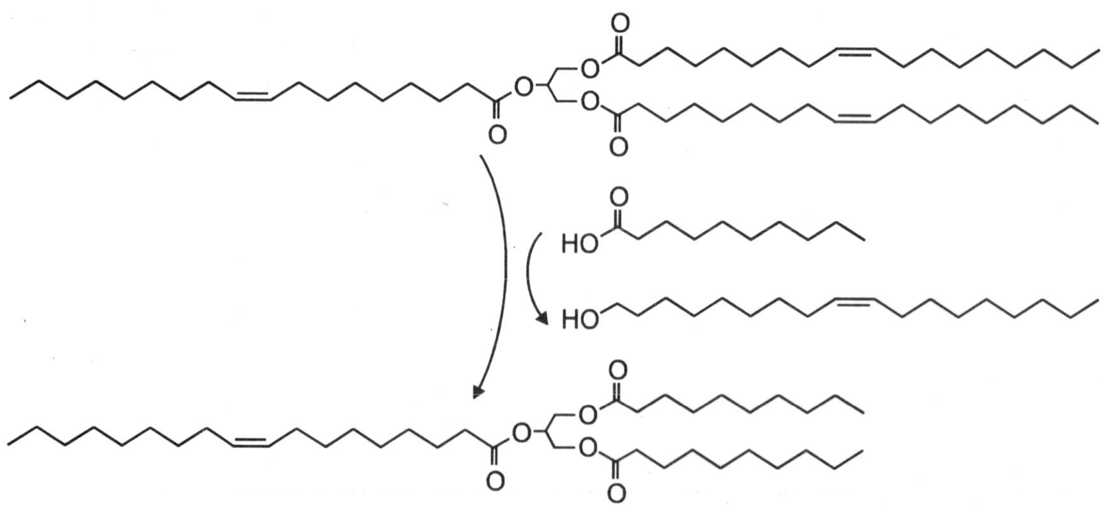

Fig. 9.1. Example of triglyceride with desirable melting properties.

One of the crucial issues for the realisation of the Unilever process was the development of immobilised preparations of lipases that exhibit the desired specificities. One such product, Lipozyme, is currently available in industrial quantities. New improved immobilised lipase preparations are being developed. Many discussions have been published that address various aspects of improved biocatalysts for the Unilever process, for example, new carrier types, chemically modified lipases soluble in organic solvents, and lipases entrapped in hydrophobic polymers. Also, several reviews discuss improvements of the actual process design by using loop reactor systems, membrane reactor types, batch reactor systems, microemulsion techniques, and packed-bed reactor systems. Packed-bed systems are currently the preferred method for interesterification of triglycerides. Collectively, the technological problems in the use of microbial lipases for triglyceride modification have been solved to the extent that interesterification of triglycerides can be considered an industrially feasible process. These technological advances, as represented by Lipozyme, must be seen as a breakthrough in industrial enzymology, since there are no previous examples of applications of enzymes in non-aqueous environments that actually generate products in industrial yields (10 tonnes product/1 kg immobilised biocatalyst). The actual application of the process for triglyceride modification is a strategic issue and will depend on the developments of competing technologies, such as plant genetics, and on the availability of natural materials that match the performance of biosynthesised materials.

The interesterification process is likely to be used first in the preparation of high value materials for use as speciality nutrients or as pharmaceuticals. It is thus conceivable that, with the aid of lipase technologies, it will be possible to make fats rich in essential fatty acids, for example, DHA (4,7,10,13,16,19- all *cis*-docosahexaenoic acid), EPA (eicosapentaenoic acid), arachidonic acid, linoleic acid, or linolenic acid. These fats would exhibit improved food properties by increasing bioavailability of essential fatty acids over naturally occurring materials such as fish oils. Triglyceride molecules that currently are expected to exhibit these highly desirable properties will carry the essential fatty acid in the 2 position and short-chain fatty acids (i.e. C_8 and C_{10} carboxylic acids) at the C-1 and C-3 positions of the triglyceride molecules. These lipids (so-called structured lipids) will be hydrolysed at an increased rate in the digestive tract because of the increased activity of the pancreatic lipase toward short-chain fatty acids and, thus, will provide an easily absorbable monoglyceride that contains the essential fatty acid. Structured lipids carrying DHA may, in the future, become of prime importance in infant nutrition whereas structured lipids containing EPA may become an important remedy for the general public for prevention of heart diseases.

STRUCTURED LIPID WITH THE AID OF LIPASES

Structured lipids manufactured with the aid of lipases are likely to become important in connection with total parenteral nutrition (TPN) of hospitalised patients. Currently, triglyceride emulsions based on sunflower oil are used extensively to supply energy efficiently in a TPN regime. Some disadvantages of the currently applied emulsions may be reduced by using structured lipids instead. These molecules are hydrolysed more rapidly than their non-structured counterparts by the 1,3-specific lipoprotein lipase that hydrolyses fat emulsions in blood vessels. Structured lipids will give rise to more hydrophilic fatty acids, which are distributed more easily in the body, whereas the essential fatty acid in the 2 position of the triglyceride molecule will be readily bioavailable as its monoglyceride after removal of the short-chain fatty acids by lipoprotein lipase. Moreover, these short-chain fatty acids can be transported directly to the liver and used there for energy generation. In addition, use of structured lipids will lessen the overload with essential fatty acids encountered by patients in connection with the use of traditional lipid emulsions over extended periods of time. Finally, lipase technologies make possible the design of triglyceride molecules with precise compositions as specified by the demand of the diseased patient.

Because of the central involvement of the essential fatty acids as precursors of prostaglandin biosynthesis, they may even be used to regulate the diseased state. Structured lipids must be expected to have an exciting future in the area of TPN and probably will be among the first examples of entirely new food molecules designed on the basis of lipase technologies.

Application of Lipases in Edible Fats and Oils

The potential application of lipases to hydrolysis of edible fats and oils has been the subject of several studies and has been found to be technologically feasible. However, this technology is subject to competition from other technologies such as steam splitting. Application of the microbial lipases to fat splitting therefore is likely to occur mainly for treatment of very valuable materials such as triglycerides that contain valuable and sensitive polyunsaturated fatty acids, since these are gaining increasing importance in the area of nutrition and as pharmaceutical products, as mentioned earlier.

In the field of synthesis of food ingredients such as emulsifiers and flavour molecules, interest in the application of lipases is increasing. This change is due not only to the superiority of lipases over traditional methods for some esterification reactions but also to the safety of the enzymatic procedures and the natural quality of materials synthesised using biocatalysts. The concept of 'green' chemistry and of 'green' products ('bioesters') appears to appeal not only to the general public but also to the authorities. Therefore, compounds manufactured with the aid of lipases are likely to gain importance in the future.

The power of lipases in the synthesis of esters of interest to the food industry is illustrated by the process shown in Fig. 9.2. This process is based on the best currently available lipase for ester synthesis—the lipase produced by *Candida antarctica*. The example shown illustrates how lipases can be applied for regioselective esterification of polyfunctional alcohols. The process shown is carried out simply by mixing the reactants in the presence of the enzyme and removing water from the reaction mixture over the course of the reaction. The process shown is also an example of how new, potentially useful chemicals become abundantly available because of the developments of lipase technologies. The glycoside esters shown are excellent surfactants and emulsifiers and can be prepared at low cost using the lipase-based process. The products are natural, since the process for their synthesis and the raw materials used are also natural.

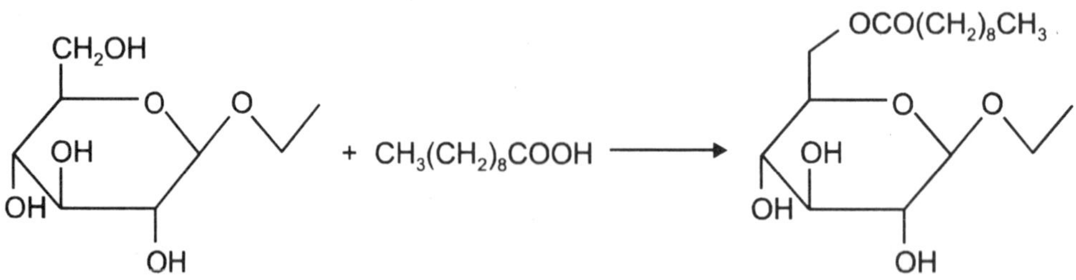

Fig. 9.2. Process of application of lipases to synthesis of esters.

To date, lipases applied in food manufacture and processing have been found in nature mainly as microbial products. Some of these enzymes have been cloned subsequently and expressed in host systems that are, as described earlier, particularly amenable to industrial fermentations. Such processes provide the lipases cost effectively in a pure state that is well suited to the various applications described.

These expression systems also lend themselves to the production of engineered lipases in which changes are made in the protein molecule to optimise its reactivity and specificity, according to the needs of the user of the enzymes. A good basis for design of such improved biocatalysts is the structure of the enzymes. Until recently, no triglyceride lipase structures had been solved to high resolution, but at least three structures are currently available. One structure is that of human pancreatic lipase, which was solved by investigators at Hofmann La Roche. Structures of Novo Nordisk lipase derived from *Mucor miehei* and Novo Nordisk Lipolase also have been solved. Several additional lipase structures are likely to become available in the next few years. Thus, a firm basis for design of improved lipases will become available. Also, new compounds amenable to preparation in lipase-catalysed processes and useful in foods are likely to be identified and are likely to play an important role in the future in connection with increasing consciousness of and knowledge about the effects of nutritional fats and oils. The application of lipase technologies in the food area is still at its beginning and is expected to become increasingly important.

Chapter 10

Oxidoreductases

INTRODUCTION

A member of a group of enzymes involved in redox reactions. Oxidoreductases were formerly known as oxidases or dehyrdogenases. In other words oxidoreductases are a class of enzymes that catalyse the reversible transfer of electrons from one substance to another (oxidation-reduction, or redox reaction).

POLYPHENOL OXIDASE

General Characteristics

Polyphenol oxidase has been given two entries in the International Union of Biochemistry (IUB) classification; as EC 1.14.18.1, monophenol monoxygenase, and EC 1.10.3.1, catechol oxidase. Common trivial names for monophenol monoxygenase are tyrosinase, phenolase, monophenol oxidase and cresolase. Common trivial names for catechol oxidase are diphenol oxidase, *o*-diphenolase, phenolase, polyphenol oxidase, and tyrosinase. Tyrosinase was the first name used for polyphenol oxidase because tyrosine was the first substrate studied. The mammalian enzyme has limited specificity, acting only on tyrosine and dihyroxyphenylalanine (DOPA).

The enzyme complexes from higher plants and fungi are virtually non-specific and oxidise a wide variety of monophenolic and *o*-diphenolic compounds. In this chapter, the enzyme will be referred to as polyphenol oxidase. Depending on its biological source, it may have monophenol monoxygenase and/or catechol oxidase activities. Monophenol monoxygenase (tyrosinase) can oxidise both tyrosine and DOPA to melanin, whereas catechol oxidase can convert only DOPA to melanin. Tyrosinase oxidises monohydroxy phenols to *o*-dihydroxy phenols and catechol oxidase dehydrogenates *o*-dihydroxyphenols to *o*-quinones. In some publications, tyrosinase is described to carry out both the hydroxylation and the dehydrogenase reaction. Robb gives the reactions as sequential: hydroxylation followed by dehydrogenation, with the final product of *o*-quinone.

Polyphenol oxidases can catalyse many reactions that involve phenolic compounds. When the monophenol *p*-cresol is the substrate, it is oxidised to the diphenol 4-methyl catechol. The oxygen for the hydroxylation comes from O_2, not from water. Thus, the enzyme acts as a monoxygenase in this reaction. Hydroxylation of monophenols is usually referred to as cresolase activity because *p*-cresol is frequently the substrate. When the *o*-diphenol catechol is the substrate, it is dehydrogenated to *o*-benzoquinone.

Polyphenol oxidase is a copper-containing enzyme widely distributed in plants, animals, and human. In this chapter, we will discuss the enzyme as it occurs in plants. Many reasons have been put forward for the lack of pigment formation in the intact cell, despite the coexistence of polyphenol oxidase and its substrates in the same organ and the same cell. Some of these reasons are separation of substrate and enzyme, substrate presence as precursor, enzyme inactivity, enzyme complexing with an inhibitor, and reduction of quinones formed.

Polyphenol oxidase is little understood in its role in higher plants. It is a plastidic enzyme that is nuclear encoded, but is not active until incorporated into the plastid. Apparently, it also exists free in the cytoplasm in degenerating or senescent tissues such as ripe or ripening fruit. Cytochemical techniques to determine the localisation of polyphenol oxidase established that the enzyme is found in root plastids, potato tuber amyloplasts, hypocotyl plastids, epidermal plastids, carrot tissue culture plastids, leukoplasts of *Aegopodium*, apical plastids, etioplasts, chiomoplasts, and chloroplasts of many different species. The function of polyphenol oxidase in plants is an enigma because, although it is localised in the plastid, most of the phenolic compounds are in the vacuole, a cellular location not juxtaposed to the plastid. The plastidic location of the enzyme might be necessary to provide a strong reducing environment to prevent the oxidation of *o*-diphenols to *o*-diquinones. The plastidic polyphenol oxidase also may be necessary to provide an electron donor for monophenol oxygenase activity. Polyphenol oxidase and chlorogenic acid have been implicated in resistance of the tomato plant. These enzymes are known to form orthoquinones in damaged plant tissue. Orthoquinones are known to alkylate amine and sulphydryl groups of proteins and amino acids. This alkylation alters solubility, digestibility, and infectivity for some pathogenic viruses.

The effects of orthoquinone alkylation on an important microbial insecticide, *Bacillus thuringiensis kurstaki* (Btk), were tested. When Btk was incubated with polyphenol oxidase and chlorogenic acid and fed to *Heliothis zea* larvae, there was a more toxic response than without the phytochemicals. Less dramatic but similar results were obtained with only polyphenol oxidase. Digestibility experiments indicated that alkylation enhanced solubilisation and/or proteolysis of the crystal protein *in vivo*.

To determine the physiological importance of an enzyme, reliable assay methods are imperative. Unfortunately, in many studies of polyphenol oxidase, attempts have not been made to differentiate it from peroxidase. In the presence of very small concentrations of hydrogen peroxide, peroxidase will form quinones from phenolic compounds. Peroxidase activity generally can be eliminated or minimised by adding catalase or alcohols to the assay solution. Catalase (EC 1.11.1.6) will remove the hydrogen peroxide and the alcohol will be oxidised preferentially to its aldehyde. Kahn discusses the use of tropolone (2,4,6-cycloheptatriene-1-one) to differentiate between tyrosinase and peroxidase activities. Tropolone is an effective copper chelator and, thus, inhibits tyrosinase. Inhibition (50 per cent) occurs at 100, 3, and 0.4×10^{-6} M tropolone when 4-methyl catechol, dopamine, and D,L-DOPA, respectively, are the substrates. In addition, tropolone can serve as a substrate for peroxidase. The resulting product is yellow and has all absorbance peak at 418 nm. The oxidation of tropolone occurs only when both peroxidase and hydrogen peroxide are present.

A commonly used polyphenol oxidase assay is the measurement of the formation of quinones and quinone polymers spectrophotometrically at 470–500 nm. Another method is the measurement of the decrease of O_2 by polarographic methods. A common dilemma for all polyphenol oxidase assay methods is the low affinity of the enzyme for oxygen. Consequently aeration of the assay solution is necessary for reliable results.

Of Mushroom

Mushroom polyphenol oxidase usually is referred to as tyrosinase. Tyrosinases catalyse the hydroxylation of monophenols, for example, tyrosine, to *o*-diphenols and convert *o*-diphenols to *o*-quinones. The mushroom enzyme can use tyrosine, catechol, or DOPA as a substrate.

Many physical characteristics have been determined for mushroom tyrosinases, including diffusion coefficients, sedimentation coefficients, frictional ratios, Stokes radii, isoelectric points, and apparent molecular weights. Mushroom tyrosinase appears to be composed of two heavy chains of 43–45 kDa each and two light chains of about 13 kDa each. The heavy chains contain the catalytic sites. Different isoenzymes apparently have different heavy chains. The role of the light chains is unknown. Although there is abundant information on the physical and catalytic properties of mushroom tyrosinase, little is known about the enzyme in developing mushrooms.

Tyrosinase activity was monitored in four stages of mushroom development: small pins (0–0.5 cm), large pins (0.5–1 cm), immature mushrooms, and mature mushrooms (classified by cap size, gill development, and veil covering). Using catechol as the substrate, there was no apparent trend in enzyme activity during development. In contrast, DOPA oxidase activity appeared to decrease with a concomitant increase in latent tyrosinase activity. The investigators concluded that tyrosinase activity and the isoenzyme content: (i) are complex, (ii) differ in preharvest development, (iii) vary from tissue to tissue, and (iv) are composed of active and latent forms. No new isoenzyme forms appear to be made during development.

Other workers investigated the effect of ascorbic acid, sodium bisulphite, and thiol compounds on mushroom polyphenol oxidase. Ascorbic acid and sodium bisulphite had little effect on initial reaction velocity whereas dithiothreitol and glutathione decreased activity by 35 per cent. By spectrophotometry, the I_{50} values were: dithiothreitol, 0.06 mM; glutathione, 0.17 mM; sodium bisulphite, 0.20 mM; and ascorbic acid, 0.24 mM. At 0.1 mM dithiothreitol, a complete loss of activity was seen after 70 minutes at 25°C, while sodium bisulphite, glutathione, and ascorbic acid caused 50 per cent inactivation after 28, 106, and 130 minutes, respectively, at 5 mM.

Of Grape

Enzymatic browning of musts and wines depends mainly on the oxidation of endogenous phenols by grape polyphenol oxidases that are released during wine making. Catecholase and cresolase activities of grape polyphenol oxidase were investigated during development and maturation of Monastrell and Airien grape berries. Cresolase and catecholase were extracted from the grapes by placing the flesh in 100 mM phosphate buffer at pH 7.3 with 10 mM sodium ascorbate. The mixture was homogenised, centrifuged, and extracted with Triton X-100. The supernatant was subjected to ammonium sulphate precipitation and, after dialysis, was used as the enzyme source. Phenolic compounds were extracted from the grape with 80 per cent ethanol and determined spectrophotometrically at 480 and 520 nm.

Catecholase activity measured below pH 5.0 increased during maturation, whereas its activity measured above pH 5.0 decreased during maturation, indicating the presence of two catecholases in Monastrell grapes. These two activities disappeared after further maturation, revealing the presence of a third enzyme. These changes in catecholase activity appear to be related to the formation of anthocyanins, because there was no shift in the optimum pH of catecholase activity in the white grape Airen, nor was there any change in the polyphenol oxidases separated by gel electrophoresis during the 7-week study. The catecholase activity of both cultivars when measured at pH 4.0 increased continuously during the 7-week study. In contrast, the cresolase activity of both cultivars increased much less. In an earlier study, partially purified polyphenol oxidase from Arien grapes also had both cresolase and catecholase activities. Catecholase had a pH optimum of 3.5–4.5 and was relatively heat stable. The apparent K_m for 4-methylcatechol was 9.5 mM. Cresolase activity had a lag phase, but the presence of *o*-diphenol abolished it. The apparent K_m for *p*-cresol was 0.35 mM and the activation constant for *o*-diphenol was 1.75 μM.

Catecholase and cresolase activities of purified Monastrell grape polyphenol oxidase were characterised. The polyphenol oxidase was purified 126-fold by ammonium sulphate fractionation. The partially purified enzyme had both catecholase and cresolase activity. The former had temperature and pH optima of 20–40°C and 3.5–5.0, respectively. The apparent K_m for 4-methyl catechol was 9 mM. Cresolase activity showed a lag phase, which was decreased by a higher enzyme concentration and increased when the substrate concentration was increased. Cresolase activity rose with a pH change from 3.5 to 7.5 without attaining a defined maximum activity. The apparent K_m for *p*-cresol was evaluated by Hanes plot and a value of 0.5 mM was observed.

Polyphenol oxidase activities of Koshu grapes were affected by storage temperature. Polyphenol oxidase activity of the filtered homogenate of grapes stored at –20°C was twice as high as when the grapes were stored at 4°C. The best harvest time for Koshu grapes, judging from soluble polyphenol oxidase activity in their juice could be determined because of the relationship between soluble polyphenol oxidase activity and Brix and total acid. There was no correlation between the insoluble fraction of polyphenol oxidase and Brix or total acid. The existence of tasteless or, in Japanese, ajinashi berries among Koshu grapes is believed to be due to a virus infection. In the ripe stage of the grapes, the polyphenol oxidase activity and total phenolic content of the ajinashi grape were about the same as those of the uninfected grape. The number of active bands of polyphenol oxidase in ripe ajinashi berries on electrophoresis was smaller than in normal berries. The intensity of the bands also indicated less polyphenol oxidase in the infected berries. An *in vitro* oxidation system was used to determine the extent of enzymatic oxidation on seven grape seed phenolics: gallic acid, (+)-catechin, (–)-catechin, and four procyanidins. There was a considerable depletion of the phenolics. In the absence of polyphenol oxidase, the decrease of phenolics was much less. There was a significant correlation between red grape browning and polyphenol oxidase content, which accounts for 74 per cent of the browning. Polyphenol oxidase only accounted for 16 per cent of the browning in white grapes. Browning of white grapes was concluded to be caused by other factors. There was no correlation between browning of red grapes and hydroxycinnamic acid content.

Of Pear

The extraction of polyphenol oxidase from plant sources is complicated by the presence of endogenous phenolic compounds. These are oxidised to the corresponding quinones or semi-quinone radicals by the enzyme. In turn, these phenolic compounds react with the polyphenol oxidase, changing its properties. Research was conducted to investigate different methods of extracting polyphenol oxidase from d'Anjou pears to determine the most efficient method of binding the phenolic compounds during extraction. A water extract of the pear tissue at pH 4.2 and 4°C lost 60 per cent of its polyphenol oxidase activity 7 hours after extraction. Visible browning of the extract occurred within 30 minutes and absorbance at 410 nm increased for 10 hours, indicating the formation of polymerised phenolic compounds. This increase in absorbance at 410 nm was accompanied by a decrease in absorbance at 324 and 280 nm. To prevent these changes, the investigators prepared a buffered extract of acteone powders, and extracts using the phenol adsorbants polyvinylpolypyrrolidone (PVPP), Amberlite XAD-4, and the ion-exchange resins BioRad AG 1-X8 and AG 2-X8. Electrophoresis of the water extract revealed three regions of polyphenol oxidase activity with 11 distinct bands of activity. In later experiments, when the phenolic compounds were removed with XAD-4, the number of apparent isoenzymes was decreased to three. Thus, endogenous phenolic substrates can react with protein to produce additional isoenzymes that are not present in the intact tissue. In the pH range 2.0–6.5, the purified polyphenol oxidase was most stable at pH 4.0 at 4°C. The enzyme was unstable at pH 6.5.

d'Anjou pear polyphenol oxidase was extracted in the presence of a phenolic binder, AG 2-X8, and the non-ionic detergent Triton X-100. After extraction, the enzyme was extensively purified by chromatography on Phenyl Sepharose CL-4B, DEAE-cellulose, and hydroxyapatite columns. An interesting discovery was that sharp peaks and good resolution were obtained only after the columns were run at room temperature. When the columns were run at 4°C, the resolution was poor. Hydrophobic chromatography, using Phenyl Sepharose CL-4B resin, provided a rapid partial purification of pear polyphenol oxidase. The enzyme was eluted as a single major peak and clearly separated from the bulk or the other 280 nm absorbing material. The purification was 24-fold. Occasionally the yield was close to 200 per cent, indicating the removal of an inhibitor. After DEAE chromatography, three peaks appeared with purifications of 30-, 12-, and 7-fold. The isoenzymes corresponding to fractions 1 and 2 were purified further using hydroxyapatite chromatography. The isoenzyme corresponding to peak 1 was purified further to 148-fold and the one corresponding to peak 2 was purified further to 103-fold. The DEAE peak was only purified 7-fold, probably because of its high protein concentration. The stability of the purified polyphenol oxidases was excellent. The enzyme of peak 1 lost 15 per cent of its activity when stored at 4°C over 15 days at pH 7.0, but the remaining activity remained at that level for 120 days.

The effect of dimethylsulphoxide (DMSO) on enzyme activity was studied using the peak 1 fraction. DMSO concentrations below 50 per cent significantly increased the enzyme activity, with a maximum of 180 per cent occurring at 5 per cent DMSO. At 80 per cent DMSO, the activity decreased to about 30 per cent. The compatibility of pear polyphenol oxidase suggests that the enzyme is active in a hydrophobic environment. The purified polyphenol oxidase is being used currently to study to the mode of inhibition by sulphite.

Of Kiwi Fruit

The polyphenol oxidase of kiwifruit was purified partially by ammonium sulphate fractionation, dialysis, and chromatography on a DEAE-cellulose column. The cresolase fraction was in the first peak and the catecholase appeared in the fourth peak when eluted from the DEAE-cellulose column. Six peaks absorbed at 280 nm. The catecholase was shown to have four isoenzymes by polyacrylamide electrophoresis and enzyme staining. These four isoenzymes (FA1P–FA4P) were characterised further. The estimated molecular masses were 15, 20, 25, and 40 kDa. The FA1P fraction was less heat resistant than the FA4P fraction. The K_m values were 50 mM and 8.7 mM for FA1P and FA4P, respectively, when, (+)-catechin was the substrate. The activation energy on (+)-catechin was 4 kcal/mol for FA1P and 7 kcal/mol for FA4P. The pH optima were not too different; pH 6.8 for FA1P and pH 7.3 for FA4P when catechol was the substrate. When (+)-catechin was used, the pH optima were 7.5 and 8.0. Experiments to determine the active site of kiwifruit polyphenol oxidase were done using Rose Bengal. The results indicated that histidine might be associated with the active site. Inhibition studies using NaCN, reduced and oxidised glutathione, ethylenediaminetetraacetic acid (EDTA), sodium diethyldithiocarbamate, and quinic acid were performed. Reduced glutathione diminished FA4P activity to 9 per cent, but had no effect on FA1P. EDTA had little effect on the activity of any of the four fractions. Sodium diethyldithiocarbamate decreased the activity of fraction FA4P to 11 per cent and the activity of fraction FA1P to 91 per cent. Quinic acid had little effect.

Of Sago Palm

Sago palm has a higher productivity of starch than either wheat or corn. In addition, it can grow on tropical acidic marshy ground that cannot be used as farmland. The plant is also somewhat salt resistant. However, sago starch is not exploited fully because of its brownish colour, which is caused by browning of the pith. To prevent browning and improve starch quality, the mechanism of browning was studied and found to be the oxidation of polyphenols by polyphenol oxidases in the pith. The oxidation of D-catechin and D,L-epicatechin in the pith by polyphenol oxidase was responsible for the browning. Two polyphenol oxidases were extracted and partially purified from sago palm pith by hydroxyapatite chromatography, DEAE-cellulose chromatography, and gel filtration. The purification procedure appeared to, remove essential copper from both enzymes, because 0.1 mM $CuSO_4$ increased the relative activity to about 300 per cent for both enzymes. Also of interest is that 0.9 mM $ZnSO_4$ increased the relative activity of both enzymes to about 140 per cent. The two enzymes showed similar behaviour with inhibitors. Among them, KCN and Na-diethyldithiocarbamate were the most potent inhibitors. Reducing agents such as Na_2SO_3 and $NaHSO_3$ were also potent inhibitors, as might be expected. The optimum pH for both enzymes was 6.5; there was little activity below pH 4 and above pH 9. One of the enzymes was surprisingly stable at pH 10. About 90 per cent of its activity remained after 18 hours at 4°C. The other enzyme was more acid stable, losing only about 10 per cent of its activity at pH 3. Both enzymes exhibited similar thermal denaturation. The activities began to decrease precipitously at about 50°C and were almost 0 at about 80°C. 4-Methyl catechol was the best substrate for both enzymes, having a relative activity of 925 per cent for one of the enzymes and 168 per cent for the other when D-catechin was set at 100 per cent. In general, both enzymes hardly oxidised monohydroxy or *m*- or *p*-dihydroxy phenols. They oxidised trihydroxy phenols to some degree. The activities of both enzymes were inhibited by an excess of D-catechin. Michaelis-Menton's formula could not be applied for either enzyme at more than 15 mM. Both these enzyme fractions had molecular masses of about 40 kDa.

Of Tea

Partially purified polyphenol oxidase from fresh tea leaves oxidises amino acids to aldehydes in the presence of (+)-catechin or other diphenols. These aldehydes contribute to development of aroma in tea. Formation of the aldehydes was directly proportional to the concentration of the amino acids, polyphenol oxidase, and (+)-catechin. However, when (+) catechin concentration was greater than 0.3 mM, production of aldehydes was reduced, possibly because of substrate saturation of polyphenol oxidase. As expected, diethyldithiocarbamate inhibited formation of aldehydes. The inhibition was reversed by Cu^{2+}. Diethyldithiocarbamate inhibited polyphenol oxidase non-competitively. Green tea leaves and crushed, torn, and curled tea (CCT) had activities of 10.9 polyphenoloxidase units per gram dry weight. Fermented tea had no detectable activity and black tea had little activity. All three polyphenol oxidase assays were done using substrates, present in tea. When the assays were repeated using pyragallol as the substrate, polyphenol oxidase activities were 18.0 for green tea leaves, 11.2 for CCT tea, 5.4 for fermented tea, and 0.9 for black tea, indicting that polyphenol oxidase was still present in fermented tea and, in a lesser quantity, still present in black tea. Depression in polyphenol oxidase during withering of tea leaf (*Camellia sinsensis*) affected the oxidative condensation of tea flavanols by forming theaflavins and, thearubigins. These compounds are associated with the briskness, brightness, and body of tea liquors. The addition of pectic substances during processing was shown to reduce the content of theaflavin and thearubigin. This reduction may have been related to the reduction of polyphenol oxidase content. Oxidation of black tea components theaflavin, theaflavin monogallates, and theaflavin digallates by polyphenol oxidase was tested in the presence of (–)-epicatechin in a model fermentation system under aerobic conditions. Oxidation of theaflavin monogallates was more rapid than that of theaflavin digallate. The variation of polyphenol oxidase activity, total polyphenol content, and catechins on the quality of black tea (*Camellia sinensis* L. O Kuntze) was studied with respect to different clones and shoot components. There was a wide variation of polyphenol oxidase in the six different clones. Optimum fermentation time and polyphenol oxidase activity of different clones showed a hyperbolic relationship. A good but non-linear relationship was found between polyphenol oxidase activity of fresh tea shoots of different clones and the quantity of theaflavins of corresponding black teas. Formation of theaflavins during fermentation of different shoot components and the quality of black tea agreed well with polyphenol oxidase activity. In another study, clonal tea was found to be better than seedling tea. One of the differences noted between clonal and seedling leaves was that polyphenol oxidase activity was higher in the former. As a result, the theaflavin content was higher in black tea made using the clone; therefore, the browning potential is greater in the clonal tea. When 0.1 per cent of a microbial polyphenol oxidase-was added to black tea during fermentation, the fermentation time was reduced. The colour of the infusion was improved, and theaflavin and thearubigin concentrations were increased significantly. Polyphenol oxidase and peroxidase activities of tea associated microflora were evaluated. Ten bacterial and seven fungal isolates were screened for these activities. In general, bacterial strains were more promising than fungal isolates. Polyphenol oxidase activity was much higher than peroxidase in the three most potent *Bacillus* strains.

In Cocoa Processing

One of the least understood factors in the quality of cocoa products is the presence of polyphenol oxidase in cocoa beans. In chocolate production, polyphenol oxidase is one component responsible for the formation of flavour precursors, beginning in the oxidative phase of the fermentation and continuing into the drying phase. Among the changes affecting flavour are reduction in the bitterness and astrigency that result from polymerised polyphenol-protein interactions. An increase in oxygen penetration into the cocoa bean mass during drying induces maximum oxidation of (–)-epicatechin and procyanidins, causing the production of melanin and melanoproteins that are responsible for the colour in brown chocolate. In an attempt to understand the role of the enzyme in chocolate flavour and colour development more fully, polyphenol oxidase was extracted from cocoa beans, then isolated and enriched. Crude polyphenol oxidase extract was prepared by grinding frozen beans and extracting them with buffer at pH 8.0. Ammonium sulphate at 65 per cent saturation precipitated the polyphenol oxidase. The enzyme was enriched by high performance liquid chromatography (HPLC) using a hydrophobic resin resulting in a 6.5-fold purification with a 25 per cent recovery. There appear to be multiple isoforms of polyphenol oxidase and no contaminating enzymes in the purified preparation. This rapid purification technique should allow for easier study of the properties of cocoa polyphenol oxidase.

Of Green Pepper

Green pepper, an important article of commerce, is obtained from unripe but fully developed berries of *Piper nigrum*. The fresh berries become black in a short time unless preserved. This blackening was found to be due to the enzyme-catalysed oxidation of 3,4-dihydroxyphenylethanol glycoside. The enzyme described was polyphenol oxidase, which was found to oxidise the aglycone of the glycoside. Damage to skin tissues resulted in a very rapid blackening due to oxidation. Among five commercial varieties of *P. nigrum* studied, there was great variation in the quantities of phenolics and polyphenol oxidase-oxidisable phenolics. However, no correlation was observed between the total or oxidisable phenolics of a particular variety and the extent of blackening during sun-drying.

This result might be explained by the findings that sun-drying resulted in a 75 per cent decrease in total phenolic content and a complete loss of polyphenol oxidase. A new compound, 3,4-dihydroxy-6-(*N*-ethylamino)benzamide, was identified in pepper as an efficient substrate for pepper polyphenol oxidase.

Of Strawberry

The red colour in strawberries comes from two main anthocyanin pigments: pelargonidin-3-glucoside (PGN) and cyanidin-3-glucoside (CYN). Anthocyanins are very unstable. Colour loss can occur after thawing of frozen fruit and during processing of strawberry products. The presence of an active polyphenol oxidase system can cause the loss of anthocyanins, but the fate of the pigment is unknown. Model systems containing PGN, CYN, D-catechin (CAT), and polyphenol oxidase were prepared. It was concluded that strawberry polyphenol oxidase played a role in the degradation of anthocyanin pigments. Direct oxidation of these pigments did not seem to be the main pathway of decolourisation, although polyphenol oxidase could oxidise cyanin at a very slow rate. The reaction of polyphenol oxidase with CAT, a flavonoid found in strawberries, occurred at a rapid rate with the formation of a polymer absorbing at 390 nm. Quinones and other intermediates formed during the oxidation of CAT by polyphenol oxidase may be responsible for the destruction of anthocyanin, either by oxidation or co-polymerisation. In an earlier study by the same group, polyphenol oxidase was extracted from strawberries using a hydrophobic Phenyl Sepharose resin. Two polyphenol oxidase fractions were isolated, both of which had very high activity with D-catechin. Studies of model systems containing CAT, CYN, or PGN indicated that anthocyanin pigments were destroyed either by direct oxidation by quinones formed from the D-catechin resulting from polyphenol oxidase action or by co-polymerisation into tannin formed by D-catechin–quinone polymerisation.

Of Potato

Dopachrome is the first visible pigment produced during adventitious enzymatic browning of plant tissue. Homogenised raw potato was used as a source of polyphenol oxidase and its substrate, DOPA. Dopachrome formation was found to be inhibited by sodium benzoate, reduced temperature, and acid pH. Results indicate that the benzoate competes with DOPA. The effect of isopropyl-*N*-(3-chlorophenyl) carbamate (CIPC) on total nitrogen content and polyphenol oxidase activity was evaluated on potatoes in India. CIPC treatment increased total nitrogen. Polyphenol oxidase activity was reduced by CIPC treatment at room temperature but not at colder temperatures. The extent of enzymatic browning at the cut surfaces of potatoes correlated with total phenolic compounds, tyrosine, and, to a lesser extent, polyphenol oxidase activity. Browning in the Atlantic cultivars was almost eliminated by a water dip. Russett Burbank was much more susceptible to browning than the Atlantic cultivars. The effects of soil applications of sodium molybdate on Katahdin potato quality was investigated. The parameters investigated were polyphenol oxidase activity, total phenolic content, chlorogenic acid level, tyrosine, and the amount of ascorbic acid. Soil applications of high levels of sodium molybdate increased ascorbic acid but decreased polyphenol oxidase activity, enzymatic discolouration, chlorogenic acid, and tyrosine.

Lower levels of sodium molybdate did not affect these parameters. One of the problems in attempting to determine polyphenol oxidase activity in potato is the inhibition of polyphenol oxidase by endogenous phenols. Endogenous phenols were removed from potato tuber homogenates by polyvinylpyrrolidone (PVP), extensive dialysis, ammonium sulphate or acteone precipitation, or Sephadex G-25 chromatography. G-25 gel filtration was superior in removing endogenous phenols. When the potato homogenate was treated in this manner, polyphenol oxidase activity was retained.

Of Pomegranate

Husk scald incidence on Wonderful pomegranate fruit during storage was related to the amount of *o*-dihydroxyphenols extracted from the husk and was controlled by inhibiting their oxidation by polyphenol oxidase. Post-harvest measures included dipping the fruit in boiling water for 2 minutes and in antioxidant solutions. The most effective control of husk scald was storage of late-harvested fruit in a low-oxygen environment (2 per cent oxygen at 2°C). However, this treatment resulted in the accumulation of ethanol, which caused development of off-flavours.

Of Persimmon

Seasonal changes of polyphenol oxidase activity in the flesh of the Japanese persimmon were measured from July to October. Results showed that polyphenol oxidase activity increased rapidly in two cultivars and there was flesh darkening. In the other two cultivars, polyphenol oxidase activity remained relatively low during fruit development and no flesh darkening was observed. Ethanol vapour treatment or attached intact Japanese persimmon quickly removed the astringency and induced flesh darkening within 3 weeks. A crude polyphenol oxidase preparation as an acetone powder from persimmon mesocarp was much more active when extracted at room temperature than at 4°C. Of the five most common substrates, 4-methyl catechol was oxidised most rapidly by the enzyme. Polyphenol oxidase activity increased sharply just after the decrease in astringency and coincided with the onset of darkening at the calyx end. High polyphenol oxidase activity was maintained during progressive darkening of the entire flesh.

Of Olive

A positive correlation between the rate of browning of the crude homogenate of the fruit and the polyphenol oxidase content of the acetone powder extracted from the fruit was seen in five olive (*Olea europaea*) varieties. An inhibitor was ruled out and it was concluded that the rate of olive browning was directly proportional to polyphenol oxidase activity. Although there were significant differences between the varieties in *o*-diphenol content and polyphenol oxidase activity, the browning could be correlated to these parameters and not to peroxidase activity.

Of Beet

Endogenous polyphenol oxidase activity is associated with the 'black ring' defect in canned beet. This black ring resulted after 5–10 minutes exposure of beet roots to live steam. A previous study implicated peroxidase and polyphenol oxidase activity as the cause. This study produced evidence that polyphenol oxidase was the primary cause of discolouration. When beet roots were dipped in 5 mM cysteine, discolouration was inhibited; when they were dipped in 5 mM dithiothreitol, it was prevented. Both sulphydryl agents are known inhibitors of polyphenol oxidase. Discolouration also was prevented by 1 mM cyanide, a known inhibitor of metalloproteins, and 1 mM tropolone, a copper chelator and peroxidase substrate. Beet slices dipped in a solution containing both tropolone and hydrogen peroxide discoloured to a similar extent as the 0.1 per cent sodium chloride control, indicating that, in the presence of peroxide, peroxidase also can contribute to the discolouration of thermally processed beet roots.

PEROXIDASE

General Characteristics

Peroxidases are distributed ubiquitously in both the plant and animal kingdoms. Perhaps the one most widely known to enzymologists is horseradish peroxidase (EC 1.11.1.7), because of its wide use as an indicator enzyme in standard spectrophotometric and immunoassay techniques. All peroxidases act on hydrogen peroxide as an electron acceptor and oxidise a multitude of donor compounds, many times to coloured end products. Horseradish peroxidase has a heme prosthetic group and is a hemoprotein. NADH peroxidase (EC 1.11.1.1) is a flavoprotein. Glutathione peroxidase (EC 1.11.1.9) is a selenium-containing protein. Chloride peroxidase (EC 1.11.1.10) is probably a heme-thiolate protein.

Studies on Horseradish Peroxidase

Thermal inactivation of horseradish peroxidase in buffer was studied by differential scanning calorimetry and residual activity determination. For both experimental systems, the apparent reaction order for thermal inactivation was 1.5. Since the four isoenzymes varied in heat resistance, the 1.5 order probably was caused by the heterogeneity of the enzyme. Sucrose at concentrations of 40 per cent or greater stabilised the enzyme against thermal inactivation. At 10–20 per cent sucrose, the thermal stability was reduced. At equimolar concentrations of glucose, fructose, or lactose, horseradish peroxidase was inactivated more rapidly than in the presence of sucrose. The thermal stability of immobilised horseradish peroxidase was tested in water-organic solvent systems.

Water-dioxane reduces and water-DMSO increases thermal stability. Results also indicated that horseradish peroxidase is stabilised against thermal inactivation by immobilisation. Aminated horseradish peroxidase was inactivated by interaction with S-Sepharose. Peroxidase that was aminated by periodate oxidation and reductive amination was purified by cation-exchange chromatography on S-Sepharose. Instead of the single peak of aminated enzyme, two distinct peaks were eluted from the column. Evaluation of the protein in each of the peaks showed that the first peak was similar in spectral properties and activity to the native enzyme. The second peak had a 3-fold reduction in extinction in the Soret region at 404 nm and had no activity.

The gene encoding horseradish peroxidase has been cloned and sequenced, so the gene is commercially available. A synthetic gene encoding horseradish peroxidase isoenzyme C was expressed in *Escherichia coli*. The non-glycosylated recombinant enzyme was made in inclusion bodies in an insoluble inactive form containing only traces of heme. The peroxidase was solubilised and the conditions under which it folded to produce active enzyme were determined. The folding was shown to be critically dependent on the concentrations of urea, Ca^{2+}, and heme, and on oxidation by oxidised glutathione. The purified peroxidase had about half the activity of the enzyme from horseradish root. Glycosylation is not essential for correct folding and activity. The overall yield of enzyme was only 2–3 per cent, but could be increased by re-processing material that precipitated during folding. More recently, a cDNA clone encoding a neutral horseradish peroxidase was isolated and characterised. The cDNA contained 1378 nucleotides excluding the poly(A) tail and the deduced protein contained 327 amino acids, including a 28-aminoacid leader sequence.

The predicted amino acid sequence is 9 amino acids shorter than the major isoenzyme belonging to the horseradish peroxidase C group (HRP-C). The sequence of the neutral isoenzyme shows 53.7 per cent identity with HRP-C but appears to have three fewer glycosylation sites. The described cDNA clone was concluded to encode a neutral horseradish peroxidase that belongs to a new, not previously described horseradish peroxidase group. Genomic DNAs encoding the two peroxidase isoenzymes *prx*C2 and *prx*C3 were cloned and sequenced. Genes *prx*C2 and *prx*C3 encoded 347 and 349 amino acid residues, respectively, including putative signal sequences at the N termini. Properties of four monoclonal antibodies to horseradish peroxidase were investigated. The antibodies were shown to be directed to different epitopes on the polypeptide chain.

Other Analytical Peroxidases

Peroxidase from *Coriolus versicolour* is now being used in analytical procedures as a substitute for horseradish peroxidase. The peroxidases from *C. versicolour* are excreted into the medium only when the carbon source is limited and veratryl alcohol is added to the medium as an enzyme inducer. In a medium with limited nitrogen, the excretion of enzyme is not dependent on veratryl alcohol. The optimum pH of the peroxidase is 4.0, and it is immunologically similar to horseradish peroxidase. A European Patent Application described a peroxidase derived from the basidiomycete *Coprinus cinereus* that also can substitute for horseradish peroxidase. A peroxidase also can be made efficiently from the callus tissue of wasibi.

Blanching of Foods

General considerations

Peroxidases damage most foods if they are not inactivated before further processing. Therefore, the inactivation and characteristics of peroxidases have been studied extensively in many fruits and vegetables, as well as in pasta and dough. Peroxidases in foods cause the formation of undesirable end products unless they are completely or at least partially inactivated. However, tomato and potato tissue cell walls apparently are not degraded by either endogenous or added peroxidase, although the application of peroxidase substrates such as indoleacetic acid, NADH, KI, and Na_2SO_3 led to an increase in weight loss of ripening tomato tissue. Exogenous peroxidase added to potato slices did not lead to tissue degradation, but added pectinase caused cell wall degradation. Blanching usually is employed to preserve most vegetables and even some fruits before further processing. Other methods to inactivate peroxidases in various food products have been studied also. Blanching, however, is still the most widely used method to inactivate peroxidases and other enzymes. Blanching of 16 vegetable varieties was done in water at 70°–100°C for various times. Peroxidase and lipoxygenase activities were determined before and after the various blanching regimes. Inactivation of the two enzymes in beans, asparagus, zucchini, and carrots showed a biphasic pattern. In all vegetables except French beans, lipoxygenase had the higher heat stability. Heat resistance was lower for soluble than for ionically bound peroxidase. A thermokinetic model of peroxidase inactivation in corn-on-the-cob during blanching was developed. Using this thermokinetic model, the average peroxidase activity retention in the kernel and outer-cob during blanching—cooling was predicted. Flavour changes that occur in unblanched frozen vegetables have been suggested to be related to enzyme activity. The activities of palmitoyl-CoA hydrolase, α-oxidase, lipoxygenase, peroxidase, and catalase were determined in 22 fresh vegetable types. Protein and ascorbic acid were also determined. All vegetable types had palmitoyl-CoA dehydrogenase and lipoxygenase activity. Palmitoyl-CoA dehydrogenase activity ranged from 0.05 (onion and beetroot) to 3.63 (potato) mmol/min/100 grams wet weight. Lipoxygenase activity ranged from 1.0 (parsley) to 44.9 (peas). α-Oxidation activity was not detected in 9 vegetable types but ranged from 0.3 (crisp-headed lettuce) to 63.6 (kale). All vegetables tested except red pepper had peroxidase activity; levels ranged from 2 (onion) to 2656 (Brussels sprouts). All vegetables except red pepper, snap pea, and tomato had catalase activity; levels ranged from 2 (cucumber) to 403 (peas). Linear regression analysis failed to find any correlation between any of the enzyme activities or between enzyme activities and ascorbic acid concentration. Palmitoyl-CoA dehydrogenase was suggested to be useful as a blanching indicator. The temperature resistance of peroxidase was studied in cabbage, leeks, carrot, celery, spinach, squash, potatoes, onions, and green beans. Peroxidase activities of cabbage and green beans were very high. In contrast, onions had very little activity. Blanching of green beans, potatoes, and squash at 75°C was not adequate for inactivating peroxidase. Inactivation also was affected by the type of vegetable and the size of the vegetable pieces.

Blanching of asparagus

The thermal stability of peroxidase in fresh asparagus tips was investigated by heating at 50, 60, and 70°C. Peroxidase inactivation followed a biphasic curve and the activation energy for the partially purified enzyme was 41.9 kcal/mol. Asparagus spears were blanched at 88°C for 1–4.5 minutes and stored at –18°C. Hexanal, ascorbic acid, free sulphydryl, and peroxidase activity were monitored during storage. Blanching for 2 minutes reduced peroxidase activity by 98 per cent. No hexanal was detectable in product blanched for 2 minutes or more. Ascorbate stability during storage increased with blanching time, but there was no direct relationship with peroxidase loss. Free sulphydryl level was sensitive to heating and also decreased during storage of unblanched asparagus. However, there appeared to be a stable fraction of free sulphydryls.

Blanching of cauliflower and cabbage

The residual peroxidase activity was determined in cauliflower after blanching and sulphur dioxide treatment. The organoleptic qualities were assessed after the cauliflower was stored for up to 1 year in sealed polyethylene bags at –35°C. The quality of the blanched product correlated inversely with the peroxidase activity. Blanched cauliflower following a brief dip in metabisulphite solution prior to freezing gave a superior product, even when stored for 1 year. The sulphur dioxide disappeared after 3 minutes of cooking in boiling water. The heat stability of purified spring cabbage peroxidase isoenzymes was determined. The anionic isoenzyme (pI 3.7) was relatively heat stable. The cationic isoenzyme (pI 9.9) was inactivated by heat more readily. Neither isoenzyme showed regeneration after heat inactivation.

Reduction of Peroxidase Activity by Other Methods

Reduction of peroxidase activity was determined in tomato, carrot, or eggplant extracts using the naturally occurring antioxidants quercetin, rutinic acid, chlorogenic acid, and α-tocopherol, alone or with heat treatment. Regeneration of enzyme at room temperature and after frozen storage (–18°C for 4 weeks) was tested also. Combining antioxidant (125 mg per cent) and heat (2 minutes at 75°C) resulted in almost complete inhibition of vegetable peroxidases. Regeneration of peroxidase in the extracts that were treated with heat and antioxidant was less than in extracts treated with heat or antioxidant alone. These same antioxidants without heat were tested as inhibitors of peroxidase in tomato, carrot, or eggplant extracts. Four concentrations of each were tested: 125, 250, 500, and 1000 mg per cent. Peroxidase activity of carrot and eggplant was inhibited more effectively than that of tomato by all four antioxidants. α-Tocopherol was the least effective and chlorogenic acid was the most effective, followed by quercetin and rutinic acid. Gibberellic acid has been tested as a post-harvest treatment for mango fruit to reduce peroxidase activity. Gibberellic acid at the rate of 100–200 mg/litre significantly delayed ripening, reduced peroxidase and amylase activities, and retarded the loss of weight, chlorophyll, and ascorbic acid content.

Of Apple

Peroxidases have been associated with flavour deterioration of fruit juice during storage. Cox's Orange Pippin apples frequently are used in the production of apple juice. The heat stability and number of peroxidase isoenzymes in the peel and pulp were determined. Soluble and ionically bound peroxidases were extracted from the peel and pulp. Most activity was in the soluble fraction of the peel. Soluble peroxidases were cationic and anionic. The ionically bound peroxidases were solely cationic. Five isoenzymes were found by isoelectric focusing. The soluble peroxidase in pulp was most heat stable. A limited regeneration of activity was noticed after heating in all instances. pH optima were 5–6. A commercial fresh apple juice contained slight peroxidase activity when assayed at pH 6.0. Apple polyphenol oxidase was inactivated by 98, 63, and 34 per cent of the initial activity after 20 minutes at 25°C at adjusted pH values of 2.00, 2.25, and 2.50, respectively. Apple peel peroxidase was purified 30-fold with 80 per cent yield using hydrophobic chromatography followed by affinity chromatography on concanavalin A-substituted agarose. Isoelectric focusing showed the presence of at least four isoenzymes in the purified extract. The pI values ranged from 4.6–9.8. The bulk or 99 per cent of apple peel peroxidase activity was due to the cationic forms. Kinetic studies with guaiacol, chlorogenic acid, and (+)-catechin indicated that peroxidase followed Michaelian kinetics with a ping-pong mechanism and that it was inhibited by high substrate concentration. Apple peel peroxidases also were able to act on the main polyphenolic compounds of this fruit catechins, hydroxycinnamate derivatives, and flavonol compounds. Apple peel peroxidase is believed to be involved in the enzymatic browning of apple fruit.

Of Orange

Flavedo peroxidase from Valencia Late oranges was purified by ammonium sulphate fractionation, gel filtration, and ion-exchange chromatography. The anionic fraction consisted of an apparently homogeneous isoenzyme with a pI of 2.5. An anionic fraction contained three isoenzymes with pIs from 9.5 to 10.5. Both fractions had similar specific activities with benzidine and indoleacetic acid; the anionic fraction had a considerably higher specific activity with eugenol and guaiacol.

Of Bean

Peroxidase activity was determined in haricot seed prior to germination, 5 days after germination, after 50 days of development, and in the mature succulent green bean. Activity increased during each of the stages until maturity, when it decreased from 190 to 55 mmol per mg protein per minute. Blanching denatured 98 per cent of the peroxidase, but 20 per cent was restored when the seed was stored for 12 months at –20°C. Peroxidase was regenerated dramatically when the seed was held at room temperature. Freeze drying also inactivated the enzyme, but air-drying showed limited inactivation. Haricot seed peroxidase was extracted and partially purified by ammonium sulphate fractionation. Optimum pH for activity was 5.4. When 4 mM hydrogen peroxide was added, the activity increased about 92-fold. At higher concentration of hydrogen peroxide, the activity was decreased. Carbonyl compounds resulting from peroxidase action were isolated as dinitrophenylhydrazones and purified by preparative thin layer chromatography. Subsequent analysis showed that acetone was the principle carbonyl compound.

Black beans (*Phaseolus vulgaris*) were studied to characterise the soluble peroxidase and to determine the effect of heat pre-treatment and storage conditions on hardening of the bean and inactivation of enzyme. The enzyme had a pH optimum of 6.4–6.8, and optimal temperature of 40°–45°C, and an activation energy of 1.19 kJ/mol. There was strong dependence on water activity and little sensitivity to Ca^{2+} ions. Thermal inactivation of peroxidase was facilitated with increasing moisture. When the peroxidase was inactivated by heat at 102°C, there was a reduction of hardening. When the temperature was raised to 105°C, there was a further reduction of hardening. Thus, heating only to inactivate peroxidase was not adequate to inactivate the hardening mechanism.

Of Grape

Soluble and ionically bound peroxidase was determined in homogenised Ohane grapes. The soluble fraction contained 87 per cent of the activity. Heat inactivation was non-linear; about 90 per cent of the activity was destroyed after 10 minutes at 80°C. There was no significant regeneration of activity. Using isoelectric focusing, six peroxidase isoenzymes with pIs from 3.5 to 9.8 were detected. Peroxidase was extracted from grape berries and purified 124-fold by ammonium sulphate fractionation and column chromatography. Four anionic isoenzymes were detected. Estimated molecular mass was about 40 kDa. pH and temperature optima were 5.5 and 40°C. Partial heat inactivation occurred at 65°C in 2 minutes; complete inactivation was achieved at 85°C in 5 minutes. K_m was 5.5 and 16 mM for guiacol and hydrogen peroxide, respectively.

Of Brussels Sprouts

Using gel filtration and ion-exchange chromatography, four peroxidase isoenzymes were isolated from extracts of Brussels sprouts. The isoenzymes differed in their substrate specificities and heat stability properties. Three showed biphasic thermal inactivation, whereas the fourth was relatively more heat labile and was inactivated in a more linear manner with time. Regeneration was not observed.

Of Barley

An endosperm-specific barley peroxidase was cloned, characterised, and expressed. A barley peroxidase (BP 1) was purified from mature barley grains. Using antibodies to this enzyme, a cDNA clone was isolated from a cDNA expression library. BP 1 is less than 50 per cent identical to other sequenced plant peroxidases. Analyses of RNA and protein from aleurone, endosperm, and embryo tissue indicated maximal expression 15 days after flowering. High levels of enzyme were found only in the endosperm. BP 1 is not expressed in leaves.

In Spices

Peroxidases are believed to contribute to the damage of spices by oxidising malonaldehyde, a product of lipid oxidation. Malonaldehyde in the presence of Mn(II), dissolved molecular oxygen, and horseradish peroxidase is oxidised with consumption of O_2 in an apparent oxygenase reaction catalysed by the peroxidase. However, results show that the peroxidase functions as a normal peroxidase. A reaction sequence is initiated by the autoxidation of malonaldehyde, generating a free radical that reacts with molecular oxygen, forming hydrogen peroxide that is acted on by the peroxidase to form more free radicals, causing more malonaldehyde oxidation, and additional oxygen consumption.

Of Potato

The cDNA of a highly anionic peroxidase from potato was cloned and sequenced. This peroxidase was strongly suggested to be involved in deposition of the aromatic domain of suberin. The deduced amino acid sequence of this anionic peroxidase showed considerable homology to other peroxidases. This anionic peroxidase was barely detectable 2 days after wounding and reached a maximal level 8 days after wounding. Using the cDNA for the anionic peroxidase as a probe, the mRNA for the enzyme was shown to be induced in suberising potato. The maximum level was reached in 3 days.

Many of the 1000 potato varieties cultivated throughout the world have similar organoleptic features, posing commercial problems for identification of different potato varieties. It has been proposed that peroxidase zymograms of tuber sap obtained after polyacrylamide gel electrophoresis at pH 7.9 could be used to identify the 50 potato varieties marketed in Spain. The peroxidase pattern obtained after electrofocusing in a pH 3.5–9.5 gradient could be used if the results are in doubt. Yams also can be identified using isoelectric focusing of peroxidase and acid phosphatase isoenzymes. This technique was suitable for the positive identification of shoot culture materials of four yam species. This method was also suitable to identify somatic hybrids following protoplast fusion and the regeneration of hybrid plants from microcalluses.

Of Cucumber

In cucumber, as in many other plants, peroxidases are a major ethylene-induced protein. Ethylene-induced peroxidases do not appear to play a major role in cell wall lignification, chlorophyll degradation or disease resistance. In normal plants, an increase in peroxidase is associated with ageing tissue. Since ethylene promotes ageing, the induction of peroxidase is thought to be the consequence of ageing. From a cDNA library of ethylene-treated cucumber cotyledons, two cDNA clones encoding putative peroxidases were isolated using a synthetic DNA probe based on a partial amino acid sequence of a 33 kDa cationic peroxidase that was known to be induced by ethylene. DNA sequencing indicated that the two clones were derived from two closely related RNAs related to published plant peroxidase sequences. An increase in the level of mRNA is evident after 3 hours ethylene exposure and plateaus after 15 hours. Antibodies to the pI-9 and pI-4 isoenzymes were used in a radial immunodiffusion assay to show that ethylene induced similar peroxidases in other cultivars of *Cucumis sativis* other species of *Cucumis* and other genera of Cucurbitaceae. Examination of ethylene-induced peroxidases by isoelectric focusing demonstrated the presence of a series of other peroxidases, most slightly acidic, with an isoelectric focusing pH of ~6. The data suggest that the induction of peroxidase isoenzymes during ethylene-induced senescence is a common response in this family of plants. It also appears that the acidic and basic peroxidases are highly conserved. Peroxidase activity in fresh cucumber fruit was highest in the skin, followed by pericarp and, finally, carpel tissues. Its optimum pH was about 7. Using hydrogen peroxide with benzidine, guaiacol, *p*-phenylenediamine, *o*-phenylenediamine, *o*-dianisidine, or 3-amino-9-ethylcarbazole as substrate, more anodic isoenzymes were seen in the skin than in the pericarp or carpel. There was no reaction using these substrates in the absence of hydrogen peroxide. The high thermal stability of cucumber peroxidase is believed to play a role in the darkening of processed cucumber products. Mechanical stress also stimulates peroxidase activity in cucumber. Moderate and severe mechanical stress caused the appearance of a new slow-migrating peroxidase isoenzyme immediately after treatment. This enzyme disappeared after 24 hours of storage, then re-appeared after 48 hours. Severe stress also stimulated the appearance of two additional moderate-migrating peroxidases 24 hours after treatment. Using peroxidase as an indicator, mechanical stress appears to induce accelerated ageing of cucumber. Peroxidase activity was determined in tissues and brines of pickling cucumbers. Peroxidase activity was relatively unaffected by the first 3 days of brining, but decreased during fermentation and storage, and after processing. $CaCl_2$ reduced the loss of activity in tissue and brine during fermentation and storage. No peroxidase, catalase, or lipoxygenase activity was detected in cucumber extracts fermented by four different lactic acid bacteria. Peroxidase activities were determined in commercially obtained pickle products. Large variations were found between products and sources of a given product. All sources of refrigerated dills, fresh pack dills, and spears had activity. Spear tissues located in the centre and bottom portions of jars had the highest peroxidase activity.

Of Pea

A simple method was developed to identify peroxidase isoenzymes in crude pea seed extract without the need for purification. Analytical thin layer isoelectric focusing on polyacrylamide gel resulted in high resolution of the isoenzymes. *o*-Dianisidine: H_2O_2 and 3-amino-9-ethylcarbazole: H_2O_2 were sensitive and specific sensing reagents. In all three cultivars, 18 isoenzymes were detected that appeared consistently. Residual peroxidase activity was determined in untreated and blanched peas to control the blanching of deep-frozen peas. Enzyme extracts were prepared by homogenising samples in phosphate buffer at pH 6 and 4°C. Activity was measured using *o*-dianisidine: H_2O_2. These tests allow the rapid characterisation of enzyme extracts relative to pH and conditions of blanching.

Peroxidase isoenzymes were purified from green peas by ion-exchange chromatography. Three isoenzymes, one neutral and two cationic, were identified and characterised. A further study attempted to optimise blanching conditions by examining the effect of various blanching treatments on the activity of peroxidase, polyphenol oxidase, lipoxygenase, and catalase, and by monitoring quality changes during storage at –23°C. Enzyme activity was concluded not to be a reliable indicator for the adequacy of blanching. The combined effects on peroxidase activity of crude extracts of green peas of heat treatment and various antioxidants were evaluated. The main conclusions were that a combination of antioxidant [butylated hydroxyanisole (BHA), butylated hydroxytoluene (BHT), or α-tocopherol] and heat treatment was more effective in inactivating peroxidase than heat treatment alone. In the presence of antioxidant, the heat treatment time could be reduced, permitting better retention of heat sensitive nutrients and organoleptic qualities.

Of Tomato

Peroxidase in tomato extract was measured using 2,2′-azino-bis-(3-ethylbenzthiazol-6-sulphonic acid) (ABTS). This substrate is much superior to guaiacol because there was only one oxidation product and the molar extinction coefficient of the oxidation product was 31100, six times higher than that of tetraguaiacol. Peroxidase activity measured with ABTS had a pH optimum of 3.3 when measured in tomato extract. The measurement could be completed in a, much shorter time: 30 seconds compared with 3 minutes using guaiacol. In suberising tomato (*Lycopersicon esculentum*) fruit, the mRNA showed induction and reached a maximum in 3 days. The data suggest that the induction of peroxidase by wounding is preceded by transcriptional activation of the peroxidase gene or by increased stabilisation of the mRNA. Activities of acidic and basic peroxidases from tomato fruit were determined at six ripening stages, from green to red-ripe fruits. The acidic and basic isoenzymes reached a maximum activity during the climacteric, at the pink stage. The relative increase of the basic peroxidase was much more pronounced. Tomatoes were harvested at three stages of ripeness and three peroxidase isoenzymes were isolated. One was soluble and two were ionically bound. All three had molecular masses of about 46 kDa and optimal pH for activity of ~6.

Of Paprika

The relationship between pigment content, peroxidase activity, and sugar composition was determined in paprika (*Capsicum annuum* L.). The pods were ripened and treated afterward with ethylene. The skin was analysed for glucose, fructose, sucrose, peroxidase, and pigment concentration. Results indicated that sugar content could be related to colour and that the method of drying had a marked effect on all the characteristics. Peroxidase activity decreased as the pods turned red and decreased about 50 per cent when dried at room temperature. Freeze drying produced inferior paprika that had a higher peroxidase content. Whether this was a causal relationship was not determined. Analysis of samples taken from different points of the band driers used for large scale processing of paprika showed that paprika is damaged by current production methods. Heat damage caused the loss of pigment as well as a considerable loss of peroxidase activity and sugar content. The loss of colour is believed to be a nonenzymatic process and caused by, among other factors, Maillard browning. There were six isoenzyme bands in red pepper extracts and eight in green fruits.

Of Peanut

Native cationic and anionic peroxidases from peanut cultures lost activity and became more susceptible to proteolysis when they were deglycosylated or their carbohydrates were oxidised with periodate. In the presence of tunicamycin, which inhibits glycosylation, the non-glycosylated form of the peroxidases was secreted. The oligosaccharides in peanut peroxidases may not be essential for their secretion but they are important in maintaining catalytically active conformation and stability. Cationic and anionic peanut peroxidases were purified to homogeneity, as shown by two-dimensional electrophoresis. The two isoenzymes appeared to be similar and only showed a small difference in their heme absorption maxima and specific activity. Antisera cross-reacted with both isoenzymes. The two major isoenzymes of peanut cationic peroxidase found in the medium of cultured cells were characterised. A monoclonal antibody for use in a solid phase enzyme-linked immunosorbent assay (ELISA) specific for an anionic peroxidase from peanut suspension cell medium was developed. The antibody had high affinity and specificity for the anionic peroxidase, had only weak interaction with α-amylase, and had virtually no reactivity with other enzymes.

In Canola

In some instances, for example, in canola, peroxidase activity is desirable because it is directly related to degreening. Peroxidase bleaches chlorophyll in the presence of hydrogen peroxide and 2,4-dichlorophenol. Peroxidase activity in isolated thylakoids from degreening canola seeds was demonstrated. Activity is inhibited initially and then stimulated by sub-lethal freezing. Therefore, inhibition of peroxidase activity following sub-lethal freezing may be responsible for the degreening of seeds. The economic value of canola is increased when there are fewer green seeds.

LACTOPEROXIDASE

General Characteristics

Lactoperoxidase is a haemoprotein that catalyses the oxidation of thiocyanate and iodide ions in the presence of hydrogen peroxide to produce highly reactive oxidising agents. These products have a broad spectrum of antimicrobial effects. They can kill viruses, gram positive and gram negative bacteria, fungi, and probably also mycoplasmas and parasites. The molecular components that are oxidised in the cells are sulphydryl groups, NADH and NADPH, and, under some conditions, aromatic amino acids. The major products responsible for these effects are the hypothiocyanate ion ($OSCN^-$), hypothiocyanous acid (HOSCN), and iodine (I_2). Lactoperoxidase is found in milk, saliva, tears, cervical mucous, and nasal glands. Human milk does not appear to contain lactoperoxidase. However, a more recent study confirmed the presence of lactoperoxidase in human colostrum. The human enzyme resembles bovine lactoperoxidase B in immunoreactive properties and has an M_r of 80000. The enzyme is a glycoprotein containing about 10 per cent carbohydrate.

The lactoperoxidase in mammalian milk has a heme very tightly bound to a single polypeptide chain of M_r 78000. Studies of the prosthetic group were carried out using the proton homonuclear Overhauser effect, a technique useful in studying paramagnetic proteins. Data indicate the presence of arginine and histidine residues in the distal pocket with stereochemistry resembling those in cytochrome C peroxidase and horseradish peroxidase. The interaction of thiocyanate ion and lactoperoxidase was studied using ^1H and ^{15}N nuclear magnetic resonance. The results were consistent with specific binding of SCN^- to the enzyme approximately 0.72 nm from the ferric ion in the heme group in the vicinity of the porphyrin periphery. Cyanide ion bound to the ferric ion at the sixth position and did not inhibit SCN^- binding to the enzyme. Iodide inhibited the binding of SCN^- and appeared to bind at the same site.

In Milk

In food processing, the most widely studied lactoperoxidase system is in bovine milk and milk products. The lactoperoxidase system is a potent bacteriostatic or bacteriocidal system that has been used to preserve raw milk without refrigeration. It inhibits cell multiplication, lactic acid production in milk, and oxygen uptake by resting cells. This system was studied by ^{15}N NMR and optical spectroscopy at different concentrations of thiocyanate (SCN^-) and H_2O_2, and different pH values in an attempt to identify the chemical species responsible for antimicrobial activity. Conditions giving optimal activity, that is, maximum inhibition of oxygen uptake by *Streptococcus cremoris*, were pH < 6.0 and equimolar concentrations of SCN^- and H_2O_2. The use of equimolar concentrations of SCN^- and H_2O_2, in this instance, 0.25 mM, for optimum effect was also found by Wang. The inhibitory power of the lactoperoxidase system also was investigated in milk on *Listeria monocytogenes* and *Staphylcoccus aureus*. The naturally occurring lactoperoxidase in milk was activated by the addition of 1.2 or 2.4 mM SCN^- and 0.3 or 0.6 mM H_2O_2 for *S. aureus* and *L. monocytogenes* studies, respectively. The lactoperoxidase system increased the predicted time to reach half the maximum attainable cfu/ml by 326 hours for *L. monocytogenes* at 10°C and 36 hours for *S. aureus*, also at 10°C. Another study in milk indicated that the lactoperoxidase should be supplied from an exogenous source and that 0.4 mM H_2O_2 was necessary for activation. A lactoperoxidase system consisting of lactoperoxidase (0.37 U/ml), KSCN (0.3 mM), and H_2O_2 (0.3 mM) delayed but did not prevent the growth of *L. monocytogenes* at 20°C in milk. In another study, it was recommended that the lactoperoxidase system in milk be reactivated periodically by adjusting SCN^- and H_2O_2 concentrations to 0.25 mM during storage of raw milk at 4°C prior to pasteurisation. Using the lactoperoxidase system in milk activated by the addition of 0.2 mM SCN^- and H_2O_2, it was found that the growth of different *Salmonella* serotypes decreased markedly in milk acidified to pH 5.3. The lactoperoxidase system, when used in milk acidified to pH 5.5, caused a reduction of *Campylobacter jejuni*. A glucose-glucose oxidase-activated lactoperoxidase system delayed the onset of exponential growth of *Salmonella typhimurium* in infant formula milk. The preservative properties of the lactoperoxidase system were evaluated in another study on infant formula. The formula was reconstituted with lake water that contained a mixed flora population and incubated at 30°C for 72 hours. The lactoperoxidase system was activated by glucose-glucose oxidase. In the control sample, proliferation of *Enterococcus* and *Pseudomonas* spp. and *Enterobacteriaceae* occurred in 24 hours. In the presence of the lactoperoxidase system, these bacteria failed to grow in the first 24 hours. However, yeasts proliferated and *Rhodotorula glutinis* was the dominant species after 48 hours.

In Milk Products

The lactoperoxidase system also has been tested to reduce the numbers of viable micro-organisms in milk products. Cottage cheese previously inoculated with *Pseudomonas fragi*, *P. fluorescens*, *E. coli*, and *S. typhimurium* was preserved by the lactoperoxidase system. The system was activated by H_2O_2 generated by glucose-glucose oxidase. Detectable levels of the test micro-organisms were not present during the 21-day test period. In French soft cheese, the lactoperoxidase system applied to the cheese surface led to the elimination of viable *L. monocytogenes* from cheeses previously inoculated with the organism. The manufacture of a pickled soft cheese (Domiati type) was not affected when the lactoperoxidase system was used to treat the milk prior to the making of cheese. The processing time was reduced by the lactoperoxidase system, which avoided the direct addition of NaCl to the milk; because the whey was unsalted it could be utilised more readily. Cow, buffalo, and mixed milks were treated with SCN^- and H_2O_2 or heat treated at 90°C for 5 minutes, then refrigerated for 2 days or held at 25°C for 1 day before being inoculated with a 2 per cent yogurt culture and incubated at 40°C for 48 hours. Rate of increase of acidity and decrease of pH was similar for all three treatments. During refrigerated storage for 7 days, there were no significant differences in chemical composition of the three yogurts. Some studies on the effects of the lactoperoxidase system on starter cultures used in the manufacture of cheeses also have been done. Thirteen strains of thermophilic lactic acid bacteria commonly used in the manufacture of cheese were tested for their sensitivity to the lactoperoxidase system by continuous pH monitoring. Inhibition was expressed as the difference in time to reach pH 5 between control and test treatments. *Lactobacillus acidophilus* and three strains of *L. bulgaricus* were inhibited in the presence of SCN^- and lactoperoxidase, which indicated their ability to produce H_2O_2. *Lactobacillus helveticus*, *Streptococcus thermophilus*, and one strain of *L. lactis* required an external source of H_2O_2 to cause inhibition. *Streptococcus faecium* and one strain each of *L. lactis* and *L. bulgaricus* were resistant to the lactoperoxidase system. In another study, a delay of 4.5 hours in coagulation was observed when a thermophilic cheese starter was inoculated into milk 2 hours after activation of the lactoperoxidase system. When milk was inoculated 6 hours after activation, there was no increase in the coagulation time but less acid was produced.

CATALASE

General Characteristics

Catalase (EC 1.11.1.6) is a haemoprotein that catalyses the dismutation of hydrogen peroxide into dioxygen and water. Several organic substances, especially ethanol, can act as hydrogen donors; ethanol is oxidised to acetaldehyde. Catalase isoenzymes with enhanced peroxidase activity were isolated from barley and mutant maize. A manganese catalase from *Lactobacillus lactis* was crystallised and its structure studied. In plants, the major function of catalase is believed to be the decomposition of potentially harmful hydrogen peroxide. When seedlings were submitted to increasing water deficit, peroxidase and catalase activities were preserved or increased as the availability of water to plant tissue was decreased. In most species tested, the two hydrogen peroxide scavenging abilities vary in a complex way in relation to the degree of water stress.

In Various Fruits and Vegetables

Catalase from the pericarp of green tomato fruit was purified to homogeneity as judged by silver staining of SDS gels. The M_r of the holoenzyme was 225 kDa and that of each sub-unit polypeptide 55 kDa. Catalase activity was optimum at pH 6.0–6.5; half-maximal activity was observed at pH 4.5 and 8.0. Tomato pericarp catalase was four times as sensitive to inactivation by silver nitrate as beef liver catalase. Tomato catalase also was more sensitive to protein-modifying agents.

Catalase from potato tubers was purified to homogeneity. The purified enzyme had a low specific activity, about 3000 U per mg protein, and a marked tendency to form aggregates. SDS-polyacrylamide gel electrophoresis showed a single 56 kDa peptide. The apparent M_r of the holoenzyme was estimated to be 224 kDa. There was some homology between the N-terminal sequences of potato (*Ipomea batatas*) and bovine liver catalase. There were time dependent increases of catalase, superoxide dismutase, and α-tocopherol when potato tubers were stored for 40 weeks at various temperatures. The increase of these antioxidants is believed to be in response to activated oxygen production.

Catalase also occurs in spices. 34 spices were tested for catalase activity. Catalase content was highest in basil, marjoram, and thyme. It was lowest in turmeric and cloves.

In Milk

Catalases from milk have been isolated and characterised. The separation and purification of catalase from bovine milk has been performed rarely and much still remains to be understood concerning its structure and mechanism of action. A study by Ito answered some of these questions. This group of investigators made a catalase antibody in rabbit using crystalline bovine liver catalase as the antigen. This IgG antibody was used to prepare an affinity matrix that was able to adsorb bovine milk catalase. The purified protein was electrophoretically homogeneous. Approximately 7 mg catalase was obtained from 10 litres of milk. The enzyme had a specific activity of 328 U per mg protein, compared with a specific activity of 0.13 U per mg protein in bovine milk. Catalase distribution in goat milk was determined. Catalase isoenzymes from goat milk were isolated and purified. About 20 per cent of the catalase was found in the cream and about 72 per cent in the skim milk. Recovery was 17 per cent in buttermilk and 55 per cent in whey. The catalase from these two sources was purified 156 and 56-fold, respectively, with yields of 6 per cent and 26 per cent, respectively. On polyacrylamide gel electrophoresis, two bands were formed that had catalase activity. The apparent isoenzymes had different pH optima.

SULPHYDRYL OXIDASE

General Characteristics

Sulphydryl oxidase catalyses the oxidation of thiols to their corresponding disulphides using molecular oxygen as the electron acceptor. The reaction requires that two moles of thiol are consumed for each mole of oxygen. A FAD-activated sulphydryl oxidase was isolated and identified from whey obtained from bovine skim milk. The enzyme also has been identified in the spores of the fungus *Myrothecium werrucaria*. More recently, the enzyme was isolated and characterised from *Aspergillus sojae* and *Aspergillus niger*.

Mammalian

Sulphydryl oxidase from bovine milk was isolated and characterised. Cream was separated from the skim milk by centrifugation. Chymosin was added to the skim milk to precipitate the casein and the precipitate was separated from the whey by centrifugation. The whey was saturated to 50 per cent with ammonium sulphate and the precipitate collected by centrifugation. The precipitate was dissolved in phosphate buffer at pH 7.0 and dialysed against the same buffer. The dialysed enzyme was concentrated to about 3 per cent protein and allowed to stand overnight at 4°C. The sedimented enzyme was centrifuged and the precipitate was dissolved in buffer, dialysed against deionised water, and lyophilised. The sulphydryl oxidase was purified further by gel filtration and differential centrifugation. The enzyme was purified about 3200-fold with a 41 per cent recovery. The specific activity was 104 μmol oxygen consumed/min/mg N.

The reaction catalysed by bovine milk sulphydryl oxidase involves oxidation of sulphydryl groups to the disulphide in the presence of molecular oxygen. The sulphydryl groups in cysteine, glutathione, or even protein can be oxidised. Glutathione was chosen as the substrate because it is more stable than cysteine and reacts more quickly than protein-bound sulphydryl groups. This enzyme can be assayed by three techniques: oxygen consumption, disappearance of sulphydryl groups, and hydrogen peroxide production. The oxygen consumption method usually is used because of its greater sensitivity and ease of use. The enzyme exhibited a sharp peak at 35°C and had only 10 per cent activity at the relatively low temperature of 50°C. It had an optimal pH of 7.0 and showed very little activity at pH 4.5 and 8.5. The K_m for glutathione was 9.0×10^{-5} at 35°C and pH 7.0.

Sulphydryl oxidase is a glycoprotein containing about 11 per cent carbohydrate, consisting of 7.2 per cent total hexose, 2.9 per cent N-acetyl hexosamine, 0.4 per cent sialic acid, and 0.44 per cent fucose. Atomic absorption spectroscopy revealed the presence of 0.030 per cent iron and 0.011 per cent copper. Apparently, 1 gram-atom of iron may be required for two enzyme sub-units, each having a molecular mass of about 45 kDa, as determined by gel electrophoresis. The molecular mass may be inaccurate since it was difficult to determine because of the tendency of the enzyme to form aggregates. As expected, 0.1 mM EDTA inhibited its activity almost 100 per cent. When the EDTA-treated enzyme was assayed in the presence of 1 μM iron, the activity was restored almost completely. Cupric ion restored its activity about 50 per cent and manganese ion about 40 per cent. Cobalt, zinc and molybdenum ions had little effect.

Only recently have data elucidating the physiological role of sulphydryl oxidase become available. Evidence has shown that sulphydryl oxidase may be involved in the interconversion of xanthine dehydrogenase to xanthine oxidase. Sulphydryl oxidase was demonstrated to catalyse the refolding of reductively denatured ribonuclease. Molecular oxygen did not appear to be responsible for the catalytic effect. Chromatographically purified sulphydryl oxidase was used to examine the refolding of reductively denatured ribonuclease. It was demonstrated that no readily diffusible oxygen species was responsible for the reaction. Reductively denatured ribonuclease A (0.040 μmol) was renatured in about 4 hours using sulphydryl oxidase (22 μg protein) at pH 7.0. Controls using boiled enzyme, ferrous sulphate, or no enzyme required 24 hours to re-nature ribonuclease A completely.

The possible interaction of sulphydryl oxidase with xanthine oxidase was first revealed when both enzymes were present in the same void volume from a BioGel A-150 M column. This result probably indicated that the enzymes were bound to each other because the xanthine oxidase has a much higher molecular mass (303 kDa) than sulphydryl oxidase (90 kDa). A purified xanthine oxidase preparation from bovine milk that had been converted to the dehydrogenase by dithiothreitol was converted to xanthine oxidase by sulphydryl oxidase. However, the xanthine oxidase lost 50 per cent of its activity in about 4 hours when it was incubated with sulphydryl oxidase. The assay method for xanthine dehydrogenase used NAD^+ as the co-factor; the oxidase was assayed in the absence of NAD^+ and the presence of oxygen. However, it was not demonstrated that the xanthine oxidase produced hydrogen peroxide and superoxide anion.

The rate of oxidation of reduced glutathione by sulphydryl oxidase was increased 5-fold with horseradish peroxidase. The degree of enhancement was directly proportional to the amount of peroxidase. Further, the enhancement was not affected by excess catalase. Therefore, it appears that same sort of direct interaction between sulphydryl oxidase and peroxidase may occur because, an initiation of the reaction in the presence of peroxidase, the absorption of the peroxidase increased at 417 nm. The wavelength of the increase in the absorbance spectrum indicates that the species formed corresponds to. HRP II, the one-electron reduced form of oxidised horseradish peroxidase. These observations are consistent with a direct transfer of an intermediate species of oxygen, presumably a hydroperoxy group, from sulphydryl oxidase to horseradish peroxidase. Sulphydryl oxidase activity also increased with the addition of peptide disulphides and oxytocin to assay mixtures. These results were interpreted as being caused by substrate binding to a second site on the enzyme. In the presence of oxidised glutathione, the expected substrate inhibition in the presence of reduced glutathione does not occur. The increased activity of the enzyme in the presence of same disulphides may be due to competitive binding at an allosteric binding site for the substrate. Sulphydryl oxidase was isolated from bovine milk in a form sufficiently pure for sequencing. Although precautions were taken during the electrophoretic elution to prevent reactions that cause N-terminal blocking, negative results when sequence analysis was attempted indicated that the N-terminus is blocked.

Microbial

Sulphydryl oxidase from *Aspergillus sojae* was purified from about 500 U/litre in the fermentation broth to about 2000 U/g. The enzyme was produced by aerobically culturing the micro-organism in a media containing cerelose, inorganic nitrogen, and essential minerals. During growth, the pH was prevented from falling below 3.5 by the addition of alkali, the preferred pH being 4.5–6.0. When a suitable biomass was attained, the culture was harvested, the cells were removed from the whole broth by centrifugation, and the cell-free broth was filtered using filter aid. The broth was concentrated by ultrafiltration followed by further concentration by diafiltration with a membrane with a 10000 molecular weight cut-off. When the intracellular and extracellular sulphydryl oxidases were recovered, the overall recovery was 40 per cent. The sulphydryl oxidase preparation contained an acid and a neutral proteinase. These were removed by chromatography using BioRad P100 resin.

The activity profile of sulphydryl oxidase from *A. sojae* appears to be similar to that of the bovine milk enzyme. In both instances, there is conversion of two disulphide groups in the presence of molecular oxygen to the corresponding disulphide with the production of hydrogen peroxide. With reduced glutathione set at 100 per cent in an oxygen depletion assay, only L-cysteine ethyl ester showed appreciable activity of about 20 per cent. Five other sulphydryl compounds had activities of less than 10 per cent compared with reduced glutathione. The enzyme exhibited a pH optimum of 6.5, a temperature optimum of 42°C, and a very sharp decrease in activity above this temperature. The enzyme showed relatively good stability to about 40°C. Above this temperature, the enzyme activity decreased sharply and at 60°C there was virtually no activity. Unlike the bovine milk enzyme, the sulphydryl oxidase from *A. sojae* does not appear to require metal co-factors because it is not inhibited by EDTA and cyanide.

The sulphydryl oxidase from *A. niger* was characterised and found to be a flavin enzyme. Stoichiometric measurements showed a relationship of 1.7 mol flavin/mol protein. This sulphydryl oxidase is of interest because it requires two reducing substrate molecules. Most flavin oxidases derive two reducing equivalents from one substrate molecule.

The procedure for the isolation and purification of *A. niger* sulphydryl oxidase involved methanol precipitation, ammonium sulphate precipitation, gel filtration, acetone precipitation, hydroxyapatite chromatography, and DEAE fractionation. The specific activity increased from 7.5 to 115 units/mg protein. Recovery was about 36 per cent.

Sedimentation equilibrium experiments indicated a molecular mass of about 106 kDa and SDS-polyacrylamide gel electrophoresis indicated two sub-units, each with a molecular mass of 53 kDa. The enzyme was a glycoprotein containing about 20 per cent hexose and 2 per cent aminohexose. The amino acid profile was unremarkable. The flavin component could be separated from the protein by boiling for 5 minutes. The apoenzyme, when treated with 5 M guanidine chloride, could be re-constituted to about 60 per cent of the original activity when incubated with excess FAD. Incubation with excess FMN re-gained no activity.

The rate of re-activation of reductively denatured ribonuclease A was enhanced by sulphydryl oxidase, demonstrating that the enzyme could oxidise sulphydryl groups associated with protein molecules. Glutathione substrate had the lowest K_m (0.3). Other thiol-containing substrates—cysteine, dithiothreitol, and 2-mercaptoethanol—had much higher K_m values: 43, 66, and 340, respectively. There was no activity with substrates of other oxidases. The substrates tested were D-glucose, L-lysine, D-lysine, benzylamine, xanthine, putrescine, catechol, NADH, and NADPH. With GSH as the substrate, pH optimum was between 5.0 and 5.5. There was no activity at pH 2.5 and very little at pH 10.0.

GLUCOSE OXIDASE

Failure in the Food Industry

Glucose oxidase (EC 1.1.3.4) has not lived up to its promise as an antioxidant in the food industry. It was modified by treatment with succinic anhydride to be much more acid stable. In fact, it had a half-life over 24 hours at pH 2.4 at room temperature. One of its major applications was seen to be the stabilisation of orange soft drinks against oxidative deterioration. This application failed because of the reversible inhibition of the glucose oxidase by the dyes in the orange soft drink below pH 3.0. There is considerable precedent for the formation of complexes between various chemical agents and the bound flavin moieties of flavoproteins. Another important reason for the lack of success in the food industry has been its inactivation by hydrogen peroxide, one of its products, when the enzyme is complexed with glucose.

Of Aspergillus niger

The production of glucose oxidase in *A. niger* was enhanced by treatment with mutagens. A number of mutants of *A. niger* affected in glucose oxidase expression are described. The overproducing mutants could be classified into seven groups. Conidia of *A. niger* strain G 13 were subjected to four stage mutagenisation using different combinations of mutagens. Each stage was treated with UV radiation and nitrosomethylurea, ethyleneimine, acryflavin, or N-methyl-N′-nitrosoguanidine. In all, 400 of 5208 colonies isolated after mutagenisation had higher glucose oxidase activity. The 9 most active showed an activity increase of almost 30 per cent.

The formation and location of the enzyme was studied also in *A. niger* that was pre-grown under citric acid-producing conditions. Glucose oxidase could be induced by shifting the pH of the medium from 1.7 to 5.5. The induction required either glucose or glucose-6-phosphate. Rapid secretion into the medium of oxidase produced in this fashion by a mechanism other than autolysis was surprising, because in *A. niger* glucose oxidase is usually an intracellular enzyme and must be recovered from the mycelium by grinding or lysis. Histochemical evidence for the peroxisomal location of glucose oxidase in *A. niger* would lead one to propose an intracellular origin of gluconic acid. When O_2-enriched air was used to increase the dissolved oxygen concentration of an *A. niger* fermentation, the glucose oxidase activity increased about 3-fold. At the same time, the production of gluconic acid increased about 2-fold. The enzyme was produced in both endocellular and exocellular forms.

Glucose oxidase from *A. niger* was cloned from cDNA and genomic libraries using oligonucleotide probes. Two yeast expression plasmids containing the entire glucose oxidase coding sequence, the untranslated region, and about 2 kb flanking gt10 sequences were transformed into yeast. One plasmid contained a yeast alcohol dehydrogenase II–glyceraldehyde-3-phosphate dehydrogenase promoter, glucose oxidase presequence, glucose oxidase coding sequence, and glyceraldehyde-3-phosphate dehydrogenase terminator inserted into the *Bam* HI site of the yeast—*E. coli* shuttle vector pAB24. The second plasmid had the yeast α-factor leader substituted for the glucose oxidase presequence. The transformed yeast was grown in yeast extract peptone medium containing 4 per cent glucose for 144 hours in shaker flasks at 300 rpm. When grown under derepressing conditions, transformants of these plasmids in yeast secrete large amounts of active glucose oxidase into the medium. Comparison of the yeast-derived glucose oxidase with purified *A. niger* glucose oxidase by polyacrylamide gel electrophoresis showed that the yeast glucose oxidase migrated more slowly than the *A. niger* enzyme. When the yeast enzyme was treated with endoglycosidase H, it migrated with about the same mobility as the *A. niger* glucose oxidase. Thus, it appears that the yeast enzyme has more extensive N-linked glycosylation. The yeast-derived enzyme had a turnover number of 17,000–20,000 min^{-1} compared with 16.200 $min.^{-1}$ for the wild-type enzyme.

Other kinetic parameters for both the yeast-derived and wild-type enzymes were similar. The hyperglycosylated yeast enzyme may be more thermostable than the wild-type glucose oxidase. The primary structure of an O-linked sugar chain derived from the glucose oxidase of *A. niger* was examined. Mannitol appeared to be the only O-linked monosaccharide detected. The O-linked sugar chains were mainly present as single mannose residues. These short O-glycosidically linked oligosaccharide units contained mannosyl-serine and mannosyl-threonine linkages. The *A. niger* glucose oxidase gene was also cloned into *A. niger*, *Aspergillus nidulans*, and *Saccharomyces cerevisiae*. Self-cloning of the gene into *A. niger* increased enzyme production; cloning into *A. nidulans* conferred a novel capacity to produce glucose oxidase. Cloning the gene into *S. cerevisiae* used a plasmid in which the enzyme gene was preceded by a yeast secretion signal. This effected the high-level production of extracellular glucose oxidase in the transformed yeast. Monoclonal antibodies to glucose oxidase from *A. niger* were made using apoenzyme as the antigen. Five of these antibodies, all of the IgG1 sub-isotype, were characterised. Glucose oxidase lost its immunological reactivity with the antibodies under direct covalent immobilisation. The carbohydrate moiety of the enzyme is not immunogenic. None of the antibodies had any detectable effect on the catalytic properties of the enzyme and all five antibodies bound the non-native enzyme coated on ELISA plates in preference to the enzyme in solution.

Of Penicillium

Aglyco- and glycoglucose oxidases from *Penicillium amagasakiense* were compared. *P. amagasakiense* secreted the aglyco version when grown in the presence of 20 mg/litre of tunicamycin. The carbohydrate content of the purified aglyco version was 2.9 per cent and the molecular mass of it and its sub-unit were 130 kDa and 70 kDa, respectively. The glycosylated version had 20.4 per cent carbohydrate and molecular mass of 150 kDa and 70 kDa. Both versions showed remarkably similar properties. For the aglyco and glyco versions, K_M values of 3.4 and 2.7 mM, V_{max} values of 320 and 270 U/mg protein, and pH optima of 5.5 and 6.0 were observed, respectively. Isoelectric focusing also showed similar patterns. Thermostability was also similar. It was concluded that the carbohydrate moiety plays no significant role in the functioning of glucose oxidase from *P. amagasakiense*. Glucose oxidase from *P. amagasakiense* was purified by fast protein liquid chromatography in combination with ion-exchange chromatography. Elution conditions were optimised to permit the simultaneous purification and separation of the glucose oxidase isoforms. Elution was performed with a mixed pH and salt gradient. Three peaks, each consisting of 1–2 isoforms, were resolved with a very flat linear gradient in 40 ml elution buffer. Three more peaks were eluted at 10, 30, and 100 per cent elution buffer. Glucose oxidase apoenzyme from *Penicillium chrysogenum* was purified by chromatography on Con A-Sepharose 4B.

Of Talaromyces Flavus

Talaromyces flavus produces a glucose oxidase that may be involved in the control of the fungal plant pathogen *Verticillium dahliae*. The extracellular glucose oxidase was isolated and purified and found to be a glycoprotein. Amino acid analysis indicated that it was similar to other fungal glucose oxidases. It was stable from pH 3.0 to 7 and the optimum pH for activity was 5.0.

PYRANOSE OXIDASE

Pyranose oxidase or glucose-2-oxidase (EC 1.1.3.10) oxidises the C-2 atoms of D-glucose, D-xylose, D-sorbose, and D-glucono-1,5-lactone to form their respective pyranosones. Hydrogen peroxide is formed as a by-product. It was studied in the 1980s as a more economical way to produce D-fructose from D-glucose. The production of D-fructose from D-glucose by glucose isomerase suffers from a drawback because the isomerisation reaction provides an equilibrium mixture typically containing only 42–50 per cent D-fructose. Pyranose oxidase typically oxidises 95 per cent of D-glucose to glucosone. D-Fructose can be prepared from the glucosone using catalytic hydrogenation. Pyranose oxidase free from glucosone-utilising impurities was obtained from strains of *Coriolus versicolour*, *Lenzites betulinus*, and *Polyporus obtusus*. The enzyme can be used to make pure glucosone for use in making high purity food sugars, for example, fructose, mannitol and sorbitol. When pyranose oxidase is prepared from *P. obtusus*, it is a tetramer with a molecular mass of about 200 kDa with a sub-unit mass of about 69 kDa. The enzyme is a flavo-protein with a sedimentation coefficient of 11.4. Its relative activities toward glucose, sorbose, xylose, and δ-gluconolactone substrates are 100:80:48:12, respectively, at pH 5.0. The enzyme can be used to make D-glucosone from D-glucose, D-xylosone from D-xylose, 5-keto-D-fructose from L-sorbose, or a mixture of 2-keto-D-gluconic acid and D-isoacscorbic acid from δ-D-gluconolactone. Pyranose oxidase can be stabilised against heat inactivation partially by amidinating it with either dimethyl adipimidate or dimethyl suberimidate. The stabilised enzyme has about three times the heat stability of the native enzyme at 65°C. When the organisms are exposed to a mutagenic agent such as UV radiation, higher amounts of pyranose oxidase can be produced. A new basidiomycete that produces pyranose oxidase has been isolated. Pyranose oxidase also can be used for the quantitative determination of glucose without the need for mutarotase, which might be needed by glucose oxidase if the reaction is expected to go to completion in a reasonable period of time. The enzyme also is useful for assaying 1,5-anhydroglucitol, a marker for the diagnosis of diabetes.

XANTHINE OXIDASE

General Characteristics

Xanthine oxidase (EC 1.1.3.22) is an iron-molybdenum flavoprotein that can oxidise xanthine, hypoxanthine, some other purines and pterins, and aldehydes. It is unique among oxidases because it can produce superoxide anion O_2^- as well as H_2O_2. Superoxide dismutase dismutates two molecules of O_2^- to H_2O_2 and water.

In Milk

Xanthine oxidase is found in milk and liver. This section addresses xanthine oxidase in milk and dairy products primarily. About 200 commercially processed dairy products were assayed polarographically for xanthine oxidase activity. Average activity of 22 of 25 fluid milks ranged from 4.5 to 139.6 mU per litre O_2 g^{-1} hr^{-1}. Most of the dried milks and evaporated milk products and all the butters had no activity. Average activity of ice-creams and yogurts ranged from 0 to 40.3 mU; 50 per cent of the 24 brands of creams had activity from 4.2 to 405 mU. About two-thirds of 111 brands and varieties of cheeses had activities ranging from 0 to 420 mU. Activity of fresh raw milk increased 3.7-fold in storage at 4°C and declined when stored for longer periods. An earlier study showed that raw milks had an average xanthine oxidase activity of 110 mU/ml. Homogenisation and pasteurisation inactivated more than 50 per cent of the enzyme and, when packaged, averaged 34 mU/ml. Raw creams had activities from 153 to 319 mU/ml. Much of the xanthine oxidase in bovine milk is localised in the lipid membrane of fat globules. Low speed agitation (5000 rpm for 20 seconds) of washed fat globules added to skim milk released xanthine oxidase. Effects of cold storage (5°C for 24 hours) and heat treatment (60°C for 5 minutes) on free and membrane xanthine oxidase were studied. Both treatments increased total xanthine oxidase activity in milk. In buttermilk, activity of membrane xanthine oxidase increased and free xanthine oxidase decreased with both treatments. When bovine milk was heated at 70°C for 30 seconds, xanthine oxidase activity generally increased, 3 to 4 fold in some samples. Pasteurisation at 85° or 90°C, however, almost completely, inactivated the enzyme.

Xanthine oxidase was isolated and purified from raw bovine milk. The milk was treated with 1–2 per cent Triton X-100, followed by a two-step fractionation using ammonium sulphate. The enzyme was then purified by chromatography on Sephadex G-75, G-100, and Sepharose 6B columns. Final purification was by chromatography on a DEAE-Sephadex A-50 column eluted with a continuous salt gradient in 0.005–0.1 M phosphate buffer at pH 8.6. The purified enzyme exhibited a single peak on gel chromatography or polyacrylamide electrophoresis. The enzyme had an average protein to flavin ratio of 2.0:4.1. The effect of isotopic substitution of hydrogen-8 of xanthine (with ^2H and ^3H) on the rate of oxidation by bovine xanthine oxidase was measured. Results suggested the presence of a rate-limiting step prior to the irreversible C–H bond cleavage step. Xanthine oxidase was converted to xanthine dehydrogenase by 1 per cent 2-mercaptoethanol. This conversion was inhibited partially by 6 M urea. When purified from fresh bovine milk, the enzyme had a specific activity of 3.59 IU/mg, contained 14.8 per cent protein N, had no lipid, was hydrophobic in nature, and had lysine as the N-terminal residue. It was indicated that xanthine oxidase could not be the source of milk riboflavin. Xanthine oxidase purified from bovine milk had a flavin protein spectral absorbance ratio of 5.0, indicating that it was a highly purified protein. Blood serum from a rabbit hyperimmunised against this enzyme fraction exhibited two precipitation lines. The second antigen was identified as bovine IgG. Commercial preparations of xanthine oxidase also contained IgG as a contaminant. When xanthine oxidase was prepared from raw bovine milk in the presence of dithioerythritol, 94 per cent of its xanthine oxidase was found as a dehydrogenase type. The enzyme was converted to an oxidase type when dithioerythritol was removed. Xanthine dehydrogenase also can be converted to the oxidase form with sulphydryl oxidase. When xanthine oxidase was purified by limited proteolysis, structural alterations results in diminished stability, sub-optimal substrate binding, and reduced catalytic efficiency.

Freshly prepared enzyme isolated from the fat globule membrane using deoxycholate yielded one zone on SDS-polyacrylamide gel electrophoresis corresponding to a M_r of 153 kDa. After storage at 4°C for 30 days, there were three zones. It was suggested that an endogenous milk proteinase, possibly of membrane origin, was causing this breakdown. After exposure of the enzyme to trypsin the enzyme retained full activity, although selective cleavage of the enzyme was occurring, as shown by gel electrophoresis. However, after 24 hours of digestion, the enzyme retained only 24 per cent of its activity and the only protein present had an M_r of 92 kDa.

The influence of various milk proteins on lipid oxidation in a related model system was assessed with purified milk metalloproteins on the oxidation of triglyceride emulsions. Xanthine oxidase had little effect on lipid oxidation in the absence of added metals, but was strongly pro-oxidative in the presence of 10 mM Cu^{2+}.

LIPOXYGENASE

General Characteristics

Plant lipoxygenases have been known for over 50 years. Several have been purified and mechanistic details have been elucidated. Lipoxygenases play an important role in the stability of various processed foods. All lipoxygenases use molecular oxygen (dioxygen) to catalyse the oxidation of polyunsaturated fatty acids and initially form fatty acid peroxy free radicals that remove a hydrogen from another unsaturated fatty acid molecule to produce hydroperoxides. The hydroperoxides are converted into acids, aldehydes or ketones, and other compounds. Lipoxygenases also can undergo an anaerobic reaction. Soyabean lipoxygenase-1 can use the hydroperoxide product as its second substrate. Although the mechanism of the reaction is simple, a complex pattern of products is formed: fatty acid dimers, oxodienoic acids and *n*-pentane. Lipoxygenases also show the phenomenon of self-inactivation. The instabilities of the lipoxygenases of pea seeds and wheat were confirmed in recent studies. In the kinetic study of the reaction of soyabean lipoxygenase with arachidonic acid as substrate, a self-catalysed destruction of the enzyme was observed that was involved in the interaction of the enzyme with the hydroperoxy fatty acid in only a minor way. Several types of lipoxygenases are classified by the Enzyme Commission. The lipoxygenase that oxidises linoleate to 13-hydroperoxyoctadeca-9,11-dienoate has been assigned the Enzyme Commission number 1.13.11.12. Other lipoxygenases that have been studied are arachidonate 12-lipoxygenase (EC 1.13.11.31), arachidonate 15-lipoxygenase (EC 1.13.11.33) and arachidonate 5-lipoxygenase (EC 1.13.11.34). Many investigators have suggested that the physiological roles of lipoxygenases in plants might be lipid oxidation with respect to seed ageing and death, regulation of metabolism in germinating seeds, production of plant hormones and inhibitory effects on pathogenic organisms. These effects are not clearly understood. Defense against insect attack was studied for the lipid oxidation products of soyabean lipoxygenases. Bean bugs were repelled by products of linoleic acid oxidation such as linoleic acid monohydroperoxides and hexanal. The insects also preferred lipoxygenase null seeds more than normal soyabean seeds. The oxidation products also repelled two species of leaf beetles that usually do not feed on soyabean seeds. Although the repellant powers of the lipoxygenase products were not great, the results suggested a defensive role for lipoxygenases. More recently, lipoxygenase was implicated in carotenoid metabolism and synthesis of abscisic acid in *Lycopersicon* and *Phaseolus* seedlings. *In vitro*, lipoxygenase cleaved neoxanthin and violaxanthin, a widely distributed carotenoid pigment formed in plants from zeaxanthin, to small ($\leq C_{13}$) fragments. Perhaps *in vivo*, any apocarotenoids formed by the specific cleavage of 9′-*cis*-neoxanthin during abscisic acid synthesis are metabolised rapidly by lipoxygenase or similar enzymes.

Of Soyabean

In the plant kingdom, the most widely studied lipoxygenase is soyabean lipoxygenase-1. This enzyme is a metalloenzyme with a non-heme iron as its prosthetic group. The iron is bound directly to the polypeptide backbone and is not readily removed by iron chelators such as bipyridyl and *o*-phenanthroline. The M_r of the enzyme is 98.5 kDa. Soyabean isoenzymes lipoxygenase-1, -2a, -2b and -2c were examined spectroscopically for the presence of covalently bound pyrroloquinone quinone (PQQ) after derivatisation by phenylhydrazine, 2,4-dinitrophenylhydrazine, and 3-methyl-2-benzothiazolinone. None of these compounds showed evidence for PQQ, contrary to a previous report. Alkyl free radicals from the β-scission of fatty acid alkoxyl radicals were detected by spin trapping in a soyabean lipoxygenase system in borate buffer and linoleic acid, linolenic acid, or arachidonic acid.

Soyabean lipoxygenase has been used since the 1930s to bleach flour to produce white bread crumb. Improvement of dough-forming properties and baking performance of wheat flour using soyabean lipoxygenase to oxidise free lipids in the flour is well known. When the free lipids are removed from the flour, this effect is diminished greatly.

The stability of foods made from soyabeans are affected by their lipoxygenase content. In addition to the oxidation of polyunsaturated fatty acids, soyabean lipoxygenase has been reported to co-oxidise fat-soluble vitamins. Although vitamins A and D do not occur naturally in plant tissues, they may be present in foods containing animal products. Lipoxygenase-catalysed oxidation of linoleic add can be accompanied by considerable losses of vitamins A, D_2, D_3, and E at neutral pH. Co-oxidation of β-carotene was also reported at pH 7.4. Chlorophyll A was bleached by soyabean lipoxygenase L1. It was concluded that linoleinyl radicals were responsible.

Among the methods studied to prevent the deleterious effect of lipoxygenase on soya-containing foods are inhibitors, inactivators, and heat denaturation. Dithiothreitol, EDTA, $CaCl_2$ and ascorbic acid decreased lipoxygenase activity but did not eliminate it. Lipoxygenase activity in soya milk was decreased by propyl gallate, which inhibited the enzyme more than BHA, BHT, and ascorbic acid. Citric acid and ascorbic acid enhanced this inhibition by propyl gallate. Turmeronol A and turmeronol B, phenolic sesquiterpene ketones derived from turmeric, inhibited soyabean lipoxygenase at IC_{50} values of 16 and 9 μ*M*, respectively. These compounds prevented the autooxidation of linolenic acid at a concentration of about 200 ppm. The prior incubation of soyabean lipoxygenase with arachidonic acid resulted in a gradual decrease of activity. Pre-incubation with linoleic acid or 11,14,17-eicosatrienoic acid had no effect on activity. A lipoxygenation product of arachidonic acid, 15(*S*)-hydroperoxy-eicosatetraenoic acid, irreversibly inhibited soyabean lipoxygenase in a time-dependent manner. Inactivation of soyabean lipoxygenase at acid pH was evaluated by grinding whole beans in water acidified to pH 3.0. Lipoxygenase was found to be inactivated completely and irreversibly at or below this pH. The acid treatment slightly reduced the functional properties of soya proteins produced from the beans. Protein solubility and extractibility were reduced but emulsifying ability was not affected. Emulsion stability was decreased and viscosity was increased. Water binding capacity also decreased. Various acids, without heat, were used to inactivate soyabean lipoxygenase with little effect on protein solubility. An ancillary benefit of the acid treatment was the reduction of urease activity to acceptable levels. Lipoxygenase-1 was stable when pre-incubated at pH 3.2–9.2, but was irreversibly denatured at or below pH 3.0. The prevention of the 'beany' and bitter flavour in soya full-fat flour, caused by lipoxygenase, during milling reportedly was prevented by a simple blanch-dry process. Soyabeans were dehulled and blanched in water at 80°–99°C. Water blanching for 3 minutes at 99°C was necessary when the cotyledons were subsequently sun dried. Other workers found that the presence of soyabean solids stabilises lipoxygenase against heat inactivation.

Breeding program to eliminate lipoxygenases in soyabean seeds is another technique to attempt to reduce the detrimental effects of lipoxygenases in processed food products. Lipoxygenase isoenzymes are localised in the cytoplasm of cotyledons of germinating soyabean seeds. Lipoxygenase appears to be absent from the mitochondria, glyoxysomes, protein bodies, lipid bodies, and other organelles. A simple test to detect lipoxygenase isoenzymes using the preferred substrate in a coupled co-oxidation and destruction of β-carotene was devised to assay for the three isoenzymes, LOX1, LOX2 and LOX3, in soyabean seeds. A simple test is necessary to determine the progress of genetic manipulations.

Lipoxygenase-free soyabean seeds were used to study accelerated ageing compared with seeds containing lipoxygenase. When stored at 40°C, 100 per cent relative humidity, for 1–10 days, germination data indicated that the absence of lipoxygenase did not increase resistance of seeds to accelerated ageing. The storage stability of lipoxygenase-free full-fat soya flours were compared with those with lipoxygenase (19–21 U per mg) activity. Free fatty acids, peroxide value, lipoxygenase activity, N solubility index, and β-carotene content were evaluated. Lipoxygenase-free flour stored better than flour with lipoxygenase. The role played by lipoxygenase in flavour quality of soyabean oil was tested by comparing oil processed from special soyabeans lacking lipoxygenase-1 with oil from normal beans. No significant differences were found in flavour quality or stability based on total volatiles. 2,4-Decadienal levels were also similar. Factors other than lipoxygenase-1 appear to affect the food quality of soyabean oils.

Soyabean lipoxygenase may have some important applications in genetic engineering. The regulatory sequence of the lipoxygenase gene is expressed vigorously in soyabean germination. This sequence may be useful for directing high levels of expression of heterologous genes in transgenic plants. The preparative synthesis of linoleic acid 13(*S*)-hydroperoxide using soyabean lipoxygenase-1 has been reported. Immobilised lipoxygenase of soyabean was used to prepare 13-hydroxy-γ-linolenic acid. The enzyme was immobilised on chitosan; product yield may reach 70 per cent at pH 9.0.

Of Potato

Potato arachidonate 5-lipoxygenase was purified and characterised. Potato tuber lipoxygenase preparations converted α-linolenic acid not only to 9(*S*)-hydroperoxy10(*E*),12(*Z*),15(*Z*)-octadecatrienoic acid but also to more polar metabolites. Potato 5-lipoxygenase also appeared to have peroxidase-like activity. It catalysed the reduction of 12(*S*)-hydroperoxy-9(*Z*), 11(*E*)-octadecadienoic acid in the presence of vitamin E. Hydroxyperoxy acids functioned as suicide substrates and irreversibly inactivated potato lipoxygenase. Pre-incubation of lipoxygenase in a haemoglobin solution abolished this inhibitory effect. The results indicate that substrates that can be converted to an epoxide intermediate act as suicide substrates.

The thermal inactivation kinetics of two lipoxygenase isoenzymes were studied in solution. Each of the two isoenzymes followed first-order kinetics with different inactivation rate constants. However, the apparent inactivation kinetics of crude and reconstituted lipoxygenases showed second-order kinetics. This was explained by the presence of two isoenzymes with different thermal properties.

With the purpose of exploring the reasons for enhanced destruction of carotenoids in irradiated potato tubers during storage at 15°C, changes in carotenoids and lipoxygenase activity were measured during storage at various temperatures. Irradiation enhanced the disappearance of carotenoids in potatoes stored at 15 and 20°C and reduced its formation at 4 and 25°–30°C. Irradiation at a sprout-inhibiting dose of 100 Gy caused an immediate decrease in lipoxygenase activity as well as in its capacity to co-oxidise β-carotene. The reduced lipoxygenase activity of irradiated tubers was seen throughout their storage, regardless of the temperature.

Irradiated tubers also had decreased protein content and increased levels of peptides and amino acids. There appeared to be no relationship between lipoxygenase activity and carotenoid destruction in irradiated potato tubers.

Of Tomato

A microsomal membrane-associated lipoxygenase from tomato was characterised. It appears to use the free fatty acids released from phospholipids. The microsomal enzyme is active over a pH range of 4.5–8.0, comprises over 38 per cent of the total (microsomal plus soluble) lipoxygenase activity in the tissue, has an apparent K_m of 0.52 mM, and a V_{max} of 0.186. The membranous enzyme also cross-reacts with polyclonal antibodies raised against soyabean lipoxygenase-1 and has an apparent molecular mass of 100 kDa. The reaction oxidising linoleic acid was inhibited strongly by *N*-propyl gallate. Lipoxygenase was extracted from tomato by an ammonium sulphate technique. Crude and partially purified preparations were stable for 1 week at 40°C and pH 7. With linoleic acid as the substrate, the enzyme had a K_m of 0.37×10^{-5} M and a pH optimum of 4.0–4.5. The effects of ammonium sulphate and EDTA on the enzyme suggested that natural inhibitors are separated during purification and trace metals are needed for activity. A large amount of 5-lipoxygenase can be extracted from the callus of tomato. The callus is induced by culturing the cut tomato stem in a medium containing nutrients, hormones, and vitamins. 5-Lipoxygenase may be useful in the production of leukotrienes.

Of Maize

Maize seeds, after 5 days of germination, contain at least two lipoxygenase isoenzymes, L1 and L2. These were extracted from acetate-buffered acetone powder containing 0.1 per cent Brij 99, a non-ionic detergent. Both isoenzymes had a broad pH optimum: 6.0–8.2 for L1 and 7–9 for L2. The pI of L1 was 6.40; isoelectric focusing of L2 resulted in two peaks with pI values of 5.55 and 5.70. When L1 and L2 were heated at various temperatures for 5 minutes, 50 per cent residual activity remained at 45 and 65°C, respectively. Substrate specificity was tested using linoleic, α- and γ-linolenic, and arachidonic acids. Using initial velocity determinations, linoleic acid was the preferred substrate for both isoenzymes. However, apparent K_m and V_{max} values showed that linolenic acids were the best substrates for L1 whereas linolcic acid was best for L2.

Of Strawberry

In strawberry homogenates, lipoxygenase oxidised unsaturated fatty acids in water but not in chloroform/methanol. The differences between ripe and unripe fruit can be detected by measuring the lipoxygenase content. 10-Hydroxyoctadeca-8,12-dienoic acid and 10-hydroxyoctadeca-8-enoic acid were detected in large quantities in the aqueous homogenates. These compounds are toxic to fungi and may be involved in protecting the plant from pathogenic fungi.

Of Canola Seed

Lipoxygenase from canola seed was purified using ammonium sulphate at 20–50 per cent saturation followed by conventional anion-exchange chromatography using DEAE-cellulose. Only a single active fraction appeared. The pH for optimal activity was 7.5; the apparent K_m was 2×10^{-4} when linoleic acid was the substrate. The partially purified enzyme had a considerably higher activity with free linoleic acid than with linoleic acid esters. The enzyme also showed a high reactivity with the lipids extracted from canola. Rapeseed free of erucic acid has both lipoxygenase-1 and -2 activities; there was great variability of the lipoxygenase activities of different rapeseed samples.

Of Bean

Two lipoxygenase isoenzymes from French bean pericarp were separated and characterised. The two isoenzymes differed in molecular weight, pI, apparent K_m, and lag phase. Both could be characterised as lipoxygenase-2 because of their acidic optimum pH, simultaneous production of carbonyl compounds and hydroperoxides in the presence of oxygen, aerobic chlorophyll bleaching ability, and inhibition by chlorophyll *a*. In the presence of preformed hydroperoxides, both isoenzymes were inhibited competitively by chlorophyll *a*. The effect of cyanide on green bean lipoxygenase, in both the crude and purified hydrophobic forms, was studied to determine if this method could be used to reduce the blanching treatments used in the vegetable industry. The results indicated that cyanide, probably in the undissociated acid form, is a weak competitive inhibitor of green bean lipoxygenase. It is possible that other non-toxic *N*-containing compounds also inhibit lipoxygenase. Green bean lipoxygenase was purified in an attempt to determine its role in food quality deterioration. Lipoxygenase was extracted from fresh and unblanched frozen beans using buffer and subsequently fractionated with ammonium sulphate. The crude enzyme was purified by hydrophobic interaction chromatography followed by isoelectric focusing, ion-exchange chromatography, and gel filteration chromatography. The results suggest three forms of green bean lipoxygenase that can be separated by virtue of their ionic charge and hydrophobicity.

The combined effect of microwave heat treatment and irradiation on lipoxygenase in fresh and dry faba beans was studied. Faba beans were irradiated at 500 krad, then microwaved at 190°C (1500 Watt, 2540 MHz) for various lengths of time. When dry beans were irradiated for 4 minutes followed by microwaving for 2 minutes, the lipoxygenase activity was destroyed completely. The inactivation of lipoxygenase was faster in fresh beans because of their higher moisture content.

Of Other Plants

Lipoxygenases from other plant sources have been studied also. The enzyme was purified from wheat kernels using ammonium sulphate precipitation and DEAE-Sephadex A-50 anion-exchange chromatography. Arachidonic acid was converted mainly to 5Δ8-hydroperoxy-6(*E*),8(*Z*),11(*Z*),14(*Z*)-eicosatetraenoic acid. Moreover, evidence was obtained for leukotriene A4 synthase activity of the wheat lipoxygenase.

γ-Irradiation at doses of 4 and 6 kGy reduced the lipoxygenase activity of intact buckwheat seeds. Avocado lipoxygenase was purified to near homogeneity by affinity chromatography. The pH and temperature optimum were 7.1 and 36°C. The K_m for linoleic acid was 7.2×10^{-2} mM and V_{max} was 432 mU per hr. per mg protein. Epicatechin was a competitive inhibitor with a K_i of 9.0×10^{-5} mM.

The thermal inactivation of asparagus lipoxygenase was studied by Ganthavorn. Inactivation followed first-order kinetics and activation energies for thermal denaturation of partially purified enzyme was 47.5 kcal/mol.

Lipoxygenase heterogeneity in two species of pea was demonstrated by antibody testing. At least seven different polypeptides were identified; five of these were identified as lipoxygenase precursors. Limited N-terminal sequence data indicated further heterogeneity when compared with sequences predicted from cDNAs. Cloning and sequencing of two cDNAs from mRNA of maturing pea seeds allowed the deduction of the complete amino acid sequence of a lipoxygenase polypeptide that is very similar to that of soyabean lipoxygenase-2. The predicted molecular mass of this polypeptide is about 97 kDa, and its sequence permits comparisons of lipoxygenase-2 and -3 isoforms from pea and soyabean.

Of Fish

Lipoxygenases are also widely distributed in the animal kingdom. In teleost fish, arachidonic 12-lipoxygenase is the predominant lipoxygenase. An arachidonic acid 15-lipoxygenase was found in gill tissue of teleost fish and is believed to help control the quality and characteristics of volatile compounds that are implicated in fish odour and flavour. Total activity of this enzyme following hydroxyapatite chromatography was significantly higher than in the crude tissue preparation, suggesting that inhibitors were removed. The enzyme was active toward polyunsaturated fatty acids present in the tissue, producing hydroxylated products from fatty acids with 18-, 20- and 22-carbon chain lengths at carbons 13, 15, and 17. The enzyme had a molecular mass of 70 kDa. The arachidonic acid 12-lipoxygenase from trout gill tissue effectively bleaches β-carotene in conjunction with the peroxidation of different polyunsaturated fatty acid substrates. The maximum velocity of bleaching was higher for trout lipoxygenase than for arachidonic acid 15-lipoxygenase from soyabeans. This difference may reflect the presence of oxidative co-factors in the trout gill lipoxygenase preparation. Trout gill arachidonic acid 12-lipoxygenase can initiate oxidation of polyunsaturated fatty acids to produce acyl peroxides. Using a gill homogenate as a model system, the ability of gill lipoxygenase depended on concentration of the enzyme and substrate and length of time. The major volatile compounds produced from the oxidation of arachidonic and eicosapentaenoic acids were 1-octen-3-ol, 2-octenol, 2-nononal, 2-nonadienal, 1,5-octadiene-3-ol, and 2,5-octadien-1-ol. Formation of oxidative flavours was related to the activity of the lipoxygenase and was inhibited by esculetin and BHA. The arachidonic 12-lipoxygenase of rainbow trout gill tissue was active at pH 7–9 and was inactivated rapidly at temperatures higher than 40°C. Inhibition studies showed that inhibition was generally non-competitive and that the most potent inhibitors were fisetin and quercetin. The presence and relative activities of arachidonic acid 12-lipoxygenase were investigated in 15 species of freshwater fish. All the species contained lipoxygenases capable of initiating the production of off-flavours by the formation of highly unstable hydroperoxides, which appear to be the direct precursors of the off-flavours. Gill lipoxygenase of rainbow trout was reactive with arachidonic, eicosopentaenoic, and docosahexaenoic acids, but had low reactivity with linoleic acid. Optimum activity was at pH 7.5 and inactivation occurred at temperatures higher than 40°C. No lipoxygenase activity was found in muscle, but activity was found in the skin of all the species, although levels varied among species. Glutathione enhanced the stability of gill lipoxygenase, indicating that thiol groups are involved in the activity of the enzyme. Lipoxygenase activity also was detected in sardine skin. A crude lipoxygenase from sardine was partially characterised. Its substrate specificity was elucidated; the enzyme was the most active on α-linolenic acid (113 per cent) when linoleic acid was set at 100 per cent.

Of Chicken Meat

Lipoxygenase-type enzymes were shown to be present in chick muscle. Examination of the oxidation products of [^{14}C]-arachidonic acid revealed the presence of 15-lipoxygenase. The enzyme was purified partially by affinity chromatography on linoleoyl-aminoethyl Sepharose. The enzyme was stable for 12 months at –20°C. Lipoxygenases were suggested to play a role in some of the oxidative changes occurring in fatty acids while chicken meat is in frozen storage.

DEHYDROGENASES

General Characteristics

In the enzyme nomenclature listing of oxidoreductases that act on the CH–OH group of donors (EC 1.1),199 dehydrogenases using NAD^+ or $NADP^+$ as acceptors (EC 1.1.1) and 3 using cytochrome as acceptor (EC 1.1.2) are listed. These oxidoreductases are categorised mostly as dehydrogenases, although 37 are categorised as reductases. This class of enzyme is very important to the metabolic functions of plants and animals but, in foods or food processing, only a few might be considered important, that is, recently mentioned in the literature. These enzymes are alcohol dehydrogenases, shikimate dehydrogenases, malate dehydrogenases, isocitrate dehydrogenases, and lactate dehydrogenases.

In Various Plants

In peach, 18 dehydrogenase isoenzyme systems were examined. Among these were 3 isocitrate dehydrogenases (EC 1.1.1.41), 3 malate dehydrogenases (EC 1.1.1.37), and 2 shikimate dehydrogenases (EC 1.1.1.25).

A shikimate dehydrogenase from cucumber was extracted and purified 7-fold by precipitation with ammonium sulphate and elution from columns of Sephadex G-25, DEAE-cellulose, and hydroxyapatite. Two activity bands were detected on polyacrylamide gel electrophoresis at the last purification step. The enzyme was inhibited competitively by protocatechuic acid. Heat inactivation at 50°–55°C was biphasic. NADPH or shikimic acid protected the enzyme against heat inactivation.

An NAD-dependent alcohol dehydrogenase was isolated and purified from parsley leaves. The M_r of the enzyme was 37 kDa by SDS-polyacrylamide gel electrophoresis and 68 kDa by Sephadex G-100. The enzyme was assumed to consist of two equal-sized sub-units. The pH optimum was 8.9; K_m and V_{max} values for acetaldehyde were 0.6 mM and 279 U/mg, respectively. The nucleotide sequence of an alcohol dehydrogenase gene in octoploid strawberry was deduced.

Studies at the International Rice Research Institute on the *N*-assimilating isoenzymes in rice showed that the shikimate dehydrogenase high-mobility alloenzyme (Sdh-12) was associated with higher protein accumulation. The Sdh-12 homozygotes had a significantly higher accumulation of oryzenin. Since oryzenin is the major component of seed storage proteins in rice, the Sdh-12 alloenzyme can be an effective marker in screening for higher seed protein content among different cultivars.

A rapid and economical method was developed to ascertain hybrid purity of tomato by testing for two variants of alcohol dehydrogenase. In the cultivated tomato, two electrophoretic variants of alcohol dehydrogenase are encoded by two different alleles that are present in the desired F_1 hybrid, open-pollinated varieties, other inbred lines, or commercial hybrids. One of the alcohol dehydrogenases is encoded by an allele in the F_1 hybrid. Therefore, testing for this enzyme will give an indication of the purity of the hybrid. A watermelon mitochondrial malate dehydrogenase (EC 1.1.1.37) cDNA was sequenced and cloned.

In barley, alcohol dehydrogenase and lactate dehydrogenase were induced by anaerobiosis in both aleurone layers and roots. Under aerobic conditions, developing seeds accumulate alcohol dehydrogenase activity that survives seed drying and dehydration. Activity of lactate dehydrogenase also increases during seed germination, but the activity in dry or re-hydrated seeds is very low. Developmental expression of alcohol and lactate dehydrogenase was monitored from 0 to 24 days post-germination. Neither activity was induced to any extent in germinating seeds; however, both enzymes were highly induced by anoxia in root tissue during development. Based on gel electrophoresis results, this increase in activity was due to the differential expression of different *Adh* and *Ldh* genes in root tissue. The increase in activity seems to be due to the *de novo* synthesis of these two proteins. The alcohol dehydrogenase gene family of barley was cloned and analysed to address the question of gene homology between barley and maize systems directly. Alleles of each of the three loci specifying alcohol dehydrogenase in barley were isolated by molecular cloning and analysed. Using the criteria of DNA and amino acid homology with the maize *Adh* genes, two of the genes (*Adh*2 and *Adh*3) have been designated as maize *Adh*2-like and the third was identified as an *Adh*1 mutant allele.

Meat Lactate Dehydrogenase

Lactate dehydrogenase may be important in meat research. There is a relationship between the lactate dehydrogenase isoenzyme pattern and different types of muscle fibres. Differences exist in muscle lactate dehydrogenase activity and in isoenzymes among different species of animal.

APPLICATION OF OXIDOREDUCTASES

Oxidoreductases are not currently large volume products for enzyme manufacturers, but they are being used in interesting specialised applications in food and associated industries. Laboratory and small scale production processes for applications other than those covered in this chapter (enzyme applications in the food industry Bottomley, decolourising annatto whey using peroxidase; Suwa deoxygenation of foods and beverages with lactate monoxygenase) indicate many more possible uses for oxidoreductases, but to date, these applications are not known to be of commercial significance.

Because of their high degree of specificity, oxidoreductases are used extensively as analytical tools for quantitative and qualitative analysis. These enzymes have been used in colorimetric test strips, in test kits with enzymes in solution, and immobilised in membranes for 'instant' read out sensors. Most of these test methods were developed originally for medical diagnostic use, but now are used frequently as analytical tools by the food industry.

Alcohol Oxidase

Alcohol oxidase catalyses the oxidation of short chain linear aliphatic alcohols to the corresponding aldehyde and H_2O_2. Because the reaction rate varies with the alcohol, the approximate alcohol composition of the system must be understood before using the enzyme. Although the reaction catalysed by alcohol oxidase should make the enzyme ideal for the deoxygenation of alcoholic beverages such as wine, wine coolers, and beer, the pH optimum for the enzyme is 7.5–8.0 with an operating range of 6.0–9.5. Alcohol oxidase also is inhibited reversibly by 5–10 per cent ethanol or methanol. With its high pH optimum and substrate inhibition, alcohol oxidase solubilised in these relatively acidic beverages is not an effective deoxygenation system, but could be used in these products if contained behind a selectively permeable cap liner.

A 1990 US patent describes a multienzyme system used to minimise or prevent the formation of off-odours in fish oil emulsions. Alcohol oxidase is one of the enzymes in this system designed to limit the formation of alcohols and aldehydes that contribute off-odours and off-flavours. Despite the fact that alcohol oxidase is not currently being used directly in food, it is being used extensively as an analytical tool for the determination of alcohol concentration. Methodologies have been established to use this reaction for the quantification of alcohol with free and immobilised enzyme.

Two interesting food analysis procedures have been developed using alcohol oxidase. The first of these uses an alcohol oxidase enzyme electrode to determine the amount of aspartame in beverages. The procedure involves the addition of α-chymosin to cleave methanol from aspartame; methanol then is measured by the enzyme electrode. The second method uses alcohol oxidase to determine the methyl ester content of pectins by measuring the quantity of methanol released from the pectin.

Catalase

In 1988, the use of H_2O_2 in North America grew by 15 per cent, the highest rate of growth of any commodity chemical. Although much of this increase was for non-food use (mostly paper and pulp), H_2O_2 producers saw considerable growth in small but growing niches. The growth of the market is caused predominantly by the perceived environmental safety of the product and by new technologies such as waste treatment and aseptic packaging. The use of catalase for removing residual H_2O_2 also has shown significant growth, despite the fact that many of these applications do not require the removal of excess peroxide.

The US Food and Drug Administration (FDA) has affirmed the use of H_2O_2 as a direct food substance with GRAS status (Generally Recognised as Safe) [Code of Federal Regulations (21 CFR) 184,1366, 1989]. The primary functional uses of H_2O_2 in the food industry are categorised by the FDA (Table 10.1) as an antimicrobial agent, an oxidising and reducing agent (as defined in 21 CFR), or as a bleaching agent. The FDA regulations stipulate functional uses of H_2O_2 for each food listed. However, the actual functional use is sometimes not exactly what appears in the regulations. For example, two of the US Department of Agriculture (USDA) approved methods for pasteurisation of egg white, the Armour and the Standard Brands methods, use a combination of heat and H_2O_2 to achieve the desired microbial reduction. 21 CFR lists the use of H_2O_2 in egg products as an oxidising and reducing agent, not as an antimicrobial agent.

Table 10.1. FDA approved uses for H_2O_2.

Food	*Maximum treatment level in food (%)*	*Function*
Milk intended for use during the cheese-making process as permitted in the appropriate standards of identity	0.05	Antimicrobial agent
Whey, during the preparation of modified whey by electrodialysis methods	0.04	Antimicrobial agent
Coloured (annatto) cheese whey	0.05	Bleaching agent
Wine, dried egg whites, and dried egg yolks	Amount sufficient for the purpose	Oxidising and reducing agent

(Contd...)

Food	*Maximum treatment level in food (%)*	*Function*
Tripe, beef feet, herring, and instant tea	Amount sufficient for the purpose	Bleaching agent
Starch	0.15	Antimicrobial agent; removal of SO_2 from starch slurry
Corn syrup	0.15	Reduction of SO_2 levels in finished syrup
Wine vinegar	Amount sufficient for the purpose	Reduction of SO_2 in wine prior to fermentation
Emulsifier containing fatty acid esters	1.25	Bleaching agent

FDA regulations also stipulate that residual H_2O_2 be removed from the product by appropriate physical and chemical means. Although this section does not define an actual method for removal of residual peroxide, many of the regulations dealing with specific products still suggest catalase as the enzyme to be used to catalyse the decomposition of H_2O_2. Catalase does provide a relatively simple and cost effective means of removing residual H_2O_2.

Two commonly used sources for catalase sold in the United States are bovine liver and *Aspergillus niger*. Each type of catalase has unique benefits that makes it useful in different situations. Beef liver catalase catalyses a 'burst effect' of rapid H_2O_2 decomposition but has a fairly limited pH range. Fungal catalase, on the other hand, functions over a wide pH range (pH 2–10) at a slower rate than beef liver catalase, but decomposes a greater quantity of H_2O_2 over time.

H_2O_2/Catalase in cheese milk

Pasteurisation (72°C for 16 seconds) is one of the methods used to ensure the microbiological safety of milk. When milk is being used for cheese production, this heat treatment adversely affects the flavour quality of the finished cheese. An alternative method permitted by the FDA is the use of up to 0.05 per cent H_2O_2 for the chemical sterilisation of milk for Cheddar (21 CFR 133.113), Colby (21 CFR 133.118), washed curd, soaked curd (21 CFR 133.136), granular and stirred curd (21 CFR 133.144), or Swiss and Emmenthaler cheese production. Sufficient catalase to eliminate residual peroxide is required when using H_2O_2. Although the FDA regulations permit the use of up to 0.05 per cent H_2O_2, the chemical generally is used at a lower level in conjunction with HTST (high temperature short time) heat treatment to destroy the bacteria in the cheese milk effectively.

If H_2O_2/HTST treatment is used for cheese milk, the H_2O_2 should be diluted to 5 per cent with water before addition to the milk, then metered into the milk just before passage through HTST equipment. The milk is then cooled to the proper temperature for the desired cheese type and pumped into the cheese vat. The residual H_2O_2 is decomposed by catalase added to the treated milk [approximately 4.2 ml beef liver catalase (100 Keil Units/ml) or 3.9 ml fungal catalase (1000 Baker Units/ml) per 1000 lb milk with 0.02 per cent H_2O_2 added]. The catalase can be added soon after the milk begins to flow into the cheese vat or it can be mixed in after the vat has been filled. About 20 minutes after catalase addition, the cheese milk should be tested for residual H_2O_2. It is important to achieve a negative test for residual peroxide to prevent the destruction of the starter culture that will be added. After achieving the negative H_2O_2 test, catalase manufacturers generally suggest a 10 minutes waiting period before starter culture addition to insure that no residual H_2O_2 remains.

The major benefit attributed to the use of H_2O_2/catalase treatment of cheese milk is the improved flavour of the finished product. The flavours of the cheeses made from H_2O_2/catalase-treated cheese milk are closer to that of cheese made with raw milk without the microbial problems associated with the use of raw milk. Care must be taken when using H_2O_2 since excessive use leads to a chemical off-flavour and a soft pasty cheese.

H_2O_2/Catalase in eggs

As mentioned earlier, two of the pasteurisation methods listed by USDA use a combination of H_2O_2 and heat (USDA). The amount of heat required to achieve proper pasteurisation of albumen by these methods is lower than that required by other approved methods. USDA regulations stipulate that, when using these H_2O_2 methods, procedures must be followed to insure that the residual peroxide is dissipated completely (negative test by appropriate method) before final packaging. One procedure uses catalase whereas the other states that 'methods other than catalase addition are used'.

As early as 1901, egg white has been shown to have catalase activity. Ball and Cotterill characterised and isolated this activity, concluding that catalase is a native component of egg white. They also stated that the amount of catalase activity varies from egg to egg. Because of this variation, the time required for the native catalase to dissipate the residual peroxide can vary. To minimise the hold time and eliminate the variability in the time required for a negative test, large scale producers of egg albumen add additional catalase to the product after heat treatment. Based on a 0.05 per cent H_2O_2 addition to the albumen, a dose of 0.005 per cent (wt./wt.) catalase (fungal catalase, 1000 Baker Units/ml) is recommended as a starting dose. The level of addition can be varied based on temperature and time constraints. Higher temperatures increase the rate of reaction but decrease the total amount of H_2O_2 decomposed by a specific quantity of catalase.

H_2O_2/Catalase in whey and other products

The FDA permits the use of benzoyl peroxide or H_2O_2 as a bleaching agent for coloured (annatto) cheese whey. Both chemicals also are allowed as antimicrobial agents, but H_2O_2 is limited to use in the preparation of modified whey by electrodialysis.

Although benzoyl peroxide is easier to work with, some companies prefer to use H_2O_2/catalase to avoid residual benzoate in their product. Both beef liver and fungal catalase are used by the whey industry to dissipate residual H_2O_2, but the fungal product is preferred when the pH is lower than 5.0 or if the product must be kosher. The FDA limits for H_2O_2 in this application are 0.04 per cent as an antimicrobial and 0.05 per cent as a bleaching agent. The procedure for removal of the residual peroxide with catalase is similar to that used for cheese milk with the exception that, as the pH of the whey decreases, the dose of beef liver catalase must be increased.

As seen in Table 10.1, the FDA permits the use of H_2O_2 in products and procedures other than cheese milk, eggs, and whey. When using catalase to dissipate residual peroxide in these systems, the procedure for cheese milk is a good starting point. Appropriate modifications to the procedure should be made based on time, temperature, and pH of the product and process.

Glucose Oxidase/Catalase

Most of the glucose oxidase/catalase used in the food industry is a semipurified enzyme preparation obtained from *Aspergillus niger* that has glucose oxidase (GO) and catalase as the predominant activities. The catalase in glucose oxidase preparations is present to prevent the build-up of H_2O_2 from the oxidation of glucose. The glucose oxidase to catalase ratios of most commercial preparations used in food applications range from 5:1 to 1:1 (titrimatric units of glucose oxidase; Baker units of catalase). The addition of glucose oxidase/catalase to products is based on the glucose oxidase units (GOU) of the product. The reactions catalysed by glucose oxidase/catalase permit it to be used to scavenge glucose or oxygen or to produce gluconic acid. Glucose oxidase without catalase can be used to produce H_2O_2.

Glucose oxidase was first identified as of potential commercial interest in 1943 by Dwight Baker. He developed the first methods of commercial type production as he sought a means of coping with the oxidative changes in beer due to the high air level in bottled beer, then averaging 5–6 ml per 12 oz bottle. The laboratory-scale trials of glucose oxidase/catalase in beer were successful but production-scale use never became very large. Having sufficient glucose in bottled beer is a problem and the bottling procedures for beer were improved, reducing the need for an oxygen scavenger.

Desugaring of egg components

The desugaring of eggs prior to spray drying is one of the best known and earliest successful uses of glucose oxidase/catalase. In this application, glucose oxidase/catalase is used to oxidise the glucose present in eggs, thereby preventing Maillard browning in the finished dry product. Powdered egg producers still remove glucose from eggs prior to spray drying, but microbial culture is now the most common procedure for desugaring. The enzymatic procedure continues to be used for some products, since it is preferred for egg products that are components of finished products with low flavour profiles. In these systems, the flavours from the culture process can interfere with the flavour of the finished product.

Glucose oxidase/catalase can be used to desugar liquid whole eggs or the separated yolk and white (albumen). A procedure for the desugaring of albumen will be presented as a guide for the enzymatic process. Pasteurisation can be performed before or after desugaring, but the desugaring process should be performed at a temperature of 45°–50°F. The pH of albumen is adjusted to 6.8–7.0 to optimise the performance of the enzyme system. The pH adjustment is made with the slow addition (to prevent localised precipitation) of citric acid (approximately 1 lb/1000 lb albumen). After acidification, about 600 ml of 35 per cent H_2O_2/1000 lb albumen is added. Once the peroxide is blended well into the albumen, add 75000 GO U/1000 lb. This dose is based on a 16-hour desugaring process. A higher or lower dose can be used to shorten or lengthen the procedure. To prevent a rate reduction caused by O_2 limitation, more H_2O_2 should be added to the albumen during the desugaring. Once the foam subsides after enzyme addition, 35 per cent H_2O_2 is added at a rate of 5 ml/min/1000 lb liquid egg white. Peroxide addition is continued at this rate until foam begins to build up (about 2.5 hours). The rate of addition is reduced to 2.5 ml/min/1000 lb. Again, when foam begins to build, the peroxide addition is reduced by half. This procedure of incremental reduction in the rate of peroxide addition is followed until the desired level of glucose is achieved (0.1 per cent glucose on a dry basis). Once the desugaring is complete, other component additions can be made prior to drying, based on the desired end product.

The desugaring procedure for egg yolk is similar to the procedure for albumen but there is no acid addition and the process is complete in about 4 hours. The procedure for whole egg does not require acidification and the enzyme dose is increased to about 1,00,000 GO U/1000 lb. At this level, the process takes about 6 hours.

Removal of glucose for the minimisation of Maillard browning also is performed commercially for other food ingredients, using procedures similar to those used for egg desugaring. In these applications, the enzymatic process is preferred over fermentation because of the minimal change in the flavour of the treated product. Two glucose oxidase/catalase desugaring processes described in the literature are glucose reduction in potatoes for minimisation of browning and conversion of glucose to gluconic acid in grape juice for the production of a low alcohol wine.

Deoxygenation of beverages and other applications

Glucose oxidase/catalase has been used for many years to deoxygenate beverages, but commercial use is limited currently to certain specific applications. As mentioned earlier, the initial commercial purification procedures for glucose oxidase/catalase came from the desire to deoxygenate bottles of beer. Successful use of glucose oxidase/catalase to deoxygenate wine was also reported by researchers but large scale use never occurred. One of the primary causes of these unutilised successes is the inability of fungal catalase to function catalatically in the presence of ethanol. Without catalase activity, glucose oxidase still deoxygenates the product but is, in reality, an effective H_2O_2 producer until either glucose or O_2 is depleted or the enzyme is inactivated. Some organoleptic benefit still can be realised with the use of glucose oxidase/catalase in ethanol-containing beverages if other compounds such as SO_2 or ascorbic acid can scavenge the peroxide formed by glucose oxidase and the oxygen permeation rate into the finished package over time is very low. Currently, there is minimal commercial use of glucose oxidase/catalase in ethanol containing beverages.

One of the best markets for glucose oxidase/catalase for deoxygenation of beverages continues to be the production of lemon or orange soft drinks sold in countries in which the product is exposed to light during distribution. In these situations, the use of the traditional oxygen scavengers and antioxidants, ascorbic and erythorbic acids, would bleach the beverages when exposed to light. To avoid a 'white' citrus-flavoured soft drink, glucose oxidase/catalase is used instead of ascorbic or erythorbic acid to scavenge the oxygen in these products. The use level of the enzyme in this application is 0.25–0.5 GO U/fluid oz beverage. Depending on the temperature, pH, and initial dissolved oxygen content, this dose can remove the dissolved oxygen within 3 hours. The headspace oxygen in the sealed bottled or canned beverage is also scavenged over time, but its rate of removal depends on the rate of diffusion into the beverage.

Soft drinks sold in the United States do not have the extreme light exposure problems prevalent in some other countries with less refined sales and distribution systems. Therefore, ascorbic and erythorbic acids can be used in citrus-flavoured soft drinks without bleaching the product. Another reason for the low level of glucose oxidase/catalase use by the United States soft drink industry is that the FD&C dyes (Red #40, Yellow #5, and Yellow #6) used as colourants in orange and lemon beverages inhibit glucose oxidase. Ironically, one of the markets targeted for glucose oxidase/catalase use because of the susceptibility of its flavours to oxidation never materialised because of the dyes used in those beverages.

The commercial use of glucose oxidase/catalase in juices with relatively high ascorbic acid content (e.g. orange or grapefruit) currently is limited to deoxygenation treatment before pasteurisation. In these products, using the same quantity of enzyme used in soft drinks, the dissolved O_2 is removed (from O_2 saturation to < 0.5 ppm at room temperature) in 30 minutes as opposed to 3 hours for beverages without ascorbic acid. The deoxygenation in these products is rapid because H_2O_2 from the glucose oxidase-catalysed reaction oxidises the ascorbic acid rather than being decomposed by catalase to water and $\frac{1}{2}O_2$. Although this system gives a relatively rapid deoxygenation of high ascorbic content juices, the oxidation of ascorbic acid by H_2O_2 can cause problems in long-term storage, especially if the packaging material allows O_2 to diffuse into the product over time. The sterile-filtered addition of glucose oxidase/catalase to aseptically packaged orange juice actually brought about more rapid browning of the product than the control during storage because of slow oxygen permeation into the package and the enzymatically formed H_2O_2 oxidisation of ascorbic acid. Despite the failure of glucose oxidase/

catalase in aseptically packaged juice, the enzyme system works well as a pre-pasteurisation treatment for juice and juice drinks. In this application, O_2 is removed from the juice prior to heating, improving the flavour of the finished product. The enzyme system is inactivated by the heat treatment, preventing the storage problem.

Work conducted by Sagi and Mannheim at Technion, Israel Institute of Technology, has shown that the glucose oxidase/catalase system can be used to increase the shelf-life of fresh-squeezed non-pasteurised orange juice. Using the glucose oxidase/catalase system to create an anaerobic environment rapidly after squeezing, the growth of spoilage organisms with concomitant off-flavour production was delayed. Currently, there is no commercial use of this process.

The use of glucose oxidase/catalase is not limited to the use of liquid enzyme for deoxygenation of liquid products. With proper pH, temperature, and substrate concentration conditions, freeze-dried glucose oxidase/catalase can be used to deoxygenate packages of non-liquid food products with a water activity as low as 0.65. A commercial application of this system is the deoxygenation of packages of shredded cheese to inhibit mould growth. Use of the dry enzyme system eliminates the spraying of mould inhibitor and the drying of shredded cheese. Instead, the enzyme and anticaking agents are added directly to the cheese coming from the shredder, and the product is packaged immediately. Other applications of this system that become commercialised will be of interest. Glucose oxidase/catalase has been used commercially since the early 1950s. Although there have been many applications of this enzyme system since that time, only a few of them have been commercially successful. Some of these failures were due to unforeseen problems with H_2O_2 formed by the reaction, whereas others were due to the economics or materials available at that time. Only a few of the many glucose oxidase/catalase applications found in the literature have been presented, since these are known to be practiced commercially.

Lactoperoxidase

Lactoperoxidase (LP) is the enzyme component in one of the naturally occurring antimicrobial systems found in bovine milk. For the LP system to exhibit antimicrobial action, lactoperoxidase, thiocyanate, and hydrogen peroxide must be present. Lactoperoxidase catalyses the oxidation of thiocyanate by H_2O_2 to hypothiocyanous acid and hypothiocyanate. The strong antimicrobial effect is due to the ability of these products to oxidise sulphydryl groups and reduced nicotinamide nucleotides. The bacteriostatic and bacteriocidal effectiveness of the LP system depend not only on the organism but also on the product conditions. Several reports provide varying data on the effectiveness of the LP system on *Listeria monocytogenes*.

Preservation of raw milk

Currently, it appears that one of the first commercial uses of the LP system in foods will be for the preservation of raw milk when refrigeration is not possible. This application relies on the lactoperoxidase naturally found in milk, supplementing it with 14 mg NaSCN/litre milk (due to the variable levels of thiocyanate in milk) and 30 mg sodium percarbonate/litre milk (to supply the H_2O_2) to activate the LP system. The use of this antimicrobial system extends the amount of time that the initial microbiological quality of the milk can be maintained without refrigeration. Depending on the initial quality and the ambient temperatures, these additions can extend the time before spoilage occurs by 7–26 hours, thereby extending the range that raw milk can be transported from collection centres to the central dairy. The LP system is not a substitute for good sanitary procedures or pasteurisation; milk treated with the LP system still must undergo pasteurisation.

This additional treatment insures not only microbiological quality but also complete removal of any residual concentrations of active oxidation products of the LP system. Successful field trials of this system have been conducted in many countries where refrigerated transport is currently not available.

Much of the development of the LP system focused on preservation of raw milk where refrigeration is not available during transportation, but research results reported by Kamau and co-workers demonstrate the potential of this system for extending the shelf-life of milk transported, treated, and held in a more typical manner. Their results show that the addition of thiocyanate and H_2O_2 to raw milk, followed by pasteurisation, significantly extended shelf-life at 10°C compared with untreated milk.

Preservation of other products

Use of the LP system to preserve raw milk uses the naturally occurring lactoperoxidase in milk. Although the LP system is not currently approved for food use, tests are being conducted in other food products such as cottage cheese and ice cream. Lactoperoxidase for use in these products is purified from non-pasteurised milk or sweet whey. Depending on the product in which the LP system is to be used, thiocyanate and H_2O_2 must be added also. Thiocyanate does occur naturally at levels as high as 600 ppm in plants and vegetables, particularly cruciferous vegetables (e.g. cabbage, cauliflower), but most products being investigated require the addition of sodium or potassium thiocyanate (currently not acceptable food additives). H_2O_2 for the reaction can be formed with the addition of glucose oxidase and glucose. Results of these tests indicate significant extension of shelf-life using the LP system, but more acceptable methods of thiocyanate enrichment need to be found to help attain regulatory approval.

Sulphydryl Oxidase

Sulphydryl oxidase catalyses the oxidation of free sulphydryls, forming a disulphide bond and H_2O_2. This enzyme has been isolated from a variety of sources such as bovine milk, bovine kidney, *Aspergillus niger*, *Aspergillus sojae*, and other microbial species. In several publications, Swaisgood and co-workers have reported the purification and characterisation of sulphydryl oxidase from bovine milk.

UHT milk treatment

Bovine milk sulphydryl oxidase was used initially to treat UHT (ultra-high temperature) milk for the elimination of the 'cooked' flavour. In this application, the sulphydryl oxidase catalyses the oxidation of the thiols responsible for the cooked flavour, thereby eliminating the off-flavour. Swaisgood report the use of an immobilised sulphydryl oxidase column for the continuous treatment of UHT milk. Starnes isolated sulphydryl oxidase from *A. sojae* and demonstrated its ability to remove the 'burnt' flavour of UHT milk, but there is no commercial use of either sulphydryl oxidase in milk at this time. Perhaps if a higher value product could be produced with this technology, sulphydryl oxidase could be commercialised.

Dough strengthening

The use of sulphydryl oxidase for the oxidation of sulphydryl groups with the formation of disulphide bonds has been speculated as a method of strengthening dough in baking applications. Using the bovine milk enzyme, Kaufman and Fennema evaluated the effectiveness of sulphydryl oxidase to strengthen wheat flour dough. Although several parameters were measured, the investigators were unable to detect significant differences. A European patent demonstrates the effectiveness of sulphydryl oxidase from *A. niger*, in conjunction with other enzymes, to strengthen dough and improve the quality of baked products made with weak flour. Further testing is necessary to demonstrate commercial utility of this process.

SECTION III

Application of Enzymes in Industry

Chapter 11

Milling and Baking

INTRODUCTION

Enzymes are involved in the formation of starch, gluten, hemicelluloses, and lipids during ripening and are the key to dough fermentation and bread making, since they degrade these polymers. The activities of enzymes in the flour used for bread making depend on the climate, the type of grain, and the harvesting conditions. Dry-harvested grain usually contains too low a level of enzyme activity. During germination, the activities greatly increase. A humid climate during the harvest favours germination and can lead to grain with too high an enzyme content, particularly of amylase. Such sprouted wheat cannot be used for bread making.

In times when bread was made by hand, the baker was still able to make good bread from flour poor in enzymes. He could, for example, prolong the action of the enzymes by extending the dough fermentation time. In modern manufacturing processes, the dough must be adapted to the baking process, which is done by adding enzyme preparations. Flours that are poor in enzymes are adjusted at the mill to a standardised activity, by adding malted flour or fungal enzymes. The ideal enzyme quantity for each manufacturing process is added by the bread-maker. In many countries, enzyme activity is reached by adding a combination of bread improvers containing enzymes. In the United States, this state can be achieved by adding enzyme tablets that are dissolved in the baking water. The use of enzymes is also of particular importance in modern baking because tremendously varied products, such as bread, rolls, cookies, and crackers, are made from the same flour. In addition, the demands for machinability of the doughs and constant quality of the baked goods are very high. The baked items must fit into pre-fabricated packaging containers and fill these optimally.

The typical constituents of wheat flour are shown in Fig. 11.1. They can vary to a limited extent. Starch and protein are the main components. The ash content depends on the milling conditions. For example, ash content of 0.55 per cent multiplied by 1000 leads to the flour type 550. The starch granules are partly damaged during milling. The proteins and pentosans are partly soluble and partly insoluble in water. Using enzymes, the properties of the polymer substances starch, protein, and pentosan can be modified. The controlled use of enzymes pre-supposes methods that determine the activity of the enzymes indigenous to the flour. Determination of the enzyme activities has been proved, in itself, to be insufficient, since no correlation exists with practical efficiency. Therefore, practical rheological methods are required for examining the dough, by means of which the enzyme activities can be evaluated. Moreover, the development of enzymes for flour treatment demands knowledge of manufacturing methods and of the practical needs of the manufacturer. Whereas, in the past, malt was the only substance available for the enzymatic treatment of flour, current microbiological techniques have made available a whole variety of fungal and bacterial enzymes with which the machinability of the flours can be influenced in a controlled manner. Even today, the enzymes are selected largely on an empirical basis. The baking test still finally determines the quality of the enzyme products. Many imponderables still remain in this complex interaction of dough constituents and enzymes with widely varying activities. However, many test results help fit more and more pieces into the 'jigsaw puzzle' of baked goods manufacturing.

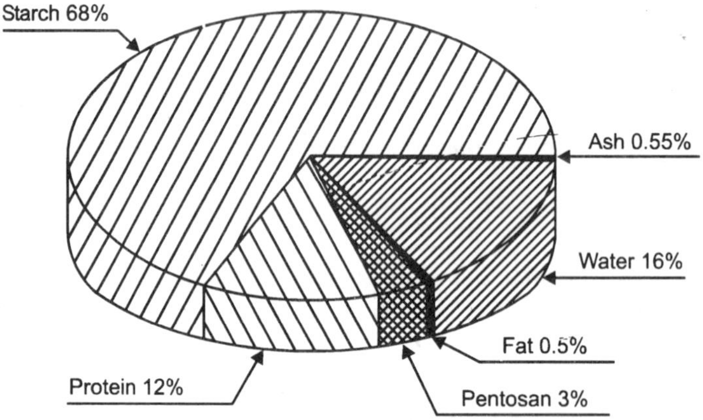

Fig. 11.1. Constituents of wheat flour.

ENZYME IN FLOUR

Flour contains many enzymes of which α- and β-amylases, proteinases, peptidases, hemicellulases, and oxidases are technically most important. These enzymes are normally present at low levels in ungerminated wheat. The low moisture content keeps them inactive during the storage of wheat and flour. On germination, the activity of most flour enzymes increases to such an extent that sprouted flour is unsuitable for baking.

The enzymes of the greatest technical importance are hydrolases, which act on natural macromolecules and degrade them ultimately to the monomers. This degradation is accomplished effectively by the combined action of endo- and exoenzymes shown in simplified form in Fig. 11.2. For branched macromolecules, there is normally also a debranching enzyme, for example, pullulanase in the case of starch enzymes. The endoenzymes change the rheological properties of the macromolecule by splitting a few bonds, whereas the exoenzymes are more important in producing sugars and amino acids, which serve as food for the yeast during dough fermentation, as flavour precursors and as crust-forming compounds.

Amylases

α-Amylases

Up to 22 multiple forms of α-amylase have been detected in wheat. Some multiple forms are a result of the genetic diversity of cereal grains. Certain multiple forms could also be artifacts of the isolation procedure. Perhaps some of the isoenzymes bind to their inhibitors. α-Amylase can be detected in immature wheat. The amounts decrease with further maturation and desiccation. On germination of the grain, the amylase activity increases, several hundred-fold over a 4–5 day period. The α-amylases of malted wheat have a pH optimum of about 5.5, pI values between 6 and 6.2, and molecular masses of 42–42.5 kDa measured by SDS gel electrophoresis. Isoelectric focusing also showed that germinated wheat and rye contain multiple forms of α-amylases that could be divided into low (pI < 5.5) and high (pI > 5.8) pI groups. Zawistoswka and others separated the two groups of α-amylases by immobilised metal affinity chromatography. The α-amylases of immature wheat also exist in multiple form, with a lower pI between 4.6 and 5.1 and higher molecular mass between 52 and 54 kDa. These enzymes are more heat labile but more stable under acid conditions than the germinated α-amylases. Germinated α-amylases adsorb to starch granules. They are preferentially adsorbed to small granules, which they degrade.

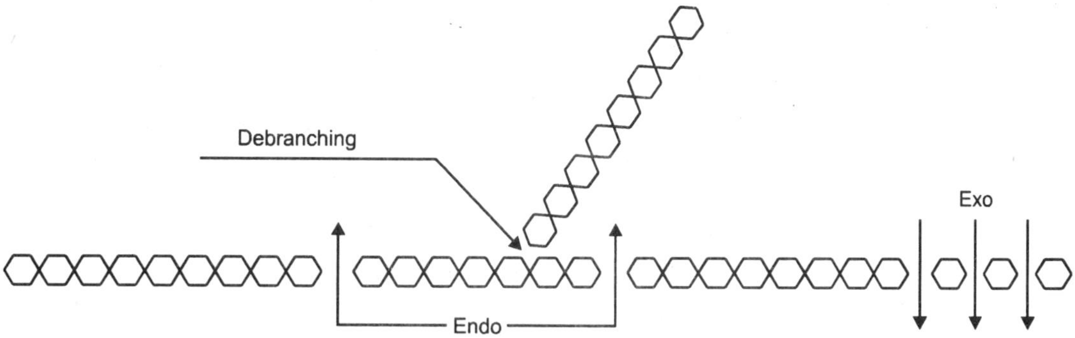

Fig. 11.2. Strategy of enzymatic degradation of natural polymers.

An endogenous α-amylase inhibitor was detected in various cereal grains by immunochemical methods.

β-Amylases

β-Amylases belong to the exoenzyme class (Fig. 11.1). They split alternate α-(1,4)-D-glucosidic linkages in starch or dextrins stepwise from the non-reducing end. The action stops in the region of α-(1,6)-D-glucosidic linkages. Branched starch can be degraded to maltose and β-limited branched dextrin, which is used in the Sandstedt, Kneen, and Blish (SKB) method for determination of α-amylase activity. Multiple forms of β-amylase result from genetic variation, but also from aggregation by disulphide bonds. Up to 80 per cent of the β-amylases are attached to glutenin and are insoluble in water. On germination, the soluble form increases but the total amount does not change significantly. By treatment with papain, insoluble β-amylase can be solubilised to five distinct species with pI values identical to species present in germinated wheat. The multiple forms have molecular masses of about 64.2 kDa, but differ slightly in their pH optima (4.5–5.5) and their pI values (4.1–4.9). β-Amylases are heat labile as is wheat α-amylase, but they are more acid stable. They do not require calcium for stability, but can be inactivated with sulphydryl reagents.

Methods for determining α- and β-amylases in flour

The amount of amylases must be known before proper compensation of enzyme deficiencies by addition of malted flour or a microbial amylase preparation can occur. It is difficult to determine α- and β-amylase separately because of their combined effect on starch degradation. The methods used, therefore, more or less determine both activities. We can assume that, in flour, β-amylase activity is always present in sufficient amounts. The limiting activity is generally α-amylase. The maltose value measures, by the ferricyanide method, the sugars liberated from 5 grams flour in 46 ml buffer at pH 4.7 by the action of α- and β-amylase within 1 hour at 30°C. Another practice-oriented method is the determination of the gassing power, in which gas production is measured after 6 hours in a fermenting dough. Both methods depend not only on enzyme activity, but also on the number of damaged starch granules in the flour, which are attacked much more easily enzymatically.

The Hagberg falling number and the Brabender amylograph method are sensitive to heat-stable α-amylase. In both methods, flour is the substrate for the enzyme and, at the same time, the enzyme preparation. The flour-water suspension is heated. At about 60°C, the starch gelatinises and the viscosity increases.

With increasing α-amylase activity of the flour, more gelatinised starch is broken down and the viscosity of the starch paste decreases. These two methods are adapted to the heat-stable wheat α-amylase, which can act on gelatinised starch at temperatures up to 90°C. Both methods are not able to determine the activity of fungal α-amylase supplements, because heat labile fungal α-amylase is inactivated at about 65°C and therefore can act on gelatinised starch only slightly. However, both methods are influenced by the properties of the starch. The falling number value of flours with nearly the same α-amylase activity increases with the rising gelatinising temperature of the starch. A modified falling number method has been described by Perten and Sproessler. The flour is mixed with gelatinised starch and the falling time is measured at 30°C. With this method all amylases can be determined.

On investigation, the nephelometric method is most suitable for determining both cereal and fungal α-amylase. The enzymes are extracted from the flour and 0.2 ml extract is added to a cloudy substrate suspension of amylopectin or, better, of β-limited dextrins. The turbidity of the suspension decreases according to the α-amylase concentration. The decrease can, for example, be determined with the Model 191 Grain Analyser within 2 minutes of incubation at 37°C. By extracting the enzyme the influence of flour components is diminished and the enzyme activity can be determined on a defined substrate.

Proteases and Peptidases

Flour proteins

The proteins that contribute to dough formation can be divided into glutenins and gliadins, according to their different solubility in a 70 per cent ethanol solution. Both protein fractions are very complex. The rheological properties of flour are determined by the quantity and ratio of gliadin to glutenin. At an increasing proportion of gliadin the resistance to extension decreases and the extensibility increases. In the presence of reducing agents and anionic detergents such as sodium dodecyl sulphate and 2-mercaptoethanol, the glutenins can be dissolved and separated chromatographically. The high molecular weight (HMW) sub-units have sizes of 95,000–1,40,000 daltons. The baking properties correlate with the pattern of HMW sub-units.

The gluten is strengthened by oxidative polymerisation of the proteins. A reduction of the disulphide bonds weakens the gluten. Proteinases weaken the gluten structure by splitting the protein chains. Peptidases form amino acids and can improve the flavour. Both types of enzymes are present in flour.

Proteases

Fox and Mulvihill have compiled the results of research on proteases and peptidases in flour. Most endoproteases are sulphydryl proteases. The carboxypeptidases of flour contain serine at the active site.

Aminopeptidases still have not been detected definitely. The protease activity is distributed irregularly throughout the cereal grain. The activity increases about 15-fold during germination. Therefore, malt contains protease as well as amylase. Too high a protease activity in relation to the amylase can, therefore, be a hindrance during bread making. Sulphydryl compounds such as cysteine and glutathione increase the protease activity; oxidants such as bromate and iodacetic acid are inhibitors. Only some of the proteases can be extracted with water.

The proteolytic activity of the extract was a maximum at pH 3.8, whereas maximal activity for whole flour was obtained at pH 4.4. The extractable enzymes are less heat stable. The thermostable non-extractable proteases are complexed with the proteins of the flour and, thus, protected against thermal denaturation. The absolute number of proteases in the flour is not yet known. Certainly, not all are determined genetically but also arise by autodigestion, microbial infection, or infestation by insects.

Peptidases

Flour extracts contain free amino acids that have been formed by the combined action of proteases and peptidases. A peptidase with a molecular mass of 59000 daltons and a pH optimum of 8.6–9.0 has been described by Kruger. The enzyme does not break down haemoglobin or gelatine. Two carboxypeptidases with pH optima of about 4.2 were isolated by Preston and Kruger. They release amino acids from the carboxylic end of haemoglobin and gelatine.

Measuring activity of proteases in cereal products

Analytical determinations of activity are carried out with protein substrates that do not occur in flour. Therefore, the values obtained fail to address the practical effect of the proteases on the baking properties. Rheological measurements are required to judge their effectiveness. Cereal chemists use the method of analytical determination. Its principle is to measure the amount of hydrolysed protein that cannot be precipitated by trichloracetic acid after enzymolysis under well-defined conditions. This determination can be made by measuring absorption at 275 nm or by using the Folin reagent. Haemoglobin is used as the preferred substrate because of its uniformity and solubility. The activities are expressed conventionally in 'haemoglobin units' (HU) per gram. Wheat, flour has activities of about 1–3 HU and malted wheat flour of about 25–50 HU per gram.

Many other substrates such as caseine, gelatine, and gluten have been used, but none of them have been satisfactory. The action of enzymes on these substrates does not correlate with their performance in bread making (Fig. 11.3).

Non-starch Polysaccharidases

Our knowledge of the indigenous cellulases, β glucanases, hemicellulases, and pentosanases is still fragmentary. Drews had stressed the importance of these enzymes and that of soluble pentosans in the making of rye bread. Non-starch polysaccharides comprise a series of substances consisting of cellulose, pentosans (hemicellulose), β-glucans, mannans, and gluco- and galactomannans. These polysaccharides form complexes with proteins and ferulic acid. They influence the rheological properties of the dough by forming complexes, by their high capacity to bind water, and by their ability to swell. These complex structures can be influenced by enzymes that hydrolyse polysaccharides, but also by proteases and oxidases. Current knowledge of the non-starch polysaccharides is reviewed by Meuser and Sukow.

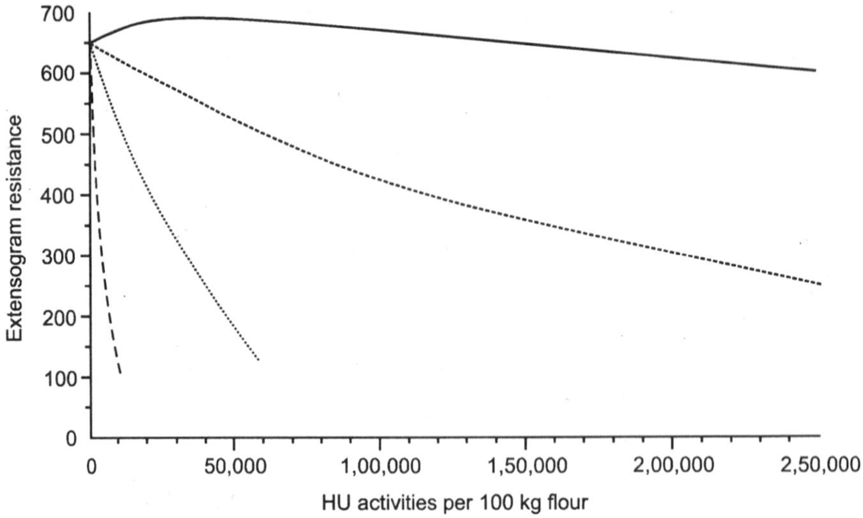

Fig. 11.3. Effect of proteases on dough resistance. The doughs prepared with proteases were fermented (135 minutes at 28°C). Bacterial protease (VERON P®),_ _ _; papain, ...; fungal protease 2 (VERON PS®), -------; fungal protease 1, ——. Resistance was measured with the Brabender Extensograph.

Wheat flour contains approximately 2–3 per cent pentosans, 25 per cent of which are soluble in water. Water-soluble, enzyme-extractable, and total pentosans were estimated in whole wheat and laboratory-milled products by Hashimoto. The enzyme-extractable pentosans were estimated after treatment with enzymes from *Trichoderma viride*. Total pentosans were determined after hydrolysis with 2N HCl at 100°C by the orcinol-HCl method. Wheat flours contain 0.7–0.83 per cent soluble pentosans, 1.66–1.86 per cent enzyme-extractable pentosans, and 1.5–2.12 per cent total pentosans. In rye flour, about 5 per cent pentosans are found, 40 per cent of which are soluble.

The water-soluble pentosans contain arabinoxylans and arabinogalactans. Arabinoxylans have an elongated structure of one long chain of β-(1,4)-linked D-xylopyranose units. Every second or third xylose unit has a side chain consisting of L-arabinose. The arabinogalactans consist of a β-(1,3)- and β-(1,6)-linked galactose chain with individual arabinose side chains. These chains are bound covalently with a peptide and thus form a highly branched structure. The high solubility of arabinogalactan is attributed to the large degree of branching within the molecule.

Pentosans contribute to the physical properties of dough mostly by non-specific interaction, with gluten proteins. The insoluble pentosans—the so-called tailings—consist chiefly of arabinoxylans that oxidatively form an insoluble network, perhaps via ferulic acid. In conjunction with a glycoprotein, a gel forms by means of oxidation that is made of 25 per cent protein, 75 per cent arabinoxylan, and small amounts of ferulic acid. Proteases cause this gel to become fluid. Using proteases, therefore, can alter the properties. The improvement of flour quality by means of oxidases surely results not only from the oxidation of the thiol groups in the gluten protein and the unsaturated fatty acid of the flour lipids, but also from the oxidative modification of the glycoproteins.

The pentosans and glycoproteins absorb a great deal of water and cause gel formation. They are not denaturated by heat and do not retrograde in the cold. They thus prolong the shelf-life and are able to retain gas in the dough. These properties make it possible to manufacture bread from rye flours.

In keeping with their higher pentosan proportion, and their major contribution to dough formation, pentosanases and β-glucanases or hemicellulases play a particularly important role in the treatment of rye flour. Bruemmer disclosed the relative effects on the viscosity of rye flour suspensions exhibited by various enzymes such as pentosanase and α-amylase. Rye dough treated with pentosanases gave softer doughs and increased bread volume. Also in treatment of wheat flour, especially of whole wheat flour, hemicellulases or pentosanases can change the swelling and water absorption properties of the dough and improve bread quality. They improve dough tolerance and ovenspring, bread volume, shape, and texture. During germination of wheat, endo-(1,4)-β-xylanase and endo-(1,3),(1,4)-glucanase activity is very low for the first 3 days but increases rapidly on the fourth and fifth day. Extracts from wheat germinated for 5 days reduce the viscosity of arabinoxylane from rye.

Gaines and Finney added several commercial enzyme preparations with protease, cellulase, and β-glucanase to cookie doughs. Cellulase preparations from *Trichoderma reesei* were effective in maintaining stability of dough consistency and cookie size when doughs were held for 2 hours before extrusion.

Oxidases

Lipoxygenase

Wheat grain contains a relatively high level of lipoxygenase activity in the germ and a low level in the endosperm. Several isoenzymes have been detected electrophoretically in various types of wheat. During storage of the wheat, the activity decreases. The temperatures that occur when mixed feed is pelletised are sufficient to inactivate the lipoxygenase almost completely. The lipoxygenases indigenous to the flour only oxidise free linoleic and linolenic acids.

This type-1 lipoxygenase only forms 9-hydroperoxide. Soya also contains type-2 lipoxygenases. These enzymes oxidise all lipids containing linoleic acid. During this reaction, peroxy radicals arise that co-oxidise the carotenoids of the flour and improve the strength of the dough and the baking properties of wheat flour.

The type-2 lipoxygenase of soya has its optimal activity at pH 6.5 and loses its activity more rapidly during storage than the technologically inactive type-1 enzyme that has a pH optimum of 9. Soya preparations used to improve flours therefore should be evaluated by means of activity determination at pH 6.5. Flour improvement using type-2 lipoxygenase probably occurs via the oxidation of low molecular weight-sulphydryl compounds and the cross-linking of protein molecules of the gluten by radicals. Flour improvement is not achieved in flours from which the lipids have been extracted but it is improved by the addition of an emulsion of linoleic acid.

Peroxidases

Wheat flour contains peroxidases that presumably can cross-link phenolic constituents such as ferulic and vanillic acid. However, the pH optimum of the wheat peroxidase is about 4.5. In the pH range 5–6 of wheat doughs, the enzyme is barely effective, which may be the reason no correlation has been found between the peroxidase content of wheat flours and dough properties. Horseradish peroxidase still has approximately 80 per cent of its optimal activity at the pH of the dough. The addition of horscradish peroxidase in combination with H_2O_2 or pyrocatechol leads to strengthening of the dough. The replacement of pyrocatechol by ferolic or caffeic acid provides similar results. H_2O_2 can be substituted by the H_2O_2-producing system glucose oxidase and glucose. Phenol oxidase from mushroom shows a comparable effect.

Lipases and Esterases

Lipase activity is found mainly in the germ and aleurone layers. Not more than one-third of the total lipase of wheat occurs in patent flours. The distinction between lipases, which only hydrolyse emulsified esters, and true esterases, which act on esters in solution, has not always been clear. In most studies on wheat lipase, esterolytic rather than lipolytic activity has been measured. The activity can be assayed on triacetin or tributyrin. Isoelectric focusing demonstrated the presence of 17 esterase fractions in a crude wheat extract. All fractions were inhibited completely by diisopropyl fluorophosphate (DFP), but not by EDTA or *p*-chloromercuric benzoate. Lipases need Ca^{2+} and bile salts for optimal activity with olive oil as substrate.

The storage of grain under adverse conditions results in an increase in free fatty acids, indicating grain deterioration during storage. Yeast has a much higher level of lipolytic enzymes than flour. The influence of these lipases and esterases on the functional properties of flour is not well known. All these enzymes are inactivated during baking. The function of lipids and lipases is described by Nierle and Mohsen. Protein/lipid interactions in the dough have been reported by Bushuk, Chung, and Zawistowska.

USE IN THE BAKING INDUSTRY

Wheat Bread and Rolls

As early as 3500 B.C., flat types of bread were baked in ovens. Because of the sour dough occasionally used even then, the flat bread rose in the oven and also stayed fresh somewhat longer. It is not known whether the Egyptians used malt as a bread improver. At any rate, they were familiar with the procedure for obtaining malt by moistening barley and allowing it to germinate. In an eulogy by Dinias (360 B.C.), Greek bread was described as beautifully white with an admirable taste that was varied by adding salt, oil, cheese, or sesame. Bread is still made in a similar way today. However, industrially manufactured yeast and the improved knowledge of the use of flour additives, make is possible to obtain a great variety of baked goods of consistently high quality.

Dough process

Basically, baked goods are prepared using the following steps. The ingredients flour, water, yeast, and additions are kneaded to a dough. The dough processing includes fermentation and leavening, dividing, moulding, and shaping.

In the straight dough process, which is used mostly in Europe, the dough is fermented 1–3 hours. In the sponge and dough process consisting of two steps, a sponge is prepared from about 60 per cent of the total flour and water quantity, yeast (2.5 per cent), and enzymes. Mixing is done for 1 minute at low speed and 3 minutes at high speed. The fermentation time is 3–5 hours at about 30°C. For dough preparation, the remaining 40 per cent flour and water, salt (2 per cent), sugar (8 per cent), and shortenings are added and mixed. The mixture is then fermented (20 minutes), divided, proofed for about 55 minutes at 42°C, and baked (18 minutes at 230°C). No-time dough is prepared by immediately dividing the dough after mixing without a fermentation time and proofing for about 55 minutes at 42°C. This short-time process requires fungal amylase and proteinase, which stabilise the dough, in addition to oxidising agents such as potassium bromate and ascorbic acid. The Chorleywood Bread Process is a special no-time dough process used in the United Kingdom that has the particular feature of using a vacuum during dough mixing.

Dough development and baking

Flour doughs contain 60–70 per cent water, depending on the water-binding capacity of the proteins, pentosans, and damaged starch. During mixing, water is bound initially to insoluble protein (glutenin). A water content of about 30–35 per cent free water dissolves soluble flour components such as gliadins. The gluten is the structure maker, gliadin acts as a plasticiser. The well-developed gluten network is filled with starch granules, admixed with pentosans and lipids. Microscopic investigations support the significance of the gluten network, filled with starch present as native, partially, or wholly gelatinised granules. Air bubbles are introduced during mixing and CO_2 bubbles by yeast fermentation. The yeast converts sugar in a sequence of 12 enzymatic reactions to carbon dioxide gas (CO_2). Theoretically, 1 gram glucose yields 0.4 gram, that is, 249 ml CO_2. In practice, this value is lowered by side reactions that yield glycerol, organic acids, alcohol, and precursors for flavour development.

The CO_2 gas expands the dough, provided that the protein network is able to retain the gas and to expand without rupture. The baking properties also depend on the distribution of the gas and the size and stability of the gas bubbles, which can be influenced by kneading and baking improvers. During the baking process, the starch gelatinises. Proteins are denatured at the same time and build a rigid structure. Water is released and picked up by the starch. In the heat of the oven, a dark crust encloses a fine, tender crumb. Figure 11.4 summarises the changes during baking.

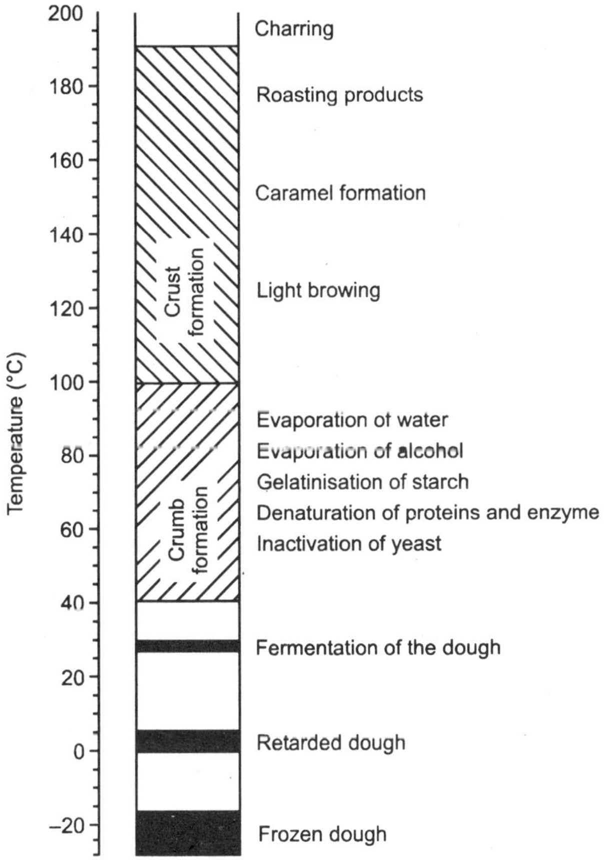

Fig. 11.4. Changes in dough and baking affected by temperature.

These changes take place in both the crust and the crumb. The browning reaction of the crust involves caramelisation of sugars and interaction between sugars and amino acids or peptides. At the same time, flavour and taste compounds are formed. Both browning and flavour forming are influenced by enzymatically produced sugars and peptides during dough fermentation and early baking. At early stages of baking, the increase in temperature enhances enzymatic activity and growth of yeast. At about 50–60°C, the yeast is killed.

Above that temperature, starch gelatinisation, protein denaturation, enzyme inactivation, and alcohol and water evaporation take place. The crumb temperature inside the loaf does not exceed 95°–100°C. The temperatures in the crust are much higher. The taste of the baked bread depends on dough composition, fermentation time, addition of components such as enzymes, and the baking process. Bread flavour is created by the combined effects of many substances. The whole bread-making process, including crumb and crust formation and flavour development, can be influenced by amylases, proteinases, pentosanases, β-glucanases, hemicellulases, lipases, and oxidases.

Effects of amylase supplementation on bread quality

A certain number of starch granules are ruptured when wheat is milled into flour. This damaged starch binds water and can be attacked in the dough by amylases and converted into dextrins, which are then hydrolysed by β-amylase to maltose or by amyloglucosidase to glucose. These reactions constantly supply the yeast with fermentable sugar. A uniform yeast fermentation and gas production in the dough is reached. The addition of sucrose or glucose results in gas production at the wrong stage of the baking process.

Amylases also improve the gas retention properties of the dough through starch modification. The addition of fungal amylase leads to doughs with better machinability and bread with a finer crumb. The crust colour is intensified by means of cereal amylase, but also by fungal amylase. The higher reducing sugar content increases Maillard browning and also gives more flavour. The combined addition of fungal amylase and amyloglucosidase increases the amount of reducing sugars, particularly in sourdough.

Microbiological amylase products usually are standardised to SKB units/g. The highly active amylase concentrates are diluted with starch or maltodextrin in such a way that they can be distributed evenly in practice in the 100 ppm range. The activity of the malted flour may vary from type to type. It is not possible to dose different exogenous amylases based on SKB activity, as is shown in Table 11.1.

Table 11.1. Dosage and inactivation of different exogenous amylases.

Commercial product	*Concentration (SKB units/g)*	*Dosage per 100 kg flour (SKB units)*	*Inactivation temperature (°C)*
Fungal α-amylase	500–20000	25000–1,50,000	65
Bacterial α-amylase	50–5000	Maximum 1000	80
Malted flour	40–80	8000–15000	70

With flours poor in maltose, fungal amylase can be added up to 1,50,000 SKB per kg flour without producing a negative effect. In contrast, malted flour may be used only up to approximately 15000 SKB units and bacterial amylases only to about 1000 SKB per 100 kg flour. The low dosage tolerance is related to the thermal stability of the enzymes. Table 11.2 shows the results of a straight-dough baking test. Amylases increase the loaf volume. Fungal amylases additionally improve the crumb score. The comparison of different fungal amylase preparations at the same SKB level gave different results (Table 11.2). The highest volume and the best dough tolerance with a very fine crumb were obtained using a fungal amylase preparation (VERON ST). This fungal amylase product not only contains different fungal amylases, but also contains β-glucanase and hemicellulase activity.

Table 11.2. Straight-dough baking test using different amylase preparations.

Preparation	*Dosage (g/100 kg)*	*α-Amylase (SKB/100 kg)*	*Loaf volume (ml)*	*Crumb grain*	*Texture score*
Control			1200	Fine	Fair
Malted wheat	100	6000	1300	Fine	Good
	200	12000	1350	Coarse	Fair
Fungal amylase A	15	25000	1370	Fine	Good
Fungal amylase B (VERON ST)	15	25000	1480	Very fine	Very good
Bacterial amylase	10	1000	1290	Fine	Fair
Bacterial amylase	30	3000	1330	Coarse	Very poor

Thermal stability of α-amylases from various sources

Differences in the heat stability of the amylases are important because of the decisive effect of enzyme action during baking when the starch begins to gelatinise at 60°–65°C and become highly susceptible to hydrolysis. The temperature in the centre of a loaf of white pan bread rises by 4°–6°C/min during baking and reaches 90°–95°C. Figure 11.5 shows the baking temperature and baking time of rolls and 1 kg white pan bread. In the case of rolls, even heat-stable amylases have only a short time to act on gelatinised starch before they are inactivated by heat. Fungal amylase is inactivated at 75°C within 10 minutes (Fig. 11.6). Malt and bacterial amylases are more heat stable and have an adequate time to act on gelatinised starch. This difference explains the high yield of soluble dextrins in the crumb. The thermostability of α-amylases from cereal, fungal, and bacterial sources in the yeast dough systems was investigated by Smith. In this model experiment with small rolls, bacterial α-amylase activity increased during fermentation, proofing, and especially during the last 4 minutes of baking. An excess of heat-stable amylases results in a sticky crumb. Such loaves cannot be sliced and bagged directly after baking. Flour of partly sprouted grain or an overdose of malt or bacterial amylase can cause such a serious problem. However, such enzymes can improve the shelf-life of the bread if they are used in a controlled manner.

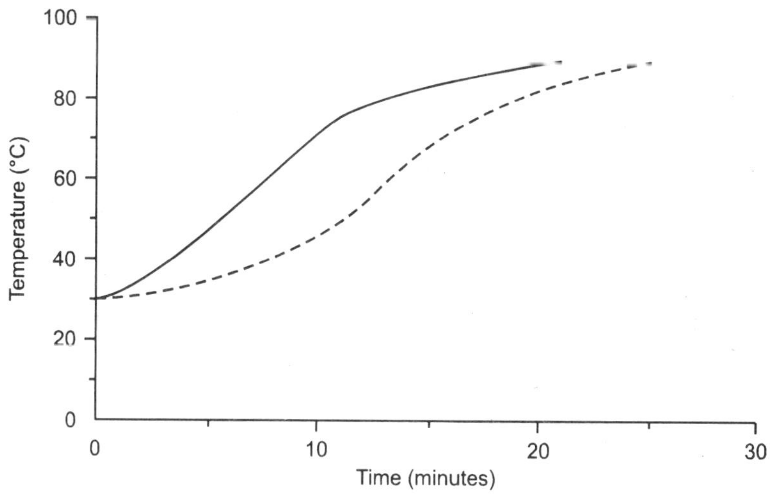

Fig. 11.5. Baking time and temperature in the crumb centre of bread (-----) and rolls (——).

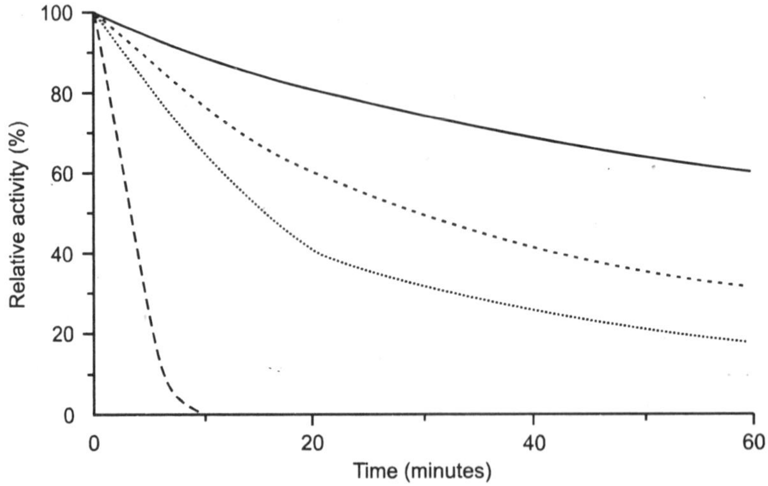

Fig. 11.6. Heat stability of different amylases at 75°C in a 1 per cent enzyme solution in water, pH 5.8, heated for different lengths of time. Enzymes tested were bacterial amylase (——), malt amylase (_ _ _), VERON F 25® (.......), and fungal amylase (------).

Proteases and dough consistency

During the preparation of dough, the proteins of wheat flour form a network. A simplified model of a gluten network with disulphide bridges is shown in Fig. 11.7. Each square represents an amino acid molecule.

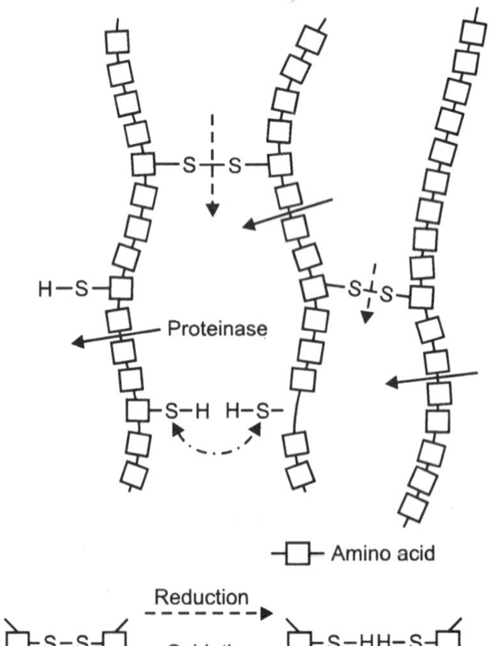

Fig. 11.7. A simplified model of gluten network and possibilities for modification.

The gluten structure gets stronger the more sulphur bridges are present. Bromate and dehydroascorbic acid strengthen the dough in this manner. The proteases break down the network by splitting peptide bonds. Therefore, the addition of proteases to a strong dough improves the elasticity and texture of the gluten and the handling properties of the dough.

The gas formed by the yeast is retained better. The gluten structure is softened and expands under the increasing gas pressure. Thus it is possible to obtain good dough rise, even with strong flours, and, therefore, make bread with high volume and good crumb structure. Fungal proteases are used in large quantities, particularly in the United States. About two-thirds of the white bread in the United States is produced with the addition of proteases. Protease enzymes relevant to the baking industry have been reviewed by Lyons. The reason fungal proteases are particularly suitable for bread making is that they soften the gluten structure of the dough specifically, and only to a certain extent. Unintentional overdosing does not lead to decomposition of the structure, as can happen easily when bacterial protease or papain is used. Figure 11.8 shows that the extension resistance of doughs measured in the extensograph drops rapidly because of the action of papain and bacterial proteases. The gentle action of the fungal proteases is, according to the most recent tests, attributed to the fact that the fungal proteases only have limited access to the compounds in the dough complex that they will break down. This limited access also may explain why many fungal pro teases in the dough show almost no effect, although they are excellent at breaking down proteins such as haemoglobin and glutenin in solution.

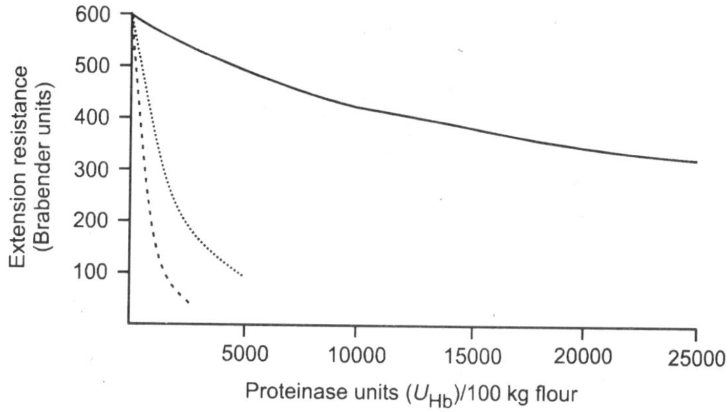

Fig. 11.8. Extension resistance of doughs after treatment with different proteinases: VERON PS®, ——; papain (COROLASE L 10®) .... ; bacterial proteinase (VERON P®), -------, dough resistance was measured with the Brabender extensograph.

The practical efficiency of various fungal proteases therefore cannot be determined analytically, for example, by the HU method, but only by means of methods similar to practical conditions. Suitable equipment is the Brabender extensograph or the Chopin alveograph. These methods are based on rheological measurement of the dough. The extension resistance and the extensibility of doughs thus can be determined. Extensograph curves of flour dough treated with fungal and bacterial proteases are shown in Fig. 11.9. Added proteases must act within the pH range of the dough (pH 6–5.5).

Alkaline or acid proteases are therefore less suitable. Further, it must be considered that flour contains protease inhibitors that affect the added protease products in different ways. Thus Gabor found that the bacterial protease thermitase and brewery enzymes can be inhibited by a serine protease inhibitor derived from wheat; VERON P®, on the other hand, is not influenced by this inhibitor.

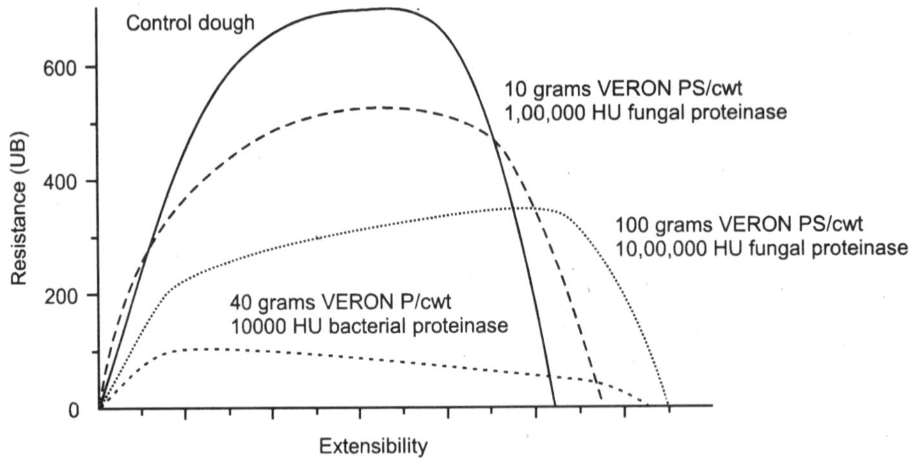

Fig. 11.9. Effect of fungal proteinase (VERON PS®) and bacterial protease (VERON P®) on Brabender extensographs of flour dough.

Bread staling

Extended shelf life has become of crucial importance to the industrial baker. At least 5 per cent of all bread produced is lost as a result of staling. We know that retrogradation of starch plays a major role, but gluten, lipids, and pentosans are also factors contributory to staling. Enzymes cannot change the quantity of these polymers but can change their properties. Bacterial amylase delays staling by increasing hydrolysis of gelatinised starch to dextrins. Dextrins of a particular size interfere with cross-links between starch and protein. The heat-stable bacterial amylase can act on the starch gelatinised at approximately 65°C for a longer time during baking. Thus, more starch is degraded and its retrogradation is delayed. It is possible to detect more dextrins in bread crumb extract with a degree of polymerisation of 6 or higher (Fig. 11.10).

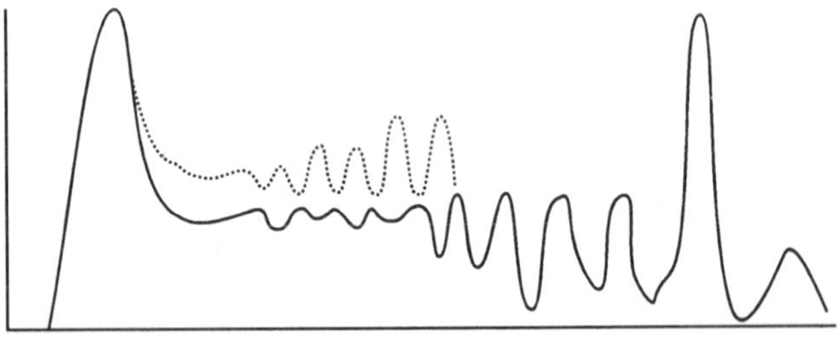

Fig. 11.10. Gel chromatographs of bread crumb extracts on Biogel P2-400 mesh (column: 200 × 2.5 cm, 65°C), carbohydrates in the eluate are detected with orcinol sulphuric acid at 420 nm. Control extract, ——; baked with VERON F 25®, ...

The properties of the starch in the bread crumb change. Bread crumb from bread made with bacterial amylase, that has been homogenised with water, shows lower gelatinisation maxima on heating in the amylogram (Fig. 11.11). It is very difficult to control the level of bacterial α-amylase. An excess of enzyme leads to a soft gummy crumb. Good results are obtained with 3 SKB α-amylase units per kg flour. Rye bread made with the addition of bacterial amylase was kept fresh for 6 months. The dosage used for this purpose was about 100 times that for white bread. At the low pH value of the rye doughs (~4.2), both the activity and the heat stability of the bacterial amylase are reduced greatly. Attempts were made to achieve the same effect as the bacterial amylase without its side effects by stabilising the fungal amylase in sucrose syrup. 'Intermediate stability' amylases with thermostability properties between those of traditional fungal and bacterial enzymes have been described by Hebeda. The debranching enzyme pullulanase improves the crumb structure when used in combination with bacterial amylase.

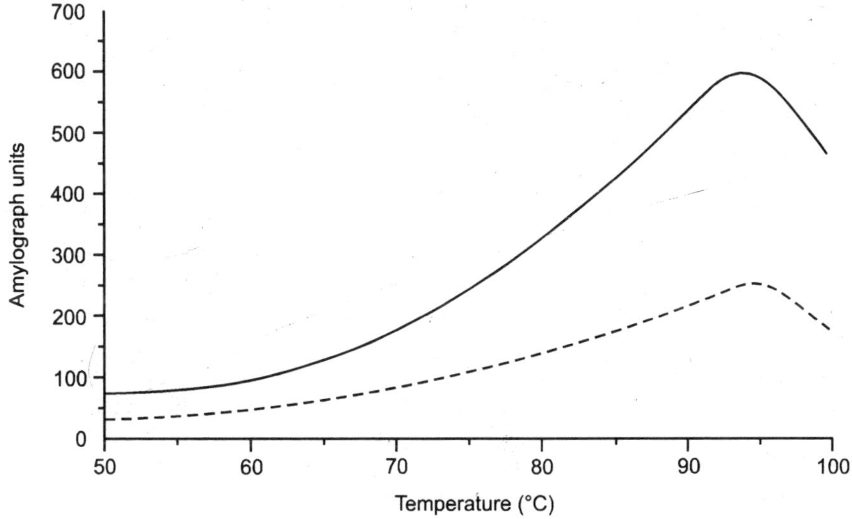

Fig. 11.11. Brabender amylograms of bread crumb homogenisation with water. Control extract ——; baked with VERON F25®, - - - -.

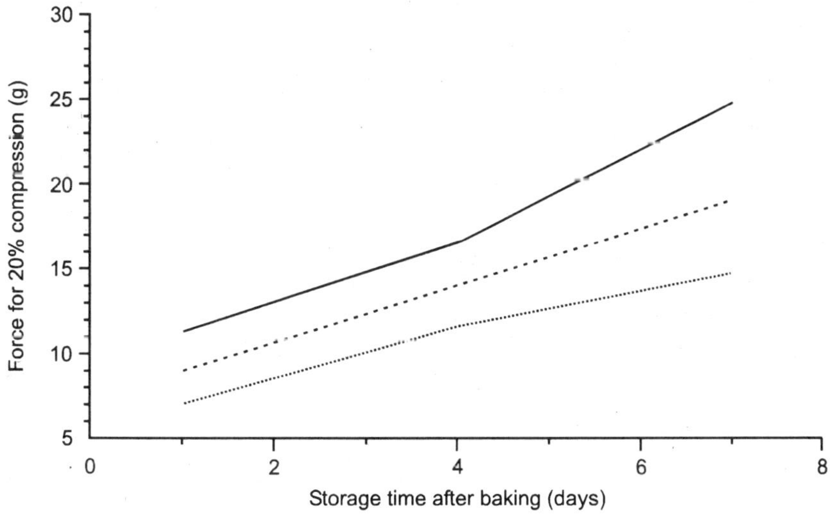

Fig. 11.12. Shelf-life studies of bread baked with emulsifier and enzymes. Breads were stored in bags at 28°C. Crumb softness was measured with compressimeter. Control bread, ——; with bacterial amylase (5 g/cwt), --------; with VERON ESL® super (5 g/cwt), ...

Fungal enzymes can be used in combination with emulsifiers without the risk of overdose to prolong the shelf-life of bread. By carefully balancing amylase, proteinase, pentosanases, and β-glucanases, and by combining them with sodium stearyl lactylate and monoglycerides, a higher volume and an improved crumb structure are obtained and shelf-life can be prolonged by 3–7 days (Fig. 11.12).

The influence of monoglycerides, proteins, and pentosans and their effects have been reviewed by Kulp and Ponte. Staleness of bread involves several sensory perceptions such as odour, flavour, mouth feel, and firmness of the crumb. All these attributes are used by a taste panel to judge the freshness of bread.

The firming of the crumb during storage can be measured by instruments such as the Baker Compressimeter and the Instron testing instrument. These instruments measure the force required to depress a plunger to a given depth into the bread crumb. The changes occurring during bread staling have been reviewed by Pomeranz. Despite considerable work in this field, bread staling is of interest because many questions remain unanswered.

The storage temperature has a strong influence on staling. Bread should be stored at room temperature or deep frozen, but should not be kept in the refrigerator. The retrogradation of the starch is lower at room temperature. Thus, it is best to store bread at room temperature under conditions that minimise moisture loss. At –22°C, bread can be stored properly for 15–20 days, but the flavour and aroma deteriorate thereafter. Freezing maintains bread at the freshest level at the time of freezing, but cannot improve the quality of stale bread.

Retrogradation is basically irreversible, but stale bread can be improved, to a limited extent by re-heating. The starch crystallinity can be measured by X-ray diffraction studies or by differential thermal scanning.

Frozen and retarded doughs

Consumers increasingly expect an oven-fresh assortment of baked goods. The frozen dough process puts deep freezing to work to preserve dough pieces at temperatures below their freezing point. The fermentation stops and the action of the enzymes ceases. In the retarded dough process, dough pieces undergo intermediate storage at temperatures between –5° and +8°C. The required storage time can be adjusted by the storage temperature and the yeast content. Fermentation is retarded, but not interrupted completely. Before baking, the dough pieces are warmed again and proofed.

Correct freezing, flours with low thermostable cereal α-amylase content, and freezing-tolerant yeast are most important. Because of its low heat stability, a fungal amylase shows no effect on gelatinised starch. Because of the prolonged dough processing times, the swelling phenomena are amplified and result in decreased water retention in the baked goods. Special fungal enzymes with amylase and balanced side activities improve dough stability and fermentation tolerance, and increase ovenspring and loaf volume.

In addition to fungal enzyme treatment, it is recommended that, flours with low natural enzyme content be used for frozen or retarded dough. Frozen dough must be fermented after thawing. Instead of freezing the dough, one can store deep frozen prebaked rolls. The rolls are baked insufficiently to gelatinise the starch, form a rigid structure, and destroy micro-organisms. Completing the baking process immediately before consumption insures a fresh product with crisp crust. Prebaked rolls are ready for use after 10 minutes baking time.

High-fibre bread

Although per capita consumption of bread has dropped in the United States, speciality breads such as whole-wheat, multigrain, and other high-fibre breads are produced increasingly. Dietary fibres have different physiological as well as functional effects. They are not digested by human digestive enzymes and therefore are not absorbed nutritionally in the small or large intestine. They can be determined analytically by the AACC Method 32–10. In fibre-rich dough, the percentage of regular wheat flour is reduced. The swelling and water absorption properties of the dough are changed, which can result in slack and sticky or excessively stiff doughs and leads to bread of inferior quality. The particle size of the added whole wheat and multigrain and the relative proportions of soluble and insoluble fibre are important, causing relatively faster or slower water absorption.

Negative effects of the hemicelluloses, β-glucans, and pentosans in the supplement can be offset by enzyme preparations containing hemicellulase or pentosanase. Straight-dough bread formulations containing fibre-rich materials can be influenced by enzyme preparations. VERON HE®, developed for rye flour treatment, hydrolyses high-polymer insoluble pentosans, breaking them down into fast-swelling fragments. VERON ST® is designed to have a stabilising effect on slack and sticky dough.

Treatment of Rye Flour

The use of rye increases the number of different types of bread in Europe. It is also possible to manufacture a variety of rye-wheat mixed breads. Rye bread consisting entirely or partly of rye flour is made using sour dough or acidifiers. A sour dough is acidified and leavened by its micro-organisms. Sour dough processing leavens the dough. Leavening of a sour dough by yeast frequently is augmented by adding baker's yeast to the final dough. Sour dough processing also develops flavour. Acidification of the dough results from the action of lactic acid bacteria. Such acidification is essential for baking rye bread. A high ratio of lactic acid to acetic acid is desirable. Acidification is retarded considerably by adding salt to the rye sour doughs. Sour doughs use a starter culture in the initial stage. The starter cultures can be bought or taken from a ripe sour dough. The culture also can be produced by spontaneous fermentation initiated by micro-organisms indigenous to the flour. The addition of fungal amylase improves the fermentation of the sour doughs. The low pH value of rye doughs influences the activity of the flour enzymes. The amylases indigenous to the flour become less active as the degree of acidification increases. By adding acid ingredients, even rye rich in amylase can be made into bread. In contrast to wheat dough, the pentosans rather than the proteins play the decisive role in dough formation. Rye flour contains more soluble and insoluble pentosans. They absorb many times their own weight in water, but produce porous and well-leavened baked products only with difficulty. The water bound by the pentosans is transferred to the starch during heating in the oven. Amounts and properties of starch are of major significance in rye bread making. The starch forms the sheetlike component. The gums, probably in cooperation with protein, form fibrils. Rye starch gelatinises at 50°–70°C. Thus, even the heat-labile fungal amylase can act, until inactivated at 65°C, on the gelatinised readily accessible starch during baking. Therefore, the addition of fungal amylase can be detected both in the amylogram and by means of the falling number. Both methods serve to characterise the quality of the rye flour and serve as a basis for treatment with amylases.

In the amylogram, the gelatinisation temperature of the starch as well as the quantity is important for the evaluation. Good baking properties are obtained above 200 Amylogram Units and 63°C gelatinisation temperature. The sooner the starch gelatinises, the greater its degree of enzymatic degradation. This characteristic can lead to a damp crumb and a weakening of crumb elasticity. The gelatinisation temperature therefore also should be taken into account. The amylogram can be lowered not only by means of amylases, but also by means of pentosanases, indicating that the pentosans also contribute greatly to the viscosity at increased temperature. In contrast to proteins, pentosans do not denature on heating. Rye flour from a dry harvest contains few enzymes, and high molecular weight starch and pentosans. The doughs are tough and subject to poststiffening. The bread has low volume and the crumb is dry and tends to crack. The addition of pentosanase solves these problems. The insoluble high molecular weight pentosans already are transformed partially into swellable fractions during kneading. Therefore, the doughs do not become tougher during fermentation. The loaves are larger, the crumb is softer, and the bread stays fresh longer.

Use of Enzymes in Cookie and Cracker Production

For cookie and cracker making, flours poor in protein with a weak gluten structure are most suitable. The dough from such flours does not shrink. Such flours are very difficult to obtain in uniform quality. By adding suitable proteases, flours rich in protein can be used also. Particularly suitable for this purpose are bacterial proteases and papain. These enzymes hydrolyse the gluten structure to such an extent that the dough loses its elastic properties, is easy to roll out, keeps its shape, and does not shrink. The energy required for kneading is clearly reduced. Proteases replace sodium sulphite, which is no longer allowed in many countries and which softens the gluten structure by reducing the disulphide bridges. The incorporation of proteases permits baked goods of superior organoleptic quality to be obtained. The enzyme dosage depends on the flour grade, recipe, and method of dough preparation. When using protein-rich flours and for recipes with a high fat and sugar content, high doses of enzymes are used. Intensive kneading or a higher water content promote the contact between the glutenin substrate and the enzyme. A higher dough temperature, for example, 40°C, accelerates the enzyme reaction. Both measures help reduce the enzyme quantity used. Gaines and Finney measured the effects of different enzymes on cookie spread and cookie dough consistency.

Wafer Production

When protein-rich flour is used in wafer making, the batter becomes thick and the protein partially forms lumps, which clogs the sieves and pouring nozzles. In contrast to cookie and cracker doughs, wafer dough contains a high proportion of water which promotes the effect of the enzyme. However, the batter is poured onto the wafer plates and baked immediately after the dough has been prepared. Bacterial protease and papain are suitable for wafer making because of their rapid hydrolysis of the gluten. However, it has been shown that these products first promote thickening and only liquefy the dough in the desired manner after a rest of 10–20 minutes.

Special enzyme products function without dough rests. They can be used without altering the manufacturing process, since the dough becomes liquid in the mixer and does not stiffen subsequently. Therefore, it is possible to obtain the advantages described by Drechsel and Ruttloff without reservation. It becomes possible to manufacture wafers from any type of flour. The homogeneous batters are more liquid. Thus, pumpable doughs with a higher flour content can be made, meaning less water must be evaporated during baking, leading to a decrease in energy use. If the flour content remains unchanged, the water evaporates more rapidly from the thin-bodied dough, leading to a reduction in baking time and an increase in production capacity.

CONCLUSION

Flour contains enzymes. Their type and quantity, however, are not optimally suited to the needs of baked goods manufacturing. This deficiency was recognised long ago and compensated for by the addition of malt flour. Microbiological enzymes such as fungal amylases, fungal proteases, pentosanases, hemicellulases, and β-glucanases have since then taken the place of malt flour.

In the new generation of microbiologically produced enzymes, special products of bread, rolls, cookies, crackers, and wafers, retarded dough processing and shelf-life extension figure ever more prominently in research efforts. This trend is likely to continue.

Chapter 12

Starches, Sugars and Syrups

INTRODUCTION

Corn is the largest and most important crop and the corn refining industry is the largest single consumer of corn in the food and industrial category.

Corn is nearly the sole source of starch, syrups and sweeteners in the United States, because of the abundant supply, low cost and ease of storage that insures a continuous source of raw material throughout the year. Elsewhere in the world, other crops such as sorghum, wheat, rice, potato, tapioca, arrowroot and sago are used in addition to corn for the production of starch and starch-based sweeteners. The selection of raw material generally is based on availability, cost and by-product value. However, refined starch products all exhibit essentially the same properties, regardless of raw material source.

STARCH PRODUCTION

Starch is produced in the corn refining industry by a process in which aqueous streams are used to facilitate separation of the individual components present in the corn kernel. Consequently, the process is called 'wet-milling', as opposed to the 'dry-milling' process that separates corn fractions by milling, aspiration and screening.

In the wet-milling process, shelled corn is first steeped or soaked, in a sulphur dioxide solution at the proper temperature and pH to initiate a lactic acid fermentation. The corn absorbs water and is softened to facilitate subsequent separation into the various components. The steeped corn is broken by coarse grinding, the germ is separated by centrifugal hydroclones and the fibre is recovered by screening. The remaining material consists of gluten (a high protein fraction) and starch. Gluten is removed by centrifugation and a starch stream containing 3–5 per cent protein is recovered and further purified by washing with water through a series of hydroclones. The relatively pure starch slurry (about 45 per cent w/w solids) is processed to a dry product or used as substrate for production of sweeteners and syrups.

Sweetener and Syrup Production

Corn sweeteners and syrups are produced by enzyme or acid hydrolysis of starch (Fig. 12.1). Degree of hydrolysis often is expressed in terms of dextrose equivalents (DE). DE is defined as the percentage of reducing sugars present, calculated as dextrose (D-glucose) on a dry solids basis. Starch DE is essentially 0 since very few reducing end groups are present in the polysaccharide. Complete hydrolysis of starch yields dextrose, that is, a 100 DE product. Therefore, an intermediate degree of hydrolysis yields a product exhibiting a DE between 0 and 100. DE is measured by a variety of techniques, including oxidation/reduction reactions, osmometry and liquid chromatography.

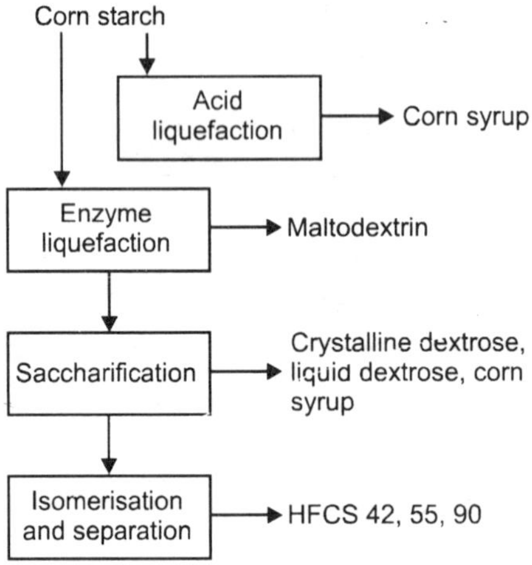

Fig. 12.1. Production of dextrose, corn syrups and maltodextrin from starch.

Initial hydrolysis (liquefaction) is conducted with a thermostable bacterial α-amylase to produce a soluble low-viscosity hydrolysate of 10–15 DE. The liquefied hydrolysate can be processed directly to a maltodextrin (maltodextrin is defined as a hydrolysis product of less than 20 DE) product or used as the substrate for conversion (saccharification) to dextrose with the enzyme glucoamylase. The dextrose hydrolysate is processed to crystalline dextrose (dextrose is defined as a product of ≥ 99.5 DE), liquid dextrose, or a high dextrose corn syrup. In addition, a portion of the dextrose can be isomerised to fructose with a glucose isomerase enzyme to produce a high fructose corn syrup (HFCS).

A wide variety of corn syrups (corn syrups are defined as products of 20–99.4 DE) can be made by acid or enzyme hydrolysis. Direct acid hydrolysis of corn starch yields straight acid corn syrups. Enzymatic saccharification of acid hydrolysate or of low DE enzyme hydrolysate produces acid-enzyme or enzyme-enzyme corn syrups, respectively.

STARCH PRODUCTION AND APPLICATION

Corn starch, as manufactured by the corn refining industry, is a 99 per cent pure product containing about 0.65 per cent lipid, 0.25 per cent protein and 0.1 per cent minerals. The starch is composed of amylose and amylopectin polymers in the form of individual granules that are 5–25 μm in size. The amylose fraction represents 25–30 per cent of the starch and is composed of dextrose units arranged in a linear fashion through α-1,4 glucosidic bonds. An amylose molecule consists of 100–1000 dextrose units. The amylopectin fraction represents the remaining 70–75 per cent of the granule and is composed of linear starch chains branched through an occasional α-1,6 glucosidic linkage. Amylopectin molecules are very large and range from 50–100 million in molecular weight.

Corn starch is used for a variety of food and non-food purposes (Table 12.1). In the unmodified form, starch is used in applications in which the ability to absorb water and swell is important in developing high paste viscosity and strong gels for thickening properties. Modification of starch alters functionality and improves properties for many applications.

Table 12.1. Food, drug, cosmetic and industrial uses of corn starch.

Abrasive paper	Cord polishing	Ore refining
Adhesives	Cork	Paints
Antibiotics	Crayon	Paper
Aspirin	Desserts	Pie filling
Baby foods	Detergents	Pharmaceuticals
Bakery products	Dispersing agents	Photographic films
Baking powder	Dressings, surgical	Plastics
Batteries, dry cell	Drugs	Plywood
Beverages	Dyes	Printing
Binding agents	Fermentation	Pre-cooked meals
Boiler compounds	Fibre board size	Protective colloids
Bookbinding	Fibreglass	Salad dressing
Briquettes	Fireworks	Sauces
Cardboard	Flours	Soaps
Chalk	Gravies	Sugar, powdered
Chemicals	Insecticides	Textiles
Chewing gum	Insulation	Tiles, ceiling
Chocolate drink	Lubricating agents	Tyres, rubber
Cleaners	Meat products	Vegetables, canned
Coatings	Mixes, prepared	Wallboard
Colour carrier	Mustard, prepared	Wallpaper
Confectionery	Oilcloth	Water recovery
Cosmetics	Oil well drilling	Yeast

Food and non-food applications account for 35 and 65 per cent of corn starch utilisation, respectively. The largest food uses are in brewing, chemicals and drugs and pharmaceuticals, each of which accounts for about 20 per cent of total usage.

Baking, confectionery and canning each utilise 6–7 per cent of corn starch produced. In non-food applications, over 90 per cent of starch is used in the paper industry to improve paper strength and appearance. Generally, unmodified starch used for this purpose is thinned enzymatically or cooked at high temperature prior to use to improve adhesive quality.

Several different procedures are used to modify starch and alter physical properties. Methods include acid treatment, oxidation, cross-linking and pre-gelatinisation. Acid modification is conducted by controlled hydrolysis of an aqueous starch slurry to produce a thin boiling product of reduced viscosity, increased gelling tendency and improved gel clarity.

Acid treatment and roasting of starch under low moisture conditions produces dextrins of various types that exhibit high water solubility and low viscosity. Oxidation with sodium hypochlorite converts free hydroxyl groups to carboxyl groups and reduces gelatinisation temperature, lowers viscosity and improves paste clarity by reducing retrogradation. Cross-linking starch with bifunctional reagents restricts swelling and provides a stable paste. Simultaneous pasting and drying of starch produces a pre-gelatinised product that develops viscosity in cold or warm water.

SWEETENER AND SYRUP PRODUCTION

Starch Liquefaction

The first step in the production of corn sweeteners is enzymatic liquefaction, or partial hydrolysis, of corn starch to provide a soluble, low-viscosity, low DE material suitable for subsequent conversion to a variety of products.

Liquefaction is conducted with thermostable bacterial α-amylases that have the ability to withstand the initial high reaction temperatures needed for rapid and complete starch gelatinisation. Enzymes that are used for the purpose of starch liquefaction include α-amylases derived from bacterial organisms such as *Bacillus stearothermophilus, B. licheniformis* and *B. subtilis*. The *B. stearothermophilus* and *B. licheniformis* enzymes are very thermostable and are used routinely at temperatures in excess of 100°C. The *B. subtilis* enzyme is less thermostable and generally is not used above 90°C, to prevent a rapid loss in activity.

The bacterial α-amylases specifically catalyse the hydrolysis of α-1,4 glucosidic linkages and act in a random but reproducible manner to reduce polysaccharide molecular weight. Saccharide compositions attained with each enzyme at DE levels of 10–40 are shown in Table 12.2. At the normal liquefaction DE of about 10.85–90 per cent of the oligosaccharides are larger than DP-5 because of the limited degree of hydrolysis. If liquefaction is allowed to proceed to 30 DE, both the *B. stearothermophilus* and the *B. licheniformis* enzyme produce high levels of maltopentaose (23–26 per cent). In contrast, the *B. subtilis* enzyme does not have a DP-5 producing ability, forming only 8–13 per cent pentasaccharide over a 20–40 DE range.

Table 12.2. α-Amylase action pattern[a].

		Saccharide distribution (% dry basis) for DP[b]							
α-Amylase source	*DE*[c]	*1*	*2*	*3*	*4*	*5*	*6*	*7*	*8+*
Bacillus stearothermophilus	10	0	2	5	1	3	11	10	68
	20	1	7	9	3	11	20	9	40
	30	4	12	12	5	26	10	3	28
	40	8	23	21	8	15	4	5	16
Bacillus licheniformis	10	0	2	6	3	4	7	6	72
	20	1	7	12	4	12	13	6	45
	30	4	12	15	5	23	7	3	31
	40	11	17	16	6	21	4	4	21
Bacillus subtilis	20	2	7	9	5	8	18	6	45
	30	6	13	12	7	13	10	3	36
	40	13	22	14	7	9	4	4	27

[a]Substrate is 20 per cent w/v acid modified starch at 70°C, pH 6, for *B. stearothermophilus* and *B. licheniformis* enzymes and at pH 7 for *B. subtilis* enzyme.

[b]DP, degree of polymerisation, where DP-1 is a monosaccharide, DP-2 is a disaccharide and so forth.

[c]DE, dextrose equivalent.

Enzymatic liquefaction requires careful control of reaction parameters such as percentage of solids, temperature, time, pH and calcium level to insure efficient hydrolysis, minimise enzyme cost and reduce the potential of downstream processing problems. Typical operational ranges for each of the liquefaction parameters follow.

Starch solids

Starch solids level generally is maintained at 30–40 per cent, preferably at 30–35 per cent. At higher solids concentrations, available water content is not sufficient to achieve complete starch gelatinisation, whereas at lower solids, increased cost of water removal impacts negatively on process economics.

pH

Starch slurry pH normally is controlled at 5.8–6.5. At a higher pH, by-product formation is increased, resulting in loss of product yield and increased refining costs. As pH is decreased below the optimum range, enzyme becomes more subject to inactivation if upsets in process control occur. The *B. stearothermophilus* amylase has a lower practical pH minimum than the *B. licheniformis* amylase (5.6 vs. 6.0); a mixture of the two enzymes may offer advantages in low pH operations.

Calcium

Calcium is added in the chloride or oxide form as an enzyme co-factor to enhance α-amylase thermostability. Generally, a calcium level of 100–200 ppm (starch dry basis) is sufficient for *B. stearothermophilus* and *B. licheniformis* enzymes. A higher calcium level of about 300 ppm (starch dry basis) is often used with *B. subtilis* enzymes.

Time/Temperature

Reaction time and temperature are balanced to provide the optimum conditions for enzyme activity and rapid starch gelatinisation. Normally, temperature is controlled to be as high as possible while still permitting sufficient enzyme activity to achieve the required degree of hydrolysis.

Typical industrial starch liquefaction processes are diagrammed in Fig. 12.2. The EHE (enzyme-heat-enzyme) process was the first enzymatic liquefaction process used when *B. subtilis* α-amylases first became commercially available. The HT (high temperature) and LT (low temperature) processes were developed to take advantage of the increased thermostability of the *B. stearothermophilus* and *B. licheniformis* α-amylases.

In the EHE process, a 30–40 per cent starch slurry containing 200–400 ppm calcium (dry basis) is adjusted to pH 6–6.5, dosed with enzyme and passed through a stream injection heater at a temperature of 85–90°C. The heater is designed to mix the starch slurry and steam thoroughly while rapidly increasing temperature to a level that provides instantaneous gelatinisation and hydrolysis. The reaction is continued for about 1 hour to liquefy and solubilise the starch. Under these conditions, however, complete starch solubilisation is not attained because of the presence of a small amount of insoluble starch-lipid complex that is not disrupted by enzyme action. If the insoluble material is allowed to persist throughout the process, the small starch particles interfere with subsequent filtration and refining operations. Consequently, a 120°–140°C, 5–10 minutes heat treatment step is included to dissociate the complex. The α-amylase is denatured during the heat treatment step, necessitating a second enzyme addition and an additional reaction for 1 hour at about 85°C. The dissociated starch polymers are hydrolysed sufficiently to prevent reformation of the starch-lipid complex and a 10–15 DE hydrolysate is produced that is suitable for subsequent processing.

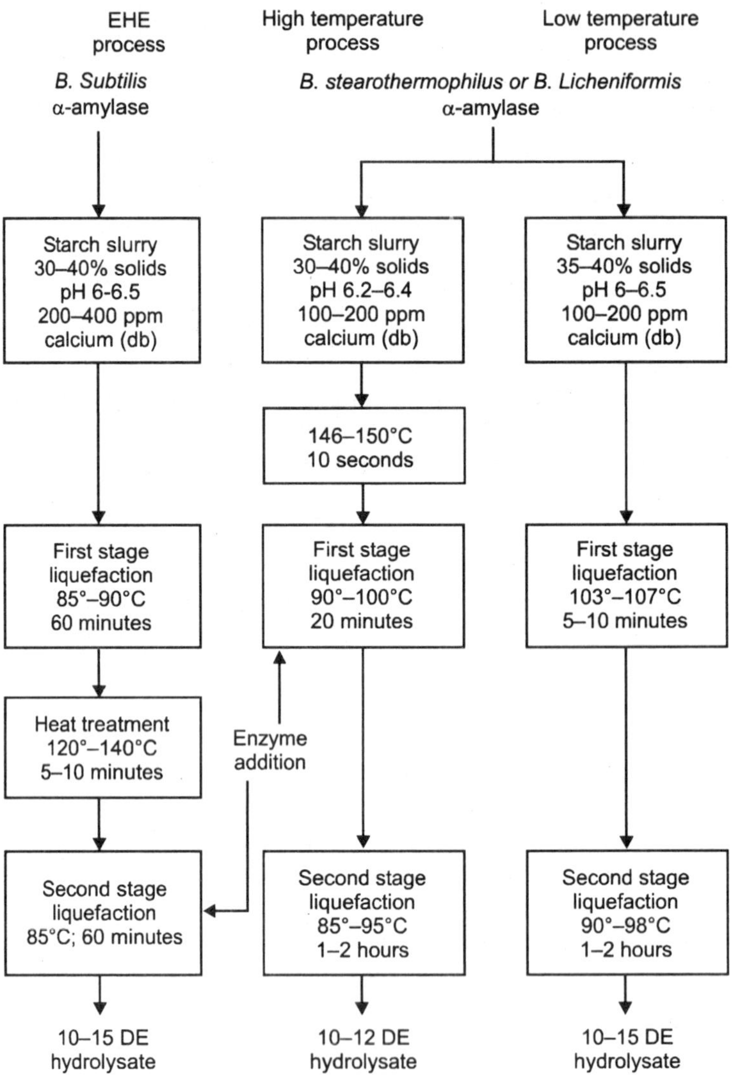

Fig. 12.2. Starch liquefaction processes.

In the HT process, the starch slurry is prepared in a manner similar to the one previously described, a thermostable α-amylase is added and initial liquefaction is conducted at 146°–150°C for 10 seconds or longer. As in the EHE process, enzyme is inactivated thermally, requiring a second enzyme addition followed by a 95°C reaction for at least 20 minutes. Further hydrolysis is then conducted at 85°–95°C for 1–2 hours to produce a 10–12 DE hydrolysate. The advantage of this process is reported to be enhanced starch gelatinisation due to the high initial temperature, resulting in improved filtration characteristics and hydrolysate quality.

The LT liquefaction technique has become the primary industrial liquefaction process used because of the relatively low temperature needed for efficient hydrolysis. In this process, a starch slurry at pH 6.0–6.5 containing 100–200 ppm (dry basis) calcium is dosed with enzyme, passed through a steam injection heater at 103°–107°C and held at that temperature for 5–10 minutes. The hydrolysate is flashed to atmospheric pressure and held at 90°–98°C for 1–2 hours to produce a 10–15 DE hydrolysate. Compared with the other described processes, the LT process has the advantages of lower temperature (i.e. reduced energy requirements) and only one enzyme addition. The initial temperature is sufficiently high to disrupt the starch lipid complex but not to cause excessive inactivation of the thermostable enzymes used for liquefaction.

Stability of α-amylase during high temperature liquefaction is important in achieving efficient starch solubilisation. In the absence of substrate, *B. stearothermophilus* and *B. licheniformis* α-amylases are 25–135 times more stable than a *B. amyloliquefaciens* (*B. subtilis*) α-amylase because of additional salt bridges that maintain enzyme conformation and minimise thermoinactivation. Presence of stability factors such as substrate or calcium are effective in increasing enzyme stability by several additional orders of magnitude, as has been shown for α-amylases derived from *B. stearothermophilus* and *B. licheniformis*.

Factors that affect α-amylase stability during liquefaction include temperature, time, pH, substrate solids and calcium level. Half-lives of α-amylases derived from *B. stearothermophilus* and *B. licheniformis* under various combinations of typical process conditions were calculated from the data of Henderson and Rosendal. Table 12.3 lists the stability of *Bacillus stearothermophilus* and *Bacillus licheniformis* α-amylases under process conditions. Lowering solids level from 37.3 to 26.5 per cent or reducing calcium level from 40 to 10 ppm (on an 'as is' basis) decreases half-life of either enzyme by a factor of 3. Reducing pH from 6.5 to 5.5 reduces stability of *B. stearothermophilus* and *B. licheniformis* α-amylases by factors of 3 and 5, respectively. Comparing all conditions tested, the *B. stearothermophilus* enzyme is 3 to 4-fold more stable than the *B. licheniformis* enzyme. Consequently, LT process changes have been proposed to take advantage of the highly thermostable nature of the *B. stearothermophilus* α-amylase. For instance, increasing the initial reaction time in the LT process to 10 minutes from the typical 6 minutes allows a reduction in enzyme dosage of 40 per cent while still maintaining efficient liquefaction.

Table 12.3. Stability of *Bacillus stearothermophilus* and *Bacillus licheniformis* α-amylases under process conditions.

Temperature (°C)	*pH*	*Calcium (ppm, as is)*	*Solids (% w/w)*	*Half-life (min)*	
				BS[a]	*BL*[b]
102	6	25	37.3	213	70
102	6	25	31.4	126	44
102	6	25	26.5	81	25
105	6.5	40	35	132	50
105	5.5	40	35	43	10
105	6.5	40	25	75	20
105	6.5	10	35	42	15

[a]*B. stearothermophilus* α-amylase.

[b]*B. licheniformis* α-amylase.

The low DE hydrolysate produced by enzymatic liquefaction is used as substrate for production of corn sweeteners or processed directly to a dry or liquid maltodextrin product.

When the final product is intended to be a maltodextrin, the reaction is conducted to a specific DE and terminated by heat treatment or pH adjustment. The hydrolysate is refined and spray dried to a moisture content of 3–5 per cent or evaporated to a syrup of 75 per cent solids.

Dextrose

Glucoamylase is used to convert (saccharify) liquefied starch to a dextrose containing hydrolysate by cleaving both α-1,4 and α-1,6 glucosidic bonds, sequentially from the non-reducing end of the molecule. The enzyme is produced from *Aspergillus* or *Rhizopus* fungal sources. Glucoamylase from *A. niger* generally is used in commercial processes because of adequate thermostability.

The rate of hydrolysis by glucoamylase depends on the particular linkage, the size of the molecule and the order in which the α-1,4 and α-1,6 linkages are arranged. For instance, an α-1,4 linkage in maltose is hydrolysed at a rate 100 times faster than an α-1,6 linkage in isomaltose. Increasing the number of α-1,4 linkages in sequence from 1 (DP-2 saccharide) to 4 (DP-5 saccharide) increases the hydrolysis rate of the first glucosidic bond by a factor of 10. In the same manner, increasing the number of α-1,6 linkages from 1 to 4 increases the hydrolysis rate of the non-reducing glucosidic bond by a factor of 23. The hydrolysis rate of an α-1,6 linkage is increased 9-fold when the adjacent linkage is an α-1,4 bond rather than an α-1,6 bond. Conversely, the hydrolysis rate of an α-1,4 linkage is reduced when the adjacent linkage is an α-1,6 bond rather than an α-1,4 bond.

Glucoamylase also catalyses the reverse reaction (reversion), in which dextrose molecules are combined to form maltose and isomaltose. Minimising this reaction is important in maximising dextrose yield during commercial operation.

The reversion of dextrose specifically involves the condensation of a β-anomer of D-glucopyranose with either an α- or a β-D-glucose molecule in the presence of glucoamylase, as shown here:

$$\text{D-Glucose} + \beta\text{-D-glucose} \longrightarrow \text{Disaccharide} + H_2O$$

Consequently, the equilibrium constant (K_{eq}) for the reversion of dextrose to maltose or isomaltose can be calculated as:

$$K_{eq} = \frac{C_m \times C_w}{C_d \times 0.63\ C_d} = \frac{C_m \times C_w}{0.63\ C_d^2}$$

where, C_m is the molar concentration of maltose, or isomaltose, C_w is the molar concentration of water, C_d is the molar concentration of dextrose and 0.63 is the decimal fraction of β-D-glucose present at equilibrium.

Rate of maltose formation during saccharification is rapid. The disaccharide reaches an equilibrium level early in the reaction before maximum dextrose is reached. In contrast, formation of isomaltose via reversion is slower than that of maltose, but the equilibrium level is much higher. Therefore, isomaltose level continues to increase during saccharification.

In commercial processes, liquefied hydrolysate at 30–32 per cent w/w solids is adjusted to pH 4–4.5 and 58°–61°C. A glucoamylase dosage is added to produce a maximum dextrose level in 1–4 days. Reaction time is inversely related to dose, so if a given dose yields maximum dextrose in 4 days, doubling or quadrupling the dose will reduce reaction time to 2 days or 1 day, respectively (Fig. 12.3). A high dose, and therefore a short reaction time, requires small saccharification tanks so the final hydrolysate can be processed rapidly to prevent excessive reversion. If large reaction tanks are available, a lower dose generally is used, since a considerable amount of time is required to empty the tank. Saccharification tanks range in size from less than 2,00,000 to several million litres.

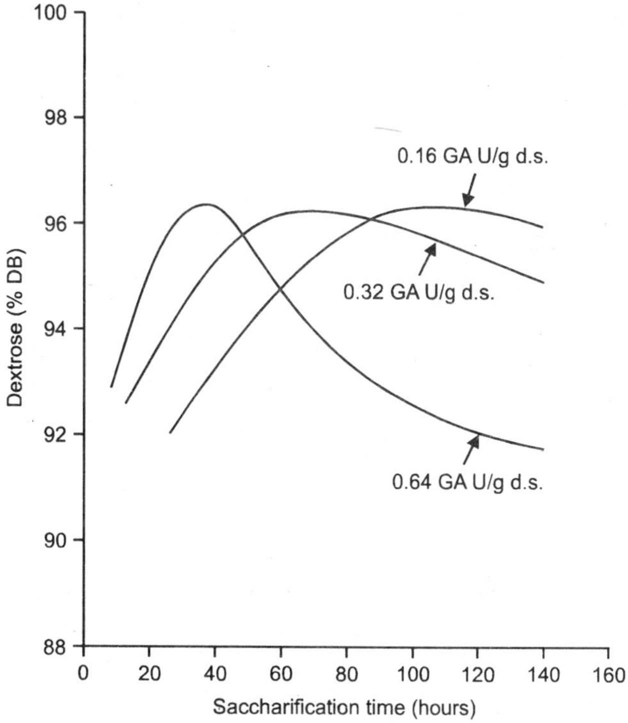

Fig. 12.3. Saccharification with glucoamylase.

Saccharide distribution attained at maximum dextrose is typically about 96.1 per cent dextrose, 1.5 per cent isomaltose, 1.0 per cent maltose, 0.2 per cent maltulose, 0.3 per cent DP-3 and 0.9 per cent DP-4 and higher saccharides. Both maltose and isomaltose are formed via reversion, that is, the maltose concentration of 1.0 per cent is at the equilibrium level dictated by the final solids level of about 33 per cent, whereas isomaltose continues to increase during saccharification. At maximum dextrose, the isomaltose concentration is about 1.5 per cent but would eventually reach 10–12 per cent if the saccharification reaction was allowed to proceed beyond maximum dextrose. Consequently, overdosing with glucoamylase or extending reaction time beyond the maximum dextrose level will result in a decrease in dextrose yield because of continued formation of isomaltose.

Maltulose concentration is normally 0.1–0.3 per cent, although much higher levels of the disaccharide are produced if the proper pH is not maintained during liquefaction. When starch slurry pH is above 5.8, maltulose precursors are formed by a mechanism in which a portion of the polysaccharide reducing end groups is converted to fructose by alkaline isomerisation. Glucoamylase does not hydrolyse the final glucosidic bond of a starch polymer if the end group is fructose. Therefore, the disaccharide maltulose is released during saccharification with an equivalent loss in dextrose yield. As liquefaction pH is increased, maltulose content is increased substantially to a level dependent on time and temperature (Table 12.4). For instance, in a typical LT process conducted with a 3-hour hold time at 98°C, maltulose level in final hydrolysate increases from 0.2 per cent at pH 6 to 2.9 per cent at pH 7.

The DP-3 saccharides generally are present only in trace quantities and are, for the most part, formed via the reversion reaction. DP-4+ saccharides are composed of highly branched material that is hydrolysed only slowly by glucoamylase. This material can be converted completely to dextrose if given sufficient reaction time, but an extended saccharification results in a loss in dextrose yield due to continued formation of isomaltose via reversion.

Table 12.4. Effect of liquefaction conditions on maltulose level in dextrose hydrolysate.

Temperature (°C)	*Time (min.)*	*Maltulose (% dry basis) at pH*		
		6.0	*6.5*	*7.0*
–	0	0.0	0.0	0.1
90	30	0.0	0.0	0.2
90	180	0.1	0.5	1.8
98	30	0.0	0.1	0.4
98	180	0.2	1.1	2.9
105	30	0.1	0.2	0.8
105	180	0.6	1.7	2.9

Maximum dextrose level can be increased by operation at lower solids, since the reverse reaction is less favoured as water concentration is increased. For instance, reducing solids from 30 to 20 or 10 per cent increases maximum attainable dextrose level from 96 to 97.5 and nearly 98 per cent, respectively. Problems associated with operating at reduced solids include the requirement for larger tanks, increased cost of water removal from final product and increased risk of microbial infection in a dilute system.

Dextrose level also can be increased when certain enzymes are used in combination with glucoamylase. Commercially available enzymes used for this purpose include a debranching enzyme derived from *B. acidopullulyticus* and an enzyme exhibiting both amylase and transferase properties that was isolated from *B. megaterium* and transferred genetically into *B. subtilis*.

Debranching enzymes are specific for the hydrolysis of α-1,6 glucosidic bonds and provide additional substrate for the action of glucoamylase. The rate and extent of dextrose production is increased when the enzyme combination is used and an additional 0.5–1.0 per cent dextrose yield is achieved compared with use of glucoamylase alone. The combined action of *B. megaterium* amylase and glucoamylase also provides a 0.5–1.0 per cent increase in dextrose yield. The mechanism for this reaction is shown in Fig. 12.4. The DP-4+ oligosaccharides that are not easily hydrolysed by glucoamylase are in the form of a highly branched structure. The *B. megaterium* amylase degrades these oligosaccharides to panosyl units that are then transferred to dextrose, forming a maltotetraose (DP-4) saccharide containing one branch point. The maltotetraose saccharide is hydrolysed easily to dextrose, so an increase in dextrose yield is achieved. Other advantages of both the debranching and the *B. megaterium* enzyme include reduced reaction time, reduced glucoamylase dose and the opportunity to operate at higher solids without a loss in dextrose yield.

Saccharified hydrolysate is clarified by vacuum filtration and refined by a combination of carbon and ion-exchange treatments. The refined hydrolysate is processed to a high dextrose corn syrup, crystalline dextrose, or liquid dextrose, or used as feed for HFCS production.

A high dextrose syrup is prepared from the refined hydrolysate by evaporating to 70–75 per cent w/w solids.

Crystalline monohydrate dextrose is produced by concentrating refined hydrolysate to 70–78 per cent w/w, seeding with crystals in a batch crystalliser and cooling from about 43°C to 20°–30°C over several days or until 50–60 per cent of the dextrose has crystallised. The crystals are separated by batch or continuous centrifugation, washed to remove impurities and dried in hot air rotary dryers to about 8.5 per cent moisture.

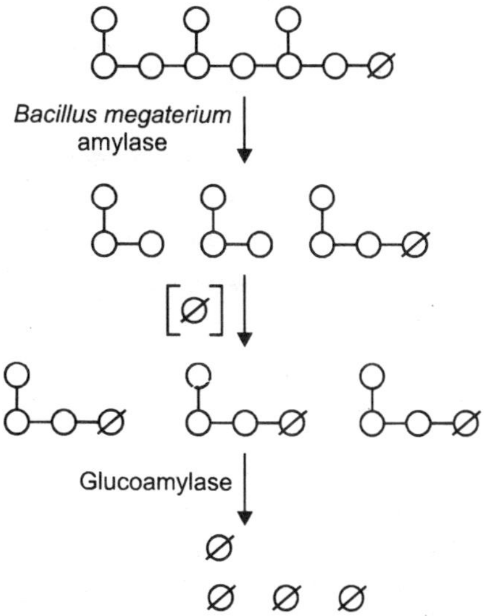

Fig. 12.4. Action of *Bacillus megaterium* amylase.

Anhydrous dextrose is produced by dissolving monohydrate dextrose in water and crystallising at 60°–65°C. Liquid dextrose is prepared by dissolving monohydrate dextrose to 71 per cent solids.

High Fructose Syrups

Immobilised glucose isomerase is used to convert a refined 93–96 per cent dextrose hydrolysate to a 42 per cent fructose product. Commercial enzymes are produced intracellularly from a variety of bacterial sources. Selected commercial enzyme products and recommended operating conditions are listed in Table 12.5.

Table 12.5. Commercially available glucose isomerase enzymes.

Supplier	Enzyme source	*Recommended operating conditions* Temperature (°C)	pH	Solids (%)
Enzyme Bio-systems	*Streptomyces olivochromogenes*	55–60	7.8–8.6	35–50
Genencor International	*Streptomyces rubiginosus*	54–62	7.6	40–50
Gist-Brocades	*Actinoplanes missouriensis*	58–60	7.5	40–45
Solvay Enzyme Products	*Microbacterium arborescens*	≤ 60	7.5	40–50
Novo Nordisk	*Bacillus coagulans*	59–61	7.8–8.3	40–45
Novo Nordisk	*Streptomyces murinus*	50–60	7.5	35–45

During the initial stages of HFCS development in the early 1960s, it became obvious that the use of a soluble isomerase for fructose production would require a high enzyme dose or a long reaction time. The high dose was economically unacceptable and the prolonged reaction time caused the formation of undesirable by-products such as mannose, psicose, colour and off-flavours that increased refining costs. Consequently, efforts were directed to the development of immobilised enzyme systems that would improve process economics.

An immobilised isomerase system was found to have many of the benefits and few of the disadvantages of bound enzyme systems proposed for other purposes. For instance, the substrate used for HFCS production is a clarified and refined low viscosity stream that is passed easily through a column of bound enzyme. Both the substrate (dextrose) and the product (fructose) are simple low molecular weight saccharides that easily diffuse through porous carriers used for enzyme immobilisation. In addition, the bound enzyme exhibits very long half-lives of several months under normal process conditions, resulting in low enzyme cost. Also, reaction time due to the high enzyme concentration is reduced sufficiently in continuous operation to minimise production of by-products. Consequently, with the exception of initial fructose production in the late 1960s, all HFCS has been manufactured using immobilised enzyme.

Two isomerase immobilisation processes are used to produce commercial products. These are referred to as 'whole cell' and 'soluble enzyme' processes.

In the 'whole cell' process, the bacterial cells containing the intracellular isomerase are recovered from fermentation broth and treated to immobilise the enzyme within the cells. In one commercial process, bacterial cells are concentrated by centrifugation, disrupted by homogenisation, cross-linked with glutaraldehyde and flocculated with a cationic flocculent. The resulting material is filtered, extruded, dried and sieved to produce a 300–1000 μm final product. In another process, a cell-gelatine mixture is cross-linked with glutaraldehyde, washed and sieved. Cellular immobilised enzyme products are discarded after use and replaced with new material.

In the 'soluble enzyme' process, cells are treated enzymatically or by physical means to break the cell wall and release isomerase. The isomerase is recovered by filtration or centrifugation and concentrated by ultrafiltration. The soluble enzyme is then bound to a carrier. One method involves recovering a purified soluble isomerase by ultrafiltration and crystallisation and contacting the enzyme for 3–5 hours with a carrier composed of DEAE-cellulose, titanium dioxide and polystyrene. The material is ground to produce a 400–800 μm final product. In another type of immobilisation, polyamine is absorbed on alumina and treated with glutaraldehyde to cross-link the amine and immobilise the isomerase by forming covalent bonds between the enzyme amino groups and the carrier aldehyde groups. In some cases, the carrier can be re-used for several cycles by removing residual enzyme and re-binding with fresh enzyme. For instance, the DEAE-cellulose carrier can be regenerated by washing sequentially with water, 2 per cent sodium hydroxide, water, buffer and rinse water to remove inactivated enzyme and accumulated impurities. Fresh isomerase is then bound by batch or column operation to achieve 95 per cent of the original capacity. Alternatively, additional enzyme can be added periodically via the feed and bound to the carrier during the isomerisation process to maintain a constant level of activity over a long period of operation.

The immobilised isomerase is loaded in a column and substrate is passed through at a rate that produces an effluent containing 42 per cent fructose. Several parameters must be controlled carefully for maximise productivity. These parameters include feed purity, solids, temperature, pH, oxygen level, magnesium content and reaction time.

Feed purity

Isomerase feed is refined hydrolysate containing 93–96 per cent dextrose. Efficient refining is required to remove insoluble material that may accumulate in the column over time and cause pressure drop problems and impurities that could cause inactivation of the enzyme. Ion-exchange treatment of feed is required to lower calcium levels to 3 ppm or less, since the metal is inhibitory to isomerase.

Solids

Feed solids level is controlled in the 40–50 per cent range, generally at about 45 per cent. If the solids level is too high, increased viscosity may reduce rate of feed diffusion within the carrier pores and cause a reduction in product yield. A solids level that is too low could result in microbial contamination.

Temperature

Temperature control is important in maximising productivity and useful life of the column. Normally, temperature is controlled at 55°–61°C. At a lower temperature, the risk of microbial contamination is increased. As temperature is increased beyond the recommended range, activity of the isomerase increases but, at the same time, long-term productivity decreases due to a faster rate of enzyme inactivation. This situation results in a reduction in the useful life of the column. Only when high productivity is required for a short period of time at the expense of long-term productivity is a higher temperature a consideration.

pH

Long-term isomerase stability is highly dependent on pH. Depending on the particular enzyme, recommended pH varies from 7.5 to 8.2. Enzyme activity increases at higher pH levels, but productivity decreases due to reduced enzyme stability. For best operation, pH should be controlled as low as possible to minimise by-product formation while maintaining maximum productivity.

Oxygen

Presence of dissolved oxygen in dextrose feed is detrimental because of possible oxidation of isomerase cysteine residues, resulting in a loss in activity. Deaeration of feed to remove excess oxygen or addition of SO_2 to reduce oxidation is helpful. Typically, 1–2 mM SO_2 as sodium bisulphite is added for this purpose.

Magnesium

The presence of magnesium is important as an enzyme co-factor for activation and stabilisation of glucose isomerase. Generally, a magnesium level of 0.5–5 mM is recommended. Magnesium counteracts the adverse effect of excess calcium and commonly is added at a molar ratio of 20 times the level of calcium present. Cobalt and iron enhance glucose isomerase activity or stability whereas copper, zinc, nickel, mercury and silver inhibit activity.

Reaction time

Flow rate through a column is controlled to yield an effluent fructose level of 42–45 per cent (dry basis). Under typical operating conditions at a temperature of 60°C, an equilibrium fructose level of 51.2 per cent is achievable (48.6 per cent fructose with a 95 per cent dextrose feed). However, the very long reaction time needed to achieve the equilibrium level reduces overall productivity significantly.

Control of these parameters is required to achieve maximum productivity. If any of these conditions falls outside the recommended range, a temporary or permanent loss in performance can be expected. Temporary loss in productivity can be caused by a moderate change in pH, low temperature, high solids, insufficient magnesium, or excess calcium. Correction of these parameters will reinstate column operation. More serious problems in productivity are caused by a temperature above 65°C or a pH considerably below or above the normal range. The one major disadvantage of column operation is that several months of potential enzyme use may be lost by an uncorrected fluctuation of pH or temperature to an unacceptable level.

If reaction parameters are controlled within the limits recommended by the enzyme manufacturer, efficient operation over several months or longer can be expected. The enzyme normally is used through several half-lives until residual activity is reduced to 10–20 per cent of the original level.

The 42 per cent fructose product containing 51–54 per cent dextrose is refined by carbon and ion-exchange treatments to eliminate colour, off-flavours, salts and other impurities and concentrated to 71 per cent solids. A storage temperature of 27°–32°C is required to prevent dextrose crystallisation.

Products containing higher fructose levels are manufactured by a combination of chromatographic separation and blending. The refined 42 per cent HFCS at about 50 per cent w/w solids is passed through a cationic adsorbent containing calcium groups. The fructose is retained and a stream (raffinate) is collected containing predominantly non-fructose saccharides. The raffinate, composed of 80–90 per cent dextrose, 5–10 per cent fructose and a higher saccharide fraction, is recycled to the saccharification or isomerisation process. Elution of the high fructose fraction with water yields an HFCS product containing 80–90 per cent fructose and 7–19 per cent dextrose. In actual practice, the separation process operates continuously in a simulated moving-bed process, providing a continuous flow of enriched HFCS and raffinate. The primary use of enriched fructose product is for blending with 42 per cent HFCS to produce a 55 per cent fructose product containing about 40 per cent dextrose.

The 55 per cent HFCS is evaporated to 77 per cent solids for storage and shipment. Crystallisation is generally not a problem due to the low dextrose level. A small amount of 80–90 per cent HFCS is evaporated to 77–80 per cent solids and sold as a non-crystallisable syrup.

Crystalline fructose is manufactured using as raw material the 90+ per cent fructose product obtained by chromatographic separation of 42 per cent HFCS. The 90 per cent fructose material is evaporated to a solids level of 90 per cent or higher and the fructose is crystallised by batch or continuous techniques over a period of several hours or days. The crystalline product is washed free of impurities and dried.

Traditional Syrups

Traditional corn syrups are manufactured by straight acid, acid-enzyme, or enzyme-enzyme hydrolysis processes.

Straight acid syrups have been produced in much the same manner throughout the history of the corn refining industry. Starch at 35–40 per cent solids is acidified with hydrochloric acid to 0.15–0.2 N and heated at 140°–160°C under pressure until the desired DE is reached; the reaction is stopped by neutralisation with sodium carbonate. Hydrolysate is clarified and refined by carbon treatment and, possibly, ion-exchange and evaporated to 83 per cent solids. Continuous and batch processes are used to produce a standard 42–43 DE product that contains an average saccharide distribution of 19 per cent dextrose, 14 per cent maltose, 12 per cent maltotriose and 55 per cent DP-4 and higher saccharides.

Lower DE products are produced occasionally. Higher DE products normally are not prepared by acid hydrolysis, since the excessive levels of colour and off-flavour formed during the reaction are difficult to remove during refining. Higher DE syrups, however, can be produced by enzymatic hydrolysis of straight acid syrups or by a process in which both liquefaction and saccharification are conducted enzymatically.

The two standard syrups manufactured by acid-enzyme or enzyme-enzyme techniques contain higher levels of dextrose and/or maltose than acid syrups. In one case, a 38–42 DE acid hydrolysate is adjusted to pH 4.8–5.2 and 55°–60°C and saccharified with a combination of glucoamylase and *Aspergillus oryzae* fungal α-amylase for several hours or days, depending on enzyme dose. The glucoamylase dose is considerably less than that used in dextrose saccharification so only a low level of dextrose is produced. The fungal α-amylase is a maltogenic enzyme, producing maltose. When saccharification is complete, a 63–65 DE product is produced that exhibits an average composition of 36 per cent dextrose, 30 per cent maltose and 13 per cent maltotriose. Products of even higher DE can be made by extending the reaction time, although the increased dextrose level may result in crystallisation problems during storage or shipment if heating is not used.

Syrups containing high levels of maltose are manufactured using only a maltogenic enzyme for saccharification. Enzymes used for this purpose are fungal α-amylase or barley β-amylase. Saccharification is conducted at 55°–60°C, pH 4.8–5.2 and a solids level of 35–45 per cent. Maltose levels of 50–55 per cent are achieved in a reaction time that depends on the dose of enzyme used. If higher levels of maltose are desired, a combination of the maltogenic enzyme and a pullulanase is used for saccharification. The pullulanase hydrolyses the α-1,6 glucosidic linkages and provides additional substrate for the production of maltose. In this manner, maltose levels of 60–80 per cent or higher are produced. The very high maltose syrups can be prepared only if a low DE substrate is used. If substrate DE is too high, maltotriose will be produced from the oligosaccharides containing an odd number of dextrose units and a high maltose level will not be achieved.

APPLICATIONS OF SWEETENERS AND SYRUPS

Corn sweeteners exhibit a variety of properties and are used in many different food and non-food applications. Functional properties include sweetness, fermentability, viscosity, flavour enhancement, osmotic pressure, humectancy, hygroscopicity, colour development, cohesiveness, nutritional quality, texture enhancement and crystallisation prevention.

Dextrose

Dextrose is used primarily for sweetening, as a source of fermentable sugars and for osmotic pressure. The perceived sweetness of dextrose depends on several factors including solids, temperature and presence of other ingredients. In general, dextrose exhibits about 75 per cent of the sweetness of sucrose and, therefore, is used in applications in which less sweetness is desired. In some formulations, combinations of dextrose and sucrose provide a synergistic effect, yielding the same sweetness as sucrose alone.

Dextrose is fermented completely by yeast and is used as a carbohydrate source in applications in which a high degree of fermentability is desired. In baking, dextrose is used as a fermentable carbohydrate source for yeast to provide the proper degree of carbon dioxide production and product volume. In addition, dextrose is important in improving bread strength for better handling and slicing, for producing desired colour and flavour via the Maillard reaction with protein and for providing good texture and crumb and crust characteristics. In beverages, dextrose is used to impart sweetness, body and osmotic pressure. In beverage powders, dextrose is used for flavour enhancement and to reduce excessive sweetness. Dextrose is also used in wine and beer for its fermentable properties. In light beer production, dextrose is used as a completely fermentable additive to reduce residual carbohydrate and caloric level.

In canning, dextrose often is used alone or in combination with sucrose in sauces, soups, gravies, fruit juices, jams and jellies to control osmotic pressure and to provide flavour, sweetness and texture.

In confections, dextrose is used in candies to control crystallisation and provide sweetness and softness. It is also used for coating, strength, hardness, colour and gloss in candy and gum.

In non-food applications, dextrose is used in a purified anhydrous form in medical applications such as intravenous solution production and tableting.

In addition, it is used as a raw material in fermentations for the production of antibiotics, citric acid, amino acids, enzymes, lactic acid and ethanol. Dextrose also is used as a raw material for production of sorbitol and polydextrose. As an additive, dextrose is used in adhesives to control flow, in library paste to increase bonding time, in wallboard to prevent brittle edges, in concrete to retard setting, in resins as a binder, in dyes as a diluent, in metal treatment as a reducing agent and in leather tanning for pliability.

High Fructose Syrups

The most important property of high fructose syrups is sweetness. A 42 per cent HFCS product exhibits about the same sweetness as sucrose, 55 per cent fructose is equivalent to medium sucrose invert and 80–90 per cent fructose is sweeter than sucrose. In general, HFCS exhibits the greatest sweetness perception under conditions of low temperature and low pH and is less sweet at warm temperatures and neutral pH.

The major use of HFCS is in soft drinks. HFCS first was used in this application in 1974, when HFCS-42 was approved for 25 per cent replacement of sucrose. In 1980, HFCS-55 was used at a 50 per cent sucrose replacement level and, by 1984, 100 per cent replacement of sucrose was approved. Currently, HFCS is used in man cases as a complete replacement for sucrose. Specifications to insure continued high quality were designed through a combined effort of the Society of Soft Drink Technologists, the Corn Refiners Association, corn wet-millers and soft drink bottlers. These specifications include solids, ash, colour, pH, fructose level, sediment, SO_2, flavour and odour.

HFCS-42 is used in canning to provide sweetness, in dairy products as a bodying agent and to improve texture and mouthfeel and in confections for sweetness, grain control and humectancy. HFCS-90 is used in reduced calorie products to lower caloric levels by 30–50 per cent.

Crystalline fructose generally is reported to be 20–80 per cent sweeter than sucrose; a sweetness synergy is often observed when fructose is used in combination with other sugars. The actual degree of sweetness perception obtained depends on several factors including temperature, pH and concentration. Pure fructose often is used in low calorie and speciality foods to provide sweetness at a reduced caloric level.

Traditional Syrups

By definition, corn syrup is any starch hydrolysis product that exhibits a DE between 20 and 99.4. Consequently, a wide range of syrups containing a variety of saccharide distributions is possible. Functional properties can vary considerably depending on composition. For instance, corn syrups are used for sweetness, fermentability, viscosity, humectancy and hygroscopic properties. In addition, corn syrups contribute osmotic pressure, control crystallisation, enhance foam stability and produce browning reactions.

Sweetness depends on the particular saccharide distribution as well as on pH, solids and temperature; a synergy often develops when used in combination with sucrose. For instance, a combination of 25 per cent 42 DE corn syrup and 75 per cent sucrose at 45 per cent solids yields the same sweetness as 100 per cent sucrose. Syrups containing high levels of fermentable saccharides, that is, dextrose and maltose, are used in brewing and baking where fermentability is needed. The high viscosity of the lower DE syrups is used to modify food texture. The humectance and hygroscopic properties of corn syrups allow moisture absorption to maintain product freshness. Corn syrup is used in baking, beverage, canning, confectionery and dairy.

In baking, corn syrups are used as a source of fermentables while improving shelf-life by maintaining a proper moisture balance and providing good crust colour. In beverages, corn syrups are used as an adjunct to provide a source of fermentables and body as well as to enhance flavour. In canning, corn syrups are used to improve flavour, colour and texture while providing body and preventing sucrose crystallisation. High osmotic pressure is important for maintaining juices in canned fruit. In confections, corn syrups, provide viscosity, mouthfeel, sweetness, texture, grain inhibition, hygroscopicity and resistance to discolouration. In some products, corn syrups are instrumental in participating in browning reactions to produce colour, give a gloss to hard candies and provide the proper texture for chewiness in gums. In dairy products, corn syrups are important for texture and smoothness by preventing crystal formation without adding excessive sweetness.

Corn syrups also are used in a wide range of non-food applications. For instance, syrups are used in colognes and perfumes to control evaporation, in air fresheners as humectants, to improve stability in adhesives and as a setting retardant in concrete. Corn syrups also are used as carriers and for sweetness in medicinal products such as lozenges. Other uses are in dyes, inks, explosives, metal plating, shoe polish, leather tanning, paper and textiles.

Maltodextrins

Maltodextrins are hydrolysis products of less than 20 DE and, therefore, are bland tasting with essentially no sweetness perception. Maltodextrin often is used in a dry form as a dispersing agent in coffee whiteners and as a bulk carrier for flavour ingredients. Maltodextrin is used in bakery, soup and sauce mixes, beverage powders, condiments, dehydrated foods, instant tea, low calorie sweeteners, marshmallows, pan coatings, nougats and snack foods. Currently, a rapidly growing application area is the use of maltodextrins as fat substitutes in a variety of food products.

Chapter 13

Dairy Products

INTRODUCTION

Dairy products are generally defined as foodstuffs produced from milk. They are usually high-energy-yielding food products. A production plant for such processing is called a dairy or a dairy factory. Raw milk for processing generally comes from cows, but occasionally from other mammals such as goats, sheep, buffalo, yaks, or horses. Dairy products are commonly found in European, Middle Eastern and Indian cuisine, whereas they are almost unknown in East Asian cuisine.

NATURAL ENZYMES OF MILK

Raw bovine milk contains about 50 native enzymes, at least 27 of which have been classified by the International Union of Biochemistry Enzyme Commission. Some, such as catalase and aldolase, may be introduced from blood serum. Most originate in the epithelial cells of the mammary gland. Many enzymes found commonly in milk are not intrinsic milk enzymes, but are in milk because of the presence of micro-organisms.

Pasteurisation eliminates most natural enzyme activity in milk. Some enzymes lose activity after being subjected to treatments as mild as 52°C for 15 seconds. Other milk enzymes are stable; plasmin, for example, loses only 60 per cent of its activity after 15 seconds at 100°C.

Some native enzymes in milk are employed for analytical or quality control purposes. For instance, destruction of alkaline phosphatase by ordinary pasteurisation temperatures is used to monitor adequacy of pasteurisation. Microbial phosphatase activity and phosphatase re-activation in high temperature pasteurised milk samples may generate unreliable results in such assays.

Another example of quality control based on enzymes is measurement of serum enzyme concentrations to detect abnormal milk. Catalase activity in milk is an indicator of elevated leukocyte levels characteristic of mastitic milk. Direct measurements of somatic cells or concentration of deoxyribonucleic acids (DNA) are preferred over the catalase test for this purpose.

Spontaneous development of oxidised flavours is common in dairy products. Several native milk enzymes are implicated in the process. Peroxidase (EC 1.11.1.7), which is stable at pasteurisation temperatures, catalyses oxidative reactions in milk and dairy products. Xanthine oxidase (EC 1.2.3.2), which catalyses oxidation of xanthine or hypoxanthine to uric acid and hydrogen peroxide, also causes oxidative flavours.

An important positive use of a native milk enzyme is the activation of lipase by homogenisation, which conditions the milk fat substrate in the manufacture of blue cheese. Homogenisation disrupts the milk fat globule membranes where lipase enzymes in milk are concentrated, making them more available for lipolysis of milk fat.

One milk protein, α-lactalbumin, is more concentrated in milk than is any enzyme. Although not an enzyme, in many respects this protein should be considered with the milk enzymes. α-Lactalbumin is one of two protein moieties essential for optimal activity of lactose synthetase; its concentration controls lactose synthesis and, hence, the colligative properties of milk. In addition to its own enzyme activity, α-lactalbumin is closely homologous with the enzyme lysozyme found in egg whites.

MILK-CLOTTING

Enzymatic Milk Coagulation

The predominant use of enzymes in the dairy industry is in coagulation of milk to make cheese. Most proteolytic enzymes clot milk but not all make acceptable cheese. The Food and Nutrition Board uses the term 'rennet' to describe all milk-clotting enzyme preparations (except porcine pepsin) used for cheese making. The Board defines rennet as 'aqueous extracts made from the fourth stomachs of calves, kids, or lambs' and bovine rennet as 'aqueous extract made from the fourth stomach of bovine animals, sheep, or goats'. Microbial rennet 'followed by the name of the organism' is the approved nomenclature for milk-clotting preparations derived from micro-organisms.

United States Standards of Identity for Cheddar cheese allow the use of 'rennet and/or other clotting enzymes of animal, plant, or microbial origin'. Milk-clotting enzymes generally recognised as safe (GRAS) are rennet and bovine rennet. Microbial rennets derived from *Cryphonectria parasitica* (formerly *Endothia parasitica*), *Bacillus cereus*, *Mucor pusillus* var. Lindt and *Mucor miehei* var. Cooney *et* Emerson may be used as secondary direct food additives. Porcine pepsin may be used to make cheese also, but is not called rennet. Chymosin (EC 3.4.23.3) from the abomasa of suckling calves remains the enzyme of choice and the standard against which all others are evaluated. Genetic engineering has been employed to produce chymosin microbially in *Escherichia coli* and in *Saccharomyces cerevisiae*. Cheese making trials comparing recombinant chymosin with calf rennet have found no significant differences between the two; recombinant enzymes are now approved for commercial use. Recombinant rennet has the added benefit of meeting some religious dietary requirements not met by calf rennet.

Milk-clotting activity is present as chymosin in the abomasum of a bovine fetus by the sixth month of development and increases in potency as the fetus approaches full term. At birth, chymosin is present in gastric mucosa at 2–3 mg/g but its production declines after 1 week. Procedures for extraction of chymosin were described by Ernstrom and Wong. Liquid rennet preparations have a pH between 5.6 and 5.8 to provide the most stable environment. Bacterial content can be reduced by filtration or centrifugation. Crude rennet extract contains active chymosin and an inactive precursor (prochymosin). Removing a peptide from the N terminus of prochymosin (36,000 daltons) converts it to chymosin (31,000 daltons). This transformation is favoured below pH 5.0. Chymosin is most stable between pH 5.3 and 6.3.

Rennet contains bovine pepsin if stomachs of older calves or cows are included with the stomachs of suckling calves. Such rennets are acceptable substitutes in cheese manufacture if their cost is lower than rennets with higher percentages of chymosin. Higher chymosin : pepsin ratios accelerate cheese ripening. When comparing five commercial milk-clotting preparations, Shaker and Brown observed that decreasing chymosin : pepsin ratios cause more general proteolysis and loss of curd to the whey.

Porcine pepsin is used as a substitute for part of the rennet in making many varieties of cheese. This enzyme is secreted by hog stomach mucosa as pepsinogen (40400 daltons). Conversion of pepsinogen to pepsin is catalysed by pepsin below pH 5. On average, 7–9 peptide bonds are hydrolysed during formation of pepsin, which releases about 20 per cent of the molecule. Cleavage of only one of these bonds is likely to be necessary to release the active enzyme. The pH of cheese milk is such that both the activity and the stability of porcine pepsin are far from optimum (pH 2.0); the enzyme is inactivated rapidly. Slow coagulation and a weak set resulting in excessive fat losses and reduced yield occur if insufficient enzyme is used, but porcine pepsin is inexpensive relative to chymosin. Mixtures of porcine pepsin and bovine rennet are used as less expensive coagulants. Such mixtures typically contain 20–25 per cent chymosin, 40–45 per cent bovine pepsin and 30–40 per cent porcine pepsin activity.

Fungal proteases are used extensively as substitutes for chymosin in milk-clotting. *Mucor miehei* rennet, a protease preparation from *M. miehei* var. Cooney *et* Emerson, is the most common fungal milk-clotting preparation. It is also most heat stable of all the commonly used milk-clotting enzymes. This enzyme remains active in the whey and is concentrated in condensed whey products, causing problems when whey products containing residual *M. miehei* rennet are mixed with casein. Therefore, nearly all *M. miehei* rennet used by the cheese industry is now modified to decrease its heat stability by treatment with hydrogen peroxide under controlled conditions. Some enzymatic activity is lost, but the modified enzyme has about the same stability as calf rennet.

Rennet from *C. parasitica* has been used with varying success for cheese making. Among the commercially used fungal rennets, *C. parasitica* rennet is the most proteolytic on α- and β-caseins and least proteolytic on κ-casein, but generally causes bitterness in the cheese product. Cheese varieties including Swiss, which are cooked at slightly higher temperatures, can be made successfully with *C. parasitica* rennet. *Mucor pusillus* var. Lindt protease has given satisfactory results as a chymosin substitute in the manufacture of a number of cheese varieties, but not all strains of *M. pusillus* var. Lindt are capable of producing acceptable cheese. The clotting activity of *M. pusillus* var. Lindt protease is more sensitive to pH changes between 6.4 and 6.8 than chymosin, but is much less sensitive than porcine pepsin.

The search for suitable chymosin substitutes has led to the investigation of a number of proteases produced by plants and bacteria. Occasional reports of success in cheese making reflect the fact that suitable coagulating enzymes may exist as part of crude mixtures containing other highly proteolytic enzymes that are detrimental to cheese making. None of these mixtures is in commercial use. Pepsin from chicken has been used for cheese making when dietary laws restrict use of bovine rennet.

Milk Coagulation Chemistry

Milk-clotting begins with enzymatic cleavage of κ-casein, destroying the ability of κ-casein to stabilise casein micelles. This cleavage is followed by a non-enzymatic aggregation of the altered casein micelles; then aggregates of casein micelles form a firm gel structure. Possibly, a separate fourth step exists during which the curd structure tightens and syneresis occurs.

Proteolytic enzymes are active in milk clotting by cleaving κ-casein to form para-κ-casein and a macropeptide:

$$\kappa\text{-casein} \longrightarrow \text{para-}\kappa\text{-casein} + \text{macropeptide}$$

Although chymosin splits κ-casein at a specific Phe-105-Met-106 bond, other milk-clotting enzymes are less specific. Splitting of κ-casein destroys the stability of the milk system and casein aggregates. Aggregation of κ-casein micelles begins long before proteolysis of κ-casein is completed. The enzymatic and non-enzymatic phases of milk clotting overlap. Under normal cheese making conditions, aggregation of micelles begins at the same time as enzymatic cleavage of κ-casein.

Both pH and temperature affect milk coagulation. Cheese starter cultures lower the pH before milk-clotting enzyme is added and continue to lower the pH during setting of the curd. Each milk-clotting enzyme has an optimal pH that must be considered in balance with culture activity to attain the correct pH at setting and cutting of the curd. The temperature coefficient (Q_{10}) for aggregation of casein micelles after rennet cleavage of κ-casein is 5–6 times that of the enzymatic step. This fact has been used by many to study the secondary phase of milk-clotting after allowing the enzymatic phase to go to completion at a lower temperature.

Measurement of Milk Coagulation

Milk-clotting activity is determined by the rapidity with which the enzyme clots milk under a set of specified conditions. This set of conditions includes the whole coagulation process, not just the enzymatic reaction. Measurement of the enzymatic reaction alone must be done with single substrates such as κ-casein or a peptide similar to the cleavage site in κ-casein. It is possible to take advantage of the different temperature coefficients to separate rennet cleavage of κ-casein and aggregation of casein micelles, but this separation does not allow separate measurement of enzyme activity.

Apparatuses for visual observation of the formation of a clot and standard substrates for comparing activities of different milk-clotting enzymes have been described by Sommer. Berridge developed a standard substrate to overcome the variability in different milk samples. Berridge substrate is made by reconstituting 60 grams low heat nonfat dry milk in 500 ml 0.01 M $CaCl_2$ and holding the solution at 4°C for 20 hours before use.

The Formagraph instrument was designed for following the progress of milk coagulation by recording the movement of small stainless steel loops immersed in milk as the milk is moved. In comparison with the visual method of Sommer and Berridge, the Formagraph has no difference in precision, but has the advantage of unattended operation. It can be used over a wide range of chymosin concentrations (0.001 to 0.16 chymosin units/ml).

McMahon monitored coagulation of Berridge substrate with a spectrophotometer connected to a computer. An initial dip in the tracing as κ-casein is cleaved is followed by a rapid rise as aggregation begins. The maximum on the curve matches the point of coagulation (τ) described by Payens. More information about the course of milk coagulation is available from this method than from any other devised to date.

Hori reported an innovative method for following the progress of milk-clotting with potential for in-plant use. A 0.1 mm × 106 mm platinum wire is immersed in milk and heated by application of 0.7 ampere DC electricity. As the milk surrounding the wire coagulates, dissipation of heat away from the wire decreases and the temperature of the wire increases. This procedure follows the progress of coagulation without disrupting the curd.

Storch and Segelcke proposed that the product of coagulation time (t_c) and enzyme concentration (E) should be a constant (k).

$$t_c E = k$$

Holter added a factor (x) to correct for the time lag between enzymatic cleavage of κ-casein and aggregation. Foltmann rearranged Holter's equation, keeping k and x separate to show that k is a constant but that x varies according to measurement conditions.

$$t_c = \frac{k}{E} + x$$

McMahon and Brown and Collinge showed that the x added to Storch and Segelcke's equation by Holter is not necessary if τ (the 'actual coagulation time' defined by Payens) is used instead of t_c.

CHEESE RIPENING

Proteolysis

Milk-clotting enzymes are added to milk to cleave κ-casein and begin coagulation of the milk, but they also have general proteolysis capabilities that contribute to cheese ageing. Chymosin is the least proteolytic of the commonly used milk coagulants. Excessive general proteolysis leads to excessive loss of fat and cheese yield and adversely affects flavour and texture.

Proteolytic activity during the ripening of cheese can be extensive. In hard cheese, 25–35 per cent of the insoluble protein of the curd may be converted into soluble protein, whereas in soft varieties such as Brie, Camembert, or Limburger, over 80 per cent of the insoluble protein can be converted to water-soluble compounds such as peptides, amino acids and ammonia. Apart from the participation of starter cultures, proteolysis depends in part on the coagulating enzyme system used in the formation of the curd. The amount of residual enzyme in curd after draining the whey depends on the choice of coagulating enzyme. Production of Cheddar cheese without starter culture results in cheese with much lower levels of free amino acids. Production of bitter flavours generally is attributed to the formation of bitter peptides. Bitter flavour results when these peptides are formed faster than they can be broken down further by proteolytic enzymes of the starter organisms.

Some rennet substitutes show high rates of proteolysis and frequently produce bitter taste in cheese. However, there are small differences between approved coagulants. When porcine pepsin and calf rennet were used at ratios that gave equivalent clotting times, proteolysis by porcine pepsin was slightly less than that by calf rennet, as determined by formation of non-protein nitrogen and the total concentration of free amino acids.

Lipolysis

Most lipolytic activity during cheese ripening is from enzymes produced by the numerous species of micro-organisms in cheese. The microflora of cheese is complex and changeable because different organisms predominate at different stages of cheese ripening. Adventitious organisms, including lactobacilli and micrococci, play significant roles in lipolysis during cheese ripening. Most hard cheeses are ripened by bacterial action in the cheese, whereas the soft varieties of cheese are ripened largely by yeasts, moulds, or bacteria growing on the cheese surface.

Milk lipases may lead to undesirable rancidity if freshly drawn milk is cooled too rapidly. This effect usually is attributed to the presence of 'membrane' lipase, which is adsorbed on the fat globules on rapid cooling and initiates lipolysis. The lipase becomes active if raw milk is homogenised or agitated or if foaming or great temperature fluctuations occur. Because of the fairly high pH range for milk lipase (6–9) and its sensitivity to heat, it does not seem to play an important role in the ripening of cheese made from pasteurised milk.

In some of the mould cheeses, such Roquefort, Gorgonzola, blue and Stilton, the lipases produced by *Penicillium roqueforti* play an important part in flavour formation.

Accelerated Cheese Ripening

Proteases and lipases or the systems producing them are used for manufacture of cheese paste products in which cheese flavour is accelerated significantly. Enzymes often are added as crude preparations made by heat treating or freeze shocking micro-organisms. Flavour enhancing enzymes have been added as pastes, as solutions, through encapsulation techniques and as soluble enzymes with salt. Addition of lactase is effective in accelerating the ripening of Manchego cheese.

In view of the extensive lipolysis of milk fat during cheese ripening, efforts have been made to supplement lipolytic activity of micro-organisms through lipase addition. Addition of pre-gastric esterases is practiced for production of Italian type cheese. Traditionally, Italian cheeses have been prepared with rennet paste made by drying the entire stomachs of calves, kids, or lambs, including the milk contents of the stomach. The paste made from kids or lambs particularly gives the cheese a characteristic piquant flavour that cannot be obtained with rennet extracts or milk free pastes. The difference is created by enzymes secreted by glands at the base of the tongue of the animals. These enzymes are extracted from the existed glands of calves, lambs, or kids. They are the major source of the characteristic lipolytic activity associated with rennet paste. These pre-gastric esterase enzymes are sometimes called oral lipases or oral glandular lipases.

LACTASE

Lactose Intolerance

The inability of some populations to catabolise lactose properly because of inadequate intestinal lactase (β-galactosidase, EC 3.2.1.23) activity is a concern for the dairy industry and for nutritionists. Lactose is the major carbohydrate of milk and milk is the only source of lactose. Use of dairy foods as dietary sources of specific nutrients such as calcium suggests the need for lactase treatment of milk and milk products.

Lactase in Dairy Processing

Use of lactase in the dairy industry is one of the most promising applications of enzymes to food processing. Lactose has poor solubility (about 15 per cent in water at 20°C) and is less than one-sixth as sweet as sucrose. Lactose hydrolysis produces a more soluble, sweeter mixture of equal proportions of glucose and galactose and small amounts of diverse polygalactosides. The hydrolysate has an osmotic pressure about twice that of lactose. Lactose is in abundant supply, particularly in cheese whey, presenting serious disposal problems in the cheese industry.

The lactose fermenting yeast such as *Kluyveromyces lactis* (formerly *Saccharomyces lactis*) and *Kluyveromyces fragilis*, or fungi such as *Aspergillus niger* and *Aspergillus oryzae* are the most frequently used sources of the enzyme lactase and commonly have been used for its preparation. Most of the work on dairy applications has been done with the yeast enzyme. Fungal enzymes with lower pH optima have proved advantageous in bread doughs.

Lactose in ice cream mixes can be hydrolysed by directly incorporating lactase into the mix and holding for 5 days at 7.5°C. Because storage of mix or skim concentrates for the periods required to allow adequate hydrolysis is impractical, such procedures are not useful for adjusting osmotic properties of mixes. Addition of lactase to mix for added sweetness, reduced lactose intolerance potential and prevention of crystallisation in the product is common.

HYDROGEN PEROXIDE AND CATALASE TREATMENT

Use of hydrogen peroxide (H_2O_2) to reduce the bacterial count of milk has been suggested since about 1900. Two obstacles have prevented extensive use of the process: unavailability of sufficiently pure and stable H_2O_2 solutions and unavailability of catalase, the enzyme used to break down any residual H_2O_2 remaining in milk after treatment.

In two situations the H_2O_2/catalase process is preferred over heat pasteurisation of dairy products. In many underdeveloped areas where cooling facilities are not available on farms and where heat pasteurisation is not feasible, addition of H_2O_2 is a practical means of milk preservation. An effective procedure is addition of 1 ml 33 per cent H_2O_2 per litre of milk at the farm, addition of 1 ml H_2O_2 at the dairy plant, heating to 50°C for 30 minutes, cooling to 35°C and adding catalase.

A second reason for the interest in H_2O_2 treatment is the relative advantage over pasteurisation in treatment of milk for cheese making. Pasteurisation destroys not only milk pathogens but also acid forming micro-organisms and inactivates some of the natural enzymes of the milk. Hydrogen peroxide treatment is effective in reducing the counts of pathogenic micro-organisms and inactivates milk catalase and peroxidase, but it permits lactic acid-forming micro-organisms to survive and lipases, proteases and phosphatases are not affected. The H_2O_2/catalase treatment cannot be substituted legally for pasteurisation, since complete elimination of pathogens is not assured.

Hydrogen peroxide is decomposed slowly by the natural catalase of raw milk. However, the amount of catalase in non-pasteurised milk is not sufficiently high to insure complete destruction of H_2O_2. Therefore, addition of catalase is essential to avoid the presence of residual H_2O_2 and a high oxidation/reduction potential, which interferes with the active growth of starter organisms in dairy products. Catalase preparations from liver and bacterial (*Micrococcus lysodeikticus*) sources have been approved for cheese manufacture. Only the liver preparation is sold because of limited demand.

MISCELLANEOUS APPLICATION OF ENZYMES IN DAIRY PROCESSING

The use of pre-gastric esterases has been discussed for manufacture of cheese, particularly of the Italian cheese varieties. These esterases also can be used to produced pronounce dairy flavours in milk fat emulsions or in whole milk concentrate substrates. Such flavours are desired for production of milk chocolate and account for the unique superiority of some milk chocolate flavours. The flavours can be produced by treating whole milk or whole milk concentrates with pre-gastric esterases and drying to whole milk powders. Such powders contain 28.5 per cent milk fat, 2.5 per cent moisture, and otherwise resemble whole milk powders in their gross chemical composition. These powders are remarkably resistant to oxidative rancidity, unlike their whole milk counterparts. They can be used in the production of milk chocolate (0.5–3 per cent based on the weight of the finished chocolate), fudge, caramel fillings, cast creams and other food applications in which a pronounced dairy flavour is desired. Low levels of lipolysed concentrates contribute a rich dairy flavour to margarines and baked goods. Basic lipolysed concentrates have been modified by fortification changes to allow their application in coffee whiteners, snack foods and imitation dairy foods. Higher levels of these products produce Italian cheese characteristics and can be used in soups, pizza and similar products where a free fatty acid character is essential.

Another lipolysed product reported for similar applications is produced by sequential growth of *Lactobacillus bulgaricus* in cream, followed by controlled lipolysis. Bacterial spoilage during cream incubation was prevented using this technique; the lactic culture flavour advantage finds application in some dairy food products.

One of the more uncommon suggestions for the use of enzymes in the dairy industry deals with the use of rennet in ice cream mixes to increase the viscosity of the mix, to eliminate the need for ice cream stabilisers and to improve the melt characteristics of ice cream. The use of small amounts of pepsin has been suggested to increase the viscosity of evaporated milk so it can be sterilised by high temperature, short time methods. The use of 0.003 ppm crystalline pepsin (1 hour at 0°–1°C) doubles the viscosity of the evaporated milk.

The possibility of future developments of commercial sources of galactose oxidase should be mentioned. The action of this enzyme is specific for galactose. It oxidises galactose to D-galactohexodialdose and H_2O_2. Although its applications are essentially analytical, the use of such a product could have as wide an application as glucose oxidase in the removal of oxygen.

Chapter 14

Pectic Enzymes in Fruit and Vegetable Juice Manufacture

INTRODUCTION

Pectic enzymes occur in higher plants and are produced by micro-organisms. Their substrates are the pectic substances that occur as structural polysaccharides in the middle lamella and the primary cell walls of higher plants and are prominent in parenchymous tissues. Endogenous pectic enzymes therefore can produce important textural changes in fruits and vegetables during ripening and storage. The necessity to activate or inactivate them often has a decisive influence on processing steps in the manufacture of products derived from fruits and vegetables. Microbial pectic enzymes serve functions in plant pathology and fermented foods but also are produced industrially as processing aids in the food industry. Pectic enzymes purified to well defined activities are used in pectin analysis, structure research of pectins, and cell wall studies.

PECTIC SUBSTANCES AND PECTIC ENZYMES

Pectic substances (pectins) are chain molecules with a rhamnogalacturonan backbone. This backbone consists of 'smooth' α-D-1,4-galacturonan regions that are interrupted to a small extent by insertion of 1,2-linked α-L-rhamnosyl residues and highly branched regions with an almost alternating rhamnogalacturonan chain. Side chains composed of neutral sugars are attached by glycosidic linkage to carbon atoms 3 and 4 of the rhamnose units and carbon atoms 2 and 3 of the galacturonic acid units, giving the rhamnogalacturonan portion of the pectin backbone a 'hairy' character (Fig. 14.1). The predominant sugars D-galactose and L-arabinose are present in complex chains of considerable length, whereas some xylose occurs in monomeric or oligomeric side chains. Commercial apple or citrus pectins, extracted for use as gelling agents, have a degree of polymerisation that usually does not exceed 500, but pectin fractions from fruits may have much higher molecular weights. Pectins extracted from most fruits under mild conditions usually have a degree of esterification (DE, percentage of galacturonic acid monomers that are methyl esterified) of over 70. Acetic acid may occur as substitute on the hydroxyl groups of galacturonic acid at carbon atom 2 or 3. Acetyl groups are prominent in sugar beet pectins, which also carry feruloyl groups linked as esters to terminal galactose or arabinose units of side chains.

Pectic enzymes are classified according to their mode of attack on the galacturonan part of the pectin molecule. Referring to the scheme in Fig. 14.2, one can distinguish between pectin methylesterases (PE; EC 3.1.11.1), which deesterify pectins to low methoxyl pectins or pectic acid, and pectin depolymerases, which split the glycosidic linkages between galacturonosyl (methylester) residues. Polygalacturonases (PG; EC 3.2.2.15 and 3.2.1.67) split glycosidic linkages next to free carboxyl groups by hydrolysis; pectate lyases (pectic acid lyases, PAL; EC 4.2.2.2 and 4.2.2.9) split glycosidic linkages next to free carboxyl groups by β-elimination (Fig. 14.3). Both endo and exo types of PGs and PALs are known. The endo types (EC 3.2.1.15, EC 4.2.2.2) split the pectin chain at random. Exo-PGs (EC 3.2.1.67) release monomers or dimers from the non-reducing end of the chain, whereas exo-PALs (EC 4.2.2.9) release unsaturated dimers from the reducing end. Pectates and low-methoxyl pectins are, therefore, the preferred substrates for PGs and PALs; the enzymes have very low activity on highly methylated pectins. Such pectins are degraded by pectin lyases (PL; EC 4.2.2.10), for which only the endo type has been described. Highly esterified pectins are, of course, also degraded by a combination of PE with PG or PAL. The existence of a polymethylgalacturonase for hydrolytic degradation of high ester pectin has never been proved and has not been found by careful screening of commercial preparations and micro-organisms in the laboratory.

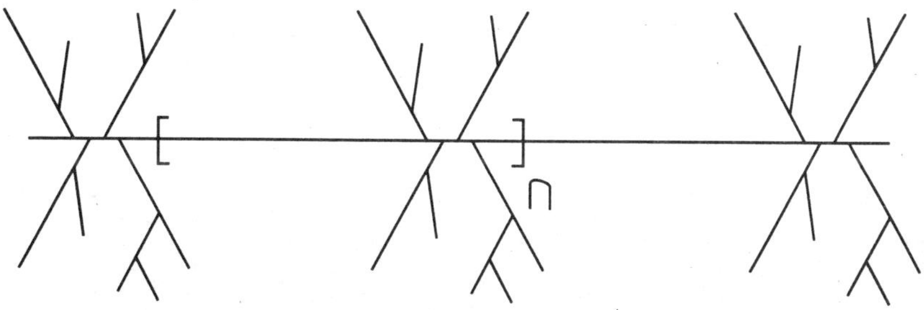

Fig. 14.1. Schematic structure of apple pectin; rhamnogalacturonan backbone with regions rich in neutral sugar side chains (hairy regions).

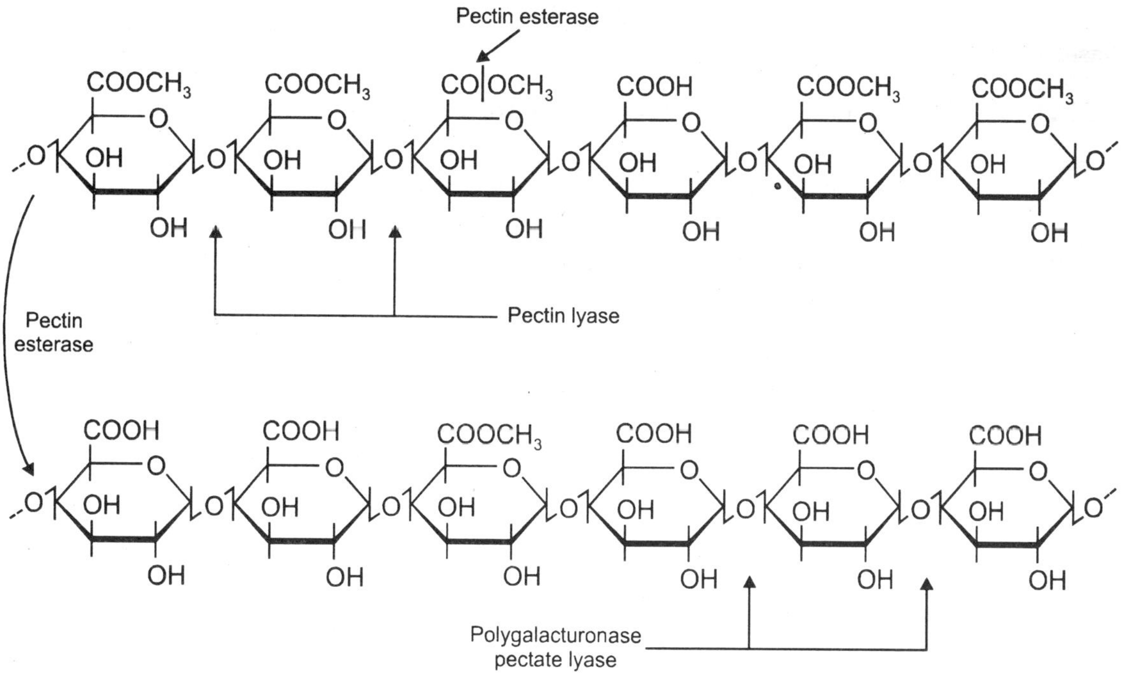

Fig. 14.2. Fragment of a pectin molecule and points of attack of pectic enzymes.

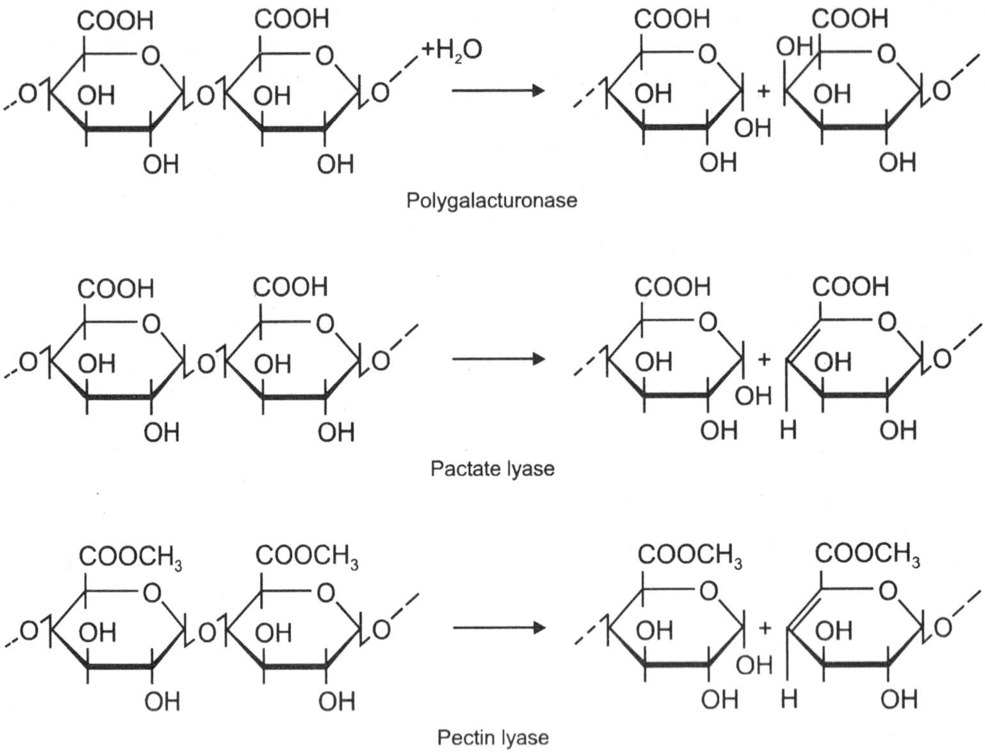

Fig. 14.3. Splitting of glycosidic bonds in pectin by hydrolysis (polygalacturonase) and by β-elemination (pectate lyase and pectin lyase).

OCCURRENCE AND MICROBIAL PRODUCTION OF PECTIC ENZYMES

Table 14.1 shows the presence of PEs and PGs in higher plants. These enzymes also are produced by micro-organisms, which produce PAL and PL as well and are given in Table 14.2.

Table 14.1. Occurrence of pectin esterase and polygalacturonase in higher plants[a].

		PG					*PG*		
Plant	*PE*	*Endo*	*Exo*	*NS*	*Plant*	*PE*	*Endo*	*Exo*	*NS*
Apple	+		+		Passion fruit	+	+	+	
Apricot	(+)			+	Peach		+	+	
Avocado		+			Pear	+	+	+	
Banana	+	+			Plums	+			
Berries	+			+	Beans	+			
Citrus					Carrots	+		+	
Lime	+				Cauliflower	+			
Orange	+				Cucumber	+	+		
Grapefruit	+		+		Leek	+			
Mandarin	+				Onions	+			

(Contd...)

		PG					*PG*		
Plant	*PE*	*Endo*	*Exo*	*NS*	*Plant*	*PE*	*Endo*	*Exo*	*NS*
Cherries	+	+			Pea	+			
Current				+	Potato	+		+	
Grapes	+			+	Radish	+			+
Mango	+	+			Tomato	+	+	+	
Papaya	+	+	+						

[a]Abbreviations: PE, pectin esterase; PG, polygalacturonase; NS, not specified.

Table 14.2. Occurrence of pectic enzymes in micro-organisms.

		PG		*PAL*		*PL*
Organism	*PE*[a]	*Endo*	*Exo*	*Endo*	*Exo*	*Endo*
Fungi						
Aspergillus (niger)	++[b]	++	++			+
Penicillium	+	+				+
Fusarium	+	+		+		
Rhizopus	+	++				
Sclerotinia	+	+				+
Collectotrichum		+	+			
Yeasts						
Kluyveromyces		++				
Bacteria						
Bacillus (polymyxa)				+++		
Clostridium	+				+	
Erwinia		+	+	+++	+	+
Pseudomonas				+++		
Arthrobacter				+		

[a]Abbreviations: PE, PG, see Table 14.1. PAL, pectate lyase; PL, pectin lyase.

[b]+++ means produced in high levels; no data means activity absent.

FOOD-PROCESSING-RELATED PROPERTIES OF PECTIC ENZYMES

Some molecular and kinetic properties are given in Table 14.3.

Pectinesterases

Plant PEs are thought to attack either at the nonreducing end or next to a free carboxyl group and to proceed along the molecule by a single chain mechanism, creating blocks of unesterified galacturonic acid that are extremely calcium sensitive. Irregularities in the galacturonan chain, such as acetylated monomers, ester groups transformed into amides or reduced to primary alcohol, and occurrence of hairy regions, inhibit PE activity. Pectinesterase is highly specific for the methylester of polygalacturonic acid. Other esters are attacked only very slowly. The methylester of polymannuronic acid is not attacked at all. The rate of pectin deesterification depends on chain length; trimethyl trigalacturonate is not attacked at all. Fungal PEs differ from plant PEs by obeying a multichain mechanism, removing methoxyl groups at random. Further properties of PEs will be discussed in the section on technological functions.

Table 14.3. Properties of pectic enzymes from some plants and food-grade micro-organisms.

Enzyme type/source	*Molecular weight*	*Isoelectric point*	*Specific activity (units/mg)*	*Optimum pH*	K_m *value (mg/ml)*
Pectin esterase					
Orange I[a]	36000	10.0	694	7.6	0.083
Orange II[a]	36200	11.0	762	8.8	0.0046
Orange III[a]	54000	10.2	≥180	8	0.041
Aspergillus niger	39000	3.9	406	4.5	3
Polygalacturonase					
Tomato I[a]	1,00,000	8.6		4.5	
Tomato II[a]	44000	9.4		4.5	
Kluyveromyces	46000	6.1	89	4.0	
fragilis	48000				
Aspergillus aculeatus	47000	4.7–4.8	67	4.0–5.0	
A. niger[a]	42000	6.0–6.4	275	4.0	
	49000				
Pectin lyase					
A. niger I	35400	3.65	17	6	5.0
A. niger II	33100	3.75	44	6	0.9
Pectate lyase					
Bacillus subtilis	33000	9.85		8.5	

[a] Multiple molecular forms (isoenzymes).

Polygalacturonases

Activity measurements by viscosity on methyl and glycol esters of pectic acid show a rapid decrease in the rate and degree of hydrolysis with increasing esterification; have shown that acetyl groups decrease the extent of degradation by lowering the affinity to the substrate molecules through the blocking of binding sites. The limitation of degradation by the presence of acetyl groups was confirmed by Rombouts and Thibault using sugar beet pectin as substrate (Table 14.4). The limiting effect of methoxyl groups is shown in the same experiment. However, Koller and Neukom have isolated a fungal PC that is fully active on 70 per cent acetylated pectin. Also, the mode of attack is known to differ for PGs of different origins. An endo enzyme may hydrolyse one bond randomly in a single enzyme-substrate encounter, followed by complete dissociation of enzyme and products. Additional scissions of this type follow, resulting in a gradual appearance of oligomers from the depolymerisation of higher oligogalacturonides. This pattern is observed with endo-PG from *Kluyveromyces fragilis*. It is characterised as a multichain attack. In the case of single-chain multiple attack, for example, with PG from *Collectotrichum lindemuthianum*, a single random hydrolytic scission is followed by a number of non-random attacks on one of the products, resulting in rapid liberation of oligogalacturonates. Many multiple forms and isoenzymes of PG have been described. Points of difference are dependence on cations and mechanism of action on oligomers. Particularly relevant to their industrial application is possible inhibition by vegetable extracts, which is noted for fungal PGs but not for PG from tomatoes.

Table 14.4. Degradation limits of sugar beet pectins before and after enzymatic demethylation and saponification.

Pre-treatment and mode of degradation	*Pectins*			
	Water soluble	*Oxalate soluble*	*Acid soluble*	*Alkali soluble*
No pre-treatment				
β-Elimination	7.2[a]	4.9	4.6	–
Endopectin lyase	8.6	4.3	4.1	–
Endopolygalacturonase	1.2	3.7	1.7	26.9
Endopectate lyase	≤1.0	4.8	1.3	23.7
After pectin esterase treatment				
Endopolygalacturonase	19.1	22.4	16.3	–
Endopectate lyase	11.8	13.0	9.9	–
After saponification				
Endopolygalacturonase	37.5	36.7	30.8	29.3
Endopectate lyase	33.4	34.5	31.0	26.4

[a]Percentages of galactosyluronic acid bonds broken.

Endopectin Lyase

The best substrate for endopectin lyases, measured at pH ≥ 7, is completely esterified pectin, as indicated by degradation limits and affinity (M^{-1}). At lower pH values, affinity for less highly esterified pectins increases and a marked stimulation by calcium and other cations is observed. This behaviour makes PL a useful enzyme for fruit processing. These enzymes need the methylester groups; they are inactive on glycol esters and on amidated pectates.

Endopectate Lyase

Endopectate lyases, on the other hand, do not differentiate between the methyl ester and the glycol ester of pectic acids. Surprisingly, pectate is not the best substrate for bacterial PALs. These enzymes show maximal activity (initial velocity and degree of degradation) on low methoxyl pectins. Figure 14.4 shows that the DE optima for PALs from *Arthrobacter* 547, from *Arthrobacter* 370, and from *Bacillus polymixa* are 21, 44, and 26 per cent respectively. This feature and their high pH optimum make them interesting for vegetable processing. Changes produced by PALs are studied easily since enzyme action can be stopped by the addition of a chelating agent because of an absolute requirement for calcium ions.

Rhamnogalacturonase

An enzyme has been described that splits the galacturonic acid-rhamnose glycosidic linkage in hairy regions of apple pectins with strongly enhanced activity when the ester groups have been deesterified and arabinose has been removed by acid hydrolysis (modified hairy regions). The enzyme was found in a commercial pectinase preparation and surely must be grouped with pectic enzymes. The end products are oligomers with alternating galacturonic acid and rhamnose units, rhamnose forming the non-reducing end. The availability of these compounds make it possible to determine the anomeric form of the rhamnose units as α using advanced NMR techniques.

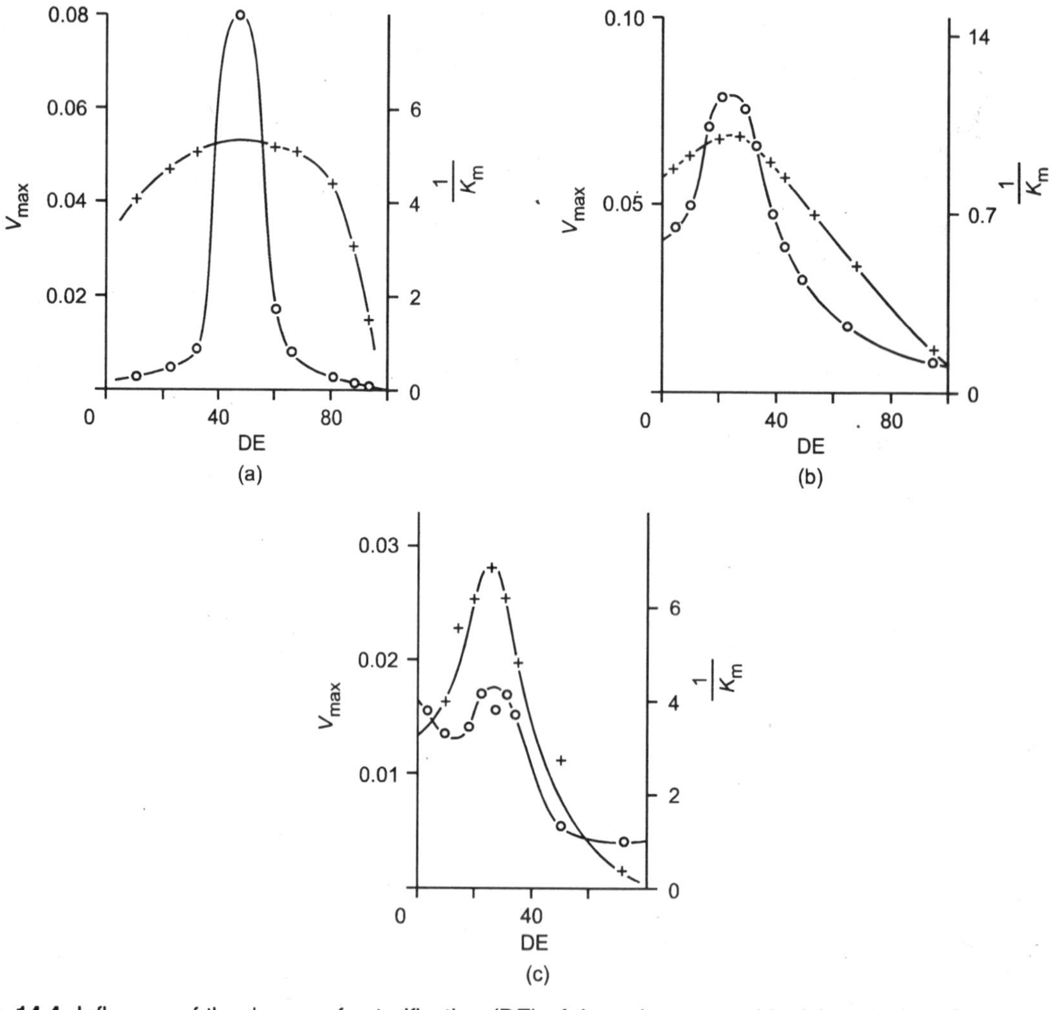

Fig. 14.4. Influence of the degree of esterification (DE) of the substrate on V_{max}(+) and $1/K_m$ (O) of pectate lyases of *Arthrobacter* 370 (A), of *Arthrobacter* 547 (B) and of *Bacillus polymyxa* (C) measured by spectrophotometry.

Commercial Pectinases

Commercial pectinases are fungal preparations, mainly from *Aspergillus* species, produced by a number of companies. These preparations are usually mixtures of PE, PG, and PL activities, hemicellulases, and endo-β-glucanases (C_x-cellulases). The activities of three different commercial preparations are shown in Table 14.5. The activities found are all from the mould used for the production of the pectic enzymes, with the exception of the C_1-cellulase (cellobiohydrolase), which is added for technological purposes. The arabinanases play a special role in fruit juice processing. Preparations containing polygalacturonase as the sole pectic activity are on the market and experimental quantities of fungal PE are available. Only one pectinase to date has been seen to contain rhamnogalacturonase activity. Gene technology surely will be used to produce desirable single-activity preparations that will open new possibilities for use alone or in optimised combinations.

Table 14.5. Specific enzyme activities of multienzyme complexes in technical fungal pectinase preparation[a].

Enzyme activity	*Substrate*	*A*	*B*	*C*
Polygalacturonase	Polygalacturonic acid	1982	3314	1878
Pectin lyase	Pectin DE 90	43	53	74
Pectin esterase	Pectin DE 65	548	448	227
Combined pectolytic	Pectin DE 75	198	290	274
C_x-Cellulase	CMC	998	180	1228
C_1-Cellulase	Avicel	99	1	22
Arabanase				
Linear arabinan	1-5-α-L-Arabinan	9	10	16
Branched arabinan	1.3; 1,2; 1,5-α-L-Arabinan	14	14	16
α-L-Arabinofuranosidase	PNP-Arabinofuranoside	35	37	333
Galactomannanase	Galactomannan	3	4	9
Mannanase	1,4-β-D-Mannan	7	0	0
Galactanase	1,4-β-D-Galactan	11	58	91
Xylanase	1,4-β-D-Xylan	4	0	2

[a]Activity is expressed in international units per gram enzyme preparation.

METHODS USED IN TECHNOLOGICAL PECTIC ENZYME RESEARCH

Assay Methods

An extended discussion by many researchers on methods using pectic enzymes from specific micro-organisms and plant cell cultures has been compiled by Wood and Kellog. This section is a more general treatment of the subject.

Pectinesterase activities

Pectinesterase activities produce free carboxyl groups and free methanol. The increase of free carboxyl groups is monitored easily by automatic titration. At pH values below 4.5, or when assessing PE action in a natural substrate not suited for automatic titration, the increase in free-methanol can be measured by gas liquid chromatography (GLC) or high performance liquid chromatography (HPLC). Head-space GLC after conversion of methanol to methyl nitrite is convenient and sensitive. Even more convenient, although less sensitive, is HPLC executed, for example, on an alkaline isopropanol extract with an ion-exchange column. One unit of PE is defined as the amount of enzyme that liberates 1 micromole free carboxyl groups or methanol per minute under specified assay conditions. Random or blockwise distribution of free carboxyl groups obtained by various PEs can be distinguished by high performance ion-exchange chromatography of the substrate or by measuring the affinity for calcium ions, which is much stronger in the case of blockwise distribution. Semi-quantitative tests are sensitive and conveniently done by observing pH shift in a neutralised pectin solution, or gelling when calcium ions have been added. Such tests are useful to check for (undesirable) residual PE activity in citrus juices after pasteurisation.

Lyase activities

Lyase activities can be measured by monitoring the increase in light absorbance at 232 nm caused by the formation of the double bond (Fig. 14.3) at the new non-reducing ends of the molecules. Differentiation between PALs and PLs is possible by choosing low methoxyl or high methoxyl pectins, respectively, as substrates. Care must be taken to use clear pectin solutions, since any change in turbidity by pectin degradation would affect the values measured. Exo and endo activities, initial velocity, and extent of depolymerisation are measured in this way. One unit lyase liberates 1 micromole unsaturated products per minute under specified assay conditions. The specific extinctions at 232 nm are 4600 M^{-1} cm^{-1} for PAL and 5500 M^{-1} cm^{-1} for PL.

Polygalacturonase activities

Polygalacturonase activities can be determined by measuring the increase in reducing end groups. The tests of Nelson-Somogyi are used preferentially, using galacturonic acid as standard. These tests measure endo and exo activities as well as the activity of oligopolygalacturonases. One unit of activity is defined as the quantity of enzyme that liberates 1 micromole reducing groups per minute under specified conditions. The substrate used is low-methoxyl pectin or pectate, which also responds to PAL, the activity of which also results in an increase in reducing end groups. PG, however, does not form unsaturated products and, thus, can be differentiated. Further, the pH optima differ widely (Table 14.3); PAL has an absolute requirement for calcium ions. It is recommended that viscosity measurements be combined with end-group (PG) and light-absorbance (lyase) measurements. This allows the calculation of glycosidic linkages split when the specific viscosity has reached half the original value. For endo activities, only a small percentage must be split whereas values as high as 40 per cent are necessary for exo activities. Information about the presence of exo activities and the breakdown patterns in general can be obtained by gel chromatography of oligomers and by peak shifts in elution diagrams from size exclusion chromatography for which versatile HPLC methods are now available. A reliable qualitative test for complete pectin degradation, for example, in fruit juice clarification, is the absence of flocculation on addition of two volumes of alcohol.

Preparation of Pectic Enzymes for the Laboratory

Pectinesterase with only weak exopolygalacturonase activity is obtained from orange pulp by extraction with a borax-acetate buffer. Polygalacturonase is found as the only pectic activity in liquid growth medium of *Kluyveromyces*.

Pectate lyases as single pectic activities are isolated from liquid cultures from certain strains of *Pseudomonas fluorescence*. There is no simple source of pectin lyase. Separation and purification of these activities is achieved by chromatographic methods. The usual steps are:

1. Desalting by gel filtration (Biogel P100).
2. Protein separation by anion exchange (DEAE Biogel A) or cation exchange (CM Biogel A).
3. Separation of pectic enzymes by cross-linked alginate.
4. Final cleaning-up step by fast protein liquid chromatography (FPLC).

Cross-linked alginate works by a combination of affinity and electrostatic effects and replaces cross-linked pectate. It is less problematic to prepare and has a greater capacity. Biogel is used because of its stability in the presence of glycosidases. Detailed procedures have been published for polygalacturonase, pectate lyase, pectin lyase, and pectinesterase. Figure 14.5 gives the scheme for the purification of pectin lyases I and II from a commercial preparation.

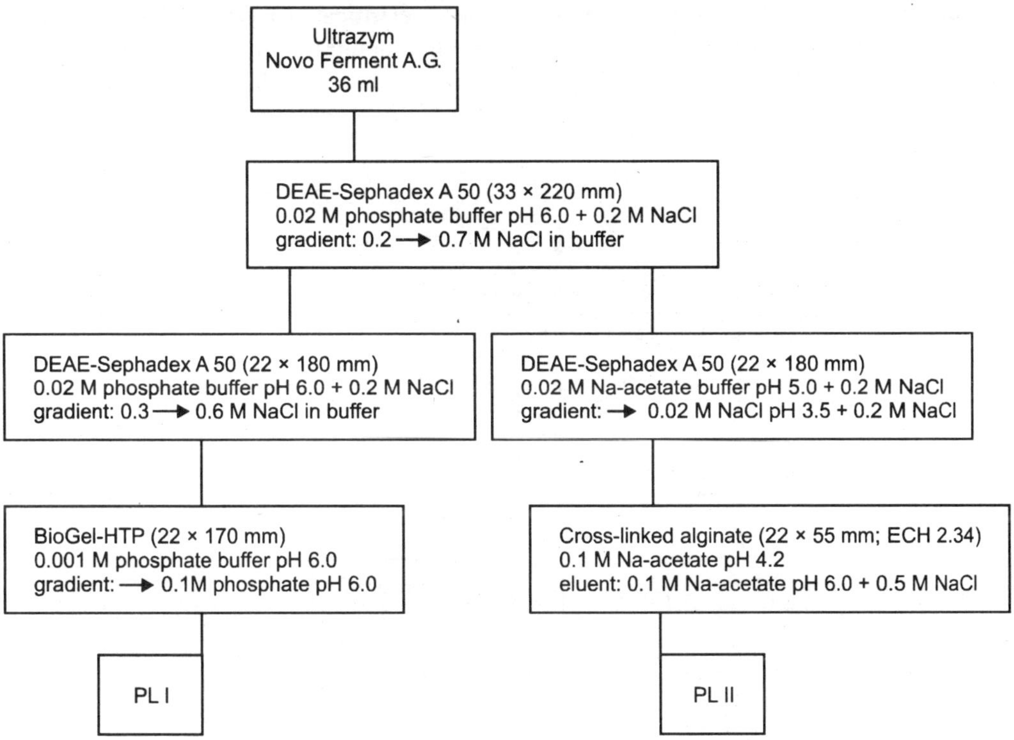

Fig. 14.5. Purification scheme for pectin lyase from a technical pectinase preparation of fungal origin.

USE OF COMMERCIAL FUNGAL ENZYMES AS PROCESSING AIDS

Figure 14.6 shows a general flow diagram for the processing of fruit and vegetable juices. The arrows indicate the point in the process at which enzymes are applied usefully.

Fruit Juice Clarification

Fruit juice clarification (Fig. 14.6, arrow 'a') is the oldest and still the largest use of pectinases, applied mainly to deciduous juices and grape juice. The traditional way to prepare such juices is by crushing and pressing the pulp. The raw press juice is a viscous liquid with a persistent cloud of cell wall fragments and complexes of such fragments with cytoplasmic protein. Addition of pectinase lowers the viscosity and causes cloud particles to aggregate to larger units ('break'), which sediment and are removed easily by centrifugation or (ultra)filtration. Indeed, pectinase preparations once were known as filtration enzymes. Figure 14.7 shows a schematic presentation of cloud particles: a protein nucleus with a positive surface charge is coated by negatively charged pectin molecules. Partial degradation of this pectin by pectinase results in aggregation of oppositely charged particles. Also, the reduction of the viscosity of the raw juice is due to pectinase action. Note that the mechanism of enzymatic clarification was not elucidated until 35 years after its introduction into the fruit juice industry. Careful experiments with purified enzymes have shown that this effect is reached either by a combination of PE and PG or by PL alone in the case of apple juice, which contains highly esterified pectin (≥ 80 per cent), whereas in grape juice, which contains pectins with a lower DE (44–65), PL alone does not perform as well. A hot clarification process saving enzyme and time and continuous clarification processes have been described. The advent of membrane processes has created the possibility for ultrafiltration of enzyme-treated juices, a process that is being introduced increasingly in the juice industry. The clarification of juices by pectin degradation is also important in the manufacture of high Brix concentrates to avoid gelling and the development of haze.

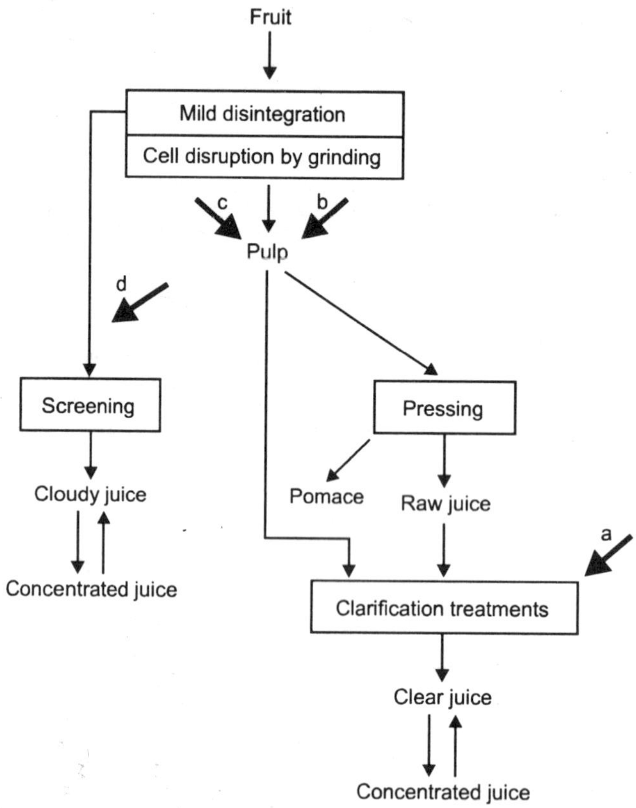

Fig. 14.6. Flow diagram of fruit juice manufacture. Arrows indicate eventual enzyme treatments by (a) pectinases for clarification; (b) pectinases for pulp enzyming; (c) pectinases and C_1 cellulases for liquefaction; and (d) polygalacturonase, pectin lyase, or pectate lyase for maceration.

Enzyme Treatment of Pulp for Juice Extraction

In the early period of use of pectinase for clarification, it was found, first for black currants, that enzyme treatment of the pulp (Fig. 14.6, arrow b) before pressing improved juice and colour yield. Pulping of such soft fruits with much soluble pectin results in a semi-gelled mass that is very difficult to press. Enzymatic pectin degradation yields thin free-run juice and a pulp with good pressing characteristics. In the early 1970s this treatment was applied to apples with poor pressing characteristics created by cultivar nature (Golden delicious) or the necessity for storage. Prevention of the inactivation of added enzymes by polyphenols is an important aspect of the process, and is achieved by aeration of the pulp. Endogenous polyphenol oxidases oxidise the polyphenols, which subsequently polymerise to high molecular weight compounds that are unable to inhibit the added enzymes. At the same time, the danger of a polyphenol haze in the finished product is diminished.

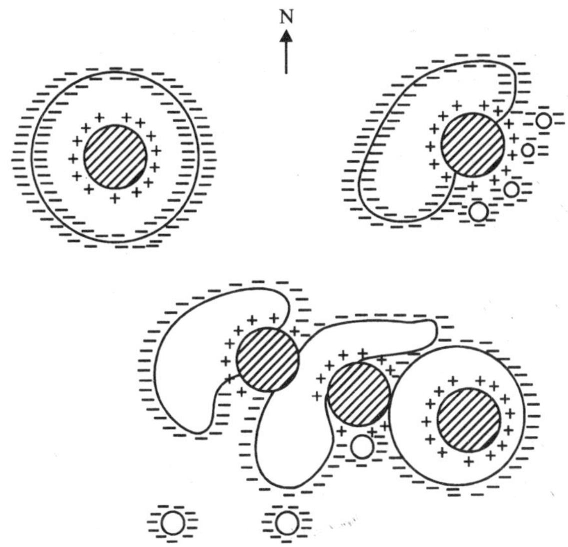

Fig. 14.7. Aggregation of cloud particles.

Addition of microbial polyphenol oxidases has been suggested. An alternative is the addition of polyvinyl polypyrrolidone, which binds the phenolic substances. Enzyme preparations performing well in juice clarification are also suitable for the enzyme treatment of pulp. In the case of apples, it has been shown that any combination of enzymes that depolymerises highly esterified pectin (DE ≥ 90) can be used successfully. The good pressability of enzyme-treated pulp is shown in Table 14.6. High yields are always connected with enzyme treatment. The better release of anthocyanins of red fruits into the juice, achieved by cell wall destruction, is another advantage of the pulp enzyme process, which has received attention from red wine manufacturers also. The process permits good yields of red grape juice, especially in combination with a mild heat treatment, which then can be fermented directly, as is the case for white wine. This technique is simpler than the classical fermentation, 'on the skins' as required for red wine to obtain the desired colour. In the United States, extraction of red grape juice from the pectin-rich concord grape (*Vitis labrusca*) by the enzymatic process is common. Enzyme treatment of pulp of olives, palm fruit, and coconut to increase oil yield has been described also.

Table 14.6. Effect of enzyme treatment and storage on apple juice yields.

Variety	*Storage*	*Enzyme treatment*	*Press yield (%)*
Treatment with 0.04 per cent pectinol flussing at 45°C for 10 minutes			
Horneburger	+	–	66
Horneburger	+	+	86
Golden delicious	+	–	59
Golden delicious	+	+	73
Golden delicious	–	–	71
Golden delicious	–	+	83
Jonathan	–	–	61
Jonathan	–	+	88

(Contd ...)

Variety	*Storage*	*Enzyme treatment*	*Press yield (%)*
Treatment with 0.004 per cent ultrazym 100 at 40°C for 20 minutes after pre-oxidation			
Golden delicious	+	+	78
Golden delicious	+	–	52
Golden delicious	+	+	75
Golden delicious	+	–	62
Karmijn	+	+	82
Karmijn	+	+	74

Liquefaction

In Fig. 14.6, arrow c indicates a process in which pulp is liquefied enzymatically so pressing is not necessary. The enzymes needed are pectinase and exo-β-glucanase (C_1-cellulase, cellobiohydrolase). The latter enzyme is derived from *Trichoderma* spp.; the so-called liquefaction process was developed when the enzyme became commercially available; Exo-β-glucanase works in conjunction with endo-β-glucanases (C_X-cellulases) which are present as *Aspergillus*-derived enzymes in most preparations. Figure 14.8 shows the viscosity decrease of stirred apple pulp during treatment with pectinase, C_1-cellulase, and a mixture of the two enzyme preparations.

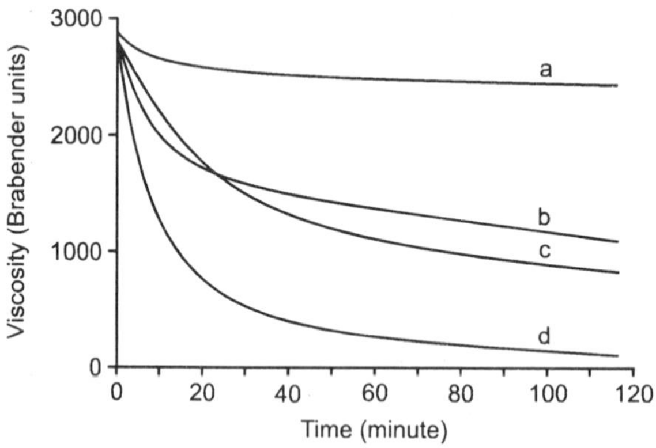

Fig. 14.8. Viscosity decrease of stirred apple pulp at 25°C. (a) No enzyme added; (b) cellulase added (C_1 + C_X); (c) pectinase added; and (d) cellulase and pectinase added.

For the mixture, a synergistic effect is seen. The low viscosity values reached correspond to complete liquefaction; at that stage, microscopic examination shows that cell walls have disappeared. The liquefied juices are almost clear (papaya, cucumber), cloudy (apples, peaches), or pulpy (carrots), depending on the accessibility of the cell wall compounds to the enzymes (lignin).

The liquefaction products can be clarified further by usual techniques; increased preference is given to ultrafiltration. The action of the enzymes can be followed chemically in enzyme-treated cell wall material by comparing the amounts of galacturonic acid and neutral sugars in alcohol-insoluble residues with the amounts produced in the control sample. Table 14.7 gives results of experiments on apple cell wall material with purified enzymes, showing that pectinase activity (PE + PG) alone releases 75 per cent of the pectic material, including the pectin sugars arabinose, galactose, xylose, and rhamnose. No glucose is released from cellulose.

Table 14.7. Neutral sugars and galacturonides released from apple AIS by pure enzymes.

Enzyme treatment	*Neutral sugars*[a,b]								
	Rha/Fuc	*Ara*	*Gal*	*Man*	*Xyl*	*Glc*	*Total*	*AGA*	*Total*
PE (citrus)	–	–	–	–	–	–	–	–	0.1
PG	4	9	5	–	4	–	3	21	8
PL	19	39	27	–	20	4	17	57	28
PE + PG	25	52	34	–	13	–	18	75	33
C_1	9	3	12	26	20	22	16	5	13
C_1 + PE + PG	58	89	69	84	71	79	78	82	80

[a]Abbreviations: AGA, anhydrogalacturonic acid; Rha, rhamnose; Fuc, fucose; Ara, arabinose; Gal, galactose; Man, mannose; Xyl, xylose; Glc, glucose.

[b]Data presented as percentage recovery, corrected for control.

Cellulase alone had little effect on pectin and solubilised only 22 per cent of cellulose. Combined cellulase-pectinase activities released 80 per cent of the polysaccharides. This synergistic effect is noticed in pulp viscosity experiments (Fig. 14.8). A similar effect has been found for grapefruit segment membranes when treated enzymatically. The breakdown products increase the soluble solids content of the juices. In this way, high yields of juice and solids are obtained, which is of special interest to manufacturers of concentrated juices. Table 14.8 compares pressing and liquefying apple juice. Liquefaction is also interesting for processing of vegetables and fruits that yield no juice on pressing or for which no presses have been developed (mango, guava, bananas). Therefore much interest is shown in this new, simple, low capital, easy maintenance technology in developing countries. Enzymatic liquefaction is attractive to all countries because costly waste disposal is virtually eliminated. Further, socio-economic developments in industrial countries make seasonal operations undesirable if not impossible. More attention will be paid to the transformation of stored apples, which are very difficult to press, so enzymatic liquefaction will have to be used.

Table 14.8. Comparison of yields from 1000 kg apples with good pressing characteristics.

Pressing
786 kg juice 12°Bx (750 litres)
131 kg concentrate 72°Bx
1000 kg 72°Bx concentrated juice from 7633 kg apples
Enzymatic liquefaction
950 kg juice 13.5°Bx (900 litres)
178 kg concentrate 72°Bx
1000 kg 72°Bx concentrated juice from 5618 kg apples

Maceration

Maceration (Fig. 14.6, arrow d) is the process by which organised tissue is transformed into a suspension of intact cells, resulting in pulpy products used as base material for pulpy juices and nectars, as baby foods, and as ingredients for dairy products such as puddings and yogurts. The enzymatic process for the manufacture of such products must be contrasted to the mechanical process in which many cells inevitably are disrupted, with negative consequences for quality because of action of endogenous enzymes that damages flavour, colour, and ascorbic acid. To limit the action of these enzymes, heat must be applied with the concomitant danger of heat damage. In any case, finisher operations to remove coarse tissue result in mechanical losses of nutritive constituents and a smooth consistency is not always achieved. Enzymatic degradation of pectin after a mild mechanical treatment often improves product properties if the process is carried out to leave as many cells as possible intact. Since the aim of the enzyme treatment is the transformation of tissue into a suspension of intact cells, pectin degradation should affect only the middle lamella pectin. This process is called enzymatic maceration. The so-called macerases are enzyme preparations with only PG or PL activity. The effectiveness of these macerases on various substrates has been compared. For maceration of vegetables, bacterial PAL is interesting because of its high pH optimum. The restricted pectin degradation solubilises rather than degrades the middle lamella pectin, which may improve the creamy mouthfeel. The process is often used for carrots. Whether maceration or cell wall breakdown has occurred can be checked by microscopy and by the absence of enzymatic browning. A very interesting use of enzymatic maceration is pulp for the production of dried instant potato mash. Enzymatic maceration prevents starch leaking out of the cells, avoiding the quality defect of gluiness of the reconstituted product. Potatoes are blanched to gelatinise starch and to inactivate endogenous PE, which in conjunction with the added PG would extend the possibilities for pectin degradation and turn the maceration process into pulp enzyming technology. Inactivation of endogenous PE is important for the maceration of many products.

Most uses of exogenous pectinases for fruit juice extraction and clarification are based on the destruction of pectin and other desirable dietary fibre components of fruits. Macerated products with only limited pectin degradation therefore may gain appeal for the consumer in future and promote the production of purified PGs and lyases. The use for pulpy products also means that immobilisation of these enzymes is of little interest.

Product-Specific Uses of Pectinases

Pulp wash

In the citrus industry, pulp wash is the product obtained by countercurrent water extraction of citrus pulp after dejuicing. Pulp wash is added back to the juice before concentration, but is often too viscous for this purpose because of dissolved pectin. The viscosity is reduced easily and efficiently by addition of pectinase. There are also claims of an increased solids yield. Also a process in citrus production has been patented to facilitate the manufacture and improve the quality of citrus fruit segments. Citrus fruit is heated to a core temperature of 20°–40°C; then peel surface is pierced to penetrate the albedo barely without penetrating the juice sections. Gas bubbles in the albedo are replaced with pectinase solution and the fruit is incubated to permit removal of peel and other membranes from intact juice segments. The use of enzymes to obtain citrus carotenoid suspensions as natural colouring matter will probably gain interest because of current consumer attitudes.

Apricot nectar stabilisation

Apricot nectar exhibits a yet unexplained phenomenon of cloud loss. The nectar is made by mixing finely screened pasteurised apricot pulp with water, sugar, and acid to contain 30–50 per cent apricot ingredient. On standing, the slowly sedimenting pulp particles form a gel that contracts, leaving clear syneresis fluid as supernatant. This separation can be avoided by grinding the pulp so finely that at least one-third of the cells is broken open. This process is difficult to control during production, but cells can be broken open by the action of pectolytic enzymes. For this purpose, an arabinanan-rich pectin fraction must be degraded. Pure PG or commercial preparations containing mainly PG are not suitable. The cell walls become thinner but do not break. If PE or an exoarabinanase is supplemented, the enzyme preparation becomes a cloud-stabilising enzyme, capable of degrading the extracted arabinan-rich pectin and of breaking open çells in the apricot pulp from which a cloud-stable product can be made. Figure 14.9 shows the elution patterns of gel filtration chromatography of the arabinan-rich pectin after treatment with various enzyme systems. Only treatments with cloud-stabilising enzymes move the uronide peak to an elution volume approximately equal to the bed volumę, indicating depolymerisation.

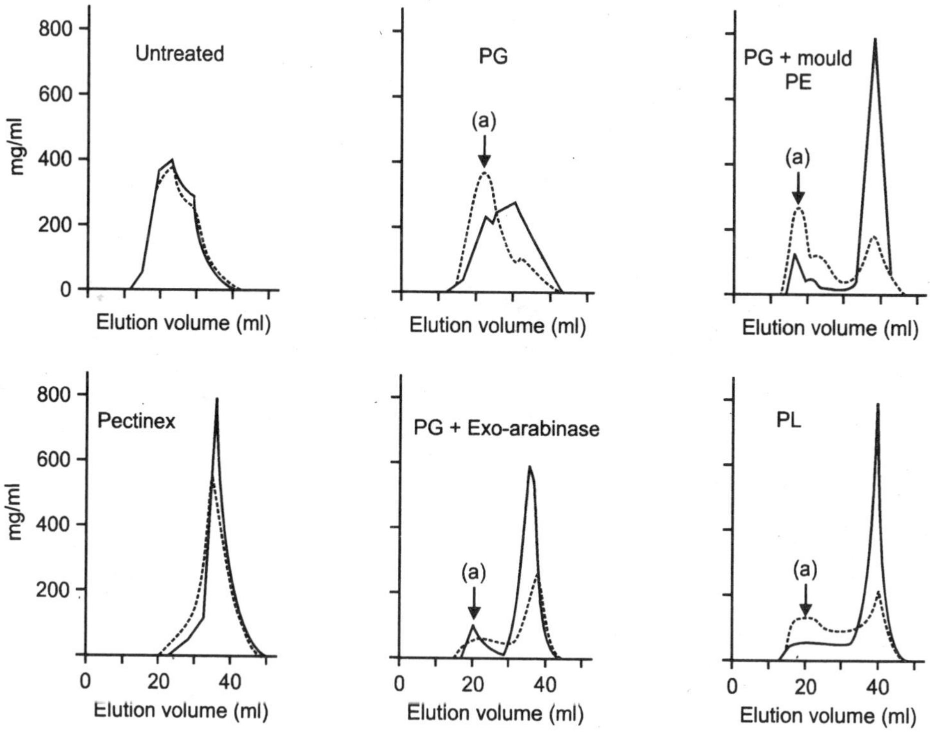

Fig. 14.9. Gel filtration chromatography of the HCl-soluble apricot pectin fraction degraded by various enzyme systems on Sephacryl S-300 column. Anhydrogalacturonic acid (AUA; µg/ml) indicated by solid line. Sugars (µg/ml) indicated by dashed line. Pectinex is a wide spectrum commercial pectinase. PG activity alone hardly degrades the pectin and does not prevent cloud loss.

QUALITY ASPECTS OF CLEAR FRUIT JUICES OBTAINED BY ENZYME TREATMENT OF JUICES OR PULP

Sparkling Clarity

The primary quality requirement of the consumer for sparkling clarity is achieved in all processes (juice clarification, pulp enzyming, liquefaction), provided pectin is degraded sufficiently (alcohol flocculation test). Nevertheless, haze has appeared in concentrated apple and pear liquefaction juices and sometimes in concentrated juices from enzyme-treated pulp. Analysis of the haze material shows it to be 90 per cent arabinose, of which 88 per cent is α-1,5-linked (Table 14.9).

Table 14.9. Sugar composition of haze material and glycosidic linkage composition for arabinosyl residues.

Sugar	*Amount (g/100 g haze)*	*Glycosidic linkages[a] arabinosyl units (%)*
Rhamnose	1.2	4.2 1-linked (terminal furanose unit)
Arabinose	90.0	0.2 1-linked (terminal pyranose units)
Xylose	0.3	2.3 1,3-linked (unbranched furanose units)
Mannose	0.7	88.4 1,5-linked (unbranched furanose units)
Galactose	1.6	3 1,3,5-linked (branched furanose units)
Glucose	0.1	1 1,2,5-linked (branched furanose units)
Galacturonic acid	3.9	
Total	97.8	

[a]Derived from methylation studies.

Structure work on nonhaze apple juice arabinans and the study of mould arabinanases indicate the haze-forming mechanism shown in Fig. 14.10. The enzyme treatment of pulp releases a branched arabinan.

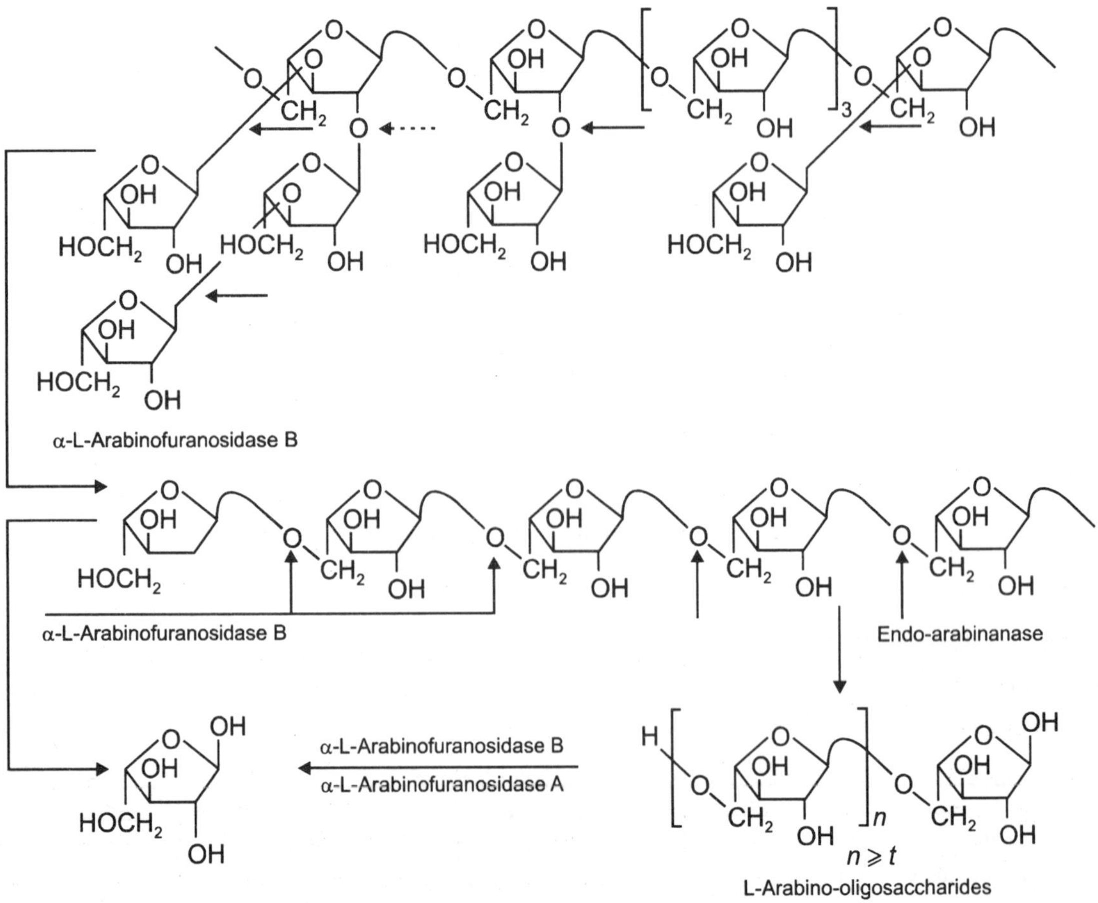

Fig. 14.10. Structure of arabinan and points of attack of arabinan-degrading enzymes.

The low uronide content indicates that the haze may come from the hairy regions of the pectin molecule. Its main chain is α-1,5 and the side chains are α-1,3. Researchers have identified and purified various arabinanases from commercial pectinase preparations. An exo-α-L-arabinofuranosidase II preferentially splits the α-1,2 linkages of the side chains, which results in debranching. Remaining chains then can retrograde under conditions such as low water activity and low temperature and crystallise, generating haze. Addition of sufficient endo-arabinanase activity to depolymerise the α-1,5 chain inhibits haze formation. Enzyme manufacturers are now providing for this effect in their preparations to minimise the problem. Table 14.10 shows the monomer analysis of freeze-dried dialysis retentate of apple juices made in the laboratory using the three technologies. The table provides information on the degree of polymerisation of the solubilised cell wall polysaccharides and on the activities of the enzyme preparations used.

Table 14.10. Amounts and characteristics of polysaccharides isolated from apple juice prepared by different processes and different enzymes.

		Pulp enzyme treatment[b]			*Liquefaction*[b]	
Analysis[a]	*Pressing*	A	B	C	C	D
Polysaccharide content (mg/kg pulp)	155	1075	1860	1075	3000	4300
Sugar composition (mole %)						
Rhamnose	3	5	3	5	4	4
Arabinose	21	38	6	52	43	54
Xylose	2	5	2	5	6	8
Galactose	19	9	4	7	6	8
Glucose	2	1	1	1	1	0
Anhydrogalacturonic acid	53	42	84	30	40	26
DM[c]	52	72	76	31	59	39
DA[d]	15	18	6	13	28	29

[a] Analysis was of dialysis retentates.

[b] Different enzyme preparations were used: A, B, and D were experimental preparations from Gist-Brocades, Delft, The Netherlands: A was a pectinase, B a pectinase rich in hemicellulase activity, and D a pectinase-cellulase mix. C was ultra SP-L from Novo Ferment, Dittingen, Switzerland.

[c] Degree of methylation.

[d] Degree of acetylation calculated from anhydrogalacturonic acid content.

Thus, liquefied juices with a high arabinan content of a high total carbohydrate retentate may be susceptible to haze formation. A simple test such as the alcohol flocculation test for pectin or the iodine test for starch must still be developed. Detection of the presence of the (partly) debranched arabinan chain involves separation by gel filtration and degradation by pure endo-arabinanase.

Uronide Content of Apple Juices

Table 14.11 shows the total uronide content of juices obtained in a laboratory study on the composition of juices prepared with the use of enzymes. The uronide content is seen to increase with intensity of pectic enzyme treatment. Free methanol from PE action is also present in quantities corresponding to about 10 per cent of the uronide content, that is, 60–450 mg/litre. Concern about free methanol is expressed recurrently but does not consider the fact that ester-bound methanol also enters the metabolism. The higher figure would be about 100 mg/kg above the range found in all kinds of beverages. Of course, in modern fruit juice technology, volatiles are stripped from juice, kept, and traded separately or mixed for use. Thus methanol content in reconstituted juices can be controlled. Depending on the PL action of the enzyme preparations used, 20–40 per cent of the uronides are present as unsaturated oligomers (Fig. 14.3). These enzyme-treated juices undergo more rapid browning than traditional juices. Figure 14.11 presents the results of model experiments from which it is seen that the unsaturated oligomeric uronides are powerful browning precursors.

The models for concentrated apple juices were: indicated quantities of oligouronides and fructose were brought to 70°Brix with sorbitol and to pH 3.5 with malic acid. In some experiments, 0.5 per cent amino acid mixture (asparagine, aspartate, and glutamate) were also added. Storage temperature was 50°C. At high temperatures, unsaturated oligomers were seen to split into saturated oligomers with one unit less than the original compound.

The unsaturated monomer reacts further to give a number of compounds, among which are the known browning precursors 2-furoic acid and 2-formyl-2-furoic acid. These findings are surely of interest in view of the large quantities of concentrated apple and pear juice shipped under non-refrigerated conditions.

Table 14.11. Uronide content of apple juice (12°Brix).

Methods of preparation	*Uronide content (mg/ml)*
Pressing and clarification	0.58
Water extraction	1.31
Methanol extraction	0.35
Enzymatic pulp treatment	2.51
Enzymatic pulp liquefaction	4.32

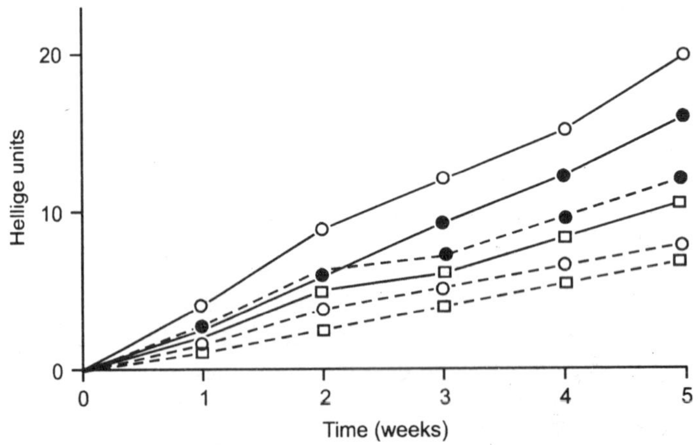

Fig. 14.11. Unsaturated oligogalacturonides (o, 0.5 per cent) as precursors for non-enzymatic browning reactions. Saturated oligogalacturonides (0.5 per cent), □; fructose (43 per cent), ●; with amino acids (0.5 per cent), —— ; without amino acids, – – –.

TECHNOLOGICAL FUNCTIONS OF ENDOGENOUS PECTIC ENZYMES

Commercial pectinases used as processing aids make certain processes possible; the presence of endogenous pectic enzymes makes certain processes necessary. PE action during fermentation of fruit is responsible for methanol in the distillates from the fruit. Distilleries pasteurise fruit pulps before fermentation to insure that methanol content of end products remains within legal limits. This release of methanol was observed by von Fellenberg and led to the conclusion that pectin is a methyl ester.

The tomato processing industry differentiates between a hot break and a cold break process. The hot break process is necessary to obtain the highly viscous juice the consumer prefers and the high consistency concentrated juice (tomato paste) used for sauces, soups, ketchup, and similar products. In this process, ingenious systems are used to heat-inactivate the abundant PE and PG activities of tomatoes for example, by crushing the tomatoes directly into circulating hot tomato pulp. If tomatoes are used for colour and flavour only, and consistency is provided by other ingredients such as starch, a more easily handled thin 'cold break' juice is the starting material for paste production. A holding time is introduced between crushing and heat treatment to insure breakdown of the pectins by combined PE/PG action. A process has been patented to macerate vegetables in fresh unpasteurised tomato pulp for the preparation of vegetable juices. A tomato cultivar has been developed in which most PG activity has been removed by gene technology; a firmer consistency of the fruit and a higher viscosity of the juice is claimed.

Citrus fruits are very active in PE; only a weak exo-PG activity as an additional pectic enzyme can be detected. Citrus PE causes one of the most intensively studied problems in food technology, the cloud loss of citrus juices. A juice protected against microbial spoilage can be seen to separate on standing into a clear supernatant and a layer of sedimented pulp. In concentrated juice, a gel is formed from which no juice can be reconstituted. These serious quality defects are caused by the action of PE, which deesterifies the native pectin which, at a certain point of deesterification, coagulates with native calcium ions just as a solution of commercial low methoxyl pectin is coagulated by the addition of calcium ions, forming flocs in diluted or a gel in a more concentrated solution. The manufacture of single strength juice with a stable cloud, or of concentrated juice that can be reconstituted to a stable cloud juice, therefore involves inactivation of PE, either by heat or by freezing. Both possibilities have disadvantages. The heat-sensitive orange juice flavour is easily damaged by heat treatments; freezing, on the other hand, is an expensive technology involving an uninterrupted line for the frozen concentrate from the juice extraction plant to the (overseas) customer. Alternatives are, therefore, investigated:

1. Inhibition of PE: Addition of polyphenols has been patented. A protein extract from the kiwi fruit (*Actinidia chinenis*) has been shown to inhibit PEs from various sources and to maintain cloud in unpasteurised orange juices. Both additions would need legal clearance and may create flavour problems. PE is known to display end-product inhibition, but the addition of pectic acid clarifies the juice, thus creating the effect one wants to prevent. However, we have been able to establish that, by degrading pectic acid, a degree of polymerisation can be found (8–10) that has the inhibitory effect of the high-polymer preparations but does not coagulate with calcium ions. The addition of such preparations will not prevent but will delay self-clarification, probably sufficiently long to be useful for short-term distribution of fresh chilled juice. No taste problems were noted and food clearance may be obtained more easily for a natural citrus juice constituent.
2. Binding of calcium ions: Figure 14.12a shows the cloud-preserving effect of adding ammonium oxalate to PE active orange juice. Similar results are obtained with chelating agents such as EDTA. This phenomenon is more interesting as an explanation of cloud loss than as an industrial practice.

3. Removal of PE substrate by addition of exogenous enzymes: Figure 14.12b shows that PG prevents cloud loss by depolymerising the low ester pectins since they are formed by PE before calcium coagulation occurs. The successful use of PL (not readily available) also has been described; in this case, the high ester juice pectin is depolymerised to low molecular weight products that are not calcium sensitive, even if deesterified, and that, in fact, might serve the function of PE inhibitor also. Such depolymerising action by exogenous enzymes also has been considered for lowering juice viscosity and, thus, facilitating concentration of orange juice to higher Brix values than ordinarily possible. Multistage evaporators that allow optimised heat treatment and aroma recovery combined with subsequent frozen storage and shipment of the concentrated juice, largely have overcome the PE inactivation-heat damage problem. This problem, however, still exists in developing countries for the manufacture of other cloudy juices that have strong PE activity and a heat-sensitive flavour, for example, guava and mango. The general trend for minimally heat processed products may bring the PE problem back, even to the modern citrus industry.

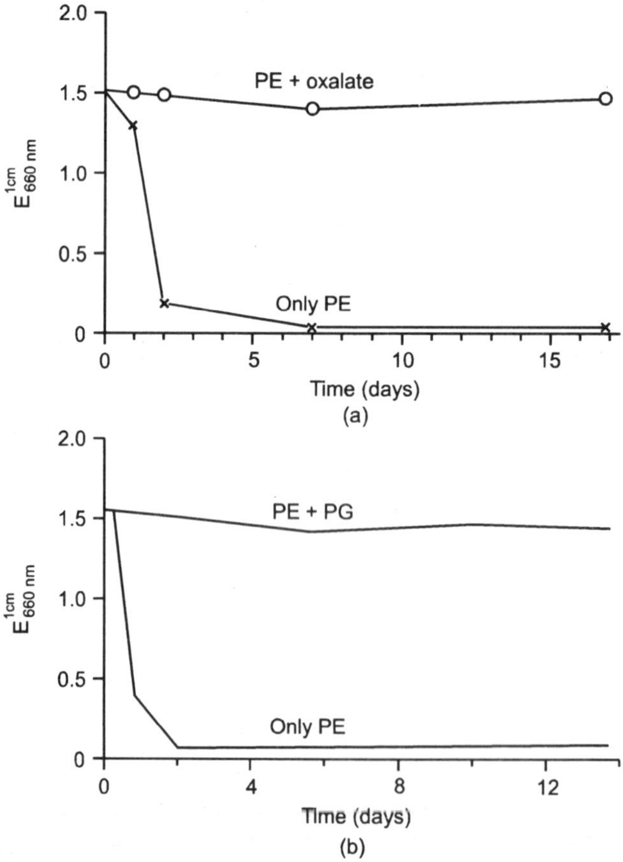

Fig. 14.12. Cloud stabilisation of pasteurised orange juice with added PE measured as extinction of supernatant after centrifugation. (a) Cloud stabilised by calcium binding as oxalate; and (b) cloud stabilised by PG. Not shown: Almost equal release of methanol with PE alone and PE + oxalate, respectively PE + PG. This proves that PE is not inhibited and cloud is stabilised as described in text.

The intensive research on citrus juice cloud and gelling problems has resulted in better knowledge of citrus PEs. Versteeg and others have isolated various multiple forms of PEs in oranges and have shown that they differ in affinity to pectins and pectates (Table 14.12).

Table 14.12. General properties of pectinesterases isolated from navel orange.

Characteristic	*PE I*	*HMW-PE*	*PE II*
Specific activity (u/mg protein)	200	≥180	762
Molecular weight	36200	54000	36200
Isoelectric point	10.05	10.2	≥11
pH optimum	7.6	8.0	8.0
K_m value (mg pectin/ml)	0.083	0.041	0.0046
K_i value (mg pectate/ml)	0.416	0.010	0.0016
V_{max} (mmol/mg min)	0.564	–	0.561
Turnover number (mmol/mmol sec)	340	–	338

They also have different heat stabilities (Fig. 14.13). It is understandable that they play different roles in cloud loss phenomena. One form, characterised by its high molecular weight, represents about 5 per cent of the total activity and has various outstanding properties. It is heat resistant and therefore constitutes the main activity responsible for cloud loss after insufficient heat treatment of pasteurised juice.

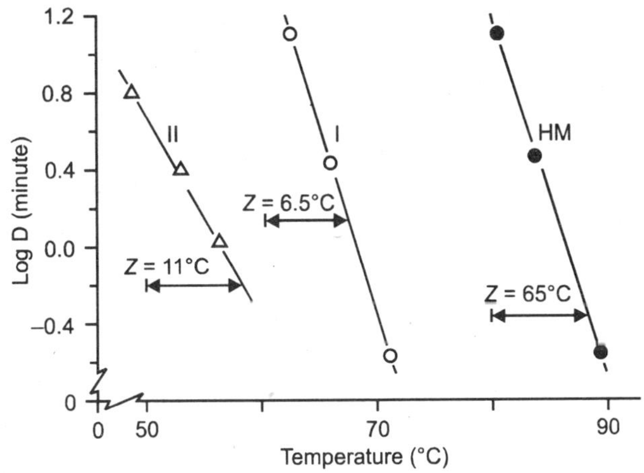

Fig. 14.13. Thermal destruction curves of pectinesterases I and II and of the high molecular weight pectinesterase (HM) in orange juice at pH 4.0.

This enzyme is also active at low temperature and low pH, and must have been the important activity in the now almost extinct process of self-clarification for clear lemon and lime juices that were obtained in holding tanks under SO_2 preservation. Such clear juices are still important commercial products and are now produced with commercial pectinases. Only form I of the two other isoenzymes in Table 14.12 has clarifying power. PE II has no clarifying power; the amount of free methanol generated is less than that generated by the clarifying enzymes. Table 14.12 shows the extremely high inhibition (K_i) by pectate that probably stops the activity at a certain point of saponification. This effect surely is enhanced by a blockwise distribution of free acid groups. Figure 14.14 shows that PE reactivation occurs when PC is added. Versteeg has noted also that the rate at which free methanol is formed in PE active orange juice is enhanced by PG. Such kinetic differences between PEs also can explain incomplete inhibition by oligogalacturonates. The different affinities to pectates can be used for separating PEs also. One PE in orange juice does not cause self-clarification, even after liberating the same amount of methanol as do the clarifying PEs. A different mode of attack must be supposed that produces low-ester pectins with reduced calcium sensitivity, probably by random distribution of free carboxyl groups. This mechanism, usually connected with fungal PEs, also has been postulated for a PE from *Aspergillus japonicus*; its production has been patented for the manufacture of low ester pectin for the preparation of low sugar jams. Such pectins are usually manufactured by an acid process. Fungal PE also has been developed by some enzyme manufacturers for clarification in the French cider industry. The two isoenzymes with orange juice clarifying properties are found in all component parts of orange fruit. All orange cultivars and citrus species tested so far contain the 12 PEs found in laboratory.

PE activity must be destroyed when peel is processed to so-called pectin pomace, the raw material for industrial pectin extraction. This modification must be done immediately after juice extraction by a blanching process, since even slight PE action creates pectins that are too calcium sensitive for normal use as gelling agents. In contrast to this application, PE is activated in the much larger operation of dried citrus pomace production cattle feed. In the so-called liming process, a slurry of calcium hydroxide is added to the peels while grinding them, bringing the pH to neutral and to alkaline levels. The high pH and the calcium ions activate the PE; rapid deesterification and calcium pectate coagulation occur. A bound liquid phase is released and easily pressed out, so only a fraction of the original water content must be removed by expensive thermal drying. The press liquor is similar in composition to the juice and is concentrated to molasses, which is also used as cattle feed. This application is probably the second largest use of a native enzyme in the food and feed industry, the largest being the use of native amylases in the cereal processing industries (malt, beer, bread). Endogenous PE is exploited also to protect and improve texture and firmness of several processed fruits and vegetables, including apple slices, canned tomatoes, cauliflower, carrots, potatoes, and peas. A low temperature long time blanching process will activate PE causing a partial deesterification of the pectins, which then react with (added) calcium ions, resulting in stronger intercellular cohesion.

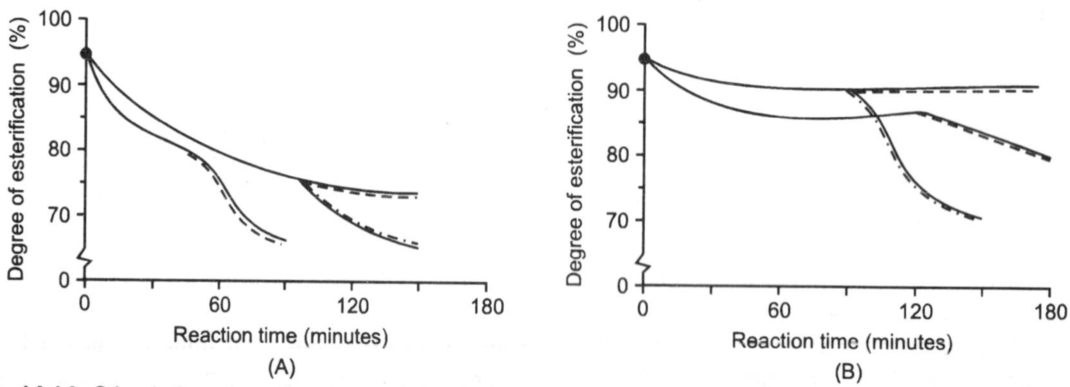

Fig. 14.14. Stimulation of pectinesterase by polygalacturonase (– · – · –·) and by pectate lyase (----). The stimulation is stronger for polygalacturonase and stronger in experiment B, in which the esterase used was inhibited more strongly by end product than in experiment A. Obviously, pectin lyase (——), which cannot degrade the deesterified regions, has no effect.

Pectolytic activities from microbial infections are, of course, of exogenous nature, but the consequences of their presence make them similar to endogenous activities. Yeast PG and bacterial PAL are involved in the softening of cucumbers and olives in brine. This quality defect can be avoided by inhibitors such as polyphenols leaching out from wine leaves into the brine, as practiced in some Mediterranean regions. Recirculated brine must, in any case, be pasteurised. *Byssochlamys fulva*, a heat-resistant mould, produces enzymes that cause decay of strawberries in syrup or jam. A highly resistant PG from *Rhizopus* spp. is responsible for texture breakdown in canned apricots. Mould infection has been described to cause clarification of aseptically filled orange juice. It is not clear whether this effect is caused by PE, PG, or both, but an increased amount of methanol was found in the clarified juices. Yeast PG can cause pectin breakdown in fermenting apple pomace, which then loses its value as raw material for the pectin industry. Desirable aspects of pectolysis by micro-organisms are seen in coffee and cocoa fermentation. During fermentation, the mucilage layer surrounding these beans is decomposed and washing of the beans before drying is facilitated greatly. Noble rot or *Botrytis cinerea* rot of overripe grapes is often considered desirable. The skin of the grapes is pierced. As a result of water evaporation, higher sugar concentrations are obtained, yielding a wine with higher alcohol content and a special flavour.

Chapter 15

Enzymes Associated with Savoury Flavour Enhancement

INTRODUCTION

Enzymes have long been exploited to improve the palatability of foods, even before their existence was known and understood. The manufacture of cheese using calf stomach as catalyst was probably the first example of enzyme use; this practice is so old that its discovery predates recorded history. With the understanding of the nature of enzymes and their catalytic potential, the use of enzymes gradually has been extended to a variety of fields. Among these fields, enzymes have found the most ready usage in the rapid evolution of the food industry. Some of the beneficial effects are flavour development, more desirable physical characteristics, greater nutritional value, and cost effectiveness. Although this entire field is of interest to many, this chapter will deal only with the production of flavour potentiating substances commonly found in savoury foods and the enzymes associated with their production.

FLAVOUR ENHANCERS

Flavour enhancers or flavour potentiators are known to be chemical substances that have little flavour of their own but, when mixed with food products, have the ability to enhance the flavour of the food. The best known flavour enhancers in use extensively worldwide are monosodium glutamate; disodium 5′-inosinate, and disodium 5′-guanylate. These compounds accentuate meaty flavour and have found applications as flavour potentiators in soups, sauces, gravies, and many other savoury products.

Traditionally, the Japanese culture has used naturally occurring sources of flavour potentiators in their food preparations. Typical sources of these flavour potentiators were sea tangle (*Laminaria japonica*), dried bonito tuna, and the black mushroom (*Lentinus edodus*). These products have been reported to exhibit a distinct savoury and delicious taste commonly referred to in Japanese as 'umami'. The important chemical compounds responsible for flavour potentiation action later were identified as monosodium glutamate (MSG), inosine 5′-monophosphate (5′-IMP), and guanosine 5′-monophosphate (5′-GMP). Kuninaka observed that 5′-nucleotides have a remarkable synergistic effect with MSG in flavour enhancement. All three potentiators that have been produced commercially are now used extensively in the food industry worldwide.

Among the more important developments in the use of enzymes is their application in the production of 5′-IMP and 5′-GMP. Of the 5′-nucleotides that exist in nature, these two show the most acceptable flavour enhancement properties. Although xanthosine 5′-monophosphate (5′-XMP) has some flavour enhancement properties, it is significantly less effective than 5′-GMP or 5′-IMP; hence, it has not been produced commercially.

Several researchers have been able to elucidate further the relationship between flavour potentiating properties and the chemical structure of the nucleotides. The nucleotides first recognised as having umami taste were those compounds that had a purine base with a hydroxyl group (or keto group, its tautomeric form) at the 6 position and a ribose moiety esterified with phosphoric acid in the 5 position. Additionally, it was implied that the hydrogen at the 2 position could be replaced by other groups such as a hydroxyl or an amino group without adversely affecting the flavour-enhancing properties. Thus, it is interesting to note that the three major purine-based 5′-nucleotides (5′-GMP, 5′-IMP, and 5′-XMP) have these common structural features which seem to be important in flavour potentiation. However, the intensity of flavour perception and enhancement was defined clearly by the specific nature of the substituents at position 2 of the purine moiety. From the standpoint of flavour enhancement, 5′-GMP has been judged to be the most effective, followed by 5′-IMP and 5′-XMP.

Both 5′-IMP and 5′-GMP occur naturally in foods. Animal muscles are generally rich in 5′-IMP. Several fresh seafoods are rich in 5′-AMP, which may be converted to 5′-IMP due to the action of endogenous enzymes released during processing. Raw vegetables have neither 5′-IMP nor 5′-GMP, yet these compounds were detected after the vegetables were cooked. Thus, it seems probable that the 5′-nucleotides found in cooked foods are the products formed following enzymatic degradation of intracellular RNA.

Several important enzymes that have applications in flavour-nucleotide production have been identified in naturally occurring products of plant, animal, or microbial origin. Best known within this group are the two enzymes 5′-phosphodiesterase and adenylic deaminase, derived from *Penicillium citrinum* and *Aspergillus* strains, respectively. This chapter primarily includes information relevant to the biochemical formation of the flavour enhancer nucleotides through the degradation of RNA by the aforementioned enzymes. Some effort has been made also to describe the characteristics of a few competing enzymes that must be minimised or eliminated for the desirable enzymes to be most effective in the flavour-nucleotide production process.

Substrate Specificity

Before addressing how different types of enzymes degrade the nuclear material in all biological systems, we must understand the structure of these macromolecules and the specific bonds that the enzymes break down during the process of hydrolysis. Like proteins, nucleic acids are informational macromolecules and therefore contain non-identical monomeric units in a specific sequence. The two major types of nucleic acids are DNA (deoxyribonucleic acid) and RNA (ribonucleic acid). Both RNA and DNA are long chain polymers made up of mononucleotides that contain three characteristic components: (i) a nitrogenous purine or pyrimidine base, (ii) a five-carbon sugar, and (iii) a phosphate group. The nitrogenous bases of DNA include ademine, guanine, cytosine, and thymine.

The composition of RNA is very similar to that of DNA with one exception. Thymine is replaced by uracil in the polymer chain. Further, RNA contains D-ribose as the 5-carbon sugar component. DNA contains 2-deoxy-D-ribose sugar. Functionally, DNA serves primarily as a repository of genetic information, whereas RNA plays several different roles in the expression of that information.

In recent years, yeast has found application as a substrate and a source of nuclear material for the production of flavour nucleotides. Although yeast contains both DNA and RNA, the latter typically makes up nearly 8–12 per cent, with less than 1 per cent DNA on a dry solids basis. Yeast extracts produced from baker's or brewer's yeast have been cleared for use in the United States by the Food and Drug Administration (Code of Federal Regulations 199010). Clearly, yeasts appeared particularly attractive as a source of nucleic acid for the production of flavour nucleotides because these micro-organisms are well-known to be edible with little risk of toxicity.

Three basic classes of ribonucleic acids are distinguished by their molecular weight and the way they function during protein synthesis. Messenger RNA (mRNA) provides the information that dictates amino acid sequence during polypeptide synthesis, transfer RNA (tRNA) directs the correct amino acids to the correct site in an elongating polypeptide chain, and ribosomal RNA (rRNA) provides the site for protein synthesis. Structurally, these polymers are formed by the linking of each nucleotide to the next through a phosphate group. Specifically, the phosphate group already attached by a phosphoester bond to the 5′ carbon of one nucleotide becomes linked by a second phosphoester bond to the 3′ carbon of the adjacent nucleoside phosphate (Fig. 15.1). The resulting linkage is called a 3′,5′-phosphodiester linkage.

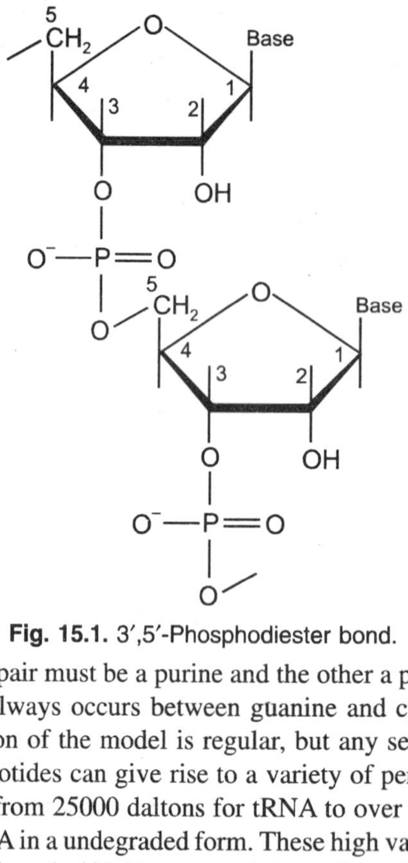

Fig. 15.1. 3′,5′-Phosphodiester bond.

One member of a nucleotide pair must be a purine and the other a pyrimidine in order to meet steric requirements. Pairing in RNA always occurs between guanine and cytosine, and adenine and uracil. The phosphate and sugar skeleton of the model is regular, but any sequence of base pairs will fit the overall structure. The four nucleotides can give rise to a variety of permutations; molecular masses of different RNA molecules range from 25000 daltons for tRNA to over 5,00,000 daltons for rRNA. It is difficult to prepare particular RNA in a undegraded form. These high values obtained for freshly isolated RNA therefore would be possible only if RNA escaped the action of nucleolytic enzymes during the purification process. Ribosomal RNA is by far the most abundant and most stable form of RNA in most cells. Usually, this RNA is referred to when speaking of 'yeast RNA' for flavour-nucleotide production.

ENZYMATIC HYDROLYSIS OF NUCLEAR MATERIAL

Several enzymes, predominantly belonging to the general group of hydrolases, are capable of digesting the nucleic acid chain, resulting in the formation of fragments of nucleotides, nucleosides, and free bases. Such a breakdown is determined primarily by the specificity of the enzymes. The enzymes that attack nucleic acid are distributed widely in yeast and in other biological materials in the microbial, animal, and plant kingdoms. The kinetics of enzymatic hydrolysis of various synthetic substrates by these nucleolytic enzymes has been the subject of a great deal of work in the past few decades.

The distribution of the four major classes of nucleolytic enzymes, that is, nucleases, phosphatases, nucleosidases, and nucleodeaminases, as found in all biological systems is outlined in Fig. 15.2. It is important for living cells to carry such an array of enzymes, both to break down the nuclear material that the cell may absorb from the environment during the growth phase and to break down the specific substrates inside that the cell no longer requires for sustenance. Under normal conditions, these hydrolytic enzymes must be sequestered carefully until actually needed by the cell, lest they digest vital cellular components that are not designed for destruction. Most of these enzymes are present in yeast and other biological materials at varying concentrations. Thus it becomes critical for the enzyme preparation used in the production of 5′-nucleotide rich extracts to have the maximum activity for the formation of the expected product, with minimal activity of the other competing enzymes. Needless to say, this characteristic is an absolute requirement for achieving acceptable flavour and optimum product yields in the process.

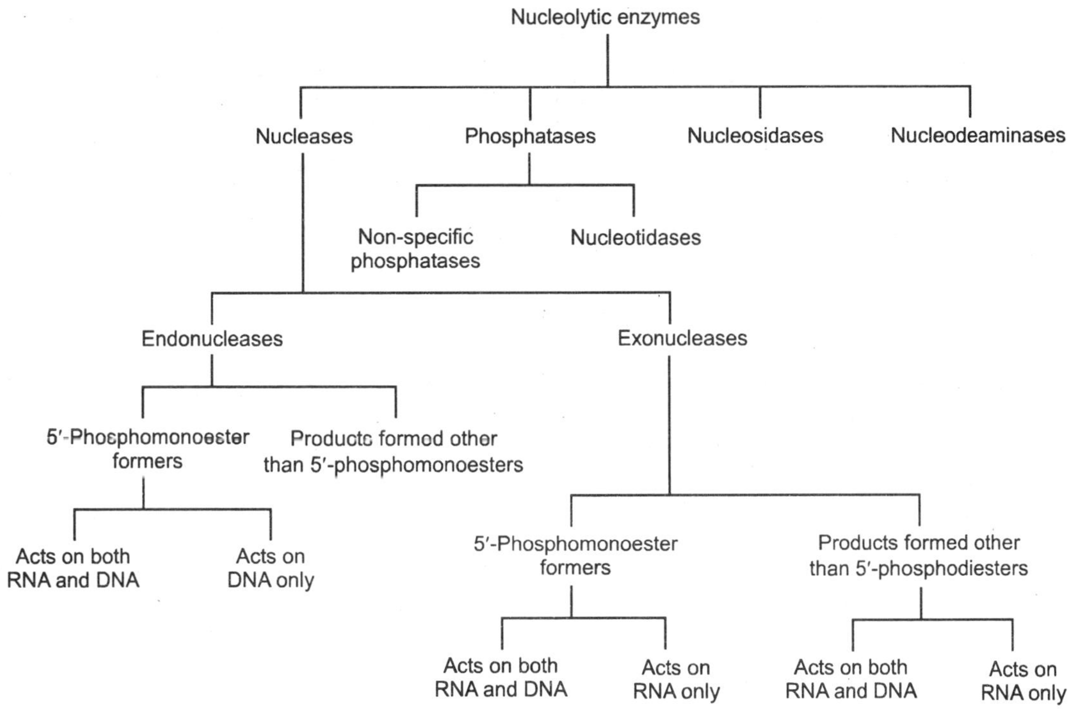

Fig. 15.2. Classification of nucleolytic enzymes.

Nucleases

Of the four major groups of nucleolytic enzymes presented in Fig. 15.2, several enzymes that are nucleases have found use in the production of flavours and flavour enhancers. All nucleases are known to catalyse the breakdown of internucleotide bridges without liberating the inorganic phosphates. Hence, these enzymes are capable of producing a wide variety of nucleotides when present in the reaction mixture. Therefore, a careful selection of the desirable enzymes is critical for the development of the desirable nucleotides for the flavour industry.

Three main features of nuclease activity can be used as the basis for classification. First is the mode of attack: polynucleotides can be attacked at points within the polymer chain (endolytically) or stepwise from one end of the chain (exolytically; Fig. 15.3). The second feature is the substrate specificity, which is based on whether the action is on RNA, DNA; or both. The third feature is the mode of cleavage of the phosphodiesterase bond. This feature depends on whether the hydrolysis of the bond occurs between 3′-OH and the phosphate group of the phosphodiester bond, to give 5′-phosphoryl end groups, or whether the attack is specific for the bond between the 5′-OH and the phosphate, which results in 3′-phosphoryl end groups. Based on these criteria, the industrially useful nucleases have been described using the classification scheme shown in Fig. 15.2.

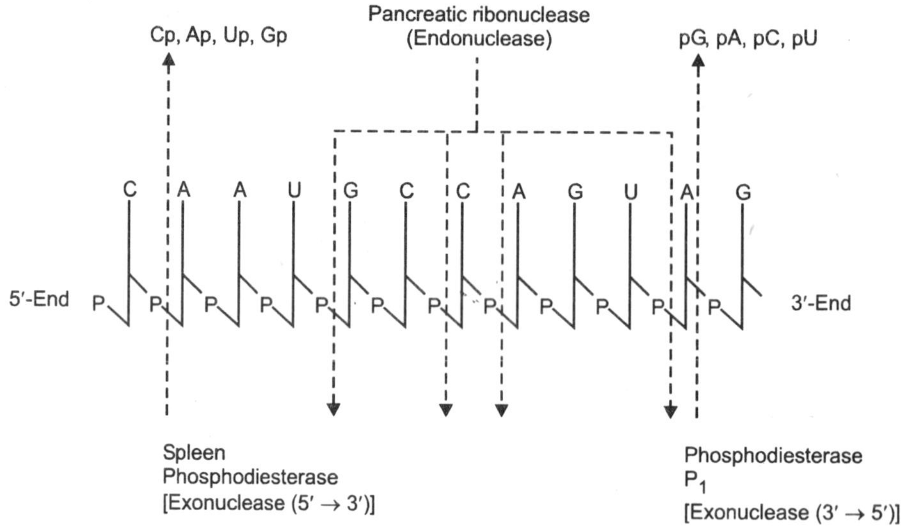

Fig. 15.3. Substrate specificity of nuclease enzymes.

Endonucleases

Endonucleases are enzymes capable of hydrolysing phosphodiester linkages within the nucleic acid chain at several points, with the simultaneous liberation of several mono- and oligonucleotide fragments. This class of enzymes can again be subdivided into two groups, namely, those enzymes that produce 5′-phosphomonoesters, some of which are capable of enhancing savoury flavour, and others that produce nucleotides other than the 5′-phosphomonoesters. The very common and widely recognised nucleotides belonging to the latter category do not enhance flavour and include 3′- and 2′-phosphomonoesters and, at times, 2′,3′ cyclic nucleotides.

Nonspecific endonucleases forming 5′-phosphomonoesters (EC 3.1.30)

The endonuclease enzymes that produce 5′-phosphomonoester compounds as a result of their activity on RNA or DNA commonly are derived from sources of plant or microbial origin. The best known enzyme belonging to this category that has found commercial use is nuclease P_1 derived from *Penicillium citrinum*. This enzyme shows a preference for cleaving phosphodiester bonds in single-stranded fragments of nucleic acids yielding 5′-nucleotides. Other less well known but nevertheless important enzymes that exhibit similar enzyme specificities are those from *Aspergillus* (*Aspergillus* nuclease S_1), *Neurospora crassa* (*N. crassa* nuclease), and Mung beans (*Phoseolus aureus* nuclease). An enzyme of similar properties also has been isolated from malt rootlets, and now has become of commercial importance in the production of flavour nucleotides. All these enzymes are capable of cleaving phosphodiester bonds of polynucleotides, regardless of the type of sugar and the base. The reaction proceeds with either denatured DNA or RNA; the extent of formation of 5′-phosphomononucleotides and oligonucleotides depends largely on the pH and the temperature of the reaction medium.

Nuclease P_1 shows a preference for certain specific linkages over others in the nucleic acid chain. For example, in the case of dinucleotide monophosphates, X-Y, nuclease P_1 showed a higher preference when X is adenosine or deoxyadenosine. Further, the base preference seems to depend largely on X rather than Y when nuclease P_1 acts on substrates that are typically of X-Y type.

It has been claimed that, for the substrate to be recognised by nuclease P_1, the critical structural requirements are the nucleoside 3-phosphate portion, with preferential breakdown of the diester bond taking place between the 3′-hydroxyl group of adenosine or deoxyadenosine and the 5′-phosphoryl group of the succeeding nucleotide (pA-Y) in the chain.

Although nuclease P_1 is placed among the endonucleases according to the most recent classification, it is also known for its simultaneous exonucleolytic activity, with the rapid generation of 5′-mononucleotides. Accordingly, the rate of appearance of mononucleotides from RNA was shown to be:

$$\text{AMP} > \text{GMP} > \text{UMP} > \text{CMP}$$

whereas the rate for denatured DNA was:

$$\text{dAMP} > \text{dTMP} > \text{dCMP} > \text{dGMP}$$

(see Table 15.1).

Table 15.1. Molar ratio of 5′-nucleotides released during partial hydrolysis of RNA and heat denatured DNA.

	Molar ratio (%)	
Nucleotide type	*RNA*	*DNA*
5′-AMP	37.7	39.5
5′-GMP	24.6	17.8
5′-CMP	17.1	20.4
5′-UMP or 5′-dTMP	20.6	22.3

Nuclease P_1 from *Penicillium citrinum* thus possesses some unique properties that differ from certain other well-studied nucleases. For example, as mentioned earlier, the enzyme is capable of hydrolysing both RNA and denatured DNA in an endo- and exonucleolytic manner, yielding predominantly 5′-mononucleotides. Although similar, the mode of action of snake venom phosphodiesterase is somewhat different because it functions only as an exonuclease, yielding only 5′-phosphomononucleotides. On the other hand, although the structural requirements necessary for nuclease P_1 and spleen phosphodiesterase are similar, the latter enzyme is known for its action at the diester bond between the phosphate and the adjacent non-specific hydroxyl component of the diester bond, thereby yielding 5′-phosphomononucleotides and 3′-phosphooligonucleotides as end products. Subsequent studies by Kuninaka and others have demonstrated that the 2′,5′-phosphodiester linkage was resistant to both nuclease P_1 and spleen phosphodiesterase enzymes, although these bonds were highly susceptible to snake venom diesterases. *Penicillium* nuclease P_1 also was noted for its ability to attack a variety of nucleic acids, regardless of the type of sugar or base, the size of the molecule, or the presence or absence of the terminal phosphate group in the nucleic acid chain. Thus it is clear that the type and the extent of the formation of phosphomononucleotides depend largely on the origin of the enzyme. Nevertheless, nuclease P_1 can be regarded as one of the most useful enzymes available on a commercial scale (or the production of 5′-mononucleotide-rich flavour extracts from substrates such as yeast that contain substantial levels of nucleic acids. Among the nucleases known to date, an endonuclease enzyme of plant origin that has been studied rather extensively is that of potato tuber. According to Suno, the structural requirements of the substrate and the cleavage sites on the nucleic acid chain for the nuclease enzyme from potato have been found to be similar to those of nuclease P_1. This enzyme is regarded as an endonuclease active on either RNA or denatured DNA to yield, almost exclusively, 5′-phosphomononucleotides and 3′-phosphooligonucleotides.

The purification of cell wall-bound nucleases from potato tuber was realised by Nguyen in small amounts that were, nevertheless, adequate for the systematic study of the nucleotides produced by its action. The results of this study indicated an estimated molecular mass of 37.6 kDa for this enzyme. The enzyme showed a pH optimum in the range 6.5–7.5 and a temperature optimum at or around 70°C. However, there appeared to be a basic difference between the cell wall-bound nuclease and the cytoplasmic nuclease derived from potato with respect to the optimum temperature of activity. Thus, according to the experimental data, the cytoplasmic nuclease performed best at 80°C, 10°C higher than the cell wall nuclease. The cell wall nuclease differed from the cytoplasmic nuclease in potato tuber with respect to its relative molecular size (37.7 kDa for cell wall; 34.2 kDa for the cytoplasmic nuclease) and its substrate specificity (the Michaelis-Menten constants, K_m, of the cell wall and cytoplasmic nucleases were 60 µg and 40 µg, respectively, when yeast RNA was used as substrate).

Table 15.2 shows the relative activity of the purified potato cell wall nuclease enzyme in the presence of various activators and inhibitors. These results have indicated a strong inhibition of enzyme activity in the presence of Zn^{2+}, Cu^{2+}, Mn^{2+}, pyrophosphates, citrate, phosphate, and EDTA. Nitrate, Mg, glutamate, and sulphite showed minimal or no inhibition of the enzyme.

Table 15.2. Effect of various ions and EDTA on nuclease activity.

Additive	*Concentration*	*Nuclease relative activity (%)*
Control (none)	–	100
Na-Glutamate	10	82
Na-Citrate	10	21
$NaNO_3$	10	104
$NaHSO_3$	10	95
NaH_2PO_4	10	51
Na-Pyrophosphate	10	0
$CuSO_4$	1	0
$ZnCl_2$	1	0
$MnCl_2$	1	20
$MgCl_2$	1	92
EDTA	1	17

Endoribonucleases forming 5′-phosphomonoesters (EC 3.1.26)

For most of the ribonucleases, the position of cleavage is such that the fragments formed predominantly terminate in 3′-phosphate groups. However, at least seven endoribonucleases have been identified that are capable of degrading ribonucleic acid chains to form fragments that terminate in 5′-phosphate groups. The best known examples of enzymes in category are ribonuclease P, ribonuclease III, and ribonuclease IV. Although these enzymes can yield 5′-nucleotides, they are of lesser importance in flavour-nucleotide production research because of their lack of availability on a commercial scale. Further, their presence in yeast has not been reported to date in the literature. Hence, they will not be discussed here in much detail.

Endoribonucleases forming products other than 5′-phosphomonoesters (EC 3.1.27)

These endonucleases preferentially hydrolyse RNA with the formation of mononucleotides and oligonucleotides that texminate in 3′-phosphate groups. Also it is now uncommon to find 2′,3′-cyclic phosphate intermediates in the reaction mixture. The best known enzymes belonging to this class of enzymes are RNase T_1, RNase T_2, and pancreatic ribonuclease. Enzymes similar in activity to these RNases have been identified in several materials of microbial and plant origin. Hydrolysis of RNA by these RNases leads almost exclusively to the formation of the four ribonucleoside 3′-phosphates.

These enzymes catalyse the hydrolysis of RNA in a two-step process. In the first step, the enzyme catalyses the cleavage of RNA to form 2′,3′-cyclic phosphate. In the second step, it catalyses the hydrolysis of the cyclic phosphate to ribonucleoside 3′-phosphates following the regeneration of the hydroxyl group at the C-2 position of the ribose unit (Fig. 15.4). The presence of these contaminating enzymes, either in the enzyme preparations or in the substrates for the production of 5′-nucleotide rich extracts, can increase the risk of possible deteriorative changes to potentially important flavour enhancing compounds and their precursors. Such contamination seriously decreases the overall quality of the final product.

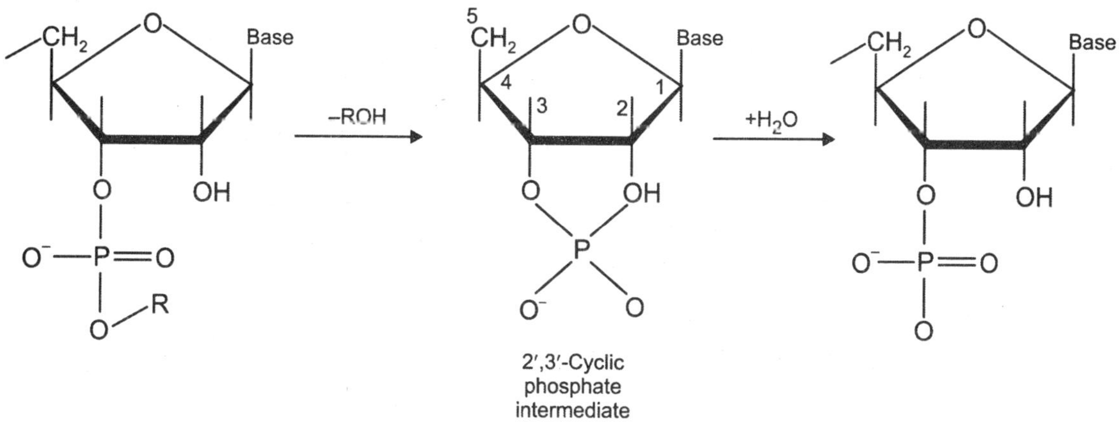

Fig. 15.4. Mechanism of hydrolysis of RNA by endoribonucleases.

The comparative properties of RNase T_1 and RNase T_2 have been reviewed by Egami. According to these studies, the molecular weights of these two enzymes were found to be 11,000 and 30,500, respectively. Their optimum pH values were determined to be 7.5 and 4.5. Both enzymes were inhibited by Cu^{2+} and Zn^{2+} ions. Studies have indicated that RNase T_1 cleaves phosphodiester bonds adjacent to the 3′-phosphate attached to a guanosine residue, yielding oligonucleotides terminating in guanosine contrast, pancreatic ribonucleases are pyrimidine specific. They cleave bonds adjacent to the 3′-phosphate linked to cytidine and uridine, resulting in a mixture of oligonucleotides terminating in either cytidine 3′-phosphate or uridine 3′-phosphate. The specificity of RNase T_2 does not appear to be as strict as that of RNase T_1. However, it is capable of producing the two-stage endonucleolytic cleavage liberating 3′-phosphomononucleotides and 3′-phosphooligonucleotides with 2′,3′-cyclic phosphates as intermediates. Similar types of enzymes that are likely to be present in 5′-phosphodiesterase preparations from either potato or malt rootlets or from substrates such as yeast must, however, be minimised or eliminated, probably by heat inactivation, to achieving best results in the production of extracts rich in purine-based 5′-nucleotides.

Exonucleases

So far we have considered the typical endonucleases that are capable of attacking the nucleic acid chain to result in the rupture of phosphodiester bonds at several points with simultaneous liberation of mono- and oligonucleotide fragments. With this information in hand, we are ready to examine the exonucleases that are common to all biological systems. These enzymes require a free 3′-hydroxyl group or a free 5′-hydroxyl group and proceed stepwise from the ends of the polynucleotide chain. These enzymatic reactions progressively increase the level of 3′- or 5′-nucleotides in the reaction medium, depending on the enzyme used in the reaction.

Nonspecific exonucleases forming 5′-phosphomonoesters (EC 3.1.15)

The best known example in this category is venom exonuclease (EC 3.1.15.1), the exonuclease cleavage reaction of which occurs in the 3′ to 5′ direction to yield the 5′-phosphomonoesters. In the purified state, this enzyme is active on both RNA and DNA. It exhibits no strict specificity with respect to the nucleotide sequence, but always yields 5′-mononucleotides. In this respect, snake venom phosphodiesterase appears to have properties similar to those of nuclease P_1 described previously. Its activity was, however, found most prominent at alkaline pH, between 7 and 10, with an optimum at pH 8.6. Although snake venom phosphodiesterase enzyme has the required properties for the production of flavour nucleotides, it is not certified as a food-grade enzyme for obvious reasons. A similar type of exonuclease enzyme has been observed in *Lactobacillus casei* cells; this enzyme degrades RNA to nucleoside 5′-monophosphates. The activity of the enzyme is stimulated by the presence of K^+ ions. Its optimum activity is at pH 8.0–8.2. Very little work has been published on this enzyme.

Exoribonucleases forming 5′-phosphomonoesters (EC 3.1.13)

These enzymes attack only RNA, providing exonucleolytic cleavage either in the 3′ to 5′ or 5′ to 3′ directions, yielding 5′-phosphomononucleotides. The best known examples of enzymes in this category are exoribonuclease II and exoribonuclease H. Although these enzymes are of interest to a flavour-nucleotide manufacturer, suitable food-grade sources have not been identified to date to produce this enzyme in a cost effective manner commercially.

Non-specific exonucleases forming products other than 5′-phosphomonoesters (EC 3.1.16)

These enzymes react with RNA and DNA, resulting in the exonucleolytic cleavage on the nucleic acid chain in the 5′ to 3′ direction to yield 3′-phosphomononucleotides. An example of an enzyme with these characteristics is spleen exonuclease. However, similar properties have been found in *Lactobacillus acidophilus* nuclease and *Bacillus subtilis* nuclease enzymes. The presence of enzymes of similar properties can lead to lower yields of 5′-nucleotides in a flavour-nucleotide production process. Typically, these enzymes have lower temperature optima and can be inactivated by simple heat treatment without destroying the more temperature-tolerant 5′-phosphodiesterase enzymes.

Exoribonuclease forming products other than 5′-phosphomonoesters (EC 3.1.14)

These enzymes have the characteristic property of hydrolysing the polynucleotide chain of RNA in a stepwise manner by successively detaching mononucleotides in sequence, starting from one end of the chain, to liberate primarily 3′-phosphomononucleotides. The best known exonuclease in this category is yeast ribonuclease. This enzyme is of particular interest to the producers of 5′-nucleotide rich extracts from yeast because of its negative effect on the yields of 5′-nucleotides if not removed completely from yeast prior to 5′-phosphodiesterase enzyme treatment. It is also important to inactivate similar enzymes that could possibly be present in the commercial grade 5′-phosphodiesterase preparations used in the production of 5′-nucleotide rich extracts. Typically, this inactivation is achieved by differential heat treatment.

Phosphotases

Phosphatases are phosphoric monoester hydrolase enzymes found extensively in almost all tissues and micro-organisms. These enzymes are able to hydrolyse the terminal phosphoryl group of nucleotides to cause the formation of free inorganic phosphates and nucleosides. These enzymes have a relatively broad specificity, so they are able to act on many compounds with common structural features. They thus may hydrolyse many different esters of phosphoric acid and are not entirely limited to nucleotides.

Placed in this group (EC 3.1.3) are 51 enzymes according to the Enzyme Nomenclature. Again these enzymes are grouped broadly under non-specific phosphatases and specific phosphatases. As the term implies, the enzymes included in the first category are known to hydrolyse all phosphomonoesters without distinguishing between the groups esterifying the phosphoryl residue. These enzymes would, therefore, hydrolyse both 3′-phosphates and 5′-phosphates. Included in this group are two sub-groups that have been known for a long time under the familiar names alkaline and acid phosphatases. Since they are non-specific phosphatases, they are capable of hydrolysing other phosphomonoesters such as glycerophosphates and phenyl phosphates, in addition to catalysing the familiar reaction with nucleic acids and their corresponding mononucleotides.

The enzymes that belong to the second category are different phosphatases with varying specificities, each being very specific in the nature of the substituent on the phosphoryl group during hydrolysis. At least 49 enzymes are grouped in this category. Among the enzymes in this series, those that are considered destructive to the processing of flavour-enhancing nucleotides are the nucleotidases, the best known of which are 5′-nucleotidases that are characterised by the position at which the phosphate group is hydrolysed on the respective nucleotides. For example, the 5′-nucleotidases are known for their activity in splitting flavour-enhancing 5′-nucleotides to their corresponding nucleosides and orthophosphates, causing a decline in the flavour-enhancing ability of the final product. Non-specific phosphatases and nucleotidases are distributed widely in gramineous seeds, potatoes, bacteria, and fungi. Use of any such ingredients or their extracts for the production of flavour nucleotides requires the prior elimination of the aforementioned degradative enzymes to insure optimal flavour-nucleotide production in the process.

Nucleosidases

The nucleosidase sub-group of enzymes under hydrolases (EC 3) is likely to be of interest to the flavour industry, particularly because of the problems these enzymes can cause by destroying the desirable flavour-enhancing nucleotides formed in the manufacturing process. These contaminating enzymes have the ability to hydrolyse N-glycosyl compounds thereby liberating free bases as a result of the hydrolysis of the β-glycosidic bond. These reactions are generally single step conversions, for example, the conversion of 5′-inosinate to hypoxanthine and D-ribose-5′-phosphate in the presence of inosinate nucleosidase (EC 3.2 .2.12).

$$5'\text{-Inosinate} + H_2O \xrightarrow[\text{nucleosidase}]{\text{Inosinate}} \text{Hypoxanthine} + \text{D-ribose-phosphate}$$

These enzymes are likely to be present in high concentrations in many of the materials employed in the production of 5′-phosphodiesterase enzymes used in flavour-nucleotide production. Further, the presence of these contaminating enzymes in yeast can be a serious concern to those who use this substrate as a source of RNA for nucleotide production. Every precaution must be taken to eliminate or minimise such competing enzymes prior to the use of yeast in the process.

Nucleodeaminase

The nucleodeaminases preferentially act on C–N bonds other than peptide bonds in cyclic amidines, liberating the free ammonia by deamination. This reaction can occur in compounds that contain purine or pyrimidine bases. In a typical reaction, adenine deaminase (EC 3.5.4.2) converts adenine to hypoxanthine and ammonia:

$$\text{Adenine} + H_2O \xrightarrow[\text{deaminase}]{\text{Adenine}} \text{Hypoxanthine} + NH_3$$

Adenosine phosphate deaminase (EC 5.4.17) is the best known enzyme in this category that has the greatest influence on the production of flavour nucleotides. Like the 5′-phosphodiesterase enzyme, adenosine phosphate deaminase, sometimes referred to as adenylic deaminase, has become an important enzyme for the production of 5′-nucleotide rich extracts. Its primary function is to convert 5′-adenosine monophosphate (5′-AMP), which offers no flavour-enhancing properties, to 5′-inosine monophosphate (5′-IMP), a product that has marked flavour-enhancing properties in food systems (Fig. 15.5).

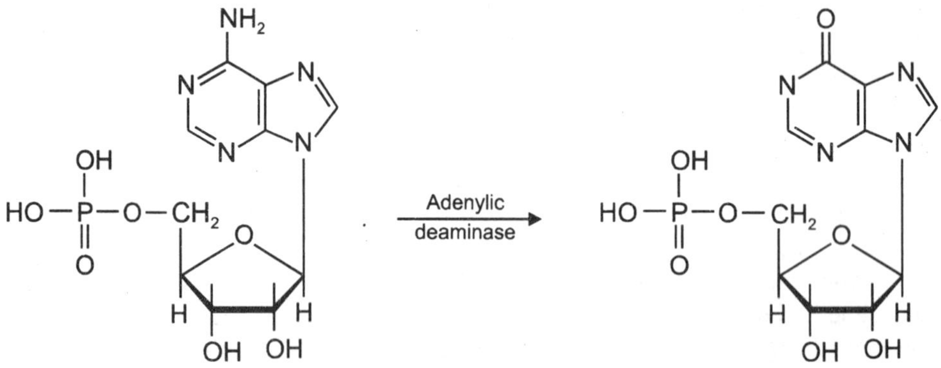

Fig. 15.5. Conversion of 5′-AMP to 5′-IMP by reaction of the enzyme adenylic deaminase.

Yeast is used as a common source of RNA for the commercial production of flavour-nucleotide rich extracts. Of the four 5′-nucleotides formed during the hydrolysis of RNA by the action of 5′-phosphodiesterase enzyme, only 5′-GMP is known to provide the flavour-potentiating properties. Nonetheless, the 5′-AMP produced, which has no flavour-enhancing properties, can be converted to 5′-IMP, which has flavour-enhancing properties like those of 5′-GMP, by applying a second enzyme, adenosine phosphate deaminase, as described. Both compounds then have flavour-enhancing properties, although guanylate is approximately twice as effective as inosinate.

Although adenylic deaminase is highly specific to 5′-AMP, it also can act on ADP, ATP, NAD^+, and adenosine, in decreasing order of activity. Adenylic deaminase has been produced commercially from *Aspergillus oryzae*.

COMMERCIAL PRODUCTION OF EXTRACTS RICH IN FLAVOUR ENHANCERS

Traditionally, the most commonly used flavour potentiators other than table salt are the naturally occurring food preparations rich in monosodium glutamate, as exemplified by Kombu or sea tangle (*Laminiarie japonica*) used by oriental cultures for centuries. This flavour potentiator, like the 5′-nucleotides 5′-GMP and 5′-IMP, generally is known to stimulate the corresponding taste buds to bring about the full savoury character in foods for maximum enjoyment.

YEAST EXTRACTS

Yeast extracts and hydrolysed vegetable proteins are commercially available products rich in glutamic acid. They have been used extensively by the food industry as flavouring agents. Yeast extract is a concentrate of the soluble fraction of yeast generally made following autolysis, which is essentially a degradation process carried out by activating the yeast's own degradative enzymes to solubilise the cell components found in the cell.

Autolysis or self-digestion of viable yeast cells takes place at an acceptable pace when the yeast slurry is maintained at 45°–55°C for 24–36 hours at pH 5–6. Lysis of these live whole cells is primarily brought about the action of β-1,3 gluconase and protease enzymes present in the cells. The β-1,6 gluconase and mannanase enzymes participate in the further solubilisation of the cell wall matrix.

More than 40 proteolytic enzymes have been identified to date in *Saccharomyces cerevisiae*, of which only a few can be described as vital to the autolytic process. A few of the important proteolytic enzymes that play a major role during lysis are proteinase ysc A, proteinase ysc B, carboxypeptidase ysc Y, and carboxypeptidase ysc S.

All these soluble proteases appear to be localised in the vacuole, whereas their inhibitors are in the cytoplasm outside the vacuole. Such compartmentalisation segregates biologically compatible components from the corresponding enzymes to allow the control and integration of the intracellular activities for the orderly functioning of the viable cell (Table 15.3).

Under autolytic conditions, the gradual breakdown of barriers in the highly compartmentalised cell matrix releases proteolytic enzymes from the vacuole into the degenerating cell matrix. These enzymes are, at first, inactivated because of the formation of complexes with corresponding inhibitors present in the immediate vicinity.

This inactivation is evident when proteinase ysc A and proteinase ysc B are extracted by autolysis at pH 7. However, incubation of the yeast at pH 5 caused reactivation of these proteinase enzymes by the proteolytic inactivation of the firmly bound inhibitors.

The vacuolar proteinases are involved in highly non-specific protein and peptide degradations. Likewise, nucleases present in the cell begin to act on the RNA and DNA, reducing them to polynucleotides, mononucleotides, and nucleosides or free bases. These processes can, however, be enhanced by initially subjecting the yeast to an external protease attack, for example, by papain, followed by treatment with a lytic enzyme such as glucanase. These enzymes are currently available commercially to improve the extract yield during yeast autolysis.

Table 15.3. Properties of yeast proteases.

Property	*Proteinase ysc A*	*Proteinase ysc B*	*Carboxypeptidase ysc Y*	*Carboxypeptidase ysc S*
Type	Acid endopeptidase	Serine endopeptidase	Serine exopeptidase	Metallo exopeptidase (Zn^{2+})
Optimum pH	2–6	6–7	4–7	7
Optimum T (°C)	35–40	45–55	45–55	60
Cell location	Vacuole	Vacuole	Vacuole	Vacuole
Solubility	Soluble	Soluble	Soluble	Soluble
Molecular weight	60000	32000–44000	61000	Not determined
Isoelectric point	3.8	5.8	3.6	–
Inhibitors	Pepstatin, etc.	Chymostatin, etc.	Hg^{2+}	EDTA
Cellular role	Protein degradation	Protein degradation	Protein degradation	Protein degradation

Glucanase enzymes have been isolated from yeast, and are known to hydrolyse the β-1,3 and β-1,6 linkages of the glycan building blocks of the cell wall. The glucanase enzymes are known to participate in the budding process.

Nonetheless, under autolytic conditions, β-1,3 glucanase enzymes, with support from the proteinases, are capable of disrupting the cell wall, thereby facilitating the release of the soluble compounds such as amino acids, peptides, mono-nucleotides, and other degraded low molecular weight materials from within the cell matrix.

The yeast proteases have a broad specificity for their substrates. They are known to exhibit different pH and temperature optima. Therefore, the process conditions used by extract manufacturers are not optimal for all proteolytic enzymes. Consequently, the variation of flavour profiles of extracts made by autolysis under different pH and temperature conditions is understandable. Thus, the control of conditions of autolysis is vital to maintaining product quality and consistency.

Glutamic acid is the most abundant amino acid in a vast majority of proteins found in foods. Proteolytic degradation during autolysis results in the release of much glutamic acid into the medium. When protein hydrolysis is incomplete, it is common to find some glutamic acid units as parts of peptides. Such bound glutamic acid generally exhibits minimal influence on the flavour. The free glutamic acid can be in equilibrium with a number of ionic forms that are dependent on the environmental pH. The predominant ionic form of glutamic acid at a pH range of 5.5–8, which is the common pH of most food systems, has a net negative charge. Under these conditions, both glutamic acid and MSG appear to be present as the glutamate ion, thus enhancing the sensory properties of the food.

Yeast extracts rich in flavour-enhancing nucleotides such as 5′-GMP and 5′-IMP also are being produced from *Candida utilis* by heating a cell suspension for 30 minutes at 80°–100°C, cooling the medium to 40°C, and treating with proteinase, 5′-phosphodiesterase, and deaminase enzymes. The suspension is then centrifuged and the supernatant rich in 5′-nucleotides is concentrated and spray dried. The resulting product is claimed to have properties particularly effective in improving the flavour of savoury meats, gravies, sauces, and soups.

Hydrolysed Vegetable Protein

Another important development in the food industry is the production of amino or peptide savoury flavourings from soyabeans, wheat, and other plant substrates by acid hydrolysis. The products made by this procedure are commonly referred to as hydrolysed vegetable protein (HVP). However, acid hydrolysis is known to bring about the destruction of certain amino acids and vitamins, resulting in a decline in the nutritive value. Additionally, HVPs made using HCl have been found to contain monochloro- and dichloropropanols (MCPs and DCPs) that are generally known as carcinogens.

One way to minimise the formation of such undesirable compounds is to provide a partial proteolysis of the protein substrate initially after which the content is subjected to a much milder acid hydrolysis. Some of the vegetable proteins selected for this process are generally high in glutamic acid. Processes that are designed to use acid hydrolysis exclusively or those that use the combination of partial proteolysis followed by acid hydrolysis are known for their higher degree of hydrolysis of proteins than the autolysis process described previously.

Hence, these products are likely to have higher levels of free glutamic acid than yeast extracts. Under physiological conditions, these hydrolysates have a higher level of glutamate ions with a higher efficacy for flavour enhancement. These products are known worldwide for their versatility and ease of the application in a variety of savoury food systems.

CONCLUSION

Trends in consumer preference for all-natural foods and a desire on the part of the food industry for a 'clear label' has resulted in an increased demand for substitutes such as yeast extracts to serve the function of the commonly used flavour enhancers in processed foods. As has been described previously, the procedure for the production of such yeast extracts requires the use of two enzymes, 5′-phosphodiesterase and adenylic deaminase.

One of the problems facing extract manufacturers is the escalating price of the enzymes. Because the current usage levels of these enzymes in the food industry are relatively low, the total amounts produced are correspondingly small. As expected, the cost of such low volume products is high and will continue to increase. A substantial improvement in the market share for these yeast extracts in the flavour industry should result in an increase in the usage volumes for these enzymes. More competition would result in more reasonably priced enzymes available for production of extracts.

However, it is not critical to operate such processes with highly purified enzymes. Indeed, one easy and certainly viable alternative is the use of crude enzyme preparations since, by adopting this procedure, the purification and stability problems could be circumvented. Further, this method is highly cost effective. However, one must bear in mind that, when operating with crude enzyme preparations, there is the risk of side reactions competing for the same substrate, thereby causing significant reduction in flavour-nucleotide yields. There is currently a shortage of crude enzymes and the yeast extract manufacturers are not comfortable with the quality and price of those crude preparations. Accordingly, there is a definite need to make these enzymes available in commercial quantities with higher purity at affordable prices.

The enzyme 5′-phosphodiesterase is known to be highly unstable after it is extracted from its native environment. The factors that influence stability most are pH, temperature, moisture, ionic strength, shear, pressure, and any combination of these effects. A more complete understanding of these effects should enable enzyme producers to produce and commercialise more active and stable 5′-phosphodiesterase enzyme for use in extract production.

Opportunities still exist to reduce the cost of enzyme use further without affecting the quality of the extracts. Use of immobilised enzymes to carry out the precise degradation of the RNA lowers costs substantially. An important feature of this method is that immobilisation enhances enzyme stability and biocatalytic recovery. The economic benefit of using this approach is based on the fact that it allows for decreased fixed costs, including those of labour and construction. Moreover, the immobilised enzymes are ideally suited for long-term continuous operation, which can bring about economic benefits to the processor. Studies have also shown that, when the two enzymes critical for production of flavour enhancers are bound to an alkylamine ceramic support using glutaraldehyde, the pH optima of both enzymes tend to shift toward the more acidic side, thereby providing improved heat stability for both enzymes. A change of this nature could allow the reaction to be carried out at a higher temperature (60°C) and a lower pH (pH 5). Although the technology is available and economically attractive, the extract manufacturers must grapple with this problem to decide when and how the immobilisation technology will be applied to commercial processing.

Current trends in biotechnology and, in particular, genetic engineering should widen significantly the scope of research in areas that directly relate to flavour enhancement. Indeed, this technology will be most useful to introduce specific genes, for example, those encoding the enzymes 5′-phosphodiesterase and adenylic deaminase, into yeast in high copy number without affecting the RNA production capacity or its growth characteristics. A successful transformation would allow extract manufacturers to produce yeast extracts high in 5′-nucleotides without relying on external sources of enzyme. The economic advantages of a change of this nature would be overwhelming.

Current commercial strains do not yield more than 10–12 per cent RNA that is of value to extract manufacture. To realise higher yields of 5′-nucleotides, the RNA content of the yeast must be increased, either by genetic means or by control of the processing of yeast. Unfortunately, genetic alterations of yeast to enhance its RNA content have received very little attention to date.

Although protein engineering is still in its infancy with respect to application to food technology, there seems to be an almost limitless potential to develop more active or more stable enzymes by making targeted alterations to the native enzyme. Such changes would permit the selected tailored enzymes to perform certain specific process tasks more efficiently. However, the future prospects for development of such superactive, stable 5′-phosphodiesterase and adenylic deaminase enzymes are not clear.

Chapter 16

Wine

INTRODUCTION

The production of high quality wines requires carefully optimised parameters such as the growing environment for a given grape variety and adapted wine technology. In the wine making process, the enzymes play a very important role. In fact, more than 10 different enzymes are involved in alcoholic fermentation. Without these enzymes, grape juice would never become wine.

In addition to these enzymes are others that act during the wine making process. Some of them are beneficial, others are detrimental to wine quality.

Therefore, it is extremely important to the enologist to know precisely the nature and the behaviour of these enzymes. The enologist must create the optimal conditions for the desirable enzymes to work and, at the same time, inhibit those enzymes detrimental to wine quality.

To accelerate certain enzymatic reactions to improve the economy and the quality of the wines, Cruess and Besone suggested the use of commercial enzyme preparations developed for the fruit juice industry. The positive results obtained led to the use of these biotechnological products in wine making. Research on grape juice and wine colloids have initiated development of new enzymatic preparations, in wine making. The first part of this chapter discusses endogenous, enzymes, while the second part is devoted entirely to the discussion of the use of commercial enzyme preparations.

ENDOGENOUS ENZYMES

This section is a brief survey of a few endogenous enzymes that are of technological relevance to the winemaker. Among the desirable enzymes, the pectolytic enzymes and the proteases are described. The enzymes involved in the formation of volatile C_6 alcohols and the polyphenol oxidases will be mentioned as detrimental to wine quality.

Pectolytic Enzymes

The pectolytic enzymes present in juice or wine have several origins: the grapes, the yeasts, and other micro-organisms such as *Botrytis cinerea*. All types of pectolytic enzymes are found, including pectinesterases, polygalacturonases, pectin lyases, and pectate lyases.

Technological relevance of pectolytic enzymes

Some pectolytic enzymes (pectinesterases) deesterify the methylated pectin and release methanol. Others, called depolymerases (polygalacturonases, pectin and pectate, lyases), hydrolyse the pectin chain into small fragments.

The depolymerases, especially the endo- type, are able to reduce the viscosity of a pectin solution in a very short period of time. The hydrolysis of a small percentage of the linkages in the main chain is sufficient to reduce the viscosity of a pectin solution by more than 50 per cent. This rapid drop of viscosity allows a faster sedimentation of the cloud particles. The benefits of such a reaction are a better juice extraction, a faster and more complete juice clarification, and, finally, an easier wine filtration.

Sources of endogenous pectolytic enzymes

Grapes

Grapes contain mainly pectinesterase (PE) and polygalacturonase (PG) activities. These activities increase during the ripening of the grapes and decrease toward the end of maturation. The activities are thermolabile; in the case of PG, an exposure at 60°C for 20 minutes reduces the activity by more than 50 per cent. PE is located mainly in the skin of the grape berries. Its activity decreases during alcohol fermentation.

Yeasts

Very little has been published on the pectolytic activities of yeast. Yeast has been reported to produce PG and, to a lesser extent, PE activities. The hydrolysis mechanisms of yeast PC have been described by Demain and Phaff.

Botrytis cinerea

Work performed by Grassin indicates that *Botrytis cinerea* produces PE and PG mainly during the growth phase. The PG level of a 'botrytised' juice is about 200-fold higher than that of a healthy juice, explaining why such juices contain much less pectin than juices, extracted from healthy grapes.

Proteases

Technological relevance of proteases

Grape proteins are considered responsible for a certain type of haze formed during the storage of wine. If this precipitation occurs after bottling, the wine becomes unsalable. The molecular mass of these 'haze forming' proteins ranges between 12600 and 30000 daltons. To prevent this precipitation, bentonite has been used as an adsorbing agent. Although this technology is successful in removing the proteins, it is not selective and removes other desirable compounds from the wine. This lack of selectivity often affects the organoleptic qualities of the wine, especially when a high dose of bentonite is applied. Therefore, the use of a more selective technology (e.g. microbial protease) is more suitable. However, the hydrolysis products of grape protein serve as nutrient for the bacteria involved in the second or malolactic fermentation.

Sources of endogenous proteases

Grapes

A protease activity has been detected in grapes, but this activity is weak and is inhibited rapidly by the alcohol formed during fermentation, as well as by juice clarification.

Yeasts and other micro-organisms

The proteolytic activity produced by yeast apparently is more stable than the one produced by the grapes. This enzyme is still active during alcohol fermentation. The enzyme activity varies greatly from one type of yeast to another. Positive correlation has been observed between the rapid start of malolactic fermentation and the ability of the yeast to hydrolyse protein and produce peptides that serve as nutrients for *Lactobacillus oenos*.

According to Lurton, the major enzymes involved in autolysis of yeast cells are proteases, whereas liberation of mannoproteins from yeast cell walls is caused by a β-1,3-glucanase. The diffusion of these macromolecular compounds into the wine has a positive effect on the organoleptic quality of the wine.

A protease produced by *Botrytis cinerea* has been identified by Heale and Movahedi.

Enzymes Involved in the Formation of C_6 Alcohols

Technological relevance

Volatile C_6 alcohol compounds such as hexanol are responsible for the grassy taste of juices and wines. Crushing of the grapes leads to a greater formation of C_6 compounds than the pressing of uncrushed grapes. Juice clarification reduces the amount of C_6 compounds in the juice.

Work done by Rapp and Crouzet has shown that the formation of C_6 compounds in juice requires the sequential action of four enzymes, namely, acyl hydrolases, lipoxygenase, a hydroperoxide-cleaving enzyme, and an alcohol dehydrogenase.

Enzyme sources

The C_6 compounds are formed in the presence of oxygen during the pre-fermentation phase, after the mechanical treatment of the grapes. During alcohol fermentation, hexanal and hexenals are transformed almost entirely into hexanol and hexenols by the yeasts. Grape and yeast enzymes therefore are involved in the formation of these unpleasant grassy-tasting compounds.

Polyphenol oxidases

The oxidation of phenolic compounds in the juice is catalysed by polyphenol oxidases. One must distinguish between the grape polyphenol oxidase and polyphenol oxidase produced by *Botrytis cinerea*.

Polyphenol oxidase of grapes

The polyphenol oxidase of the grapes, also called tyrosinase, is mainly bound to insoluble cellular compounds such as chloroplasts. The enzyme can be removed from the juice by a clarification (natural sedimentation of the cloud or centrifugation). This enzyme is not stable under wine making conditions (pH, SO_2, alcohol). Therefore, this polyphenol oxidase can be controlled easily by clarification and SO_2 addition.

In hyperoxygenation technology, this enzyme is used advantageously to reduce the level of oxidisable polyphenols in white wines. The influence of this technology on the varietal flavours of white wines remains to be investigated more closely.

Polyphenol oxidase of Botrytis cinerea

The polyphenol oxidase produced by *Botrytis cinerea* is called laccase. This enzyme is about 30 times more active than the tyrosinase. However, it is pH and SO_2 resistant and stable under wine making conditions. Because of its characteristics, this enzyme is considered detrimental by winemakers. Laccase is responsible for the loss of colour and the browning of red wines. Therefore it is of greatest importance to understand and control this polyphenol oxidase. Inactivation usually is performed by heat treatment and/or SO_2 addition. Laccase can survive for some time in the wine. The presence and activity level of this enzyme therefore must be controlled from time to time, especially before the wine is exposed to air (racking). A selective and simple colourimetric method to determine laccase activity has been developed by the Wine Research Institute of Bordeaux.

USE OF COMMERCIAL ENZYME PREPARATIONS IN WINE MAKING

Commercial enzymes have been developed to reinforce the endogenous enzymes present under wine making conditions. In some cases, the suitable enzymes are missing; exogenous enzymes are used to fill that gap.

Two types of commercial enzymes are used: the pectolytic enzymes and the beta glucanases. The pectolytic enzymes were introduced into wine making in the early 1970s. The beta glucanases were introduced in 1984. Pectolytic enzymes are used to improve juice extraction or the extraction of other valuable compounds, as well as to improve clarification and filtration. The beta glucanases were developed to solve specific clarification and filtration problems caused by the presence of Botrytis beta glucan.

Use of Pectolytic Enzymes

The potential uses of pectolytic enzymes in wine making are illustrated in this section, using a few examples.

White wine technology

Juice extraction

Pectin is a structural compound of the cell wall. It is responsible for the firmness of the grape berries. A low free-run juice yield and a long pressing time usually are correlated with the pectin level of the grapes. The influence of enzymatic pectin hydrolysis on the juice yield is shown in two examples.

The first example illustrates the results obtained on Kadarka, a Hungarian grape variety that has a high pectin content. The enzyme preparation is added continuously to the crushed mash while filling the press. The contact time of enzyme and mash was 1–2 hours. The mash temperature was 14°C. The amount of juice obtained in the different fractions—free-run juice including the first and second press and press juice (third to fifth press)—is indicated in Fig. 16.1. As we can see from the figure, the first quality juice fraction (free-run juice and first and second press) is much higher for the enzyme-treated batch than for the control. Another trial, performed on grapes containing less pectin than Kadarka (Table 16.1), shows similar results. This trial was performed on Spanish grapes. Table 16.1 indicates the yield obtained for the different press fractions, as well as the total juice yield. The juice yield obtained for the first quality juice (free juice and first press fraction) is about 66.3 per cent for the enzyme-treated batch and 57.6 per cent for the control. The overall juice yield is 82.5 per cent for the enzyme-treated batch and 79.1 per cent for the control.

Table 16.1. Influence of enzyme treatment (Rapidase CX) on juice yield.

	Control	*Enzyme treated*
Batch size	4000 kg	4000 kg
Free run juice and first press fraction	2305 litre	2650 litre
Second and third press fraction	630 litre	500 litre
Fourth and fifth press fraction	230 litre	150 litre
Total yield	3165 litre	3300 litre

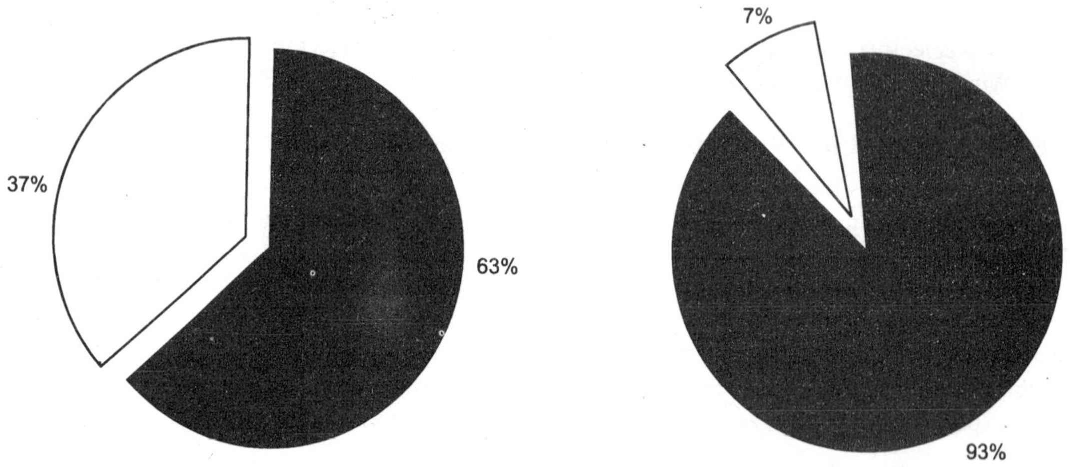

Fig. 16.1. Influence of an enzymatic mash treatment on the free-run juice (including first and second press; ■) and press fractions (third to fifth press; □).

These two examples demonstrate very clearly the positive influence of an enzyme treatment. In addition to the positive effect on juice extraction, the enzymatic mash treatment also improved the extraction of aromatic compounds such as terpenic substances. Table 16.2 indicates the results obtained by a mash treatment using Ultrazym. The enzymatic maceration of the mash took place at 15°C over 20 hours.

Table 16.2. Influence of enzyme treatment (Ultrazym 100) during maceration on level of terpenic substances in wine (Semillon).

Terpenic substance	*Control*	*Enzyme treated (5 g/hl)*
Linalol	2.5	3.1
Terpineol	0.5	0.9
Citronellol	5.3	8.3
Geraniol	6.7	7.5
Total	15.0	20.0

Juice clarification

The effect of an enzyme treatment on the juice clarification and especially on the compactness of the cloud sediment is indicated in Fig. 16.2. To illustrate this effect, a trial on Chasselas grape juice (low pectin level) was chosen. The differences obtained between treated and untreated samples are considerable. The effect would have been greater in a juice of high pectin content. As we can see in Fig. 16.2, the sedimentation of the cloud is much faster after enzyme treatment. Sedimentation of the control required 4–5 times more time without obtaining the same compactness of sediment.

In another trial performed on Chasselas, the degree of clarification of the juices was followed for 20 hours. The turbidity of the juices (supernatant) was determined after centrifugation (10 ml juice centrifuged at 3000 grams for 5 minutes). The results obtained are indicated in Fig. 16.3. The clarification effect obtained after enzyme treatment is quite spectacular. Even after 20 hours, the control samples do not reach the same degree of clarification obtained by the enzyme-treated juice after 1 hour.

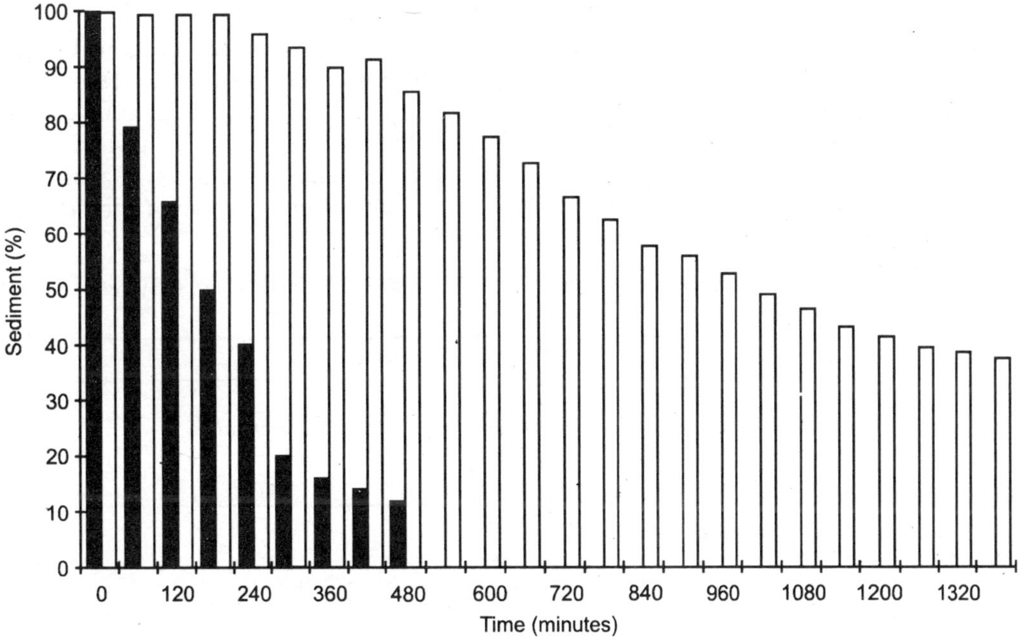

Fig. 16.2. Influence of enzymatic treatment of the juice on the compactness of the sediment. Control, open bars; enzyme treated, shaded bars.

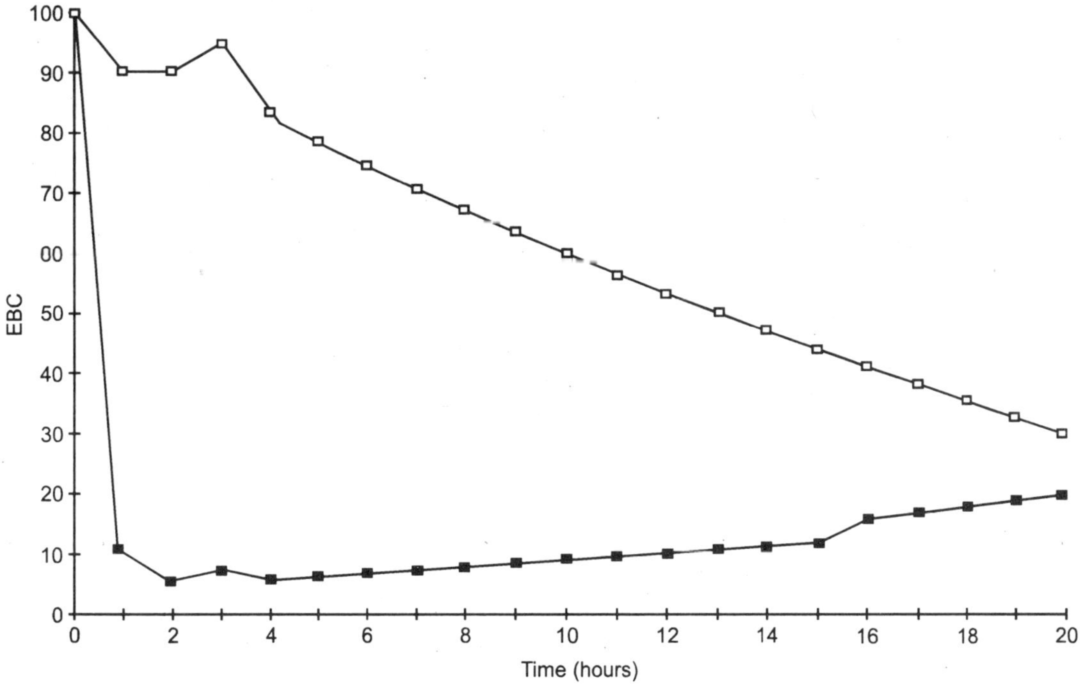

Fig. 16.3. Control, □; enzyme treated, ■. Influence on enzyme treatment of clarification of juice.

Wine filtration

To illustrate the effect of enzyme treatment on wine filtration, we have chosen a trial performed with Sylvaner grapes (medium pectin level). The enzyme treatment was done on the mash before pressing.

No further enzyme treatment took place between pressing of the grapes and filtration of the wine. Filtration was carried out with a 10 m^2 Kieselgur filter. The filtration behaviour of both wines is indicated in Table 16.3. As we can see from the table, the enzyme treatment allowed filtration of about 2.5 times more wine per filtration run. The economy realised by such an enzyme treatment (filtration material, filtration time, wine losses) is considerable.

Table 16.3. Effect of enzymatic treatment (Vinozym, NOVO NORDISK FERMENT) of a Sylvaner grape mash on filterability of the wine[a].

	Filtration behaviour of the wine			
	Enzyme treated		*Control*	
Filtration time (hr.)	*Differential pressure (bar)*	*Volume filtered (litre)*	*Differential pressure (bar)*	*Volume filtered (litre)*
0.5	3.0	3500	3.0	3500
1.0	3.0	7000	6.0[b]	6500
1.5	3.0	10500	–	–
2.0	3.0	13500	–	–
2.5	6.0[b]	15500	–	–

[a]Filtration over Kieselgur (10 m^2).
[b]Filtration surface clogged; filtration run stopped.

Red wine technology

The classical application of pectolytic enzymes in red wine technology is the treatment of press wines to improve clarification and filtration. The results obtained are generally on the same order as or better than those reached for white wines.

Another enzyme application in red wine making is colour extraction. Experimental trials on a laboratory scale indicate that pectolytic enzyme preparations with a broad spectrum of activities (hemicellulases, cellulases) give better results than standard pectinases. Also it has been observed that the contact time during maceration must be about 1 week to extract valuable polyphenols that subsequently stabilise the red colour compounds. Thus, colour intensity is not correlated with colour stability. Well-coloured wines can be obtained after 2–3 days maceration with enzyme treatment, but experience has shown that such wines generally lose a great amount of colour after malolactic fermentation. Therefore it is necessary to perform the maceration for 8–10 days to achieve good colour stability. More work is necessary to understand and master the colour stability problems in red wine technology.

Use of β-Glucanases

Wines obtained from botrytised grapes (grey mould or noble mould) are always difficult to clarify and to filter because of the presence of a high molecular weight β-glucan (in the past, erroneously called dextran) produced by *Botrytis cinerea*.

The presence of Botrytis glucan can be detected by the following specific and very simple test. Of the wine, to be tested, 10 ml is poured into a test tube. Then, 5 ml 96 per cent alcohol (industrial or denatured alcohol) are added to the wine. The solution is shaken lightly just to insure thorough mixing of the two liquids. If more than 15 mg glucan per litre are present in the wine, a filamentous precipitate will become visible within minutes.

To detect lower quantities (3 mg/l and above), the following test has been developed. In a conical centrifuge tube, 5 ml 96 per cent alcohol are added to 5 ml wine. The contents of the tube are mixed and left to stand at room temperature for 30 minutes. The mixture is centrifuged at 3000 rpm (700–800 grams) for 20 minutes. The supernatant is decanted carefully. The sediment is dissolved in 1.0 ml water, 0.5 ml 96 per cent alcohol is added. The formation of a filamentous precipitate indicates the presence of small amounts of Botrytis glucan. This test is somewhat more labourious than the first one but is, nevertheless, highly recommended, since filtration problems already start with the presence of 3–5 mg glucan per litre.

Specific clarification and filtration problems cannot be mastered by centrifugation, fining, or addition of pectolytic enzymes. The development of a specific β-glucanase has solved all glucan-related problems in a spectacular way. The effects obtained by this kind of preparation are shown in the next two examples.

A Bordeaux sweet white wine containing 22 mg glucan per litre was divided into two portions. The first portion was treated enzymatically (Glucanex) whereas the second portion was kept untreated as control. The filtration took place over Kieselgur (filtration surface, 4 m^2). The filtration conditions were the same for both wines. The results obtained are summarised in Fig. 16.4. As we can see from this trial, the enzyme treatment allowed the filtration of almost 10 times more wine per filtration run.

The second example shows the filtration behaviour over filter pads. In this case, both wines (the enzyme-treated and control wines) were centrifuged prior to filtration. The results obtained are indicated in Figs 16.5 and 16.6. This trial demonstrates that centrifugation does not remove the glucan and, consequently, does not solve the filtration problems.

Glucanase treatment solves glucan filtration problems in a spectacular way. It permits better filtration at lower pressure and reduces the filtration costs drastically. The organoleptic properties of the wine are not affected by such a treatment.

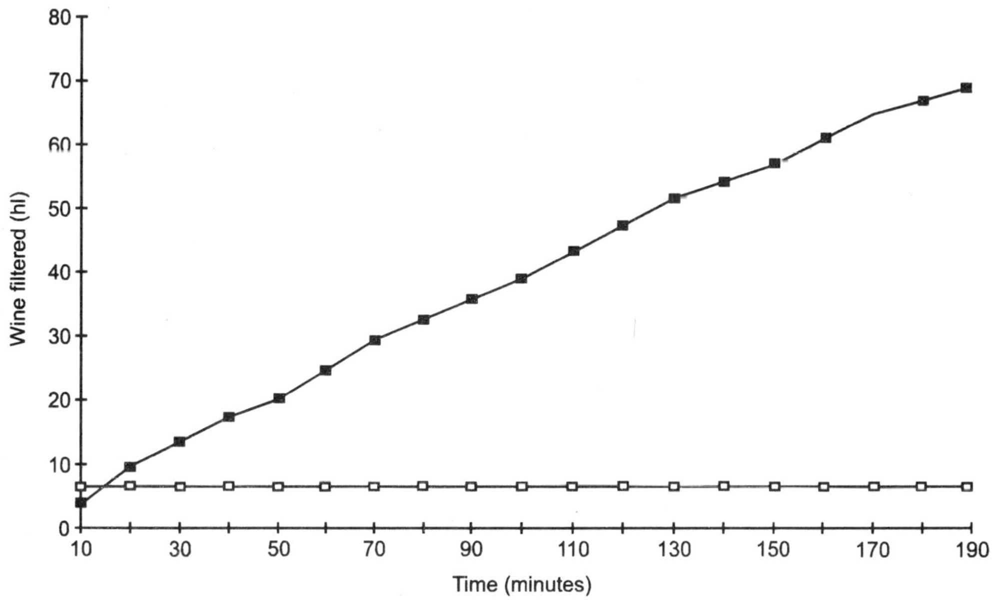

Fig. 16.4. Influence on enzymatic β-glucan degradation on wine filterability. Control, □; and enzyme treated, ■.

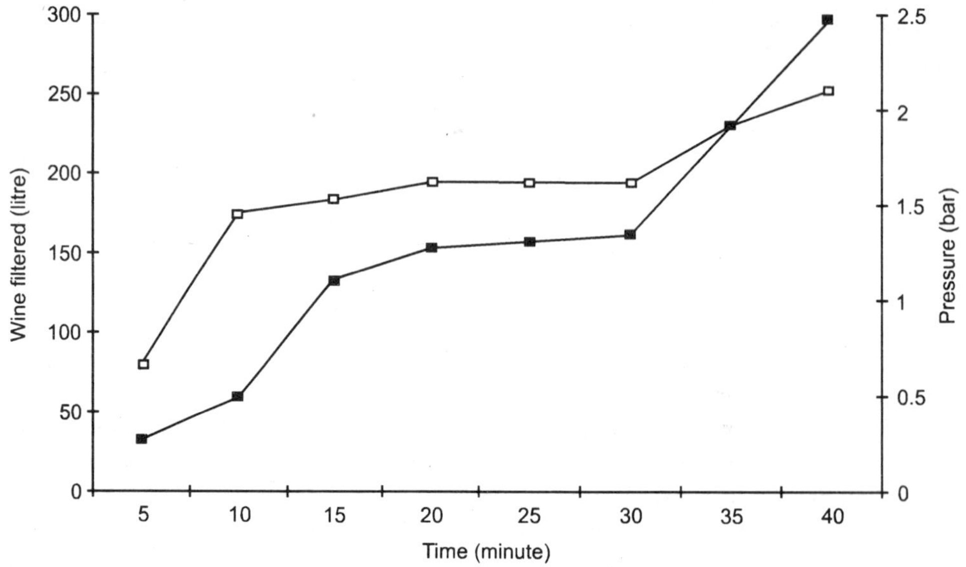

Fig. 16.5. Filtration (□) and pressure (■) curves of the control wine.

FUTURE DEVELOPMENTS

The development of pectolytic enzymes with a broad spectrum of activities (hemicellulases, cellulases) for colour extraction in red wine technology is certainly one of the projects that will be carried out in the near future.

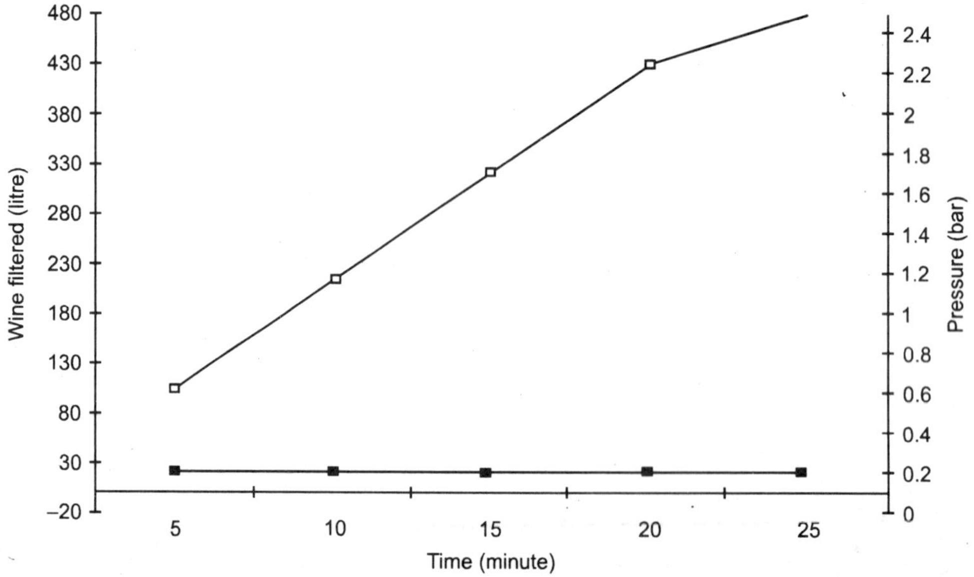

Fig. 16.6. Filtration (□) and pressure (■) curves of the enzyme treated wine (Glucanex, 1 g/hl).

As already mentioned, the replacement of bentonite by a microbial protease for the stabilisation of white wines against protein precipitation remains a desirable goal. Considerable research work is presently being done in this field. Another interesting subject is the hydrolysis of bound terpenes (terpenes bound to sugar) to reinforce the aromatic profile of certain wines. This treatment can be done using different glucosidases in a sequential order. All the microbial glucosidases tested to date are inhibited by the presence of glucose. Therefore, such enzymes cannot be used for the treatment of grape juices. In the wine treatment, the majority of enzyme reactions take place after alcohol fermentation. Several research groups are presently working on this project.

It is possible, by means of glucose oxidase/catalase treatment of grape juice, to produce a high amount of gluconic acid in the wine. Since gluconic acid is not metabolised by wine yeasts into alcohol, the fermentation of such juices can lead to the production of low alcohol wines. In this case, the excess of gluconic acid must be removed. If gluconic acid is left in the wine, the wine will be used as an 'acid reserve' to blend with add deficient wines, allowing production of better balanced wines. Glucose oxidase/catalase-related technologies are not optimised to date and must be developed further.

Chapter 17

Brewing

INTRODUCTION

Beer is a beverage made by alcoholic fermentation of a sugar solution derived from grain. Most carbohydrates in grain is present in the form of starch, which cannot be fermented by brewer's yeast. Before beer can be produced, the starch of the grain must be transformed into fermentable sugar by enzymatic digestion.

The brewing of beer is an ancient enzymatic process that was developed, as were other traditional processes, without knowledge of the existence and nature of the enzymes involved. This chapter begins with an examination of the conventional brewing processes established long ago and reviews the enzymatic reactions involved.

By the beginning of the twentieth century, enzymes were recognised as part of yeast, malt, and other living organisms. Many enzymes became available in commercial quantities. The application of enzymes from non-traditional sources made possible many changes in the brewing industry. In 1911, Leo Wallerstein demonstrated that, by adding papain from papaya to beer, the beer could be kept clear during shipment over long distances and during storage for long periods of time.

This discovery was critical to the establishment of larger brewing companies that market their packaged beer over expanded geographical areas.

One of the most significant recent developments in the brewing industry is the use of microbial enzymes for the production of lower calorie or 'light' beers. The success of enzymatically produced 'light' beer has led to changes in marketing strategy throughout the industry and has encouraged research into the development of other new products.

Other applications of enzymes not native to the brewing process have been developed and have found industrial application. However, none have had as dramatic an effect on the industry as the applications just described. The use of enzymes to process cheaper or more readily available raw materials has allowed their incorporation into beer in place of more expensive traditional barley malt. Other enzymes have been found that facilitate filtration of the cloudy beer liquid immediately after fermentation to yield a clear final product. Enzymes that react with the small amount of oxygen found in packaged beer have been used to extend the shelf life.

MALT AND ITS REACTIONS DURING BREWING

Malt is grain that has been moistened and allowed to germinate under controlled conditions and is then dried to a stable form that is rich in enzymes but similar to the original grain in external appearance.

Malt is stable; it can be shipped over long distances and stored for long periods of time. The most commonly malted grain, and the primary grain used in brewing, is barley. Barley is preferred for many reasons, two of the most important of which are that barley has an adhering husk that protects the malt and acts as a filter aid during separation of the grain from the liquid after brewing and that barley has a good ability to form enzymes during germination.

During germination of the barley, several important hydrolytic enzymes are produced or released. Some enzyme hydrolysis of the grain occurs during the several days of germination. Phytase digests much of the phytin of the grain, releasing phosphates and associated metallic minerals that are important mineral nutrients for yeast during fermentation. Hemicellulases are produced that degrade much of the cell wall structure of the endosperm tissue, which holds most of the starch in the grain. The chief component of the barley endosperm cell wall is $(1\rightarrow3)$ and $(1\rightarrow4)$-β-D-glucan. A number of enzymes generally referred to as β-glucanases are involved in degradation of cell wall hemicellulose. Usually, degradation of the cell walls is extensive, leading to the production of a malt that is called 'modified', that is, less rigid in structure than the original grain and capable of releasing soluble components relatively easily when used for brewing. A variety of proteases is produced also that begins breaking down protein in the grain. Insoluble storage protein such as hordein is converted partially to large soluble peptides and to smaller peptides and amino acids. The larger peptides are important in brewing as a source of protein, an important component of the foam formed when a beer is poured. The smaller digestion products are an important source of nitrogen for growth of the yeast during beer fermentation.

Starch digestion is very limited during germination. Barley α-amylases are formed and digest the starch to a limited degree. The bound β-amylase found in ungerminated barley is released by the action of proteases during germination, but the β-amylase is not active on the native starch granules found in the germinating barley.

During germination, enzyme hydrolysis is minimised to obtain sufficient modification of the malt. The definition of sufficient modification varies depending on the type of malt required. Ale malts of the type classically used in England are very thoroughly modified because the traditional ale brewing equipment offers little of the flexibility required for extensive enzyme hydrolysis when the malt is used for brewing. Extended hydrolysis during germination results in higher levels of sugars and amino compounds that are converted to coloured products, melanoidins, by browning reactions when the malt is dried in a kiln. Traditional ale malt gives a relatively dark-coloured beer. The popular Pilsner style of beer, which is as pale as possible, requires much less colour development in the malt. The malts used to make pale lager and Pilsner style beers are, therefore, less modified during germination. Thus, they require a more intense enzymatic hydrolysis during brewing, which is brought about by a more complicated time-temperature schedule.

The extent of germination also influences the amount of enzyme present in malt. Pale malts with low modification intended for Pilsner production have lower enzyme levels than a darker distiller's malt. Many of the enzymes in the germinated barley or green malt, particularly the α- and β-amylases, survive the kilning process and are activatable in the brewing process.

Brewing beer begins with making a mixture of hot water and grain, called a mash. The mash usually contains malt as the major grain; the malt acts as a source of enzymes for brewing. The mash undergoes a programme of varied time and temperature to accomplish the desired hydrolysis and solubilisation of the grain components. At the end of the mash programme, the liquid contains dissolved sugars and other compounds derived from hydrolysis of starch and other components of the grain. The liquid, called wort, is separated from the remaining undigested or spent grain by filtration, using the hulls of the malt to form the filter bed. A gravity filter, called a Lauter bed, or a special plate-and-frame wort filter that operates with pressure provided by a pump may be used. The wort is boiled with flavouring hops, cooled, and fermented after the addition of yeast and oxygen. The mash programme determines some of the basic properties of the beer.

To produce a Pilsner style beer from pale malt alone, at least three temperature stages are used (Fig. 17.1). The mash begins at the lowest temperature, often called the 'mashing in' or 'proteolytic' temperature. As the name suggests, the most important function of this step is proteolysis of the malt, to break down insoluble protein in the malt to soluble peptides of varied molecular weights and even to free amino acids.

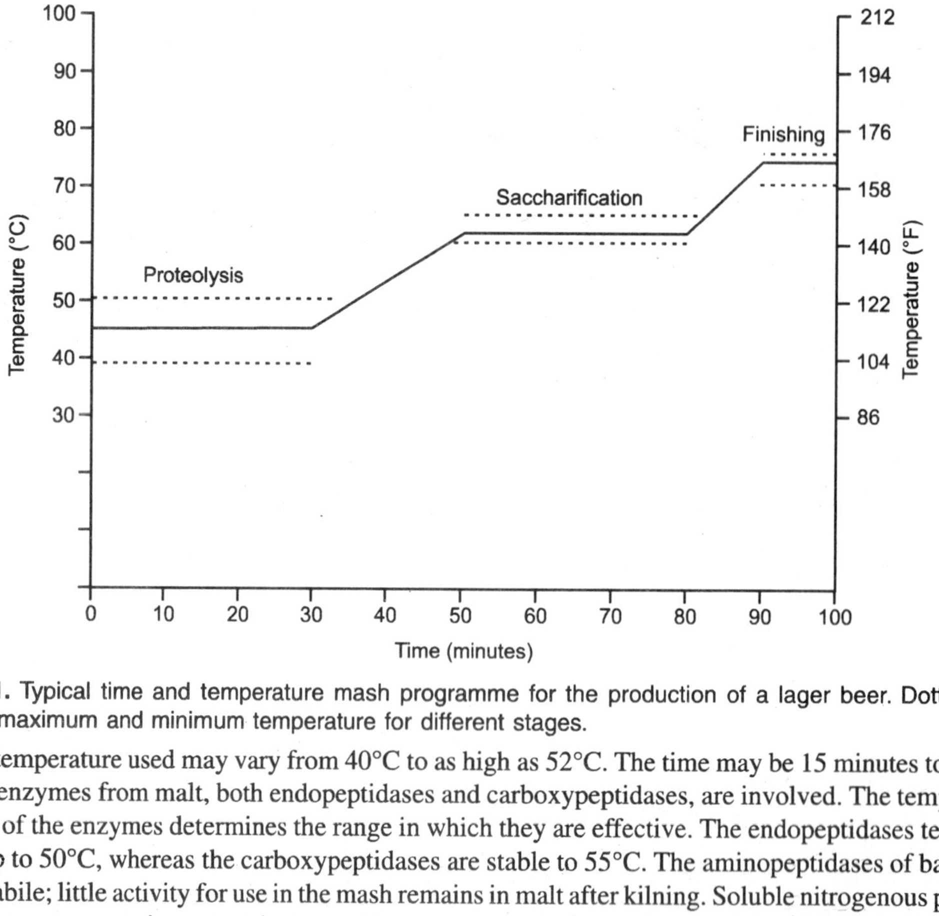

Fig. 17.1. Typical time and temperature mash programme for the production of a lager beer. Dotted lines indicate maximum and minimum temperature for different stages.

The temperature used may vary from 40°C to as high as 52°C. The time may be 15 minutes to 1 hour. Several enzymes from malt, both endopeptidases and carboxypeptidases, are involved. The temperature stability of the enzymes determines the range in which they are effective. The endopeptidases tend to be stable up to 50°C, whereas the carboxypeptidases are stable to 55°C. The aminopeptidases of barley are thermolabile; little activity for use in the mash remains in malt after kilning. Soluble nitrogenous products of these enzymes are important in supporting yeast growth during fermentation as well as in forming beer foam. Considerable amounts of soluble nitrogenous compounds will have been formed already during germination of the barley; only about one-third of the total nitrogen in the wort is released during this stage of the mash. For beers made with malt as the only grain, or made with a high percentage of malt, this stage of the mash is not very critical and may be abbreviated or run at the high end of the temperature range, given. Occasionally, malts with a low degree of modification may require additional action of β-glucanase or phytase during, the mashing process. These malt enzymes are more heat labile than the proteases, and their reactions are best carried out at the low end of the temperature range shown (40°C; Fig. 17.1). Normally malt will have completed most of the reactions that occur at this temperature to a sufficient degree earlier during germination of the barley. In the early twentieth century, barley varieties did not have as great an enzyme forming capacity as present day varieties, so low temperatures and long mashing times were common. In some breweries, the practices have not changed although the malt has.

After completion of the proteolytic stage, the mash is raised to a higher temperature. Increase in temperature can be accomplished by removing part of the mash, bringing it to a boil in another vessel, and returning it to the original vessel. The resulting rise in temperature can be controlled accurately using rudimentary equipment. The method is called decoction mashing and is a very old approach, predating the availability of thermometers for temperature control. More commonly, the temperature of the entire mash is raised at a controlled rate using steam heat or even direct flame on the vessel in a process called programmed infusion mashing. Regardless of the method used to increase the temperature, the mash is brought to the next stage, commonly referred to as the saccharification stage (Fig. 17.1), at a temperature of 60°–70°C.

Prior to the saccharification stage, the starch in the malt remains largely intact because it has not yet been gelatinised and native starch is not hydrolysed readily by enzymes. The starch of barley malt gelatinises at a temperature of 60°C. Once gelatinised, the starch is digested rapidly by the α- and β-amylases present in the malt. In the saccharification temperature range, both types of amylase are active. Their combined action results in production of large quantities of fermentable sugar, especially maltose, which is the only sugar produced by β-amylase; hence, this stage receives the name saccharification stage. Smaller amounts of glucose and maltotriose, which are also fermentable by yeast, are formed from starch by α-amylase, in addition to larger non-fermentable dextrins.

The saccharification stage seldom lasts longer than 30 minutes. At temperatures above 60°C, which are required to obtain gelatinisation of malt starch, the malt amylases, especially β-amylase, are not very stable. During saccharification at 65°C, β-amylase is inactivated almost completely; within 30 minutes (Fig. 17.2). The α-amylase also is deactivated significantly (Fig. 17.2), but some of this enzyme is still present at the end of saccharification.

Some of the malt starch usually remains ungelatinised at the saccharification temperature. The saccharification stage represents a compromise between the need for higher temperatures to obtain gelatinisation and lower temperatures to preserve enzyme activity. The temperature chosen results in optimal sugar formation but incomplete starch degradation. Starch degradation must be finished by raising the mash temperature again so gelatinisation can be completed. The finishing temperature is at least 70°C, but below 80°C. Again, the exact temperature is a compromise selected to be as high as possible while still permitting significant activity of the malt α-amylase (Fig. 17.2). The small amount of α-amylase still present is sufficient to hydrolyse starch to soluble dextrins, but not to form much more fermentable sugar.

The basic programme for an infusion mash outlined here is modified at different breweries to provide different types of beers. By adjusting the saccharification temperature, smaller or larger proportions of fermentable sugar will be formed in the wort, giving beers with different ratios of alcohol to non-fermented dextrins. Intermediate steps may be used also. A complicating factor in understanding and predicting the rates of enzyme reactions in a mash is, as Narziss points out, that the enzymes are present inside the structure of the malt and time is required for them to become soluble and able to act on their substrates. Soaking at a lower temperature will affect reaction rates at a higher temperature. Brewers' experience suggests that 63°C is optimum for fermentable sugar formation and that a temperature as high as 73°C can be used for simultaneous saccharification and finishing to give a wort with the lowest percentage of fermentable sugar. By changing the temperature schedule of a mash, the percentage of fermentable sugar in wort can be varied from about 50 per cent to 75 per cent of total dissolved solids using the malt enzymes. This range will give beers of different alcohol contents. Most beers are fermented from worts with 65–70 per cent fermentable sugar. The presence of some preformed sugars in the malt makes it difficult to reduce the percentage of fermentable sugar in a wort and, therefore, to make a very low alcohol beer.

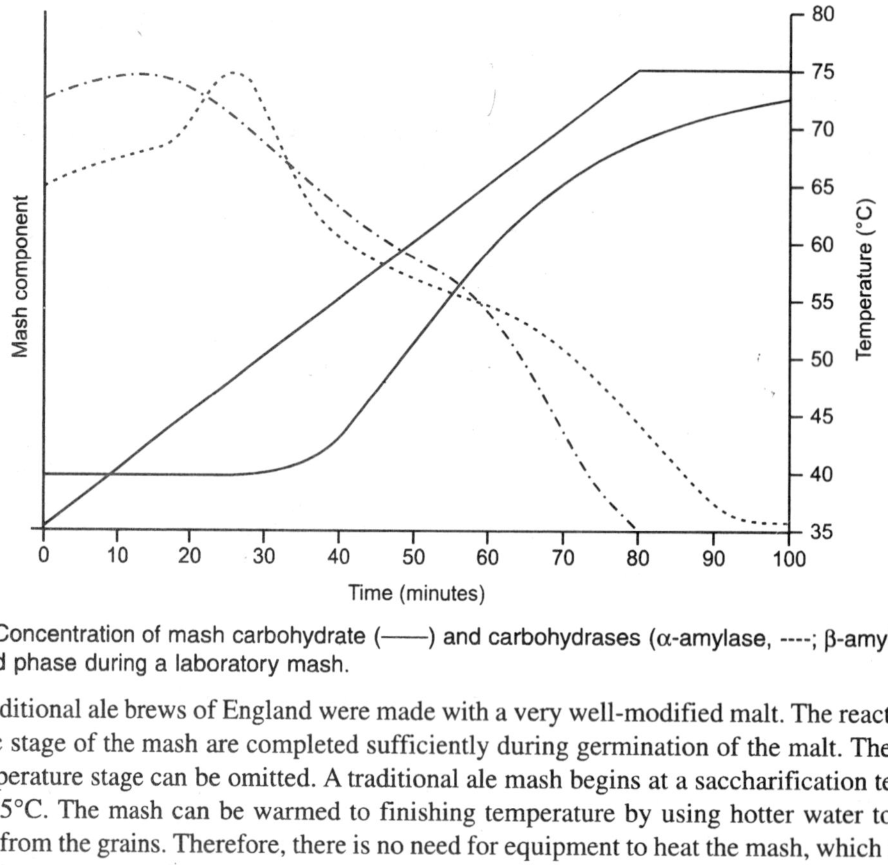

Fig. 17.2. Concentration of mash carbohydrate (——) and carbohydrases (α-amylase, ----; β-amylase, – · · –), in the liquid phase during a laboratory mash.

The traditional ale brews of England were made with a very well-modified malt. The reactions of the proteolytic stage of the mash are completed sufficiently during germination of the malt. Therefore, the lower temperature stage can be omitted. A traditional ale mash begins at a saccharification temperature of about 65°C. The mash can be warmed to finishing temperature by using hotter water to sparge or rinse wort from the grains. Therefore, there is no need for equipment to heat the mash, which means that ale breweries can be simpler and less expensive to build.

USE OF BREWER'S ADJUNCTS

The preceding discussion of brewing reactions describe reactions that occur in a brew made with malt alone. Historically, beers were made using only malted grain, but brewers found that at least small amounts of unmalted grain could be used also. By the end of the nineteenth century, brewers determined that malt was the source of whatever converted the starch of other grain to fermentable sugar. The malt sugar-producing enzyme diastase was one of the first enzyme systems recognised. The ability of malt to convert unmalted grains, called adjuncts, was recognised clearly in the United States. The varieties of six-row malting barley grown in most of the country have high protein levels and produce malts with enzyme capacities much higher than those necessary to process the malt starch alone. The malts are well suited to use with adjuncts because of this excess enzyme. The process of making whiskey from corn using a small amount of malt as an enzyme source, commonly used in the United States, suggested that corn could be used to make beer also. Anton Schwarz popularised and promoted the use of adjuncts, especially corn and rice, in brewing beer. Handke described production of a beer in Milwaukee for which corn was used to provide 45 per cent of the extract or soluble material. This amount is essentially the same as the practical limit on the percentage of adjunct material used with malt today. By the end of the nineteenth century, the use of adjuncts digested by malt enzymes in making beer had been perfected.

An important aspect of the use of unmalted grain as an adjunct for brewing is that the most commonly used grains, corn and rice, have a higher gelatinisation temperature than barley malt. From the preceding discussion of malt reactions in brewing, it is obvious that malt starch gelatinises at a temperature just low enough to allow the amylases of malt to digest the starch. The gelatinisation temperatures of corn, about 70°C, and rice, 70°–80°C, are so high that malt β-amylase cannot provide sufficient activity for normal sugar formation as the starch gelatinises. Malt α-amylase will, however, be active for a short time when corn and even rice starch gelatinises. The activity of malt α-amylase permits the use of these grains as adjuncts. Because of their higher gelatinisation temperatures, corn, rice, and sorghum must be boiled to be used completely and efficiently. Boiling is carried out in a separate vessel, a cooker. A mash containing a small amount of malt as a source of enzyme and the ground unmalted grain, preferably degerminated, is made and heated to a temperature at which the adjunct, starch just begins to gelatinise. Malt α-amylase hydrolyses the starch so excess thickening does not occur; enzyme action is facilitated by holding this temperature for a short period of time, perhaps 10 minutes. The temperature is then increased to boiling to complete gelatinisation of the enzymatically thinned mixture. After thorough boiling, the cooked adjunct mash is added to and mixed with a separate mash containing most of the malt which is proteolysing at a lower temperature. The combined mashes are sized most conveniently so the complete mixture is at saccharification temperature. Otherwise temperature adjustment will be necessary. The amylases present in the enzyme-rich six-row barley malt are sufficient to complete digestion of the starches of both malt and adjunct grains at the same temperature as an all-malt brew.

Malt α-amylase is not the ideal enzyme for thinning the starch solution that develops as adjuncts are cooked. The amylase is stable enough to work fairly well with corn starch. However, when using rice, which usually has a higher starch gelatinisation temperature, sufficient thinning becomes difficult to obtain; production problems are not uncommon. Rice for brewing is milled rather finely before it is cooked to make the starch more accessible to the amylase, but the enzyme is barely stable enough to digest the gelatinising starch sufficiently to keep the adjunct mash from being too viscous to handle. The availability of amylases from the bacteria *Bacillus subtilis* and *Bacillus licheniformis* offers the brewer an alternative source of amylase for thinning the cooking adjunct. These enzymes, especially the thermostable *Bacillus licheniformis* enzyme, are more stable than the malt enzyme and are more effective in reducing viscosity build-up. Some, but not most, breweries have adopted one of the bacterial enzymes for use in adjunct cooking. Commercial literature on these enzymes suggests use of 0.5–1 kg standard enzyme solutions for 1000 kg adjunct grain. At this rate, these enzymes are both economical and effective.

Cooked adjuncts such as corn and rice are not subjected to digestion by the malt at temperatures suitable for protease activity, so they have insignificant amounts of soluble nitrogen compounds. As a result, these starchy adjuncts contribute no soluble nitrogen compounds to the wort. Wort and beers made with these adjuncts will have lower nitrogen content than beers made from malt only. This characteristic can be an advantage in preserving clarity of the beer after it is packaged because higher molecular weight beer peptides tend to form hazes as the beer ages, and haze formation becomes a problem. If the malt provides sufficient nitrogen for the yeast to grow during fermentation, a good tasting beer can be made with adjunct.

A few grains, such as barley and wheat, contain starch with a gelatinisation temperature low enough that the grains can be converted by malt enzymes without prior boiling. The grains are mixed with malt in the mash and can be converted to sugar by the malt enzymes using the temperature that would be used for an all-malt mash. The problem with using unmalted wheat and barley arises in filtration to separate wort from grains or yeast from beer. The hemicelluloses of the unmalted grains tend to obstruct filtration. If more than 10 per cent of these grains are used as adjuncts, an additional enzyme such as glucarase must be added to digest hemicellulose.

BREWING WITH VERY HIGH ADJUNCT LEVELS

The practical limit for the amount of starchy adjunct that can be used to make a beer is determined by the failure of common adjuncts to contribute nitrogen to the wort. Experimental work has shown that the amylase content of malt is sufficient to convert the starch of adjuncts to normal amounts of sugar, even when only one-third of the sugar comes from malt. However, the practical minimum is 50–55 per cent from malt. Malt is a source of the nitrogen and minerals that the yeast requires for growth during a normal fermentation. Because the starchy adjuncts contribute only sugars and dextrin to the wort, the nitrogen and minerals become diluted by the adjunct. The nutrient that first becomes critical is the nitrogen. The flavour that a yeast imparts to a beer depends, among other factors, on the type of amino acids the yeast uses for growth. When crucial amino acids are in short supply, the yeast begins to form higher quantities of off-flavours caused by compounds such as diacetyl or hydrogen sulphide. At some level, the flavour change in the beer becomes unacceptable. The critical point and the degree of flavour change depends on the yeast strain used, but a level of free amino nitrogen of about 120 mg/litre in wort before it is fermented is necessary to produce an acceptable beer flavour. To provide a margin of safety, a level of 150 mg/litre is prudent. This level is reached when 50 per cent of the solids in wort come from a high-nitrogen, high-enzyme six-row barley malt.

To use more than 50 per cent adjunct, more nitrogen than can be provided by malt and its enzymes must be supplied. The malt contains more protein than is released in the mash; only 40 per cent of the total protein becomes soluble. Addition of protease from another source is one method that has been used to provide more nitrogen from the protein present in the malt. Plant sulphydryl proteases, ficin and papain, have been added to the mash successfully to increase the levels of soluble nitrogen in the wort. Although the major flavour problems of making a very high adjunct beer are solved by this approach, the application has never become common commercial practice. Extreme dilution of some minor flavour components that come from the malt seems to be a problem.

Another procedure for making beer that contains only a small percentage of malt was worked out in detail by Nielsen and Weig. Unmalted barley is used to replace a major portion of the malted barley. The barley provides some of the flavour normally contributed by malt. The unmalted barley must undergo many of the enzymatic changes that occur during several days of germination in a short period of time in a mash for this approach to be successful. The major changes that occur during formation of the malt are the degradation of the endosperm cell walls, the digestion of protein to form lower molecular weight nitrogen compounds, and the formation of α-amylase and release of β-amylase, which are needed for sugar formation. The major and most efficient enzyme involved in cell wall degradation is endo-β-glucanase.

Alone, this enzyme significantly digests barley cell walls without the full complement of glucanases found in germinating barley. Bacterial α-amylase from *Bacillus subtilis* reacts with starch at the same temperatures, or slightly higher, as the malt α-amylase and is a suitable substitute for this malt enzyme. Barley possesses a high level of β-amylase in bound form. The bound enzyme can be released by proteolytic action. Any of a number of proteases are suitable for this reaction. Since *Bacillus subtilis* produces both endo-β-glucanase and α-amylase, a protease also produced by this organism is the most efficient choice. *Bacillus subtilis* actually produces two different proteases, an alkaline protease and a neutral protease. The alkaline protease is susceptible to an inhibitor present in barley and is not useful in brewing. The neutral protease is very effective in digesting barley protein and is suitable for use in brewing. Thus, all the necessary enzymes can be produced by the same organism; they can be produced simultaneously in the same bacterial fermentation.

The final mixture of enzymes contains all the basic enzyme activities for making a malt substitute from barley. The mixed enzymes have been successful in producing beer from a mash containing as much as 90 per cent unmalted barley. Further experience has shown that some varieties of barley are more amenable to enzymatic digestion than others. The enzymatic process has been accepted more in areas of the world where these barley varieties are grown. Beers made with barley as an adjunct have many of the minor constituents found in malt, for example, polyphenols, but objective evaluation still finds that their flavour is not quite the same as that of beers made with malt as the major component. Brewing with barley and enzymes is still a rather rare procedure.

KEEPING BEER CLEAR WITH CHILLPROOFING ENZYMES

The major use of non-malt enzyme in the brewing industry is as enzyme that is added to fermented beer to 'chillproof' the beer. Beer is fermented and aged under chilled conditions. Almost all packaged beer is filtered while cold to achieve clarity. In spite of this filtration, the beer becomes cloudy after it is packaged, distributed to customers, and chilled again for serving. The cloudiness that develops is caused by formation of haze particles, termed 'chill haze'. The haze that develops results largely from the interaction of beer peptides with polyphenolic procyanidins. Other material, such as carbohydrate or metal ions, also may be incorporated into the haze. The amount of haze varies with factors such as the type of beer and increases with age, exposure to oxygen, and agitation during shipping. Eventually the beer becomes hazy even at warm temperature and a sediment may develop. The haze is not desirable for aesthetic reasons: it resembles cloudiness produced by microbial spoilage.

In a series of patents, Leo Wallerstein proposed the use of proteolytic enzymes to prevent development of chill haze in beer. Wallerstein looked back at the response to and development of his very successful patented process exactly 50 years after the patents had been awarded. By the early 1900s, advances in bottling and the introduction of pasteurisation created the potential for greatly increased sales of bottled beer. Packaged foods in general were becoming popular. Unfortunately, the beer could not be shipped great distances or stored for long periods of time because of problems with chill haze.

The brewing industry needed a solution to this problem. In 1911, at the Second International Brewers' Conference in Chicago announced that addition to the beer of 'a proportion of proteolytic enzymes active in slightly acid media sufficient to modify the proteids contained in the beer in such a manner that they will not be precipitated upon chilling subsequent to pasteurisation, the beer being rendered chillproof in the sense that it is capable of remaining brilliant even when kept upon ice for a considerable time'.

After some initial skepticism, the treatment with protease suggested by Wallerstein was tried with great success. Belgian scientist Effront reported of United States' beer that the Wallerstein treatment was applied to 'the greater part of the beer made in that country'. The original patents described the use of bromelin, papain, and pepsin for chillproofing. Further experience with the process led to improvement in commercial preparations. Although data on the composition of proprietary products normally are not published, papain (the extract of papaya that contains mixed proteases) is tar superior to the other enzymes used for chillproofing. Similar plant sulphydryl proteases such as bromelin and ficin are not as effective. Purified papain and chymopapain components of commercially available papain have been shown to be effective in chillproofing beer by Cayle. Virtually all the enzyme used for chill stabilisation has been and still is papain.

As new proteases were discovered, they were tested for effectiveness as chillproofing agents for beer. Only the proteases produced by some *Streptomyces* species have been effective enough to have some commercial use. However, adding a small amount of α-amylase from *Aspergillus oryzae* to commercial papain-based chillproofing preparations has become common. The amylase is thought to help keep the beer clear during storage by hydrolysing any large dextrins that may be present in the beer and that would tend to become insoluble with age. Experience has shown that the amylase often does help keep beers clear.

Wallerstein's original patents described the action of proteases in chillproofing as modifying the proteins contained in the beer 'in such a manner that they will not be precipitated upon chilling subsequent to pasteurisation'. He stated that the enzymes became active during the heating of treated beer to pasteurise it. Despite almost a century of industrial application of the enzymatic chillproofing process, the exact mechanism has never been determined. The amount of proteolytic hydrolysis in the beer is very limited, so no significant increase in amino nitrogen is found in treated beer. A decrease in trichloroacetic acid-precipitable protein is detected. Whether the chillproofing activity of papain is one of protein digestion alone is uncertain. Hebert and others also have detected an increase in coagulable protein after papain treatment and have suggested that papain-treated haze precursors coagulate very quickly, preventing further reaction with polyphenolics to form larger amounts of haze during chilling.

Whatever the reaction of papain with beer haze precursors, the affinity of papain for the precursor protein must be extremely specific because so little hydrolysis is involved in effective treatment and no other enzyme is as effective as papain.

MAKING NEW AND SPECIAL TYPES OF BEER WITH ENZYMES

The amount of alcohol in a beer is limited by the amount of solids, or extract, dissolved in the wort and by the percentage of that extract that is fermentable sugar. The percentage of fermentable sugar in wort is controlled by the extent of digestion of starch by amylases in the mash. The upper limit of fermentable sugar is 75 per cent. The amount of fermentable sugar that can be formed by α- and β-amylase is limited by the temperature stability of these enzymes and by the fact that both enzymes hydrolyse only $\alpha(1 \rightarrow 4)$ bonds of starch, leaving the $\alpha(1 \rightarrow 6)$ branch points intact.

Over the years, brewers often have attempted to make special beers with a higher alcohol content than standard beer. Alcohol can be increased either by increasing the total percentage of solids or extract in the wort or by increasing the percentage of the extract that can be fermented by yeast. Increasing the fermentability requires the application of enzymes to hydrolyse more of the non-fermentable dextrins.

The term 'malt liquor' has no strict definition and can be applied to any beer. In practice, however, the name usually is applied to a beer that is higher in alcohol than a regular beer, not very bitter, and fermented to give a relatively high ester content. Generally, the higher alcohol level is achieved by increasing the percentage of fermentable sugar in the wort. In making a malt liquor, therefore, the percentage of fermentable sugar usually is increased by a combination of increased enzyme action and addition of a completely fermentable sugar, such as pure glucose, as an adjunct.

A simple traditional method of increasing fermentable sugar during fermentation is adding wort that was separated from the mash at an early stage while the temperature was still low to cooled, fermenting wort. Some of the first 'malt liquors' produced were made using this technique. The malt extract has a full complement of malt α-amylase and β-amylase as well as a small amount of a thermolabile starch debranching enzyme found in malt. Even at the low temperature of fermentation, 15°C or less for a lager beer, the enzymes are capable of significant hydrolysis of dextrins during the several days of fermentation. The percentage of fermentable sugar in the wort can be increased by 10 per cent using this method, elevating total fermentable extract from 65 per cent to 75 per cent. Unfortunately, a variety of bacteria can be introduced into the fermenting wort with the enzyme when using this method, so it is not very popular. A more recent version of this method is adding commercially available barley β-amylase to the fermenting wort. Malt α-amylase is not very active in fermenting wort because the pH drops during fermentation, usually to pH 4.5 or less. The malt debranching enzyme is present only in small quantity, so at the pH of fermentation the only malt starch-degrading enzyme that is very active is β-amylase, which is quite acid resistant. Addition of barley β-amylase gives essentially the same results as addition of a crude malt extract.

Commercially prepared α-amylase from the mould *Aspergillus oryzae* was developed originally by Takamine as a substitute for malt diastatic enzyme. This enzyme has been added to brewer's wort during fermentation to increase the level of fermentable sugar by hydrolysing dextrins. Although the enzyme is an α-amylase, it produces a large amount of fermentable sugar from dextrin, and it is stable at the low pH of fermentation. The enzyme is capable of increasing the percentage of fermentable sugar in the wort to 75 per cent. The amount of enzyme that must be added is about 1500 SKB units per litre of wort.

In the late 1960s, a new type of beer was developed using enzyme to increase the percentage of fermentable extract in the wort. The beer type is lower calorie beer, commonly called 'light' beer. Most of this beer is made using microbial enzymes.

In a regular beer, about one-third of the wort extract material is soluble dextrins derived from starch. These dextrins include the α(1→6) branch linkages of starch. Dextrins are not fermented by the yeast, so they are present in the finished beer, to which they contribute about one-third of the caloric content. By converting these dextrins into fermentable sugar, a beer with significantly higher alcohol content and very little residual carbohydrate can be produced. If this high-alcohol beer is diluted so the alcohol content is reduced to that of a regular beer, the resulting beer has fewer calories than regular beer and very little residual carbohydrate. The diluted beer is sold as 'light' beer or, in some countries, as low carbohydrate beer.

The enzyme originally applied to beers to give near complete breakdown of the dextrins normally found in beer is glucoamylase from *Aspergillus niger*. This enzyme is an exoamylase capable of hydrolysing both α(1→4) and α(1→6) bonds in the dextrins. The application of this enzyme to increase the amount of fermentable sugar in wort was investigated by Saletan soon after the enzyme became commercially available. Under optimum conditions, glucoamylase is capable of transforming at least 95 per cent of dextrins in beer to glucose that is fermented by the yeast. Since compounds other than carbohydrate, such as peptides of various sizes, are present in beer; the percentage of the total wort extract fermented is 85 per cent. The enzyme is added to fermenting wort. Different units are used by different manufacturers to measure glucoamylase activity in commercial preparations, but the most common units are comparable in magnitude. The amount of enzyme added to fermenting wort for optimal reaction is 5–10 units per litre. A unit may be gram glucose produced from soluble starch per hour at pH 4.2 and 60°C. The assay conditions presented are quite different from conditions in fermenting wort. The pH is close to 4.2, the optimum for the enzyme, but the temperature will not be above 15°C. Therefore, the enzyme is used quite inefficiently during fermentation. However, the marketability, of the beer produced justifies the inefficiency. Usually, the enzyme is not used at such a high level that excess glucose accumulates during fermentation. Conversion of dextrin to glucose is the rate limiting step. After the sugar that is formed during brewing of the wort is used by the yeast, fermentation continues for an additional 1–2 days as the dextrins are digested by the enzyme and the products almost immediately fermented by the yeast. The original procedure for producing light beer was developed by Owades at Narragansett Brewery and gradually spread to other breweries, and has been improved over subsequent years. The increase in business that occurred as a result of the popularity of light beers made coping with some of the production problems that arise from the use of glucoamylase worthwhile. The biggest problem is that, if beer containing the enzyme is mixed with regular beer with a full complement of dextrin, the enzyme will begin digesting the dextrin in the regular beer, resulting in production of large amounts of glucose and an abnormally sweet tasting beer. Most beer is pasteurised at a temperature of 60°C to prevent spoilage. This temperature is optimum for glucoamylase activity; beer pH is the optimum pH for the enzyme. Pasteurisation of regular beer mixed with as little as 5 per cent enzyme-containing light beer will change the taste of the beer significantly by making it sweet. Extreme vigilance and control is required to insure that this does not occur, especially since it is common practice to filter two different types of beer in sequence through the same filter or to fill tanks with new beer when they are nearly but not completely empty.

Because of the potential problem of accidentally mixing glucoamylase-containing beer with regular beer, the prospect of using glucoamylase early in the brewing process, in the mash where starch from grains is made soluble and hydrolysed by malt enzymes, is of great interest. All enzymes, from malt or other sources, present in the mash are inactivated completely during the subsequent boiling of the extracted wort, Conditions in the mash during starch gelatinisation and digestion include temperatures above 60°C and a pH in the range of 5.2 to 6.0. The glucoamylase of *Aspergillus niger* begins to be inactivated at temperatures above 60°C. It is most stable and most active in a pH range of 4.0–4.5. Conditions in a brewers' mash, therefore, are not favourable for activity of this enzyme. Nevertheless, by using several-fold more enzyme than would be needed to obtain sufficient hydrolysis during fermentation, by adjusting the pH of the mash to the low end of the practical range, and by holding the mash at relatively low starch saccharification temperature (63°C) for some time, almost as much conversion of starch to fermentable sugar, mainly glucose, occurs when enzyme is added during fermentation. Because of the advantages of absence of glucoamylase from the beer during later stages of the process, some brewers prefer to use the enzyme in this manner.

Production of light beer with glucoamylase is a good source of examples of the importance of side activities in commercial enzymes. *Aspergillus niger* is capable of producing many enzymes other than glucoamylase that can have an effect on beer. The acid protease produced by this organism is a good example. In beer, the ability to form foam when poured is an important characteristic. Foam formation depends on the presence of high molecular weight peptides, often called proteins, in beer. The acid protease of *Aspergillus niger* is ideally suited for hydrolysis of peptide bonds at the acid pH of beer. If too much protease is included in glucoamylase preparations, beer treated with the enzyme is likely to be rejected because of poor foam formation.

However, the protease side activity can have a positive effect in fermentation of light beer. Many yeast strains have the ability to form flocs after they finish growing during fermentation. The yeast cell surface changes so the cells adhere to each other on contact. Hundreds or thousands of cells will adhere to each other to form a floc, which allows the yeast to settle to the bottom of the fermenting tank much more quickly than single yeast cells would. The change at the cell surface must involve protein because a variety of proteases are able to change yeast cells from the floc-forming state to the non-floc-forming state. Fermenting beer with a floc-forming yeast is often advantageous because the yeast settle out of the beer quickly after fermentation is over, making the beer easier to clarify by filtration. In production of light beer, the yeast must stay suspended for a longer period of time to ferment all the glucose released by the glucoamylase. A small amount of protease present in glucoamylase can be just enough to prevent the formation of flocs for 1–2 days. The extra time is enough to allow fermentation to be completed. When batches of glucoamylase containing no protease are used in brewing with a floc-forming yeast strain, the yeast is able to settle out of the beer too quickly and the beer does not ferment to the extent that it should. These brewers must use a glucoamylase that contains some protease, but not so much that it reduces foam quality.

Other problems occurred when glucoamylase was first applied to production of light beer. A side activity, possibly β-galactosidase, which can hydrolyse galactosyldiglycerides found in malt, caused severe problems in foam production, especially in breweries in which the equipment was known to allow higher than normal amounts of lipid to be carried into the beer. Glucoamylase produced by certain enzyme-producing plants tended to induce haze formation in beer treated with that enzyme. Beer is expected to remain clear after it is filtered. The tendency to form haze was linked to protease levels, but the exact cause of the haze was never determined. As the techniques of enzyme production evolved, purer and more concentrated glucoamylase became available and many of the earlier problems of beer production disappeared.

The glucoamylase produced by *Rhizopus* species has attracted some interest for use in brewing because it is inactivated at ordinary beer pasteurisation temperatures (60°C) and should not cause problems of excess sweetness if enzyme-treated and untreated beers become mixed. This enzyme is capable of essentially complete degradation of dextrins to fermentable glucose during fermentation of a wort in a manner similar to that of the *Aspergillus* enzyme. However, side activities or impurities in commercially available preparations lead to problems of haze formation and poor foam characteristics in beer produced with this enzyme. Therefore, this enzyme has never been used for commercial production of a light beer.

IMPROVING FILTRATION WITH ENZYMES

Filtrations are critical to the brewing process. Before fermentation can begin, the liquid wort must be filtered from the grains; after fermentation, almost all beers are filtered to a high degree of clarity. In addition, some beers are filtered almost to the point of sterility if they are to be packaged and sold without pasteurisation. The (1→3) and (1→4)-linked β-D-glucan of the barley cell wall has been identified as a particular source of problems in filtration. Enzymes that digest this polysaccharide, identified simply as β-glucanases, have been employed as an aid to filtration.

The separation of liquid wort containing the soluble products of hydrolysis of the grains from the insoluble portion of the grain, or spent grain, is critical from the standpoints of time required and of the amount of material, called extract, recovered from the grains used. Ordinarily a gravity filtration, called lautering, requires about 2 hours whereas a pressure filtration requires a little less time. A difference in time of only a few minutes can affect significantly the throughput and efficiency of equipment operating on a 24-hour basis. The amount of extract recovered from the grains obviously affects production cost. The two factors are interrelated. If wort filtration is slow, time can be saved by washing the grains with less 'sparge' water, but this adjustment leads to less efficient extraction. Therefore, factors that affect the rate of wort filtration raise considerable interest. Incompletely digested cell wall fragments or undigested small starch granules in the grain are important elements that can impede passage of liquid through the grain. Malt that has not undergone adequate digestion of cell walls during the germination process, referred to as undermodified malt, often gives poor wort filtration when it is used for brewing. Addition of β-glucanase enzyme from *Bacillus*, *Aspergillus*, *Penicillium*, or *Trichoderma* sources has been found to improve filtration when undermodified malt or even unmalted barley is present in the mash. This effect is sometimes attributed to viscosity reduction by the β-glucanase but, at the high temperature of the mash, the soluble β-glucans in the wort do not increase the viscosity of the wort significantly. Careful study has determined that blockage by small insoluble pieces of grain is more likely to be the source of filtration difficulty. Addition of enzyme also may increase the amount of extract recovered from the grain, but the mechanism for the increase has not been determined carefully. Increased filterability due to the enzyme may allow more thorough washing of already soluble material from the grain. Alternatively, the enzyme actually may allow more components of the grain to become soluble.

The soluble β-glucan that is extracted from malt into the wort remains in solution during fermentation and is still present in the fermented beer. Although poorly modified malts will contribute more β-glucan, even the highest quality well-modified malts still have some soluble β-glucan content. This β-glucan becomes part of the beer that needs to be filtered to achieve clarity and, perhaps, near-sterility. For many years it was known that poorly modified malts produce beers that are difficult to filter; it was assumed that much of the problem was due to the β-glucan in the beer. Wagner and others rigorously have correlated high molecular weight β-glucan content of beer with filtration characteristics of the beer. Smith and others have shown that filtration difficulties are caused by gel formation by the β-glucans, which depends on factors such as temperature and the shear developed as cold beer passes through a filter.

Addition of β-glucanase to beer during wort production or during storage after fermentation is effective in reducing the content of high molecular weight β-glucan and in making beer easier to filter, especially when the untreated, beer is difficult to filter. Enzyme from *Trichoderma* is especially effective in making beers easier to filter. The preparations contain a cellulase that hydrolyses the $\beta(1\rightarrow4)$ bonds of the glucan as well as hemicellulases that may aid in glucan breakdown.

EXTENSION OF SHELF-LIFE OF PACKAGED BEER

Beer is extremely sensitive to oxygen and oxidative staling. Oxygen disappears during fermentation because it is taken up by the yeast, and remains below 0.2 mg/l during processing in a well-operated plant in which the atmosphere in tanks and equipment is pure carbon dioxide. Levels of air in packaged beer are usually no more than 1 ml per package; some breweries operate below 0.5 ml. Once the development of enzymatic chillproofing made it practical to package beer and ship it over long distances while maintaining its appearance, further developments in packaging technology that reduced the air content in the package helped make beer flavour more stable as well as improved the clarity of the beer. Further improvement in packaged beer stability came with the development of antioxidants that are added to beer to retard oxidation, a development also credited to Wallerstein and his company.

Antioxidants commonly used in beer include sulphite salts, sodium hydrosulphite ($Na_2S_2O_4$), and sodium erythorbate.

A continuing interest in the application of enzymes that react with oxygen to reduce oxygen in beer is the result of the importance of oxidation in beer flavour maintenance. The most readily available enzyme for this purpose is glucose oxidase from *Aspergillus niger*. A combination of glucose oxidase, which produces peroxide from glucose and oxygen, and catalase from the same organism, which produces oxygen and water from the peroxide, will after a number of cycles through peroxide intermediates eventually remove all oxygen from a water solution containing the enzymes, glucose, and water. Blockmans applied the glucose oxidase-catalase system to beer. Although glucose is present in unfermented wort, it is used completely by yeast during fermentation so insufficient glucose is present to support complete removal of the oxygen trapped in a sealed beer package. These investigators found that addition of 0.1 g/l glucose to the beer allows complete reaction of the oxygen normally present in the package with glucose oxidase. Unexpectedly, however, beer treated with glucose and glucose oxidase-catalase enzyme tasted more stale after ageing than did untreated beer. The next discovery was that, in the presence of ethanol, catalase does not convert peroxide to oxygen and water but instead acts like a peroxidase, catalysing the oxidation of beer components by the peroxide intermediate. To eliminate the peroxide, it is necessary to add free sulphite to the beer. The sulphite is oxidised preferentially by the peroxide, providing a pathway for oxygen removal that does not adversely affect beer flavour. By adding glucose, enzyme, and sulphite to beer these individuals finally were able to achieve an improvement in flavour after ageing.

Subsequent experimentation with glucose oxidase flavour stabilisation by Goossens used glucose oxidase immobilised with glucose in the bottle crown inlay. These investigators were able to remove 90 per cent of the oxygen in the headspace gas above the beer at a rate dependent on design parameters of the crown inlay. Further experimentation with oxygen removal by enzymes is necessary to determine whether removal of the low level of oxygen present in a commercial package results in significant flavour stabilisation and whether the system is economically feasible.

Chapter 18

Fish Processing and Meat Industry

INTRODUCTION

The importance of enzymes in fish processing has, until recently, mainly been restricted to endogenous enzymes naturally present in the fish tissues. The deliberate use of added enzymes in fish processing is not common. However, the last several years, interest in controlling or aiding traditional processes that rely on endogenous enzymes by using added enzymes has increased. The use of enzymes as 'specific' tools for fish processing operations has appeared also. Deskinning of fish, membrane removal, and roe purification are examples of such processes.

TRADITIONAL PROCESSES

Production of Fish Sauce

Fermented fish sauce is a traditional product of the Far East. The products are important protein sources and are used mainly as a condiment for rice dishes. The products are consumed by people in Southeast Asia. The production procedure consists of mixing small uneviscerated fish with high concentrations of salt and storing the mixture in sealed vessels at ambient temperatures for a period of 6–18 months. During the process, hydrolytic enzymes slowly degrade the fish tissues. Massive proteolysis occurs, with concomitant formation of peptides and amino acids. Other nitrogenous compounds such as ammonia and trimethylamine are formed also. The final product is a clear liquid that is separated by filtration or is decanted off. The liquid has a high content of soluble nitrogenous compounds and a salt concentration in the range of 20–30 per cent. The remaining residue consists mainly of bones and scales and commonly has been used as a fertiliser.

The hydrolysis of the tissues appears to be mainly an autolytic process by endogenous fish enzymes. Orejana and Liston report that trypsin or trypsin-like enzymes are the principal agents of proteolysis in patis (a fish sauce) production. The pH during fish sauce fermentation is usually in the range of 5–6. Bacteria or their enzymes are only of minor importance during the digestion, because of the inhibitory effects of the high salt concentration. It appears, however, that bacteria play a role in forming the distinct flavour and aroma that, in turn, determine the quality of the fish sauces. Fish enzymes and/or atmospheric oxidation also have been reported as possible contributors to the aroma of the fermented products.

In recent years, there has been some interest in accelerating fish sauce processing to reduce production cost by shortening the digestion or fermentation period. Amano and van Veen state that pineapple juice is sometimes used as a source of proteolytic enzymes to hasten the process. Beddows and Ardeshir found that the plant enzymes bromelain, ficin, and papain could digest fish tissues in a short period, thus producing hydrolysates with distribution and concentration of nitrogenous compounds similar to that of fish sauce, but lacking in aroma. Ooshiro came to a similar conclusion when using the enzyme papain; the added enzyme accelerated proteolysis but the aroma typical of traditionally produced fish sauce was not observed. Fish hydrolysates manufactured by the use of plant proteolytic enzymes can be added to traditionally produced fish sauce without affecting the nutritional quality. However, on their own, the artificially produced fish hydrolysates are inadequate since they lack aroma, and aroma and flavour are the most important factors in consumer acceptability of the traditional fermented fish sauces.

It has been reported that a good quality fish sauce can be prepared from capelin (*Mallotus villosus*) with the help of squid hepatopancreas. These investigators conclude that the amount and type of amino acids and other non-amino acid nitrogen compounds are important for the acceptability of capelin fish sauces. Addition of proteinases such as fungal, pronase, trypsin, and chymotrypsin to minced capelin increased the initial rate of protein hydrolysis, but did not yield a product with a much higher content of free amino acids in comparison with sauces without added enzymes.

It may be possible to accelerate fish sauce processing using halophilic bacteria. Smith found that the addition of halophilic *Bacillus* strains accelerated the production of fish sauce considerably. No great differences in free, amino acid profile were observed between samples with or without the added bacteria.

Acid digestion has been suggested as a possible means to accelerate the processing of fish sauce. The disadvantage with hydrolysates thus produced is the lack of aroma and flavour. The acid hydrolysates can, however, be neutralised and added to traditionally prepared fish sauces, to increase the volume and thus reduce production cost. Gildberg found that fish sauce hydrolysis could be accelerated considerably by autolysis at pH 4 and low salt concentrations, followed by subsequent neutralisation and salt addition. The final product, although initially lacking in flavour, developed flavour characteristics similar to those, or traditionally produced fish sauce during storage.

In this context, the production of fish silage can be mentioned. The process has been used commercially for some years, mainly in Denmark, to produce animal feed. The raw material is mostly fish waste and trash fish, which is homogenised and acidified by inorganic and/or organic acids, thus creating a suitable environment for the proteolytic enzymes in the fish to degrade the tissues and form a liquefied mass. During fish silage production, no salt is used since the low pH prevents microbial spoilage. The fish silage thus produced is not suitable for human consumption because of the bitter tasting hydrophobic peptides that are formed during the enzymatic hydrolysis. These bitter peptides apparently are not observed in the production of traditional fish sauce, possibly because of the high salt concentration.

Many aspects of traditional fish sauce manufacture are still unclear. Better understanding is needed of the nature of the enzymatic hydrolysis and especially of the development of the flavour and aroma to obtain the best results with added commercial enzymes.

Salting of Herring

Endogenous fish enzymes are also of importance in traditional salt-curing of herring (*Clupea harengus*). The process probably originated in Scotland in the eighth century. Over the years, it has been changed and improved by the various countries that produce salted herring. The salting of herring is carried out by all countries that fish in the North Atlantic and the North Sea. The salting or processing of other species such as sprat (*Sprattus sprattus*) or anchovies (*Engraulis encrasicholus*) will not be addressed in this discussion.

Salting of herring is carried out by mixing salt and fresh herring (whole or gutted) and packing it in barrels. In some countries, a salt brine is added before the barrels are closed. The barrels are stored at low temperatures (–5° to 10°C) for a few weeks or up to several months until the herring is ripe.

The final salt concentration of the products varies (approximately 4–18 per cent), depending on the recipes used. Sugar, spices, and preservatives (potassium nitrate, sodium nitrite, potassium sorbate, and sodium benzoate) are sometimes used to produce certain varieties of ripened herring. If the recipes are followed closely during the manufacture of salted herring and the raw material is of good quality, the final product obtained will be characteristic of a well-ripened herring with a tender consistency and a pleasant taste and odour.

The process consists of two stages, that is, salting and ripening but the latter stage is caused mainly by enzymes naturally present in the fish. The salting stage is characterised by salt penetration into the fish muscle and ends when the concentration of salt in the tissues is equal to that of the surrounding brine. During the ripening stage, biochemical and chemical processes occur that change the characteristics of the fish tissues. Proteolysis causes an increase in soluble nitrogenous compounds, such as peptides and amino acids, with concomitant change in the tissue structure. Both endo- and exopeptidases are considered to be of importance during ripening. Luijpen found that trypsin was the major contributor to endopeptidase activity at the pH normally found in herring brines (pH 5.5–6.5). Various exopeptidases are believed to be important, especially during development of the sensory characteristics.

Although the salting of herring has been practiced for centuries, the nature of the ripening process is not known in any detail. Experience shows that herring that has been thoroughly cleaned of intestines ripens slowly and does not acquire the characteristic taste and odour. Luijpen, investigating the salt curing of matjes herring, concluded that pyloric caeca were responsible for the sensory properties of the finished product. Enzymes from muscle tissue also may be of importance, although it has been shown that some muscle enzymes, for example, cathepsins, may lose activity at relatively low salt concentrations. Micro-organisms and their enzymes are believed to be of minor importance during the ripening of herring.

Proteolytic enzymes from the internal organs, particularly the digestive enzymes, are of primary importance during the ripening of herring. However, the activity of the endogenous digestive enzymes vary with the season and is at its highest before spawning. Spent herring is considered unsuitable for salting because of the bad appearance and the lack of the characteristic odour and taste in the finished salted product. In effect, the best products are obtained from herring salted during the feeding stage, which lasts for only a few months, when the endogenous proteolytic activity is high and the fat content is high as well. Considerable effort has been put into developing a manufacturing process that relies more on added enzymes than on endogenous enzymes, so salt-ripened herring can be produced throughout the year. Acceleration of the ripening process also has been considered.

It has been reported that an acceptable ripened product can be obtained from eviscerated lean herring by the addition of a homogenate prepared from the appendices pyloricae of well-fed herring. A process for the manufacture of ripened herring has been patented that is based on an enzyme preparation from herring viscera. The mixture is reported to be suitable for ripening herring caught during non-feeding periods (both fresh and frozen) as well as herring fillets. According to Eriksson, enzyme concentrates from cod fish also can be used to control the ripening of herring. Opshaug has described a process based on the addition of homogenised herring viscera to accelerate the ripening of herring or herring fillets (fresh or frozen). The ripening takes place during a period of 1–5 days and the resulting product is described as having a smooth consistency and a good aroma. However, the use of added herring viscera enzymes to control the ripening of herring does not appear to be widely used at the industrial level; the enzyme preparations are not readily available as commercial products.

Ritskes and Ruiter found that a mixture of commercial proteolytic preparations (from beef organs) containing low amounts of lipase activity had a favourable effect on both the flavour and the texture of salted matjes herring. A process for the manufacture of salted matjes herring based on a proteolytic mixture, containing mainly trypsin and chymotrypsin, has been patented. The enzyme mixture should, according to the patent, be substantially free of pepsin and papain, but also preferably free of lipolytic and amylolytic activities. The addition of cod trypsin to eviscerated brined herring has been shown to cause a more rapid accumulation of soluble protein than addition of bovine trypsin. The accumulation of free amino acids increased with trypsin supplementation of eviscerated herring samples, although the amino acid profile was different from that of conventionally prepared samples. Moreover, trypsin-supplemented herring samples were ranked lower than conventionally produced herring samples by sensory panelists. Kiesvaara found that an incubation of herring fillets in a solution of pancreatic trypsin increases the release of peptides whereas, in traditional barrel ripening, peptides do not accumulate since the natural fish enzymes seem to cause further proteolysis. However, it has been reported that cod trypsin and an enzyme preparation from squid liver containing catepsin C (dipeptidyl peptidase I) activity accelerated the formation of free amino acids when supplemented to a brined eviscerated squid. An informal taste evaluation indicated that the product prepared with squid liver catepsin C had a better taste than other samples.

Today, products are available on the European market that are ripened by adding commercially available enzymes or enzyme mixtures. The quality, especially the taste quality, of herring products ripened this way is likely to be inferior to that of traditional salt ripened herring. The commercially available enzymes or enzyme mixtures that have been used to affect the ripening of herring seem to contain mainly endopeptidase activity, whereas the natural herring enzymes contain both exo- and endopeptidase activity. The difference between the activity of added enzymes and that of the endogenous enzymes may affect the quality of the final product, because the amount and type of free amino acids formed appears to be important in development of the flavour of ripened herring.

Although some success has been obtained with artificial ripening of herring, more work is needed, especially to understand in detail the nature of the traditional enzymatic ripening process and to select the enzymes that have properties and specifications similar to those of the endogenous enzymes.

NEW APPLICATIONS OF ENZYMES IN FISH PROCESSING

Some new interesting applications of added enzymes for fish processing operations have emerged, partly because of increased knowledge of enzymes and their uses but also because many enzymes including some fish enzymes have become commercially available. In this section, some uses of enzymes in fish processing are discussed.

Deskinning

The removal of skin from various fish species has been investigated by several workers. Fehmerling described a process for deskinning of fish species, such as tuna, that are difficult to skin by manual or mechanical means. Briefly, the process consists of pre-heating tuna fish (e.g. skipjack tuna) to 60°C with steam and placing the fish in a warm water bath (50°–60°C) containing mainly proteolytic enzymes for approximately 10–90 minutes, following the immersion, most of the skin can be removed by water jets.

A process has been reported for the deskinning of skate (*Raja radiata*) or, more specifically, skate wings. This skate species has proven to be difficult to deskin manually or by skinning machines because of the spiked skin. The main structural component of the skin is collagen, which gives the skin physical strength. The skinning is accomplished by a gentle heat denaturation of the skin collagen in warm water, followed by immersing the skate wings at low temperatures (0°–10°C) in an enzyme bath for several hours. The final step of the process consists of rinsing the dissolved skin from the skate wings and trimming the wings. A suitable enzyme solution should contain not only proteinases but also carbohydrases. The carbohydrases are apparently important during the enzymatic skinning, although not essential. Although the process is promising, it is as yet not used commercially.

Joakimsson described a process for the deskinning of whole herring (*Clupea harengus*). The skin collagen is first denatured and the scales removed by a treatment with 5 per cent acetic acid. The skin is solubilised by a cod pepsin preparation at low pH.

Solubilisation is reported to require 30–120 minutes at 15°–20°C. The final step is removal of skin and enzymes by water jets. However, this process has not yet been applied at an industrial level.

A method for deskinning and tenderising squid using enzymes has been patented by Raa and Nilsen. The skinning of squid is necessary to improve the appearance and to prevent discolouration, off-flavour, and odour of products for human consumption. However, mechanical or manual skinning is normally the bottleneck during processing. The enzymatic method can be accomplished by immersing the squid at low temperatures in a salt solution (5 per cent NaCl) containing the enzyme papain. The squid also can be skinned by immersing the tube and tentacles in a salt solution at 45°C for 10 minutes, whereby the endogenous enzymes of the squid are activated. According to the patent, the outer and the inner skins of the squid tube and the skin from the tentacles are all removed during the process. The enzymatic process is being used at the industrial level and has proved successful. The outline of the industrial process, described by Wray, indicates that the outer pigmented skin of the squid is removed by skinning machines prior to enzymatic treatment.

Deskinning of shark fin with enzymes has been described. Shark fin is a delicacy in Chinese cookery, used either for soups or as an appetiser. The preparation of shark fin is a traditional process that is both time consuming and labour intensive. Using a bacterial proteinase, the fin can be prepared in a shorter time. In the enzymatic process, the shark fin is immersed in a warm solution containing the enzyme until the skin can be loosened by rubbing. The enzyme activity is destroyed by transferring the fin to a solution with 1 per cent hydrogen peroxide.

Removal of Membranes

Membrane removal can be a problem during fish processing. The production of canned cod liver is an example. The raw material for canning, cod (*Gadus morhua*) liver, is sometimes infested with sealworm (*Phocanema decipiens*), which must be removed during processing. Removal is accomplished by removing a thin membrane that surrounds the liver since the sealworms infest the membrane. Suitable machinery is not available. Therefore, the membrane is removed by hand, but only with moderate success and at considerable labour cost. Steingrimsdottir developed a method in which this membrane is dissolved by proteolytic fish enzymes or bromelain. The method was patented in Iceland.

Membrane removal is also a problem during the production of salted cod swim bladders. The product is manufactured in small quantities (30–50 tonnes) for markets in Southern Europe. The fresh swim bladder used for the process is enclosed in a black membrane which must be removed before salting. The membrane is tightly bound to the bladder itself and is virtually impossible to remove by manual or mechanical means. However, a simple process has been developed to dissolve this membrane using enzymes. The enzymatic process increases the rate of handling of swim bladders during membrane removal at least eight-fold. The method is now used by all producers of salted cod swim bladder with a high degree of success.

Fish Roe Purification

A method of fish roe purification has been patented based on the use of proteinase to hydrolyse both the supportive and the connective tissues that envelop individual salmon and trout roe eggs and the roe sack. The product yield is claimed to be more than 85 per cent, whereas it is 50–60 per cent using conventional methods. According to the patent, any proteinase may be used, for example, proteinases active under acid, neutral, or alkaline conditions (pH 3–10). Fish pepsins are reported to be useful during the production of salmon and trout roe caviar. According to Gildberg the fish pepsins split the linkages that adhere the egg cells to the connective tissue of the roe sack without affecting the egg cells. The enzymatic process is used in Scandinavia and Canada. An enzymatic process is also reported to be used in Scotland during the production of trout roe caviar.

Miscellaneous Applications

Fehmerling described processes to remove intestines from shellfish such as clams, scallops, and mussels by the use of enzymes. The process consists of immersing the shellfish, after removing the shells, in a warm solution containing hydrolytic enzymes (mainly proteinases, lipases, and glycosidase hydrolases) until the intestines can be removed by spraying with water jets. Enzymes also may be of use in loosening or removing the shells from shellfish.

The production of fish protein hydrolysate by proteinase treatment is a means to transform cheap pelagic fish and fish offal into protein concentrates with better functionality than fish meal. The manufacture consists of mixing ground fish with proteinase (e.g. ficin, papain, or microbial proteinases) that hydrolyses part of the fish protein into peptides and amino acids. After hydrolysis, the enzyme is inactivated, the suspension is sieved, and usually it is concentrated and spray dried. Fish protein hydrolysates are soluble and can be used to supplement the nutritional value of liquid foods. They have been produced commercially in France as animal feed, mostly for use in milk replacers. However, fish protein hydrolysates are not suitable for human consumption because of the bitter peptides that are formed during hydrolysis and because of their fishy flavour.

A method has been described in which an emulsified protein mixture is produced from fish using enzyme hydrolysis. The raw material is fish or shellfish. The fish is eviscerated and skinned. In some cases, bones and shells are removed before further processing. The raw material is then ground into particles of 100 μm or smaller. A proteinase is added and the proteins are broken down so that coagulation will be minimal when heated. After hydrolysis the mixture is heated to inactivate the enzyme and heat sterilised to insure good storage life. The paste-like product thus produced is a creamy textured fish protein that has a moderate fish taste. It is claimed that the product can be used for various purposes as a food ingredient, feed (after drying), or fertiliser.

Seafood flavourants from various sources of raw material can be produced by enzymatic hydrolysis. For the process, mainly by-products from fish and shellfish are used. The process consists of liquefaction of the raw material by enzymatic hydrolysis, thermal inactivation of enzymes, separation of bones and shells, filtration or centrifugation, and, finally, concentration of the flavours naturally occurring in the raw materials. The purpose of the enzymatic hydrolysis is to liquefy but not produce any flavour or taste. Exogenous enzymes are used but endogenous enzymes are also of importance for the hydrolysis. A process has been patented based on the addition of an extract of crude papaya latex to salmon during canning. The enzymatic activity is claimed to reduce the amount of heat-coagulated soluble proteins (curd) formed during the canning process. According to the patent, other proteinases are also useful, for example, pepsin, pancreatic proteinases, ficin, and bromelain. A bacterial proteinase has been used to hydrolyse protein in stickwater during fish meal processing. The enzyme is claimed to reduce the viscosity of the stickwater and therefore generate considerable energy savings.

The combined action of the enzymes glucose oxidase and catalase in a solution of glucose has been suggested to extend the storage life of fresh fish. The formation of gluconic acid with the concomitant drop in pH at the fish surface is claimed to inhibit spoilage micro-organisms. Field found that the enzymes could extend the storage life of both fillets and whole winter flounder. However, Shaw found that the enzymes glucose oxidase and catalase or gluconic acid alone had little potential in extending the storage life of Atlantic cod fillets.

ENZYMES IN THE MEAT INDUSTRY

Sigmund Schwimmer titled his detailed discussion of the biochemical processes that follow the death of an animal, the termination of its blood circulation and cessation of oxygen supply, 'muscle to meat'. He describes the effect of enzymatic as well as non-enzymatic reactions on the texture, colour, and taste of meat. Not all factors determining meat texture are currently known.

The most important properties that the meat consumer demands are juiciness, good chewability without loss of firm texture, colour, and taste. Native meat enzymes, the cathepsins, play a special role in tenderising meat by controlled ageing. Many ethnic populations have long used meat tenderisation as a common kitchen practice. Wrapping meat in papain leaves or dipping or immersing in papaya or pineapple juice is a standard practice in tropical countries.

Technical meat tenderising enzymes have been used since about 1940. Papain in different preparations is the most frequently used enzyme. Health considerations are not an issue, because the papaya fruit itself is a food. Incomplete data are available about the tenderising mechanism, but many of the factors that influence this process are known. In addition to pH, temperature, and the enzyme quantity used, the delivery of the proteolytic enzyme into the tissues and its distribution therein is important. For this, several methods exist.

Surface Treatment

Usually, meat slabs are sprayed with a liquid papain solution, or are dipped into enzyme solutions. Distribution within the meat is assured by needle piercing; in the home this is done by piercing the meat with a fork. Generally, 5–10 ppm of a commercial papain preparation is used.

Injection Treatment

Two methods are employed: antemortem and postmortem injections. In commercial meat tenderising, the latter is performed with a multi-injection system in which the enzyme solution is injected under pressure into the meat slabs.

Antemortem injection

1. A papain solution is injected into the animal before slaughter. The incubation period prior to slaughter should be equal to 60 per cent of the time required for one blood circulation cycle.

2. A reversibly inactivated papain solution is injected prior to slaughter. In this case, papain is inactivated with hydrogen peroxide-catalase, which causes oxidation of the free sulphydryl groups. Reactivating the enzyme is accomplished during the initial phase of heating the meat.

Meat Tenderisers

There are many patents and publications describing enzymes that are used in meat tenderisation. In addition to papain, bromelain, ficin, pancreatic proteases, pancreatic elastase, collagenase, and microbial proteases are described. Papain preparations are the most widely used.

Various compounds are added to papain for spraying or dusting the preparations onto meat. Most preparations contain sodium chloride and sodium glutamate. In addition, polyphosphate, protein hydrolysate, glycerin and oil emulsions may be added.

The appropriate enzyme quantity required when dusting the preparation is 0.1 gram of a technical papain preparation (having an activity of 100-gram tryosine units per gram of protein) for each kilogram of meat. For meat injection, approximately 2–3 per cent (v/wt) of 0.005 per cent papain solution is required.

Chapter 19

Textiles and Laundry Detergents

INTRODUCTION

Weaving textile fibres exposes the fabric's warp and weft strands to substantial mechanical stress. To prevent breakage of these strands, they are almost always reinforced with sizing. Sizing can consist of various materials. Gelatine, guar gum or carbo bean meal, polyvinyl alcohol, methacrylate, water-soluble cellulase derivatives, and last—but not least—starch have been employed as sizings. The essential requirement to qualify as a suitable sizing material is good adhesion to the threads and easy removal after weaving.

DESIZING ENZYMES

Despite the many advantages of synthetic sizings, starch remains the sizing of choice because of its low cost. In Europe potato starch is used predominantly, in the United States cornstarch, and in the Far East, rice starch. For imported raw fabrics that still contain sizing, the substances used for sizing must be identified. Sometimes combinations of starch with plant gums or egg white have been used, which can present unexpected difficulties during desizing.

Coating the fibres with sizing causes changes in the fabric. Thus, the sizing must be removed prior to further processing because subsequent treatment such as dyeing, bleaching, or finishing can be compromised.

Many sizings, primarily synthetic ones, can be removed with warm water. Sizing with starch must be executed in such a way that the sizing is water-soluble. This can be achieved by steam heating with sodium hydroxide or by treatment with oxidants. Both processes, however, strongly attack the cellulose fibres and contribute to the waste-water burden.

The safest and environmentally least hazardous process of desizing starch is enzymatic. Earlier, malt α-amylase or pancreatic amylase were primarily employed. The drawback of amylases is their limited temperature stability. To obtain a rapid starch degradation, starch must first be gelatinised. For this, the fabric is briefly boiled in water. When treating with malt or pancreatic amylase, the fabric is cooled to 60° or 50°C.

Generally, the bacterial amylases will permit starch degradation above the gelatinisation temperature, at 85°–90°C, or even higher, at 105°–115°C, which will accelerate the desizing process. Table 19.1 summarises the applications of desizing textiles.

Table 19.1. Enzymes used in desizing textiles.

Enzymes	*pH range*	*Maximum operating temperature (°C)*	*Activators, stabilisers*
Malt α-amylase	4.5–5.5	55–65	Ca^{2+} ions
Pancreatic amylase	6.5–7.5	45–50	NaCl, Ca^{2+} ions
Fungal α-amylase	4.5–5.5	55–65	Ca^{2+} ions
Standard bacterial α-amylase	5.5–7.5	75–85	NaCl, Ca^{2+} ions
Thermostable bacterial α-amylase	5.0–7.0	90–105	NaCl, Ca^{2+} ions (with very soft water)

The maximum operating temperature can be increased by 5°C under special conditions. The system then operates at optimal activity but also with the least enzymatic stability.

Desizing Procedures

The process can be divided into four steps:

Pre-wash

The pre-wash with boiling water and surfactants promotes acceleration of the entire process, especially with very heavy fabrics. This operation removes lipids, waxes, and non-starch sizings. Surfactants must be selected for enzyme compatibility. This is the case for most non-ionic surfactants, but anionic wetting agents must first be tested for enzyme compatibility.

Soaking

During the pre-wash or during soaking, the fabric should absorb 90–100 per cent of its weight in water. Before the enzyme is added, the saturation bath is prepared by dissolving the wetting agent and the substances that stabilise and activate the enzyme in water and control the pH. When a standard bacterial α-amylase is used, about 300 grams of sodium chloride, 50 grams of calcium chloride, and about 50 grams of non-ionic surfactant are dissolved in 100 litres of water at 65°–70°C. Then about 100–200 grams of an amylase preparation with an activity of 3000 SKB/g is mixed into the brine. If the thermostable α-amylase is used, the temperature can be increased to 70°–80°C. A 100-litre batch should contain 30 grams of calcium chloride and 400 grams of sodium chloride with a pH of 6–8.

Starch degradation

Degradation commences with the soaking step and lasts, depending on the process, from 2 minutes to 16 hours. The variable parameters are the enzyme, enzyme concentration, temperature, and reaction time. Short reaction times can be obtained with a thermostable amylase at high concentration and high temperature. To prevent the fabric from drying, which would halt starch degradation during batch desizing, the rollers are covered and rotated. Progress in starch degradation is monitored by the reaction of iodine with starch. A 0.1 N iodine/potassium iodide solution is diluted with water (1:19) and used as the test reagent. Degradation is complete when the iodine test is negative, indicated by a light yellow spot remaining on the fabric after a drop of the solution is added.

Rinsing

The degradation products are removed by rinsing with water at temperatures of 95°–100°C. The rinse is supplemented by surfactants or detergents and conducted with vigorous mechanical agitation. Unusually heavy particulates are removed with a 5–10 grams NaOH/litre rinse.

Desizing Processes

Desizing in a pad roller unit or Foulard

In this process (Fig. 19.1) the pre-wash and soaking steps proceed continuously. For starch degradation, the material is put on a slowly rotating roller or alternatively, put into a trough. Hydrolysis is carried out with a standard bacterial α-amylase (e.g. 250 grams Degomma HB [RM] or 250–500 grams Aquazyme 120 L [NO] per 100 litres of soaking bath) at 70°–75°C for 2–4 hours, or at temperatures <50°C for 10–14 hours. A rinsing process follows.

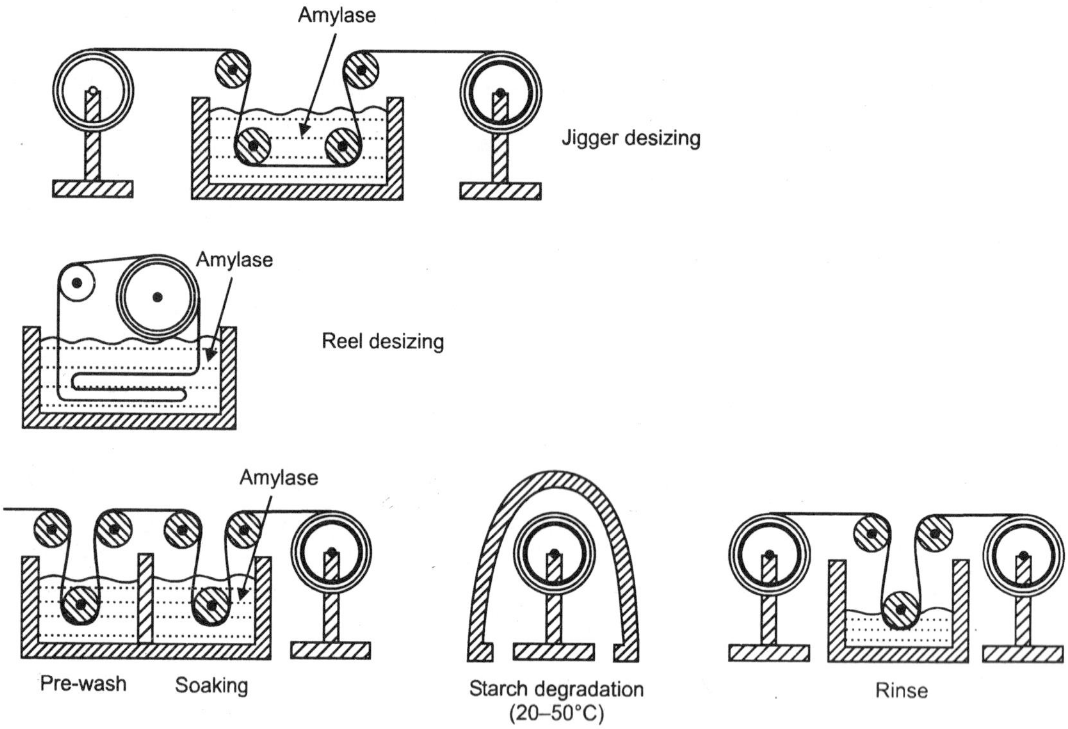

Fig. 19.1. Desizing in a pad roller unit or Foulard.

Jigger desizing process

This is a simple process requiring only two rollers and a trough. The material passes through the bath, which must be changed for each step.

For the pre-wash, two passages in boiling water containing a mild surfactant capable of emulsifying the oils and waxes are required. For soaking, 50–150 grams of enzyme with an activity of about 3000 SKB units/g of a standard bacterial α-amylase (e.g. Optisize [SOEG]) is required to hydrolyse the starch. Starch degradation at 65°–70°C requires four passages. For the last two passages, the temperature is increased to about 90°C.

Reel desizing process

This is also a relatively simple process in which the bath must be changed as well. Pre-wash and post-wash proceed in the same way as that in the jigger process.

The amount of enzyme needed is about 25–100 grams of a preparation with an activity of about 3000 SKB units/g (e.g. Aquazyme [NO]).

Continuous desizing process with a standard bacterial α-amylase

The process requires approximately 30 minutes. The material is, as described earlier, continuously pre-washed and soaked. Starch hydrolysis proceeds continuously in a J-tank, which permits controlled dwell times at elevated temperatures. The starch, degradation process is concluded by continuously passing the material through a steam chamber. Finally, a vigorous hot-water rinse containing 20–30 grams of sodium hydroxide per litre follows ending with cold water rinses (Fig. 19.2).

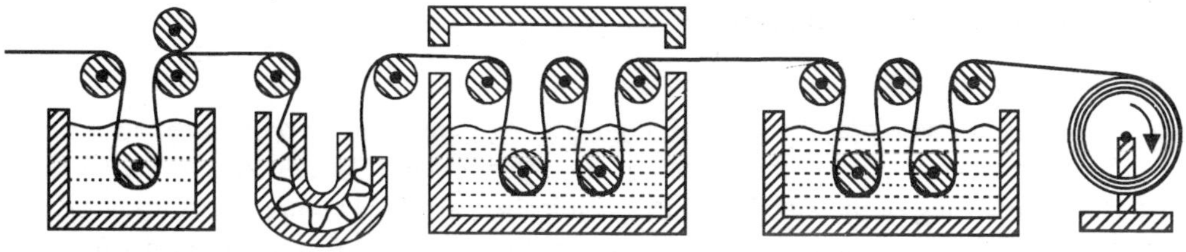

Fig. 19.2. Continuous desizing process.

A 100-litre soaking solution consists of the following:

1. 300 grams of sodium chloride.
2. 200 grams of calcium chloride.
3. 50 grams of surfactant.
4. 250–500 grams of a standard bacterial α-amylase (e.g. Degomma HB 20 [RM]).

Continuous desizing process with a thermostable bacterial α-amylase

The process time is reduced to 2–5 minutes. Removal of sizing is performed in a steam chamber at 105°–115°C requiring only 15–120 seconds. A vigorously boiling alkaline wash significantly improves the cleansing process.

A 100-litre soaking solution consists of the following:

1. 400 grams of sodium chloride.
2. 300 grams of calcium chloride.
3. 50 grams of surfactant.
4. 50–300 grams of thermostable bacterial α-amylase with an activity of 2600 SKB units/g (e.g. Thermamyl [NO]).

Stone wash with amylase and cellulase

Recently stonewashing was introduced as a means of giving blue jeans a worn, rugged appcarance. To enhance the stoning effect, bacterial amylase were introduced. About 1–1.5 litres of a standard bacterial α-amylase with 2000–2500 SKB units/g (e.g. Aquazyme [NO]) is added to a 1000-litre batch.

Since 1992, it has become standard procedure to use cellulases for finishing jeans. Cellulases are especially suitable for the stonewashing of denim dyed with indigo, as this dyestuff hardly penetrates the fibres, but stays mainly on the surface of the yarn. Depending on the process, neutral/alkaline or acid cellulases from *Humicola insolens* [NO] or *Trichoderma reesei* [MJ] are used, also in the combination with bacterial amylase.

The photographic industry produces large amounts of gelatine-silver salt emulsions. To remove the silver from exposed films, such as exposed X-ray films, the gelatine layer must be separated from the carrier material. This allows recovery of both the silver and the carrier, such as celluloid, which can be recycled. Bacterial protease as well as pancreatic protease (e.g. Degomma PNL [RM]) are used to remove the emulsions. For gelatine, which is difficult to separate, film fragments are briefly pre-soaked at 90° or 70°C for several hours. The gelatine is then digested with a dilute protease solution, such as 2–3 grams of Degomma PNL in 100 litres with non-ionic surfactants at 20–45 for 1–2 hours. With colour film, the process requires 24–28 hours in an agitated reaction bath. The treatment may have to be repeated to remove any residual colour layer.

Enzymes for Degumming Silk

The raw-silk filament has a gray-yellow-coloured coating consisting of protein called sericin or silk gum, which envelops and cements the fibroin of the silk filament. Sericin must be partially removed prior to further processing of the silk filament. Chemically, sericin is composed of 90 per cent protein and 5 per cent wax, fats, salts, and ash. It must be softened to permit unwinding of the filament into a continuous thread. Raw silk consists of about 20 per cent sericin and 80 per cent fibroin. Only about 1 per cent of the sericin is removed because some silk gum is required for further handling of the delicate filament. Earlier, degumming was accomplished with alkaline solutions and soap.

Today, degumming is performed with solutions of alkaline bacterial protease. A typical degumming solution includes:

1. Alkaline bacterial protease from *B. licheniformis* with 1 AU/g, ~1.0–2.5 g/l or pancreatic protease with 2 AU/g, 0.5–1.5 g/l.
2. Sodium bicarbonate, 5.0 g/l.
3. Non-ionic surfactant, 0.5–1.0 g/l.

The temperature for protease from *Bacillus* can be 55°–60°C. Pancreatic protease should be heated only to 55°C. The degumming process requires about 0.5–1 hour with gentle agitation. For manufacturing crepe de chine, a non-enzymatic wash at 95°C for 30–60 minutes is recommended as a first step in the degumming process. Weight loss of the silk after this process should be about 16–20 per cent. Further degumming is performed with alkaline bacterial protease, bleaching with hydrogen peroxide follows, and finally, the silk is thoroughly rinsed.

ENZYMES IN LAUNDRY DETERGENTS

Enzymes have been used in laundry soaps for more than 75 years. The foundation for using proteases in laundry detergents was laid by Otto Röhm in 1913. His patent, on 'Reinigung von Wäsche' (process of laundry cleaning) initiated the development of incorporating proteases into detergents that today are currently used throughout the world. Röhm used porcine pancreatic proteases. As early as 1919, 65 tonnes of porcine pancreases were processed for the proteases that were incorporated into the 'Burnus' detergent. The success of these products stimulated the production of enzyme-containing laundry detergents so that, in some markets, 70 per cent of the market has been tilled.

Laundry Detergents

The purpose of detergents in laundering is to facilitate the removal of soil from clothes. Dirt and stains in clothes may consist of inert soil particles, the dust of inorganic and organic substances, lipids, and carbohydrates such as starches or pectic substances. In addition, proteins, tannins, and pigments cause stains that cling with varying degrees of tenacity to different textiles. Removal of dirt with water is bolstered by the laundry detergent. Washing starts with the soaking phase: the penetration of water into the fabric, wetting the fibres. Subsequently, the actual laundering processes involving emulsifying and suspending the dirt particles begins.

Laundry powders include non-ionic and ionic detergents. They contain, in addition to soaps, sodium silicate or sodium bicarbonate, and sodium tripolyphosphates ('builders' or chelating agents) that bind calcium and magnesium ions, thereby eliminating soap scum, which results from the formation of insoluble salts of fatty acids during the wash cycle. Most recently, the phosphates have been replaced by other chelating agents to reduce phosphate in waste-water effluents. Laundry detergents may also contain bleaches, especially peroxyborate that act by liberating hydrogen peroxide when the wash water temperature is >50°C. Chlorine-containing components such as hypochlorite are also used as bleaches. Other laundry detergent additives are colloidal substances, including carboxymethylcellulose, optical brighteners, and perfumes.

Protein Stains

Household laundry soil consists primarily of lipids and carbohydrates that readily dissolve in an alkaline or highly alkaline pH range; these and other stains are removed by surface-active substances (surfactants), which break up and emulsify the dirt particles; surfactants also act as wetting agents which help to remove the soil from the fabric. Protein-containing stains are more difficult to remove. There are several reasons for this; proteins from milk, whey, meat juices, and blood can coagulate on the fabric and denture. These reductions are initiated not only by elevated temperatures but also by alkaline pH or oxidants. One such oxidation reaction occurs on contact of protein with laundry agents that contain perborate at temperatures >50°C, which stabilises the protein. Glycoproteins present in bronchial mucus, or the fibrillar serum proteins, form solid films on textile fibres on drying and are very difficult to solubilise and remove during laundering. Protein residues must be completely removed from the textile fibre, because such protein films may also include pathogens that would, if not removed, be protected and stabilised on the fabric. In this manner, they could survive subsequent laundry disinfection. In the two-step wash cycle initially developed by Röhm GmbH, the laundry is first pre-soaked with an enzyme-containing laundry detergent (Enzymolin) at 35°C for 30 minutes. The second step involves only detergent at temperatures as high as 100°C. The warm pre-wash cycle, still practiced today, avoids setting blood, milk, and other protein stains.

Protease Requirements for Laundry Detergents

Laundry enzymes must meet the following requirements:

1. Possess high efficiency and little specificity.
2. Exhibit activity at pH 9–11 for one hour at temperatures reaching 95°C.
3. Be stable in the presence of chelating agents, perborates, and surfactants.
4. Retain activity within the detergent without noticeable loss of activity for at least one year.

These requirements are best met by the microbial serine proteases, although some laundry detergent components affect these enzymes' activity and stability adversely. For example, chelating agents, which are important as water softeners, reduce enzyme stability at high temperatures. The reason for this is that by complexing the Ca^{2+} ions they remove the ions that would otherwise act as enzyme stabilisers. Loss in enzymatic activity can be expected from perborate or other oxidants.

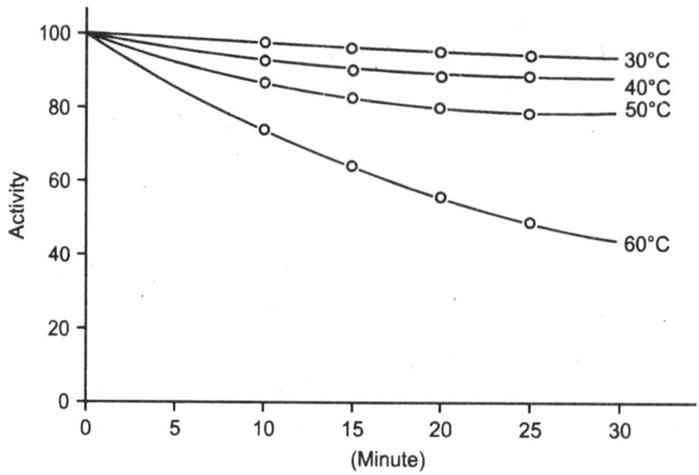

Fig. 19.3. Sodium perborate stability of a laundry detergent protease (0.3 per cent Na-perborate, pH 9.5, enzyme Optimase [SOEG]).

Such agents can oxidise a methionine residue near the active site to a sulphoxide, thereby generating an activity-lowering steric blockage.

Figure 19.3 illustrates Na-perborate stability of an alkaline serine protease (Optimase [SOEG]) at various temperatures. A significant drop in activity can be observed during the wash cycle at 60°C.

An internationally approved test for demonstrating the cleansing power of a given enzyme is its ability to remove common stains from cotton fabrics. Test kits are commercially available under the name EMPA 116 (available from Swiss Material Test Institute, Switzerland). The test sample is composed of blood, milk, and soot. Stained fabric samples are subjected, along with other laundry, to the standard wash cycle of a regular washing machine. After drying, the cleansing effect is determined by measuring the degree of light reflection. The tests are conducted in a standard wash system with the Terg-O-Tometer or the Launder-O-Meter. Dambmann designed a laundry detergent standard for comparing the cleansing power of enzymes:

Enzyme	*Cleansing power (per cent)*
Sodium lauryl sulphate (40 per cent)	5.00
Alkyl phenol ethoxylate (11 EO)	3.00
Soap	3.00
Na-carboxymethylcellulose	1.60
Na-EDTA	0.18
Na-tripolyphosphate	38.00
Na-perborate $4 \cdot H_2O$	25.00
Na-metasilicate	1.00
Na-sulphate	≤100.00

Packaging Proteases into Detergents

Laundry detergent proteases are prepared by fermentation of *B. licheniformis*, *B. amyloliquefaciens* or *Bacillus* sp. After completion of the fermentation step, the medium is purified by centrifugation or filtration; the purified culture medium is concentrated, if necessary, by thin-layer evaporation. Such enzyme concentrates are then stabilised for use in liquid laundry detergents. Stabilisation is achieved by lowering the water content in detergent preparations by adding glycerol, glycols, or sorbitol. Many patents describe various stabilising mixtures of glycerol, sugar, sugar alcohols, polysaccharides, proteins, polycarbonic acids, alkyloamines, ionic and non-ionic surfactants, and inorganic salts.

Further concentration to a dry protease preparation is done by precipitating the enzyme with alcohol or salts, preferably with sodium sulphate. Dust from protease concentrates is particularly aggressive and readily causes inhalation allergies. Procedures for dust control have been developed.

Initially, binding the protease to the surface of an inert material, such as sodium tripolyphosphate, was attempted because the latter was already present in most detergents. For this and other procedures, it is important that the particle size of the enzyme-containing granules be similar or equal to the particle size of the other detergent components, to prevent separation of the components during transport or laundering. A disadvantage of the early granulations was frequent detachment of the enzyme from the surface of its carrier. Binding the enzyme to particle surfaces has been replaced by the prilling method, in which the protease is mixed with the molten detergent components and sprayed onto rotating disks; the droplets then solidify in the cold airstream into relatively homogeneous spherical granules that can be sifted for size. Another method of packaging enzymes is by extrusion granulation (Table 19.2). The enzymes, mixed with detergents and other inert substances, are extruded, and the extrusion strands are then cut into cylinders and shaped into spheres with a 'Marumeriser' (a machine with rapidly rotating corrugated disks), to receive granulated powders.

Table 19.2. Procedures for incorporating proteases into detergents.

Prilling	*Extrusion granulation*	*Mixed granulation*
Basic ingredients and enzymes	Basic ingredients and enzymes	Basic ingredients and enzymes
↓	↓	↓
Melting	Mixing	Mixing
↓	↓	↓
Prilling	Extrusion	Drying
↓	↓	↓
Sieving	Marumerisation	Wax coating
	↓	↓
	Fluidised-bed drying	Cooling
	↓	↓
	Sieving	Sieving
	↓	
	Wax coating	
	↓	
	Cooling	
	↓	
	Sieving	

After drying and sifting, the granules are coated with wax. Other types of granules can be prepared with fluidised-bed granulation. This procedure is especially suited for more thermally labile enzymes.

Some enzyme producers granulate laundry detergent enzymes in systems that are also used in preparative food technology.

Wash Cycle

The wash cycle in Europe usually consists of a pre-wash at 20°–30°C for 10–20 minutes duration. The main wash cycle follows at 40°–50°C or higher for 30–60 minutes. This process is more enzyme-friendly than one-step cycles in which hot (50°–60°C) water is added at the onset of the wash cycle. The need to save energy promoted the development of enzymes that have optimum activity at temperatures of about 40°C. Recent experimental products have shown such properties.

Typical compositions of enzyme-containing pre-wash laundry detergents are:

Components	*Composition (%)*
Anionic components	10–15.00
Non-ionic components	2–3.00
Na-tripolyphosphate	20–45.00
Na-carboxymethylcellulose	0.5–1.00
Alkaline protease with 1.5 AU	0.5–1.00
Na-sulphate	≤100.00

Liquid Laundry Detergents

Liquid detergents present no dust problem and can be dispensed and dosed more easily than powder. Liquid detergents have already achieved a market share of more than 25 per cent in the United States. Liquid detergents containing powerful enzymes are also of interest in stain pre-treatment. A pre-soak with concentrated liquid detergents has become a customary household procedure today. The application of enzymes as spot-removal agents has been practiced in dry cleaning for years.

The enzyme is stabilised by reducing the water component in the formulations. Liquid detergents contain about 30–50 per cent of anionic and non-ionic surfactants and 10–25 per cent of builders; the overall water content is 45 per cent. According to Weber-Meyer, proteases in liquid detergents can be stabilised with glycerol, propylene glycol, and a number of other polyfunctional alcohols.

Other Enzymes for Detergents

Bacterial amylases

These enzymes are often mentioned in formulations for pre-soak or pre-wash detergents. While these additives have little importance in laundry detergents, they are of interest for dishwashing detergents. The effects are significant when removing sticky, starchy, food residues, such as those of noodles and oatmeal. The thermally stable bacterial α-amylases maintain their activity beyond pH 9.0. It would be interesting to integrate amylases into hand soaps; however, there are problems with enzyme stability in such applications.

Cellulases

An unexpected discovery found cellulases to have the ability to soften recycled cotton fabric without measurably damaging the fibre. Employing cellulases in laundry detergents has been described in a Japanese patent and a European patent by Murata listing 23 bacterial strains and 126 fungal strains as suitable sources for the cellulases.

An enzyme presently in use is the cellulase from *Humicola insolens* [NO]. This enzyme is active under mild alkaline conditions (pH 8.5–9.0) and at temperatures of ≥ 50°C. The softening optimum is pH 8.0. In addition to the softening effect, cellulase can impart colour freshness to coloured, frequently laundered cotton fabrics. This phenomenon is related to the fibre-straightening effect of cellulase.

The amount of enzyme required represents only 0.4 per cent of the total detergent; this, however, still constitutes a considerable expense.

Lipases

It has been argued for some time that the use of lipases in the laundry industry may have little value because surfactants at higher washing temperatures adequately remove lipids.

Until recently, no lipases were available that exhibited the necessary activity and stability in the pH range 9–10. However, if the quantity of surfactants could be reduced and the cleansing conducted at lower temperatures, then the new lipases could be of commercial interest. A fungal lipase (Lipolase [NO]) produced by recombinant DNA techniques is effective under alkaline conditions (pH ≤ 12) and temperatures of ≤ 60°C. Numerous cleansing tests were carried out with Lipolase in the Terg-O-Tometer with various basic laundry detergent formulations. Significant effects were reported for European formulations under cold-wash conditions as well as for US laundry detergents at higher temperatures (40°C). Lipase (0.2–0.6 per cent) with 100 KLU[3]/g was added to the detergent.

Cornelissen and others have described the use of lipases from *Humicola languinosa* or *Thermomyces languinosus* in laundry detergents. Also, patent rights were filed for the use of a bacterial lipase from *Chromobacter viscosum*, which exhibits special stability in the presence of bleaching agents.

Thus, recombinant DNA technology has already produced suitable lipases for detergent application. Another lipase was isolated from *Pseudomonas fragi* and its gene transferred to *E. coli*, a more acceptable source organism.

Cutinases can also act as general lipases having a useful high pH optimum. A cutinase from *P. putida* has a high lipase activity within a pH range of 6–11 and, therefore, is also a candidate for detergents. Presently, genetically engineered lipases are being actively studied in the detergent industry.

An enzymatic system that could hydrolyse tannins and plant phenols present in fruit stains, which are still extremely difficult to remove by laundering, would be particularly valuable.

Chapter 20

Pulp and Paper

INTRODUCTION

Pulp and paper are manufactured from wood, non-wood fibres and recycled fibres. The primary raw material used in paper manufacturing is cellulose fibre, which constitutes 50 per cent of wood. The pulp and paper industry is both huge and small in size, technically diverse, operate on a wide variety of manufacturing processes on a large number fibre types. There are two main pulping methods, mechanical and chemical. The yield of the pulp by mechanical process is as high as 90–95 per cent, but the quality of the pulp is of low grade, highly coloured and contains short fibres. In the chemical process, the wood chips are cooked with appropriate chemicals, in an aqueous solution, at an elevated temperature and pressure so as to break the chips into a fibrous mass. The yield of the pulp by this process is about 40–50 per cent of the original wood material. Chemical pulping is carried out in two media: alkaline and acidic.

To obtain a stronger fibre, various methods have been devised, including thermomechanical and chemi-thermomechanical methods. There are two main chemical processes, i.e. Kraft and Sulphite process. In Kraft process, the wood chips are cooked in a solution of sodium hydroxide (NaOH) and sodium sulphide (NaS_2) which results in the release portions of ligneous material (30–45 per cent), saccharinic acid (25–35 per cent), formic and acetic acid (10 per cent) and extractives (3–5 per cent). In sulphite process, the wood chips are cooked in a mixture of sulphurous acid (H_2SO_3) and bisulphide ions (HSO_3^-) to dissolve lignin. After, pulping the product is screened, and the chemical washed out is recovered. The resulting pulp is frequently bleached to increase the brightness of the finished paper product.

Each of these stages produces solid, gaseous and liquid waste in variable amounts. In the pulping and bleaching operation, the quantity and quality of the waste-water depends on the characteristics of the wood and the industrial process used. After cooking and pulping, the black liquid remaining in the pulp is extracted by washing, which makes one of the waste-water streams from the production of cellulose fibres. The black alkaline liquid released by paper mills constitutes only 10–15 per cent of the total waste-water, but contributes to approximately 95 per cent of the total pollution load of pulp and paper mill effluents. The black liquid that originates in the Kraft and soda processes is usually burnt to recover pulping chemicals and produce energy, thus considerably diminishing the environmental impact associated with the pulping stage. However, for small pulp mills, chemical recovery is generally not an economically feasible process, therefore, black liquid may represent an important source of pollution. In chemical pulping, the use of sulphur-based reactants results in the emission of foul-smelling sulphur compounds. The foulest odours produced by the pulp and paper industry are usually from Kraft pulp mills. Both the chemical processes produce a fairly dark pulp that has to be bleached. The bleach generally used in both processes is chlorine-based (chlorine gas, chlorine dioxide or hypochlorite), which reacts with the organic matter in the pulp to produce lignosulphonic, chlorinated resin acids, chlorophenols and dibenzo-*p*-dioxins collectively known as adsorbable organic halides (AOX). These can have a profound impact on the environment.

Numerous physico-chemical methods have been used for the treatment of bleach plant effluents. Biotechnological methods have the potential to reduce the problems associated with physico-chemical methods. But major organochlorinated chemicals formed by reaction mechanisms in the effluent and colour-containing compounds are not degraded in the effluent. Research on black liquid treatment is needed to understand the aerobic and anaerobic treatment of *ex situ* and *in situ* bioremediation of semi-chemical pulping and bleaching effluents. The prospects of enzyme technology indicate possible removal of recalcitrant compounds and recovery of products from lignocellulosic residues for bioprotection and biomaterial. Therefore, the research is primarily based on search methods for the removal of contaminants from the effluent of pulp and paper mills for the bioprotection of the environment and recovery of value-added products from residue and biocatalysts released during degradation of lignocellulosic biomaterials.

PROCESS AND PRODUCTION OF PULP FOR PAPER

The main steps for papermaking are raw material production, exploitation and transportation, raw material storage and preparation, pulping, bleaching, papermaking and the transportation of products, by-products and waste products.

Raw Material Production and Exploitation

The objective of pulping is to separate wood into discrete fibres. About 30 per cent of paper production is based on recovered paper and the balance uses virgin raw materials from developed countries. Wood is the main fibrous raw material used to produce pulp and it accounts for more than 90 per cent of the production. Non-wood fibres are an important source of raw material, especially in developing countries. They can be by-products of agriculture, such as bagasse and cereal straws or industrial crops like bamboo, esparto, sisal, flax and sorghum. After being transported to the plant, the raw materials are processed so that they can be transformed into paper. These operations involve inputs and outputs to the process. They also require space and consequently they can compete with other uses for the site.

Raw Material Storage and Preparation

Generally, pulp and paper plants have a large capacity and they need important stocks of raw material in the plant or in the production area. These stocks can be a source of pollution or diseases (particularly in forest areas). Preparation of the raw material is necessary before it enters the papermaking process. Wood, for instance, first needs to be debarked usually by mechanical or hydraulic processes, and is then disintegrated, generally by chipping into particles of adequate size. Debarking is the process of removing the bark from cut logs. If the bark were left on the wood, it would not only increase papermaking costs, but it would also negatively affect its appearance. There are four types of debarkers: drum, ring, cutter head and hydraulic jet. After the debarking process, the round wood must be cut into chips. There are three types of chippers: disc, v-drum and drum. The chips are then put through a screening process. Uniform size and thickness of the chips is necessary to ensure good quality of pulp. There are two ways to screen chips: vibrating and gyrating or disc type. After the screening process, the chips are cleaned to remove other contaminants. As the chips enter the blow line, the pressure drop causes the fibres to separate into pulp. The pulp is carried by the digester pressure to the atmospheric diffusion washer/cleaner. The purpose of the diffusion washing/cleaning process is to remove spent cooking chemicals and dissolved organic material, from the pulp blown from the digester.

Pulping

All raw materials used in papermaking, except chemicals, fuels and additives, contain fibres, but cellulose fibres constitute approximately only 50 per cent of the dryweight of raw materials. The goal of pulping is to separate the fibres from the material. The remaining components (hemicellulose and lignin) must be treated and they constitute the major potential sources of pollution in chemical pulping. Basically, there are three different groups of pulping methods:

Mechanical pulping

This method consists of grinding raw material against an abrasive surface so as to defibre the raw material (more generally softwood) without any lignin dissolution. A high yield is obtained by this method, generally more than 95 per cent of the dry-weight of the wood. Some methods use high temperature and pressure to increase the efficiency of the process. Mechanical pulping generates very low polluting effects, but is an energy intensive process, as the non-cellulosic wood components are not available (as in chemical pulping) for energy recovery. Various mechanical pulping processes in use are stone ground wood pulping, refiner mechanical pulping and thermo-mechanical pulping.

Chemical pulping

This method separates the fibres from the raw material by making all the non-cellulosic components soluble in cooking liquid, at high temperature and pressure. Chemical pulping gives better paper quality, but is potentially the cause of greater environmental pollution. Chemical pulping mainly uses chemical energy to separate fibres from raw material. Generally, more than 90 per cent of the lignin is removed from the wood, resulting in lower yields (from 40 to 55 per cent). There are two principal pulping processes: acid (sulphite) or alkaline (Kraft). Kraft pulping (KP) process is one of the most developed. Chemical agents used to dissolve lignin and separate fibres are Na_2S and NaOH. Now-a-days, most pulping installations use continuous digesters. High yield sulphite pulping (HYSP) derived from basic sulphite pulping is a variation involving less steam pollution. It associates chip cooking and the use of a refiner, like in mechanical processes. The raw acid SO_2 dissolved in water, is generally mixed with combined SO_2. The pH of the liquid varies from 2 to 5. Delignification is made by sulphonation and hydrolysis, which lead to lignosulphonates production. The wood chips are placed in a digester with the chemical mixture. There are two types of digesters: batch and continuous. In the digester, heat and pressure is added to dissolve the wood's lignin, the natural glue holding the wood fibres together. The brown fibre is then separated from the spent chemical/lignin solution, in preparation for the bleaching process. Water is a key requirement in a pulp mill. Large volumes of water are used for every ton of pulp produced. The quality of the water used to produce pulp and paper products is of vital importance. Therefore, capital and operating costs are higher than those of mechanical pulping. The yield of chemical pulping is about 50 per cent of the dry weight of raw material.

Recovery is a process that is economically important to the pulp mill. This process allows chemical pulp mills to recover a large amount of the cooking chemicals, generate about half their own power requirements and protect the air and water quality. Pulp mills generate their own power through a process known as co-generation, in which steam is generated and then converted to electrical energy.

Chemi-mechanical pulping

A wide range of processes, combining mechanical and chemical methods, are available, but they represent a small part of the worldwide paper production. They generally involve chemical pre-treatment of the raw material, before mechanical treatment, to liberate the fibres. The yields of these processes are between those for mechanical and chemical pulping. The principal methods (of chemi-mechanical pulping) are chemi-thermo-mechanical pulping, neutral sulphite semi-chemical pulping and Kraft semi-chemical pulping.

Biomechanical and biochemical pulping

The use of micro-organisms for the treatment of wood chips, prior to mechanical or chemical pulping, is called biopulping. In mechanical pulping, the aim is to reduce energy demand and increase paper strength. In chemical pulping, biopulping should help to reduce the amount of cooking chemicals, to increase the cooking capacity, to enable extended cooking, resulting in lower consumption of bleaching chemicals. When applied to mechanical and chemical pulping, biopulping will help to further decrease the negative environmental impact of pulp and paper production.

Organosolvent pulping

This pulping method has been receiving increased attention in recent years. Solvent pulping offers a number of potential advantages over conventional pulping techniques, such as relatively low chemical and energy consumption, coupled with low capital costs and low environmental impact. Aqueous organic solvents, such as methanol and ethanol are used for delignification to produce a bleachable pulp which can be bleached with non-chlorine chemicals. Pilot-scale tests have given high pulp yields with strength properties similar to sulphite and Kraft pulps.

Pulp Bleaching

Viikari reported that endoxylanases decrease chemicals needed for bleaching Kraft pulp. Since then many researchers have confirmed their results and the technology is being utilised on commercial scale today. Hemicellulase treatments are effective on both hardwoods and softwoods. Although promising results were obtained initially with mannanases, they are not as effective as xylanases. Rossari offers *Rossapase*, a blend of xylanases, cellulases and hemicellulases for effective bleaching of pulp.

There are at least two explanations for how *Rossapase* enhance pulp bleaching. The first model suggests that it increase access of bleaching chemicals to pulp fibres by removing precipitated xylan. The uncoated fibres are then more susceptible to bleaching chemicals and lignin extraction. Essentially, this model proposes that xylan physically entraps lignin and chromophores in the pulp matrix.

The second model suggests that *Rossapase* release chromophores and lignin from the cellulosic pulp matrix by breaking covalent linkages between the hemicellulose and lignin moieties. Rather than postulating a physical entrapment, it holds that residual lignin and chromophores are chemically bound in the pulp.

Recent evidence supports the role of xylanase in breaking lignin-carbohydrate bonds. Hemicellulose hydrolysis release the bound chromophores and lignin, but xylan removal, *per se*, is not good because it decreases the pulp yield and if carried to an extreme, xylan removal can decrease pulp strength. So the objective is not to take out the xylan, but rather to remove the chromophores and lignin. The reason for using enzymes is to do that in a more specific, economical and environmentally benign manner.

Enzyme Characteristics

Several features determine which xylanases are effective for bleaching:

1. It is desirable to have a low molecular weight xylanase that is able to penetrate the fibre. An enzyme must have access to the substrate if it is going to release chromophoric groups and lignin.
2. It must have an appropriate pH optimum. Kraft pulps are alkaline and the enzyme must be able to function at the ambient pH.
3. The enzyme should have an alkaline isoelectric point (pl) in order to bind to negatively-charged fibre surfaces.
4. The enzyme must be thermal stable at the temperature prevailing in the pulp stock.
5. It should possess appropriate substrate specificity. The objective is to release chromophores and extract residual lignin—not remove the bulk xylan.

Enzyme penetration

Enzyme penetration throughout the cellulose fibre is essential in order to fully access and extract residual lignin. Complete penetration might not be achieved in a single enzyme application. Enzyme penetration is determined in part by molecular weight, but other factors, such as substrate binding, may be more critical. High substrate affinity may be a positive factor when enzymes are applied to pulps in a dilute suspension (because they would keep the enzyme in the pulp rather than in free solution), but at the relatively high consistencies used in bleaching plants (>10 per cent total solids), they could work against enzyme diffusion to the interior of the fibre.

Alkaline pH optimum

The second important characteristic is that the enzymes should have an alkaline pH optimum.

Alkaline isoelectric point

A third feature that's useful is an alkaline isoelectric point. With an alkaline pl, the enzyme binds more readily to the fibres under the pH prevailing in the pulp and this is important for determining its ability to attack the substrate. Pulp fibres are negatively charged due to the presence of sugar acids and when an enzyme has an alkaline pl and hence a positive charge, it binds more effectively.

Thermal stability

Most microbial xylanases are stable at 50°–60°C and this is within the range of prevailing temperatures in pulp mills. The combination of alkaline and thermal stability, however, is more difficult to come.

Classification of xylanases

Xylanases can be classified in three ways:

Based on molecular weight and pl

Information for this sort of classification can readily be obtained during purification and initial characterisation.

Based on crystal structure

This can be derived by a determination of DNA sequence. Xylanases can be structurally classified into family F (now known as glycosidase family 10) and family G (now known as family 11).

Based on kinetic properties, substrate specificity or product profiles

Virtually all xylanases are 'endo' acting, as readily determined by chromatography, but the more detailed determination of kinetic properties—measuring the relative reaction rates on various substrates and determining the kinetics of intermediate product formation—is much less common. Family 10 xylanases occasionally exhibit endocellulase activity; they generally have a higher molecular weight and they occasionally possess a cellulose binding domain. Members of family 10 act on both PNP-xylobioside and PNP-cellobioside however, the overall catalytic efficiency on PNP-xylobioside is about 50 times higher.

This suggests that family 10 enzymes act mainly on xylan. Even though all xylanases are endo acting, they can show variations in their product profiles. Some enzymes form predominantly xylose and xylobiose and others predominantly (or exclusively) form xylotriose and other higher oligosaccharide products. This difference appears to result from the number of substrate-binding subsites on the enzyme surface. The number of pyranose rings that the enzyme will bind effectively determines the nature of the oligoproducts. Family 11 xylanases are true xylanases. They don't have cellulase activity; they consistently exhibit a low molecular weight and they can have either a high or low pl.

Chromophore release

Some xylanases release chromophores from pulps in rather significant amounts, e.g. that from Streptomyces. The crude enzyme mixture releases material that absorbs very strongly in the UV and also into the visible region. The latter absorptivity shows up as a colour that one sees in the pulp. Chromophore release correlates with enzyme dose over the effective range of enzyme addition. In fact we can also see that the kappa number following enzyme treatment and alkali extraction decreases with the enzyme dose. Absorptivity is much greater in the UV region than in the viable region, so monitoring UV absorptivity is a sensitive means to determine the extent of enzyme action on pulp. In short, treatment with *Rossapase* has two indirect effects:

1. It is possible to wash out more lignin from the pulp.
2. The pulp becomes more susceptible to bleaching chemicals. This is called 'bleach boosting' and significantly reduces the need for chemicals in the subsequent bleaching stage.

Thus *Rossapase* helps to achieve the desired level of brightness of the finished pulp using-less-chlorine.

Papermaking

Paper mills can be integrated with a pulp mill or be independent. In the latter case, the pulp is generally dried before processing. Pulp preparation includes slashing, mechanical pre-treatments, blending of pulps of different qualities and addition of fillers and other additives to make paper. The type of process differs for the paper quality required. At the end of the pulping process is the machine room. The machine room (houses the pulp finishing equipment), which processes liquid pulp stock into multibale units of dry pulp sheets. The stock preparation system blends, screens, cleans, thickens and mixes the pulp stock to the appropriate consistency for sheet forming. The stock preparation system begins at the blend chest and ends at the machine chest. Bleached hardwood or softwood pulp is pumped from the appropriate bleached high-density storage tank to the blend chest. In the sheet forming and pressing processes, the pulp stock is formed and pressed into a continuous wide pulp sheet that is dried by gravity, vacuum and mechanical means. The sheet forming and sheet pressing processes are carried out by the forming section and the press section, respectively. In the sheet cutting process, the dried full-width continuous sheet is cut into stacks of smaller sheets ready for baling. Sheet cutting is carried out by the cutter-lay boy. Wrappers are used to wrap the pulp sheets. The bale finishing process involves conveying, weighing, pressing, wrapping, tying and stacking bales of pulp, weighing approximately 1/4 ton. The bales are then stacked or tied together into larger units, ready for immediate shipping by rail or storage (in the warehouse). The most traditional paper-forming equipment is the Fourdrinier Table. Prior to this machine, paper was made in single separate sheets. On a Fourdrinier machine, a wide jet of dilute furnish is applied from a head box onto the surface of a continuously moving wire screen. It is during this process that the pulp begins to transform into continuous paper. Water is removed by gravity and with the help of hydrofoils (or alternatively, table-rolls), vacuum boxes and vacuum applied at the couch roll. Reeling is a continuous process that is only interrupted by reel changes when the parent reel diameter reachs the target value.

Most mills in India are integrated pulp and paper mills, employing sulphate process, with the exception of a few, wherein sulphite process is used and its effluents are found to be coloured, high in pH and COD and moderate in BOD. They also contain a high concentration of suspended solids which are the most objectionable constituents from the pollution point of view. Chemicals are used to bleach pulps to make the paper white. The bleaching process has hydrogen peroxide brightening, chlorine bleaching, TCF (totally chlorine free) bleaching and ECF (elemental chlorine free) bleaching. This alters the chemical structure of lignin, lightening its colour, but leaving it in the pulp. The most widely used is chlorine. Ozone bleaching offers the option of producing totally chlorine-free pulp. To ensure the lowest evaluated cost of supplying a mill with ozone, a total mill impact study must be conducted to consider the overall oxygen and ozone requirements and fibre line constraints. Oxygen alkali extraction is one stage in a pulp bleach plant where oxygen is used together with alkali to lower the consumption of other more expensive bleaching chemicals.

Enzyme Applications in Pulp and Paper Processes

Pulping starts with the conversion of wood or agricultural materials into a flexible fibre that can be made into paper. The basic steps in paper recycling are pulping, washing, screening and flotation (to remove ink and other contaminants), bleaching and waste-water clean-up. Enzymes can be useful in the steps of each of these processes. Enzymes used in pulping can increase the yield of fibre, lessen further refining energy requirements or provide specific modifications to the fibre. Enzyme pre-treatments, using cellulase, hemicellulase and pectinase have been shown to enhance the Kraft pulping of sycamore chips and other pulp sources. This enzyme mixture allows for better delignification of the pulp and savings in bleaching chemicals, without altering the strength of the paper. Other enzymes such as laccase and protease have been reported to reduce energy requirements in mechanical pulping. Laccase treatment has the additional benefit of increasing fibre bonding, which enhances the strength of the paper. Kraft pulp has been treated with cellulases and xylanases and both enzymes have reduced the energy required for further refining. The cellulases must be used carefully so that they do not reduce the strength of the fibres. Xylanase treatments are more effective on unbleached pulps, than they are on bleached kraft pulps. Deesterifying the soluble O-acetyl-galactoglucomannans of Norway Spruce, using an acetyl esterase, was shown to precipitate the galactoglucomannans onto the fibre and increase the yield of fibre from the process. Endoglucanase treatment decreased the viscosity and chain length and increased the reactivity of a pulp made from eucalyptus and acacia. The endoglucanases were more efficient at hydrolysing the pulp than were cellobiohydrolases at the same protein dosage.

The pulp and paper industry is applying new, ecologically sound technology in its manufacturing processes. Emerging technologies, based on sustainable use of renewable resources hold promise for the rejuvenation and growth of the pulp and paper industry. An enzyme technology based on microbial xylanases has helped to achieve this goal by reducing or even eliminating the need for chlorine in the manufacture of elemental chlorine free (ECF) and totally chlorine free (TCF) printing and writing paper grades. Xylanase treatment can improve lignin extraction, alter carbohydrate and lignin associations or cleave re-deposited xylan. There are some reports of direct brightening with xylanase treatments, but most of the direct bleaching with enzymes is done using oxidative enzymes that directly attack the colour-producing compound in the lignin. The enzymes that attack the lignin components of the fibre are oxidative. Peroxidases use hydrogen peroxide and laccases use oxygen and eventually react with the lignin-derived moieties. Lipases can control the accumulation of pitch during the production of paper from pulps with high resin content, such as sulphite and mechanical pulps from pine.

The potential applications of xylanases and the need for enzymes that might be more appropriate for use in pulp bleaching has caused many researchers to look for xylanases in different organisms and environments. Since the Kraft process results in pulps that are alkaline and at higher temperatures, enzymes that do not require adjustment of temperature or pH are better suited to this process. Thermophilic and alkaline sources for these enzymes have been reviewed recently. A substantial amount of work has been devoted to isolating and cloning new xylanases from all sources. Using xylanase as a search term in scientific bibliographic databases, from 1991 to 2000 revealed an average of 188 papers per year. Though not all these papers were about new xylanases, a better indicator may be the number of xylanase-encoding DNA sequences that have been entered into data banks. However, a proper search for novel xylanases is underway.

Variations in Other Enzymes Used for Pulp Bleaching

Xylanase removes reactants with the bleaching agent or obstacles to the bleaching action. A similar effect was shown for cellulase treatment and colour removal in recycled yellow pages. Enzymes can be used to enhance dewatering (drainage) rates, facilitate contaminant removal and increase bond strength in recycled fibres. Drainage resistance of secondary fibres adversely affects sheet formation, slows operation of the paper machine and increases drying energy. Deinking and contaminant removal, likewise can benefit from enzyme treatments that facilitate separation of fibres in the washing and flotation processes.

Refining and Drainage

Endoglucanases can enhance the drainage rates for recycled fibres beyond what can be attained by polymer addition. Enzyme treatments require additional retention time and sufficient mixing to affect the surface structure of the fibre. At optimum pH, temperature and consistency and with appropriate mixing, successful enzyme treatments can be achieved.

Enzyme addition after refining increases freeness, thereby enabling the operation of the paper machine at a faster rate, enables lower head box consistency for better paper formation, improved strength, decreased basic weight and increased use of recycled fibres. In this case, care must be exercised to keep treatments at a low dose, because endoglucanases that attack amorphous cellulose also cause rapid loss of fibre strength.

Deinking

One of the main applications of enzymes to recycled fibres has been to remove print. More than 70 per cent of paper consists of uncoated papers that are printed with copy and laser printer toners, which are often difficult to remove by conventional, alkaline deinking processes. Cellulases are particularly effective in facilitating the removal of toners from office waste papers. Both cellulase and a non-ionic surfactant are required for effective deinking. Recently, it has been shown that chemically modified cellulases are more effective for recycled fibres than unmodified cellulases. These researchers added polyethylene oxide co-polymer derivatives and maleic anhydride to the amino groups of the enzyme and then used the modified enzyme to treat recycled fibres. Paper made from fibres treated with modified cellulase had better physical properties, such as tensile strength and internal bond formation, than fibres made from unmodified cellulase. The modified cellulase was more effective in separating ink particles from fibre, during flotation deinking. The treatment is more effective with alkaline sized papers, after adjusting the pH to 3.0 and the addition of a hydrocarbon surfactant.

Despite greater availability of MOW (mixed office waste), its conversion to higher grade pulp is much restricted owing to difficulties in removing toner inks, which result into bigger visible specks in the final product. Deinking of non-impact printed papers (xerographic/laser) is very difficult, even after several cleaning/screening cycles in chemical deinking processes. Current deinking methods used in recycling MOW mostly involve extraction under alkaline conditions (Fig. 20.1).

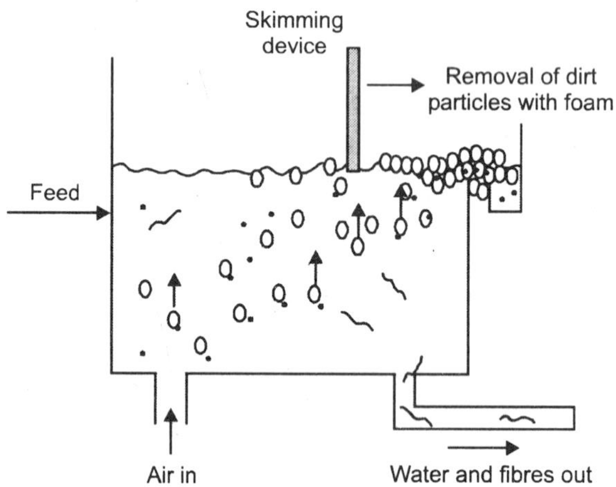

Fig. 20.1. Deinking of recycled paper.

Conventional deinking

Conventional deinking processes use alkali and detergents. The recovered paper is first dissolved in water and separated from the non-fibre impurities. The fibre is then progressively cleaned by flotation process, where air is blown into the solution. The ink adheres to air bubbles and rises to the surface from where it is separated. After the ink is removed, the fibre is bleached with hydrogen peroxide. Flotation deinking is the most important method of recycling paper worldwide.

Enzymatic deinking

Enzyme enhanced deinking processes show promise for avoiding excessive use of alkaline chemicals and surfactants and yield better quality paper in terms of brightness or non-yellowing. Several research studies have shown that regardless of ink types, enzyme treatment tends to reduce ink particle size when coupled with mechanical pulping, compared to conventional deinking.

In addition to removal of ink, it also helps improve the strength of paper sheets and freeness. It reduces fine content of the recycled fibre.

A variety of enzymes are being used for deinking, viz. cellulases, hemicellulases, lipases, esterases, pectinases, amylases and lignolytic enzymes. Rossapase is a blend of all the useful enzymes.

Hemicellulases, cellulases and lignolytic enzymes in Rossapase alter the fibre surface or bond in the vicinity of ink particles, thereby freeing the ink for removal by washing or flotation. The toner particles adhering to the paper surface released by the enzymatic treatment are further separated from the pulp by flotation.

Deinking with enzymes at acidic to neutral pH reduces the overall chemical requirements and minimises the yellowing of reclaimed paper normally associated using alkaline deinking. It can reduce the pulping time, thereby saving energy and potentially increasing production. Besides decrease in the ink particle size, milder alkaline conditions prevalent in enzymatic deinking bears a positive impact on control of stickies at mill scale. Last, but not the least, reduction in quantity of non-ionic surfactants, which biodegrade slowly in effluent treatment plants (ETP) is an additional positive environmental effect.

Enzymatic deinking: FAQ

With acid-sized (sulphite) paper, good results are obtained using a *Trichoderma reesei* acid cellulose. With alkaline sized papers, one has to adjust the pH within the activity range of the enzyme. It is more economical to use a neutral cellulase enzyme (*Humicola insolens*) in order to save alkali. Recycled fibres contain large amounts of calcium carbonate and its alkaline pKa generally necessitates the use of a neutral to alkaline enzyme.

We get high CMCase (carboxymethylcellulase) enzyme from bacteria and FPase (filter paper activity) from fungi. This mixture of bacterial and fungal cellulase prove a best choice for deinking. Synergism between endo- and exo-acting cellulases generally gives the highest overall activity.

Similarity between the mechanisms in enzyme deinking and enzyme detergency: Principal enzymes used in detergents are proteinases and lipases. Cellulases are also used to reduce 'pill' formation in cotton fabrics. However, cellulases may also help to release stains from amorphous cellulose or cellulose microfibrils on the fabric surface. The mechanism for enzymatic deinking of paper fibres is similar, but it depends, in part, on the types of contaminants being removed. The inks do not usually stain the fibres themselves, but rather form particles or micro particles that are either embedded in the paper fibre or intertwined and adherent with the fibres.

In the case of small particles that are ground into the fibres, the cellulases could release them by removing the microfibrilar matrix in which they are embedded. This can be done fairly easily. Once the microfibrils are gone, the fibres would have a lower specific surface area and would release water (and particles) more readily. In the case of larger particles (such as toner particles), which are trapped in the plastic matrix (rather than the particles trapped in the fibre), either the fibres break off from the toner particle or the presence of the enzyme allows the fibre to slip free from the adherent toner. The mechanical action of the pulper is synergistic and important to remove toner particles by cellulases.

RECENT ADVANCES IN THE APPLICATION OF ENZYMES FOR PAPER AND PULP

Some of the recent advances in the application of enzymes for paper and pulp are:

1. Effect of peroxidase and phenol oxidases on bleaching of kraft pulp.

2. Use of catalase in hydrogen peroxide bleaching of kraft pulp.
3. Effect of enzymic pre-treatment on the diffusion of sodium hydroxide in wood.
4. Application of hemicellulolytic enzymes from extremely thermophilic bacteria for pulp bleaching.
5. Enzyme pre-treatment on conventional kraft pulping.
6. Enhancing the freeness of pulp made from secondary fibre by using a mixture of cellulase and pectinase enzymes to the pulp.
7. Bleaching of kraft pulp with lignin degrading enzymes.
8. Bleachability of recycled fibres deinked with enzyme preparations.
9. Use of cellulase to improve surface smoothness in the production of single face glossy paper.
10. Deinking of mixed office paper, old newspaper and vegetable oil based ink printed paper using cellulases, xylanases and lipases.
11. Enzyme deinking of non-impact printed white office paper with α-amylase.
12. Deinking of laser printed paper using enzyme mixtures of cellulases and xylanases.

Enzyme and Micro-organism Uses in the Pulp and Paper Industry

Deposit removal

Water used for the paper process often contains nutrients and is kept in conditions conducive to microbial growth. Microbial growth or deposits can be a problem when mills have closed their water loops, providing the possibility of concentrating organisms and nutrients. The micro-organisms that cause these problems can be many different species, both aerobic and anaerobic. Slime formation on equipment is often the result of such conditions. Slime can cause many problems, such as pump failure and breaks and blockages in paper machines. The first approach to this problem is to prevent slime formation by adding biocides. Once slime has formed, its removal is an important operation in the paper-making industry and often requires shut-down and caustic boil-outs to properly clean the slime. Much of the slime or biofilm, is composed of carbohydrate and protein polymers. Enzymes such as amylases and proteases have been found useful in enzymatic boil-out of slime-encrusted equipment. Organisms have been found that could degrade these polymers, depending on the polysaccharide and organism used to make it. A similar approach was taken with colloidal material that could be reduced by culture filtrates of a fungus *T. versicolor* grown on the colloidal material. Lipase has been used to remove lipid-based materials in these suspensions for a number of years. Removal of extractives by laccase-based catalysis of polymer formation and lipases provided an additional removal of organic materials.

Fibre modifications

Interest has been growing in the enzyme-assisted modification of fibres. Enzymes such as peroxidases and laccases have been used to polymerise or co-polymerise materials with wood-based fibres. Attachment of guaiacol sulphonate, using laccase made lignin more water soluble. The attachment of 4-hydroxy-phenylacetic acid was also demonstrated, but no differences in lignin solubility could be distinguished. High kappa pulps were modified with 4-hydroxyphenylacetic acid in the presence of laccase. This modification increased the carboxylic acid groups, water retention, tensile strength and burst strength of the resulting paper. The same treatment with tyrosine and guaiacol sulphonate had only minimal effects. Laccase catalysed the polymerisation of acrylic compounds onto lignosulphonates. Peroxidases and other oxidases have been used to modify polymers, as well. While these enzymes are able to catalyse reactions efficiently, they operate by free radical addition and the bond formation can be predicted, but not necessarily controlled and directed. This is especially true when the polymer being used is complex, like lignin and woodbased fibres. Cellulases can increase fibre flexibility and thereby, improve pulp properties. Cellulase and xylanase treatments can increase the density of the resulting hand sheets made from Kraft pulps.

Compost

Composting is defined as the decomposition of organic waste material, such as animal waste, crop residues, garbage, sludge and other human activities wastes, carried out by successive microbial populations, combining both mesophilic and thermophilic micro-organisms of bacteria, actinomycetes and fungi in a warm, moist environment. Because composting is a biological process, environmental factors influencing the activity of the organisms determine the speed and the course of the composting cycles. The most important factors are particle size of the material, moisture content, aeration, temperature and initial carbon to nitrogen ratio. The optimal conditions for these factors are 35°C/55°–60°C for mesophilic/thermophilic decomposition phases and 25–30:1 for C/N ratio. The delignification of plant residue severely increases the formation of compost, prepared by slurry of pulp and paper mill effluent treated by fungi and applied for reclamation of chromium and chlorinated phenolic compounds.

Degradation of Recalcitrant Compounds

Some enzymes also seem to have the potential to remove colour and AOX from pulp and paper mill effluents. Peroxidases, laccases, etc. are the most important. The use of microbial or enzymes-based treatment offers some distinct advantages over physical and chemical AOX precipitation methods, in that only catalytic and not stoichiomatric amounts of the reagents are needed. It has been reported that lignin type peroxidases, secreted by white rot fungi are involved in the degradation of a whole range of organic pollutants, including PCB, DDT and linden, chlorinated anilines, polychlorinated phenolics, including PCP and chlorolignins and many more.

PROSPECTS OF GENETIC ENGINEERING IN BIOCONVERSION

Protein Engineering

Enzymes that attack cellulose and xylan often have portions (domains) of the protein that bind to cellulose. Binding domains specific for xylan have been identified, but a single amino acid change can alter this specificity from xylan to cellulose. Xylanase catalytic sub-units with CBDs (catalytic binding domins) as part of their overall tertiary structure are common. The biochemical role of a CBD is to keep the enzyme catalytic unit close to the substrate surface. The need for the CBD in fibre processing might depend on the uses of the enzyme. The role of CBDs in the hydrolysis and modification of chemical pulp has been examined. The presence or absence of CBDs did not affect the action of enzymes against soluble substrates. However, enzymes with a CBD enhanced fibre beating to a greater extent, than did enzymes without the CBD. Since beating is thought to increase micro-fibrillation, the CBD could play a role in opening up the cellulose crystallite structure.

Modifying Enzymes to Attain Activity under Specific Conditions

The cloning of specific enzymes from described organisms and DNA sequences is becoming a routine. Many new xylanases, cellulases and other fibre modifying enzymes have been recently cloned, produced and characterised as per their activities. Often these enzymes can be produced without ever having to deal with the natural host. With a plethora of genomic sequences being discerned, many new enzymes might be available by searching the new sequences for regions that code for specific enzymes. This directed approach has provided new potential enzymes for use. Degenerate primers coding for conserved regions of enzymes have been used in PCR (polymerase chain reaction) studies to produce gene fragments encoding these enzymes. This technique has been used to produce novel enzymes from unknown organisms present in the environment. Thermophilic organisms must have active enzymes for all their cellular functions. The nature of the environment in which these organisms are found dictates that these enzymes must be thermostable. Many enzymes have been cloned from the DNA of these organisms. Although a thermophilic enzyme can tolerate higher temperatures, it acts slower at lower temperatures. Thus, a thermophilic enzyme should be sought when a process has to remain hot or requires the use of solvents, but many other alternatives exist when the process does not have to be heated and is at a non-denaturing pH. In fact, using a thermophilic enzyme might cost more for a given treatment, if the fibre has to be specifically heated for the enzyme to be effective. When an enzyme has been studied to the extent that the crystal structure is known, some specific planned changes can be made and the response in the activity of the protein predicted. The introduction of disulphide bonds through site-specific mutations has increased the thermo stability of a *B. circulans* xylanase. All of the designed changes provided, increased thermostability. However, only one out of the eight changes, provided an increase in the optimal temperature of activity. Site-specific changes were also introduced into the MnP of *P. chrysosporium*, which allowed the enzyme to tolerate higher concentrations of hydrogen peroxide than the wild type. The increased tolerance was somewhat offset by a decrease in the affinity of the enzyme for hydrogen peroxide. Site-specific changes require a knowledge of the sequence of the proteins and some concept of how changes can be imparted. Comparisons of the sequence of *Streptomyces lividans* xylanase to thermostable xylanases from the same family provided a road map to increasing the thermostability of the *S. lividans* enzyme. The expression of xylanases in *E. coli* and a screening assay for thermo stability also helped to identify more thermostable variants of the *S. lividans* xylA xylanase. The first 29 amino acids of the *Thermomonospora fusca* xylanase A enzyme imparted increased thermostability to *S. lividans* xylanase B and also *T. reesei* xylanase II. An increase in optimal operating temperature was found in variants of *S. lividans* xylanase B by further random gene shuffling, using random fragmentation of the *T. fusca* and *S. lividans* genes. The thermostability and thermal activity of the modified *T. reesei* enzyme has been incorporated into the BioBrite HB60C, which shows improved activity in ECF bleaching. Other domains from the family 10 xylanases have also been shown to have thermostabilising effects. Domains from *C. fimi*, *Thermomonospora alba*, *Neocallimastix frontalis*, *Streptomyces olivaceoviridis* and *Fibrobacter succinogenes*, have all been studied. The results from these studies indicate that the domains often convey their function to new chimeric proteins. This could be the ability to bind to cellulose, bind to other proteins, change active site domains and change substrate binding domains. The shuffling of modules or domains can again produce proteins with altered characteristics. An example of a modified enzyme would be the combination of cellulase activity with that of (1–3,1–4)-β-glucanase activity. This multienzyme was able to degrade the specific linkages present in barley glucan. The altering of an enzyme, using random mutations is powerful and does not require detailed structural information. The *Coprinus cinereus* peroxidase was modified, using both site directed and random methods. The manual combination of individual site directed mutants resulted in an enzyme with 110 times the thermal stability and 2.8 times the oxidative stability. However, when *in vivo* shuffling of mutations through yeast homologous recombination was used, an enzyme with 174 times the thermostability and 100 times the oxidative stability was obtained. The directed evolution of an enzyme can provide a powerful tool to attain enzyme optimisation. The technique requires a useful host organism and a screen to select desired clones. Directed evolution increased the alkaline activity and thermostability of a *Neocallimastix patriciarum* xylanase. Error prone PCR, mutator strains, DNA shuffling and domain swapping can all be used when the selection is powerful and specific. *In silico* design of an enzyme may lead to further development in this area. Intimate structural knowledge of an enzyme's active site is required for these types of modifications. With an enzyme designed by *in silico* methods, the principles of protein folding, bond angles, hydrophobic cores, hydrogen bonding salvation and other forces can provide us with so many possibilities (many of which have not been produced by nature) that they cannot be all tested.

Commercial development of these applications requires a better knowledge of enzyme mechanisms, then action on fibres, development of improved processes for their use and changing enzymes to function better under existing operating conditions. In addition to the cost implementing enzyme technology, the changes require thermo-tolerant and alkalophilic enzymes or altered down stream processing, all of which might create expenses. Today, these traditional means are available, and in addition, recombinant DNA has allowed the cloning of enzymes from known producers, difficult-to-culture micro-organisms and even unculturable organisms. These cloned enzymes can be modified for temperature, pH and stability using similar techniques. Random mutagenesis, gene shuffling, directed evolution and site-specific changes in the active site or supporting structure, have all allowed the biotechnologist to change the natural enzyme to fit the need. The advent of *in silico* design adds another dimension to the possibilities by allowing the testing and design of enzymes that are not presently found in nature.

Biotechnology has the potential to increase the quality and supply of feedstocks for pulp and paper, reduce manufacturing costs and create novel high-value products. Novel enzyme technologies can reduce environmental problems and alter fibre properties. Because the pulp and paper industry is capital-intensive, with facilities specific to the tasks, new technology must either reduce expenses or fit easily into the existing process design. Pulp and paper companies are reluctant to build or expand plants when the overall industry has enough capacity to satisfy market demands. Nevertheless, the industry has embraced enzymes for use in the paper making process and has responded to environmental and economic concerns. The industry has greatly reduced chlorinated aromatic by-products that can be formed during pulp bleaching, first by reducing the amount of residual lignin in pulps and second by turning to other bleaching agents.

Chapter 21

Tanning Industry

INTRODUCTION

Since pre-history humans have obtained part of their nutrition from animals and processed the animal skins into products for their daily needs. The skins served as clothing and footwear, as a cover against rain and cold, and as protection against injury. Animal hides were also crafted into saddles and reins, belts and bags, and boats and tents. However, these articles were made of raw skins and were not very durable. Raw hides rot when wet as a result of rapid bacterial growth; alternatively, during periods of dryness and heat, the skins lose their malleability and become hard and stiff. Soon it was learned that when the hides were rubbed with fats or oils, they remained supple and became water-repellent. Further, smoking fresh hides could prevent their rapid disintegration. Treating the skins with salts and plant extracts produced an effect that resembled what is now called tanning, which gave the hides some resistance to microbial attack.

Illustrations and sculptured reliefs from ancient Egypt provide evidence of the early technology of leather manufacture. Wall paintings from burial chambers in Thebes show Egyptian tanners at work.

Some leather articles found in burial chambers in the Valley of the Kings are of excellent quality and remained well preserved over the centuries because of the arid climate. Impressions of floral ornamentation and the names of kings were found on strips of fine-coloured leather. Painted and gilded leather straps were found with the chariots in the tomb of King Tutankhamen. Scribes used a thin parchment-like leather in addition to papyrus. That careful preparation and tanning of hides occurred can be concluded from the condition of these finds. Simple methods for pre-treatment and cleaning of hides and pelts, using materials that were quite original and readily available but hardly pleasant (excrements, urine), were employed during Babylonian and Assyrian times. Therefore, it can be assumed that the vessels and the wall paintings were used for brine curing, soaking, liming, and bating hides. For the subsequent tanning an extract from the pods of the Nile acacia, or *Acacia nilotica*, was used.

Even the most ancient cultures along the Euphrates and Tigris rivers knew how to manufacture leather, but little is known about the methods they employed. Minoan art objects show persons wearing leather clothing and semi-high boots.

The practice of employing feces from animals such as dogs or pigeons as bating material (puering) to loosen hair and soften skin was still employed early in the twentieth century. Later, attempts were made to replace this 'enzymatic' process with chemicals as bating agents. It was later found that the active material in the fecal matter was trypsin. Wood ash was used to loosen hair. The German term *Äscher* is derived from this activity. Lime and sodium sulphide were added for dewooling and dehairing, as well as for disrupting the hides, collagenous fibre network. However, further conditioning (bating) was required for cleaning the skins from residual epidermis, hair, and pigments.

Wood and Law at Merck GmbH tested pancreatin as a bating agent; however, the results were unsatisfactory. In 1904, Otto Röhm became interested in bating methods as part of the tanning process. Initially, he tried to replace the feces with a dilute ammoniacal solution; subsequently his interest turned to the proteolytic enzymes.

Historically, the birth of the application of enzymes in industry commenced with Dr. Röhm's dairy entry on May 22, 1907, asking whether extracts from the pancreas whose enzymes affect the digestion of foods could also act as bating agents.

To employ pancreatic protease, it was essential to first empirically determine the effect of the reaction milieu and additives (activators). Addition of ammonium salts proved particularly favourable in that they could be adjusted for optimal effectiveness and also exhibited a deliming effect. Röhm was granted a German patent for an enzymatic liming agent. This development involved the enzymatic removal of hair and wool along with their roots.

The process is highly efficient and has been employed continuously; it is still being used with modifications at the present time. Its significance is borne out by the fact that the chemical environmental burden from leather manufacturing is virtually nil.

SKIN STRUCTURE

Animal skin consists of three different layers:

1. The epidermis, which constitutes about 1 per cent of the total skin. It includes hair roots and sebaceous (oil) and sudoriferous (sweat) glands. It is subdivided into three additional layers: the *stratum corneum*, the outer epidermal layer or horny layer, followed by the *stratum granulosum*, or granular layer, and near the dermis or *cutis vera*, the *stratum mucosum*, or mucosal layer.
2. The dermis, *cutis vera*, or *corium* makes up 85 per cent of the skin. It is composed of the *stratum papillare*, or papillary layer, making up 20–80 per cent of the dermis, with the mass of collagen fibres. The *stratum reticulare*, or reticular layer, consists largely of connective tissue.
3. The *tela subcutanea*, subcutis, or hypodermis makes up about 15 per cent of the skin, and is a loose fibrous layer containing interstitial fat, connective and muscle tissue, blood vessels, and nerves.

Chemically, the skin consists of 60–65 per cent water, about 30–32 per cent protein, approximately 10 per cent fat, and 0.5–1 per cent minerals. Figure 21.1 illustrates the main components of mammalian skin. Leather is made from approximately 150 different skins and pelts. Each has a specific surface texture that the tanner calls grain.

ENZYMES FOR LEATHER PROCESSING

A large number of different proteases that have, over time, been used in the beamhouse in one way or another are cited in the literature. In addition to the pancreatic proteases, acidic, neutral, and alkaline bacterial and fungal proteases, as well as papain and bromelain of plant origin, are used.

Many experiments have been conducted to analytically define and standardise the enzymatic methods that could be correlated with the practical procedures in the beamhouse. Because this is not yet possible, the tannery chemists have agreed to a traditional definition of proteolytic enzymatic activity. This is the Löhlein-Volhard unit (LVU).

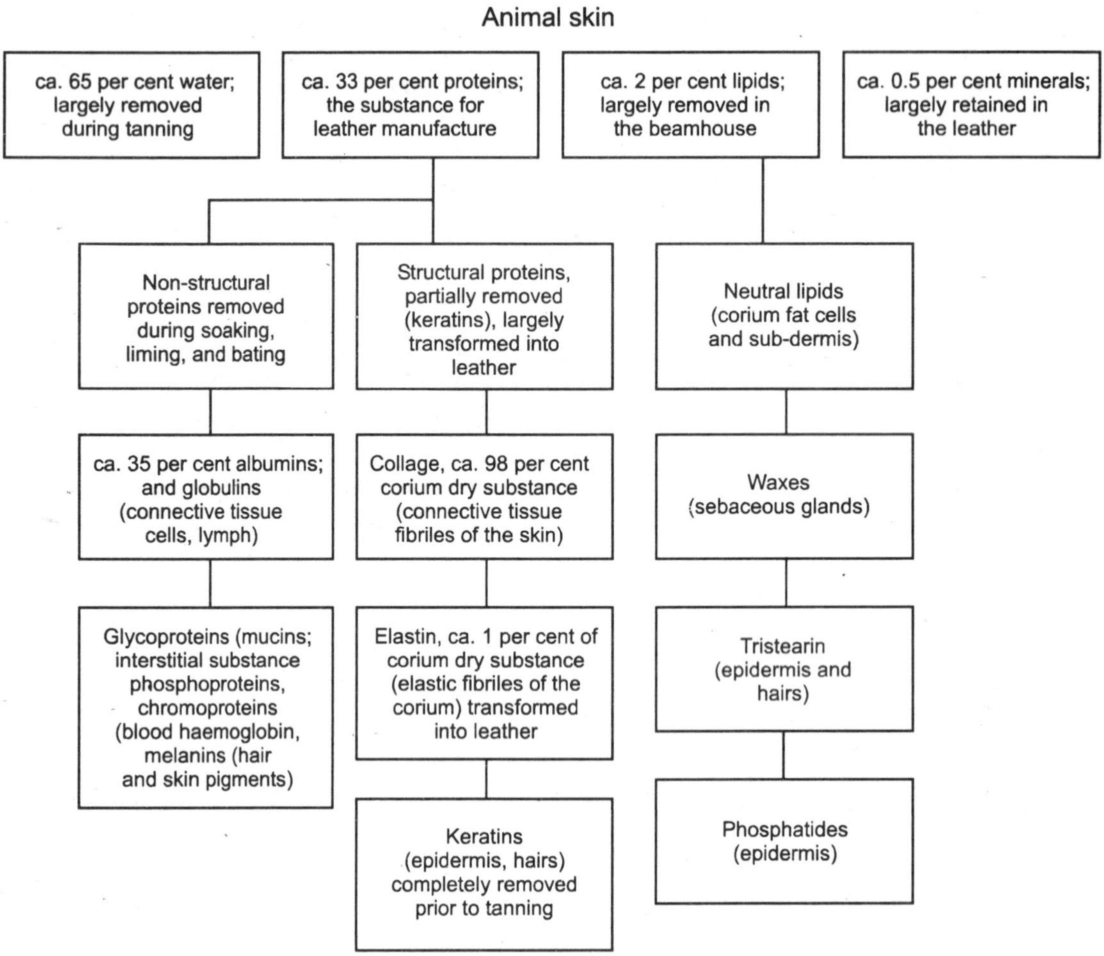

Fig. 21.1. Chemical composition of the skin.

One Anson unit (AU) of a pancreatic protease preparation corresponds to about 1,16,000 LVUs. This is the specific activity of 1 gram of enzyme preparation. Pancreatic trypsin with 3 AU/g corresponds to about 3,50,000 LVU/g.

Table 21.1 shows the relationship between the LV units and the data obtained from proteolysis with other methods and substrates. A comparison of the enzymes, standardised to 1,00,000 LVU/g protein; shows that the bacterial enzymes exhibit their highest activity with haemoglobin and gelatine substrates. The results obtained with the Kunitz method show large variations in activity between the various bacterial proteases. This is due to the different pH optima of the enzymes, a fact that is not addressed in the LVU method.

Leather Manufacture

The manufacture of aniline leather is shown in Fig. 21.2.

Table 21.1. Comparison of various proteases using different analytic methods.

	Proteases[a]						
	Bacterial			*Fungal*		*Pancreatic*	
Method (Substrate)	*I*	*II*	*III*	*I*	*II*	*I*	*II*
Anson haemoglobin	1.3	1.0	0.75	0.8	0.85	0.85	0.65
Modified Kunitz test							
pH 8 (PU method)[b]	190	90	180	185	250	100	170
pH 7	120	180	240	–	–	–	–
Gelatine viscosity test	18000	–	25000	3000	3800	7500	4500
Willstätter casein	26	21	30	–	–	20	–
Gelatine-formaldehyde titration	–	–	–	200	105	–	–
Determination of soluble N_2 after precipitation	29000	19000	26000	22000	30000	24000	22500

[a]Enzyme source and preparation number from various microbial or pancreatic fractions; enzymes standardised to 1,00,000 LVU/g protein (LVU = Löhlein-Volhard unit).

[b]Protease unit.

Raw-stock preservation

Because skins and hides rapidly decompose after slaughter, they are immediately air-dried, or dehydrated and cured by packing with dry salt. During this first step of the hide-to-leather conversion process, the raw stock is extremely susceptible to microbial attack.

Microbial decomposition of hide proteins results in a significant loss of leather quality. Adequate initial brine to curing of the hides is especially important because the shipping distances for raw for stock have increased over the years.

Soaking

On arrival in the tannery, the hides are re-hydrated in the so-called beamhouse. During the pre-tanning process, blood, dirt, salt, and fat residues are removed from the hides in an initial soaking operation. The main soak serves to restore the original moisture and hydration state of the hides. In addition, non-fibrous proteins such as globulins and albumins are removed during the main soak.

To improve and efficiently operate the complex soaking process, additives are introduced. They include surfactants (surface-active agents), reducing agents, preservatives, and enzymes.

Proteolytic enzymes effectively facilitate the soaking process. In addition, lipases have also been recommended for the soak. Most of the technical enzyme preparations can be combined with non-ionic detergents.

The synergistic effect observed with this combination is clear if one considers that the skin proteins are surrounded by lipids that must be removed in order to make the substrate (i.e. proteins) accessible to the proteolytic enzymes.

An important result of enzymatic soaking is the removal of undesirable (scud) matter from the skin. This consists of pigment cells and sebaceous and sweat glands. Such cleaning is a pre-requisite for manufacturing distinctly grained high-quality aniline-dyed leather.

It should be mentioned that removal of lipids with surfactants during pre-soak and main soak is preferred to removal during the subsequent liming step. Penetration of the lime into the hide is greatly facilitated by previous enzymatic soaking, a procedure that eliminates the use of chemicals that may subsequently present an environmental pollution problem. Soaking will also stabilise the hides against microbial growth. This can be done quite economically by adjusting the pH to 9. In cases of extreme microbial contamination, a combination of preservatives and surfactants are recommended.

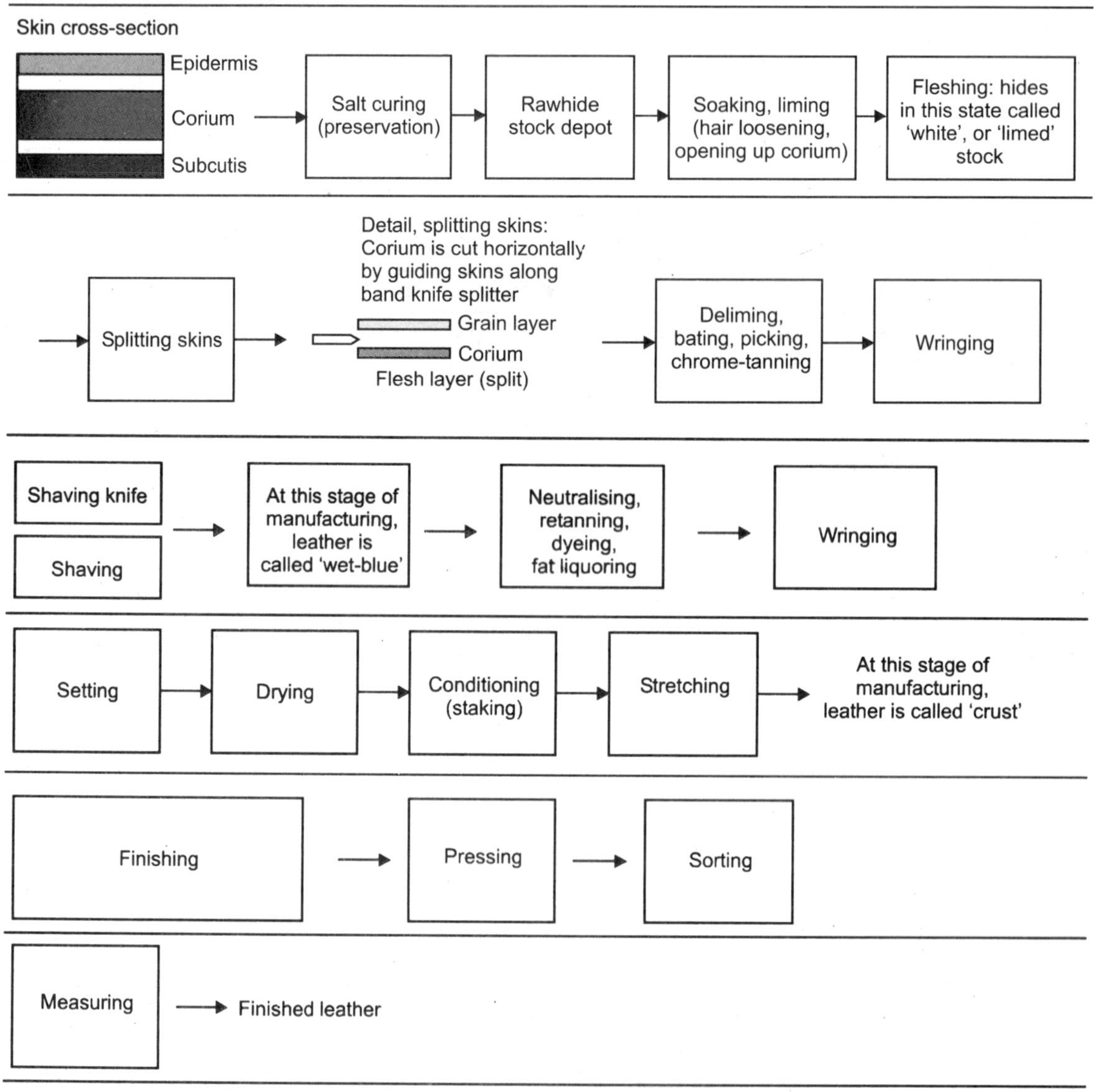

Fig. 21.2. The process of manufacturing aniline leather.

Enzymes for the soaking process

Pancreatic protease and pancreatin are employed along with neutral and alkaline bacterial proteases; in special cases, alkaline fungal proteases and lipases may also be used. In this case, as in other beamhouse operations, it becomes important that these proteases only the non-collagenous proteins, leaving the collagen essentially intact. Indeed, highly substrate-specific proteolytic enzymes can selectively remove some proteins while leaving others unaffected.

Generally, only minimal enzymatic activity in the soak is required to produce satisfactory results; for instance, 50–200 grams of a protease with 1,00,000 LVU/g protein is adequate for 1 tonne of hide. Pancreatic enzymes appear to be the enzymes of choice that can produce a satisfactory amount of skin structure (corium) loosening with a minor effect on collagen (Table 21.2).

For barrel soaking, the volume required ranges from 100 to 400 per cent, the pH from 7.5 to 8.5, and the temperature from 18° to 22°C. The duration of soaking is 6 hours for a salt-cured raw stock, and 8–12 hours for flint-dried hides.

Table 21.2. Skin loosening and opening up corium with various proteases.

Bating enzymes	*Löhlein-Volhard (units/g protein)*	*Loosening (pH 8.5)*	*Opening Corium (pH 8.5)*
Pancreatic	500	228	95
Pancreatic	500	245	117
Bacterial	500	62	204
Bacterial	500	63	118
Bacterial	500	52	34
Fungal	500	77	120
Fungal	500	68	79

Example: Enzymatic soak for upper-footwear leathers

1. Raw stock:
   (a) Salted cattle hides with a weight class of 30–39.5 kg.
2. Pre-soak (vat):
   (a) 150 per cent water at 32°C; duration 2 hours at 4 rpm.
   (b) Batch is drained.
3. Main soak (vat):
   (a) 150 per cent water at 28°C.
   (b) 0.5 0.7 per cent soaking agent from *B. subtilis* with an activity of 1000 LVU/g (Pellvitt F [RM]).
   (c) 0.4 per cent soda (calcined); duration 4 hours at 26°C.
   (d) Density: 3.0°Bé (Baumé).
   (e) Batch pH 9.8.

The hides re-hydrated during soaking can be limed in the same batch. The percentage data shown are based on the weight of salt-cured hides.

Example: Enzymatic soak and dehairing of goat skins

1. Presoak:
   (a) 600 per cent water, 4 per cent soda (sodium carbonate), 0.1 per cent preservative, and 0.8 per cent nonionic surfactant.
   (b) Duration: 12–18 hours, pH 9.3–9.6 at 27°C.
   (c) Batch is drained.

2. Enzymatic soak:
   (a) 500 per cent water at 28°C.
   (b) 0.5 per cent nonionic surfactant.
   (c) 0.6 per cent soaking agent from *B. subtilis*, *A. sojae* and pancreas with a total activity of 4000 LVU/g (Erhazym C [RM]).
   (d) 1.5–2 per cent soda (calcined); agitated for 2 hours, pause 1 hour; then alternating cycle of 2 minutes agitation and 60-minute pause.
   (e) Duration 12–18 hours, pH 9.3–9.6 at 27°C.
3. Trimming.
4. Stretching.
5. Weighing.
6. Dehairing (vat):
   (a) 250 per cent water at 28°C.
   (b) 1.7–2.3 per cent enzyme mixture of bacterial protease and protease from *A. sojae* with a combined activity of 6000 LVU/g (Arazym M 100 [RM]).
   (c) 2.5–3.0 per cent soda (calcined).
   (d) 0.5 per cent non-ionic surfactant.
   (e) 0.05–0.1 per cent preservative that must be enzyme-friendly.
   (f) Agitation for 60 minutes, 60-minute pause; then alternating cycle of 2-minute agitation with 2-hour pause.
   (g) Duration 16–20 hours, pH 9.3–9.5 at 27°C.
   (h) Batch is drained and dehaired.

Liming

The soak is followed by processes in which the hair and epidermis are removed and the corium is made accessible to tanning agents with an alkaline solution that causes the hide to swell. In addition, these processes also remove residual non-collagenous hide proteins after the soak. This is accomplished with a mix of substances containing lime, Na-hydroxide, and reducing agents. Otto Rôhm had developed special enzyme preparations consisting of alkaline fungal proteases for this liming process. The proteases used for the liming process must be especially stable in alkaline solutions.

Example: Liming mix for furniture leather

Liming solution (vat)—in a soak batch that contains protease, the composition may be as follows:

1. 0.4 per cent protease from *B. alkalophilus* with 500 LVU/g (Pellvitt KAB-P [RM]).
2. 1.0 per cent sulphide-free liming agent (Erhavit F [RM]).
3. 1.0 per cent of 72 per cent Na-sulphydrate.
4. 1.0 per cent Ca-hydroxide; agitated at 4 rpm for 45 minutes; then pause for 30 minutes.
5. 0.8 per cent of 60 per cent Na-sulphide.
6. 2.0 per cent Ca-hydroxide: agitate at 4 rpm for 1 hour, then every hour for 2 minutes at 2 rpm.
7. Duration 20 hours; final agitation for 15 minutes.
8. Then the batch is drained.

Dehairing and dewooling

The proteins of the epidermal basal membrane and the pre-keratins of the hair root shaft are easily lysed by alkalies and sulphides. Therefore, traditionally, the dehairing and dewooling operation was conducted with lime and Na-sulphide. Pastes of lime and Na-sulphide are rubbed, mostly by hand, into the flesh side of the hides. The reaction is fast and the removal of the hairs complete. The hair roots remain in the skin. In the lime drum the hairs are 'gelled'; that is, they are largely dissolved by the sulphide. Subsequent alkaline treatment hydrates the skin, and the loose hair is discharged into the waste-water, thus giving rise to the following problems of today's tanneries:

1. The large biological oxygen demand for processing tannery effluents is largely due to the quantity (60–70 per cent) of solubilised keratin.
2. The energy costs for oxidising organic matter in waste-water is very high.
3. Sulphides are eliminated by either oxidation or precipitation.
4. Worldwide, serious accidents have been reported with hydrogen sulphide liberated from waste-water when acidified, such as by action of deliming agents. Also, some tanneries have been shut down because of the foul odour of hydrogen sulphide.

Presently, no ideal enzymatic dehairing process has been developed for all types of hides and skins; however, enzymatic procedures have been worked out that permit use of a fraction of the chemicals that were required earlier.

One important consideration is the speed with which the enzymes penetrate the skin, where they impact the mesodermal hair papillae. The limiting factor during this process is enzyme diffusion, which can be influenced by pH changes, specific salt concentrations, and temperature regimes. Another consideration is the high level of enzymatic activity that has to be sustained at high pH. New proteases effective in the very high alkaline range (pH 10–11) could resolve this problem. By employing such enzymes, the amount of lime required can be cut in half. The amounts of enzymes needed depend on the type and quality of the hides.

Dehairing, however, is not exclusively the result of protease action. Combinations of proteases with carbohydrases are described in the patent literature. Proteoglycans present in the basal membrane area that surrounds the hair roots could be responsible for the synergistic effects reported for proteases in combination with carbohydrases.

Today one distinguishes between two dehairing processes: one in which the hair is preserved and one in which it is destroyed. Table 21.3 provides an overview.

Table 21.3. Various dehairing processes.

Method	*Dehairing agent*	*Hide or skin*	*Effect on hair*
Cleansing and liming	Sulphite <1 gram Na_2S/litre Ca-hydroxide	Goat, sheep, cattle, calf	Intact
Liming	Sulphide >1 gram Na_2S/litre Ca-hydroxide	All skins	Burned off
Sweating	Bacterial hydrolysis (bacterial proteases)	Goat, sheep, cattle, calf	Intact
Enzymatic dehairing	Technical proteases	Goat, sheep, cattle, calf	Intact
Dewooling	Technical proteases	Sheep skins	Intact
Debristling	Technical proteases	Pig	Intact
Pulling	Lime	Pig	Intact
Shearing	Mechanical	Sheep pelts	Intact

Conventional reductive dehairing solubilises proteins by complete reduction of all disulphide bridges in an alkaline milieu. Other reducing agents such as hydrosulphite, thioglycolic acid, mercaptoethanol, and amines can be used in concert for dehairing processes where hair is preserved.

Typically, the process of dehairing cattle hides proceeds as follows. For the initial treatment step, the hides are soaked in a bath of 1.5–10-fold water (150–1000 per cent) with 1.5 per cent soda and 0.1 per cent dehairing enzyme, a neutral bacterial protease. For the second treatment step, dehairing is further promoted by the addition of a fungal protease at pH 10. Treatment time is about 24 hours; subsequently, the loosened hair is mechanically removed.

This process can be significantly accelerated if an enzyme solution is forced under pressure into the hide from the flesh side.

Deliming and bating

Prior to tanning, the alkaline hydration of the skin has to be reversed. To accomplish this, the dehaired hides or skins must be delimed prior to or during bating. This requires partial neutralisation during which the pH is reduced from 12 to 8. Commonly used deliming agents are acid salts such as ammonium sulphate, ammonium chloride, or organic acids.

Bating means to macerate, or to soften; it is done to purify the collagenous network from other proteins present, as it loosens the collagen fibres. Bating always involves the degradation of non-collagenous proteins such as the glycoproteins and proteoglycans of the interfibrillar matrix, which are undesirable substances in leather processing.

Individual collagen fibrils should become accessible to the tanning process that requires their cross-linking. This process also seeks to eliminate other skin components such as hair roots, pigments, and residual epidermis. Clean hides or pelts, ready for tanning, are an absolute requirement for aniline-dyed leathers.

Bating enzymes

Otto Röhm produced bating compounds based on pancreatic proteases. It is important for satisfactory skin loosening that the complete pancreatic enzyme complex be available; thus, this requirement must be ascertained during its technical preparation. Other enzymatic activities from peptidases and carbohydrases also influence the bating results.

A comparison of enzymatic efficacy of various enzyme preparations has shown that trypsin is least able to digest collagen. Stirtz and Schröder have shown that chymotrypsin can cause disruption of the cross-linkages between collagen molecules.

Today, pancreatic enzymes are combined with neutral and alkaline bacterial and fungal proteases. One problem encountered when bating with highly active alkaline bacterial proteases is that the bating effect is exaggerated; this will affect the grain and yield rough surfaces.

In addition to enzymatic bating in the alkaline range, bating with acid fungal proteases in the weak acidic pH range is also practiced. This bating mix is used predominantly for soft leathers.

Example: Bating mix of dehaired pork hides used in clothing leather

1. Partially processed hides:
   (a) Limed, washed, and delimed hairless pork skins.
2. Bating mix (vat):
   (a) Deliming series (60 per cent water).

(b) +60 per cent water at 32°C.
(c) +2 per cent pancreatic proteases with 1200 LVU/g Oropon OO [RM].
(d) 60-minute agitation; check skin cross-sections with phenolphthalein indicator.
(e) Drain batch when skin section remains colourless.

(Dose percentages are correlated with the weight of the hides.)

The white stock (hairless hides, skins, or pelts) must be free of scud; the processed stock should be soft and pliable.

Enzymatic one-step process

The objective of this process developed is the optimisation of tannery beamhouse operations. Soak, hair loosening, liming, and bating are streamlined into one operation. Such processing requires an intensive enzyme treatment in which hair loosening is achieved in a maximum of 12 hours without damaging the grain.

Comparison of the conventional with the one-step process (Table 21.4) shows that treatment time and water consumption can be reduced by 50 per cent (Fig. 21.3).

Table 21.4. Basic outline for the one-step process.

Washing: Water: Salt ratio 1:1 (wt/wt at 28°C)
One-step process: mixer, barrel or vat, tanning drums
50–100 per cent water at 28°C
0.2–0.3 per cent enzyme preparation
0.2–0.3 per cent Na-hydroxide 1:10 (wt/v), 20 minutes agitation/hr. for 4 hours
3.0–5.0 per cent Ca-hydroxide
1.2–2.0 per cent liming aids
0.3–0.5 per cent 95 per cent Na-sulphydrate
0.3–0.5 per cent 60 per cent Na_2S
0.3–0.5 per cent Na-hydroxide 1:10 (wt/v), 16 hours, initial run 2 hours, with short intermittent agitation

Loosening wet-blue stock

The starting material is chrome-tanned grain layer hides, chrome-tanned split (flesh-layer or blue drop) hides, or chrome-tanned pelts.

The objective of the treatment is to loosen or soften the chrome-tanned stock for a more supple feel; better tensile strength, resiliency, and porosity; and a more even dyeing.

1. Partially processed hides:
   (a) Washed, chrome-tanned grain layer hides.
2. Loosening and softening:
   (a) 150 per cent water at 40°C.
   (b) 2–3 per cent enzymatic loosening (or softening) agent derived from fungal protease with an activity of 1000 LVU/g (Eropic WB [RM]).
   (c) 2-hour agitation, then alternating a 30-minute pause with 5-minute agitation.
   (d) Duration 12–16 hours, pH 3.8–4.2.
   (e) Then batch is drained.

Processing of tanning wastes

Tanning waste, consisting primarily of the so-called glue stock, cannot be deposited at waste dump sites but must be completely processed. To accomplish this, all proteins are solubilised at high-alkaline pH with bacterial proteases in order to partition the fat from the hydrolysate. After concentration of the protein solutes, the protein is dried and used as a nutritional supplement in animal feed or technical processes.

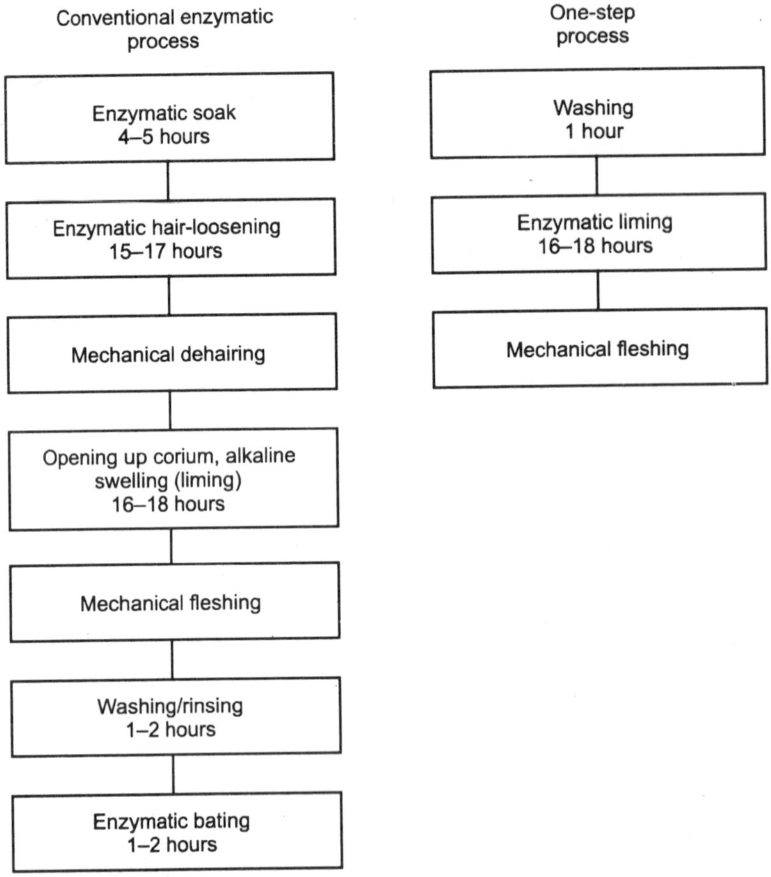

Duration	48 hours	24 hours
Water consumption	900 per cent	400 per cent
Process steps	7	3

Determined by hide weight

Fig. 21.3. Comparison between conventional and one-step enzymatic processes.

Table 21.5 reiterates the individual steps of enzyme-operated leather manufacture but focuses on economic considerations.

Table 21.5. Individual steps in leather manufacture.

Process	*pH*	*Substrates*	*Objectives*	*Enzymes*	*Alternatives*
Pre-soak (re-hydrating)	7	Albumins globulins	Degrading proteins	Proteases, bacterial proteases	Surfactants Cl ions
Main soak	9	Non-fibrillar proteins, prekeratins, lipid cell membranes	Solubilising proteins, digesting cells loosening hair, opening up corium	Fungal, bacterial, and pancreatic proteases; papain	Surfactants, weak reducing agents, amines
Dehairing	10	Keratin	Removing hair and epidermis	Keratinases, proteases	Inorganic and organic sulphides; strong reducing agents, OH ions
Liming	12.5	Non-collagen protein, mucopolysaccharides	Separating fibres, opening up corium	Bacterial and fungal proteases, elastases	OH ions, hydroxylamines, ΔT
Deliming and bating	8–9	Pigments, sebaceous and lipid cells, hair-root-associated proteins	Cleaning grain surface, increasing grain elasticity	Pancreatic, fungal, bacterial, and plant proteases	Salts: $(NH_3)_2SO_4$ NH_3Cl; surfactants
Acid pickling	5–6	Collagen	Softening skin, increasing leather elasticity, improving dyeing	Fungal proteases, pancreatic amylases, papain, etc.	Prolonged acid hydrolysis, hydrotropic agents, ΔT
Wet-blue stock softening	6	Chrome- or vegetable-tanned leathers	Degrading proteins, softening skin, separating fibres	Fungal, bacterial proteases, special pancreatic extracts	Treatment with hydrotropic agents
Scrap leather processing	11	Glue leather	Hydrolysing proteins for lipid separation	Bacterial proteases	ΔT, ΔP

Chapter 22

Clinical Analysis

INTRODUCTION

Recently, there has been a rapid growth in the use of enzymes in two areas. Firstly, as labels in enzyme immunoassay and, secondly, their incorporation into ever more sophisticated 'dry reagent systems'.

The most common routine use of purified enzymes in clinical analysis is still in the measurement of the concentration of substrates in either plasma or serum. Originally these analyses were measured using end-point or equilibrium assays. This has been encouraged by improvements in the performance of automated analysers coupled with the need for increased throughput, which is one of the main advantages of kinetic assays. The activity of many enzymes is difficult to determine, but in certain cases the addition of other enzymes simplifies their measurement. These analyses require the addition of relatively large amounts of pure enzymes. The use of purified enzymes for large numbers of analyses can be expensive, and this was one of the principal reasons for the incorporation of re-usable immobilised enzymes into clinical analyses. However, the potential cost savings were not realised in practice. Nevertheless, there has been rapid growth in the use of immobilised enzymes in dry reagent systems, which is now driven more by convenience for the manufacturer and the consequent simplification of assay procedures for the end user. There has also been much research into the development of biosensors, or bioanalytical probes. Many of these use immobilised enzymes to generate and multiply the signal. Unfortunately very few enzyme-based commercially available products have so far been produced. The other important growth area is in enzyme immunoassay. Originally, the main use of these assays was in the determination in blood or urine of the concentrations of a variety of drugs, but it is now realised that they have many other potential uses, including the measurement of hormones and the amount of enzymes, as distinct from their activity. Enzymes are also incorporated into dry reagent immunoassay systems as reliable labels.

The rationale for using enzymes in clinical analysis comes from their two best-known properties, catalytic efficiency and specificity. As they are extremely good catalysts, the reaction should proceed rapidly, even under mild conditions of temperature and pH, and they do not affect the equilibrium position of the reaction. Their specificity ensures that the rate of a particular reaction, or of a limited range of related reactions, is enhanced. This enables the rapid analysis of particular compounds that occur in complex media, such as serum, plasma and urine, to be performed without prior purification steps. In the case of enzyme immunoassays the essential feature of specificity comes from the specific binding between antigen and antibody, while the enzyme provides an easily detected label which also generally amplifies the signal.

The progress curve for a typical enzyme catalysed reaction is shown in Fig. 22.1. This indicates the two parts of the curve from which data are obtained. The equilibrium method, which can only be used for the determination of substrates, makes use of data collected when the concentrations of product or substrate are time-independent. Those methods where the data are obtained when the concentration is time-dependent are kinetic methods, in which case the data are generally collected from the early, linear part of the curve, which is called the initial velocity (or rate).

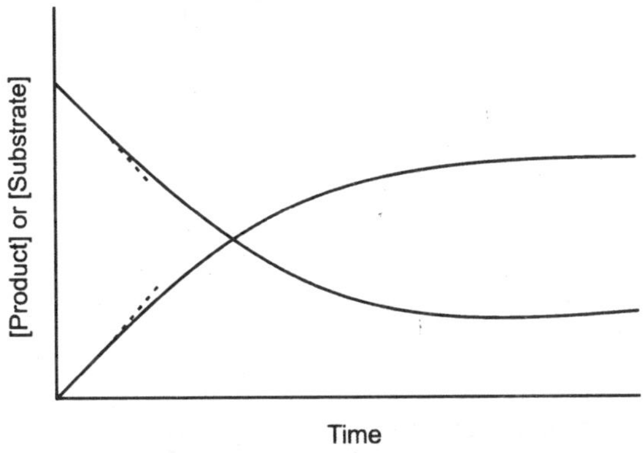

Fig. 22.1. Progress curves for a typical enzyme-catalysed reaction. The reaction can be followed by the loss of substrate or the formation of product. The initial velocity is determined from the tangent(s) to the curves (- - - - -) at zero time. The progress curves have a steady value when the reaction reaches equilibrium.

The simplest enzyme-catalysed reaction involves one substrate and one product:

$$\underset{\text{Enzyme}}{\mathrm{E}} + \underset{\text{Substrate}}{\mathrm{S}} \underset{k_{-1}}{\overset{k_1}{\rightleftharpoons}} \underset{\substack{\text{Enzyme-substrate}\\ \text{intermediate}}}{\mathrm{ES}} \xrightarrow{k_2} \mathrm{E} + \underset{\text{Product}}{\mathrm{P}}$$

The way in which the rate of this reaction is dependent on the substrate concentration is usually described by the Michaelis-Menten equation:

$$v = \frac{V[S]}{K_m + [S]} \qquad \text{... (22.1)}$$

where, v is the velocity (rate) of the reaction, $V = (k_2[E_0])$ is the maximal velocity (rate) of the reaction, $[S]$ and $[E_0]$ are the initial substrate and enzyme concentrations, K_m is the Michaelis constant for the substrate, and k_1, k_{-1}, and k_2 are the reaction rate constants. This equation describes a rectangular hyperbola as illustrated in Fig. 22.2.

Different substrate concentrations are suitable for different purposes. For the determination of substrate by a kinetic method it is necessary that the substrate concentration should be rate limiting, and therefore it should not be much higher than the K_m of the enzyme. When the activity of an enzyme in serum is to be measured, then the enzyme should be rate limiting, and therefore high concentrations of substrate are necessary.

The rate of enzyme-catalysed reactions is very dependent on pH and temperature, and it is essential that these two factors are rigorously controlled in all kinetic methods. Many compounds can either reduce (inhibitors) or increase (activators) the rate of an enzyme-catalysed reaction. The presence of these compounds will affect the accuracy of kinetic methods for more than an equilibrium method.

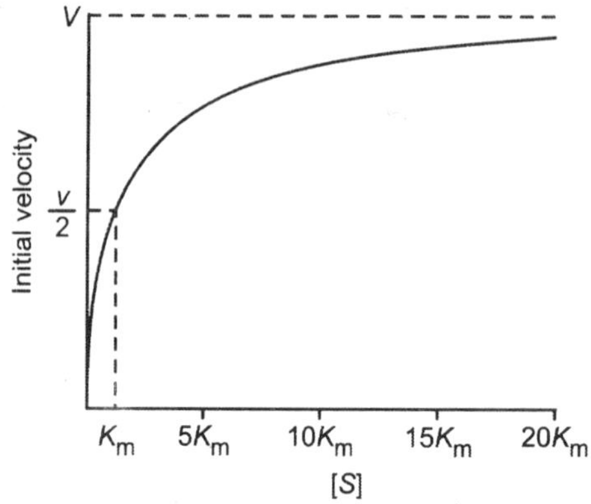

Fig. 22.2. Theoretical Michaelis plot.

MEASUREMENT OF SUBSTRATE CONCENTRATION WITH ENZYMES

Many substrates, or metabolites, can be measured with the aid of enzymes. Some co-enzymes, such as adenosine triphosphate (ATP) and nicotinamide adenine dinucleotide in its oxidised (NAD^+) or reduced form ($NADH + H^+$), can be considered as co-substrates in this context. The substrates most frequently measured are urea and glucose, but several others are determined in clinical laboratories.

Principles of Equilibrium Methods

These assays are simple to perform, but the position of equilibrium and the time needed to reach this point require consideration. Consider the equation:

$$S + C \underset{K_{eq}}{\overset{\text{Enzyme}}{\rightleftharpoons}} P + Q$$

where, S is the substrate to be measured, C is a coenzyme or coreactant (or cosubstrate), and P and Q are the products of the enzyme-catalysed reaction.

The greater the proportion of substrate that is utilised, the more accurate the assay. This proportion is determined by the relative initial amounts of coreactant (C_0) and substrate (S_0) and the equilibrium constant (K_{eq}) for the reaction. Table 22.1 shows the calculated ratios of C_0:S_0 required to give 99 per cent, 99.5 per cent, and 99.9 per cent conversion. As the equilibrium constant increases, so the relative excess of coreactant needed is reduced. The very high ratios required at low K_{eq} are likely to pose practical problems, for example insolubility of coreactant, possible inhibition of enzyme by coreactant, or high initial background absorbance. If the reaction does not approach completion a calibration curve that covers the anticipated range of substrate concentrations is essential.

Table 22.1. Calculated ratio of initial [coreactant]: initial [substrate] required for stated percentage conversion.

	Values of $C_0 : S_0$			
Conversion	K_{eq}			
(%)	*1*	*5*	*10*	*50*
99	100	20	10	3
99.5	200	40	20	5
99.9	1000	200	100	20

This also overcomes the non-linearity of calibration curves that is most pronounced with low K_{eq}. Another solution to this problem is to change the equilibrium position. This can sometimes be done by altering the pH of the reaction or by using a reagent that combines with one of the products.

Factors that alter K_{eq} should be kept constant during the assays; this is particularly important when the reaction has a low K_{eq}. The K_{eq} should not be affected by the presence of enzyme, so long as the molar concentration of enzyme is very low compared to S_0. K_{eq} is however affected by temperature and pH. Therefore these two parameters must be controlled during the assay. The time needed for an enzyme-catalysed reaction to reach equilibrium can be calculated from the integrated form of the Michaelis-Menten equation.

$$t = \frac{2.3K_m}{V} \log_{10} \frac{S_0}{S_t} + \frac{(S_0 - S_t)}{V} \quad \text{... (22.2)}$$

where, V and K_m have the meanings given above, t is the time of reaction, S_0 is the initial substrate concentration, S_t is the substrate concentration after time t, and $(S_0 - S_t)$ is the product concentration after time t.

If we accept 99 per cent conversion of substrate to product as adequate for estimation, then $\log_{10}(S_0/S_t) = 2$ and S_t can be ignored relative to S_0. Then eq. (22.2) becomes:

$$t = 2.3 \times 2\frac{K_m}{V} + \frac{S_0}{V}$$

If $S_0 \leq K_m$, then the maximum value of S_0/V becomes K_m/V.

$$t \simeq 5.6\frac{K_m}{V}$$

For the reaction to be completed within 5 minutes, K_m/V should be slightly greater than 1 min/ml. This is achieved if $K_m = 10^{-3}$ mol/l (which is 1 μmol/ml) and V is 1 U of enzyme ml^{-1}. The international unit, U, is the amount of enzyme required to convert 1 μmol of substrate to product per minute. Then,

$$\frac{K_m}{V} = \frac{1\ \mu\text{mol/ml}}{1\ \mu\text{mol/min.}} = 1\ \text{min./mol}$$

This type of calculation allows the amount of enzyme required per ml of reaction mixture to be calculated when K_m and V, which is proportional to E_0, are known.

Certain factors, which may alter the time required for the assay, are ignored in this calculation. If the product formed in the reaction causes product inhibition or a reverse reaction with a significant rate is established, V would be reduced. However, these possibilities should usually be unimportant if $S_0 \leq K_m$ as assumed above. If $S_0 > K_m$, the time required to reach equilibrium will be increased considerably, as will the potential problems caused by product formation. The presence of activators in the biological specimen will reduce K_m/V and hence the time required for the reaction to reach completion. Alternatively, the presence of inhibitors will increase K_m/V, and thus the time required to reach equilibrium will be increased.

Typical progress curves were represented in Fig. 22.1. However, it is possible in practice to obtain curves where there is a 'linear drift' after the rapid consumption of substrate. This is shown in Fig. 22.3(a). The most likely cause is an impure, non-specific enzyme which is continuing to catalyse another slow reaction. If there is 'linear drift' a correction should be made by extrapolation to the zero time as indicated. The problem is avoided if pure, specific enzymes are used. Figure 22.3(b) shows a similar curve except that here there is a reaction before the addition of enzyme. The cause is generally the presence of endogenous enzymes reacting with their substrates. This can usually be overcome by use of a relatively long (for example, 5 to 20 minutes) pre-incubation before the addition of the enzyme specific for the analyte. An alternative, common type of curve is shown in Fig. 22.3(c). Here there is a steady side-reaction, such as the oxidation of reduced NAD^+ by dissolved oxygen in the presence of the respiratory chain enzymes. In these cases an extrapolation to zero time is necessary unless the side-reaction can be prevented.

All the situations shown in Fig. 22.3 can be overcome, but this is only possible if data are collected at several time points, which increases the length of operator, or machine time required per assay.

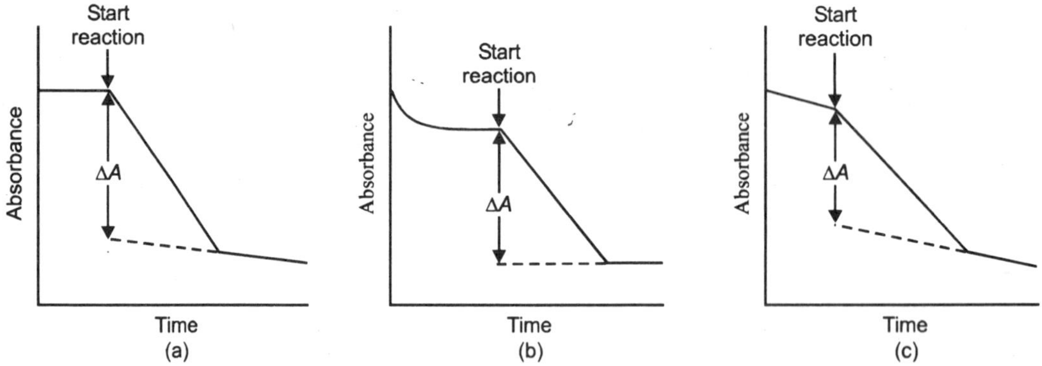

Fig. 22.3. A typical progress curves and methods for correction. (a) Shows a slow reaction continuing after the main reaction is completed. This can be corrected for by extrapolation to the time when the reaction was started by the addition of specific enzyme (↓). (b) Pre-incubation is often necessary to allow endogenous reactions to go to completion before starting the reaction of interest. (c) A continuous side-reaction can be corrected for by extrapolation to the time when the reaction was started.

Principles of Kinetic Methods

The increasing workload of clinical biochemistry laboratories in the early 1960s was partly solved by the introduction of continuous flow analysers. Part of this increase in the number of analyses involved the determination of the activity of several enzymes. Single time point assays were originally used, but as the limitations of this type of assay were appreciated it became more common to follow the linear, early part of the progress curve to determine the initial velocity (Fig. 22.1). Originally this was a slow and tedious manual process which, since the early 1970s, has been simplified by the introduction of a variety of discrete kinetic analysers. These machines follow the progress of the reaction generally at a number of fixed time intervals, check the linearity of the curve, and print out the initial velocity. The availability of free time on these sophisticated kinetic analysers encouraged the development of kinetic analyses for substrates.

The Michaelis plot (Fig. 22.2) indicates that the initial velocity is a non-linear function of substrate concentration. The sensitivity to increasing substrate concentrations decreases markedly when $S_0 > K_m$, so that $S_0 = K_m$, is about the upper limit for the use of kinetic methods. A calibration curve over a range of substrate concentrations is essential to account for non-linearity, although if substrate concentrations below $K_m/20$ are used the deviation from linearity is theoretically <3.3 per cent.

In a kinetic method the reaction velocity is measured over a short time period; but as a direct consequence of this, the change in the parameter measured, usually absorbance, is very small. For this reason electronic noise can be a significant source of error, although this can be partly overcome by increasing the amount of enzyme added in the assay.

Any factor which affects the initial velocity will cause inaccuracies with kinetic methods. The presence of either activators or inhibitors of enzymes in biological samples is difficult to detect, but they can markedly alter the results of a kinetic assay. For an accurate assay, temperature control within ±0.1°C is essential as is good control of pH, particularly if the assay uses more than one enzyme with different pH optima.

Comparison of Equilibrium and Kinetic Methods

The characteristics of equilibrium and kinetic methods are presented in Table 22.2. Equilibrium methods are generally to be preferred, but they have the major drawbacks of relatively high cost of enzyme and co-factors, and the length of time required per assay. These are the very points where the sophisticated kinetic analysers have their main advantages. However, kinetic methods are prone to large errors if enzyme activators or inhibitors are present, but undetected, in the biological material. The value of K_m, relative to the normal concentration of substrate in the assay, may be important in deciding which method of assay is most suitable.

Table 22.2. Comparison of equilibrium and kinetic methods for the assay of substrates.

	Equilibrium method		*Kinetic method*	
	Rating	*Comments*	*Rating*	*Comments*
Change followed, e.g. A	+++	Relatively large and accurate measurement. More sensitive	++	Very small but within ability of sophisticated kinetic analysers. Electronic noise can be a problem
Change in pH	++	May alter K_{eq} but unlikely to cause error unless K_{eq} is small	–	Any change from optimum pH may cause low values—especially if more than one enzyme used in assay
Change in T	+++	Will alter K_{eq} but no serious error unless K_{eq} is small and T change is several degrees	+	Good temperature control (±0.1°C) is essential
Presence of inhibitors	++	Lengthen assay time check reaction terminated	–	Low values
Presence of activators	+++	Equilibrium reached more quickly	–	High values
Influence of K_m	++	Low K_m reduces time required for assay	++	The higher the K_m the higher S_0 that can be assayed
Maximum S_0	++	$S_0 < K_m$	++	$S_0 < K_m$

(Contd ...)

	Equilibrium method		Kinetic method	
	Rating	*Comments*	*Rating*	*Comments*
Linearity with S_0	++	Linear but may get deviation at low K_{eq}	+	Non-linear, therefore calibration curve needed
K_{eq}	+	Problems with low K_{eq} of solubility, inhibition or background absorbance	+	With low K_{eq} assay may be too insensitive
Time required	+	Several minutes	+++	Generally <1 minute
Cost of enzyme	–	Relatively high to complete assay in reasonable time	++	Relatively low. Require sufficient to give adequate signal to noise ratio
Cost of co-factors	–	Can be higher than cost of enzyme especially if K_{eq} is low	++	Relatively low. Concentration must not be rate-limiting
Product formation	++	May cause inhibition or reversal of reaction	+++	Too little to cause effect
Unstable product	+	Cannot measure product	++	No problem if assay is rapid
Turbid or coloured sample	+	Blank required	+++	Should not affect reaction rate
Equipment	+++	Sophisticated equipment required but dedicated equipment being produced	+	Requires sophisticated equipment but this may be used for other analyses

Attempts have been made to extend the usable range of kinetic methods to include substrate concentrations that greatly exceed K_m. One approach is to include competitive inhibitors in the assay since they cause an apparent increase in K_m. An alternative approach has more general applicability. Hamilton and Pardue have used non-linear regression analysis to fit data for absorbance (A) and rate (dA/dt) versus time to the Michaelis-Menten equation. This gave linear calibration plots for concentrations from well below K_m to $3.5 \times K_m$.

Common Indicator Species Used in Routine Clinical Analysis

Only a limited number of compounds are frequently measured in routine clinical analyses. These indicator species and the properties which make them suitable for sensitive, specific, and yet diverse clinical analyses are considered below.

Nicotinamide adenine dinucleotides

Nicotinamide adenine dinucleotide occurs in the oxidised (NAD^+) and reduced (NADH) forms. Nicotinamide adenine dinucleotide phosphate also has two forms, $NADP^+$ and NADPH. Figure 22.4 shows their structure and interconversion. Only the reduced form, NADH, absorbs light at 339 nm where it has a large molar extinction coefficient ($\varepsilon = 6.31 \times 10^{-3}$ litre/mol cm). NADP$^+$ and NADPH have very similar properties to NAD$^+$ and NADH. Table 22.3 shows the molar absorbance of NADH at particular wavelengths, and indicates how the value changes slightly with alteration in pH, temperature, and ionic strength. The wavelength of maximum absorbance also changes slightly with temperature. The values actually obtained are of course dependent on the type of spectrophotometer used, and especially on the bandwidth of the monochromator.

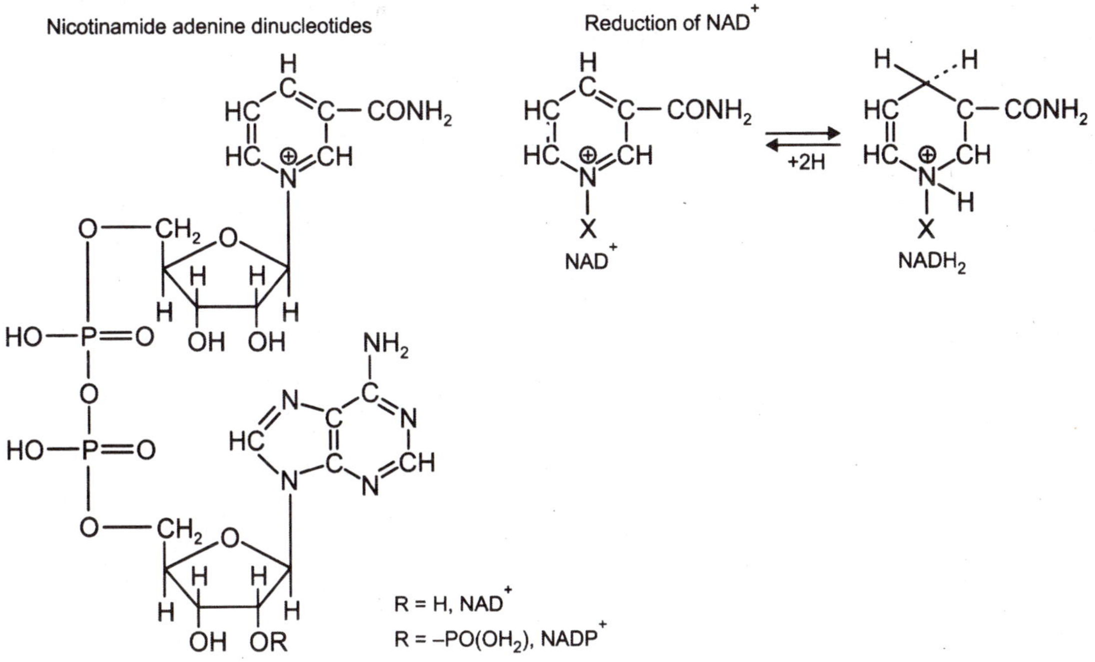

Fig. 22.4. Structure of NAD(P)$^+$ and the oxidation-reduction reaction.

Table 22.3. Molar extinction of NADH.

Wavelength (nm)	*NADH at 25°C (1/mol cm)*
334	6180
339	6310
366	3500

Decreases (i) by 16 per pH unit from 8.8 to 6.8; (ii) by 24 per 10°C from 25 to 35°C; (iii) by 60 as ionic strength (NaCl) increases from 0 to 1.0 M.

The wavelength of maximum absorption decreases from 340 nm at 0°C to 33.8 nm at 25°C and to 337.5 nm at 40°C.

A simple reaction involving these compounds is the determination of pyruvate using lactate dehydrogenase:

$$\text{Pyruvate} + \text{NADH} + \text{H}^+ \underset{\text{dehydrogenase}}{\overset{\text{Lactate}}{\rightleftharpoons}} \text{L-lactate} + \text{NAD}^+$$

The loss of NADH is followed, generally by measuring the decrease in absorbance at about 340 nm.

Since $K_{eq} = 2 \times 10^4$ at pH 7.0, the reaction will go to completion so long as [NADH] > [pyruvate]. The total change in absorbance can then be related to the original amount of pyruvate by use of the Lambert-Beer law ($A = \varepsilon \times c \times l$), where A is the absorbance, c is the concentration (mol/l), l is the pathlength (cm), and ε is the molar absorption coefficient (l/mol cm).

The same enzyme-catalysed reaction can also be used to measure lactate. To do this the reaction is done in the reverse direction in a buffer with pH between 9 and 10 which neutralises the H^+ liberated. Alternatively the pyruvate produced can be either trapped with hydrazine or removed enzymically with another enzyme, alanine transaminase. In either case the reaction is now in favour of using lactate with an equivalent amount of NAD^+. It is essential that the enzyme can function under the altered conditions of high pH and in the presence of other compounds.

In the pyruvate to lactate reaction, NAD^+ and NADH are two of the reactants, and therefore this reaction enables direct measurement of these compounds. With many analytes this is not possible, but by using two or more linked enzyme reactions a change in NADH or NADPH can be produced, which is equivalent to the concentration of the original substance. The measurement of glucose with hexokinase and glucose 6-phosphate dehydrogenase illustrates this point.

$$\text{Glucose} + \text{ATP} \xrightleftharpoons{\text{Hexokinase, Mg}^{2+}} \text{Glucose 6-phosphate} + \text{ADP}$$

$$\text{Glucose 6-phosphate} + \text{NADP}^+ \xrightleftharpoons[\text{dehydrogenase}]{\text{Glucose 6-phosphate}} \text{6-Phosphogluconate} + \text{NADPH} + \text{H}^+$$

Hexokinase has broad specificity for hexoses. Therefore fructose and other naturally occurring carbohydrates are also converted to their phosphates in the first reaction. However, the second enzyme-catalysed reaction, which involves $NADP^+$, is very specific for glucose 6-phosphate, thus allowing accurate and specific measurement of the original glucose. However, this particular measurement has three disadvantages.

Firstly, the coenzyme, $NADP^+$, is more expensive than NAD^+. Secondly, the introduction of two or more enzymes adds to the cost of analysis. Thirdly, there will normally be a lag phase during the initial part of an assay involving two enzymes which requires that there should be a short delay if a kinetic method is used.

A recently developed novel assay is worthy of note because it provides a way of measuring K^+ enzymically. In the first stage endogenous NH_4^+ is eliminated from the serum sample by a glutamate dehydrogenase-catalysed reaction:

$$\text{NH}_4^+ + \text{2-oxoglutarate} + \text{NADPH} + \text{H}^+ \xrightleftharpoons[\text{dehydrogenase}]{\text{Glutamate}} \text{Glutamate} + \text{NADP}^+$$

After 5 minutes, the reagents for the second stage are added which causes the following reaction:

$$\text{Tryptophan} \xrightleftharpoons[\text{Pyridoxal 5-phosphate}]{\text{Tryptophanase, K}^+} \text{Indole} + \text{pyruvate} + \text{NH}_4^+$$

The NH_4^+ produced by this step is now measured by the glutamate dehydrogenase catalysed reaction by following the decrease in NADPH which is linearly related to $[K^+]$. The main advantages of this coupled enzyme assay for K^+ are:

1. Unlike other methods it does not require the presence of an ionophore.
2. The reaction can be made independent of $[Na^+]$ and is linear up to 7 mM (K^+).
3. The method is easily automated with a total assay time of about 10 minutes.

An alternative method for measuring NADH or NADPH is to react them with an oxidising agent, thereby reforming NAD^+ or $NADP^+$. When a tetrazolium salt is used a very intensely absorbing coloured formazan is generated and can be measured. Since this chemical reaction only occurs slowly another enzyme, for example diaphorase, is normally added to catalyse the formation of formazan. This removal of NAD(P)H gives another way of effectively shifting the equilibrium of the reaction in the direction of the formation of reduced form.

Oxygen and hydrogen peroxide

Oxygen is used, and hydrogen peroxide formed, in three of the more common enzymic estimations performed in clinical laboratories:

$$\text{Cholesterol} + O_2 + H_2O \longrightarrow \text{Cholest-4-}n\text{-3-one} + H_2O_2$$

$$\text{Glucose} + O_2 + H_2O \longrightarrow \text{Gluconic acid} + H_2O_2$$

$$\text{Uric acid} + O_2 + 2H_2O \longrightarrow \text{Allantoin} + CO_2 + H_2O_2$$

Hydrogen peroxide is commonly measured colorimetrically by using peroxidase to catalyse the reaction below:

$$\text{Chromogen} + H_2O_2 \xrightarrow{\text{Peroxidase}} \text{Dye} + H_2O$$

Chromogens currently used are *o*-dianisidine or diammonium 2,2′-azino-*bis*(3-ethylbenzothiazoline-6-sulphonate). Alternatively, oxidative coupling of two compounds can be used. These include phenol with either 4-aminophenazone or 4-amino antipyrine, or the alternative *N*-ethyl-*N*-(3-hydroxyethyl)-*m*-toluidine and 2-hydrazone-2,3-dihydro-3-methyl-6-sulphobenzothiazole. Another system for measuring hydrogen peroxide uses the enzyme catalase.

$$CH_3OH + H_2O_2 \xrightarrow{\text{Catalase}} HCHO + 2H_2O$$

$$HCHO + 2\ \text{acetylacetone} + NH_3 \longrightarrow \text{3,5-diacetyl-1,4-dihydrolutidine} + 3H_2O$$

Methods which use peroxidase are subject to interference by many compounds including ascorbic acid, uric acid, glutathione, and acetylsalicylic acid, all of which decrease the value obtained. Conversely, the presence of peroxides increases the value. The interference by these compounds is minimised by diluting the plasma samples with a large (>200-fold) excess of reagents.

The possibility of interference and the cost of second enzyme can both be reduced by measuring one of the other common components in the three reactions. The utilisation of oxygen can be measured with an oxygen electrode.

This can be done when all the components are in solution. However, the introduction of an immobilised enzyme, glucose oxidase, has produced a more economical, semi-automated process suitable for routine analysis. More recently an H_2O_2-sensitive electrode utilising immobilised glucose oxidase has come into routine use. These electrodes are discussed in more detail later.

MEASUREMENT OF ENZYMES

Enzymes are measured in body fluids as an aid to diagnosis, and also for monitoring the effect of treatment. The rationale for this is that certain enzymes are located solely or predominantly in a tissue. Also it is known that some enzymes occur in multiple forms, isoenzymes, each of which has a different tissue distribution. When particular tissues are damaged owing to disease, their tissue enzymes and isoenzymes are liberated and their enzyme levels in serum increase.

A more specialised application of enzyme measurements is to determine if a particular enzyme is defective, generally having no or little activity. For these purposes a tissue sample is normally required. These assays are used to establish the cause of an inborn error of metabolism.

Enzymes are measured by kinetic assays, similar to those described for substrates, except that the enzyme, rather than substrate, is rate limiting. Originally, attempts to measure the amount of enzyme by immunoassays, or by titration of the active site, were originally too complex for routine use. More recently, simpler (semi-)automated immunoassays for proteins, including enzymes, have become available.

These immunoassays measure inactive enzyme protein as well as active enzyme. A major advantage is that they should not be affected by the presence of activators or inhibitors in the same way as inevitably occurs with a kinetic assay.

Principles of Enzyme Assay Using Coupled Enzymes

With very few exceptions enzyme assays available in kit form, which include other enzymes as analytical tools, use the $NAD(P)^+/NAD(P)H$ end point. The components and conditions used for this type of assay are summarised in Table 22.4.

Table 22.4. Components and conditions for enzyme assays using coupled enzymes.

Enzyme sample (e.g. serum)	Rate-limiting Typical A_{340} = 0.7–4.6 (serum)
Coenzyme/cosubstrates	Optimal concentration
Coupled enzyme	> 40-fold excess
$NAD(P)^+/NADPH$	NAD(P)H > 0.2 mmol/l
Buffer	pK close to pH optimum of enzyme
Temperature (e.g. 25°C, 30°C, 35°C, 37°C)	Controlled to ±0.1°C
Specific substrate	Start reaction by addition of optimum concentration of specific substrate
Mix	Follow conversion of $NAD^+/NADH$ at 340 nm

Serum contains many proteins, some of which will be enzymes, only one of which is to be measured in a particular assay. The activity of this enzyme in the reaction cuvette must be rate-limiting. In the case of a serum enzyme this usually involves dilution of the serum by between 1 in 10 and 1 in 30. This dilution also reduces the absorbance at 340 nm due to serum proteins.

Most enzymes have two substrates, and both of these are added at 'optimum concentration'. In practical assays the concentration of substrate may be limited by the solubility of the substrate or by the possibility of inhibition by high S_0. If S_0 of $19 \times K_m$ can be achieved, then substitution in the Michaelis-Menten equation [eq. (22.1)] gives:

$$v = 0.95V$$

that is, the initial velocity, v, is 95 per cent of the maximum velocity. Since eq. (22.1) describes a rectangular hyperbola (Fig. 22.2) there is little change of v with small changes in S_0 at this level. Therefore, this concentration of substrate should give consistent assays so long as the velocity is measured before 10 per cent of the substrate has been converted to product. This should ensure that errors associated with any reverse reaction or inhibition by product are negligible. This concentration of substrate is also sufficient to allow a proportion to be used up in secondary reactions which frequently occur because of the presence of other enzymes in the serum. These reactions should be given time to go to completion, which occurs when the endogenous substrates in serum are used up, before the actual enzyme assay is started by addition of the substrate for which it is more specific.

Aspartate aminotransferase (AST) was the first enzyme to be measured with a coupled enzyme, in this case malate dehydrogenase (MDH). The system used was

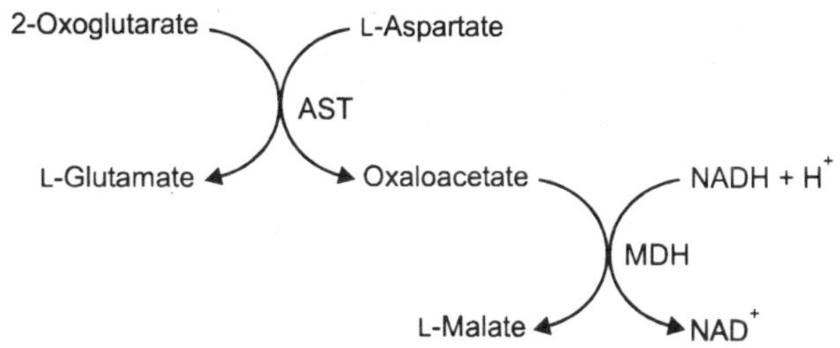

The oxaloacetate formed in the first reaction has to be utilised as soon as it is formed by the reaction catalysed by MDH if measurement of the rate of NADH conversion to NAD^+ is to be an accurate assay for AST. Thus the reaction catalysed by MDH has to operate under the unfavourable condition that the concentration of one of its substrates is virtually zero. To achieve this the total activity of the coupled enzyme must be at least 40-fold greater than the total activity of the enzyme being assayed. Even then there will be a lag phase as the concentration of oxaloacetate increases to its low steady-state level. This lag phase and the measurements of the velocity should be completed before the reaction deviates significantly from the true rate owing to substrate depletion. If there is more than one coupled enzyme, the lag phase will be increased, and higher total activities of the other enzymes are needed to ensure accurate assays.

The inclusion of large amounts of other enzymes in coupled enzyme systems requires that they are pure, and in particular they must not be contaminated with activity of the enzyme being measured. The use of these purified enzymes is a major contributor to the cost per assay.

The concentration of NAD(P)H present at the start of an assay should not exceed 0.2 mmol/l. At this concentration the absorbance at 340 nm is 1.26 in a 1 cm pathlength cuvette. This, added to other absorbances for other light-absorbing species including serum proteins, is close to the upper limit before large errors are likely to be caused by stray light. Enzyme assays should be carried out at their optimum pH. At this pH small deviations in pH have less effect on the velocity than at other pHs. The optimum pH of an enzyme is dependent on the temperature and concentration of the substrates, and should be determined under the actual conditions to be used in the assay. The components of buffer systems usually inhibit enzyme activity at high concentrations. The concentration of buffer can be minimised if the pK of the selected buffer is close to the pH used for the assay.

Enzymes do not have an optimum temperature, which accounts for the range of temperatures that have been recommended (Table 22.4). It is important that all components of the assay mixture should be kept at the temperature selected for the assay, and it is essential that the serum enzyme should not be inactivated at this temperature or during any pre-incubation step. Changes in activity of enzymes of 5–10 per cent occur per 1°C. Therefore for accurate assays, temperature control to within ±0.1°C is essential.

Before the reaction is started it is necessary to mix together and incubate all the reaction components, other than the substrate specific for the enzyme to be measured. During the incubation, endogenous substrates of other enzymes present in the serum react. For instance, in the assay of aspartate aminotransferase, any pyruvate present in serum is removed by lactate dehydrogenase which is also in the serum.

$$\text{Pyruvate} + \text{NADH} + \text{H}^+ \xrightarrow[\text{dehydrogenase}]{\text{Lactate}} \text{L-Lactate} + \text{NAD}^+$$

The assay of aspartate aminotransferase should not start before the above reaction is completed. Some samples of serum contain so much pyruvate that it is now recommended that purified lactate dehydrogenase be routinely included in the assay mixture to hasten its removal.

IMMOBILISED ENZYMES FOR MEASURING SUBSTRATE CONCENTRATIONS

During the 1970s much attention was focused on the substitution of soluble enzymes by their immobilised counterparts in enzyme-based clinical analyses. This was because of the limited stability of some enzymes under routine operating conditions, and the relatively high cost of many analytically useful enzymes. The exploitation of immobilised enzymes apparently offers the clinical chemist the advantages of re-usability, greater stability, and increased convenience. However, relatively few of these analytical techniques have been commercially exploited. The reasons for this are mainly that the potential cost benefits have not always been realised in practice, partly because the re-use of enzymes over a prolonged period requires considerable care.

The potential utilisation of immobilised enzymes in clinical chemistry may be divided into three categories: enzyme reactor tubes, biosensors or bioanalytical probes, and dry reagent products.

Immobilised Enzyme Reactor Tubes

The first problem has how to incorporate immobilised enzymes into automated analytical systems. Early attempts used packed bed enzyme reactors, which were subsequently packed into auto analyser mixing coils. These were satisfactory for systems based on non-segmented flow analysis. For the more common air-segmented continuous flow systems an alternative approach was necessary. Hornby developed the first open tubular heterogeneous enzyme reactors (OTHERs) in which the enzyme was covalently bound to the inner surface of a narrow tube. These were typically a nylon tube of 1 mm bore with one or more enzymes coated onto the internal surface.

The main advantage in using OTHERs was their reusability. However, OTHERs did not result in an economic saving, when compared to the use of soluble enzymes, unless very large numbers of samples were analysed routinely. There was also the problem that OTHERs usually had to be washed separately from the rest of the analytical system to prevent denaturation of the immobilised enzyme. The OTHERs could not be allowed to dry out and were best stored in buffer containing an antimicrobial agent. Because of these problems the market for OTHERs has decreased in the last decade. Their demise has been quickened by the replacement of many air-segmented continuous flow systems by discrete automated systems where sample and reagents are mixed together in the reaction cuvette and there is no obvious place to put immobilised enzymes. On the other hand 'dry reagent systems' have expanded greatly.

Dry Reagent System

The first dry reagent system was produced by Smith and others by adsorbing and drying the analytical reagents on to a paper (cellulose) pad. The dried paper changed colour when wetted with a sample containing the analyte glucose. The concept was quickly developed under the name 'Clinistix' by the Ames Division of Miles Laboratories and was widely used for the detection of glucose in urine. The dried reagent pad is cemented to a plastic strip (Fig. 22.5).

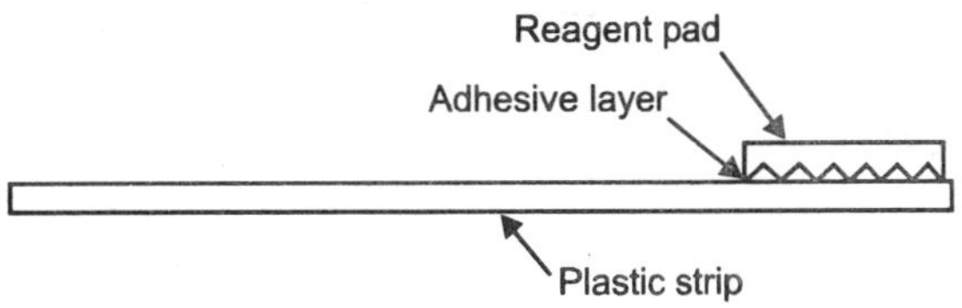

Fig. 22.5. Diagram of a dry reagent test strip.

In the subsequent 30–40 years this simple concept has been developed commercially in several ways. Compared dry reagent systems with biosensors, the two most important changes have been:

1. The change from qualitative urinary glucose analysis to greatly simplified but accurate quantitative tests on 10–30 μl of serum or blood for a range of metabolites, enzymes and proteins, and drugs. Most of these analyses use enzymes as reagents.
2. The development of a range of instruments which can be used by the diabetic patient at home right through to a fully automated discretionary analyser capable of over 500 tests per hour for use in routine analytical laboratories.

Three different formats are used for the preparation of dry reagents. Firstly, the matrix is dipped in a reagent and allowed to dry. This is now done in a continuous process such that the sequential layers of reagent are built up on individual cellulose fibres, with the last reagent in the reaction being coated on first, etc. Uniform coating is essential if the small reagent pads cut from the original large sheet of paper are to give consistent results. The pads are fixed to a reflective support since quantitative analysis generally uses reflectance photometry. Although the development of this system has been strongly influenced by the large market for blood glucose analysis by diabetic patients, a benchtop machine, the Seralyser is used.

This measures glucose, cholesterol and uric acid using adsorbed enzymes and chromogens. The activity of the enzymes creatine kinase, and aspartate and alanine transaminase are measured using coupled enzyme systems. Homogeneous immunoassays using enzymes as labels are available in dry reagent format. The first drug monitored routinely with the Seralyser was theophylline.

A second format based on impregnated paper is the Reflotron® system developed by Boehringer Mannheim. In this benchtop system a whole blood sample is placed on a mesh which spreads the sample over a layer of glass fibres. These separate red blood cells from plasma, which collects in a reservoir layer within 30 seconds. Only then is the reaction zone pressed into the plasma reservoir, the reaction starts and the measurement is made. On the reverse side of the reagent carrier is a magnetic strip with information on the type of test, incubation times, wavelength for measurement, and the mathematical factors for calculating the final result. Enzyme-based tests for glucose, creatinine, cholesterol, HDL-cholesterol, triglycerides, urea and uric acid are available. The enzymes, creatine kinase and aspartate and alanine transaminase can be measured with coupled enzyme systems. The assay of α-amylase does not require further enzymes, but antibodies to salivary amylase are incorporated into the plasma separation layer, which inhibits the salivary isoenzymes, so that the assay is specific for pancreatic α-amylase.

In the Reflotron® system the coupled enzyme assays are different from those described earlier in this chapter, where the end-point was the conversion of NADH to NAD^+. In this system a peroxidase-catalysed conversion of an indicator is used, as illustrated below for the assay of alanine transaminase (ALT). The third format has been developed by the Eastman Kodak Corporation, and is based on their experience of producing photographic film. Although photographic chemistry is not involved, Kodak have adapted their photographic film-spreading technology to produce dry reagent slides for their 'Ektachem' system.

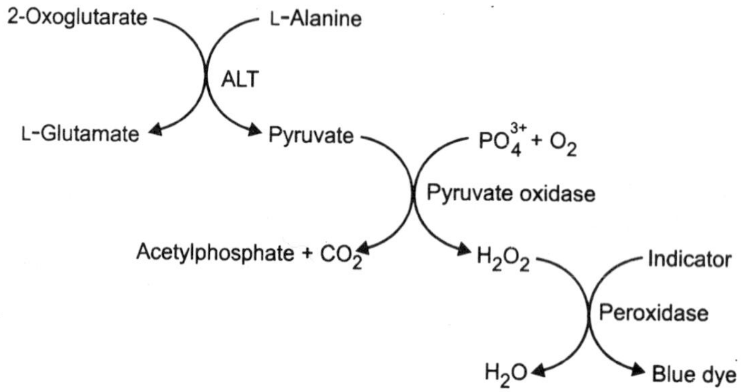

Figure 22.6 shows a cross-section of a slide for the analysis of urea. When a drop of serum touches the analysis slide it comes into contact with the porous layer which spreads the sample uniformly and rapidly by capillary action. Below the spreading layer is the reagent layer, which for blood urea determination contains urease and a buffer (pH 8.0) bound in gelatine. Urease catalyses the hydrolysis of urea to ammonia and bicarbonate. Ammonia then diffuses through a semi-permeable layer of cellulose acetate butyrate which is interposed between the reagent and indicator layers to minimise diffusion of hydroxyl ions to the dye later. A leuco indicator in a cellulose acetate binder provides the colour-forming layer and is coated directly onto a transparent support. After colour development the colour intensity is measured by reflectance through the transparent support against the white spreading layer background. Similar principles, using other appropriate enzyme systems, are used for glucose, uric acid, triglyceride, cholesterol and creatinine. Coupled enzyme systems for creatine kinase, and aspartate and alanine transaminases, linked to NADH to NAD^+ are available. Enzyme-labelled immunoassay systems for theophylline, phenytoin and phenobarbital have also been developed.

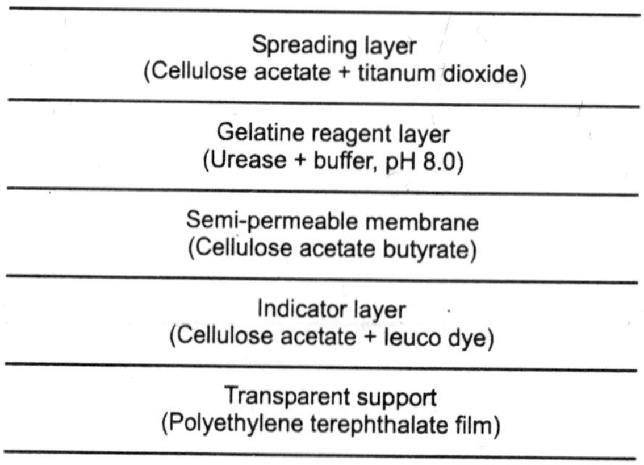

Fig. 22.6. Layer representation of Kodak Ektachem dry phase slide for urea assay. The layers are not drawn to scale.

The quality of the results obtained with dry reagent systems depend in part on the user. In the routine laboratory, results are very good and correlation coefficients with comparison to reference methods for enzyme-based analyses are all better than 0.97. However, when the tests are done outside the laboratory less satisfactory results were obtained by doctors and nurses and by patients measuring blood glucose. The reason is that many instructions need to be understood and followed meticulously. For instance, failure to replace the cap on a bottle containing reagent strips can lead to their rapid deterioration because the dessicant cannot function. For these reasons it is essential that instrument and operator performance are regularly checked, preferably by using suitable quality control material.

The advantages of dry reagent systems are not dependent on enzyme immobilisation and re-usability. However, they are convenient and simple to use, avoid waste of liquid reagents, and the solid waste from fully automated systems is easily and safely disposed of by incineration. They also save time, on make-up and dispensing of reagents, by allowing analyses to be done literally 'at-the-bedside', and the start-up and close-down procedures are very rapid. In addition they allow safe handling of high risk samples since patient samples do not make contact with the automated equipment.

Biosensors and Bioanalytical Probes

Biosensors were defined by Pickup as, either a small probe-like device which, without the addition of a reagent, gives a rapid and specific signal in response to an analyte of biological interest, or it is confined to such a device only when a biological component such as an enzyme or antibody is needed for its action. Thus biosensors usually have immobilised biological molecules attached to the surface of a transducer that allows an electronic or optical signal to be converted into an appropriate signal (Fig. 22.7). There has been an enormous amount of research effort in biosensors in the last ten years.

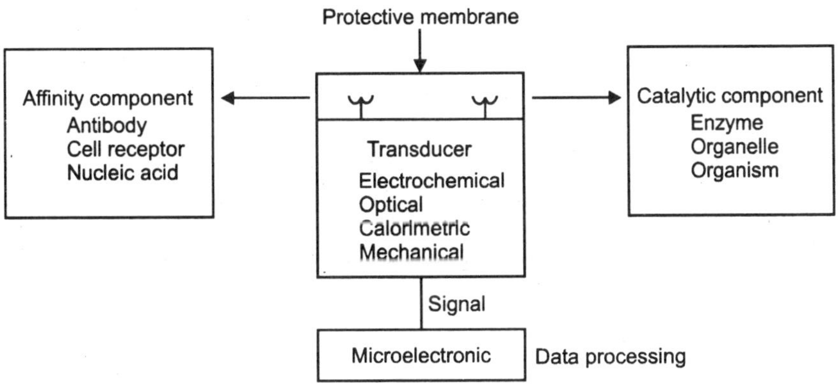

Fig. 22.7. Diagrammatic view of a biosensor showing the various biological components used and the range of transducers.

The main potential uses of biosensors are given below:

1. They can be used as rapid devices with an inexpensive disposable part which will allow clinicians, nurses and patients to perform simple tests. These devices are really in competition with dry reagent stick tests. A successful enzyme-based pen-sized device for measuring blood glucose in 30 seconds is marketed.
2. Biosensors can be used as similar devices for continuous monitoring and bioprocess control. These devices have application in monitoring fermentation processes. A major aim is to develop

a system which will continuously monitor *in vivo* glucose concentrations in diabetic patients and then activate an alarm system if the concentration goes outside specified limits. The ultimate aim is to use such a system to control glucose concentrations *in vivo*, i.e. an artificial pancreas. However, there are many problems which still need to be solved, including immune rejection and/or protein coating of implanted materials. The devices that have been tried so far have been enzyme based.

3. Biosensors can also be used as re-usable devices where interaction between the biological component and the analyte needs to be reversible. Enzyme-based systems which rely on calorimetry or conductance have been developed. The former measure milli-degrees centigrade changes which are [substrate] related. The latter can be used to measure enzyme-catalysed reactions where there is a net change in concentration of ionised species. Sample pH and buffer capacity can affect the response obtained. The specificity of response for both these types of biosensor is determined by the enzyme used. Although such devices can be useful in certain circumstances, they are unlikely to compete with the automated analysers that are currently in routine use and are more versatile in their range of analytical methods.

The following section is presented to illustrate the current state of commercially produced enzyme-based bioanalytical probes, or enzyme electrodes. There are other potential uses of enzymes, particularly in immuno-based biosensors, where the enzyme acts as a label, which will be mentioned later in this chapter.

Enzyme electrodes

An enzyme electrode is a device which consists of a combination of an electrochemical sensor in contact with a layer of immobilised enzyme, which is used to measure the concentration of a metabolite. The outer surface of the enzyme layer is exposed to the sample, which is usually stirred or agitated to minimise concentration gradients. The analyte diffuses into the enzyme layer where it is converted to an electroactive compound which is measured by the transducer, either when steady-state conditions exist or during the initial stages of the reaction. The potential or current of the electrode is measured, relative to a reference electrode, and the concentration of the substrate determined. The relationship is logarithmic for a potentiometric electrode and linear for an amperometric electrode.

Semi-permeable membranes are generally placed between the immobilised enzyme and the solution, and also sometimes between the enzyme and the transducer. These help to increase specificity and to minimise the effect of interfering substances.

Enzyme electrodes for the determination of urea, glucose, lactic acid, ethanol, penicillin, cholesterol, amino acids and many other substances have been developed. The utility of many of these devices is severely limited in practice, particularly in their application to physiological fluids. This can be illustrated by the analysis of urea using an ammonium ion-sensitive electrode; the electrode is surrounded by immobilised urease which hydrolyses urea molecules as they diffuse into it:

$$(NH_2)_2CO + 2H_2O + H^+ \xrightleftharpoons{\text{Urease}} HCO_3^- + 2NH_4^+$$

$$NH_4^+ \rightleftharpoons NH_{3(gas)}$$

The electrode responds to one of the products of the reaction, NH_4^+. However, this transducer also responds to all other monovalent cations including H^+, Na^+ and K^+. Although this particular problem of interfering ions can be eliminated by the use of an ammonia gas-sensitive electrode, this suffers from very slow response and poor washout characteristics. Thus it is unlikely that an enzyme electrode for urea will compete successfully with the simple and rapid colorimetric assays for urea that are already available.

The glucose analyser, produced by the Yellow Springs Instrument Company, was the first commercial instrument to use an enzyme electrode. It is still widely used. The glucose probe (Fig. 22.8) contains a thin layer of glucose oxidase immobilised on glutaraldehyde resin particles and held between two membranes, one of which is placed in contact with a hydrogen peroxide sensitive electrode (that is, a platinum electrode polarised at 0.7 V which acts as the anode of an electrochemical cell). The sample is injected from a special calibrated micro-syringe into a chamber of buffered reagent containing the glucose probe. Glucose which diffuses to the probe from the solution is converted to gluconic acid and hydrogen peroxide by the immobilised glucose oxidase, and the hydrogen peroxide is determined amperometrically at the electrode. The success of the instrument owes much to the nature of the membranes, which not only retain the immobilised enzyme particles but also minimise the amount of interfering substances reaching the electrode. The outer membrane is of a polycarbonate material with a pore size large enough to allow passage of glucose, oxygen, and other small molecules, but too small to allow cells and macromolecules to pass through. The inner membrane is of cellulose acetate with a much smaller pore size (cut-off 100 Daltons) which excludes glucose, uric acid, ascorbic acid and most other potentially interfering substances from reaching the electrode, while still allowing some molecules such as hydrogen peroxide to pass through. However, users of this instrument should be aware that some blood preservatives and certain drugs do reach the electrode where their oxidation produces erroneous results. The usefulness of enzyme reagents is further exemplified by another feature of the Yellow Springs analyser. The reagent solution contains catalase, in a soluble form, to scavenge 'stray' hydrogen peroxide which may diffuse back into the probe, causing drifting of the anodic current.

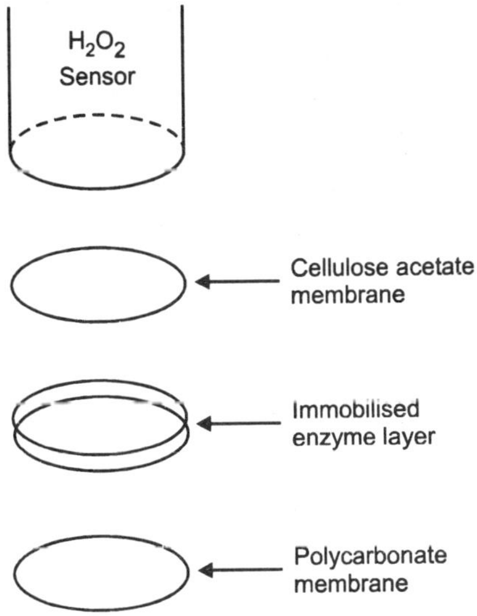

Fig. 22.8. Exploded view of the Yellow Springs Instrument Company glucose probe.

In the laboratory the Yellow Springs instrument is most frequently used to analyse urgent blood glucose specimens that arrive at irregular times both day and night; and also as a means of producing rapid results in an outpatient diabetic clinic. The immobilised enzyme-membrane disk is routinely changed at monthly intervals, typically having been used for between 600 and 700 assays. When stored at 4°C the disks have a shelf-life of about 1 year.

Other commercial glucose analysers may also be modified for use with immobilised enzyme preparations. For example, Sokol attached glucose oxidase to a derivatised Teflon membrane which they placed over the tip of the Clark-type oxygen electrode of a Beckman glucose analyser. These authors claimed that the immobilised enzyme modification gave similar results to the soluble enzyme normally used in this instrument, and offered the advantage of simplicity, economy of enzyme and linearity over a greater range of concentrations. One disadvantage of this design is that the electrode cannot be used with whole blood since the electrode is competing with haemoglobin for the oxygen needed for the enzyme reaction.

Two other enzyme-based electrodes have been developed commercially. A cholesterol analyser, which utilised cholesterol oxidase, was marketed for a few years but is no longer available, whereas a lactate analyser, based on lactate oxidase, has recently been launched.

Currently the most successful commercial glucose biosensor, the ExacTech, is marketed by Medisense. This has a disposable enzyme electrode strip which is attached to a small metre. The metre is used to time the assay, 30 seconds, and to display the result. The device works on a sample of whole blood which remains in contact with the electrode for the whole time. A novel aspect of this assay is that a mediator, ferrocene, is incorporated into the electrode. This regenerates oxidised enzyme and passes the electron to the electrode as illustrated in the scheme below:

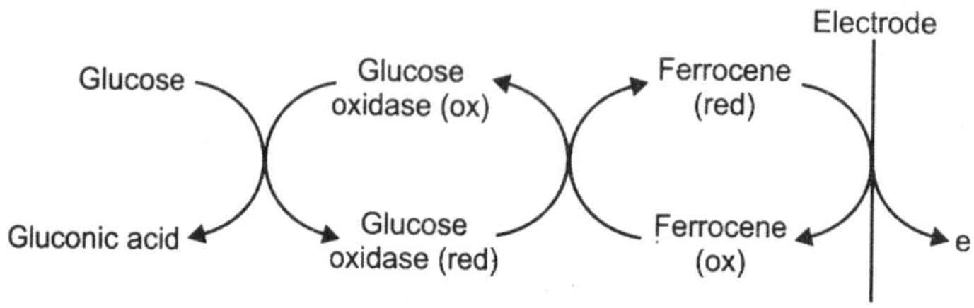

The mediator speeds up the reaction, allows the measurement to take place in the absence of oxygen, and ensures that hydrogen peroxide is not produced. These are all features that will be required if a successful *in vivo* glucose electrode is to be produced.

ENZYME IMMUNOASSAY (EIA)

The earlier literature on this topic was comprehensively reviewed by Carlier. He cited over 600 references, nearly all of which referred to the 1970s. They tabulated the uses of EIA for the quantitative estimation of serum proteins, blood clotting factors, immunoglobulins, antibodies to hormones, a wide range of drugs, viruses, bacteria and parasites. The range of EIA formats was reviewed by Gould and Marks, together with their advantages and disadvantages.

The principle of competitive immunoassays is shown in the equation below:

$$\text{Antibody (Ab)} \quad \begin{array}{c}\text{Antigen (Ag)}\\ + \\ \text{Labelled Antigen (*Ag)}\end{array} \rightleftharpoons \begin{array}{c}\text{Ab–Ag}\\ + \\ \text{Ab–*Ag}\end{array}$$

Free labelled (*Ag) and unlabelled antigen (Ag) molecules compete for a fixed but limited number of specific binding sites on antibody molecules. After an incubation period the free and bound antigen are normally separated from each other. Then the amount of labelled antigen in either or both fractions is determined. When the quantity of unlabelled antigen is increased, fewer labelled antigen molecules will be bound by antibody. Therefore a standard curve can be produced and, by using this, the concentration of antigen in biological samples can be determined. Radioactive labels, hence radioimmunoassay (RIA), have been used since the late 1950s. Enzyme labels became available in the early 1970s.

RIA allows the direct measurement of small quantities of a wide range of compounds present in complex biological samples without prior separation. However, radio labels have disadvantages. Preparation of labelled antigen may involve a health hazard. The shelf-life of the labelled product may be short, for example when γ-emitting isotopes such as ^{125}I are used. Radioactive and organic wastes have to be disposed of safely. The equipment needed to measure radioactive samples is expensive.

These disadvantages encouraged research into alternative labels, of these enzymes are now the most frequently used. Table 6.5 lists the advantages and disadvantages of enzyme labels. Gosling wrote a comprehensive review of immunoassay methodology. This deals with all types of labels used in immunoassays. At present about one-third of all new immunoassays use enzymes as labels.

Table 22.5. Comparison of enzyme—with radio labels in immunoassays.

Advantages of enzyme labels
No radiation hazards occur during labelling or disposal of waste
Labelled product can have a long shelf-life
Equipment can be inexpensive and is generally available
Homogeneous assays can be completed in minutes
Disadvantages of enzyme labels
Plasma constituents may affect enzyme activity
Assay of enzyme activity can be more complex than measurement of some types of radioisotopes

There are two types of quantitative EIA, heterogeneous and homogeneous. In the former the bound label that is finally measured must be separated from the free label, as is the case with RIA. In homogeneous EIAs the final activity of the enzyme label is altered as a result of interaction of the antigen/hapten with antibody. Therefore, separation is no longer necessary, since it is the change in enzyme activity that is determined. Homogeneous assays have the advantage, when this is necessary, of rapidity, and they lend themselves to easy automation. But they are also the assays most susceptible to interference by plasma constituents.

Preparation of Enzyme Labels

Ideally the enzyme labels, whether they are on antibody, antigen or hapten, should be stable. Also the chemical reactions used to produce these enzyme labels should not affect the activity of the enzyme. Equally the specific interaction between antibody and antigen or hapten should not be altered by the combination of antibody with enzyme. Haptens are small molecules, for example, drugs or steroids, which are not usually immunogenic but which become so when coupled to a larger carrier protein.

The reactions used to cross-link enzymes with antibody, antigen, or hapten are similar to those used to form immobilised enzymes.

The most common reactions used to form enzyme-labelled antigens or antibodies are:

1. Glutaraldehyde method: The dialdehyde links proteins through their amino groups. The main disadvantage is the self-linkage of both types of protein. This is reduced if the reaction is done in two stages instead of one.
2. Periodate method: This method is limited to glycoproteins, for example peroxidase. The periodate is used to oxidise the carbohydrate residues of the enzyme; the aldehyde produced can then react with the amino groups of the antigen protein.
3. Maleimide method: This method joins proteins by combining with sulphydryl groups on both proteins or with an amino group on the antibody. The latter reaction can reduce the amount of self-polymerisation.

Enzyme-labelled haptens are formed by a limited number of reactions:

1. Mixed anhydride method: A peptide bond is formed between a carboxylic acid residue on the hapten and an amino group on the enzyme.
2. Carbodiimide method: This reaction also forms a peptide bond between the hapten and enzyme. If necessary, carboxylic acid residues can be introduced into haptens containing hydroxyl groups.
3. Bifunctional imidate method: A cross-link is formed between amino groups on the hapten and enzyme.

Choice or Enzyme Label

Enzymes function as labels because of the reaction that they catalyse. This results in an amplification effect, so that each molecule of active enzyme causes the conversion of many molecules of substrate to product. Ideally, the enzyme label should have the properties listed in Table 22.6. In practice the enzymes used have limitations, and these need to be considered for each assay. Most of the enzymes are of bacterial or plant origin.

Table 22.6. Properties of an ideal enzyme label.

High specific activity at the optimum pH for antibody-antigen binding
Presence of reactive groups through which enzymes can be covalently linked to antibody, antigen, or hapten with minimal loss of enzyme and immune activities
Stable enzyme-labelled conjugates under storage and assay conditions
Absence of enzyme activity and factors affecting the activity of the enzyme from the tested fluid (for homogeneous EIA)
Cheap, accurate, and sensitive assay method, preferably colourimetric
Availability of soluble, purified enzyme at low cost
Absence of health hazards attributable to enzyme, substrate(s), or cofactor(s)

The enzymes most commonly used in kit form for heterogeneous assays are listed in Table 22.4. The relevant properties of these enzymes have been thoroughly discussed by Ishikawa and Yolken. Peroxidase is the most commonly used of these enzymes. It is the cheapest and the smallest (40 kDa) of the enzymes listed, and also has the highest turnover number. Since it is also a glycoprotein, it can be conjugated by the periodate method or by the normal protein methods. It is prone to inactivation by preservatives, and by oxidising and reducing agents. The enzyme is detectable in femtomole amounts by colourimetry. Several chromogens have been used to visualise the enzyme, and they each have their own disadvantages.

The most sensitive, *o*-phenylenediamine, is photosensitive; 2,2′azino-di-(3-ethylbenzthiazoline-6-sulphonate) is expensive, and the product is unstable; 5-aminosalicyclic acid has low sensitivity; *o*-toluidine and *o*-phenylenediamine have been found to be carcinogenic or mutagenic in laboratory animals. At present tetramethyl benzidene is frequently used as it is not thought to be carcinogenic and gives very sensitive assays. Alkaline phosphatase (100 kDa) and its conjugates are very stable. The calf intestinal enzyme can be detected in femtomole amounts, but the purified enzyme is expensive. Alkaline phosphatase is also in the first stage of the two commercially available amplifying systems which can be employed to improve the sensitivity of heterogeneous enzyme immunoassays. Urease produces a strong purple colour as its end-point. It is claimed that this allows for easier visual identification of positive results. Quantitative analyses as sensitive as those with other enzyme labels are also claimed. β-Galactosidase can also be detected colourimetrically in femtomole amounts, and it has a potential advantage over peroxidase and alkaline phosphatase in that it is absent from animal cells. This accounts for it being the only one of these enzymes that is also used for homogeneous immunoassays. Lysozyme was the first enzyme label used in these assays. It is still used for the semi-quantitative assay of a number of drugs present in urine in the mg/litre range. More sensitive quantitative assays in serum are possible with glucose 6-phosphate dehydrogenase as the enzyme label. The enzyme is obtained from *Leuconostoc mesenteroides*, as this enzyme specifically uses NAD^+ as one of its substrates. This overcomes the possibility of interference with the equivalent human red cell enzyme which is specific for $NADP^+$. The ease of removal and subsequent re-addition of FAD, the prosthetic group of glucose oxidase, is the main reason for this enzyme being used in the PGLIA system. A few urine assays use the more expensive but more sensitive malate dehydrogenase. This enzyme would be susceptible to interference in homogeneous immunoassays of serum components.

Homogeneous EIA

The first commonly used homogeneous EIA was the 'enzyme multiplied immunoassay technique or EMIT© (Syva Maidenhead, UK).

The principle of this assay is illustrated in Fig. 22.9. It depends on the occurrence of inhibition of the enzyme, to which the hapten is attached, when hapten-specific antibody binds. Free hapten, in the standards or samples, competes for the antibody and relieves this inhibition. Therefore enzyme activity is proportional to the amount of free hapten in the samples. The inhibition of enzyme activity by antibody may be due to: (i) prevention of access of substrate to the active site by steric hindrance, (ii) causing a conformational change in the enzyme, and (iii) prevention of a conformational change necessary for enzyme activity. A different mechanism has been found with a thyroxine assay. In this case the thyroxine-malate dehydrogenase conjugate is enzymically inactive, but on binding to thyroxine antibodies it is activated. The combination of enzyme with thyroxine must inhibit the enzyme by changing the active site. The antibody reactivates the enzyme by binding with thyroxine and reversing the change. The EMIT system is easily automated on a reaction rate analyser or a centrifugal analyser. EMIT was originally applied only to small haptens, such as drugs, but the assay range was extended to include proteins.

An alternative homogeneous EIA system uses substrate-labelled hapten. The principle of this 'substrate-labelled fluorescent immunoassay' (SLFIA) is illustrated in Fig. 22.10. The substrate of the enzyme is covalently linked to the hapten to be measured, for example umbelliferone-gentamicin which can be hydrolysed by the enzyme, which is β-galactosidase, to yield a fluorescent product. Hapten in the sample and substrate-labelled hapten compete for the limited number of sites on the hapten-specific antibody. Antibody-bound molecules of substrate-labelled hapten are inactive as substrate. Therefore, both the rate of production of fluorescence and the total fluorescence are proportional to the amount of free hapten in the sample or standard. This type of assay does not make use of the amplification due to enzyme activity found in other EIAs. This reduces its sensitivity, but the use of a sensitive fluorimetric detection procedure adequately compensates for this drawback. Despite the apparent complexity of SLFIA, this format has been converted to a dry reagent system.

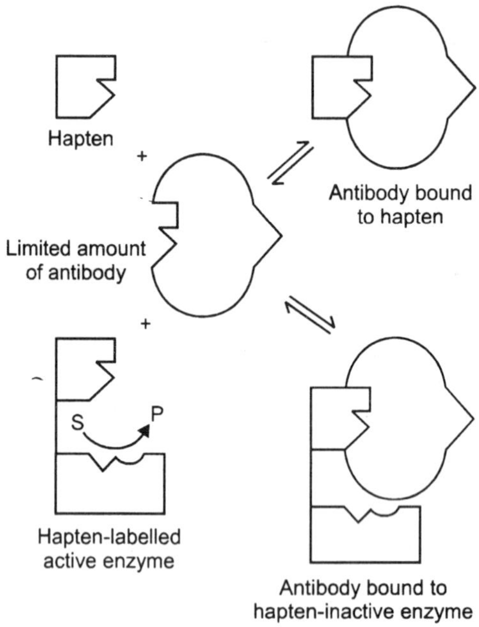

Fig. 22.9. Principle of the commercially available homogeneous EMIT system.

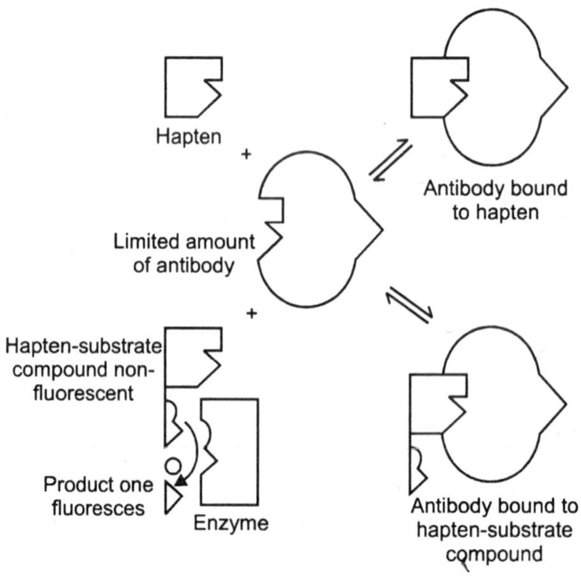

Fig. 22.10. Principle of the commercially available homogeneous SLFIA system.

Another type of homogeneous assay, which relies on the recombination of a prosthetic group with its apoenzyme to reform active holoenzyme, was described by Ngo and Lenhoff. They used glucose oxidase from which the prosthetic group, FAD, is easily removed. The principle of this 'prosthetic group-labelled-immunoassay (PGLIA)' is illustrated in Fig. 22.11. Free hapten competes with hapten-labelled prosthetic group for the limited amount of antibody. Only the remaining free hapten-labelled prosthetic group is then able to combine with apoenzyme to form active glucose oxidase. This PGLIA assay has been adapted for use as a dry reagent system.

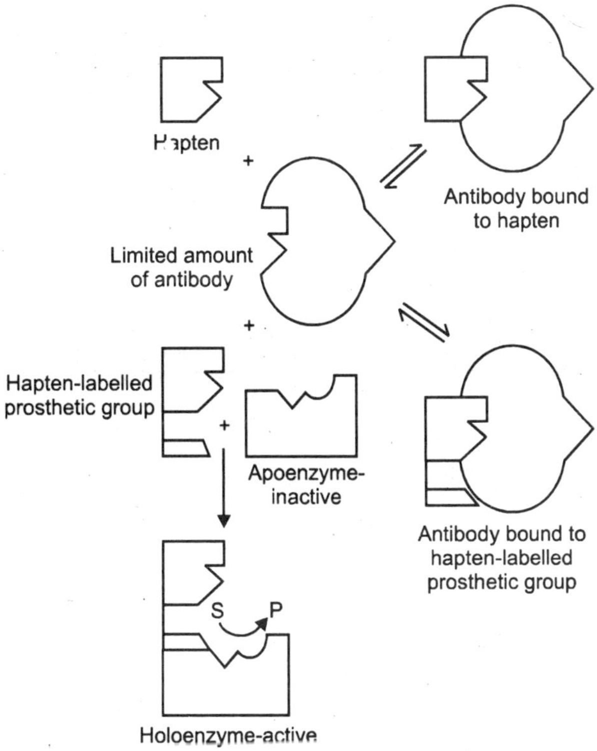

Fig. 22.11. Principle of the commercially available homogeneous PGLIA system.

A more recent development is the CEDIA^{TM} system, which is illustrated in Fig. 22.12. For this assay two separate polypeptide fragments of β-galactosidase have been synthesised using recombinant DNA technology. These fragments are: enzyme donor (ED, about 95 per cent of the protein) and enzyme acceptor (EA, about 5 per cent of the protein). These fragments can spontaneously combine to generate active enzyme. In CEDIA technology, excess antibody reacts first with hapten or standard. In the second stage, the excess of antibody is reacted with ED-labelled hapten. Finally, the remaining ED-labelled hapten is made to react with EA, to give active b-galactosidase. This assay is easily automated and the signal increases as the amount of hapten in the sample increases.

Heterogeneous EIA

Engvall and Perlmann pioneered the use of solid-phase supports for heterogeneous assays. These are now the most frequently used method of separation, and are used in all the commercial heterogeneous systems. The solid phases are made from a variety of plastics and are in the form of pre-moulded multiple sample wells, beads, cuvettes, or tubes.

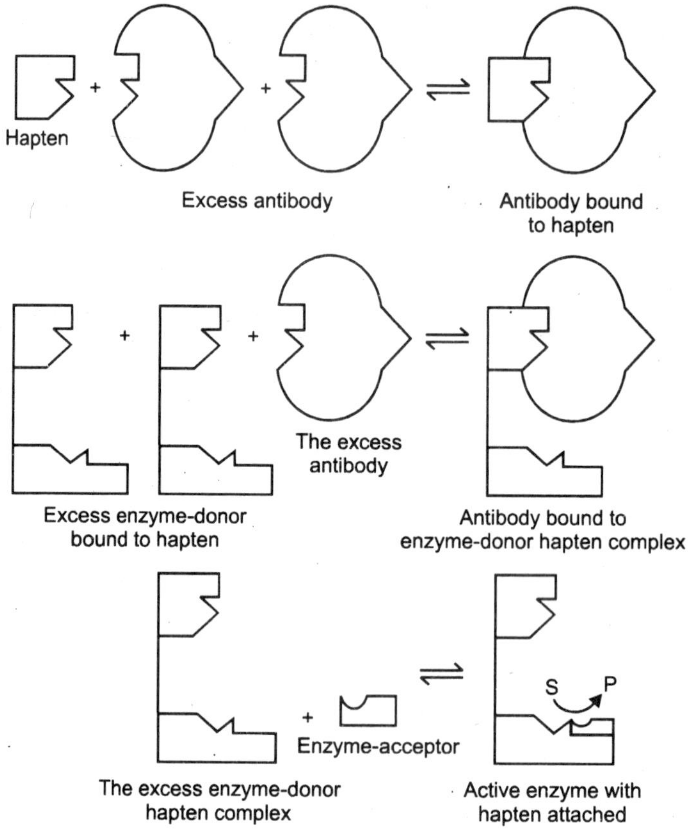

Fig. 22.12. Principle of the commercially available homogeneous CEDIA system.

These types of assay are frequently referred to as enzyme-linked immunosorbent assays (ELISA). In these assays an antigen or antibody is coated onto the surface of the solid phase by adsorption. It is essential that this coating process is reproducible and some workers therefore recommend covalent attachment.

Several types of heterogeneous EIA have been developed. The principles of a representative selection of systems that are in routine clinical use are described below.

Figure 22.13 illustrates principles of competitive EIA for antigen. This method is analogous to classical competitive RIA. The first stage involves competition between solid-phase antigen and free antigen, in the sample or standard, for the limited quantity of enzyme-labelled antigen-specific antibody. Excess free antigen and antigen bound to antibody are removed by washing.

The amount of enzyme-labelled antibody bound by solid-phase antigen is inversely proportional to the concentration of free antigen present in the sample. The enzyme activity remaining bound to the solid phase is determined. This method can be used for hapten measurement if hapten molecules are used in place of antigen molecules.

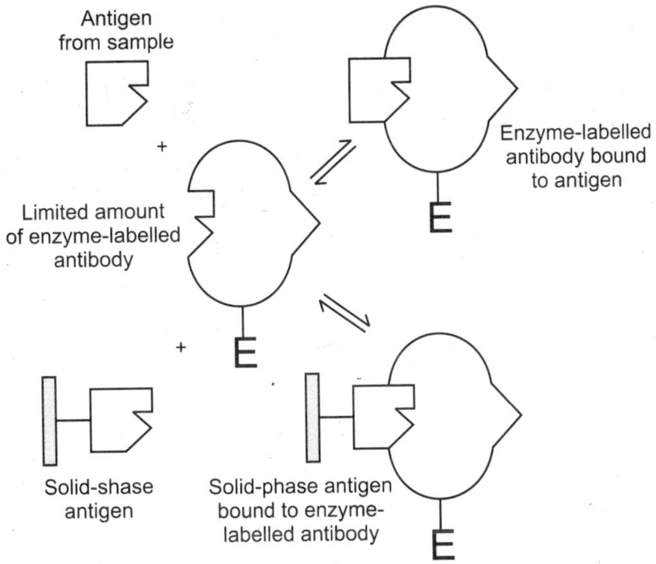

Fig. 22.13. Principle of a solid-phase competitive EIA for antigen.

Figure 22.14 illustrates the principles of the non-competitive or sandwich or capture EIA for antigen. In the first stage antigen, in sample or standard, is incubated with excess solid-phase antigen-specific antibody. After washing, excess enzyme-labelled antibody is added, the excess enzyme being eliminated by washing before the enzyme activity is determined. In this method, the enzyme activity bound to the solid phase is directly proportional to the amount of antigen in the sample. This type of assay can only be used for antigens which possess more than one antigenic determinant.

Figure 22.15 illustrates the principles of the non-competitive or direct sandwich assay for antibody. In the first stage excess solid-phase antigen is incubated with sample antibody. After washing, excess enzyme-labelled second antibody is added.

This second antibody is raised against immunoglobulins of the animal species in which the first antibody was raised. Excess enzyme-labelled second antibody is eliminated by washing, and the bound enzyme activity is measured. This activity is directly proportional to the amount of antigen-specific antibody present in the sample.

Competitive Versus Non-competitive EIA

The essential difference between these two types of assay is that the competitive assay uses only a limited amount of antibody while, in the non-competitive EIA, an excess of solid-phase antibody is available. This difference gives rise to the following consequences:

1. With the competitive ELISA the signal decreases with increasing antigen concentration. The opposite is the case with a non-competitive assay.
2. Although the competitive ELISA produces its greatest signal at low concentrations of antigen, the non-competitive assay gives lower detection limits.
3. The non-competitive assay offers the advantage of greater specificity since two separate epitopes are recognised when two different monoclonal antibodies are used. In this case the antigen and labelled antibody can be added at the same time, thus reducing assay time since one incubation step is eliminated.

4. In the non-competitive assay, the affinity of the capture antibody is less important than in a competitive assay, since excess antibody is used.
5. The accuracy of pipetting of the capture antibody is less critical since it is no longer a limiting factor as it is in the competitive format.

For these reasons the non-competitive assay is characterised by being more rapid, more reliable and more rugged than competitive assays, with the added advantages of improved sensitivity and specificity.

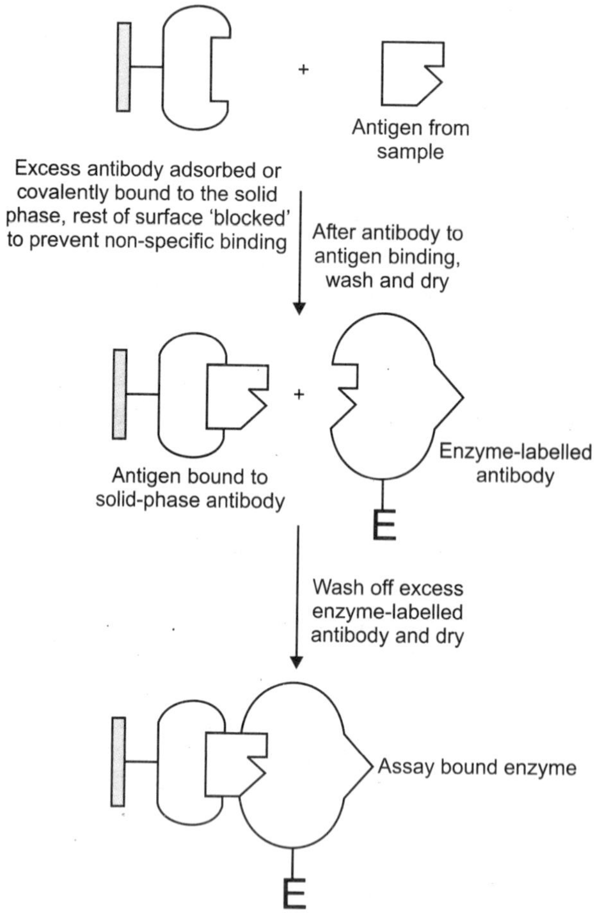

Fig. 22.14. Principle of a solid-phase non-competitive (or capture or sandwich) EIA for antigen.

Assay in EIA

In the assay of enzymes in heterogeneous EIAs, the reaction is stopped at a particular time, for example by the addition of acid or alkali, within the early linear part of the progress curve (Fig. 22.1). The whole process of incubation, washing, and enzyme assay is easily automated with the equipment that is now available. Individual assays generally take up to a day, but with microtitre plates 96 wells are processed at one time.

For homogeneous EIAs a kinetic assay is normally used. An individual assay can usually be done within two minutes, although extra time is required to produce a calibration curve. The procedure can be automated, on standard kinetic analysers or centrifugal analysers, and about 30 assays can be performed at one time in about 20 minutes.

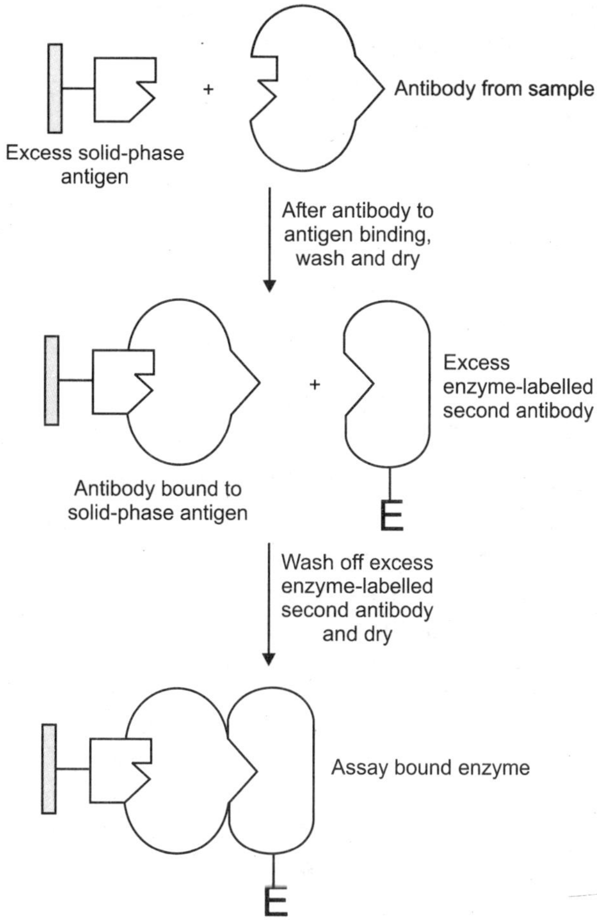

Fig. 22.15. Principle of a solid-phase non-competitive (or capture or sandwich) EIA for antibody.

The determination of enzyme activity is mainly by photometric methods, often using the same end-points as were discussed previously. β-Galactosidase is measured either colourimetrically or fluorometrically; the latter assay can be about 100 times more sensitive. The assay of lysozyme is more unusual; it is assayed by turbidimetry.

Practical approaches to increased immunoassay sensitivity using enzymes

There are three main ways of exploiting the potential for more sensitive EIAs offered by non-competitive assays discussed earlier. These are:

1. To increase the molar ratio of enzyme to antigen in the final step: This can be achieved by using the multiple binding sites of avidin or streptavidin for biotin. Alternatively, enzymes entrapped in liposomes can be disrupted at the enzyme assay step and thus used to amplify the signal.
2. To amplify the signal produced: A system based on enzymic recycling was patented and used by Self and others and is available as a kit. Alkaline phosphatase converts $NADP^+$ to NAD^+ in

the first stage (Fig. 22.16). The second stage involves two further enzymes, alcohol dehydrogenase which produces NADH, and diaphorase which regenerates NAD^+, for recycling, and a coloured formazan end-product which accumulates and is measured. Typically a 40- to 100-fold increase in absorbance for the same amount of alkaline phosphatase can be achieved.

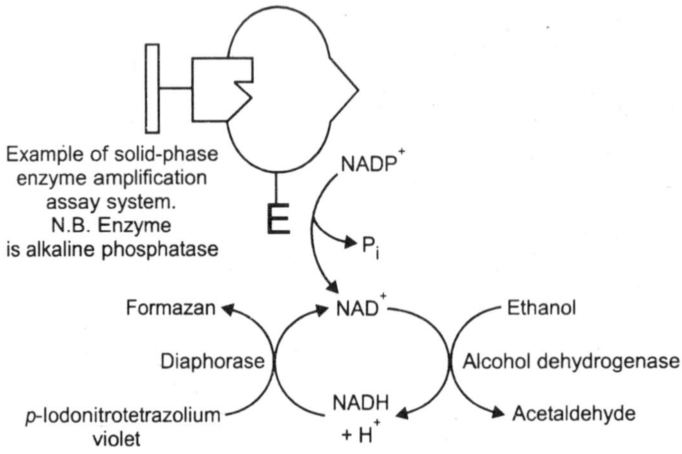

Fig. 22.16. Principle of an enzymic amplification system based on recycling between NAD^+ and NADH.

Another system, this one based on the conversion of inactive apo-enzyme to active holo-enzyme, is illustrated in Fig. 22.17. Again the solid-phase enzyme label is alkaline phosphatase. This is used to dephosphorylate a 3′-phosphorylated derivative of FAD to form FAD which then converts apo-D-amino acid oxidase to the holo-enzyme. Another probable candidate enzyme is glucose oxidase. The holoenzyme reacts with its substrate, D-proline is used, and the H_2O_2 produced is visualised with a peroxidase-catalysed assay linked to suitable dyes. The apo-enzyme is available commercially from Calzyme Laboratories, and the amplification method has been patented by Rabin. The system can detect down to 0.03–10 attomoles of alkaline phosphatase but can be detuned to quantitate higher levels of enzyme if required.

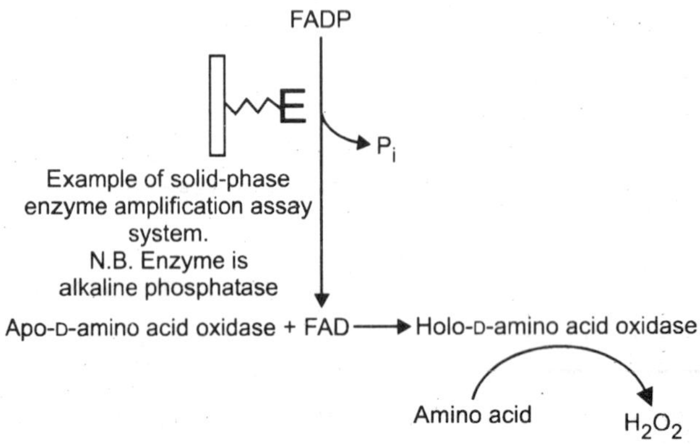

Fig. 22.17. Principle of an enzymic amplification system based on the conversion of inactive apo-enzyme to active holo-enzyme.

3. To employ a more sensitive detection method. For instance fluorimetric assays are available for β-galactosidase and alkaline phosphatase, and reactions catalysed by horse radish peroxidase can be measured by luminometry.

Non-competitive assays for haptens

Haptens, being small molecules, generally only have one epitope. This means that they cannot be assayed by the non-competitive assay formats discussed above, which all require the binding of two different antibodies. However, Freytag devised a novel non-competitive technique which only required a single antigenic site and could therefore be used with haptens. The technique uses $F_{(ab')}$ fragments, which are produced from IgG by using pepsin to form two identical parts, each with one binding site. These are conjugated with enzyme. Excess enzyme-labelled $F_{(ab')}$ is incubated with sample hapten. The mixture is then passed through an affinity column containing immobilised hapten in large excess. This binds the excess enzyme-labelled $F_{(ab')}$ but allows the enzyme-labelled $F_{(ab')}$-hapten complex to pass through and be collected. Enzyme activity, which is directly proportional to the amount of original hapten in the sample, is measured. Using this methodology, Freytag was able to measure 10 pmol of digoxin in 7 minutes. This assay could be combined with the practical approaches outlined already to produce very sensitive assays for haptens.

Immunochromatography

A novel way of measuring EIA in a dry reagent format was developed by Zuk and others. They utilised a dry paper strip containing excess immobilised antibodies to measure antigen/hapten concentrations in blood. The strip is placed in an aliquot of liquid containing the enzyme glucose oxidase, a fixed amount of peroxidase-labelled antigen and a measured volume of blood containing the antigen (theophylline was used initially) to be measured. The mixture migrates to the top of the paper strip in about 10 minutes. The distance travelled by the peroxidase-antigen conjugate increases with increasing amounts of sample antigen since they both compete for binding to the immobilised antibody. However, the glucose oxidase is evenly distributed throughout the strip. When the solvent front has reached the top of the strip it is immersed in a reagent containing glucose and a chromogen. The function of the glucose oxidase is to generate H_2O_2 which serves as a substrate for the peroxidase which is visualised by conversion of the chromogen into an insoluble blue product. The total test takes about 15 minutes, which includes reading the length of the coloured zone with a ruler, which is provided in the commercial version. The length of colour is proportional to the sample antigen concentration. A conversion table is provided with each kit since the results obtained are very dependent on the consistency with which antibody is immobilised. Kits for several drugs are available.

Simultaneous assay of two haptens

A potential advantage of EIA is the relative ease of measuring two haptens simultaneously. This possibility was noted by Wisdom, but the first working system was published by Blake. The advantages of this system, compared to separate single-hapten methods, include savings on reagent cost, sample volume, and a reduced overall assay time.

Simultaneous measurement of two haptens using two iodine isotopes, ^{125}I and ^{131}I, was achieved by Mitsuma, but the usefulness of this type of assay is restricted by the short half-life of ^{131}I and the less than ideal distinction between the γ-emissions of the two isotopes.

The first simultaneous EIA of two haptens measured thyroxine and triiodothyronine. Alkaline phosphatase, conjugated to thyroxine, was detected at 540 nm by the liberation of phenolphthalein from its monophosphate. The second enzyme was β-galactosidase which was conjugated to triiodothyronine and assayed at 420 nm by the formation of *o*-nitrophenol from its substrate, *o*-nitrophenyl-β-galactosidase. The two end-points allowed each hapten to be measured in a single tube.

FUTURE

Four aspects of enzymes in clinical analysis have been reviewed. Two of these, the measurement of metabolites and the measurement of enzymes, are well established. The basic methodologies have changed comparatively little, but the availability of increasingly sophisticated machines has considerably increased the number of assays that can be performed both on an individual specimen and in a given time. More sophisticated machines can do more tests and they are discriminatory, enabling them to do only the tests requested by the physician. The third section of this review dealt with immobilised enzymes. Their impact has been disappointing. Their re-use in automated systems has not progressed. Bioanalytical probes and biosensors have been the subject of much research but as yet the only commonly used systems measure glucose. The main problems with biosensors for *in vivo* use are to enable them to remain stable for a reasonable length of time in the hostile environment of the body. Their use *in vitro* is generally in direct competition with dry reagent systems. The recent commercial development of optically based biosensors could lead to simple devices, allowing a wide range of assays in the near future. However, few of these are likely to be enzyme based. Dry reagent technology has been commercially attractive for over 35 years. In this time a wide range of assay formats have been developed and improved. These range from simple reliable systems that can be used by patients themselves through to sophisticated automated analysers for use in routine or emergency laboratories. Hand-in-hand with such developments we should also expect to see the introduction of small portable, simple-to-operate machines for use in the physician's office; this concept is particularly attractive in the USA, or by the patient's bedside. The final part, on EIA, also described a powerful, versatile technique with considerable potential. The range of available EIAs is increasing, as is the supply of sophisticated dedicated equipment for the automation of both homogeneous and heterogeneous assays. Once again many of these assays have been produced in a form that can be used by the physician, pharmacist or patient. A major concern with these simplified assays is to ensure that there is good quality control, otherwise incorrect results may lead to inappropriate action. The use of enzymes in clinical analysis will continue, but the range of assays using them will expand because of their convenience, specificity, versatility and reproducibility.

Chapter 23

Enzymes in Organic Synthesis

INTRODUCTION

In the past decade, an abundance of material has appeared in the literature concerning the structure and properties of a whole range of enzymes, and their isolation and exploitation. The use of biological systems (including isolated enzymes, whole cells such as yeasts, novel organisms generated through recombinant DNA technology, and catalytic antibodies) in organic synthesis has been subject to extensive review. The purpose of this chapter is to offer some insight into how the structural and mechanistic properties of the major classes of enzymes currently exploited on a synthetic scale (sub-divided into hydrolytic, oxidative-reductive, and carbon-carbon bond-forming enzymes) facilitate the exploitation of these enzymes, and to indicate some of the most recent applications of enzyme technology in synthesis including possible future applications. Wong and Whitesides have emphasised that the area of enzymatic catalysis in organic synthesis is, as are most areas or practical catalysis, fundamentally a process chemistry. Therefore this review seeks to describe the process of the exploitation of biocatalysis rather than to discuss individual applications in detail. For this reason, the emphasis is on enzymes which have been (or are in the process of being) isolated, and perhaps stabilised or genetically modified, for use in organic synthesis, and on features which make these biocatalysts attractive for use in process chemistry. It is hoped that an idea of the current limitations of enzymatic catalysis will also be presented, and some indications as to how industry and academia are working to circumvent these limitations.

HYDROLYTIC ENZYMES

Enzyme Structure and Function

The most widely exploited enzymes to date, in the laboratory and in industry, are the hydrolytic enzymes: lipases, esterases, proteases, amidases, etc. Their function *in vivo* is to catalyse the formation or hydrolysis of ester and amide bonds, such as those present in proteins, lipids and sugar conjugates. Mechanistically, the hydrolytic enzymes can be divided into four categories:

1. Serine hydrolases (e.g. chymotrypsin, subtilisin, lipase, penicillin acylase and pig liver esterase) possess a triad of catalytically active residues in the enzyme active site serine, histidine and (usually) aspartate—which catalyse the formation of an *o*-acyl intermediate covalently bound to the enzyme via the serine residue. Nucleophilic displacement of this is intermediate then releases the product. The nucleophile may be water, as in the hydrolysis of amide bonds by proteases, or an alcohol, amine or thiol group.

2. The catalytic mechanism of the cysteine hydrolases (such as papain) is similar, except that a cysteine, rather than a serine, residue serves as the initial nucleophile.
3. Metalloproteases (e.g. thermolysin and carboxypeptidase A) employ a divalent zinc ion as a Lewis acid, which coordinates to the nucleophilic water molecule and to the carboxyl group of the substrate, and no covalent intermediate is formed.
4. The mechanism of aspartyl proteases (e.g. pepsin and HIV protease) is a general acid/general base catalytic process involving carboxylic acid residues in the active site, and may or may not involve the formation of a covalent intermediate.

The use of hydrolytic enzymes to generate enantiomerically enriched materials by techniques such as the kinetic resolution of racemic chiral alcohols or esters, the desymmetrisation of esters of *meso*-diols or diacids by selective hydrolysis of a single ester function (or acylation of a single hydroxyl function), or the selective coupling of amino acids or peptide fragments is now well established. Stabilised forms of many of the more frequently employed enzymes (such as immobilised lipases) are commercially available and can readily be used 'off-the-shelf'.

The use of water-immiscible organic solvents as media for serine or cysteine hydrolase-catalysed transesterifications is now well established. In such a system, a chiral secondary alcohol or a *meso*-diol acts as the nucleophile in the hydrolase-catalysed dissociation of an organic ester (the acyl donor), and undergoes kinetic resolution (Fig. 23.1) or desymmetrisation as a consequence. Such reactions often prove inconvenient or unsuccessful in aqueous media, owing to the typical low solubility of the substrates and the occurrence or competing hydrolytic reactions. Use of an organic solvent system offsets these difficulties, frequently leading to improved yields and enantio-selectivities, and also often conferring improved stability to the enzyme (which remains encased in a hydration sphere). However, in order to avoid the establishment of complex equilibria in such systems it is necessary that the acyl donor be a highly activated species (such as the ester of an oxime or trihaloacetic acid), or else (for instance) an enol ester (so that when the acylenzyme intermediate is formed, the released enol rapidly tautomerises and thus makes the reaction irreversible).

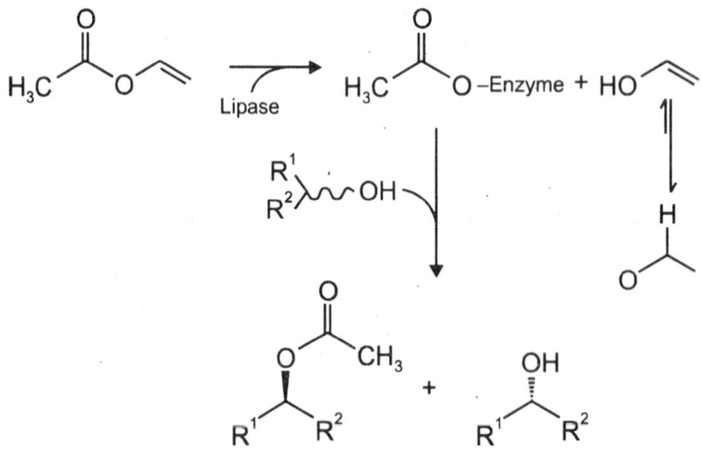

Fig. 23.1. Kinetic resolution of a secondary alcohol by irreversible enzymic acylation.

Lipases, which are among the most widely used of all the hydrolytic enzymes, catalyse the cleavage of esters of long-chain fatty acids. Lipases from a variety of sources have found wide applicability in organic synthesis, and have been extensively used for kinetic resolutions of chiral alcohols and desymmetrisations of *meso*-diols in organic solvents. *In vivo*, lipases function at the interface between the enzyme and a large lipid droplet; in aqueous solutions with soluble substrates, lipases display little activity, but activity increases very markedly once the substrate concentration exceeds the critical micellar concentration. Sih and his colleagues isolated two forms of the lipase from *Candida rugosa*, formerly *Candida cylindracea* (CRL) which differed significantly in their enantioselectivity towards chiral carboxylic acids [resolutions of racemic chiral carboxylic acids using this lipase had consistently demonstrated poor enantioselectivity, though resolutions of racemic chiral alcohols with excellent enantioselectivity had been observed] and also demonstrated that the poorly enantioselective form of the enzyme could be converted to the more enantioselective form by treatment with sodium deoxycholate followed by precipitation and repeated purification. Kazlauskas later showed that a simple treatment of the crude lipase with propan-2-ol was sufficient to increase significantly its activity and enantioselectivity. The reason for these observations is that lipases tend to exist in two conformational forms, distinguished by the position of a helical 'lid' above the hydrophobic crevice which contains the active site of the enzyme. In the 'closed' conformation the lid partially covers the active site (though not completely in the case of CRL, so that accessibility of the active site is restricted, but not entirely blocked), whereas in the more hydrophobic 'open' conformation (which is thus favoured by treatment with a less polar solvent than water, such as propan-2-ol) the active site is entirely exposed; moreover, the oxyanion-stabilising residues in the active site are re-orientated slightly, so as to form an 'oxyanion hole' which binds the carboxylic acid substrate in a very precise orientation, thereby conferring high enantioselectivity to the reaction.

The crystal structures of several lipases are known, and a great many of them possess a helical 'lid'. The equilibrium between 'closed' and 'open' forms of the lipases is responsible for the phenomenon of 'interfacial activation'. Above the critical micellar concentration for catalysis, the enzyme is adsorbed on the surface or a lipid droplet; the environment close to the enzyme surface is therefore hydrophobic, and the more hydrophobic 'open' conformation, which allows entry of lipid molecules to the active site and generates an 'oxyanion hole' therein, is favoured. Interfacial activation was not, however, observed with crude CRL, presumably because the 'lid' does not close off the active site completely to the substrate, and (slow) diffusion of lipid into the active site is possible even in the less active 'closed' form of the enzyme.

Active Site Models

Even when the X-ray structures of enzymes are known, exact prediction of the stereochemical outcome, or the enantioselectivity, of a given reaction is seldom straightforward. Some hydrolytic enzymes have demonstrated unusual lack (or even reversal) of expected enantioselectivity among a series of closely related substrates. In order to rationalise such behaviour, empirical models for the active sites of a number of enzymes have now been described, based on a summation of the known substrate selectivity data. One of the most detailed is that proposed by Jones for pig liver esterase in which a cubic space model of the active site is drawn, comprising two hydrophobic regions and two more polar regions. This active site model has successfully accounted for the enantioselectivity of the enzyme towards a huge range of substrates, and was found to have significant value as a tool for predicting enantioselectivity. Active site models for other hydrolytic enzymes, including chymotrypsin and *Mucor miehei* lipase, have also been proposed and 'Kazlauskas' rules' (Fig. 23.2) have been proposed to account for the enantioselectivity of CRL.

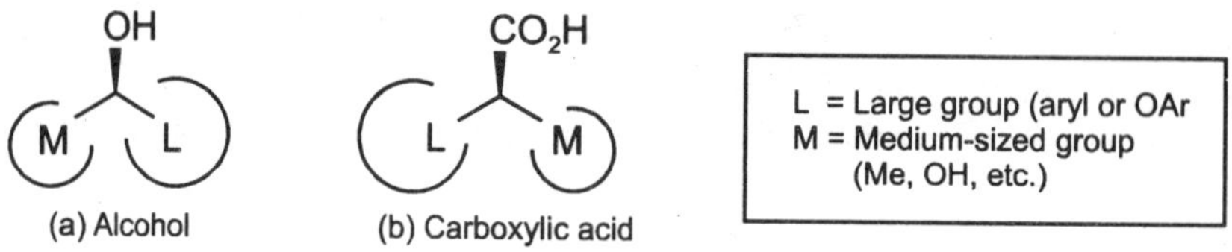

Fig. 23.2. Kazlauskas, rules for *Candida rugosa* lipase.

Hydrolase Enzymes for Novel Chiral Intermediates

Hydrolase enzymes have been extensively used in the generation of enantiomerically enriched (or pure) chiral alcohols, diols, esters, amides, peptides, etc., many of which are valuable synthetic intermediates in the production of pharmaceuticals, materials or fine chemicals. New applications for hydrolase enzymes in the production of chiral chemicals are constantly being discovered, as are enzymes which catalyse novel and/or highly selective transformations. A few such recent applications are reported below.

Epoxides such as glycidaldehyde diethyl acetal (1) are extremely valuable chiral building blocks in organic synthesis. In addition to chemical methods such as the Sharpless asymmetric epoxidation reaction, several enzymatic methods exist for the synthesis of chiral epoxides. One of the most versatile methods is the cyclisation of chiral secondary alcohols derived from the kinetic resolution of racemic 2-acetoxy-3-substituted acetals and, ethers using porcine pancreatic lipase (PPL) or LP-80 lipase (Fig. 23.3).

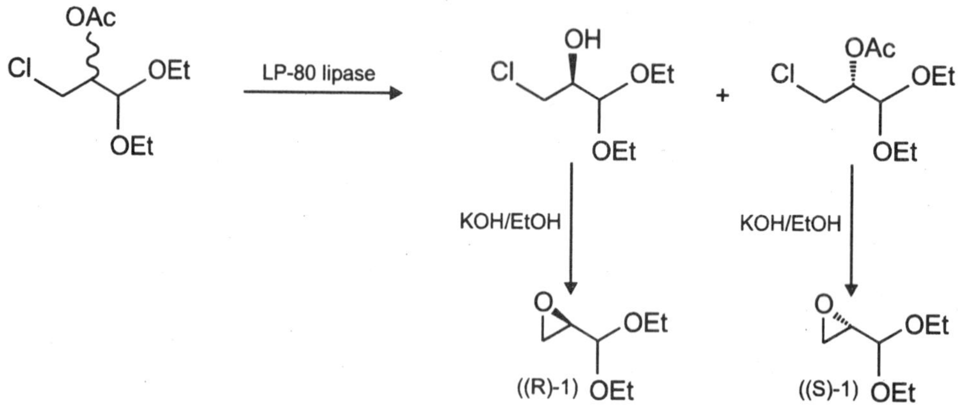

Fig. 23.3. Kinetic resolution route to enantiomerically pure epoxides.

Recently Furstoss's group has reported the use of the epoxide hydrolase activity of the micro-organisms *Aspergillus niger* and *Beauveria sulfurescens* as a route to enantiomerically enriched chiral aromatic epoxides and 1,2-diols from a racemic mixture of the appropriate epoxide. *A. niger* and *B. sulfurescens* appeared to be enantio-complementary in the hydrolysis of racemic styrene oxide, hydrolysing the (*R*)- and (*S*)-epoxide respectively. For other aromatic epoxides, the substrate specificity of *B. sulfurescens* appeared to be wider than that of *A. niger*. It was also noticed that during the hydrolysis of racemic *cis*-1-phenylpropene oxide, both enantiomers were hydrolysed with different regioselectivities, leading to an almost enantiomerically pure diol product in an overall preparative yield of 85 per cent (Fig. 23.4). These observations have enabled active site models for the fungal epoxide hydrolase enzymes to be proposed.

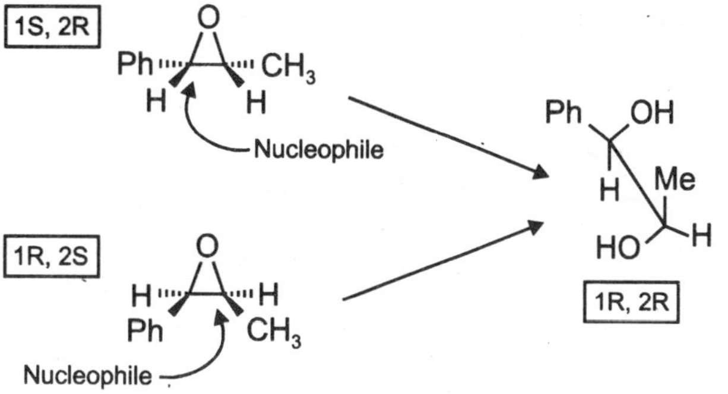

Fig. 23.4. Selectivity of *Beauveria sulfurescens* epoxide hydrolase.

Nitrile hydrolase enzymes are also finding wide applicability in organic synthesis. Nitrile hydrolysis typically involves a nitrile hydratase which converts the nitrile to an amide, followed by an amidase which further hydrolyses the amide to a carboxylic acid; either or both may show selectivity. A recent synthesis of the lactone pharmacophore of the hypocholestemic agents mevinolin and compactin utilised the nitrile hydrolase activities of two micro-organisms: first, a selective hydrolysis of the pro-*S* nitrile moiety of 3-(benzyloxy)glutaronitrile using *Brevibacterium* R312, and later a hydrolysis using immobilised *Rhodococcus* sp. SP361 (Fig. 23.5). Other applications of these enzymes include the industrial preparation of acrylamide from acrylonitrile utilising the nitrile hydratase activity of *Pseudomonas* sp.

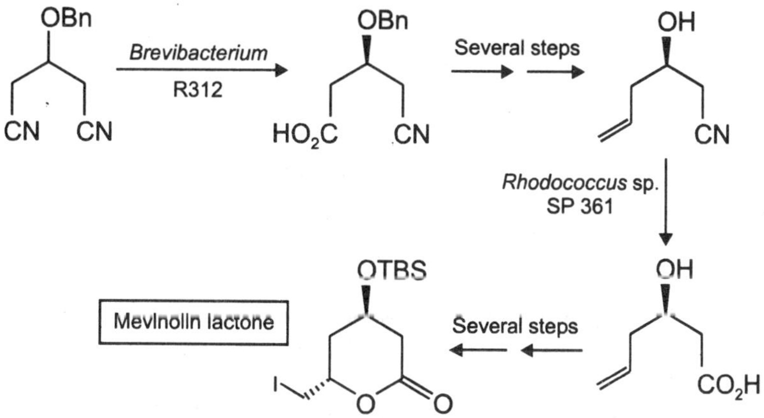

Fig. 23.5. Chemoenzymic synthesis of mevinolin lactone utilising nitrile hydrolase enzymes.

Both epoxide hydrolase and nitrile hydrolase systems presently necessitate whole-cell systems rather than isolated enzymes. However, some progress has been made in the isolation of epoxide hydrolase and nitrile hydratase. Whether these isolated enzymes will be of such general use as, for instance, isolated lipases remains to be seen.

Enzymic Protecting Group Strategies

Another area where hydrolytic enzymes are beginning to come to the fore is that of protecting group chemistry in the synthesis of highly functionalised molecules such as peptides, saccharides, and their conjugates. The use of lipases and proteases to acylate selective hydroxyl groups of carbohydrates, and of lipases, esterases and proteases to achieve selective unmasking of multiply acylated carbohydrates, has been reviewed by Roberts. Depending on the enzyme and solvent systems employed, selectivity for the primary or for a particular secondary hydroxyl group can often be attained in good yield. Thus, for example, a naturally occurring rhamnopyranoside of *p*-hydroxybenzaldehyde was synthesised in good yield using selective acetylation catalysed by *Pseudomonas fluorescens* lipase (PFL) in the final step (Fig. 23.6).

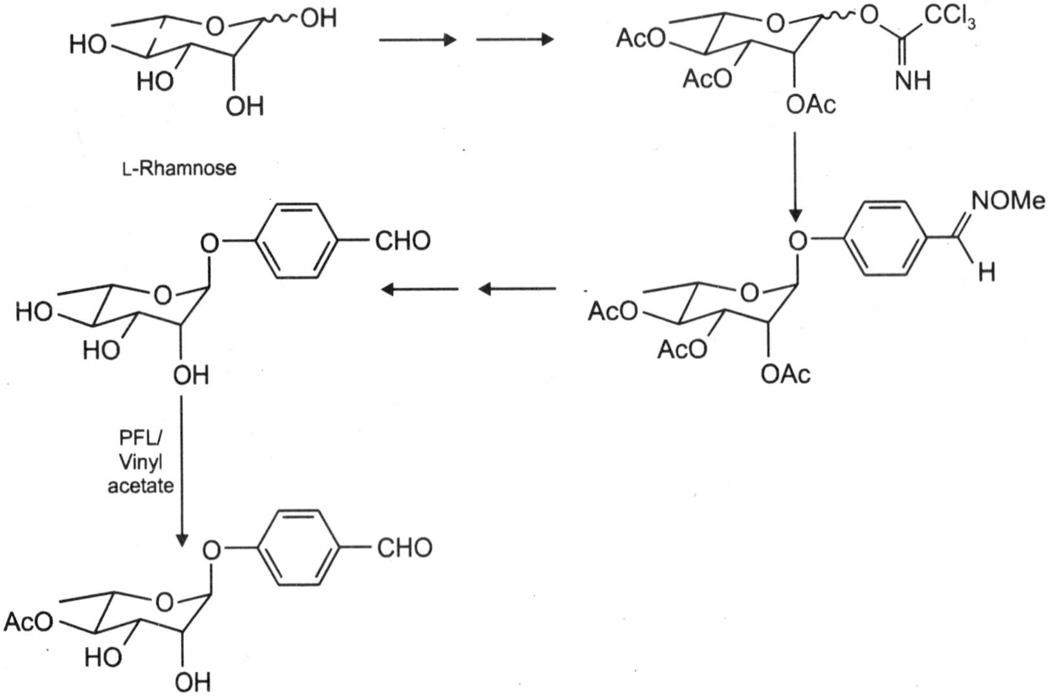

Fig. 23.6. Selective protection in carbohydrate synthesis.

The successful construction of peptide conjugates requires extremely mild conditions. Even mildly basic conditions can result in the cleavage of fatty acid, nucleotide or carbohydrate moieties from the peptide chain, while under acidic conditions glycosidic bonds are prone to anomerisation or cleavage. Therefore assembly of the peptide chain requires the mildest conditions, which are not always possible with conventional chemical synthesis. The *Ras* proteins, which play a crucial role in the regulation of cell growth, illustrate the difficulty. The C-terminal lipohexapeptide (**2**) associated with human N-*Ras* protein possesses an acid-labile *S*-farnesylated cysteine methyl ester and a base-labile *S*-palmitoylated cysteine residue; therefore conventional peptide synthesis using acid or base-labile protecting groups for the amino acid side chains is impossible. The use of protecting groups which can be cleaved under the mildest conditions using hydrolytic enzymes can therefore be essential to a successful synthesis of such molecules, and a range of protecting groups and deprotection methodologies have been developed. A range of such protecting groups are listed in Table 23.1.

—Cys(Pal)—Met—Gly—Leu—Pro—Cys(Far)OMe

(**2**)

With the recent enormous growth of interest in methods for the synthesis of peptide conjugates and oligosaccharides using solid-phase 'combinatorial' methodology, the challenge of developing linkers which can be cleaved under mild conditions has arisen. This is another area where enzyme technology may be of great importance, although to date relatively few examples of enzyme-cleavable linkers are reported in the literature.

Table 23.1. Enzyme-cleavable protecting groups for peptide synthesis.

Protecting group	*Use*	*Cleavage conditions*	*Comments*
Phenylacetamido (PhAc) (Waldmann)	N-terminal protection	Penicillin acylase	Stable to acid, base, hydrogenation. Can lead to some racemisation in conventional peptide synthesis
p-Acetoxy-benzyloxycarbonyl (AcOZ) (**3**) (Waldmann and Nägele)	N-terminal protection	Acetyl esterase, acetylcholinesterase, *M. miehei* lipase	See Fig. 23.7 for cleavage mechanism. Used in first synthesis of N-*Ras* hexapeptide (2)
2-Bromoethyl (EtBr) (Waldmann and Sebastian)	C-terminal protection	Lipase	
Heptyl (Hep) (Waldmann and Sebastian)	C-terminal protection	Lipase	
2-(*N*-morpholino)-ethyl (Waldmann and Sebastian)	C-terminal protection	Lipase	
Choline (Cho) $RCO_2(CH_2)_2NMe_3^+$ (Schelhaas)	C-terminal protection	Cholinesterase, chymotrypsin, papain	Enhances peptide solubility. Can be used as activating group in protease catalysed peptide synthesis

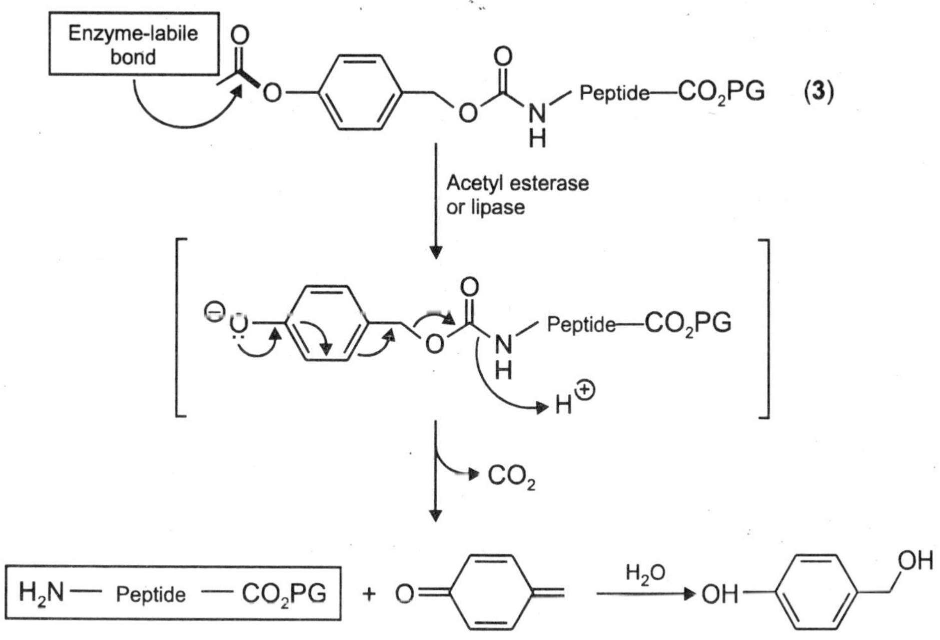

Fig. 23.7. Enzymic deprotection of the *p*-benzyloxycarbonyl (AcOZ) group.

Elmore and others have devised a linker containing a phosphodiester moiety for use in solid-phase peptide synthesis, wherein the phosphate group may be hydrolysed, and the peptide released from the support, by the use of phosphodiesterase. Wong employed a phenylalanyl ester group, recognisable by chymotrypsin, in the solid-phase synthesis of a sialyl Lewis^{X} (SLe^{X}) glycopeptide. It is highly likely that similar applications of enzyme technology will be reported in the future.

Process Chemistry

The transfer or enzyme technology from laboratory bench to industrial process is fraught with challenges. Economy of time, resources and finance are essential to produce a viable industrial process, and any or all of these factors can be a considerable encumbrance to the application of enzyme technology in industry. The costs of substrates, enzymes (and cofactors if necessary) and reactor operation and design (if a new reactor has to be devised for the process) can all be limiting factors. In fact, hydrolytic enzymes possess a number of advantages: they are co-factor-independent, many of them are readily available in large quantities and can be immobilised or otherwise stabilised for use in versatile chemical reactors, and they are often easy to remove and recycle once the process is finished.

A wide range of industrial processes making use of hydrolytic enzymes are now well established. One example is the thermolysin-catalysed synthesis of aspartame (**4**) from *N*-benzyloxycarbonyl (*Z*)-protected aspartic acid and phenylalanine methyl ester. Thermolysin is a robust metalloprotease which operates at neutral pH and retains significant activity even at elevated temperatures, and is therefore well suited to use in an industrial process. Moreover, it is entirely selective for L-amino acids (which means that cheap racemic mixtures of amino acids can be employed in the reactor), and for the α-carboxylate group of aspartic acid, and no racemisation is observed in the process. Thermolysin is also of interest in the context of combinational chemistry and enzyme inhibition (Fig. 23.8).

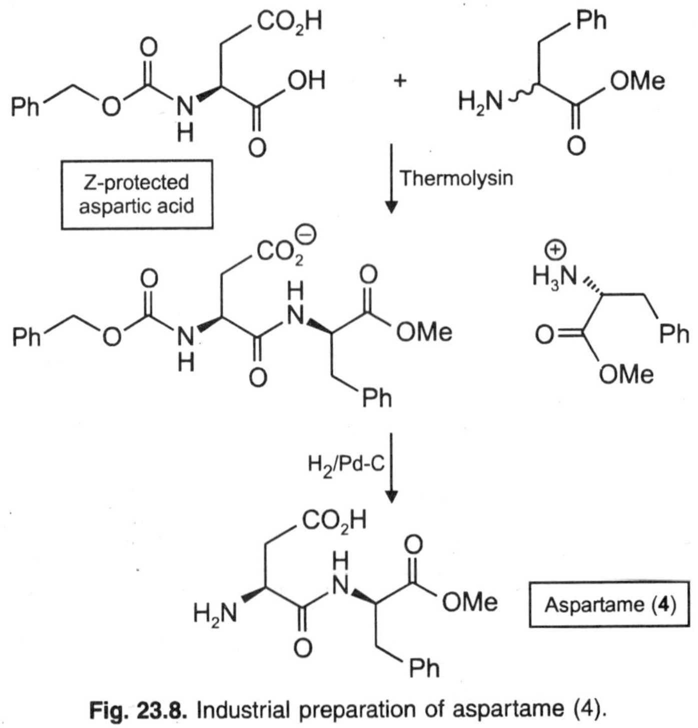

Fig. 23.8. Industrial preparation of aspartame (4).

Due to the cost of enzyme purification, crude enzyme preparations are most commonly used in industrial applications. However, several disadvantages are inherent in this approach, most notably the fact that contaminants in the enzyme preparation can reduce the enzyme lifetime, complicate downstream processing, and even lead to inconsistent or contradictory results in the process itself (for instance, the presence of other hydrolases as impurities in crude cell-free preparations of lipases, etc. can lead to a lower enantioselectivity than expected, and to unpredictability in the effect of alterations in reaction conditions such as temperature and pH. Efficient, cheap method for the purification and stabilisation of enzymes can therefore be essential to the operation of a viable biocatalytic process; they may additionally confer other advantages such as allowing the enzyme to be recycled. Enzyme immobilisation, i.e. attachment of the enzyme to a water-insoluble solid support, is the most common method of stabilising an enzyme, allowing the use of continuous packed-bed reactors and enabling the catalyst to be recovered (often by the simple process of filtration) and recycled. Immobilised enzymes, however, do tend to lose activity over a period of time, and can degrade quite substantially by the physical process of stirring. Moreover, the specific activity of the immobilised enzyme is often considerably lower than in the crude cell-free preparation, as the amount of enzyme in an immobilised preparation is typically less than 5 per cent by weight. An alternative to enzyme immobilisation is the use of cross-linked enzyme crystals (CLECs), which can be obtained by means of a simple purification step followed by crystallisation and treatment with a cross-linking agent such as glutaraldehyde.

The resulting biocatalyst consists of a rigid, close-packed network of molecules, which are often water-insoluble, highly active, and significantly more stable to extremes of heat and pH and to the presence of exogenous proteases and water-miscible organic solvents than the crude enzyme preparations. Cross-linked crystals of thermolysin have been shown to be highly efficient catalysts in the preparation of aspartame, and crystals of CRL were shown to have significantly superior enantioselectivity to the more often used crude enzyme preparation in the hydrolysis of a range of esters of 2-arylpropanoic acids, including methyl ibuprofen (**5**), which was efficiently resolved on a 100 grams scale [recrystallisation of (*S*)-ibuprofen (*S*)-**6** gave material of >99 per cent enantiomeric excess (e.e.)]. Given the low enantioselectivity of CRL in the resolution of chiral acids (*vide supra*), this represents a significant improvement in catalyst performance. Similar results for a range of hydrolases, indicate that CLECs are potentially highly viable catalysts in large-scale industrial processes.

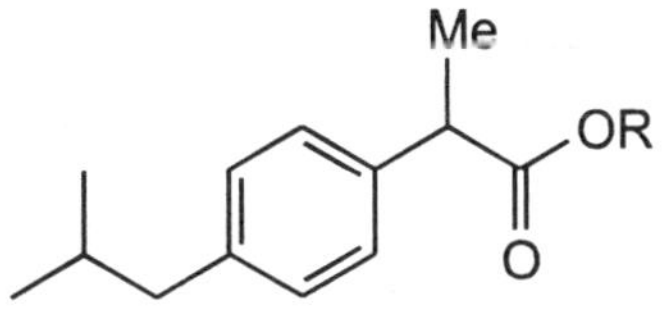

R = Me; **5**
R = H; **6**

Chemical modification of enzymes can also be invaluable in improving their performance as catalysts. For instance, pig liver esterase modified at surface lysine residues by chemical attachment of polyethylene glycol monomethyl ether units has demonstrated significantly improved activity in water-immiscible organic solvents over the native enzyme (which is all but inactive in such systems). The use of site-directed mutagenesis techniques to produce genetically modified enzymes may in future also enable the generation of biocatalysts with enhanced stability or activity in organic solvent systems, though the exploitation of such materials on a wide scale is not yet a reality.

Where immobilisation or cross-linking has not proved effective in the production of stabilised biocatalysts, an alternative is to utilise a homogeneous system with the enzyme enclosed within a membrane. The pioneering work of Wandrey and colleagues has enabled the development of enzyme membrane reactors (EMR) for the industrial-scale preparation of a number of valuable chiral chemicals, e.g. the resolution of racemic *N*-acetyl-DL-amino acids using acylase I from *Aspergillus oryzae* enclosed in an ultrafiltration membrane through which substrate and product molecules can flow as necessary. The process can be applied to the synthesis of non-proteinogenic amino acids such as α-aminobutyric acid as well as proteinogenic amino acids, and has the advantage that the recovered *N*-acetyl-D-amino acids can be readily racemised, reducing chemical waste to a minimum.

Dealing with 'waste' materials is of great importance to the viability of a chemical process. In an enzymatic kinetic resolution, only 50 per cent of the desired material can be obtained, and requires separation from the remaining 50 per cent. Often a use can he found for the unwanted material, which will also be enantiomerically enriched. Where an unwanted enantiomer is not directly useful, the ideal situation is for it to be converted to its enantiomer, or else racemised.

Thus, the by-products from the acylase-catalysed resolution of amino acids described above can readily be racemised by acetic anhydride in alkaline solution, or by using a racemase enzyme, and returned to the reactor. Even better is to use conditions which allow racemisation of the unreacted enantiomer *in situ*, allowing a so-called 'dynamic resolution' of the starting material, with a theoretical maximum yield of 100 per cent. Dynamic resolution of amino acids using hydantoinase has been exploited on an industrial scale, and can be performed in tandem with a second hydrolytic enzyme (carbamoylase) to release the free amino acid of desired configuration, as both D- and L-hydantoinases are known, this process can be used for the synthesis of either D- or L-amino acids (Fig. 23.9). More recently, an application of the Lipozyme®-catalysed dynamic resolution of oxazolin-5(4H)-ones (azlactones) derived from racemic *N*-benzoylamino acids to the synthesis or enantiomerically pure amino acids [including the valuable chiral intermediate L-*tert*-leucine (**7**)] has been reported by Turner (Fig. 23.9). Both hydantoins and azlactones racemise rapidly under the conditions employed in the biotransformation. An interesting new form of dynamic resolution process has recently been described for the resolution of chiral allylic alcohols using acetylcholinesterase or PFL in combination with a palladium (II) catalyst which racemises the allylic acetate starting material rapidly in aqueous buffer (Fig. 23.9). Simultaneous enzyme and transition metal-catalysed reactions are unusual, and the scope for their wider implementation as tools in process chemistry is considerable.

OXIDATIVE-REDUCTIVE ENZYMES

Enzyme Structure and Function

Redox enzymes have not been so widely exploited as hydrolytic enzymes in organic synthesis. A few large-scale processes exist, mostly involving whole-cell preparations of yeasts or micro-organisms rather than isolated enzymes; exploitation of free enzymes has often been hampered by the fact that a number of useful species are membrane-bound, and difficult to isolate, and all require redox-active co-factors, which have to be supplied in the reaction medium and recycled when isolated enzymes are employed. Those enzymes which have been exploited fall into two classes: metalloenzymes and nicotinamide-dependent oxidoreductases. The metalloenzymes are involved with the oxidation of relatively redox-inactive functionalities, e.g. alkanes, carbon-carbon double bonds or heteroatoms, whereas the nicotinamide-dependent species take part in the oxidation and reduction of ketones and alcohols. The former have great potential significance in biomediation.

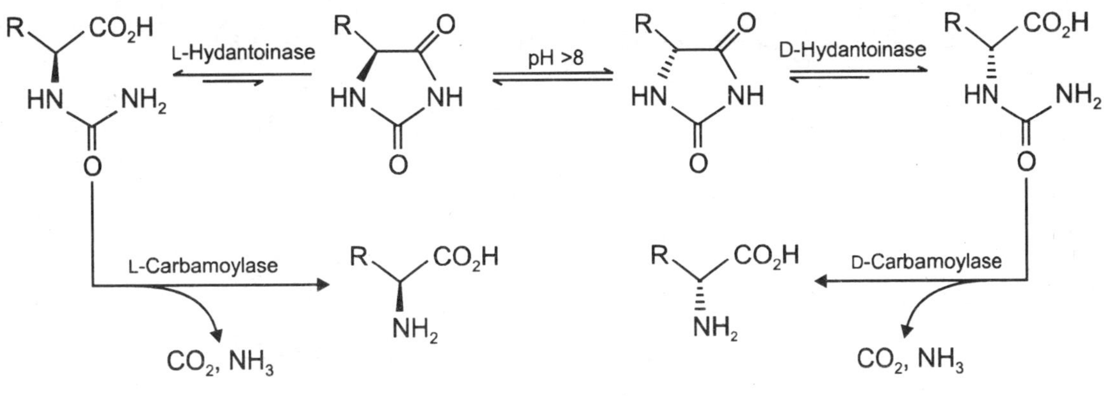

(a) Amino acid synthesis using hydantoinase/carbamoylase.

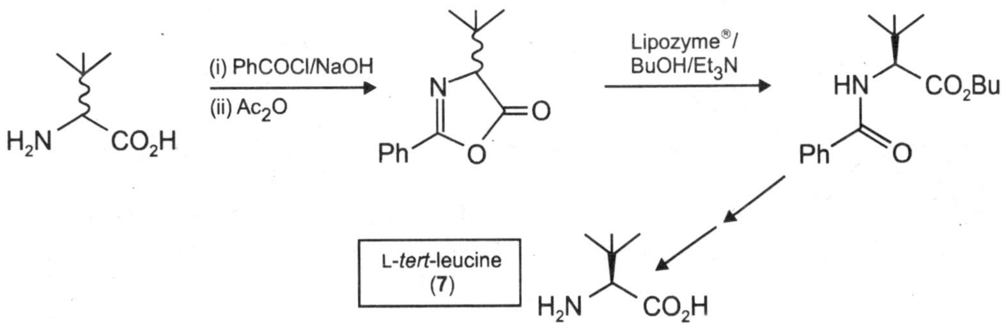

(b) Lipozyme®-catalysed dynamic resolution of azlactones.

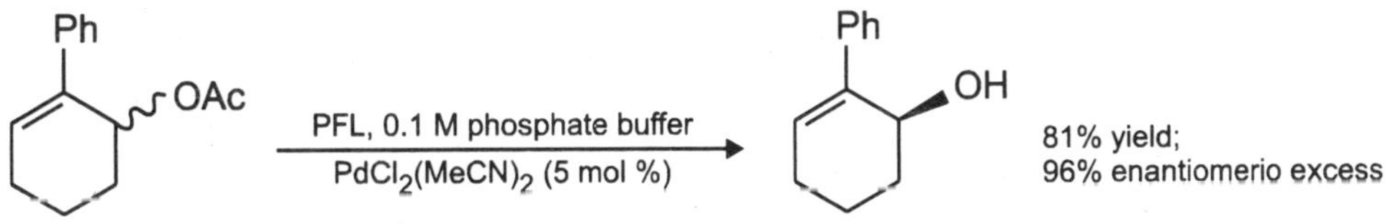

(c) Hydrolase/palladium dynamic resolution.

Fig. 23.9. Dynamic resolution processes.

Metallo-oxidases of various kinds are known. The cytochrome P-450 monoxygenases contain a heme group, and catalyse the oxidation of alkanes (with oxygen insertion), carbon-carbon double bonds (forming a chiral epoxide), and heteroatoms (e.g. oxidation of sulphides to chiral suphoxides) by molecular oxygen, with a high-valent $Fe^{V} = O$ intermediate as the active species. The P-450 monoxygenase from *Pseudomonas putida* involved in the oxidation of camphor (P-450_{cam}) has been overexpressed in *Escherichia coli* and isolated, though as yet the use of the isolated enzyme as a synthetically useful catalyst has not been reported. *Pseudomonas oleovorans* monoxygenase possesses a non-heme iron atom in the active site, and also function: via an $Fe^{V} = O$ active intermediate. The proposed catalytic cycle for these mono-oxygenases is shown in Fig. 23.10. Chloroperoxidase from *Caldariomyces fumago* also makes use of a non-heme $Fe^{V} = O$ intermediate, though in this case the actual oxidant is not molecular oxygen but the hydroperoxide ion.

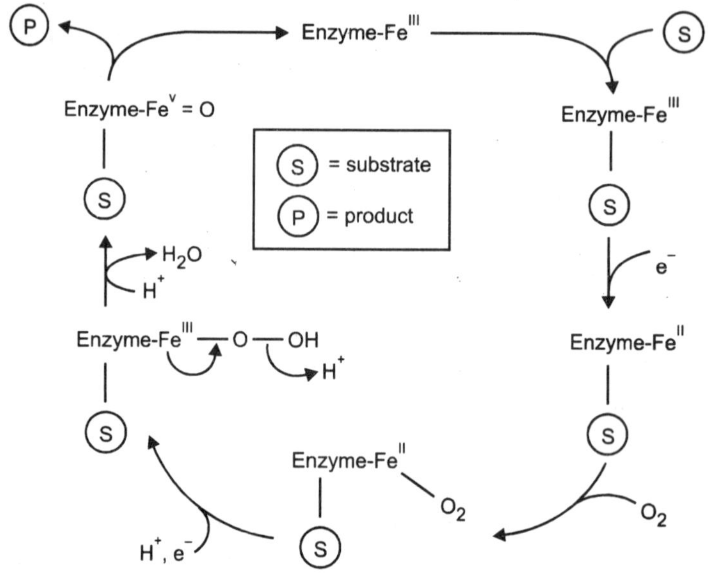

Fig. 23.10. Catalytic cycle for iron-dependent monoxygenase [the mechanism of the substrate oxidation step(s) is not clear].

The enzyme has been overexpressed and is commercially available, and has been found to be useful in the enantioselective epoxidation and halohydration of alkenes and the oxidation of sulphides; a radical mechanism has been proposed. Some success has been achieved in the modification of substrate specificity of enzymes of the P-450 class by site-directed mutagenesis techniques. Other metallo-oxidases include the Cu^{II}-dependent galactose oxidase, the Fe-dependent isopenicillin-N synthetase, and the arene dioxygenases (which add two oxygen atoms to each molecule of substrate, with molecular oxygen as the source from *P. putida* which perform regio- and enantioselective dihydroxylation of aromatic rings, generating *cis*-cyclohexadienediols which are valuable intermediates in the synthesis of a huge range of chiral molecules, including sugar analogues.

The alcohol dehydrogenase enzymes are the most well-known examples of the nicotinamide-dependent oxidoreductases. Their function is dependent on the presence of nicotinamide adenine dinucleotide (NAD^+, **8**) or its 2′-phosphorylated derivative ($NADP^+$, **9**). The nicotinamide ring system readily accepts a proton and two electrons or a hydride ion, to form the 1,4-dihydronicotinamide derivatives NADH (**10**) and NADPH (**11**). The reversible hydride transfer from a reduced substrate to NAD^+ or $NADP^+$, or from NADH or NADPH to an oxidised substrate, is stereoselective, the nature of the selectivity varying from enzyme to enzyme. Most alcohol dehydrogenases, including those from horse liver (HLADH), yeast (YADH) and *Thermoanaerobium brockii*, follow 'Prelog's rule' (Fig. 23.11), transferring the pro-*R* hydride from the nicotinamide ring to the *re* face of a carbonyl substrate.

Alcohol dehydrogenases have been successfully employed in oxidative mode in the synthesis of α-hydroxy- and α-amino aldehydes with kinetic resolution of racemic 1,2-diols and of chiral lactones by desymmetrisation of *meso*-diols, and in reductive mode in the synthesis of chiral alcohols by reduction of the corresponding ketone. The enzymes have also proven useful in the synthesis of deuterium- or tritium-labelled alcohols and α-hydroxyacids, utilising the precise nature of the deuterium or tritium transfer from a suitably generated nicotinamide cofactor (typically prepared through reduction of the oxidised species using labelled propan-2-ol). A cubic space model of the horse liver enzyme active site has been proposed by Jones and shown to be valuable in predicting the stereochemical outcome of HLADH-catalysed reactions with new substrates. Another class of NAD/NADP-dependent redox enzymes are the 'Baeyer-Villigerases', monoxygenases which catalyse the regio- and enantioselective Baeyer-Villiger rearrangement of ketones. The ultimate oxidant in these systems is dioxygen, and the enzymes utilise a tightly-bound flavin group in addition to the nicotinamide co-factor; a flavin hydroperoxide adduct serves as the initial nucleophile, with the migrating C—C bond of the intermediate, the peroxidic (O—O) bond and a lone electron pair on the carbonyl oxygen of the substrate all adopting an antiperiplanar arrangement. Whole-cell cultures of *Acinetobacter* strains NCTMB 9871 and TD63, and of *Pseudomonas putida* strain NCIMB 10007 have exhibited 'Baeyer-Villigerase' activity with striking regio- and enantioselectivity utilising a range of bi- and monocyclic ketones, the *Acinetobacter* and *Pseudomonas* strains being basically enantiocomplementary in their selectivity. A cubic model of the active site of *Acinetobacter* monooxygenase has been proposed by Furstoss to explain the selectivity observed with bicyclic and monocyclic ketones.

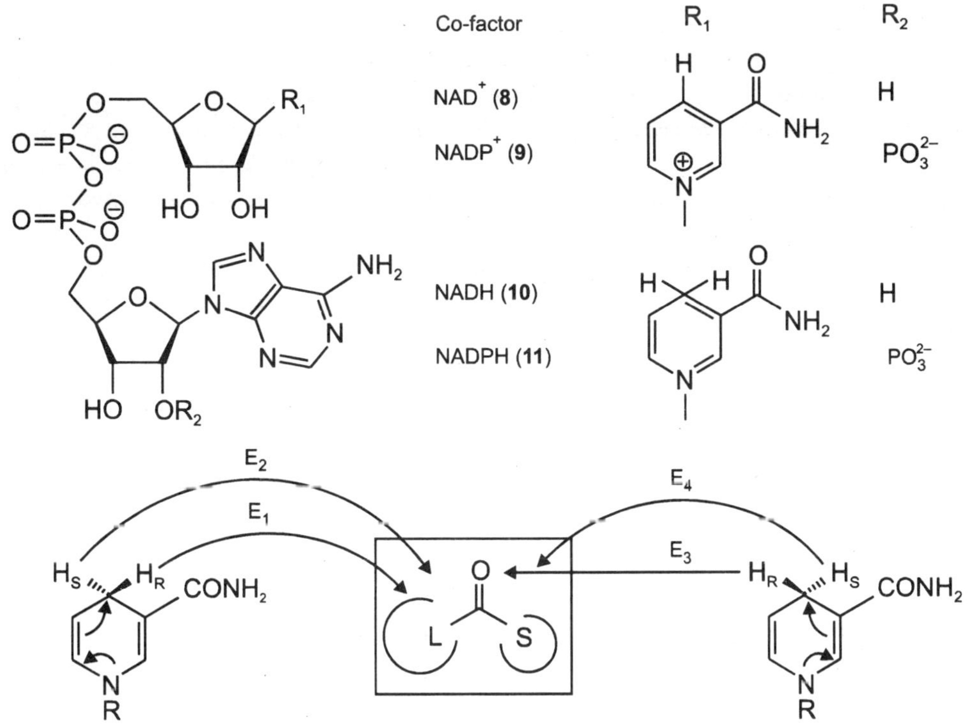

E_1 (Pro-*R*/*si* face): Alcohol dehydrogenase from *Pseudomonas* (NADH) or *Lactobacillus kefir* (NADPH)
E_2 (Pro-*S*/*si* face): Alcohol dehydrogenase from *Mucor javanicus*
E_3 (Pro-*R*/*re* face): Alcohol dehydrogenase from yeast, horse liver and *Thermoanaerobium* sp. ('Prelog's rule')
E_4 (Pro-*S*/*re* face): unknown

Fig. 23.11. Selectivity of alcohol dehydrogenase-catalysed ketone reductions.

Co-factor Regeneration

The majority of biocatalytic reductions are performed using whole-cell systems. Yeasts especially are organisms which are adept at performing reductions on systems as diverse as ketones, aromatic nitro compounds, and carbon-carbon double bonds. Whole-cell systems can also be employed effectively in biooxidations such as Baeyer-Villiger reactions. Whole-cell biotransformations suffer from a number of disadvantages, however: notably the fact that the presence of other active enzymes within the cells can lead to side reactions and correspondingly poor yields of product. Isolation of the product from cell debris after the process can also be troublesome. One significant obstacle to be overcome in the harnessing of isolated redox enzymes for use in biocatalysis is the need to recycle cofactors. The nicotinamide cofactors are far too expensive to be used stoichiometrically on a synthetic scale; therefore, methods have been developed to regenerate these co-factors with high efficiency once consumed. Cofactor recycling can, in fact, impart advantages to a process. Most importantly, it can drive reactions at equilibrium (or thermodynamically unfavourable reactions) to completion by coupling with a favourable regeneration process. It also eliminates product inhibition by the cofactor by-product and can simplify considerably the reaction work-up.

At present, the best recycling systems are those which regenerate the co-factor via a second enzymatic reaction. NADH, for instance, can be regenerated from NAD^+ by oxidation of propan-2-ol using an NAD^+-dependent alcohol dehydrogenase, or by oxidation of formate using formate dehydrogenase (Fig. 23.12). The formate dehydrogenase system is cheap and efficient, but enzyme activity is low, and the enzyme is specific for NAD^+. An alternative system utilises glucose and glucose dehydrogenase from *Bacillus* sp., which is a highly active, robust enzyme; however, the by-product (gluconate) may complicate product isolation. The glucose dehydrogenase system can also be used to recycle NADPH, as can the propan-2-ol *Thermoanaerobium brockii* alcohol dehydrogenase system; however, most alcohol dehydrogenases currently available are NAD^+-specific.

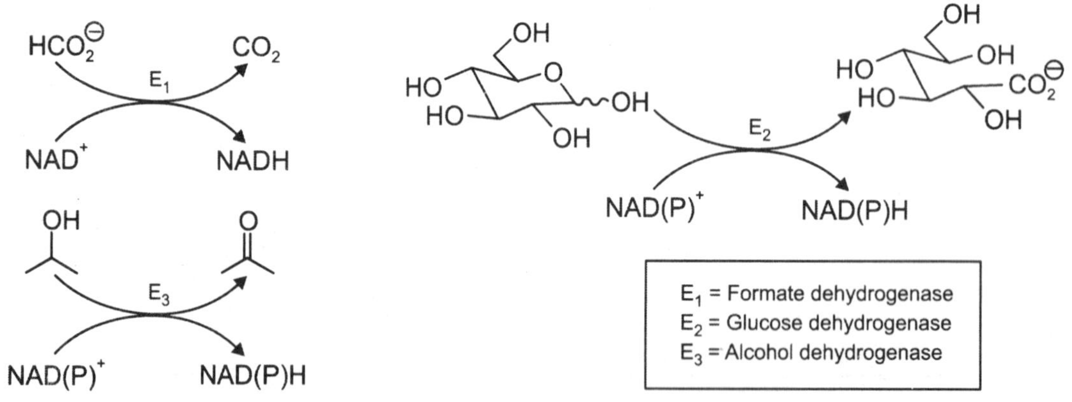

Fig. 23.12. Some recycling systems for NAD(P)H.

Enzymic methods for the recycling of NAD^+ or $NADP^+$ are shown in Fig. 23.13. The most commonly used systems are α-ketoglutarate with glutamate dehydrogenase (glutamate in the reaction mixture can complicate product isolation) and pyruvate with lactate dehydrogenase (NAD^+-specific). Other systems are also known.

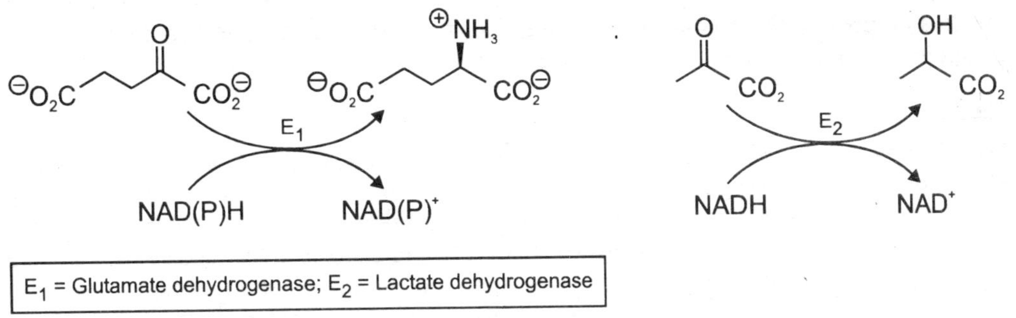

Fig. 23.13. Some recycling systems for $NAD(P)^+$.

Process Chemistry

The usefulness of redox enzymes as catalysts for large-scale processes is hampered by a number of issues, notably enzyme availability, co-factor recycling and the susceptibility of the enzymes to denaturing and product inhibition. However, as the scope of these enzymes to perform transformations with often unique selectivity (as in the case of the Baeyer-Villigerases for instance) is realised, a wide range of methods is currently in development for improving the synthetic utility of these enzymes.

New enzyme sources are constantly under investigation. Enzyme overexpression is making more and more potentially useful enzymes available, as shown recently in the case of the P-450_{cam} monooxygenase. Isolated enzymes from synthetically useful whole-cell systems are also under extensive investigation, for instance the two purified Baeyer-Villiger mono-oxygenases (MO1, NAD^+-dependent and MO2, $NADP^+$-dependent) from *P. putida* NCIMB 10001, which were found to have very different substrate selectivities (MO1 being a highly efficient catalyst for the oxidation of bicyclic ketones, MO2 being more efficient for monocyclic ketones).

A number of industrially useful enzymatic redox processes are in operation. Commercial syntheses of the amino acids L-leucine and L-*tert* leucine, for instance, consist of the reductive amination of α-ketoacids using the NADH-dependent enzyme leucine dehydrogenase, with formate/formate dehydrogenase to regenerate NADH. The enzyme is enclosed within a membrane, and in the case of L-leucine the cofactor is also retained within the membrane via attachment to soluble polyethylene glycol, enabling the reactor to be operated continuously for over three months.

CARBON-CARBON BOND-FORMING ENZYMES

General Observations

The stereocontrolled formation of carbon-carbon bonds is of the foremost interest to research chemists, and although the technology exists for a wide range of asymmetric coupling reactions (such as stereocontrolled aldol condensations) to be performed in the laboratory, the high precision of the enzymes which catalyse carbon-carbon bond formation and their usefulness under mild conditions makes them potentially attractive catalysts in the synthesis of natural products, sugars and other highly sensitive molecules. Although the use of such enzymes in the laboratory-scale synthesis of a vast range of chiral molecules has been extensively documented, most are not yet widely exploited on a large scale. The fact that many of these enzymes require, expensive co-substrates and can frequently experience severe product inhibition has to date limited their application. However, recent advances in biotechnology have made a number of these enzymes available in large quantities for the first time, making their application in large-scale processes an area of current interest.

Aldolases

Aldolases belong to a class of enzymes known as lyases. Their function *in vivo* is to catalyse the degradation of metabolites (sugars in the case of aldolases), but the reactions which they catalyse are reversible and so the enzymes can be used *in vitro* for the synthesis of carbon-carbon bonds. Aldolases, as the name implies, synthesise C—C bonds by means of an enzymatic aldol condensation between two substrates, the nucleophilic donor (usually a ketone) and the electrophilic acceptor (invariably an aldehyde). Type I aldolases, mainly found in animals and higher plants, operate by forming a Schiff base between an active site lysine residue and the donor, which subsequently adds stereospecifically to the acceptor; type II aldolases, mostly found in micro-organisms, are metallo enzymes utilising a divalent zinc ion in the active site as a Lewis acid co-catalyst (Fig. 2.14). Although most aldolases will accept a wide range of aldehydes as electrophilic substrates, their specificity for the nucleophilic component of the aldol condensation is fairly rigid. Table 23.2 lists the four main classes of aldolase, according to the preferred donor, with examples from each class (this is not an exhaustive list). Many aldolases have been exploited successfully in synthesis and some representative examples appear below.

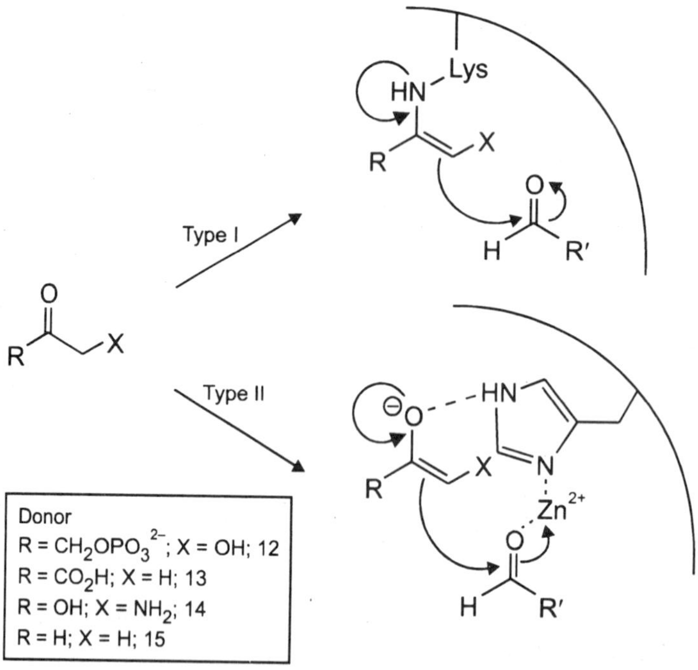

Fig. 23.14. Mechanisms of aldolase-catalysed aldol condensation.

The dihydroxyacetone phosphate (DHAP, **12**) dependent aldolases have to date been the most widely exploited in synthesis. The FDP aldolase from rabbit muscle (RAMA) is commercially available; overexpressed FDP aldolase from *E. coli* has also been employed in synthesis and microbial sources of the other three DHAP aldolases listed in Table 23.2 have also been reported. These four DHAP aldolases show effectively complementary diastereoselectivity (Fig. 23.15). The specificity of these aldolases for DHAP as the donor substrate is relatively rigid—only a small number of DHAP analogues were found to be (poor) substrates for RAMA—though a mixture of dihydroxyacetone and inorganic arsenate can serve effectively as a replacement for DHAP and more recently the thio analogue of DHAP has been shown to be a synthetically versatile donor. The acceptor specificity is more flexible, a wide range of aldehydes being accepted as substrates (though species which are poorly soluble in aqueous buffer, such as aromatic aldehydes, tend to be non-substrates). FDP aldolase is highly stereoselective, with the two new chiral centres formed in the aldol reaction possessing the [3*S*, 4*R*] or 'D-*threo*' stereochemistry in nearly every case (though one recent report refers to a relaxed selectivity under thermodynamic control; Gijsen and Wong). Fuc1P and Rha1P aldolase similarly generate 'D-*erythro*' and 'L-*threo*' diols, though a degree of relaxation in the selectivity of Rha1P aldolase with aliphatic aldehydes has been reported. TDP aldolase, by contrast, also showed unwanted FDP aldolase activity, which has to date limited its usefulness in synthesis; however, a recombinant enzyme was recently reported with significantly suppressed FDP aldolase activity which may be synthetically useful.

Table 23.2. Some synthetically useful aldolases (the bold bond indicates the carbon-carbon bond formed by the enzymic aldol reaction).

Donor	*Examples*	*Natural substrate/product*
Dihydroxyacetone phosphate (DHAP) (**12**) 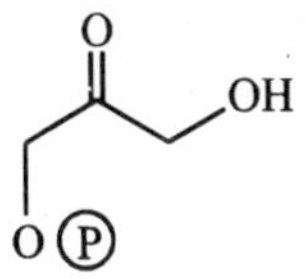	Fructose 1,6-diphosphate (FDP) aldolase [E.C.4.2.1.13]	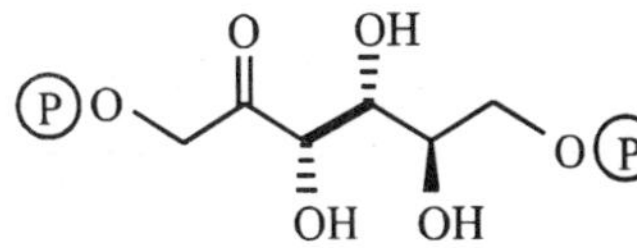
DHAP (**12**)	L-Fuculose 1-phosphate (Fuc1P) aldolase [E.C. 4.1.2.17]	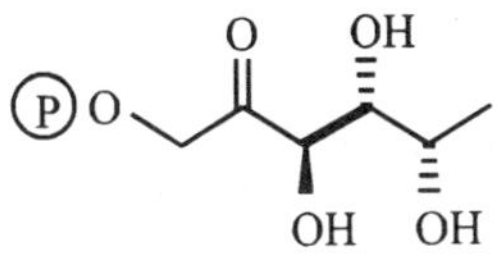
DHAP (**12**)	L-Rhamnulose 1-phosphate (Rha1P) aldolase [E.C. 4.2.1.19]	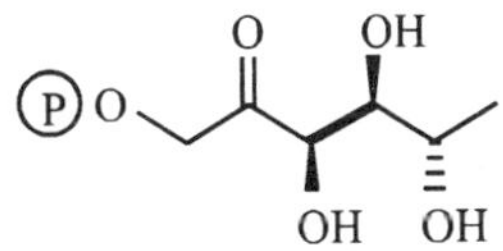
DHAP (**12**)	Tagatose 1,6-diphosphate (TDP) aldolase	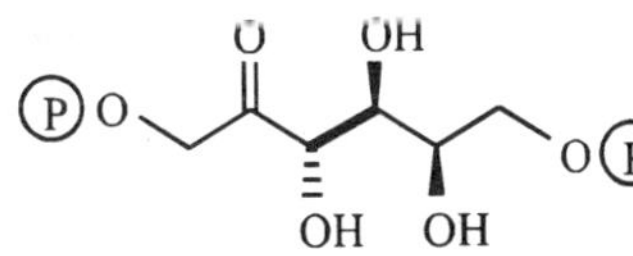
Pyruvate (**13**) 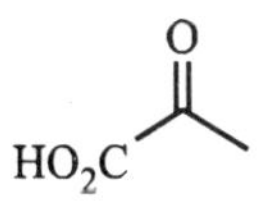	*N*-acetylneuraminic acid (Neu5Ac) aldolase [E.C. 4.1.3.3]	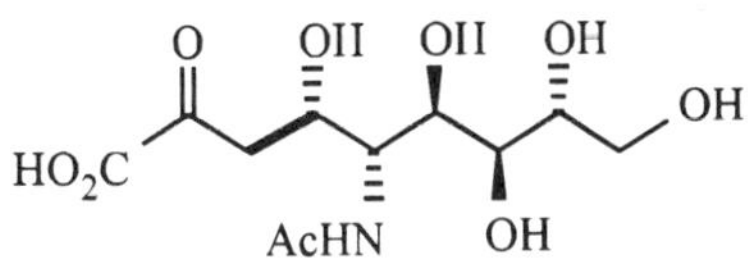
Pyruvate (**13**)	3-Deoxy-D-mannooctulosonate (KDO) aldolase [E.C. 4.1.2.23]	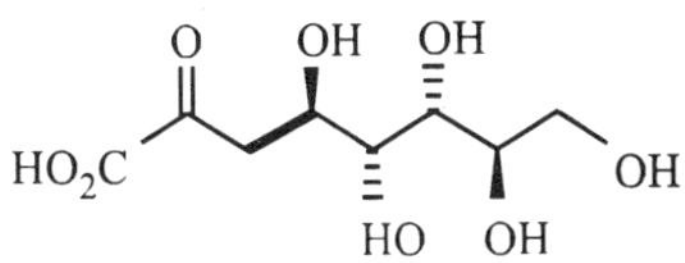

(*Contd...*)

Donor	*Examples*	*Natural substrate/product*
Pyruvate (**13**)	2-Keto-3-deoxy-6-phosphogluconate (KDPG) aldolase [E.C. 4.1.2.14]	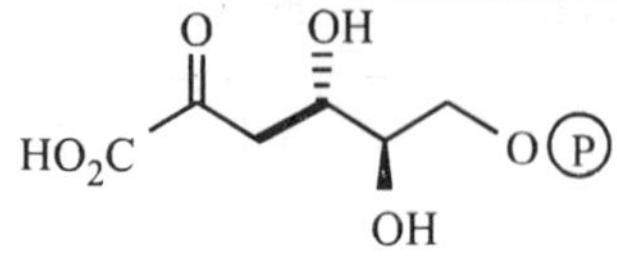
Pyruvate (**13**)	2-Keto-3-deoxy-6-phosphogalactonate (KDPGal) aldolase [E.C. 4.1.2.21]	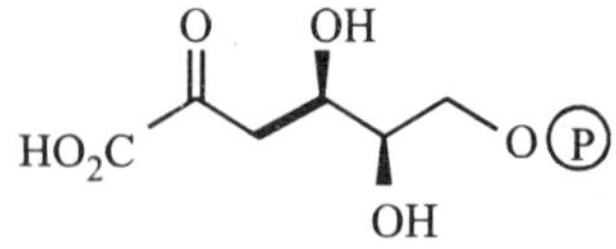
Glycine (**14**) 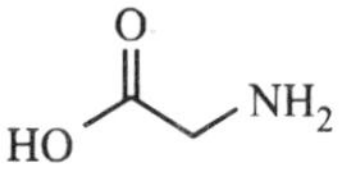	L-threonine aldolase [E.C. 4.1.2.5]	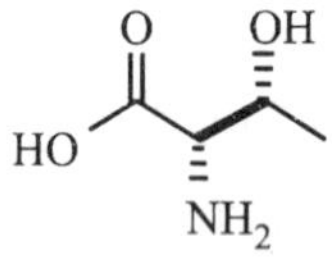
Acetaldehyde (**15**) 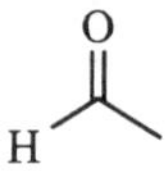	2-Deoxyribose-5-phosphate aldolase (DERA) [E.C. 4.1.2.4]	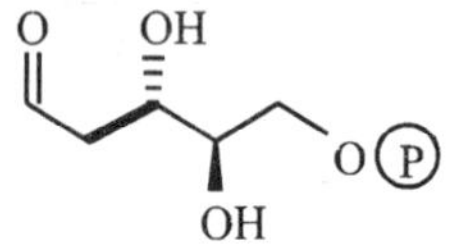

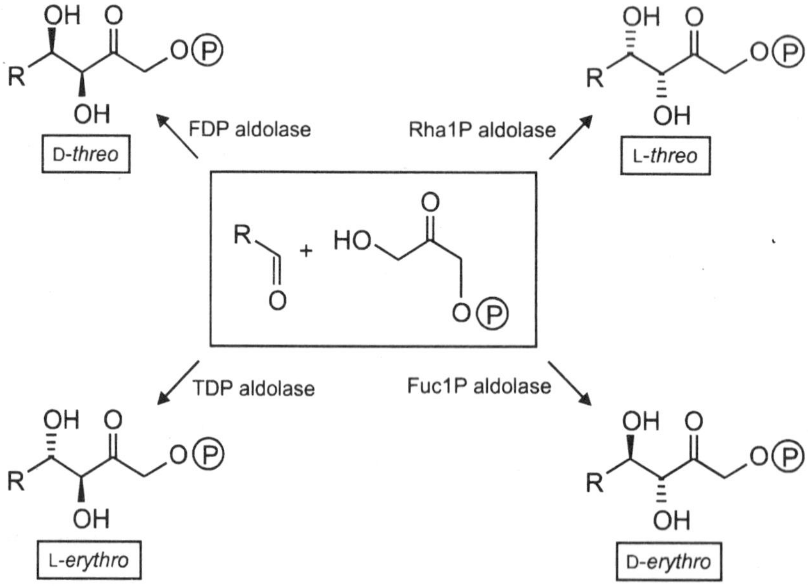

Fig. 23.15. Stereochemical preference of DHAP aldolases.

The DHAP aldolases show varying degrees of selectivity for chiral centres in the acceptor substrate. Fuc1P and Rha1P aldolases show a high preference for the L-enantiomers of racemic mixtures of α-hydroxyaldehydes, allowing kinetic resolution of the D-(R)-enantiomers. The presence of a negatively charged residue (such as phosphate or carboxylate) at least five bonds away from the aldehyde group of the acceptor was shown to have a profound influence on the kinetic diastereoselectivity of RAMA for α-hydroxylated aldehydes the formation of an ion pair between a negatively charged residue on the substrate and an active site lysine residue effectively anchoring the substrate in position and resulting in high kinetic selectivity [cf. the 'oxyanion hole' in *Candida* lipase (vide *supra*)]. Thus, under kinetic control, RAMA accepted D-glyceraldehyde-3-phosphate (**16**) with a 20:1 selectivity over the L-enantiomer, though with the unphosphorylated sugars little selectivity was observed. Thermodynamic differentiation between the enantiomers of a racemic aldehyde can also occur, if the products can reversibly cyclise to form stable hemiketals. When the reaction reaches equilibrium, a mixture of aldehydes and the two possible products will be present, and if one cyclic product is significantly more stable than its diastereomer, e.g. if there are fewer 1,3-diaxial interactions) it will predominate in the equilibrium mixture (Fig. 23.16 for an example). However, where there is little or no difference the stabilities of the cyclic products, an ≈1:1 mixture of products will predominate.

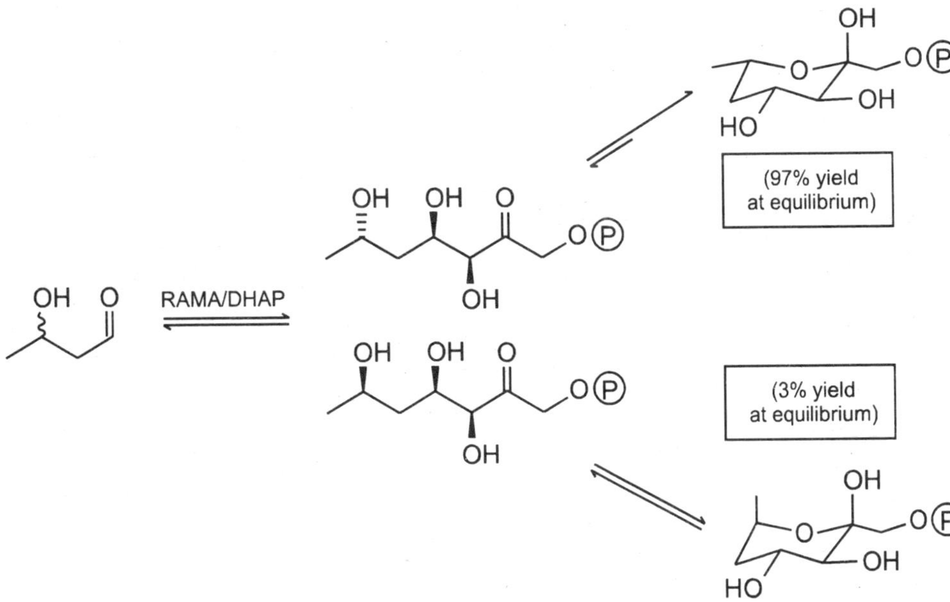

Fig. 23.16. 'Thermodynamic resolution' using RAMA.

To date, the most extensively studied of the pyruvate aldolases has been *N*-acetylneuraminic acid aldolase, a type I aldolase which catalyses the reversible aldol reaction of *N*-acetylmannosamine (ManNAc) **17** and pyruvate **13** to form *N*-acetylneuraminic (or sialic) acid (Neu5Ac), **18**, an important component in a number of biologically significant oligosaccharides and glycoconjugates such as sialyl Lewis^x (SLe^x). The enzyme is specific for pyruvate, but accepts a range of 4-carbon and higher sugars as substrates. A free hydroxyl group at C-3 of the acceptor is necessary for reaction; the stereochemistry at C-2 is also critical, as ManNAc is an efficient substrate but *N*-acetylglucosamine (GlcNAc), **19**, its C-2 epimer, is not. A wide range of configurations and substitution patterns at the remaining chiral centres of the substrate is tolerated. The stereochemical course of the reaction can follow one of two paths (Fig. 23.17): the normal mode involves *si*-face attack of the enamine intermediate on the acceptor, with the formation of an (*S*)-chiral centre as in the *in vivo* reaction. The 'inverted' mode, which is observed partially or exclusively with some substrates [e.g. with D-arabinose **20** to form 2-keto 3-deoxy-D-*manno*-octulosonic acid (KDO), **21**, as well as its expected C-4 epimer **22**)], involves re-face attack. Although the *si*-face attack is the kinetically preferred mode, under thermodynamic control the products of re-face attack can accumulate if the resulting (*R*)-product in its cyclic form is the more stable diastereomer. An active site model of the enzyme proposed by Wong suggests that in the kinetically disfavoured re-face attack, the substrate is forced into a 'boatlike' conformation at the active site.

Other pyruvate aldolases have also been exploited in organic synthesis, though less widely. KDO aldolase catalyses the condensation of pyruvate with D-arabinose 20 on the re-face to yield 21 (Fig. 23.17); several other sugars were found to be substrates, but the reactions were often under thermodynamic control and product mixture resulted. 2-Keto-3-deoxy-6-phosphogluconate (KDPG) aldolase, which catalyses the condensation of pyruvate and D-glyceraldehyde-3-phosphate **16** to form KDPG 23, has been isolated from three microbial sources and shown to have a wide acceptor specificity, even accepting pyridine-carboxaldehydes as substrates (though simple aliphatic and aromatic aldehydes are not accepted, nor are aldehydes with bulky sidechains. The enzymes from *E. coli* and *P. putida* were highly selective for D-α-hydroxyaldehydes, and all three strains exhibited solely kinetic diastereoselectivity.

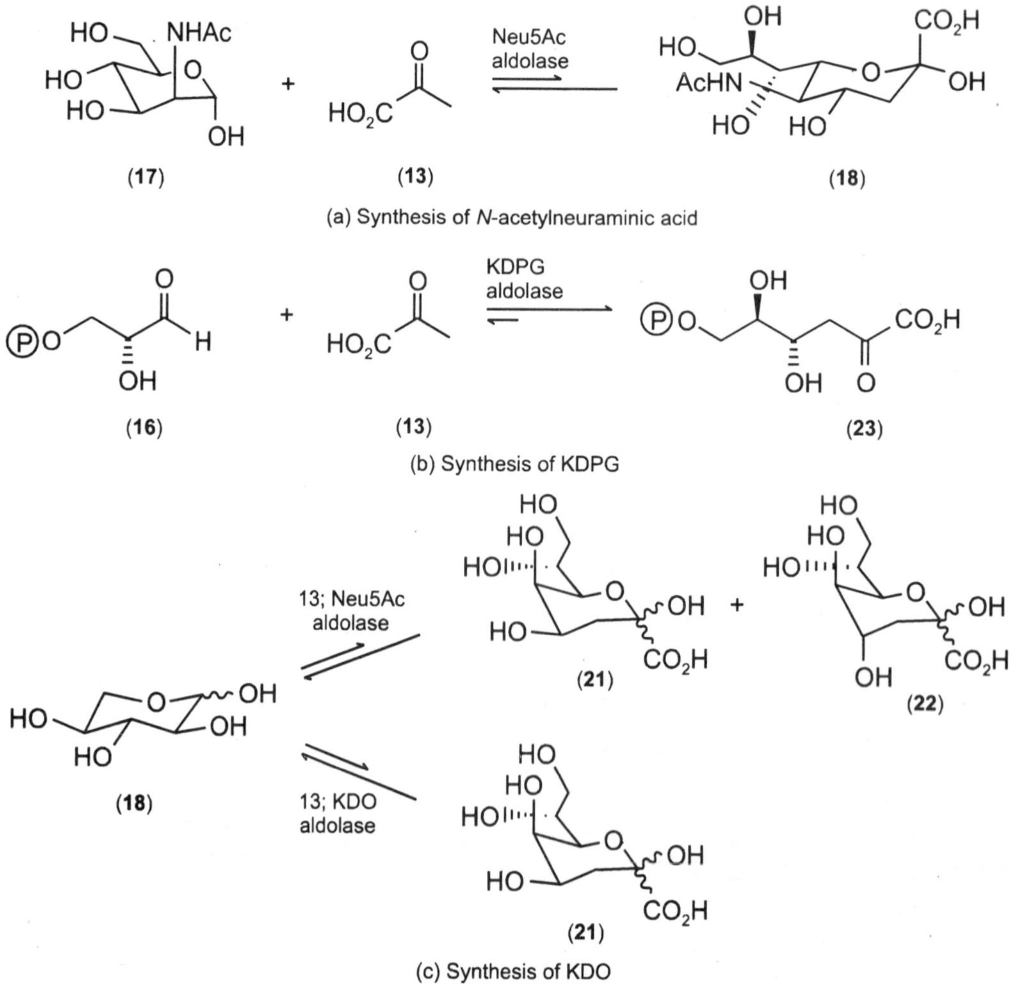

Fig. 23.17. Reactions catalysed by pyruvate aldolases.

Two other aldolases which have recently emerged as useful biocatalysts are L-threonine aldolase and 2-deoxyribose-5-phosphate aldolase (DERA). Threonine aldolase uses glycine (**14**) as the donor, forming α-amino-β-hydroxy acids with high stereocontrol at the α-centre but a highly variable degree of control at the β-centre. DERA is unusual as it is the only known aldolase which utilises an aldehyde (acetaldehyde, **15**) as the donor, forming a chiral centre of (*R*)-configuration. When an α-hydroxy-aldehyde is used as the substrate, the preference is for the (*R*)-enantiomer. The structure of the products is such that they can undergo sequential aldol reactions *in vitro*, the reactions eventually ceasing when a stabilised cyclic product can form (Fig. 23.18).

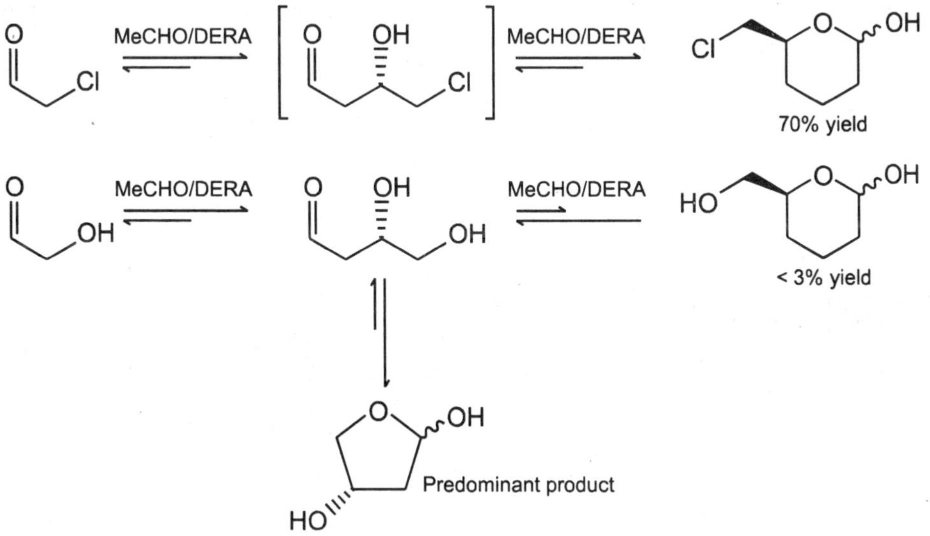

Fig. 23.18. Sequential aldol additions catalysed by DERA.

Transketolase and Transaldolase

In living cells, the link between the metabolic pathways of glycolysis, which degrades sugar molecules with the generation of chemical energy stored in the form of adenosine triphosphate (ATP), and the pentose phosphate pathway, which is used by the cell to provide small-molecule precursors for biosynthetic pathways and reducing power in the form of (reduced) NADPH, is mediated by the enzymes transketolase and transaldolase. These two enzymes provide a reversible link by catalysing the interconversion of sugar phosphate molecules. Transketolase (TK) is responsible for the transfer of a two-carbon ketol unit, $C(O)CH_2OH$, from a donor ketose such as D-xylulose 5-phosphate **24** to an acceptor aldose, e.g. ribose 5-phosphate **25** or erythrose 4-phosphate **26** (Fig. 23.19). The enzyme requires magnesium ions and the co-factor thiamine pyrophosphate (TPP) (**27**) for its activity. The enzyme has been isolated from a variety of sources, and transketolase from yeast, spinach and *E. coli* have all been employed in synthesis.

Transaldolase (TA) catalyses the transfer of the C1-C3 aldol unit from D-sedoheptulose 7-phosphate **28** to D-glyceraldehyde-3-phosphate **16** to produce D-fructose 6-phosphate **29** and D-erythrose 4-phosphate **26**, functioning via a Schiff base intermediate. Although TA is available commercially, its substrate specificity has not been studied in detail.

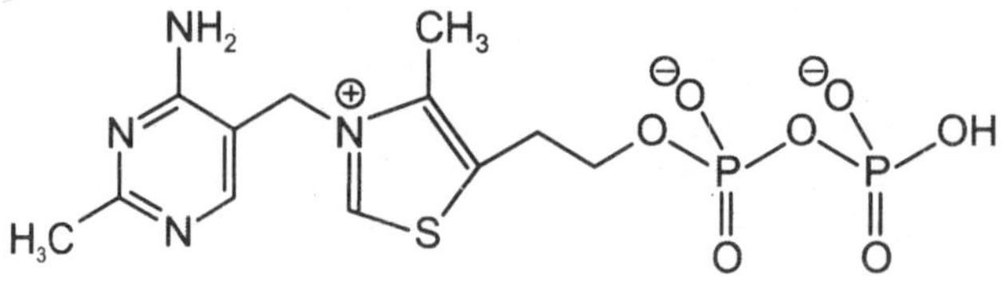

(27)

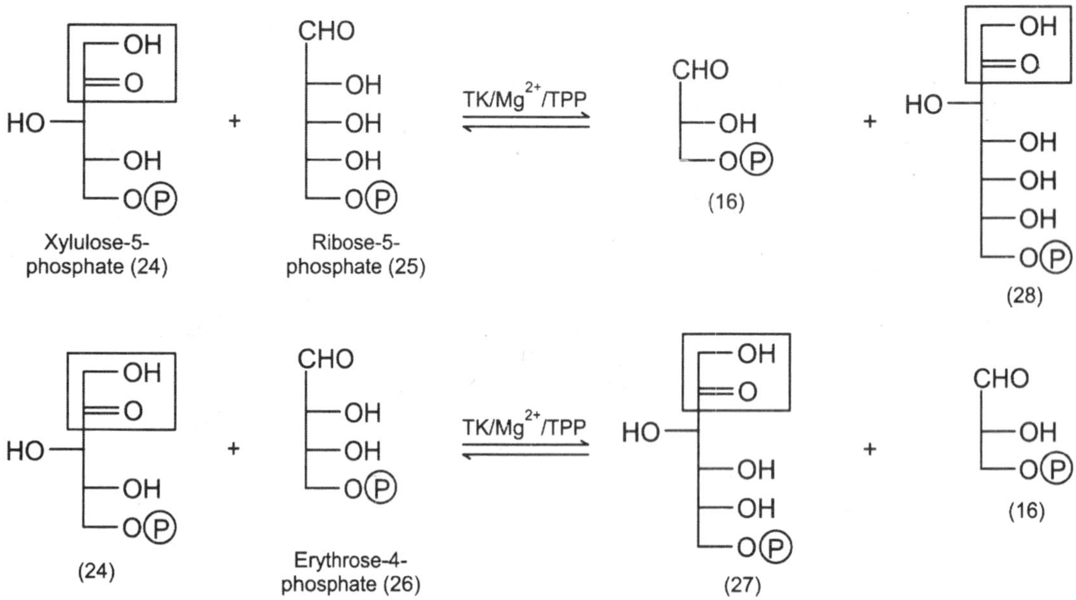

Fig. 23.19. Reactions catalysed by transketolase *in vivo*.

Oxynitrilase

Oxynitrilases or hydroxynitrile lyases, catalyse the reversible addition of hydrogen cyanide to an aldehyde. (*R*)-Oxynitrilase from bitter almonds and (*S*)-oxynitrilases from sorghum and *Hevea brasiliensis* have been isolated and employed in the stereospecific synthesis of carbon-carbon bonds. The cosubstrate, HCN, tends to add non-stereospecifically to the substrate when reactions are performed in aqueous solution reducing the enantiomeric purity of the isolated cyanohydrins; however, the extent of the non-enzymic process can be minimised by a choice of low operational pH and temperature. The enzymes can also be employed in solvents such as ethyl acetate and diisopropyl ether in the presence of minimal amounts of water, thereby suppressing the non-enzymic reaction. Another alternative is to use acetone cyanohydrin as the HCN donor; a transcyanation reaction, rather than a direct addition of HCN, then occurs, and the problem of non-enzymic addition is avoided.

The substrate specificities of oxynitrilases have been examined for a small range of aldehydes. Almond (*R*)-oxynitrilase will accommodate a range of aliphatic and aromatic aldehydes, forming the (*R*)-cyanohydrins in high enantiomeric excess. The selectivity for chiral centres at remote positions on the side-chain appears broad, but a near-total loss of selectivity was observed with (*R*)-2-phenyl-propionaldehyde as the substrate. (*S*)-Oxynitrilase from *Hevea* also appears to have a wide substrate specificity, though only aromatic aldehydes are tolerated by the enzyme from sorghum. The enzymes have been employed in the synthesis of chiral α-hydroxy carboxylic acids and the (*R*)-oxynitrilase has been employed in the kinetic resolution of (*S*)-ketone cyanohydrins and the formation of enantiomerically pure substituted heterocycles. A recent report also describes the use of (*R*,*E*)-2-hydroxy-3-pentenenitrile, available in ≈200 mmol batches from an (*R*)-oxynitrilase-catalysed addition, in the synthesis of enantiomerically pure α-hydroxy-β-amino acids, including *N*-benzoyl (2*S*, 3*S*)-3-phenylisoserine **30**, the amino acid component of the sidechain of Taxol (Fig. 23.20).

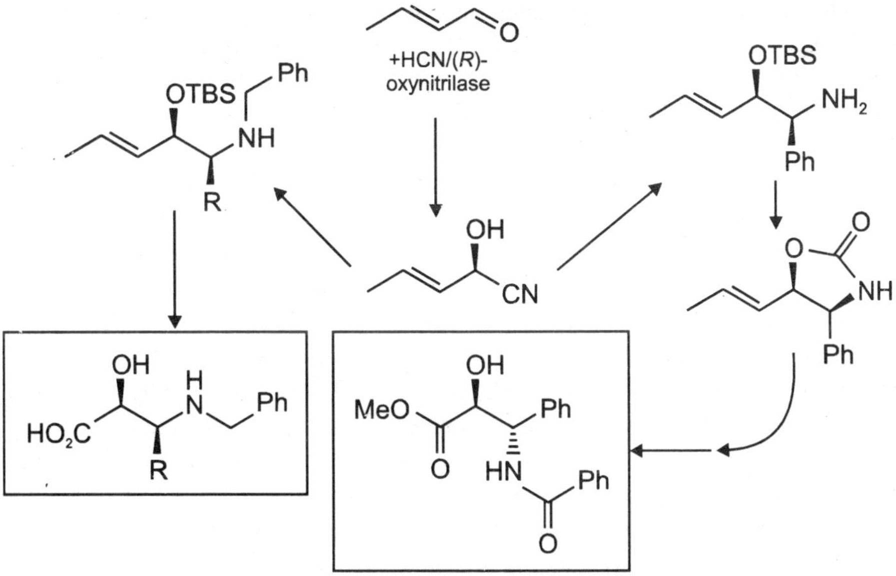

Fig. 23.20. Synthesis of α-hydroxy-β-amino acids using oxynitrilase.

Anabolic Carbon-Carbon Bond-forming Enzymes

An enormous class of carbon-carbon bond-forming enzymes are involved in the biosynthesis of fatty acids, polyketides, terpenes and numerous other secondary metabolites. The majority of these synthase enzymes function by repetitive condensation of small molecule building blocks (such as acetate, malonate or propionate) activated in the form of coenzyme A (CoA) thioesters; the resulting carbon chains are then modified by reduction, cyclisation, etc. The mechanisms of action of these synthases, many of which consist of huge multienzyme complexes with each part of the complex catalysing a particular biosynthetic step, have been extensively investigated, as has the arrangement of the polyketide synthases on a genetic level. A number of organisms possessing genetically engineered polyketide synthases for the production of novel antibiotics have been recently reported, and the sequential arrangement of the synthase genes in the microbial genome has even made possible the selection of genes to produce new polyketides (a concept which one reviewer has colourfully described as 'combinatorial biosynthesis). Isolated enzyme systems are not yet exploited; however, a number of obstacles will have to be surmounted, e.g. the recycling of acyl CoA units, before any of these systems are usable as synthetically useful isolated enzyme systems.

Process Chemistry

The efficiency of carbon-carbon bond-forming enzymes in laboratory-scale synthesis has, on the whole, not yet been transferred to the large-scale industrial process. The fact that most syntheses employing

DHAP-dependent aldolases are carried out under thermodynamic conditions means that reaction times are typically long (five days being not uncommon for mmol-scale laboratory transformations) with repeated additions of enzyme being necessary; the fact that mixtures of products often form under these conditions (*vide supra*) can further complicate product isolation (chromatography is normal in laboratory-scale syntheses). An additional hindrance is the expense of the co-substrate DHAP. Chemical methods now exist which can produce multigram quantities of high-purity DHAP *in situ* enzymatic methods (such as the cleavage of FDP in the presence of RAMA with subsequent isomerisation of glyceraldehyde-3-phosphate **16** using triosephosphate isomerase (TIM), or the oxidation of L-glycerol-3-phosphate using glycerolphosphate oxidase (GPO) are often more convenient, but suffer from their own disadvantages, such as the expense of starting materials and the presence of extra components in the eventual equilibrium mixture. However, the versatility of the DHAP aldolases remains their greatest asset as potential process catalysts. The recently reported use of cross-linked crystals of RAMA, which display exceptionally high stability and catalytic efficiency even in relatively high concentrations of water-miscible organic solvents, may indicate new opportunities for the application of this catalyst beyond the research laboratory.

The pyruvate aldolases may prove to be even more versatile than the DHAP aldolases as process catalysts. Pyruvate addition to an aldehyde produces a densely and differentially functionalised carbon skeleton (four different oxidation states of carbon being present in four contiguous carbon centres). The high commercial value of many of the products obtained from these aldolases, particularly Neu5Ac aldolase, has led to the development of efficient reactors for their synthesis. [*N*-acetylneuraminic acid derivatives have found application as anti-influenza compounds, as well as in the synthesis of new glycoconjugates many of which have significant biological activity.] Wandrey's enzyme membrane reactor for the multigram synthesis of Neu5Ac utilises an epimerase to convert the cheap *N*-acetyl glucosamine, **19** (a non-substrate for Neu5Ac aldolase) into the substrate ManNAc, **17**, with an excess of pyruvate to drive the aldol condensation equilibrium in the direction of Neu5Ac (Fig. 23.21). The process was successfully scaled up by Glaxo-Wellcome, who have used simple solvent treatment and crystallisation steps to obtain highly pure Neu5Ac on a multi-kilogram scale. To date, the Glaxo-Wellcome process represents the most successful large-scale enzymatic carbon-carbon bond-forming process. Other pyruvate aldolases, notably KDPG aldolase, however also have characteristics which may make them suitable large-scale process catalysts, including high stability in organic cosolvents and entirely kinetic, rather than thermodynamic, patterns of reaction.

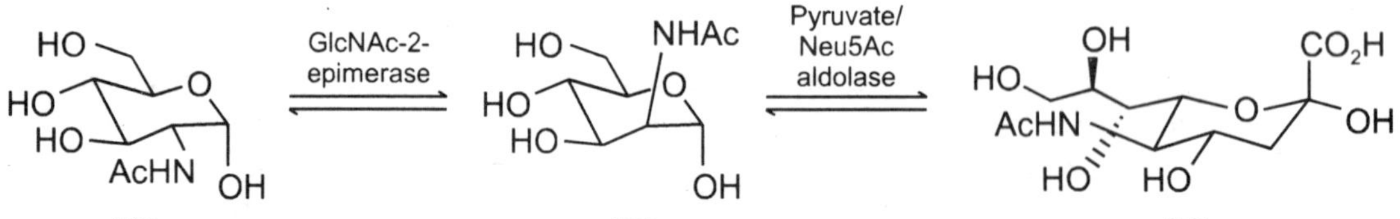

(19) (17) (18)

Fig. 23.21. Production of Neu5Ac in the enzyme membrane reactor.

Oxynitrilases also have considerable potential as catalysts for large-scale processes. The enzymes lend themselves readily to immobilisation and have been shown to be highly effective in packed-bed reactors at high substrate concentrations and in an enzyme membrane reactor. The co-catalyst (HCN) is inexpensive and does not unduly complicate downstream processing. At present the major limitation on their exploitation is catalyst availability; the availability of recombinant oxynitrilases may well provide the key to their successful implementation as process catalysts.

Thus, the organic synthesis has experienced strongly the impact of enzyme chemistry in the past twenty years. Several features stand out significantly. First, enzyme-catalysed steps and chiral intermediates produced by enzyme-catalysed reactions are now common in synthesis. Second, the range of useful and accessible enzymes has expanded greatly beyond hydrolysis and redox reactions into selective carbon-carbon bond formation. Third, major attention has been paid to the requirements of processes using enzymes. There is no reason to believe that the heights of scientific and technical practicality have been reached.

Chapter 24

Enzymes and Bioremediation

INTRODUCTION

Technological developments have had their own consequences, and pollution is one of the major resulting problems. As people become conscious about the problems of pollution, laws are being enforced to clean up hazardous wastes. Biological methods or 'bioremediation' promises to provide an ecofriendly solution to this menacing problem. Biodegradation refers to the breakdown of substances or molecules by biological system into smaller units. This, essentially is a part of normal cellular physiology of all living beings; for instance, glucose is broken down to yield carbon dioxide and water by a process known as cellular respiration, and so on. However, the term 'biodegradation' used in this chapter would refer to the degradation of undesirable and harmful compounds into harmless or more tolerable products. It is difficult to draw a dividing mark between desirable and undesirable compounds. For instance, cellulose is important and desirable for fabric and paper, while it becomes a nuisance in the environment; many similar examples can be illustrated. Almost all natural organic compounds and many synthetic compounds are biodegradable, although some persist for a longer time than the others, or require some special conditions to be degraded. A variety of micro-organisms possess the ability to degrade the undesirable or xenobiotic compounds. These organisms include bacteria such as *Mycobacterium, Bacillus, Pseudomonas, Coryneforms*, etc. fungi such as *Cladosporium, Cunninghamella, Phanerochaete*, a few yeasts, algae and protozoans. The rates at which xenobiotics are degraded are usually far too slow compared with the rates at which pollutants are discharged into the environment, and become toxic to most life forms. The slow rates of degradation may be attributed to many physio-chemical factors or the genetic make-up of the organism present in that environment. It is important to realise that in this field of enzyme technology and chemistry, success in applications depends not only upon effective organic chemistry but also upon physico-chemical properties such as mass transport and in bioremediation, above all the viability of the micro-organisms responsible for the desired degradation.

Biodegradation involves a series of enzymatic reactions. We shall try to elucidate the concept of these enzymatic transformations with regard to hydrocarbons for two main reasons. First, it would be too extensive to discuss biodegradation of all known xenobiotic compounds; however, a brief description of these compounds is presented in the following sections. Second, it is convenient since many enzymes which are involved in the degradation of various hydrocarbons have been characterised and their modes of action are known.

Discussion of their scope of action and genetic regulation provides good examples of the basic mechanisms and the complexity of the sequences of transformations involved in bioremediation. It is noteworthy that virtually all of the metabolism of xenobiotic compounds begins with oxidation of a carbon hydrogen bond.

XENOBIOTIC COMPOUNDS

In a broader sense, the term xenobiotics refers to chemicals or to groups of chemicals which are often alien to the existing biological system. In view of the capability of micro-organisms to adapt to the changing environment at remarkably rapid rates, it is difficult to class any chemical or group of chemicals as alien in absolute terms. Every year there is addition to an already existing enormous list of chemicals that are required to undergo some sort of environmental impact assessment. Many such compounds are either toxic, carcinogenic or have the potential to pose similar hazards. For instance, heavy metals like lead and mercury, which are present in petrol or in seed dressings, interact with enzymes or their co-factors and render them inactive, causing serious metabolic disorders. As mentioned earlier, there can be no clear-cut distinction between toxic and non-toxic compounds, as the tolerance limits for different compounds vary with the organism in question as well as the nature of the compound. Apart from the heavy metals, some of the most common xenobiotics are grouped under the categories of polychlorinated biphenyls (PCBs), dioxins, pesticides, detergents and hydrocarbons.

Polychlorinated Biphenyls (PCBs)

These are a group of biphenyls containing one to ten chlorine atoms per molecule. Their properties include high viscosity, boiling and melting point, and resistance to chemical attack. They possess low electrical conductivity and find application as plasticisers in polyvinyl polymers, insulators and coolants. PCBs are also quite stable to biological attack, and their stability increases with the degree of chlorination. It is observed that most of the tetrachlorobiphenyls and more chlorinated derivatives of PCBs have an extremely slow rate of degradation. However, mono-, di- and trichlorinated biphenyls are biodegradable, as evidenced by the degradation of these by *Alcaligenes* and *Arthrobacter*, and can serve as source of carbon and energy to the cells.

Dioxins

These are a group of compounds which arise primarily as by-products of pesticides industries and PCB manufacture. They result either from the combustion of chlorobenzenes and other chemical wastes or during bleaching of paper. Chlorinated dibenzo-*p*-dioxins like 1-chlorodibenzo-*p*-dioxin, 2-chlorodibenzo-*p*-dioxin and 2,3,7,8-tetrachlorodibenzo-*p*-dioxin (TCDD) have been recognised as contaminants in the manufacture of various pesticides and herbicides derived from chlorinated phenols.

The worldwide use of these chlorophenols in industry and agriculture is of environmental concern as these compounds are highly toxic, e.g. TCDD is a proven carcinogen in guinea pigs. These chlorinated compounds are eliminated from the natural environment by reductive dehalogenation. The reaction is initiated in the presence of ultraviolet radiation and an organic hydrogen donor. Metabolism of these compounds by micro-organisms was initially explored by Klecka and Gibson who reported co-oxidation of dibenzo-*p*-dioxin to *cis*-1,2-dihydroxy-1,2-dihydrodibenzo-*p*-dioxin and 1,2-dihydroxybenzo-*p*-dioxin by *Pseudomonas* sp. (Fig. 24.1). In *Beijerinkia* sp., co-oxidation was observed under certain conditions: whole cells, grown on succinate with biphenyl as inducer, oxidised dibenzo-*p*-dioxin and several mono-, di- and trichlorinated dibenzo-*p*-dioxins.

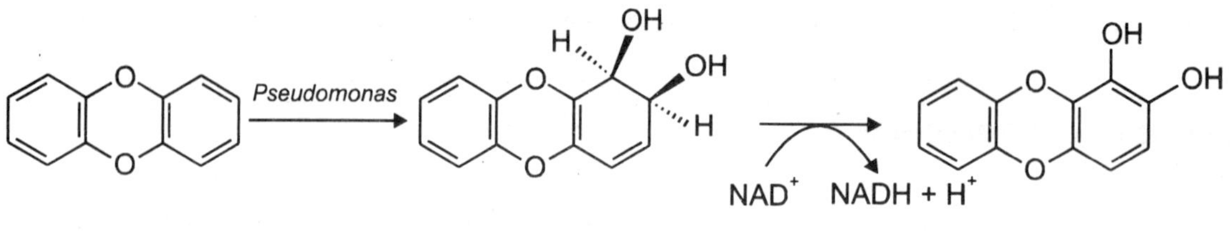

Fig. 24.1. Metabolism of dibenzo-*p*-dioxin in *Pseudomonas* sp.

Pesticides

Organic pesticides include chlorinated hydrocarbons, methyl carbamates and organophosphate compounds. Methyl carbamates and organophosphates are readily biodegradable by hydrolysis whereas chlorinated hydrocarbons [e.g. caldrin, heptachlor, chloridane, 1,1,1-trichloro-2,2-bis(*p*-chlorophenyl)ethane (DDT), 1,1-dichloro-2,2-bis(*p*-chlorophenyl)ethylene (DDE)] resist rapid metabolism. However, many micro-organisms are capable of degrading pesticides. *Pseudomonas putida*, for instance, degrades Dolapon and can use it as sole source of carbon. Attempts have been made to modify the chemical nature of a pesticide to improve its biodegradability. *s*-Triazines, for instance, were structurally modified to yield a degradable product, cyanazine. The bioaccumulation of DDT derivatives by substitution of various functional groups in the *p*-position has provided information on biodegradability. The degradation pathway of DDT analogues like methoxy-methichlor, methyl-ethoxychlor, chlormethylchlor, etc., have been studied in insects. It is observed that in general, addition of electron-donating functional groups such us CH_3-, CH_3S-, CH_3O- and C_2H_5O- makes the chlorinated hydrocarbons more easily biodegradable, as would be expected if oxidation is the primary route. Schematic representations of some the chlorinated hydrocarbons are illustrated in Fig. 24.2 and at the *meta*-position of the ring makes the compound more resistant to biodegradation.

Detergents

Commercial detergents contain approximately 20 per cent surfactant, while the rest consist of polyphosphates, silicates, sulphates, carboxymethyl celluloses, alkanamides, dyes and perfumes. The surfactant is a mixture of alkylsulphonates [$CH_3(CH_2)_nSO_3Na$ where $n = 9$ to 13]. The partial oxidation of various compounds leads to eutrophication of lakes and rivers into which effluents containing detergents are discharged. The microflora of such natural waters have acquired the ability to transform xenobiotic compounds. In these micro-organisms the alkyl side chains of alkyl sulphonates are degraded by the enzymes of lipid metabolism. Naturally occurring sulphate esters have been shown to act as inducers of sulphatases that initiate the catabolism of detergent alkyl sulphates by hydrolysis.

Hydrocarbons

Although a great majority of the hydrocarbons are biodegradable, there are a few which can be termed recalcitrant as they persist in the environment for a longer time. Such hydrocarbons are present in crude oil and industrial effluents. Various fractions of petroleum oil are alkanes, aromatics and asphaltanes. Asphaltanes contain phenols, fatty acids, ketones, esters, porphyrins and resins. Recalcitrance is seen maximally with asphaltanes, high-molecular weight polycyclic aromatic hydrocarbons (PAH) and also branched-chain alkanes. We shall consider the degradation of hydrocarbons separately in detail. Often, xenobiotic compounds themselves are non-toxic but, during the process of biotransformation, the intermediates that are produced by the biological system may be far more toxic than the parent compound. It is, therefore, essential to know the routes by which these compounds are degraded and the nature of those intermediate metabolites.

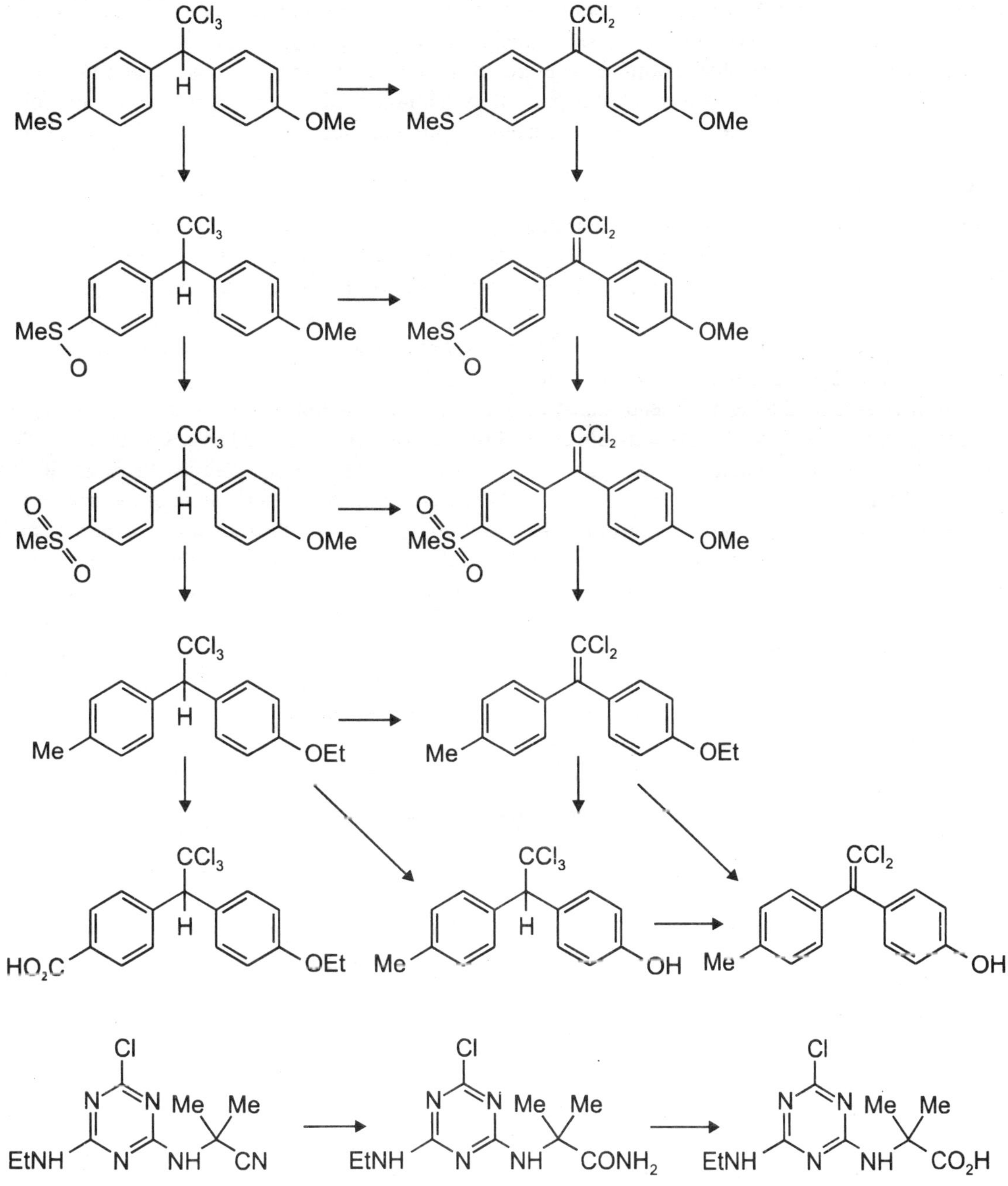

Fig. 24.2. Metabolism of chlorinated hydrocarbons.

After having classified the principal xenobiotic compounds simply, one may wonder why these compounds resist biological degradation, especially when we have stated above that almost all natural organic compounds and many synthetic compounds may be degraded biologically. The reasons are many, and range from physiological factors present within the biological system to physico-chemical factors prevailing in the environment. We shall quickly review some of the major factors that influence biodegradation. Firstly, we shall examine the toxic effects of xenobiotic compounds with reference to hydrocarbons, and the adaptive mechanisms present in the biological system (especially the microbial system), which not only reduce the toxic effects but also use the xenobiotics as source of carbon and energy.

TOXICITY

Higher homologues of hydrocarbons, both aromatic and aliphatic, are primarily hydrophobic in nature. All biological systems, with the exception of certain viruses are enclosed within a lipid membrane. Lipid molecules present in these membranes consist of two parts, the hydrophilic part and the hydrophobic part. The orientation of these parts of lipid molecules is such that the hydrophilic moiety is able to interact with the surrounding milieu directly (which invariably is water), while the hydrophobic moiety stays away from the predominantly aqueous environment.

In bacteria, the lipid bilayer is encapsulated within a reasonably tough coat. In Gram-negative micro-organisms this external coat lodges a certain class of protein known as porins. These are fairly unbiased towards the type of molecule that can pass through them. Besides these proteins, there are highly specialised transporters present in the membrane which are specific towards the kind of molecule to be transported. This uptake process can either be energy intensive (active transport) or driven by a concentration gradient between the inside and outside of the cell so that there is no need to spend energy in order to get certain molecules (passive diffusion). Most xenobiotic compounds (hydrocarbons) are believed to enter the cell through passive diffusion. As both the intracellular and extracellular milieu are aqueous, these hydrophobic (or lipophilic) xenobiotic compounds tend to accumulate in the lipid bilayer. This accumulation can be associated with serious consequences if no action is taken by the biochemical machinery of the cell, such as excretion, metabolism, etc. Some of the major effects of accumulation of lipophilic compounds in the lipid bilayer are as follows.

The accumulation of lipophilic compounds in the bilayer results in an increase in the volume of the membrane. Since the distribution of lipid molecule is heterogeneous within the membrane, the accumulation of lipophilic compounds is not uniform and causes bulging of the membrane at those points where accumulation has taken place. Depending on the extent of accumulation, the cell may lyse at those points.

Even if lysis does not occur, in order to maintain the membrane integrity, the side chains of the hydrophobic moiety of the lipid must interact with each other. Small lipophilic molecules compete with this interaction and the result is loss in membrane integrity, which has serious consequences. Ion gradients across the membrane drive many essential cellular processes. Loss of membrane integrity results in the loss of such ion gradients and thus the cellular process is severely affected.

The membranes are host to many enzymes of metabolic importance as well as to transport systems. Many membrane-bound enzymes have two domains: the hydrophilic domain faces either the intracellular spaces or the external surrounding, while the hydrophobic domain is imbedded in the lipid bilayer. The lipophilic compounds may interact with the hydrophobic domain of the enzyme. This interaction may bring about the change in the overall conformation of the enzyme, eventually leading to loss of activity. Besides membrane-bound enzymes there are other membrane-bound proteins which traverse the membrane through their hydrophobic domain. Interaction of lipophilic compounds with these may result in loss of their respective functions. These include essential functions such as signal transduction and ion transport.

ADAPTIVE RESPONSES

In view of the toxic effect caused by hydrocarbons it is intriguing that certain micro-organisms can survive and are able to tolerate high concentrations of xenobiotic compounds in their vicinity. The following are the general mechanisms by which the micro-organisms adapt themselves to such compounds. It is interesting that several enzymes now in industrial use for the manufacture of chiral organic compounds must have originated from adaptive responses of micro-organisms to high concentrations of xenobiotics.

Ordering of the Lipid Bilayer

Various lipophilic compounds induce changes in the fatty acid composition. In general, it is observed that polar solvents that are miscible with water cause an increase in the synthesis of unsaturated fatty acids. As the content of unsaturated fatty acids increases the membrane becomes increasingly hydrophobic, thereby reducing the fluidity of the membrane.

Isomerisation of the Lipid

Certain solvents such as benzene cause isomerisation of fatty acids from the *trans*-form to their *cis*-forms. Such an isomerisation results in increased ordering of the membrane lipids thereby reducing the fluidity, preventing the lipophilic molecules from entering. There are solvents that cause the *cis*-form to isomerise into the *trans*-form.

Other Factors

Some other factors that result in the better adaptation of micro-organisms are decreased hydrophobicity to prevent accumulation of lipophilic compounds in the lipid bilayer, active excretion of the lipophilic compounds to reduce the toxic effects, and directed evolution in the micro-organisms to modify the existing enzymes to degrade the lipophilic compounds or to transform them into another form which is less toxic to the cell. Such changes in the characteristic biochemical capability of micro-organisms can be studied by genetic methods which examine the genes associated with the enzymes that catalyse the degradation reactions.

PHYSICO-CHEMICAL FACTORS INFLUENCING BIODEGRADATION

The rate of biodegradation is greatly influenced by various physical and chemical factors. Factors that regulate the biodegradation process are detailed below:

Solubility or Bioavailability

Present conjecture is that those compounds that are soluble in water or those that have been solubilised through some means serve as a substrate for catalytic degradation. The hydrophilic phospholipids present on the cell wall of the bacteria prevent any interaction between the lipophilic molecules and lipid membrane. The question of the uptake of such molecules by bacterial cells remains to be answered. Most hydrocarbons dissolve in water at around 25°C in trace amounts. This dissolved quantity then becomes a limiting factor for biotransformation as the dissolved compounds may be transformed rapidly.

Dissolution rates generally increase at higher temperature with the increase in solute surface area and in the presence of bacterial growth. Higher rates of dissolution in the presence of bacterial growth may be attributed to production of surface-active compounds (biosurfactants) as products of the metabolism of the micro-organisms themselves. Many species of bacteria have been reported to possess the ability to produce a biosurfactant, e.g. *Pseudomonas*, *Bacillus*, *Acinetobacter*, etc. The production of a biosurfactant by micro-organisms can occur during exponential growth or during the stationary phase of growth when the nutrient limiting conditions start prevailing in the medium. In the case of growth-associated biosurfactant production, there exists a parallel relationship between substrate utilisation, growth and biosurfactant production. In *Acinetobacter calcoaceticus* RAG-1, emulsan or emulsan-like precursors accumulate as capsular material during the exponential growth phase but are released into the medium during the stationary phase of growth.

Temperature

In general, increase in temperature increases solubility of solutes; as more and more lipophilic molecules get dissolved in water, they are readily available for the cells to degrade them. It is observed that between 30–40°C the rates of degradation increase by two- to three-fold. It is difficult to draw any straightforward conclusion from such an observation as temperature may also influence the microbial growth and other physiological processes.

pH

As with temperature, pH plays a crucial role in microbial growth besides influencing the solubility of lipophilic compounds. Depending on the type of micro-organisms the optimum pH for growth varies. Similarly, the effect of pH on solubility of lipophilic compounds varies with the type of compound under question.

Oxygen

Apart from its requirement in general aerobic microbial metabolism, oxygen is involved in initiating the degradation process. The first step in the biotransformation is introduction of an oxygen atom or atoms by an enzyme, a monoxygenase or dioxygenase. The first step is not only critical in biodegradation but also rate limiting. A few exceptions to this generalisation are encountered in case of anaerobic degradation of hydrocarbons where instead of oxygen, sulphur acts as terminal electron acceptor.

Nutrients

These are required basically to maintain the microbial population in a physiologically active state. For example, rates of crude oil degradation by natural microbial flora increase by simply spraying the sites with a nitrogen source such as urea.

Moisture Content

This is an important factor when degradation is considered in soil. Besides supporting the microbial population, moisture content determines the extent of solubility of lipophilic compounds.

HYDROCARBON DEGRADATION

Hydrocarbons include both aliphatic and aromatic compounds and are widespread xenobiotics as a result of all economic activity concerned with fuel and petrochemicals. The mechanism of the degradation of the two classes of hydrocarbon and the enzymes involved differ remarkably; therefore we shall discuss them separately. Organic chemicals are metabolised aerobically to carbon dioxide and water, and anaerobically to methane and carbon dioxide. The degradation of hydrocarbons may be considered as a two-step process. Hydrocarbon uptake is an essential first step for the bacteria to degrade both straight-chain (labile) and ring (semi-recalcitrant) compounds, with the latter requiring more stages via a series of catalytic reactions. Extracellular surfactants may assist in the uptake process. The initial steps in the catabolism of aliphatic, cyclic and aromatic hydrocarbons by bacteria and fungi involve the oxidation of the substrate by oxygenases, for which molecular oxygen is required. Aerobic conditions are, therefore, necessary for this route of microbial oxidation. Once the hydrocarbon constituents have been catabolised to smaller molecules they are utilised in metabolism to obtain the energy required for microbial growth and division. Such an approach ensures, if the process is permitted to go to completion, that the hydrocarbons are completely degraded (mineralised). After the hydrocarbon in question has been degraded, the viable population density of the added or enriched bacteria declines rapidly in response to the removal of the most prevalent carbon sources (the polluting hydrocarbons) and the indigenous micro-organisms predominate once more.

Mechanism of Hydrocarbon Uptake

It might be expected that there would exist a pathway for the uptake of hydrocarbons from the outside of the cell membrane to the locus of the enzyme responsible for the primary attack on hydrocarbons in a micro-organism. However, the mechanism of hydrocarbon uptake has not been explained so far. The first step by which micro-organisms utilise liquid hydrocarbons as carbon sources is by the emulsification of the hydrocarbon and water. The hydrocarbon emulsified by micro-organisms may then be absorbed easily by the cells. Three possible pathways for hydrocarbon uptake by micro-organisms have been proposed by Nakahara:

1. Direct contact of the cells with large oil drops.
2. Direct contact of cells with fine droplets.
3. Uptake of dissolved hydrocarbons in the aqueous phase.

Hydrocarbon uptake may be carried out by free cells, either cells which absorb sub-micron droplets or cells which are attached to large oil drops to form flocks. Some investigators have reported that yeast cells grow mainly at the interphase between oil and water and that the oil phase interfacial area limits microbial growth, while others suggest that sub-micron oil droplets are available to cells for growth. Interfacial area and amount of sub-micron droplets are thus important factors for hydrocarbon uptake. Other than these essentially physical phenomena, hydrocarbon utilisation by micro-organisms requires several adaptive modification of the cells which concern the breakdown of the substrate.

Some microbes, e.g. *Acinetobacter calcoaceticus* are known to produce biosurfactants. The main physiological role of biosurfactant is to permit the micro-organism to grow on water-immiscible substrates by reducing surface tension at the phase boundary, thus making the substrate more readily available for uptake and metabolism. The molecular mechanisms of the uptake processes of these substrates are, however, still not clear. The direct uptake of hydrocarbons dissolved in the aqueous phase, direct contact of cells with large hydrocarbon droplets as well as interaction with solubilised droplets (emulsion) have been described as being the major processes involved in the uptake of hydrocarbon substrates. The studies on the phenomenon of hydrocarbon fermentation and the mechanisms of hydrocarbon transport and uptake are still in infancy, and detailed research is still needed to understand thoroughly the processes involved in these two phenomena.

Biodegradation of Alkanes

Polluting alkanes include the entire range of straight-chain (*n*-alkanes), branched alkanes and cycloalkanes. Alkanes with a chain length of up to C_{44} have been reported to be biodegraded. Biodegradation of *n*-alkanes proceeds via monoterminal attack yielding a primary alcohol as one of the products which is oxidised further to corresponding aldehyde and monocarboxylic acid. The monocarboxylic acid is further metabolised via series of β-oxidations typical of fatty acid metabolism. Subterminal oxidation of certain alkanes leads to the oxidation or a secondary alcohol which is further oxidised to a ketone. Highly branched alkanes such as pristane, undergo diterminal γ-oxidation resulting in the formation of a dicarboxylic acid. Extensive methyl branching increases the resistance of the hydrocarbon to microbial degradation. Methyl substitution at the α-position, for instance, blocks β-oxidation and therefore alternate routes such as α-oxidation have to be adopted.

Sub-terminal alkane oxidation pathway

In bacteria that utilise alkanes as sole source of carbon and energy, subterminal oxidation is a minor alternate pathway to monoterminal alkane oxidation. On the other hand, in bacteria that co-metabolise alkanes, as observed in certain *Bacillus* sp., *Streptomyces* sp., *Arthrobacter* sp. and *Chlorella*, the analysis of degradation products indicates that alkanes in these micro-organisms are metabolised via a subterminal oxidation pathway. Many fungi (moulds) and some yeasts, exclusively oxidise alkanes to ketones. Isolation of these intermediates has been the major criterion for establishment of subterminal oxidation pathway for alkane metabolism operating in these micro-organisms.

Diterminal alkane oxidation pathway

Micro-organisms unable to discriminate between the ends of *n*-alkanes, oxidise both the terminal methyl groups without change in the carbon skeleton which results in the formation of dicarboxylic acids. Extensive lists of micro-organisms with the metabolic potential for conversion of *n*-alkane to dicarboxylic acids have been compiled by Rehm and Reiff. Among them were primarily yeasts and moulds while only a few bacteria are listed in the compilation. Yi and Rehm demonstrated that dicarboxylic acids could be formed from *n*-alkanes via the α,ω-diol, i.e. diterminal alcohol. Alternatively, dicarboxylic acids could be formed by the oxidation of fatty acid, resulting from monoterminal oxidation. These are further converted into ω-hydroxy fatty acid and then to a dicarboxylic acid. The monocarboxylic acids are probably hydroxylated by a non-specific alkane hydroxylase. In fact, the cytochrome P-450 alkane hydroxylase system hydroxylates fatty acids at much higher rates than alkanes.

ENZYMOLOGY OF *n*-ALKANE OXIDATION

As already outlined in this chapter, in the majority of the alkane-oxidising micro-organisms, the initial metabolism of *n*-alkanes takes the route of monoterminal oxidation. The enzymes required for such a pathway are alkane hydroxylase, alcohol dehydrogenase and aldehyde dehydrogenase. Overall, the characteristic feature of alkane oxidation is the conversion of *n*-alkanes to the corresponding carboxylic acids, which are then channelled into the β-oxidation pathway.

Mechanisms of Alkane Oxidation

The hydroxylation reaction is catalysed by a multifunction oxidase (monoxygenase) which, in the presence of molecular oxygen and certain cofactor reductant, converts the alkane into the corresponding primary alcohol. One atom of molecular oxygen is incorporated into the alkane while the other atom is reduced to water:

$$RCH_2CH_3 + O_2 + NAD(P)H + H^+ \rightleftharpoons RCH_2CH_2OH + NAD(P)^+ + H_2O$$

In contrast, hydroperoxidation involves the incorporation of molecular oxygen into the alkane by a dioxygenase to yield an *n*-alkyl hydroperoxide, which is subsequently reduced to the primary alcohol.

$$R-CH_2-CH_3 + O_2 \longrightarrow R-CH_2-CH_2COOH$$

$$\xrightarrow{NAD(P)H} R-CH_2-CH_2OH + NAD(P)^+ + H_2O$$

This mechanism of oxidation is not well understood because so far the isolation of alkyl hydroperoxides in alkane-oxidising micro-organisms has not been achieved. The incorporation of molecular oxygen into an alkane during its oxidation by micro-organisms is an indispensable pre-requisite for its mineralisation. From the above equations, the distinction between the mechanisms of hydroxylation and hydroperoxidation is not clear, as overall stoichiometry of both the oxidation reactions is ultimately the same.

Enzymes Involved in Alkane Oxidation

The oxidation of alkanes is catalysed by a complex enzyme system that appears to be linked to one of several different electron transport systems. Since the monoxygenase involves both electron transfer and an oxygen insertion step, extensive work has been done to study the interactions between these two events at the molecular level and the nature of the activated oxygen species generated by these enzymes. Rubredoxin and cytochrome P-450 systems represent well-defined systems at the enzyme level and serve to provide a comparative basis for the study of alkane oxidation by different micro-organisms relevant to bioremediation. The range of enzymes known or implicated is very wide, as described in this section, and a significant question regarding the genetic regulation of alkane oxidation thus arises.

Rubredoxin-dependent alkane hydroxylase

This enzyme was reported by Peterson in *Pseudomonas putida*, catalysing the hydroxylation of either alkanes or fatty acids in the presence of oxygen and NADH. It is a three-component monoxygenase consisting of rubredoxin, NADH-rubredoxin reductase and P-hydroxylase. These three proteins have been studied in detail with respect to β-hydroxylation of alkanes and fatty acids.

The first component of the enzyme system is a membrane-bound monoxygenase, which catalyses the addition of one oxygen atom to the terminal methyl group. The second oxygen atom is reduced by the electron transferred from NAD(P)H to rubredoxin (a red iron-sulphur protein) by an FAD-containing enzyme, rubredoxin reductase. In other bacterial alkane-oxidising enzyme systems, studied in less detail, a similar mechanism of alkane oxidation has been observed as proposed for *P. putida* and *A. calcoaceticus*. Isolation and characterisation of rubredoxin from *A. calcoaceticus* supported the similarities between *P. putida* and *A. calcoaceticus*. Van Eyk and Bartels separated crude extracts of heptane-grown *P. aeruginosa* into three separate proteins which, though not characterised, appeared to be rubredoxin, NADH-rubredoxin reductase and hydroxylase. Examination of the substrate specificity of the alkane hydroxylase of heptane-grown *P. aeruginosa* showed broad substrate specificity. Mckenna and Coon also found similar characteristics for the *P. putida* hydroxylase. Similar proteins are also known to occur in other hydrocarbon-oxidising bacteria and have been demonstrated in *P. desmolytica*. In *Corynebacterium* sp. Matsuyama described a three-component enzyme system derived from *P. aeruginosa* catalysing the conversion of hexadecane to hexadecanol.

Cytochrome P-450-dependent alkane hydroxylases

The yeast cytochrome P-450 system involved in alkane-metabolism has been reviewed by Kapelli and Muller. Initially, this enzyme was characterised in yeast as an alkane monoxygenase and was absent from most of the bacterial strains that had been traditionally investigated except for certain strains of *Acinetobacter* and *Corynebacterium* sp. The isolation of cytochrome P-450 from these bacteria provided convincing evidence that there might be more cytochrome P-450-dependent alkane hydroxylases existing in bacteria.

Among the prokaryotes, besides *Corynebacterium*, *A. calcoaceticus* EB 104 is the only bacterium in which a cytochrome P-450 system has been identified as an alkane monoxygenase, based on the following evidence: (i) cytochrome P-451 was induced by the unoxidised alkane and not by an intermediate oxidation product, (ii) inhibition of cell-free alkane oxidising activity by CO, typical of cytochrome P-450-dependent reactions, (iii) concomitant induction of cytochrome P-450 and alkane mono-oxygenase activity by the non-aliphatic hydrocarbon biphenyl, and (iv) reconstitution of alkane monooxygenase activity by the isolated proteins cytochrome P-450, ferredoxin and ferredoxin reductase.

Alcohol dehydrogenases

In alkane-grown bacteria, most of these enzymes are constitutive, soluble and require NAD^+ or $NADP^+$ as co-factors. A few of these enzymes have been purified and are relatively non-specific for primary aliphatic alcohols within a certain range of chain lengths (short-, medium- or long-chain). Alkanol dehydrogenase oxidised alkanols to alkanals with concomitant reduction of $NAD(P)^+$ to NAD(P)H. These enzymes are related to the alcohol dehydrogenases that have found wide applicability in organic synthesis.

P. aeruginosa contains a variety of soluble, constitutive pyridine nucleotide-dependent alcohol dehydrogenases with broad substrate specificity. In addition, *P. aeruginosa* contains at least two $NAD(P)^+$-independent alcohol dehydrogenases induced by growth on primary alcohols, alkanes or α,β-diols. Growth on octane or higher alcohols or diols induced a second, similar membrane-bound enzyme the K_m of which decreased with increasing substrate chain length. The introduction of a polar group into the molecule reduced its oxidisability, suggesting the presence of two binding sites for the substrates, one a polar site binding the hydroxyl group, the other a hydrophobic site binding the alkyl chain. Hydrophobic interaction between the substrate and membrane-bound enzyme was important in oxidation of hydrophobic alkanes. Other types of alcohol dehydrogenase have also been discovered. Nucleotide-independent alcohol dehydrogenases were found to use phenazine methosulphate, as electron acceptor.

A. calcoaceticus strains have distinct NAD^+ and $NADP^+$-dependent, soluble, specific alcohol dehydrogenases. As in *Pseudomonas*, the K_m of the enzyme decreased with increasing substrate chain length. Singer and Finnerty reported the presence of three alcohol dehydrogenases ADH-A, ADH-B and HDH, in the same bacterium. Alcohol dehydrogenase-A, ADH-A, was seen to be induced by ethanol and was essentially an NAD^+-linked ethanol dehydrogenase, while the other two had mid- to long-chain specificity. ADH-B was constitutive, $NADP^+$-dependent and functioned optimally with octanol as substrate while hexadecanol dehydrogenase (HDH) was an NAD^+-dependent, inducible enzyme. Fox, in *A. calcoaceticus* HOI-N, however, found a constitutive, NAD^+-dependent alcohol dehydrogenase that showed a preference for decanol while the $NADP^+$-dependent enzyme showed a low K_m, for tetradecanol. Alkane-grown *A. calcoaceticus* also contained an $NAD(P)^+$-independent alcohol-oxidising activity detected in crude cell extracts by hexanol- or octanol-dependent reduction of dichlorophenol indophenol (DCPIP). This activity was not observed in ethanol-grown cells. No bound co-factors such as flavin have been demonstrated in these enzymes.

Aldehyde dehydrogenase

As with alcohol dehydrogenases, bacteria also possess a multiplicity of aldehyde dehydrogenases. Most of the aldehyde dehydrogenases described in earlier studies were either NAD^+ or $NADP^+$-dependent or both. In alkane-grown *Pseudomonas aeruginosa* and *Acinetobacter*, soluble or membrane-bound, $NAD(P)^+$-dependent aldehyde dehydrogenase enzyme activities were detected. Such enzymes were not detected in non-hydrocarbon grown cells such as acetate, palmitate or yeast extract-grown cells. $NAD(P)^+$-dependent aldehyde dehydrogenase activity was induced by long-chain alcohols and aldehydes and not palmitic acid or acetate. There are a number of reports on nucleotide-independent aldehyde dehyrogenases in *Acinetobacter* sp., and also in *Pseudomonas* sp. Recently Fox detected the presence of a novel nucleotide-independent aldehyde dehydrogenase which used Wurster blue as an electron acceptor. This enzyme was strongly induced by the long-chain carbon source and its activity could perhaps allow oxidation of aldehydes without affecting the $[NAD^+]/[NADH]$ or $[NADP^+]/[NADPH]$ ratios in the cell.

The biochemical versatility of bacteria in developing pathways of alkane metabolism is thus remarkable. Aliphatic alcohols and aldehydes are the intermediates of initial alkane oxidation, evidence for which in many cases is based on the abilities of cells to grow on or oxidise alcohol and aldehyde. The enzymes aliphatic alcohol dehydrogenase and aldehyde dehydrogenase exist in multiple forms within a single organism, as has been documented in a number of alkane-utilising bacteria. These enzymes require nicotinamide nucleotides as their cofactors. Membrane-bound, nucleotide independent dehydrogenases have also been reported. The enzymes utilise typical organic dyes such as phenazine methosulphate (PMS) or Wurster blue as artificial electron acceptors. In general, the long-chain alcohol and aldehyde dehydrogenases are inducible and membrane-bound but constitutive enzymes are also found to be involved in alkanol and alkanal dehydrogenation.

Alcohol and aldehyde dehydrogenases in yeasts

In alkane-grown yeasts all the alcohol and aldehyde dehydrogenases described are NAD^+-dependent. These activities are localised in the cytoplasm, mitochondria, peroxisomes and the cytoplasmic membranes. Most of these dehydrogenases exhibit higher specific activities in alkane-grown cells than in cells grown on non-hydrocarbons. In general, they oxidise longer-chain alcohols at faster rates than shorter-chain substrates, while soluble constitutive alcohol dehydrogenase oxidised alcohols from ethanol to hexadecanol at about the same rate. A few of these enzymes have been purified or characterised from alkane-grown yeasts.

Chenko and Tsfasam isolated and characterised an aldehyde dehydrogenase from *Torulopsis candida*, cultured on hexadecane. Kinetic properties of this enzyme showed that it possessed strict specificity for NAD^+ as a cofactor and virtually no activity with $NADP^+$, and its K_m was higher for lower alkanals. Some of the recent reports documented the presence of alkane induced long-chain (fatty) alcohol oxidases, instead of dehydrogenases, in the peroxisomes of different *Candida* sp. grown on alkanes. Similar long-chain fatty alcohol oxidases have also been reported in filamentous moulds.

GENETIC REGULATION OF ALKANE OXIDATION

The study of genetics of hydrocarbon-utilising micro-organisms has essentially occurred in the past two decades. Apart from the intrinsic scientific interest in understanding how nature organises the oxidation of alkanes selectively — a reaction most difficult to accomplish by standard organic chemistry — knowledge of the genetics of alkane oxidation may lead to the development of organisms with improved properties for bioremediation. This is a direct analogy of the study of the biosynthesis of polyketides in terms of the enzymology and genetics so that novel antibiotic molecules might be synthesised by cells. The discovery that the enzymes for the degradative pathway of salicylate are encoded on a plasmid in *Pseudomonas putida* stimulated a widespread investigation into the involvement of plasmid-encoded enzyme systems in hydrocarbon degradation. So far, plasmids have only been detected for simple hydrocarbons, short-chain alkanes and monocyclic and dicyclic compounds. Recent research showed certain genes encoding enzymes for hydrocarbon degradation to be located on the chromosome and not on a plasmid. It has been suggested that catabolism of more complex hydrocarbons is in general chromosomally encoded.

As already discussed several enzymes are required for alkane oxidation and assimilation of the products into metabolism. Further, several of the enzymes involve several separate protein sub-units. An effective genetic organisation for such systems would therefore be expected to include a regulatory gene activated by the inducing alkane, together with several functional genes encoding for the enzymes and associated proteins themselves. The operons containing all of these components will be crucial in the genetic make-up of an organism for it to be able to survive on a 'diet' of alkane. Such organisation of the genes for a particular biosynthetic or catabolic pathway is, of course, well-known.

The initial monoterminal oxidation of alkanes to corresponding alcohols is alkane-inducible and requires the participation of a three-component monoxygenase system. Subsequent conversions of alcohols to fatty acids are mediated by an $NAD(P)^+$-dependent alcohol dehydrogenase and an $NAD(P)^+$-linked aldehyde dehydrogenase. The biochemistry and enzymology of short-chain *n*-alkane oxidation was first elucidated in *Pseudomonas oleovorans*.

Chakrabarty demonstrated that the ability to grow on octane, a strain designated by the genetic shorthand *oct*$^+$, was plasmid encoded in stain PpG6; the *oct* plasmid encoded both the octane-inducible hydroxylase and an octane-inducible dehydrogenase which was redundant for the chromosomal dehydrogenase activity. Growth of the PpG6 on hexane, octane or decane (C_6–C_{10}) supported a genetic model for a single induction system, which was responsive to the entire range of alkanes (C_6–C_{10}). Putative alkane hydroxylase mutants (alkane$^-$, alcohol$^+$) were examined for alkane-oxidising activity in the above system.

Phenotypic analysis of point mutations in PpG6(OCT) and in cured strains distinguished between chromosomally coded and oct-coded functions for alkane oxidation. These studies showed that genes for two of the three proteins of alkane hydroxylase were clearly carried on the *oct* plasmid, which were designated *alk* A and *alk* B. At least three genetic loci are involved in alcohol oxidation (including *alc* A and *alc* B) and two in aldehyde oxidation. For the latter, *ald* A and *ald* B were used to designate the chromosomal loci coding for long- and short-chain alcohol and aldehyde respectively. Figure 24.3 illustrates the regulatory circuit operating in *Pseudomonas* sp.

Although *oct* contained the code for a competent hydrocarbon degradation pathway it was not capable of restoring β-oxidation in mutants that were unable to utilise fatty acids, demonstrating that the chromosomal location of at least one locus is necessary for the metabolism of all straight-chain fatty acids and is independent of alkanes. For an organism to be useful as an agent in bioremediation, ideally it should also be competent to degrade fatty acids. It is also important to establish whether there is a relationship between the inducer molecule and the potential of the organism to degrade a range of substrates. This can be approached by investigating the control mechanisms surrounding the activity associated with OCT. Regulatory mutations in OCT examined in *alc* A backgrounds by Fennewald and Shapiro demonstrated that substrates that supported growth were effective inducers of the hydroxylase genes, *alk* A, *alk* B and *alk* C activities. Further, all compounds that were inducers and supported the growth were effective inducers of *alk* A, *alk* B and *alk* C activities and all compounds that were inducers supported the growth of revertants. These results confirmed that inducer specificity determined the range of alkane substrates utilised by PpG6 for growth. Any mutation that blocked induction also altered inducer specificity or led to a constitutive phenotype that influenced *alk* A, *alk* B and *alk* C activity in a similar fashion. This indicated that both the activities (hydroxylase and alcohol dehydrogenase) were under a common regulatory control and hence a complete degradative pathway could be relied upon.

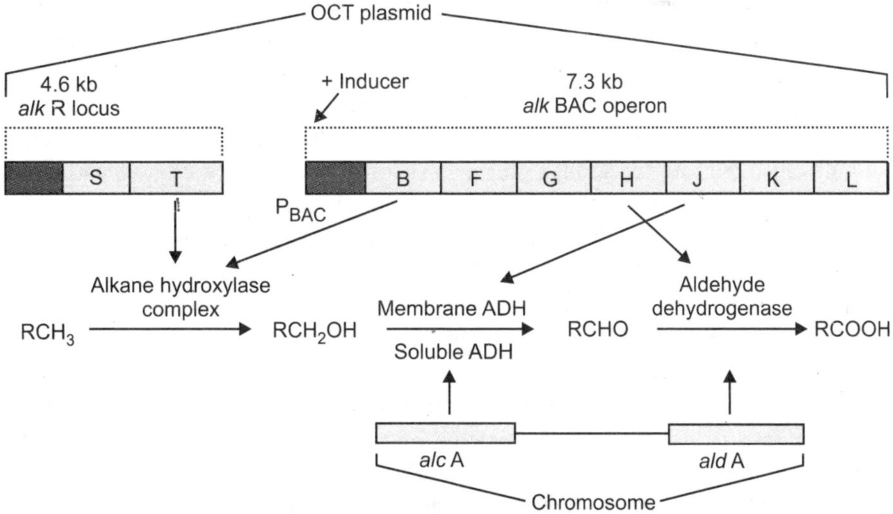

Fig. 24.3. Genetic regulation of alkane oxidation in *Pseudomonas oleovorans* involves the genes present on OCT plasmid and chromosomal genes. The *alk* genes on OCT plasmid are arranged in two transcriptional units, called operons, containing regulatory (*alk* S-T) and structural (*alk* B-L) genes. Alkane, hydroxylase complex is encoded by *alk* T and *alk* B, while genes coding for alcohol and aldehyde dehydrogenase reside both on plasmid (H and J respectively) as well as on chromosome (*alc* A and *ald* A). The product of *alk* S gene directly regulates the expression of structural genes.

The relationships between the various genes that code for the key enzymes of the pathway can be helpfully represented diagrammatically. Mapping of *alk* loci on the *oct* plasmid showed the clustering of the genes for alkane metabolism in two operons (Fig. 24.3). The *alk* BAC operon encodes the structural genes and *alk* R is the regulatory region. All the polypeptides encoded by the *alk* systems have subsequently been identified; based on this, the *alk* BAC operon has been re-named as the *alk* BFGHJKL and the genes associated with each protein can be identified. The first of these is a 45.7 kDa protein and *alk* B gene which is the membrane-bound component of the alkane hydroxylase. The *alk* A region encodes at least one of the soluble components of the alkane hydroxylase complex. Its nucleotide sequence revealed the presence of three open reading frames as *alk* F, *alk* G and *alk* H encoding respectively 14-, 18- and 52-kDa peptides previously identified as rubredoxin reductase. Among these, *alk* G is essential for alkane hydroxylation and *alk* H weakly complemented a *P. oleovorans* aldehyde dehydrogenase mutation. The *alk* C region encodes the other three peptides identified by Eggink. The corresponding cistrons are designated *alk* J (58 kDa), *alk* K (59 kDa) and *alk* L (20 kDa).

On the OCT plasmid, these structural genes are encoded on the 7.5 kb *alk* BAC operon, which is positioned on a 16.9 kb-*Eco*R1 fragment. The *alk* R region encodes two peptides (*alk* ST) and is situated on a 4.9 kb *Sal* fragment. *Alk* S and *alk* T encoded 99 kDa and 48 kDa proteins respectively. It has been shown that expression of the alk BFGHJKL operon is controlled by *alk* S and that *alk* T is not required for *alk* B expression. *Alk* T turned out to have an enzymic rather than a regulatory function and is essential for a fully functional *alk* system. It is a component of an alkane hydroxylase complex.

The variety of metabolic organisation in micro-organisms means that it is not possible to generalise from one species to another, even within a relatively narrow range of biochemistry such as alkane oxidation. Thus, for any application including bioremediation, each organism must be studied on its own merits.

In contrast to the genetic model of *P. oleovorans* described above in other alkane-utilising micro-organisms such as *Acinetobacter*, the genetic information for longer-chain alkane metabolism is found to be localised solely on the chromosome; data on genes encoding alkane assimilation are so far only preliminary. Singer and Finnerty, by chemically induced point mutations isolated alkane negative mutants (C_{12}–C_{16}) from strains HO1-N and BD413 (ADP1) that were able to grow on the fatty acid palmitic acid. The mutants clustered into the groups on the basis of transformation crosses and from intragroup crosses. This study indicated the presence of unlinked loci (*alk* X and *alk* Y) necessarily for alkane oxidation by *Acinetobacter*. Similarly, *Acinetobacter* mutants unable to grow on dodecane have been generated by Geissdorfer which were shown to be complemented by genes encoding proteins similar to rubredoxin and rubredoxin reductase. These genes are localised in close proximity on the chromosome and the lack of apparent promoter and transcriptional terminology structures flanking these genes suggest that they may be part of a larger operon.

Most of our knowledge on the genetics of short- and long-chain alkanes metabolism is contributed by studies using *P. oleovorans*, harbouring *oct* plasmid, and *Acinetobacter* showing chromosomal location of alkane-degrading genes. Little information is available on the genetic regulation of alkane degradation in other micro-organisms. Also, further studies on the genetics of long-chain alkane oxidation are required. So far, no reports are available on the involvement of plasmid(s) harbouring genes specific for degradation of long-chain alkanes and other complex hydrocarbons. Some of these have been found to be located on chromosome. The mapping of whole structural and regulatory genes on chromosome is yet to be elucidated.

BIODEGRADATION OF AROMATIC COMPOUNDS

Aromatic hydrocarbons are ubiquitous in the environment. They are released into the environment by combustion of organic materials, crude oil spills, petrogenic industrial effluent, oil refinery, waste, paper and pulp industrial effluent, coal mines and gasification plant. Aromatic hydrocarbons are pollutants of concern because of their toxic, mutagenic and carcinogenic potentials. Complete removal of aromatic hydrocarbons from the environment by physico-chemical methods is neither feasible nor cost effective. Bioremediation is in principle the most environmentally sound and least intrusive, promising and effective technology. There are many micro-organisms, isolated from soil or aquatic environments, that have been demonstrated to degrade variety of aromatic compounds. However, some aromatic hydrocarbons are toxic to organisms and are recalcitrant to biological degradation.

Nevertheless, the oxidation of complete and recalcitrant aromatic hydrocarbons by specialised bacterial strains has provided he examples which demonstrated that bacteria may break down these problematic compounds and obtain carbon and energy for their own cellular processes. This versatility allows them to be present in many environments as natural autochthonous microflora with a high potential for bioremediation of pollutants. We shall concentrate in the following sections on the degradation of monocyclic aromatic compounds, as mechanistic details are available in these systems. Micro-organisms also have the capability to degrade polycyclic aromatic compounds but the details as compared with the degradation of monocyclic aromatic compounds are scanty.

A common strategy employed by bacteria that degrade aromatic hydrocarbons is to convert the aromatic compounds into a small number of intermediates which are channelled into the central metabolic pathway, the Krebs cycle. The entire process of biodegradation of aromatic compounds can then be divided into two pathways designated 'upper pathway' and 'lower pathway'. The series of metabolic reactions leading to conversion of aromatic compounds into catechols or substituted catechols is known as the upper pathway. This may involve oxidation of side chains, opening of rings (if the compound under question happens to contain more than one benzene ring), etc. The products of the upper pathway are channelled into the Krebs cycle. This conversion may occur by *meta*-cleavage or by the β-ketoadipate pathway, in which the C—C bond next to one of the hydroxyl group present in the catechol or substituted catechol is broken as shown in Fig. 24.4; the bonds cleaved are shown by arrows. Alternatively, the C—C bond between the two hydroxyl groups can be cleaved using an intradiol dioxygenase catalysed *ortho*-cleavage. The end-product of this pathway is succinate, which enters the Krebs cycle.

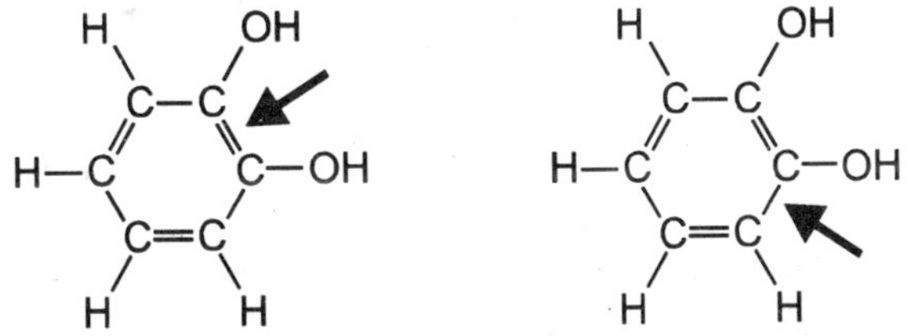

Fig. 24.4. Pathways of monocyclic aromatic hydrocarbon degradation. Bonds cleaved during the degradation are indicated by arrows.

The genetic information needed for the degradation of aromatic compounds is very often clustered in operons which may be present either on a plasmid or the bacterial genome. The plasmids acquire their name from the compound which they degrade: for instance, the plasmid responsible for the degradation of toluene is known as TOL plasmid; those degrading salicylate are called SAL; NAH plasmids are responsible for degrading naphthalene, etc. The TOL, SAL and NAH plasmids are autonomously replicating self-transmissible plasmids, belonging to P-9 group of compatibility. Their self-transmissibility could play an important role in the evolution of the aromatic degradative pathways in micro-organisms. TOL and many of its variant forms are harboured by a number of *Pseudomonas* sp. The naturally occurring TOL plasmids are extremely large, their size ranging between 110–280 kb among different variant strains. Only a small portion of this plasmid, isolated from natural environment (approximately 40 kb), is responsible for the degradation of toluene. We shall see the degradation pathways coded for by some of these plasmids which will help us to understand various possible degradation routes that monocyclic aromatic compounds follow in microbes. Apart from their ability to encode for the enzymes involved in the degradation of aromatic hydrocarbons, many fascinating aspects of these plasmids remain unresolved.

Degradative Plasmids

Many of the micro-organisms capable of metabolising aromatic hydrocarbons harbour large degradative plasmids. Some of the well-known catabolic plasmids include TOL, CAM, OCT, NIC, SAL and NAH, for the degradation respectively of toluene, camphor, octane, nicotine and salicylate. These plasmids encode all the genes required for the complete degradation of their respective compounds. Their widespread occurrence in nature can be attributed to their ability to undergo self-transmission. Studies indicate that most natural TOL plasmid transfer rates are high enough to maintain the plasmid in a natural environment even in absence of selective pressure. The plasmid transfer rates are closely related to the specific growth rates of bacteria, higher rates of transfer being observed at high specific growth rates of the donor cells.

Toluene-metabolising genes, the so-called *xyl* genes, of the TOL plasmid occupy approximately a 40 kb region. They are flanked by transposable elements. Detailed genetic analysis of the structure of these transposable regions and their relationships between species have been undertaken. From the retention or otherwise of the ability to oxidise toluene in different organisms and strains it has been suggested that the insertion of *xyl* genes into chromosomes or mobilisation of chromosomal genes during conjugation might have important consequences in the adaptability of bacterial strains to different environmental conditions. This is clearly important for the identification of effective micro-organisms in bioremediation.

The extrachromosomal gene cluster known as SAL plasmid is capable of conjugal transfer among the gene molecules of the same species. The genetic basis of biodegradation of salicylate was first explored by Chakrabarty. *Pseudomonas putida* has a NAH7 plasmid (80 kb). This plasmid has two positively regulated operons, NAH operon and SAL operon. The NAH operon with catabolic genes such as NAH ABCDE and F, catalyse the degradation of naphthalene to salicylate, while the SAL operon with genes as NAH GHIJK is specified for enzymes to convert salicylate to Krebs cycle intermediates. Thus the upper pathway requires two operons to degrade its substrates to succinate. Some of the enzymes produced from SAL operon (NAH GHIJK) have been studied in great detail. The expression of both the operons is coordinately induced by salicylate. These operons are not catabolite repressed since neither glucose nor succinate affect induction of expression of the two operons. These operons are under positive control of regulator gene NAH R. The NAH plasmid was first reported from naphthalene and salicylate-utilising *P. putida* strain 97 by Dunn and Gunsalus. This plasmid was analysed in detail in the derivatives of *P. putida* strain GI (ATCC 17453) and was re-named NAH 7.

MECHANISM OF MONOCYCLIC HYDROCARBON DEGRADATION

The biochemical reactions coded for by the *xyl* genes of TOL plasmids fall into two categories: (i) upper pathway genes, and (ii) *meta*-cleavage genes or the β-ketoadipate pathway. Reactions of *ortho*-cleavage are coded for by chromosome-located genes and shall not be discussed in this chapter; however, they involve similar steps except for the type of bond that is cleaved. The genetic information required for the degradation of toluene and similar compounds is grouped into four different clusters. Two of these contain structural genes and the other two contain regulatory genes. One of the two structural genes cluster codes for enzymes required for the upper pathway conversions. It contains five genes namely *xyl* CMAB and *xyl* while the other cluster contains *xyl* XYZLTEGFJQKI and *xyl* H. Regulatory genes *xyl* S and *xyl* R are presented as the regulatory gene cluster. Figure 24.5 shows the organisation and regulatory circuit thought to be operative between these operons in the TOL plasmid. Similar regulatory mechanisms might be operative in other hydrocarbon-degrading plasmids.

Regulation of Toluene Degradation

As with many other degradative pathways, the reaction sequence encoded for by TOL plasmids is inducible at the transcriptional level. Toluene and substituted toluenes act as inducers for the upper pathway genes, while benzoate and substituted benzoates act as inducers for the lower pathway or the β-ketoadipate pathway genes. Transcription of the upper pathway is regulated by the P_u promoter, while transcription of the lower pathway is regulated by the P_m promoter (Fig. 24.5). Transcription of the regulatory genes, *xyl* R and *xyl* S, is under the control of the P_r and P_s promoter respectively. Both the physical properties of the nucleotide sequence and the binding of proteins that alter DNA topology have an effect on the final outcome of promoter activity.

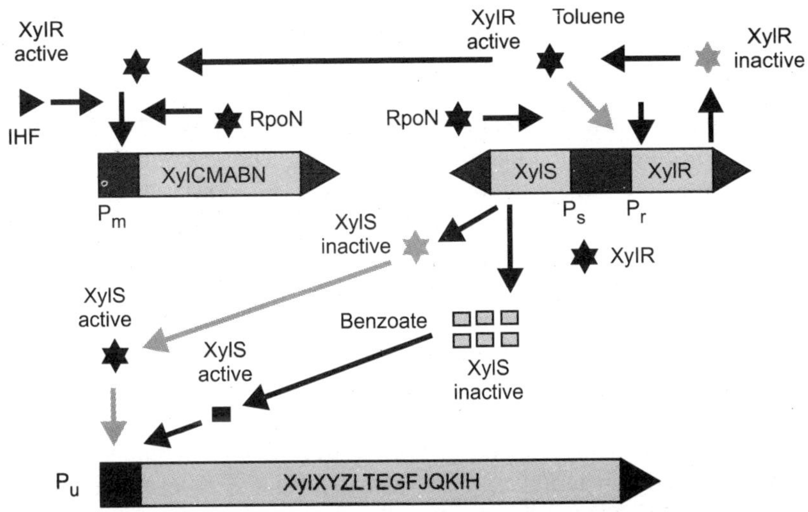

Fig. 24.5. Molecular interactions and genetic regulation during toluene degradation. Protein encoded by xylR is constitutively expressed in an inactive form which becomes activated in the presence of toluene. The activated XylR protein autoregulates its own synthesis and drives expression from P_s, P_u and P_m promoters, coding for XylS protein, upper pathway and *meta*-cleavage enzymes. Expression from P_m requires two additional cellular factors, Integration Host Factor (IHF) and RNA Polymerase N (RpoN; σ^{54}); s^{54} is also required to drive the expression from P, promoter.

Transcription of *xyl* R is constitutive and keeps taking place even in the absence of any inducer molecule. Some understanding of how inducers act at the molecular level has been achieved in these cases. An induced molecule causes certain conformational changes in the proteins that are synthesised from the uninduced P_r and P_s transcripts. This mechanism is very similar to the operation of the *Lac* operon with which most biochemists are familiar.

Induction by toluene

The *xyl* R protein interacts with toluene and undergoes conformational changes which increase the affinity of the P_u and P_s promoters for *xyl* R. The activated *xyl* R protein down-regulates its own transcription from P_r promoter while increasing the transcription from P_u and P_s promoter. *Xyl* R shows a broad effector specificity and can recognise even non-metabolisable compounds such as *o*-xylene. *Xyl* R also causes increased expression of *xyl* S which is activated by benoate, the end-product of the upper pathway for the oxidation of toluene. Therefore, the expression of *xyl* S is under control of both upper pathway as well as active *xyl* R.

Enzymes of the Upper Pathway

The primary task of this pathway is to convert substrates such as toluene and xylene into a limited number of intermediates, catechols or substituted catechols, which can in turn be channelled into the central metabolic pathway. This pathway consists of three enzymes: xylene monoxygnesase, benzyl alcohol hydrogenase, and benzaldehyde dehydrogenase, coded for by *xyl* MA, *xyl* B and *xyl* C, respectively (Fig. 24.6).

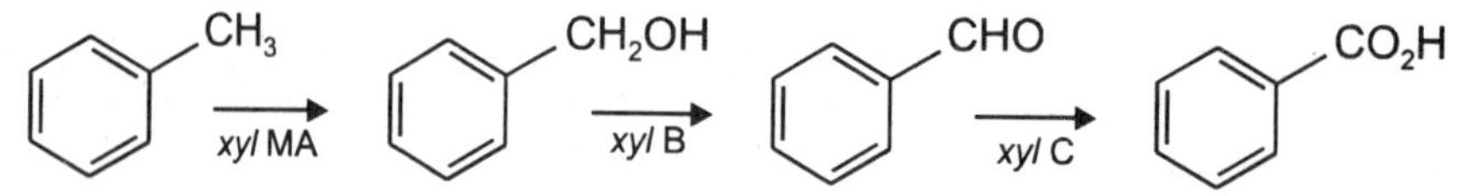

Fig. 24.6. Upper pathway for toluene metabolism in TOL plasmid.

Xylene monooxygenase

Catalysing the first step in toluene metabolism, this enzyme is a flavoprotein monoxygenase that introduces an atom of molecular oxygen into the benzene ring. The reaction involves formation of an oxygenated enzyme intermediate involving an enzyme-bound dihydroflavin in a reaction that has no good laboratory mimic. The enzyme is coded for by two genes designated *xyl* M and *xyl* A. *Xyl* A codes for a protein (40 kDa) which acts as an electron transfer protein or as the NADH: acceptor reductase. *Xyl* M codes for an enzyme (35 kDa) which acts as terminal hydroxylase. As a result of xylene monooxygenase activity, a hydroxyl group is introduced into the methyl side chain or toluene, converting it into benzyl alcohol. The proposed electron transfer sequence or the hydroxylation reaction is shown in Fig. 24.7. The terminal reductant, NADH situated at the C-terminus of the enzyme coded by *xyl* A, reduces the flavin ring of FAD generating a semi-quinone FAD or dihydroquinone FAD (the FAD is present as a prosthetic group on the hydroxylase). The electrons from this reduced prosthetic group are then transferred to the ferredoxin moiety [2Fe-2S] present at the N-terminal region of the *xyl* A hydroxylase. This in turn reduces the Fe^{III} present at the active centre of the terminal monoxygenase. The terminal monoxygenase deals with the interaction between molecular oxygen and the substrate. Thus, the electron delivered by the *xyl* A protein to Fe^{III} converts the iron into the reactive Fe^{II} oxidation state, leading to the activation of molecular oxygen. The hydroxyl group introduced into toluene comes from water while oxygen acts as the terminal electron acceptor. Sequence analysis of *xyl* A amino acid reveals that its amino-terminal is a chloroplast-type of ferredoxin reductase and the carboxyl terminal bears homology with NAD(P)H ferredoxin reductase. This suggest that *xyl* A protein would function as an electron transfer protein donating electrons to *xyl* M, the terminal hydroxylase form of NAD(P)H. These predictions have been confirmed experimentally. The ability of these proteins to organise spatially the interactions between mutually incompatible chemical species is remarkable. Without the intervention of the protein and the electron transport chain, no control of oxidation could be exerted whatsoever and the degradation of hydrocarbons would be left to random autoxidation. This exquisite control of chemistry attracts scientists to apply enzymes to demanding tasks such as bioremediation and justifies the detailed study of the genetic make-up of organisms with the ability to oxidise hydrocarbons.

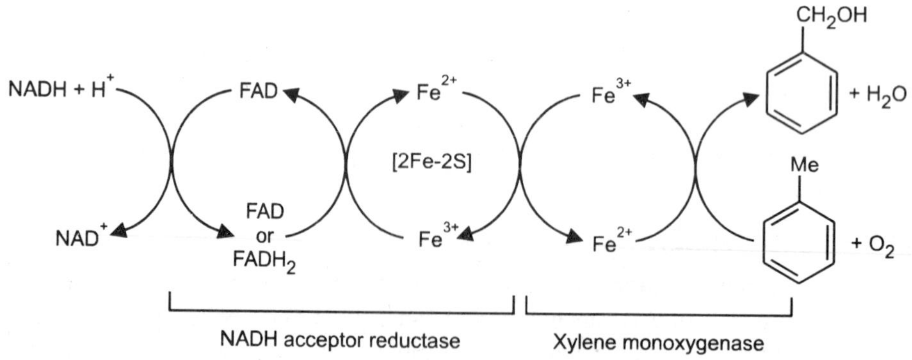

Fig 24.7. Mechanism of xylene monooxygenase action.

Benzyl alcohol dehydrogenase

This converts benzyl alcohol to benzaldehyde reversibly and is a nicotinamide-dependent enzyme. It is a single polypeptide chain coded by *xyl* B and belongs to the dimeric (2 × 42 kDa) alcohol dehydrogenase family. Many such enzymes have been used in organic synthesis. It contains two Zn dications per subunit, one of which is present in catalytic site perhaps bound by Cys^{60}, His^{61}, and Cys^{169} and the other Zn^{2+} is present in a non-catalytic site bound by Cys^{90}, Cys^{93}, Cys^{96} and Cys^{104}. Its activity is stimulated by monovalent cations like K^+, Na^+ and NH_4^+. The kinetic and mechanistic studies performed with this enzyme show that it undergoes a general ordered Bi-Bi kinetics in low H^+ acceptor concentration. Enlargement of the small substrate-binding pocket of benzyl alcohol dehydrogenase might broaden its substrate specificity and increase its potential for bioremediation. Such a change can be effected by site-directed mutagenesis, provided that the structure of the active site is known from crystallography or another reliable method.

Benzaldehyde dehydrogenase

This is encoded by *xyl* C and is responsible for the conversion of benzaldehyde into benzoic acid. It is a single polypeptide enzyme (57 kDa) and is a member of the aldehyde dehydrogenase family. Its substrate specificity is also quite broad and similar to that of benzyl alcohol dehydrogenase. Catalytic efficiency of benzaldehyde dehydrogenase (BZDH) is observed to increase in the presence of electron-withdrawing groups in the substrate. The proposed mechanism of BZDH activity consists of a nucleophilic attack on the carbonyl carbon of the substrate to form a hemithioacetal which is oxidised at the active site by NAD^+. This addition/oxidation mechanism is consistent with the effect of electron withdrawing groups and the presence of a conserved Cys (Cys^{302}) at the active site. The thiolester bond between Cys and the carbonyl carbon is then hydrolysed to produce an acid. An active-site glutamic acid (Glu^{286}) is also conserved and may be involved in this stage of the catalysis. Many enzymes like aldehyde dehydrogenase, glycerol dehydrogenase and esterase undergo similar mode of action. Consistent with this view of the mechanism, a few BZDHs exhibit esterase activity.

Metabolism of Naphthalene

Naphthalene (**1**) metabolised to salicylate by a six-step pathway, the sequence of reactions being as follows. Naphthalene is converted into *cis*-1,2-dihydroxy-1,2-dihydronaphthalene or simply naphthlene dihydrodiol (**2**, Fig. 24.8) by an enzyme, naphthalene-1,2-dioxygenase, which introduces two hydroxyl groups into naphthalene. The product, naphthalene dihydrodiol is quickly dehydrogenated by a dehydrogenase to yield 1,2-dihydroxynaphthalene (**3**). The dihydroxynaphthalene may either be oxidised to 1,2-naphthaquinone (**4**) or enzymatically cleaved to yield 2-hydroxy-4(2′-oxo-3,5-cyclohexadienyl)-buta-2,4-dienoate (**5**). This ring fission is catalysed by a dioxygenase and the product is unstable; it can either spontaneously re-aromatise to *cis*-*o*-hydroxybenzylidine pyruvate, CHBPA (**8**), or isomerise to 2-hydroxychromene-2-carboxylate, HCCA (**6**). Some investigations suggest the formation of the *trans*-isomer of *o*-hydroxybenzylidine pyruvate, THBPA (**7**). THBPA is acted upon by an enzyme, 2-hydroxychromene-2-carboxylate isomerase, to convert it into CHBPA. CHBPA undergoes spontaneous oxidation to yield 2-oxo-4-hydroxy(2′-hydroxyphenyl)-butyrate (**9**) which is cleaved by an aldolase to give salicylaldehyde and pyruvate. Pyruvate can enter the Krebs cycle directly while salicylaldehyde is oxidised by an NAD^+-dependent salicylaldehyde dehydrogenase to salicylic acid. The enzymes catalysing the first three steps in this pathway are similar to those involved in the metabolism of monocyclic aromatic hydrocarbons such as toluene. Subsequent reactions of salicylate formation differ from those of monocyclic aromatic hydrocarbon degradation because the instability of 1,2-dihydroxynaphthalene, the ring cleavage product, leads to a rearrangement followed by the formation of a hemiacetal product 2-hydroxychromene-2-carboxylate. The enzyme isomerase and hydratase aldolase represent classes of enzymes that are unique to the biodegradation of aromatic hydrocarbons having two or more rings.

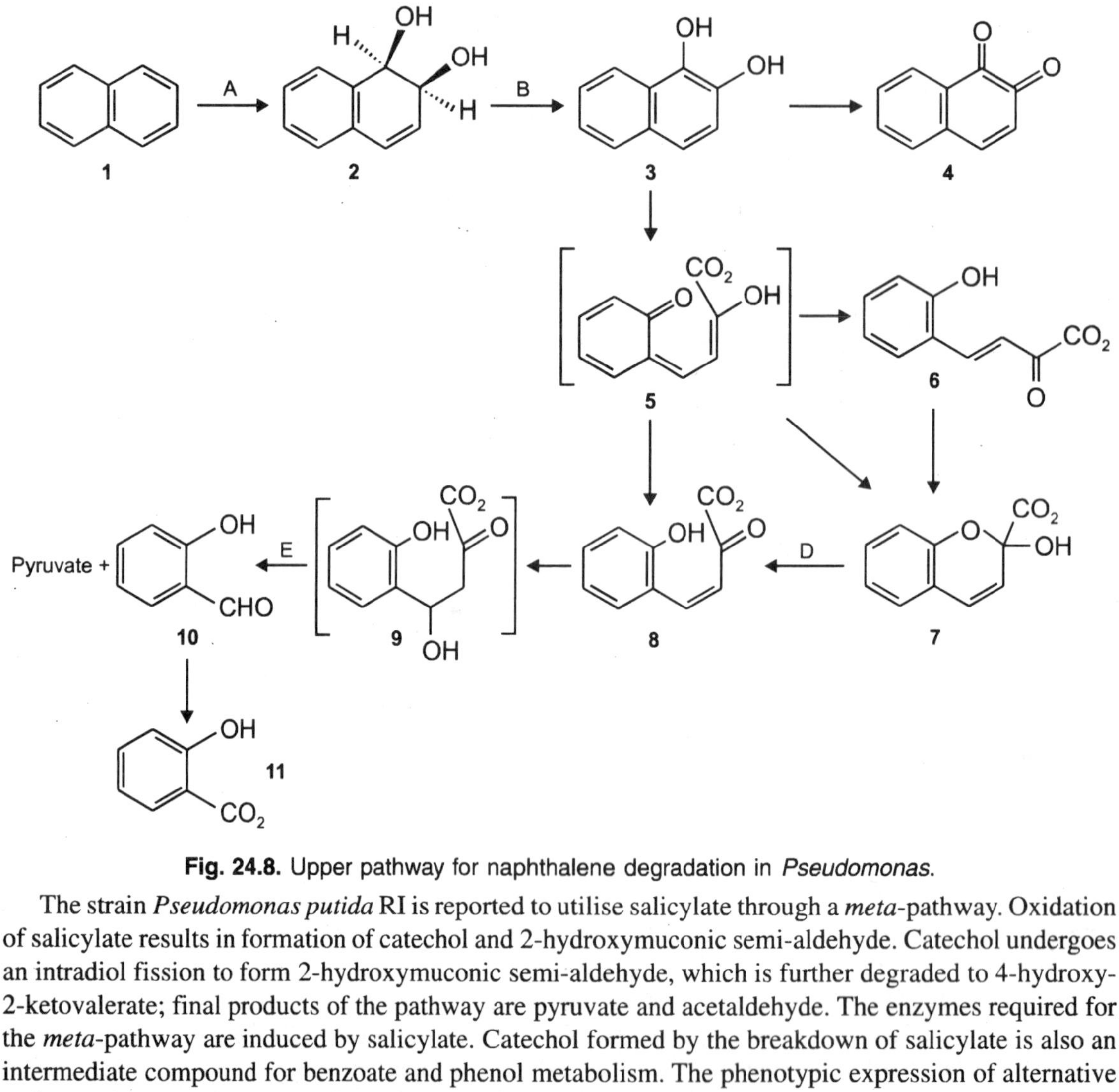

Fig. 24.8. Upper pathway for naphthalene degradation in *Pseudomonas*.

The strain *Pseudomonas putida* RI is reported to utilise salicylate through a *meta*-pathway. Oxidation of salicylate results in formation of catechol and 2-hydroxymuconic semi-aldehyde. Catechol undergoes an intradiol fission to form 2-hydroxymuconic semi-aldehyde, which is further degraded to 4-hydroxy-2-ketovalerate; final products of the pathway are pyruvate and acetaldehyde. The enzymes required for the *meta*-pathway are induced by salicylate. Catechol formed by the breakdown of salicylate is also an intermediate compound for benzoate and phenol metabolism. The phenotypic expression of alternative pathways (*meta*-pathway or *ortho*-pathway) is determined by the chemical nature of the aromatic catechol precursor with which the organism is grown. For example, it has been shown that in naphthalene-degrading *Pseudomonas* sp., the enzymes for the *meta*-cleavage pathway are induced after growth with naphthalene and salicylate, but the enzymes of the *ortho*-cleavage are induced when grown with benzoate. The genetic manipulation of reactions of this type has become industrially significant in the production of substituted benzene-*cis*-glycol derivatives using strains of organism in which catalysis of the metabolism of the diol has been blocked by deletion of the gene coding for the required enzyme.

ENZYMES OF THE LOWER PATHWAY

After the side chain is oxidised by the enzymes of the upper pathway, the products are then said to enter the lower pathway (Fig. 24.9). It is here that aromatic ring cleavage occurs to yield products that can be readily fed into the Krebs cycle. The genes encoding for the required conversions are clustered into one operon, the *meta*-cleavage operon, and are regulated transcriptionally by an upstream P_m promotor. The physical organisation of the *meta*-operon genes are: *xyl* XYZLTEGFJQKI and *xyl* H. Detailed genetic analysis of the *Pseudomonas meta*-cleavage pathway reveals that the gene order is conserved downstream from the gene for a ferrodoxin-like protein, *xyl* T, and the equivalent genes show high levels of homology. The identity of genes has been demonstrated using their nucleotide sequence. Co-linearity of DNA hybridisation has been established for *meta*-pathway operons of naphthalene-degrading plasmids *nah* NLOMK, toluene-degrading plasmids *xyl* WJKQI, and phenol-degrading plasmids *dmp* EFGH.

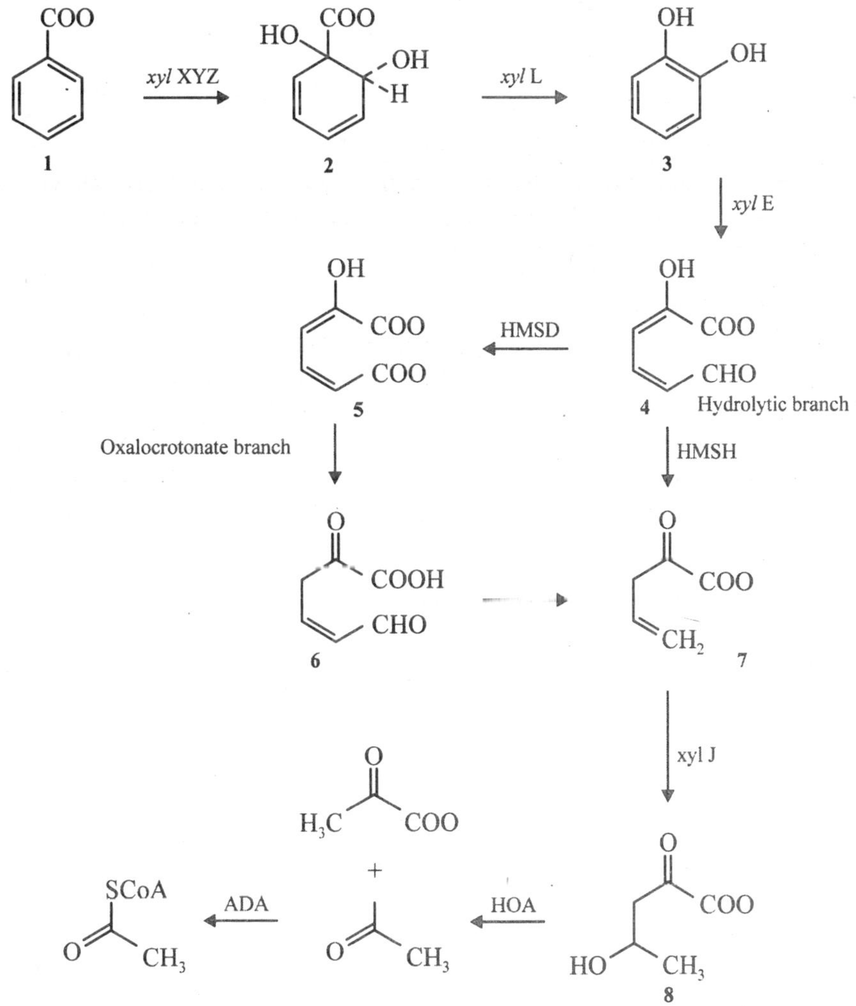

Fig. 24.9. Lower pathway for benzoate metabolism in TOL plasmid.

Formation of Catechol from Benzoate

Benzoate, which is the end-product of the upper pathway reactions, is acted upon first by a multi component enzyme followed by a step of oxidative decarboxylation. In the first step, two hydroxyl groups are introduced into the aromatic ring by benzoate 1,2-dioxygenase which is encoded by *xyl* XYZ genes. The hydroxylation reaction requires an equimolar quantity of NADH. The product of this reaction, 1,2-dihydroxy-cyclohexa-3,4-diene carboxylate (Fig. 24.9, **2**), is transformed to catechol (**3**) by the enzyme *cis*-diol-dehydrogenase and is coded for by *xyl* L gene of the *meta*-operon. *Cis*-dioldehydrogenase generates catechol by concomitant reduction of 1,2-dihydroxycyclohexa-3,4-diene carboxylate, releasing carbon dioxide; it is therefore said to catalyse an oxidative decarboxylation reaction. *Cis*-dioldehydrogenase is a 28 kDa protein belonging to the short-chain alcohol dehydrogenase family of enzymes, and shares significant homology with its counterparts in the NAD^+ binding region.

Formation of Catechol from Salicylate

Salicylate enters the *meta*-pathway by action of salicylate hydroxylase. This enzyme is encoded by *nah* G. There exists homology with *p*-hydroxybenzoate hydroxylase from *P. putida* which catalyses a similar reaction, this enzyme converts salicylic acid to catechol by introducing a single hydroxyl group. The enzyme contains 1 mole of FAD per mole of enzyme protein. The reaction also requires NADH and oxygen. The reaction is believed to involve the reduction of an enzyme–FAD–salicylate complex by NADH. Subsequent oxygen fixation by the complex results in the liberation of carbon dioxide with formation of catechol and oxidised enzyme.

Catechol-2,3-Dioxygenase

In the metabolic sequence of toluene degradation catechol-2,3-dioxygenase catalyses the first step in which the C—C bond of the aromatic ring is cleaved. Many organisms that degrade the aromatic compounds harbour this enzyme, although the primary sequence may be different, which accounts for their altered substrate specificity. For instance, *Acinetobacter* shows preferred activity with 3-methylcatechol while the enzyme present in *Pseudomonas* shows better activity with catechol as substrate. In the TOL plasmid the enzyme is coded by *xyl* E, the native enzyme being a tetramer with a molecular weight of 32 kDa per sub-unit. The enzyme is believed to be activated by a protein encoded by *xyl* T of the lower pathway operon; however, the exact mechanism of activation remains uncertain.

Catechol 2,3-Dioxygenase (C2,3O) from SAL Plasmid

The distinction between *nah* H and *xyl* E for 3-methylcatechol oxidation has been shown to be due to His^{250} substitution by Gln. The hybrids with N-terminal region of *nah* H gene showed that His^{250} substitution by Gln reduces the k_{cat} value by two-fold. The hybrids with N-terminal region of *nah* H gene showed that neither His^{250} nor Gln^{250} residue of the protein is disadvantageous, but the catalytic role of residue 250 is influenced by the surrounding residues.

Native C2,3O is a 70 kDa protein in a dimeric form and has an acidic pH optimum. The C2,3O gene of *Pseudomonas aeruginosa* has been expressed under the simian virus 40 or Rous sarcoma virus promoter in mammalian cells; it was found that the gene could be used as a reporter for the study of gene expression. Since *Lac* Z and C2,3O (from *xyl* E) activities are inherently absent in mammalian cells, they are suitable as reporters. The enzyme is convenient because its activity is easily detected by the yellow colour of 2-hydroxymuconoic acid-6-semi-aldehyde (2-HMS) (**4**), the product: formed from catechol by C2,3O. There are two enzymes which act on 2-hydroxymuconic semi-aldehyde:

1. 2-Hydroxymuconic semi aldehyde hydrolase catalyses a hydrolytic release of formate; however, this activity appears negligible.
2. 2-Hydroxymuconate semi-aldehyde dehydrogenase causes dehydrogenation of its substrate to form 4-oxalocrotonate (**5**); this dehydrogenase is found to be NAD-dependent and encoded by *nah* I gene.

The existence of further divergence has been shown within the *meta*-cleavage pathway by TOL, NAH and biphenyl-degrading plasmids. The ring cleavage product, 2-hydroxymuconic semi-aldehyde, is subjected to two different pathways of degradation, the two pathways eventually rejoining at 2-oxopent-4-enoate (**7**). The enzyme hydroxymuconic semi-aldehyde hydrolase hydrolyses 2-hydroxymuconic semi-aldehyde, the product of ring fission, directly to yield 2-oxopent-4-enoate and therefore this route of degradation is known as the 'hydrolytic' branch. The second branch, known as the 'oxalocrotonate' branch, involves the formation of an enolic form of oxalocrotonate; namely 2-hydroxyhexa-2,5-diene-1,6-dioate by an NAO^+-dependent enzyme, 2-hydroxymuconic semi-aldehyde dehydrogenase (HMSD). Two additional enzymatic steps are involved in the conversion of oxalocrotonate (the enolic form) to 2-oxopent-4-enoate; these are shown in Fig. 24.9. For a compound to enter the oxalocrotonate pathway, the essential criterion is the presence of an oxidisable aldehyde, e.g. 2-hydroxy-6-oxohepta-2,4-dienoate.

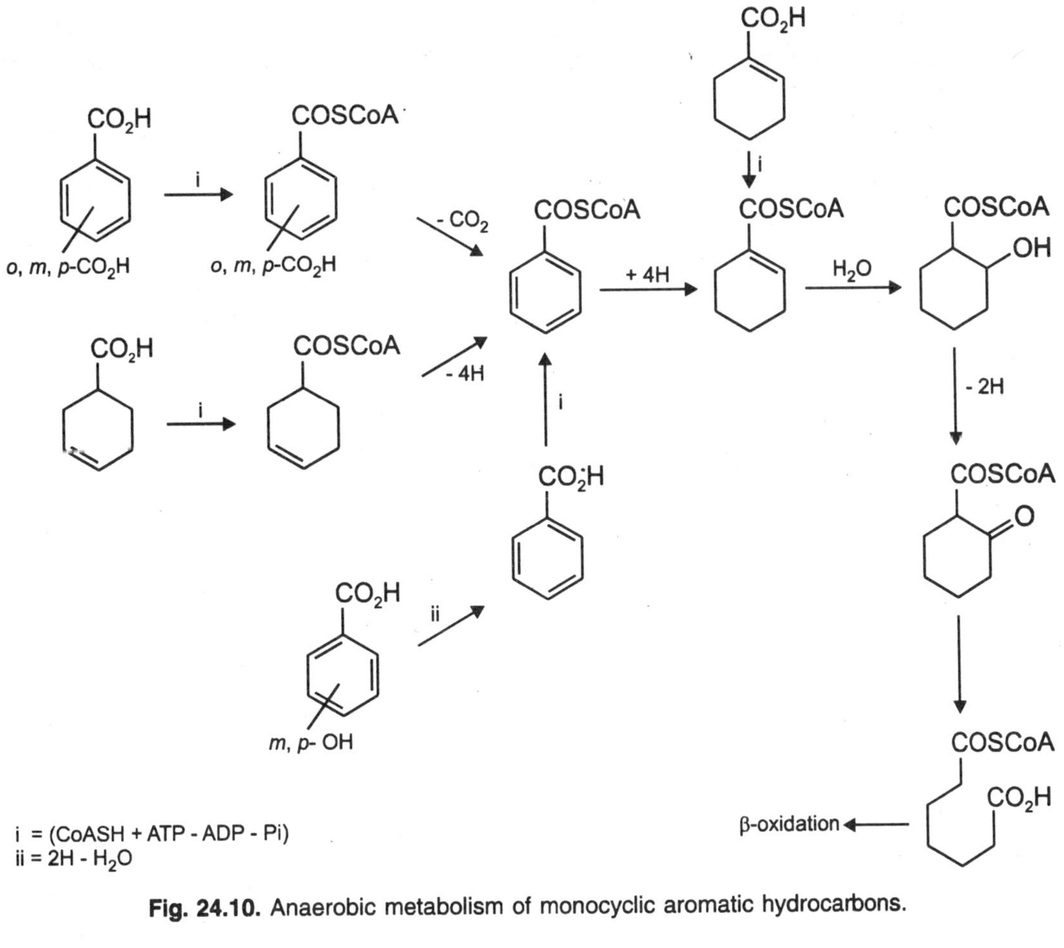

Fig. 24.10. Anaerobic metabolism of monocyclic aromatic hydrocarbons.

The cleavage product of 3-methylcatechol lacks an oxidisable aldehyde and, therefore, it is converted to 2-oxopent-4-enoate directly by the hydrolytic enzyme hydroxymuconic semi-aldehyde hydrolase (HMSH). 2-Hydroxymuconic semi-aldehyde, the product of catechol ring fission, can enter either branch, although, it is preferentially degraded by oxalocrotonate branch. This may be attributed to high levels of HMSD expression and low affinity of hydroxymuconoic semi-aldehyde hydrolase towards them.

This would imply that at low substrate concentration 2-hydroxymuconic semi-aldehyde would be metabolised via the oxalocrotonate pathway while at higher substrate concentration it can be metabolised via the hydrolytic path as well as by the oxalocrotonate branch. 2-Oxopent-4-enoate, which is the end-product of both hydrolytic branch and the oxalocrotonate branch of the *meta*-cleavage pathway is converted into 4-hydroxy-2-oxovalerate (**8**) by 4-hydoxy-2-oxovalerate aldolase, the product of *xyl* J gene. The subsequent metabolism of the intermediates to acetyl CoA and pyruvate is identical in the degradation of naphthalene, salicylic acid, phenol and other compounds.

4-Oxalocrotonate Decarboxylase

The gene *nah* K encodes for the decarboxylase that converts 4-oxalocrotonate into carbon dioxide and 4-hydroxy-2-oxovalerate. The activity of this enzyme can be stimulated by Mg^{2+} or Mn^{2+} ions and is inhibited by EDTA. In TOL plasmids the enzymes oxalocrotonate decarboxylase and enoate hydratase are physically linked together to form an enzyme complex. This physical association might help in channelling the metabolites in an efficient manner. Two enzyme activities associated with the one protein is a well-known phenomenon in biosynthesis.

2-Oxo-4-Hydroxypentanoate Aldolase

The gene *nah* M codes for an aldolase enzyme. It converts 4-hydroxy-2-oxovalerate into the final products of pathways, acetaldehyde and pyruvate. All the enzymes operative in this pathway are inducible except for 4-hydroxy-2-oxovalerate aldolase. The enzyme 4-hydroxy-1-oxovalerate aldolase (HOA) and acetaldehyde dehydrogenase (ADA), coded for by *nah* O, are also physically associated, as evidenced by their co-purification. The complex has a molecular weight of 148 kDa [(2 × 32.5) + (2 × 39)]. Aldolase activity, has a pH optimum of between 8.5–9.0, while the dehydrogenase activity has a pH optimum of 6.5–8.5. Aldolase activity is stimulated by Mn^{2+} and inhibited strongly by Zn^{2+} in the purified protein; however, the aldolase activity in crude extract is stimulated by the addition of Mg^{2+}. The activity of HOA is found to be enhanced by NAD^+ addition, which may be due to the result of its binding to an allosteric pocket in the enzyme.

There are two major advantages of the HOA and ADA association. First, the aldehyde being toxic to the cell, is immediately metabolised to yield acetyl-CoA which is required for other cellular processes. The aldehyde can also be converted into acetate and its subsequent acetylation would generate acetyl CoA. However, by physically associating ADA activity with HOA activity, acetyl-CoA is transferred to acetaldehyde in a reaction that does not require ATP, thus saving on ATP (energy) requirements.

Anaerobic Degradation of Polycyclic Aromatic Hydrocarbons

Bacterial strains have also been isolated for their ability to degrade monoaromatic compounds under anaerobic conditions. These micro-organisms utilise similar pathways as do homocyclic monoaromatic-degrading denitrifying bacteria and non-sulphur purple phototrophic bacteria. Monocyclic phenols can be degraded either using nitrate, carbon dioxide, or sulphate as terminal electron acceptor. This degradation proceeds via a reductive pathway in which the aromatic ring is reduced to yield cyclohexane followed by the ring fission to generate *n*-caproic acid which then undergoes β-oxidation (Fig. 24.10). The initial step in the metabolism of phthalates and benzoic acid is the formation of CoA ester of these compounds and subsequent decarboxylation to form benzyl-CoA. This is in contrast to the aerobic degradation where benzoic acid and phthalate are degraded by different routes. In these pathways, monocyclic aromatic compounds are metabolised via aromatic acids such as nitrobenzoate and hydroxybenzoate and subsequently they are converted into benzoates. The benzene ring is then reduced and cleaved to yield aliphatic acyl products such as pimelate. An important step of these pathways is the addition of co-enzyme A (CoA) residues to the carboxylic group of aromatic acid by CoA ligase. In this reaction carboxyl group reacts with the thiol group of CoA to form an acyl-CoA. The first step or the reaction is the formation of an acyl-adenylate intermediate that binds to the protein while pyrophosphate is released from the enzyme. In the second step, the acyl group is transferred to the thiol group of the COASH residue and AMP is released.

$$\text{Enzyme} + \text{MgATP} + \text{PhCO}_2\text{H} \longrightarrow [\text{Enzyme-PhCO-AMP}] + \text{Mg PPi}$$

$$\text{CoASH}\downarrow$$

$$\text{PhCO-SCoA} + \text{enzyme} + \text{AMP}$$

A thiol ester is more reactive than an oxyester, which is explained by the lower degree of delocalisation of the non-bonded electrons of sulphur compared with oxygen in the respective carbonyl groups. After the formation of the co-enzyme A ester, subsequent degradation reactions such as β-oxidation, which does not require molecular oxygen, are favoured. Thiol ester formation early in a degradation pathway is thus an obvious metabolic advantage for an anaerobic organism. Aromatic acids having side chains can also be metabolised as fatty acids. They are degraded to phenyl acetate (for the odd number of side chain carbon atoms) or benzoate (for the even numbers) by γ-β-oxidation. Hence the addition of CoA by CoA ligase is an important step of this metabolism as it is for the degradation of fatty acid. It has also been suggested that the CoA thiol ester residue exerts an electronic effect upon the phenyl ring which perhaps makes it easier to oxidise.

Thus, bioremediation is a potential application of enzyme technology for the cleaning of contaminated sites. It has a very low environmental impact compared with physical and chemical decontamination methods, as it encourages natural processes. The problems are not with the identification of suitable organisms, nor their adaptation to metabolise the pollutants, as the studies described above illustrate. Present information suggests that the microbes in the natural environment possess genetic information for the specialised enzyme systems and metabolic pathways involved in the degradation of the xenobiotics.

The importance of these genetic studies concerns not only the potential application of micro-organisms in decontaminating adversely impacted ecosystems, but also their industrial exploitation. Recently, a number of publications have appeared showing utilisation of structural and regulatory genes of OCT plasmid for the bioconversion of alkanes to alkanols and alkanoic acids by recombinant bacteria. Of course, many microbial isolates have been engineered and provided with structural and regulatory genes for application in the impacted sites. These studies have provided information into the molecular and regulatory mechanisms responsible for the evolution of the degradative pathways. The problems with large-scale applications reside in several features that will take more experience and research to solve. The first concerns the scale and speed at which bioremediation can work. Bioremediation is not a very rapid process and measurable impacts occur at least a couple of weeks after the application of the microbes. The application of genetically engineered micro-organisms to impacted sites is limited as an alternative to application of other bioremediation measures, because it is believed that these engineered microbes have too narrow substrate specificities. They will, therefore, be ineffective decontaminants in the real world where the variety of substrates and the presence of other perhaps better adapted organisms would dominate. It is difficult to see how the specific selection pressure required to give dominance to the engineered microbe could be ensured in a waste environment. Finally, there are always concerns about the potential adverse effects associated with any genetically modified organism in the environment; microbes designed to improve the environment do not escape the controls. It could well be, therefore, that the most effective micro-organisms in bioremediation will be naturally occurring species identified by strain selection using conventional microbiological methods and used in mixtures. Such a strategy avoids risks associated with regulatory problems and copes with multiple substrates in the environment. The most effective deployment of such cocktails will benefit from a knowledge of the composition, of the pollutants and a knowledge of the biochemical pathways available to the micro-organisms-knowledge that has developed as a result of studies in the field described in this chapter.

SECTION IV

Instrumental Techniques Used in Enzyme Analysis

Chapter 25

Instrumental Techniques in Enzymatic Analysis

INTRODUCTION

The course of enzyme-catalysed reactions may be monitored by a variety of techniques, of which the most common at the present time are spectrophotometric, spectrofluorimetric and electrochemical procedures. Radiochemical procedures may also be used, but although they are extremely sensitive, they do not permit the rate of the reaction to be monitored continuously. Dry-reagent techniques are of increasing importance.

Fully automatic procedures of enzymatic analysis may involve fixed-time methods (either discrete or continuous-flow analysis), fixed-concentration methods or continuous-monitoring methods. The last-mentioned type includes the use of discrete, rate-monitoring analysers and centrifugal analysers, both of which may be coupled to computers to allow automatic processing of results.

PRINCIPLES OF THE AVAILABLE DETECTION TECHNIQUES

A variety of detection procedures are used or have been used in the past, for enzymatic analysis by either end-point or kinetic methods; they may also, of course, be used for the investigation of enzyme characteristics. The most important ones are discussed below.

In most cases, the instrument detecting the changes which take place as the reaction proceeds can be coupled to a recorder, enabling the course of the reaction to be followed in detail without requiring constant human attention. Also, the enzyme involved may be in an immobilised form to allow them to be recovered at the end of the procedure. However, these refinements, whose use largely depends on the finances available, do not affect the principles of the detection techniques.

Manometry

If any of the substrates or products of an enzyme-catalysed reaction is a gas, the course of the reaction may be followed by manometric techniques. The basic system must be air-tight and consists of a reaction vessel attached to a manometre. The sample and reagents are put into different compartments in the reaction vessel and mixed together at zero time; then, as the reaction proceeds, the uptake or evolution of gas is indicated by the movement of the fluid in the manometre tube.

However, it will be realised that as the manometre fluid moves towards or away from the reaction vessel, the level of liquid in the two arms of the manometre will no longer be the same; this will introduce pressure changes which will affect the measurement of the progress of the reaction. Various features are therefore added to the basic system to ensure that all readings are made under identical conditions.

In constant volume manometry, the manometre is usually connected to a reservoir, so that fluid may be added to or removed from the manometre as required. The amount of fluid in the manometre is adjusted prior to taking a reading so that the level in the limb nearest the reaction vessel is at a fixed point. The difference between the fluid levels in the two limbs of the manometre gives an indication of the pressure in the reaction vessel, which changes as gas is taken up or evolved and so may be used to follow the course of the reaction.

In constant pressure manometry, the volume of the reaction chamber is changed by measurable amounts (e.g. by connecting it to a micrometre, as in the Gilson differential manometre) sufficient to keep the manometre fluid level at a fixed point as the reaction proceeds. The amount of fluid in the manometre is not changed, so this procedure ensures that the levels in both arms remain fixed and no pressure changes take place. The course of the reaction may thus be followed in terms of the volume changes required to maintain constant pressure.

Manometric methods may be used to investigate reactions where a gas is consumed, e.g. that catalysed by glucose oxidase, where molecular oxygen is a substrate. They may also be used when a gas is produced, e.g. to monitor CO_2 production in reactions catalysed by decarboxylases. Reactions where an acid is produced (including all reactions involving NAD^+ reduction to $NADH + H^+$) may be investigated by manometry by adding bicarbonate to the reaction mixture and measuring the evolution of CO_2.

However, such techniques usually involve a margin of error greater than 5 per cent, so they are rarely used today for enzymatic analysis, since more accurate and reliable procedures are now available.

Spectrophotometry

If a substrate or a product absorbs light at a characteristic wavelength, then the reaction can be monitored by following the changes in absorbance at this wavelength. The absorbance, A, is given by:

$$A = \log_{10}\left(\frac{I}{I_0}\right)$$

where, I_0 is the intensity of the incident light and I is the intensity of the light transmitted through the sample. The absorbance is related to the concentration of the light-absorbing substance in the sample by the Beer-Lambert law:

$$A = \varepsilon c l$$

where, ε is the molar extinction coefficient, c is the molar concentration and l is the length of the light-path through the sample (usually measured in centimetres).

This is generally true for any wavelength of visible light (wavelength range 400–700 nm) or ultraviolet light (200–400 nm). The wavelength required can be selected from the spectrum obtained when light is passed through a prism or a diffraction grating. In simpler and cheaper instruments (colorimetres), a waveband, rather than a particular wavelength, is selected by passing the light through a coloured-glass filter; such instruments are less reliable than spectrophotometres and cannot be used in the ultraviolet region.

For enzymatic analysis, the temperature of the reaction mixture should be maintained to within ±0.2°C, which necessitates the use of a jacketed cell-holder connected to a circulating water bath.

Procedures which involve the calculation of initial velocity should preferably be ones where an increase (rather than a decrease) of absorbance with time is observed. They will then be operating in the most sensitive part of the absorbance scale (the lower part) during the crucial early stages of the reaction.

Such procedures are best followed by instruments which give direct absorbance readings rather than by null-point instruments which require separate adjustments for every reading. Direct reading instruments may be coupled to recorders so that the whole course of the reaction may be traced out. Double-beam instruments, which automatically compare the absorbance of the reaction mixture to that of a reagent blank at each point, are preferable in these circumstances, since changes of absorbance with time of the blank may be taken into account without having to re-set the instrument during the analysis.

Spectrophotometric methods are particularly useful for monitoring reactions which involve NAD^+ or $NADP^+$ as coenzyme or ones which can be coupled to such reactions.

There are two main problems associated with spectrophotometric methods. Firstly they can only be used satisfactorily over the lower part of the absorbance range. Since at high absorbance values there is, in addition to reduced sensitivity, often some departure from the Beer-Lambert law because of interactions between solute molecules in concentrated solution. Secondly, errors may be introduced when there is a high background absorbance or when the sample contains material in suspension (e.g. blood cells or micro-organisms), thus scattering the incident light. For this reason, crude samples may often have to be partially purified (e.g. by dialysis, selective precipitation and/or centrifugation) before one of its components may be investigated by spectrophotometric procedures.

Spectrofluorimetry

Compounds are said to be fluorescent when they absorb light of one wavelength and then emit light of a longer wavelength. At low concentrations, the intensity of fluorescence (I_f) is related to the intensity of the incident light (I_0) of appropriate wavelength by the relationship:

$$I_f = I_0 \times 2.3\ \varepsilon c l q$$

where ε is the molar extinction coefficient, c the molar concentration, l the length of the light-path and q the quantum efficiency (i.e. the number of quanta fluoresced divided by the number of quanta absorbed). The fluorescence is usually measured at right angles to the incident beam, to avoid detection of any transmitted light. The wavelength of the fluorescence required for measurement, together with the appropriate wavelength for the incident beam, are usually selected by the use of diffraction gratings.

In general, spectrofluorimetry is several times (theoretically up to about a hundred times) more sensitive than spectrophotometry. However, fluorescence effects vary with temperature, so it is particularly important to maintain a fixed temperature throughout. NADH and NADPH exhibit fluorescence, absorbing light at 340 nm and re-emitting it at about 460 nm. Therefore, any reactions utilising these as co-enzymes may be followed by the use of sensitive spectrofluorimetric procedures. Spectrofluorimetry is also widely used for the investigation of hydrolytic reactions, using synthetic non-fluorescent substrates which are esters of highly fluorescent alcohols or amines: for example, Guilbault and Kramer devised an assay for triacylglycerol lipase based on the following reaction:

$$\underset{\text{(non-fluorescent)}}{\text{Dibutyryl fluorescein}} \xrightarrow{\text{Lipase}} \underset{\text{(Fluorescent)}}{\text{Fluorescein}}$$

In a similar way, fluorescein diacetate may act as a substrate for esterases and derivatives of highly-fluorescent 4-methylumbelliferone are often used in the assay of esterases, glycosidases, phosphatases and sulphatases.

Tyrosine and tryptophan fluoresce at 330–350 nm and since residues of these are usually present in enzymes, they would give a high background emission in the ultraviolet region. Hence, it is advisable to restrict the use of spectrofluorimetry in enzymatic analysis to the detection of substances which emit light in the visible region. Apart from this, the main problem associated with spectrofluorimetric techniques is quenching, where the extra energy possessed by the excited molecules after absorption of a photon of light is transferred to another molecule or group (e.g. to iodide or dichromate ions, if these are present) instead of being released as another photon. Concentration quenching may also occur, due to some net absorption of either the incident or the emitted light taking place.

Electrochemical Methods

Many different types of electrochemical procedures have been used in enzymatic analysis. Often these involve potentiometric techniques, where an electrical potential is generated which is dependent on the concentration and properties of substances in the test solution; the change in potential as the reaction proceeds can be taken as a measure of the rate of the reaction. It should be noted that a potential, which is caused by a transfer of electrons between an electrode and a solution, cannot be measured in isolation: the half-cell consisting of the electrode dipped in the test solution must be connected to a reference half-cell, of fixed potential, to enable the potentials of the two half-cells to be compared.

Of particular importance are techniques involving ion-selective electrodes, where the potential generated can be related to the concentration of a specific substance. The best known ion-selective electrode is the glass electrode used in the pH metre: in this, an internal electrode (e.g. silver wire coated with silver chloride) is dipped in a solution containing HCl at known pH. This is separated from the test solution by a thin glass wall and, in a way not fully understood, a potential is generated which is dependent on the pH difference across the glass. The potential of the entire half-cell, which includes the contribution of the internal electrode as well as that resulting from the pH difference, may be measured by linking the internal electrode to an external reference electrode (e.g. a calomel electrode) and completing the circuit by means of a salt bridge (Fig. 25.1).

Ag | AgCl | Glass | Test sample || Saturated KCl acting as a salt bridge | External reference electrode |

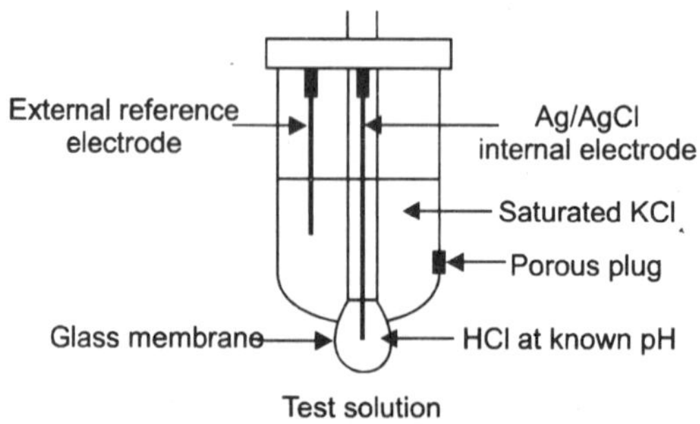

Fig. 25.1. Diagrammatic representation of a combined glass electrode and external reference electrode, as used in the pH metre.

The complete cell consists of the following interacting compartments. The potentials produced at each of these junctions will be constant for a particular instrument, the one exception being the important one between glass and the test sample. Thus, a suitably calibrated instrument may be used to determine the pH of the test sample.

Many enzyme-catalysed reactions involve the production of an acid or base from a neutral compound (or *vice versa*), so the course of such reactions could be followed by measuring the change in pH as the reaction proceeds. However, this is rarely done because a change in pH would cause the activity of the enzyme to change. The best way of monitoring the course of such reaction is by the use of a pH-stat. This usually consists of a pH metre coupled to a motor-driven burette syringe and a circuit-breaker designed to switch off the burette motor when the pH of the reaction mixture is at the original pre-set value. As the reaction proceeds and the pH of the reaction mixture starts to change, the motor switches on and forces alkali (or acid) from the burette into the reaction vessel, with continuous stirring, to restore the pH to its original value, whilst a recorder automatically makes a trace of the amount added from the burette in relation to time. Thus the pH is maintained at an approximately constant value throughout and the rate of addition of alkali (or acid) required to achieve this is a measure of the rate of reaction. Such techniques may be used, for example, to monitor the hydrolysis of esters.

Other types of electrochemical procedure include those based on polarography or voltammetry, where an increasing voltage is applied between two electrodes immersed in the test solution, whose composition determines the current which flows at each instant.

If a fixed voltage is applied between the electrodes, the current depends on the electrical conductance of the solution, which is the sum of the contribution of all the ions present (since these must be the carriers of the current through the solution); it is also the reciprocal of the electrical resistance between the electrodes. For those enzyme-catalysed reactions where there is a change in the number of charged species as substrates are converted to products, the course of the reaction may be followed by determining the change of electrical conductance with time; this is termed conductometry. Such procedures usually make use of alternating rather than direct current, to avoid electrolysis taking place, which would tend to reduce the applied voltage. Since the reaction mixture must be buffered, which will contribute a background conductance, a low concentration of a buffer of low intrinsic conductivity (e.g. Tris) is usually employed.

An example of a reaction which is ideally suited to enzymatic analysis by conductometric procedures is that catalysed by urease:

$$\underset{\text{urea}}{O = C(NH_2)_2} + H_2O \rightleftharpoons 2NH_4^+ + CO_3^{2-}$$

An uncharged molecule is converted into ions as the reaction proceeds, so the conductance must increase. This has been used for the analysis of the substrate, urea and gives a linear response up to a concentration of about 22 μmol l^{-1}, above which there is interaction with buffer ions; samples containing urea at higher concentrations than this should be diluted prior to analysis.

A specific electrode which is based on similar principles is the oxygen electrode. In the type developed by Clark there is a central cathode of gold or platinum surrounded by an annular silver anode and separated from it by an epoxy resin casing (Fig. 25.2). Oxygen dissolved in the reaction mixture (which should be in a thermostatted, air-tight container and stirred continuously) can freely diffuse to the electrodes through a membrane (e.g. Teflon or polypropylene) and an electrolyte solution (usually buffered KCI). A constant voltage of between 0.5 and 0.8V is applied across the electrodes, providing a supply of electrons at the cathode which can reduce oxygen atoms:

$$4e^- + O_2 + 2H^+ \longrightarrow 2H_2O$$

whereas at the anode there is a corresponding transfer of electrons to the electrode:

$$4Ag + 4Cl^- \longrightarrow 4AgCl + 4e^-$$

Thus a current flows between the electrodes which is proportional to the pO_2 of the sample. By monitoring how this current changes with time, the course of a reaction involving oxygen production or consumption (e.g. that catalysed by glucose oxidase) can be followed. This is a far more sensitive procedure than manometry, a possible alternative.

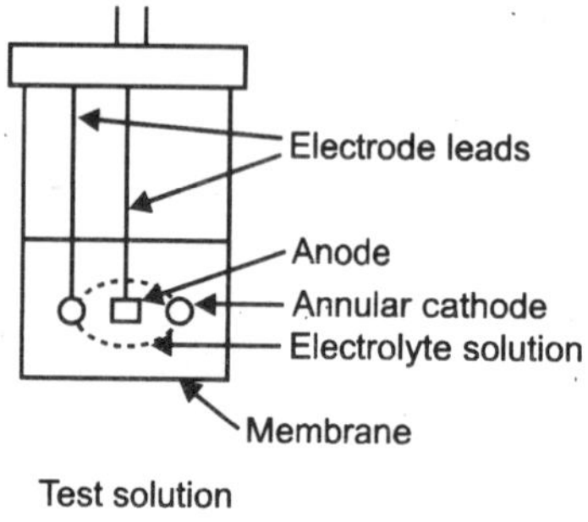

Fig. 25.2. Diagrammatic representation of a typical oxygen electrode.

As a refinement, the oxygen electrode (or other specific electrode) may be coupled with an immobilised enzyme to form an enzyme electrode for the analysis of the appropriate substrate. For example, Hicks and Updike trapped glucose oxidase in polyacrylamide gel and layered it over the membrane of an oxygen electrode, thus constructing an instrument which could be used to analyse D-glucose by a sensitive, specific and convenient procedure. Such enzyme electrodes form a major category of what have been termed biosensors. In general, electrochemical methods have an advantage over certain other procedures (e.g. spectrophotometry) in that contaminating substances can be present in suspension without affecting the analysis.

Enthalpimetry

Enzymatic analysis by enthalpimetry (microcalorimetry) is of potential importance because of the sensitivity, freedom from interference and almost universal possibilities of application of these techniques. In most reactions, heat (enthalpy) is gained or lost as the reaction proceeds. In enthalpimetry, the reaction is carried out under controlled conditions in a calorimeter and the temperature change is monitored. The main reason why these procedures have been little used in the past is that the temperature change is small (usually 10^{-2} to 10^{-4} °C) and well within the range of possible fluctuations in the ambient temperature. Hence extremely accurate thermostatting or excellent insulation is required, together with very sensitive temperature sensors. Suitably accurate temperature measurements may be obtained by the use of thermistors (temperature-sensitive resistors consisting of sintered mixtures of metal oxides in glass or plastic casings) or of thermopiles (thermocouples connected in series).

Secondary reactions with the buffer can be used to increase the signal and thus improve the sensitivity of the procedure: for example, a proton produced by the primary reaction can be trapped by buffer with evolution of heat; Tris is very useful in this respect, for its heat of protonation is large. McGlothlin and Jordan used enthalpimetry to estimate blood glucose by means of a hexokinase-catalysed reaction in Tris buffer at pH 8:

$$\text{Glucose} + \text{ATP} \longrightarrow \text{glucose-6-phosphate} + \text{ADP} + \text{H}^+ \qquad \Delta H = -28 \text{ kJ mol}^{-1}$$

$$\text{Tris} + \text{H}^+ \longrightarrow \text{Tris} - \text{H}^+ \qquad \Delta H = -47 \text{ kJ mol}^{-1}$$

Note that the secondary reaction liberates more heat than the primary reaction. However, such procedures have yet to be used routinely.

Radiochemical Methods

The use of a radioactively-labelled substrate can be valuable in enzymatic analysis. The isotopes most commonly used for labelling purposes are hydrogen-3 (tritium), carbon-14, phosphorus-32, sulphur-35 and iodine-131. All of these isotopes emit β-radiation (electrons) as they decay.

After the enzyme-catalysed reaction has progressed for a specified period, it is terminated. The substrate is then separated from the product, usually by chromatography or electrophoresis and the product concentration is determined indirectly by measuring the radioactivity of the product fraction.

High-energy β-emission, e.g. from ^{32}P or ^{131}I, may be detected and counted directly using a Geiger-Müller counter. Low-energy β-emission, e.g. from ^{3}H, ^{14}C or ^{32}S, is usually monitored by liquid scintillation counting.

Some organic solvents fluoresce when bombarded with β-radiation, giving out light of very short wavelength; this is not suitable for efficient detection, but it may be converted into light of longer wavelength by the fluorescence of one or more flours or scintillants (e.g. 2,5-diphenyloxazole, PPO) and detected by a photomultiplier.

A typical example of enzymatic analysis by a radiochemical procedure is the method of Reed and colleagues involving the cholinesterase-catalysed hydrolysis of [^{14}C]-acetylcholine:

$$\underset{[^{14}C]\text{-acetylcholine}}{^{14}CH_3.CO_2CH_2CH_2N(CH_3)_3} \longrightarrow \underset{[^{14}C]\text{-acetic acid}}{^{14}CH_3.CO_2H} + \underset{\text{choline}}{HOCH_2CH_2N(CH_3)_3}$$

After incubation, the unreacted substrate and the choline may be removed by an ion exchange resin and the radioactivity of the acetic acid fraction determined.

In general, radiochemical procedures are extremely sensitive (far more sensitive than photometric ones). They can be used in a straightforward fashion for enzyme assay and, by means of isotope dilution techniques, for other forms of enzymatic analysis also. However, they have certain disadvantages when compared to other possible techniques. These include the health hazard inevitable in the handling of radioisotopes and the separation steps which have to be performed. Also, it is impossible to use them to monitor continuously the course of a reaction and any estimate of the initial velocity which is obtained can only be based on a limited number of points (possibly only one). A further possibility of error can result from a quenching of the radiation between source and detector: this is particularly likely in the case of tritium, whose emission is so weak, it can be absorbed by paper.

Dry-Reagent Techniques

For a number of years, test strips have been available for certain routine analytical procedures. The strip, now usually made of porous plastic, incorporates a pad at one end impregnated with the reagents, which may include enzymes. This end of the strip is dipped into the test solution and after a specified time the colour produced in the pad is compared with those on the chart provided. A semi-quantitative assessment of the concentration of a reactant in the test solution is thus obtained. Such a procedure is widely used to give an indication of the D-glucose content of urine.

Systems of this type have been developed to make them fully quantitative. For example, the Kodak Ektachem system for blood urea analysis involves placing a drop of blood on a special slide. The sample then diffuses down through a layer of reflective titanium dioxide to a reagent layer impregnated with urease, which converts urea to ammonia; this in turn diffuses down through a semi-permeable membrane to a layer containing leuco dye reagent, where a coloured product is formed. The intensity of the colour is measured by reflectance densitometry through the transparent plastic base of the slide, reflected light from a suitable source being detected by a photodiode. Similar slides for different tests can be used with the same instrumentation, including enzyme assay as well as the use of enzymes as reagents. The advantage of this and other dryreagent or non-liquid reagent systems (e.g. the Roche Reflotron) is simplicity, for no reagents have to be prepared and the materials are disposable.

AUTOMATION IN ENZYMATIC ANALYSIS

As discussed already, most of the detectors used to monitor the course of enzyme-catalysed reactions can be coupled to automatic recorders. However, if each sample still has to be mixed with reagents and introduced to the detector by hand, the overall process can only be said to be semi-automatic. Here in this section, we are concerned with fully automatic procedures, where simultaneous or sequential analysis can be carried out on a batch of samples without any manual steps being involved other than in sample preparation and in loading the batch on to the analyser and possibly in the calculation of the results from recorder traces. Many analysers are now linked to computers, to enable the results to be calculated and assessed automatically. As always, the choice of instrument and of its range of facilities is limited by financial considerations.

Most fully automatic techniques used for enzymatic analysis give a result based on an estimate of the initial velocity of the reaction. In the case of relatively simple instruments, this may be based on just one or two readings; more complex instruments monitor the reaction throughout and give the best statistically-derived estimate of initial velocity. Regardless of this, the analysis must always be carried out at constant pH and temperature.

When an instrument allows only a limited number of readings to be made for each analysis, it is advisable, before using it for a particular application, to perform preliminary experiments to ensure that the progress of the reaction is linear with respect to time over the period of interest.

Under suitable conditions, most enzyme-catalysed reactions quickly reach a linear, steady-state phase which is then maintained for a few minutes before the rate of reaction starts to drop. However, some reactions (e.g. that catalysed by creatine kinase) have quite an appreciable lag-phase (i.e. one or more minutes) before the steady-state phase is reached; even the others may, in practice, take seconds to reach the linear phase, because complete mixing of the sample and reagents cannot be achieved instantaneously. The duration of the linear-phase depends mainly on the relative concentrations of enzyme and substrate. Hence, by preliminary investigations, it should be possible to choose analytical conditions which ensure that a two-point procedure gives a reasonable estimate of initial velocity provided the test substance falls within a specified concentration range (Fig. 25.3). However, ideal conditions for a one-point procedure, which assumes linearity from zero time to the time of the reading, can never be achieved if there is any appreciable lag-phase. Although linearity over the period of interest should be aimed at in both one- and two-point procedures, reasonable results may nevertheless be obtained even if there are slight departures from linearity, provided suitable standards of known composition are included among each batch of test samples. These standards may be solutions made up from solids or they may be samples which have been analysed previously by a reliable method.

Fixed-Time Methods

In fixed-time methods of analysis, the change in experimental reading (e.g. absorbance) over a fixed time interval is recorded. If the reaction has linear characteristics over this time interval, then the observed change is directly proportional to the initial velocity, and hence can be related to the concentration of the substance being analysed.

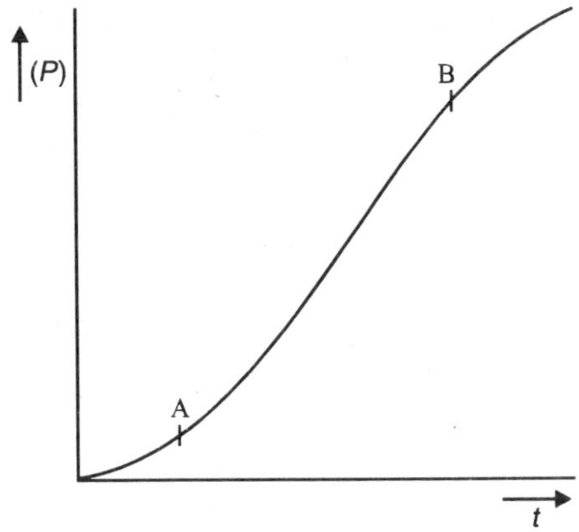

Fig. 25.3. Typical plot of product concentration against time for an enzyme-catalysed reaction. Conditions for a two-point procedure should be chosen by preliminary experiments so that the readings fall between A and B.

Such analysers may be termed discrete, when each sample is mixed with reagents by means of automatic pipettes and passed into a detector at an appropriate time, according to the programmed instructions. Discrete analysers (e.g. the Hitachi, the Technicon RA-1000 and the Roche Cobas Mira systems) can work efficiently on large batches for single tests (e.g. analysing glucose in perhaps 200 samples per hour) or can be programmed to carry out 12 or more different tests on each sample with a throughput of up to 1000 tests per hour; an even greater throughput is possible on high-capacity instruments such as the Hitachi 747 system. The stress on automatic pipetting systems is considerable and early systems, around 25 years ago, were susceptible to frequent mechanical breakdown. However, such problems have now been overcome and, at the present time, discrete analysers are employed in most high-throughput laboratories. Detection is usually by spectrophotometry, but ion-selective electrode modules may be fitted to some types. Most modern discrete analysers are selective, i.e. more tests can be done on some samples than on others, as considered appropriate. Another development is to have a random access facility, allowing the operator, through the linking computer, to be in control throughout. For instance, samples added while the instrument is already in operation can be given priority, if required. The computer may also be able to arrange the quickest way of carrying out all the required tests on the samples.

An alternative approach, widely used since the 1960s, involves continuous-flow methods (as in the Technicon Autoanalyser SMA and CHEM 1^{+} systems). If a solution is pumped through a tube at a fixed flow-rate, each molecule will take the same time to pass between two given marks on the tube. In continuous-flow analysis, a mixture of reagents is forced continuously through a tube, by means of a peristaltic or other type of pump and, at regular time intervals, one of a series of samples is automatically introduced into the reagent line. Suitable mixing coils and constant-temperature heating-baths ensure that each sample is mixed with the reagent and that, as they are pumped along the tube, a reaction takes place under controlled conditions. Eventually, the reaction mixture reaches a detector and the progress of the reaction is determined. The set-up ensures that a fixed-time elapses between the mixing of each sample with reagent and the arrival of the resulting mixture at the detector. Often air (or other gas) is pumped into the reaction tube along with the sample and reagents, producing large bubbles which segment the reaction mixture into lengths of no more than a few centimetres each. These pass along the tube separately, with gas in between and so diffusion between reaction mixtures from different samples is prevented. Alternatively, tubing of a very narrow bore may be used to minimise diffusion.

A typical set-up is shown in Fig. 25.4(a): this is simplified slightly in that it shows only one reagent line, whereas it is common for several different reagents to be pumped separately until they are mixed together with the sample; also, the mixing is usually staggered slightly so that no more than three lines ever come together at the same point. In some applications, the concentration of a product of the primary, enzyme-catalysed reaction is determined by a separate, indicator reaction. For example, in an assay of alkaline phosphatase involving the use of disodium phenyl phosphate as substrate:

$$\text{Disodium phenyl phosphate} \xrightarrow[\text{pH 10}]{\text{Alkaline phosphate}} \text{Phenol + sodium phosphate}$$

the phenol liberated must be analysed after the primary reaction has been stopped because of the nature of the indicator reagents. This can be done satisfactorily if a one-point procedure is being employed, whereas such a coupled system could not be used for monitoring continuously the course of the primary reaction.

In the King and Armstrong method, the phenol is made to react with Folin-Ciocalteu reagent and the absorbance determined at 680 nm. However, this reagent also reacts with the phenolic amino acid, tyrosine, which is present in a bound form in alkaline phosphatase, so the enzyme must be removed prior to treatment with the reagent. In a continuous-flow analysis procedure, this may be achieved by placing a dialyser at point Y [of Fig. 25.4(a)], allowing the phenol, but not the protein, to pass into the system of indicator reagent.

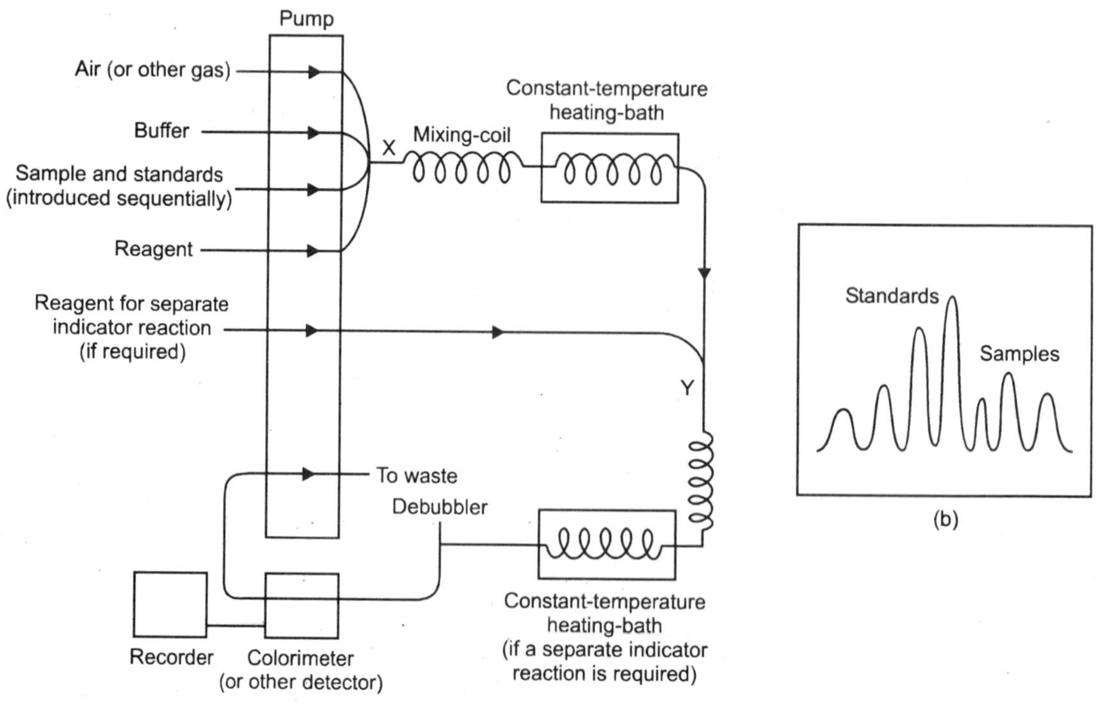

(a)

Fig. 25.4. (a) Simplified representation of a set-up for continuous-flow analysis; and (b) typical recorder tracing from a continuous-flow analysis procedure.

In the King and Kind method, on the other hand, the phenol is made to react with a reagent (4-amino antipyrine) which is insensitive to phenolic amino acids. Therefore, without any need for a dialyser, the antipyrine reagent may be introduced at Y, stopping the primary reaction and initiating the indicator reaction, whose absorbance is subsequently read at 520 nm.

In either case, the procedure may be calibrated by placing phenol standards of known concentration amongst the batch of samples: typical results would be as indicated in Fig. 25.4(b). The incubation time of the enzyme-catalysed reaction is effectively the flow-time from X to Y [of Fig. 18.4(a)]. Provided no dialyser is present at Y, this may easily be determined by introducing a coloured dye into the sample line and observing its movement. However, it is usually considered unnecessary to know the precise incubation time, particularly since it is often in the order of 10 minutes, meaning that the reaction is likely to have proceeded beyond the linear phase. Hence it is best to calibrate the procedure for each analysis by the use of standards of known alkaline phosphatase activity, this being a relatively stable enzyme. Such standards (e.g. standard sera) may be obtained commercially, but of course have to be stored carefully to avoid loss of enzyme activity.

Many continuous-flow procedures have an advantage over the ones discussed above in that there is no need to perform a separate indicator reaction. The reaction catalysed by glucose oxidase, for example, can be monitored by coupling with a peroxidasemediated dye formation, both reactions taking place simultaneously. Hence the flow circuit is simpler than the general one shown in Fig. 25.4(a), having no reagent line joining at point Y and no second heating-bath. This is also true for enzyme-catalysed reactions which do not need coupling to others; for example, the assay of alkaline phosphatase by the use of colourless *p*-nitrophenyl phosphate as substrate involves the formation of a product (*p*-nitrophenol) which is yellow in alkaline conditions, the absorbance being read at 410 nm. The system may be calibrated with *p*-nitrophenol standards and the incubation time determined as the time taken for a coloured dye to travel from the point of mixing (point X) to the colorimeter; however, as before, the system may be calibrated by the use of alkaline phosphatase standards, eliminating errors due to non-linearity.

In procedures where enzymes are being used as reagents (e.g. the determination of blood glucose by the use of the glucose oxidase/peroxidase coupled method mentioned above), a dialyser is often included at a point in the sample line before it merges with the enzyme line [i.e. before the point X in Fig. 25.4(a)]. This ensures that the only protein in the reaction mixture is the enzyme being used as a reagent, since no protein from the sample would be able to pass through the dialysis membrane. Calibration of such procedures is easier than for enzyme assay, since there is not usually any problem in obtaining standards of reliable, precisely known composition. The enzyme reagent may also be used in an immobilised form in an enzyme analytical reactor, which would replace the mixing coil in the system. For example, immobilised glucose dehydrogenase has been used to determine blood glucose, NAD^+ being reduced in the reaction.

Single-point continuous-flow procedures of this type can analyse in the region of 60 samples per hour, which is low compared to the performance of a discrete analyser. However, the sample stream may be split and identical fractions pumped simultaneously into a variety of reagent streams. Thus each sample may be analysed for several different factors at the same time, with little extra labour involved. However, such a system is wasteful of reagents and each test will be carried out on each sample, whether necessary or not.

Two-point continuous-flow systems may give slightly more accuracy than one-point systems for enzymatic analysis, since it is easier to ensure that the reaction is being investigated only in the linear-phase. The first detector is placed in the flow-line at a point immediately after the first mixing coil, while the second detector is situated after the heating-bath. The two detectors would preferably be components of a single instrument, e.g. two electrodes of an electrochemical set-up or two sensors of a differential-spectrophotometer. Ideally, the flow-time between the two detectors should be sufficiently short (e.g. 30 seconds) to keep within the linear phase and the sample should be pumped for a sufficient length of time to enable it to occupy (together with the reagent) the whole of the system. When this has been achieved and the system stabilised, the difference between the two detector readings will be a measure of the rate of reaction.

For example, Blaedel and Hicks used a two-point continuous-flow system with a differential-spectrophotometer as detector to determine glucose concentrations by a method involving glucose oxidase/peroxidase in which 15 samples per hour could be analysed.

Since the incubation time is short, the difference in readings is very small, so very accurate detection techniques are required for such procedures. Since they have neither the convenience of the one-point methods nor the comprehensive qualities which come from continuous monitoring, two-point continuous-flow methods are not widely used at the present time.

Fixed-Concentration Methods

In fixed-concentration methods of analysis, the time for an experimental reading (relating to the concentration of some substance participating in the reaction) to change by a specified amount is measured automatically. As with fixed-time methods, the readings should ideally be restricted to the linear-phase of the reaction, as indicated by preliminary experiments. Under these conditions, the time taken is inversely proportional to initial velocity. Procedures based on this principle must involve discrete rather than continuous-flow analysis. However, they have only been used to a very limited extent.

Methods Involving Continuous Monitoring

Several automated rate-monitoring systems, specifically designed for enzymatic analysis, include the LKB Kinetic Analysis System and the Beckman Enzyme Activity Analyser. Although they differed in design, all were discrete analysers which could handle in excess of 40 samples per hour, working sequentially. The main difference from other types of discrete analyser is that when a sample was mixed with reagents, the ensuing reaction was monitored continuously (or almost continuously, readings being taken at no more than 3 second intervals) by means of a spectrophotometer, the results being fed directly into a computer. Constant temperature was maintained throughout.

Any lag-phase reaction could be eliminated from the calculations either by imposing a delay before the first reading is taken or by programming the computer to recognise the various phases of the reaction. When the linear-phase had been observed for 15–20 seconds, enough readings would have been taken for the computer to calculate its slope and give a statistical assessment of the results, after which the reaction mixture could be ejected from the detector and a new one introduced. Warnings were often printed if there was any significant departure from linearity over the period of interest or if the optical limits of the instrument were being exceeded. However, the use of such specific systems has been overtaken by more general developments in automation.

With the development of microprocessor technology, several spectrophotometers now have kinetic attachments. For example, the kinetics analysis facility of the Beckman DU-70 spectrophotometer can make up to 1200 readings per minute on a single sample or can monitor simultaneously the reaction in a series of samples, making readings every few seconds. The absorbance against time curve curves can be visualised on a screen and linear regression analysis carried out on selected portions of the curves.

The more recent Beckman DU Series 500 spectrophotometers include kinetics in the standard applications software package.

Some routine discrete analysers now also have a kinetics capability. An example is the Beckman Synchron CX7 system, which has a throughput of up to 825 results per hour and a random-access facility.

Centrifugal analysers can also be used for enzymatic analysis, and some, e.g. the Roche Cobas Bio and the Baker Centrifichem systems, were widely employed around 10 years ago, although their use has since declined. In such systems, 15–30 samples are analysed simultaneously in cuvettes on a spinning, horizontal centrifuge plate. The cuvettes are on the outer portions of the plate and each is aligned with specific depressions on the inner portion which originally contain sample and reagents (Fig. 25.5). The inner plate can be removed for ease of loading and to allow one batch of sample to be loaded while another is being analysed.

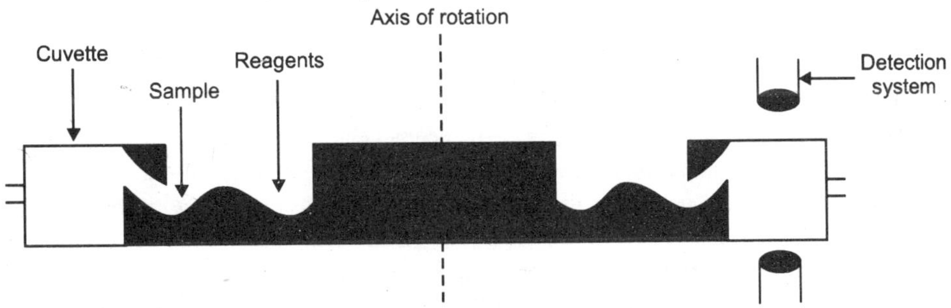

Fig. 25.5. Diagrammatic representation of a centrifugal analyser set-up (side view).

When centrifugation commences, each sample is mixed with its reagent as they are carried together by centrifugal force into a cuvette, thus starting a reaction. As the plate revolves, each cuvette passes through a detecting system (usually a spectrophotometer) linked to a computer, so readings can be taken and stored for each reaction mixture on each circuit (i.e. at about 100 ms intervals). Also on each circuit, a dark current reading is taken, when the light beam is interrupted by a solid part of the rotor. An oscilloscope may be linked to the detector, to enable the progress of the reaction in each cuvette to be visualised. Regardless of this, a suitable computer can process the readings to give the best estimate of the initial velocity of each reaction and print out the results. A trap may then be opened at the bottom of each cuvette and the contents emptied by centrifugal force, preparing the instrument for the next batch of samples. It is possible to analyse in the order of 200 samples per hour by such a procedure.

Later models, such as the Roche Cobas Fara, incorporate a variety of detecting devices, including ion-selective electrodes, while the Baker Encore has research modules which enable it to be used for the study of enzyme kinetics.

Chapter 26

Enzyme Thermistors for Food Analysis

INTRODUCTION

Biocalorimetry is the most general of all analytical techniques that exploit biocatalysts for detection. Almost all bioprocesses are exothermic, and enzyme reactions are usually accompanied by a negative enthalpy change in the range of 20–100 kJ/mol. In recent years, suitable instrumentation for routine analysis has been developed as an alternative to the more complicated traditional microcalorimeters. The enzyme thermistor system measures the temperature change associated with the reaction using a thermistor as temperature transducer. The thermistor is virtually driftless, which is of special value in monitoring concentrations over long time periods, and its high temperature resolution makes it possible to obtain adequate sensitivity with rather simple electronics. Additional advantages include immunity to electrochemical and optical interferences generated by the reaction or by the sample matrix. The obvious inherent disadvantage of calorimetry, non-specific temperature signals, in practice presents fewer problems than might be expected and can normally be avoided by careful sample preparation. If this is not sufficient, a reference column can be introduced, which usually eliminates all non-specific effects.

In spite of its many attractive features, the thermal biosensor concept has not become as widespread as might have been expected. The poor impact on the market has limited the development of instrumentation as well as of methodology. Existing equipment is still rather expensive and complex, about the same as an HPLC detector. Recently, however, several laboratories have proposed miniaturised and much simplified calorimetric constructions that may change this situation. It appears that many problems, such as thermal common-mode rejection or variations in the background temperature, are reduced when the dimensions of the device are cut down or with the introduction of alternative transducers. As with most miniature biosensors, the object here has been to develop portable and even implantable monitoring devices, especially for glucose in diabetes control. This development can, however, be expected to spill over to other analytical fields as well, resulting in much cheaper instrumentation.

Hitherto, only a few food analytical applications of thermal biosensors have been demonstrated. Previous work has been dominated by biomedical measurements and by bioprocess monitoring and analysis. Many assays from these areas can, however, easily be adapted to food analysis and are therefore included in this chapter. Previous experience with the enzyme thermistor point to its general applicability where the sensitivity is adequate. Even if this overview in many cases only indicates possibilities, the prospects for success in food analytical applications are good since the concentrations are usually comparatively high in food samples. Even if the sample matrices could be complex, the extensive dilution possible should make them less problematic than in other complex samples, such as blood serum or fermentation broth.

MEASURING PRINCIPLE

The enzyme thermistor measures the heat developed in a small adiabatic column containing an immobilised biocatalyst placed in a continuous buffer stream by monitoring the temperature of the effluent of the column. Almost all bioprocesses are exothermic, and enzyme reactions in particular are usually accompanied by an enthalpy change in the range of –20 to –100 kJ/mol. For instance, the reactions catalysed by β-lactamase (in Tris buffer), catalase, and glucose oxidase all produce heat, on the order of 100 kJ/mol, and urease, 60 kJ/mol. If the enthalpy change of the primary reaction is low, the situation can be saved by introducing one or more additional enzymes to act on the product of the first enzyme. Since the sum of all enthalpy changes is detected, such sequentially operating enzymes can be used to increase the total sensitivity.

Determinations of concentrations are usually made by flow injection analysis. A small sample volume is injected in a comparatively rapid buffer stream (Fig. 26.1). A short sample pulse results in a temperature peak that is proportional to the concentration of the analyte (the enzyme substrate) over a linear operating range of concentration that can be from micromolar to several hundred millimolar.

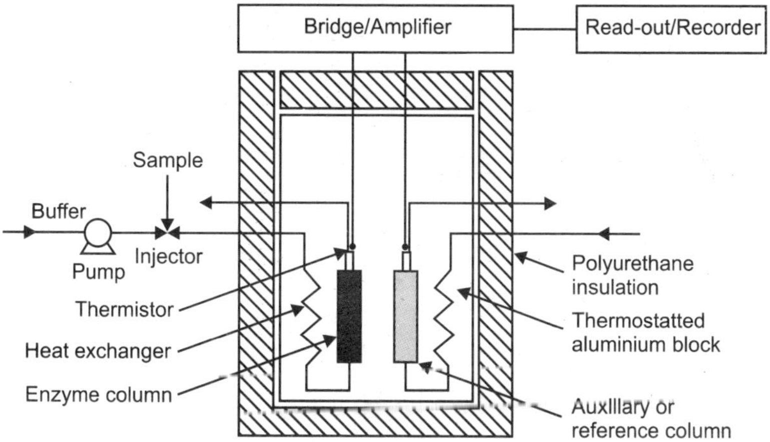

Fig. 26.1. Instrumental set-up for flow injection analysis using the enzyme thermistor.

Normally, the thermograms are evaluated by peak height determination, which is simple to perform both manually from the recorder diagram and automatically by an integrator or computer. The area under the peak and the ascending slope of the peak are also linearly related to the analyte concentration. Continuous sample introduction leads to a shift of the temperature signal to a new steady state level. Since this shift is also proportional to the analyte concentration, the instrument can easily be used for continuous monitoring.

The system shown in Fig. 26.1 contains two identical flow lines, which can be independently used for two different enzyme columns for different analyses. Alternatively, one of the columns can be used as a reference column to compensate for any non-specific reactions. In normal operation, however, the temperature of the enzyme column is registered versus the temperature measured with a reference thermistor mounted in the heat sink of calorimeter (Fig. 26.2).

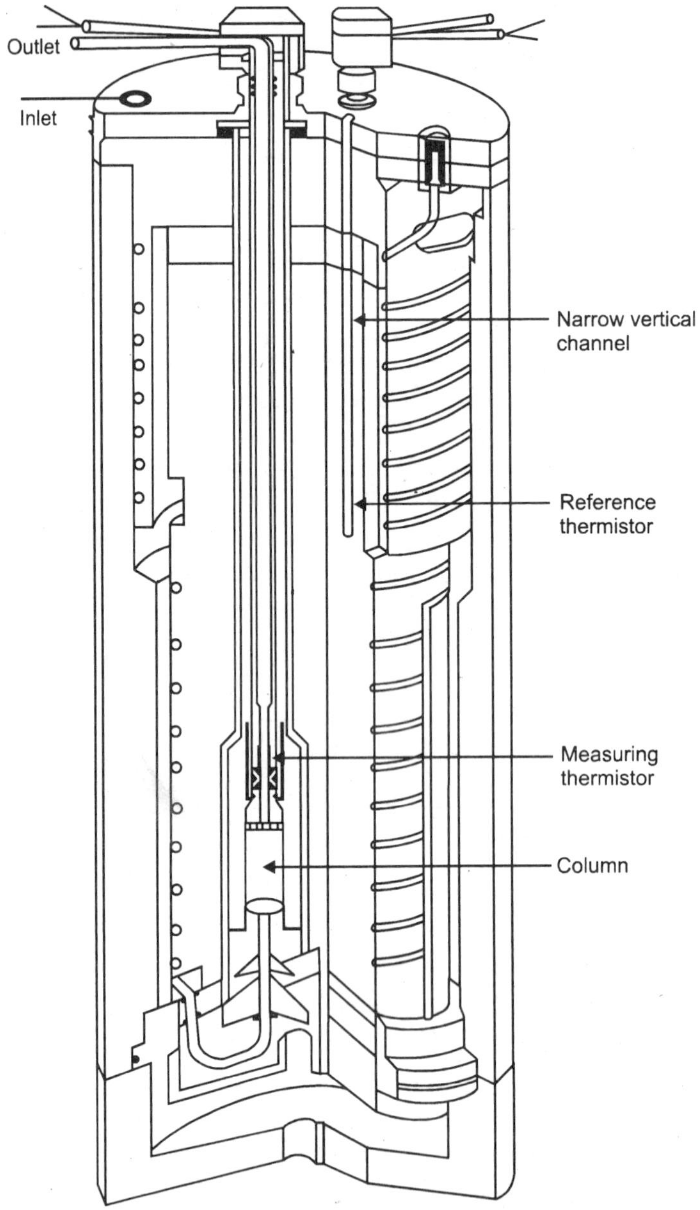

Fig. 26.2. Cross-section of an enzyme thermistor calorimeter.

INSTRUMENTATION

Most of the work described here has been done with the two-channel calorimeter shown in Fig. 26.2. This calorimeter consists of an outer aluminium cylinder (80 × 250 mm) thermostated at 25, 30, or 37°C and placed in a container insulated with polyurethane foam. Inside the thermostated jacket is a second aluminium cylinder with two column ports and a pocket for a reference thermistor. The solution to be analysed is passed through a thin-walled acid-proof steel tubing (0.8 mm ID) with two-thirds of its length in close contact with the thermostated jacket. The last third is in close contact with the inner cylinder, which acts as a heat sink. With this arrangement, short-term temperature variations of the solution entering into the column will be exceedingly small. The columns (4 or 7 mm ID with a volume up to 1 ml) are attached to the end of Delrin tubes, by means of which they are inserted into the calorimeter. These tubes also contain the temperature sensors (closely matched thermistors; 2.8 kohm ±0.25 per cent at 25°C, temperature coefficient –3.9 per cent/°C), which are mounted with heat-conducting epoxy on a short piece of gold capillary through which the column effluent passes. The resistance of the thermistors and thereby the temperature is measured with a Wheatstone bridge of DC type build with precision resistors with low temperature coefficient and a chopper-stabilised operational amplifier. The maximum sensitivity of this bridge is 100 mV/0.001°C.

A common measuring range is 0.01°–0.05°C. To give an impression of what sensitivity is needed, it can be mentioned that a reaction enthalpy change of 80 kl/mol (such as for the glucose oxidase reaction) should be expected to give a temperature change of 0.01°C for 1 ml of a 1 mm sample at a flow rate of 1 ml/min., which are rather typical parameters for this type of analysis.

A continuous flow of buffer (0.5–5 ml/min.) is produced by a low-pulsation pump, usually a good peristaltic pump. Samples are usually introduced with an injection valve using 0.1–0.5-ml sample loops. Sample volumes less than 25 µl can be injected undiluted, but larger samples should be diluted at least 5–10-fold to avoid non-specific heat effects from dilution or solvation. Such effects can also be eliminated by the split-flow technique mentioned above, in which the flow is divided into two equal streams, one through the enzyme column and the other through a similar but inactive column. Since the temperature is measured differentially between the columns, non-specific heat effects should cancel.

The most common enzyme support has been propylamino-derivatised CPG with a pore size in the range of 500–2000 Å and a particle size up to 80 mesh (0.18 mm), loaded with a large excess of enzyme, often 100 units or more, immobilised with glutardialdehyde. CPG offers high binding capacity; good mechanical, chemical, and microbial stability; and relatively simple coupling procedures, but Eupergit C (oxirane acrylic beads from Röhm Pharma, Weiterstadt, Germany) and VA-Epoxy Biosynth (Riedel-de Haën, Seelze, Germany) are good alternative support materials. The major limiting factor for column life is usually mechanical obstruction. If, however, the solutions used, as well as the samples, are filtered through a 1–5 µm filter and if microbial growth in the solutions and the flow lines is prevented, good operational stability with unchanged performance for large series of samples (thousands) can be obtained and the column may function for several months.

APPLICATIONS

Very satisfactory results have been obtained with biothermal analysers in many bioanalytical fields, including clinical chemistry, process control, fermentation monitoring and other analyses within biotechnology, and environmental control. The analysis can be carried out in various ways depending on the column filling. For the determination of enzyme substrates it is most common to use one or more immobilised enzymes (Table 26.1), but immobilised cells can also be used, especially if they are induced for the substrate in question. Immobilised enzymes or cells can also be used in studies of reversible or irreversible inhibitors or activators, such as pesticides or vitamins. A rapid on-line ELISA technique with thermometric detection (TELISA) has been devised in which the column is loaded with an immunosorbent. In addition, the enzyme thermistor device described here is, in fact, a flow calorimeter, which can be used with an empty column or with a column filled with an inert porous material for studies of cell suspensions or for the assay of enzyme solutions.

Table 26.1. List of substances of food analytical interest that have been determined with thermometric biosensors[a].

Substance	*Enzyme(s) used*	*Concentration range (mm)*
L-Ascorbic acid	Ascorbate oxidase	0.05–0.6
Cellobiose	β-Glucosidase + glucose oxidase/catalase	0.05–5
Ethanol	Alcohol oxidase/catalase	0.002–1
D-Galactose	Galactose oxidase	0.5–10
D-Glucose	Hexokinase	0.1–25
D-Glucose	Glucose oxidase/catalase	0.001–0.8
L-Lactate	Lactate-2-monoxygenase	0.005–2
L-Lactate	Lactate oxidase/catalase	0.002–1
Lactose	β-Galactosidase + glucose oxidase/catalase	0.25–5
Oxalate	Oxalate oxidase	0.005–0.5
Sucrose	Invertase	0.1–100
Urea	Urease	0.01–200

[a]The concentration range given is the linear range.

Assays Made with Immobilised Enzyme Columns

Glucose

The biosensor literature is dominated by studies involving glucose. Glucose is, of course, one of the most common clinical analytes and a metabolite and substrate of major interest in biotechnology, but considerable efforts are also made to develop implantable or portable glucose sensors for diabetes control. Fortunately, there are several enzymes available for glucose determination, which makes it easy to apply the entire arsenal of sensors — amperometric, optical, potentiometric, calorimetric, etc. In the latter case two different enzymes have been employed: hexokinase and glucose oxidase.

Glucose oxidase is invariably co-immobilised with catalase, which has three effects. As mentioned above, catalase approximately doubles the enthalpy change (and thereby the sensitivity), it restores half of the oxygen consumed in the glucose oxidase reaction, and it eliminates the hydrogen peroxide formed in this reaction, which otherwise could damage the enzyme. As a consequence the stability of a glucose oxidase/catalase column is excellent. It can be stored several years and used with thousands of samples. The main drawback of this system is caused by the poor solubility of oxygen in water (about 0.25 mm at 25°C), which limits the linear concentration range to <0.7–1.0 mm when larger sample volumes are applied. The limit of detection, however, is as low as 1 μM. The assay is usually carried out in 0.1 M sodium phosphate buffer, pH 7.0.

This method is used for the determination of glucose in blood plasma or serum and in biotechnological samples with good correlation with conventional methods as well as for glucose monitoring in fermentation broths. High sensitivity, specificity, and operational stability are the most notable features of this assay. It could be expected that this method should work well in food analysis where the glucose concentrations usually are very high and the sample matrix problems are likely to be less than in fermentation broth. To prove this point, the glucose amounts in some food samples were measured with an enzyme thermistor equipped with a glucose oxidase/catalase column in comparison with photometric determination using hexokinase and glucose-6-phosphate dehydrogenase. The agreement in results was good.

The useful working range can be considerably increased, at least to 75 mm, by using a more soluble electron acceptor than oxygen, such as benzoquinone. The feasibility of the system was demonstrated by glucose determinations in food samples such as beverages. We are currently investigating an alternative procedure using ferrocene as electron acceptor with electrochemical re-oxidation and thermometric detection. The linear range for the glucose assay based on hexokinase developed by Bowers and Carr for clinical use lies within 0.1–25 mm. With respect to sensitivity and simplicity this assay should also be useful for glucose determinations in food analysis, but to date there are no published results from this area.

Disaccharides

The glucose/oxidase/catalase-based assay can be used together with disaccharide-splitting enzymes for the determination of disaccharides containing glucose, such as cellobiose, lactose, and maltose. This approach is preferred because the enthalpy change associated with the hydrolysis of disaccharides normally is too low to be directly useful in thermometric assays except for very high concentrations. Mattiasson and Danielsson used β-galactosidase for lactose and β-glucosidase for cellobiose hydrolysis, resulting in a useful concentration range of about 0.1–10 mm.

The use of two enzymatic steps in the determination of disaccharides is clearly disadvantageous, even if this is the common practice in enzymatic analysis, regardless of detection mode. The hydrolysis of disaccharides is generally not accompanied by any easily measurable physical changes other than optical rotation. If glucose is present in the sample, which is common in food samples, the glucose concentration has to be determined separately, which doubles the analysis time. Furthermore, if the glucose concentration is several times higher than that of the disaccharide, which is not unusual, the precision of the disaccharide measurement will be poor. These problems were treated by Danielsson in a separate study on cellobiose determination at cellulose degradation. A valve arrangement for switching the β-glucosidase in and out of the flow stream simplified the stepwise determination of glucose and cellobiose with the same glucose oxidase/catalase column.

In contrast to other disaccharidases, invertase (β-fructosidase) generates enough heat for calorimetric measurements. Thus sucrose can be directly determined in the range of 0.1 mm up to at least 100 mm in the presence of glucose and other carbohydrates. This makes the invertase-loaded enzyme thermistor a unique biosensor constellation. It should be very useful in food analysis, but it has hitherto been used only with biotechnological samples.

Ascorbic acid and galactose

The concentration of ascorbic acid (vitamin C) in food samples is sufficiently high for enzyme thermistor determination with L-ascorbate oxidase, which provides a linear range of 0.05–0.6 mm in 0.1 M sodium acetate buffer, pH 5.5. The operational stability was acceptable for practical purposes. By using concanavalin A-conjugated Sepharose as enzyme support, the enzyme could readily be changed *in situ*. In the same study an assay for galactose was described. The enzyme D-galactose oxidase gave an operating range of 0.1–10 mm. The column stability was rather low but could be improved to one week by storage in buffer (0.1 M Tris-HCl, pH 7.4) containing Cu^{2+}.

Ethanol

Ethanol can be measured with alcohol oxidase from *Candida boidinii* with a linear range of 0.01–1 mm. Coimmobilisation with catalase has a profound effect on the stability of the enzyme column, making it stable for several months as shown by Guilbault and co-workers. This assay is useful for samples from beverages and for blood samples. A preliminary test with beer, wine, and some stronger liquors using the alcohol oxidase from *Pichia pastoris* showed good agreement with the photometric reference method. This enzyme has also been used with excellent results by Rank and others in ethanol monitoring during fermentation. It should be noted that methanol is a better substrate for the alcohol oxidases than ethanol and also that other low alcohols are oxidised.

Furthermore, enzymes from different sources may differ in calorimetric response depending on differences in substrate specificities. An example is given in Fig. 26.3, which displays calibration curves for different alcohols with alcohol oxidases from *Candida boidinii* and *Pichia pastoris*. The assay buffer was 0.1 M sodium phosphate, pH 7.0, with both enzymes.

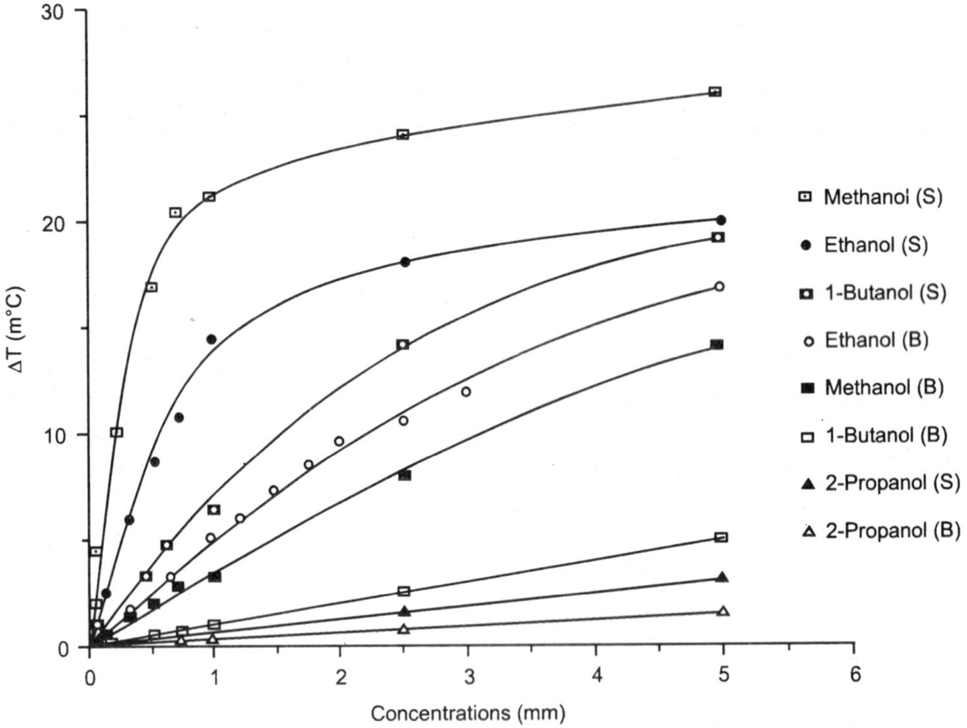

Fig. 26.3. Calibration curves for different alcohols with alcohol oxidase from *Candida boidinii* (B) and from *Pichia pastoris* (S) coimmobilised with catalase. The flow rate of the buffer stream (0.1 M sodium phosphate, pH 7.0) was 1 ml/min., and the sample volume was 0.5 ml.

Lactate

Two different enzymes have been tested for the determination of L-lactate with excellent results. With 50 units of lactate-2-monoxygenase (*Mycobacterium smegmatis*) applied to a CPG column, linear response was obtained in the range of 0.005–1 mm. Alternatively, 25 U of lactate oxidase from *Pediococcus pseudomonas* co-immobilised with 100,000 U of catalase on a CPG column resulted in a linear response in the range of 0.002–1 mm. Both enzymes yield columns that are stable for several months and can be run in 0.1–0.2 M phosphate buffer, pH 7.0. The assay should be quite useful for food analysis but has hitherto been used only for serum and biotechnological samples.

Oxalate

Oxalate determination with the enzyme thermistor, on the other hand, has been tested with food samples. Two different enzymes can also be used here: oxalate decarboxylase, for which a working range of 0.1–3 mm has been obtained, and oxalate oxidase, providing a linear range of 0.005–0.5 mm. The former enzyme is only available in preparations of rather low specific activity, whereas the oxidase can be readily prepared at reasonable purity from barley seedlings. It is also commercially available in qualities of satisfying specific activity, yielding enzyme columns with good stability. A suitable assay buffer is 0.1 M sodium citrate, pH 3.5, containing 2 mm EDTA and 0.8 mm 8-hydroxyquinoline. Winquist found that this method was suitable for the determination of oxalate in beverages and other food samples as well as in urine. In the latter case it was found to be advantageous to pass the sample through a C_{18} cartridge to remove interfering substances.

Lipids

Practically useful thermometric methods for the determination of the main blood lipid classes have been developed. It is likely that these methods could find some use in food analysis as well. Triglycerides can be determined with lipoprotein lipase (EC 3.1.1.34) immobilised on CPG with a larger pore size than usual (2000 Å). A suitable assay buffer is 0.1 M Tris buffer, pH 8.0, with 0.5 per cent Triton X-100 as detergent. Using a split-flow apparatus, a linear response was obtained for 0.05–10 mm tributyrin and 0.1–5 mm triolein.

Phospholipids can be determined with the use of three consecutively acting enzymes. Satoh added phospholipase D directly to the sample, which was then injected into the buffer stream (0.1 M Tris-HCl, pH 8.0, containing 15 mm $CaCl_2$ and 0.5 per cent Triton X-100). The enzyme thermistor column was loaded with co-immobilised choline oxidase and catalase.

The linear range, 0.03–0.19 mm, was adequate for the direct determination of phospholipids in 10-fold diluted serum samples. The choline oxidase/catalase column showed good stability, but phospholipase D is so unstable that it has to be applied in soluble form, which reduces the practical usefulness of the assay.

The stability of immobilised cholesterol oxidase also needs to be improved before a really useful system for cholesterol determination can be designed. Co-immobilisation with catalase improves the sensitivity. If the samples are treated with cholesterol esterase, preferably in a pre-column, cholesterol esters may also be measured. The assay buffer, which has to contain a detergent, was phosphate buffer, pH 6.5, supplied with 12 per cent (v/v) ethanol and 8 per cent (v/v) Triton X-100.

At present, the triglyceride determination is the only one of the thermometric lipid methods tested that seems promising for food analytical applications. In all cases a detergent system has to be employed, which puts special demands on the thermal approach in order to avoid non-specific temperature signals. Differential measurement involving the split-flow technique appears to solve this problem. Alternatively, the measurement can be carried out in an organic solvent.

Urea

Immobilised jack bean urease yields very stable columns, provided that it is protected against oxidation and heavy metals by the addition of 1 mm reduced glutathione and 1 mm EDTA to the assay buffer (0.1 M sodium phosphate, pH 7.0). A clinically useful method with a precision better than 1 per cent and a linear range of 0.01–200 mm was proposed by Danielsson. Of interest to food analysis are the attempts to develop methods for urea determination in alcoholic beverages, especially those produced by fermentation. This is important, because urea and ethanol may combine to form a carcinogenic compound, ethylcarbamate. An acid urease from *Lactobacillus fermentum* was used in these experiments.

Xanthine and hypoxanthine

Hypoxanthine and xanthine have been determined in the range of 0.1–0.5 mm using xanthine oxidase (XOD):

$$\text{Hypoxanthine} \xrightarrow{\text{XOD}} \text{Xanthine} + H_2O_2 + H_2O + O_2$$

$$\text{Xanthine} \xrightarrow{\text{XOD}} \text{Uric acid} + H_2O_2 + H_2O + O_2$$

Although the enzyme column was rather unstable (useful during one week or with >150 samples), the assay could be practical for smaller series of food samples, for instance for milk analysis.

Measurements in Organic Solvents

Measurements in organic solvents are interesting because the lower heat capacity should lead to temperature changes up to three times larger than in water at the same enthalpy change. Many enzymes have been found to work well in organic solvents, and it should be of interest to take advantage of this possibility to develop assays for substrates that are difficult to dissolve in aqueous solvents. As an example, we found that triglycerides can be determined directly in cyclohexane, in which they are readily soluble. The enzyme lipoprotein lipase functions well in this solvent, and the response is about 2.5 times higher than in water. With other enzymes the increase in sensitivity may be even higher. Despite the obvious hygienic disadvantages of using organic solvents, this approach could be acceptable with samples prepared by extraction with organic solvents. This is a common situation, for instance, in lipid analysis.

Chemical and Enzymatic Amplification

In order to obtain as large a temperature signal as possible in relation to background variations it is always important to find optimal operating conditions. Thus the temperature response can be increased by the addition of sequentially acting enzymes. This is usually necessary with hydrolytic enzymes because of their low reaction heat, even if at the same time it makes the assay more complex and increases the risk of interference. As mentioned above, lactose determination, for instance, has to be carried out with β-galactosidase followed by a glucose oxidase/catalase column.

Oxidases are generally combined with catalase or peroxidase for several reasons: the total enthalpy change is approximately doubled, the damaging effects and side reactions of hydrogen peroxide are prevented, and half of the oxygen consumed in the oxidase reaction is restored by the addition of catalase. If a proton is involved in the enzymatic reaction, as with proteolytic enzymes, the total heat production can be increased considerably by using a buffer with a high protonation enthalpy, such as Tris buffer (–47.7 kJ/mol).

Substrate or coenzyme recycling as a means to amplify the sensitivity is especially efficient in thermometric analysis because all the reactions involved may contribute to a greater total temperature change. Scheller and co-workers obtained a more than 1000-fold enhanced sensitivity for lactate or pyruvate by co-immobilising lactate dehydrogenase and lactate oxidase/catalase. L-Lactate is oxidised by lactate oxidase to pyruvate, which is reduced to lactate again by LDH and so on. The total enthalpy change of this system is as high as –225 kJ/mol, permitting lactate (or pyruvate) concentrations as low as 10 nm to be determined.

Process Monitoring and Control

The enzyme thermistor has been successfully applied to bioprocess monitoring and control. Such techniques may not be of immediate use in food analysis, but a few observations may be of interest. Determinations of substrates such as various sugars as well as of products such as ethanol and β-lactams are of immediate interest for process control. Rank obtained excellent results in fermentation monitoring of these metabolites over 1–2 weeks without changing the column. Sterile samples were withdrawn with a filtration probe inserted into the fermenter. Other sampling techniques, a field of vital importance in bioprocess monitoring, have been described by Mandenius and Danielsson. The monitoring can be performed either continuously by continuous sample introduction or semi-continuously by repeated flow injection analysis with sample injection as frequently as required. The latter technique has several advantages and is usually preferred in bioprocess monitoring where the changes in concentration are slow. The two techniques have been compared by Mandenius and co-workers in a study concerning control of an immobilised invertase reactor. With the enzyme thermistor described here, two different analytes can be measured simultaneously. Wehnert designed a four-channel enzyme thermistor by which four different substances—for instance, different sugars—can be monitored.

Enzyme Activity Determinations

Enzyme activities can be monitored by calorimetry even in very crude and complex solutions by mixing the enzyme solution with a suitable substrate to the enzyme of interest followed by measurement of the heat of the reaction. The enzyme thermistor apparatus can be used for such measurements with a small modification of the flow system and with an empty or inactive column (preferably Teflon) as reaction chamber. The sample stream and a buffer stream containing the appropriate substrate in excess are passed through the heat exchangers, thoroughly mixed, and introduced into the column after passing through a short heat exchanger to eliminate mixing and solvation effects. Even a reaction time of <1 minute is enough for enzyme activity determinations with a sensitivity down to 0.01 U/ml. This technique, which has general applicability, was reviewed by Danielsson and Larsson. It can be used with both continuous sample introduction and pulsed sampling for the monitoring of chromatography and other enzyme purification processes with a sensitivity comparable to that of UV monitoring at 280 nm.

Measurements on Cell Suspensions and Immobilised Cells

In the same way as with enzyme solutions, the enzyme thermistor apparatus can be used for studies of the total metabolic heat from a cell suspension by mixing the cell stream with a substrate stream or simply by letting a flow stream containing cells, for instance from a fermentor, pass through an empty column in the enzyme thermistor. This technique is useful as a general method to follow the state of microbial processes and in studies of metabolic effectors (such as vitamins) and toxic compounds.

If the pure enzyme is too unstable or difficult to isolate or if the desired reaction involves complex enzyme sequences, it may be advantageous to use immobilised whole cells. Since coenzyme regeneration is 'automatically' handled by the cells, no addition of expensive coenzymes is necessary. Besides, for measurements of substrates, immobilised cells can be used in assays based on the metabolic: effect of an agent, such as a vitamin or a poison. Nicotinic acid was sensitively determined with gelatine-entrapped *Lactobacillus plantarum* that had been cultivated in a medium deficient of the vitamin nicotinic acid. A disadvantage of this type of assay is that the column often can be used for only one (positive) experiment. If large amounts of the sensing cells are available, the assay can, however, be designed in a continuous-flow fashion where the sample is mixed with a continuous cell suspension stream.

Environmental Control

At least three different approaches have been suggested to employ the calorimetric detection principle in environmental control analysis. The heat of the conversion of a pollutant or toxic compound by an enzyme or a cell may in some cases be sufficient for measurement, at least in process streams or in sewage where the concentrations are high. The enzyme rhodanese has been used for the determination of cyanide (in sewage water from marzipan manufacture), and there are enzymes available that can use pesticides and chlorophenols, etc. as substrates.

The second alternative, which is much more sensitive, is to measure the inhibitory effect of a pollutant. For example, a highly sensitive technique for the determination of heavy metals (Hg^{2+}, Cu^{2+}, and Ag^{+}) based on the inhibition of urease has been developed. The column can be restored by washing with iodide and EDTA and reused many times. Sensitivities in the ppb range or lower are achievable. In a similar fashion, cells can be used for the detection of environmental poisons.

The third approach exploits the fact that many enzymes require metals in their active sites to be active. It is then usually possible to obtain an inactive apoenzyme by removing the metal ions with strong chelating agents. The enzyme can be reactivated through exposure to a sample containing the same metal ion. Satoh has used this technique for the determination of Co(II), Cu(II), Zn(II), etc., in amounts down to nanomolar concentrations.

Each one of these approaches could be useful in food analysis, especially in the food industry. Detection of antibiotics in milk, which is very important in dairies, may serve as an example of analyses that could be performed as inhibition assays. Thermometric assays for β-lactams, such as penicillin G and V, using β-lactamases have been very successful in biotechnological analysis. The useful linear range is at least 0.01–100 mm, and the enzyme columns are very stable and useful for several months or for thousands of samples. The sensitivity is, however, not sufficient for the detection of penicillin in milk. Instead it should be possible to exploit the inhibitory action of penicillin on carboxypeptidase Y or on microbial cell growth. For instance, the growth of *Lactobacillus* could be sensitively followed with the substrate-recycling assay for lactate described above.

Thermometric Enzyme-Linked Immunosorbent Assay (TELISA)

For rapid and sensitive determination of larger molecules, such as hormones and antibodies present in, for instance, fermentation broth, an automated flow-through thermometric ELISA has been designed. The determination and monitoring of the production and release of human proinsulin by genetically engineered *E. coli* cells was reported by Birnbaum and co-workers. Insulin concentrations in the range of 0.1–50 μg/ml were determined, in good agreement with the results of conventional radioimmunoassay. Standard curves were reproducible over a period of several days. The response time of the automated assay is 7 minutes after sample introduction, and a single assay cycle is completed in 13 minutes. This method has been found to be very suitable for bioprocess monitoring and in other situations where rapid determinations on a limited number of samples are desired.

Chapter 27

Enzyme Electrodes for Food Analysis

INTRODUCTION

A large variety of electrochemical sensors are described in the literature on food analysis. Some are suitable for the determination of specific compounds such as glucose, ethanol, or lactic acid, and others have been developed for the determination of more complex parameters such as 'quality' or 'freshness'. The devices range from fully automated FIA systems requiring no sample pre-treatment to disposable sensors fabricated by methods developed in the microelectronics industries. Additionally, some enzyme analysers originally designed for medical applications have been applied to food analysis.

In most applications no sample pre-treatment is required other than dilution of the sample or extraction of the analyte from a solid sample (fish, tea, etc.). Relative standard deviations of up to 5 per cent are given for all established methods.

This chapter provides an overview of the possibilities and limitations of biochemical analytical devices based on amperometric determination, with examples of a variety of setups and formats applied to food analysis.

There are even more examples of such sensing systems described in the literature, but they are based on the principles explained here.

The first biosensors described in the literature were based on the combination of glucose oxidase with the electrochemical determination of O_2 and H_2O_2. Since then, this principle has been extended to the development of sensors for other analytes using the electroactivity of not only H_2O_2 and O_2 but also NADH and other compounds, so-called mediators, for electron transfer from the enzyme to the electrode. The principles of these biosensors are described by Equations 27.1 to 27.3.

Biosensor based on an oxidase

Enzymatic reaction:

$$\text{Substrate} + O_2 \rightarrow \text{Product} + H_2O_2 \qquad \text{... (27.1)}$$

Electrochemical reaction:

$$O_2 \text{ electrode: } O_2 + 4H^+ + 4e^- \rightarrow 2H_2O$$

$$H_2O_2 \text{ electrode: } H_2O_2 \rightarrow O_2 + 2H^+ + 2e$$

Biosensor based on a dehydrogenase

Enzymatic reaction:

$$\text{Substrate} + \text{NAD} \rightarrow \text{Product} + \text{NADH} \qquad \text{... (27.2)}$$

(In some reactions, NADP or the covalently bound PQQ is required as cofactors.)

Electrochemical reaction:

$$NADH \rightarrow NAD + H^+ + e^-$$

$PQQH_2$ can be re-oxidised by mediators:

$$PQQ(red) + Fe(CN)_6^{3-} \rightarrow PQQ(ox) + Fe(CN)_6^{4-}$$

Electrochemical reaction:

$$Fe(CN)_6^{4-} \rightarrow Fe(CN)_6^{3-} + e^-$$

Biosensors based on mediators as artificial electron acceptors

Enzyme (oxidase) reaction:

$$\text{Substrate} + \text{med(ox)} \rightarrow \text{Product} + \text{med(red)} \qquad \text{... (27.3)}$$

Electrode reaction:

$$\text{med(red)} \rightarrow \text{med(ox)} + e^-$$

Much effort has also been put into the optimisation of these sensors with respect to fabrication technology, sensitivity, stability, and reproducibility, and some enzyme electrodes that can be used for food analysis are commercially available.

FUNDAMENTALS OF AMPEROMETRIC DEVICES

Most of the electrochemical biosensors used for food analysis are based on amperometry, and therefore only this principle is described in this chapter. For amperometric measurements a defined potential is applied at a working electrode with respect to a reference electrode. This results in a current that can be related to the concentration of an electroactive substance in the solution. At low current densities, the most elementary electrochemical setup of only two electrodes (the working and reference electrodes) is sufficient.

At higher current densities the potential of the reference electrode may change with the current, and reproducible determination of the analyte is not possible. To avoid this problem, a third electrode, the counter electrode, is required. In this three-electrode setup the current is measured between the working and counter electrodes, whereas the potential is measured with respect to the reference electrode. The current is directly related to the rate of the overall process in the electrochemical cell. This process consists of several steps.

The essential step is the heterogeneous electron transfer between the electrode and the electroactive substance in the solution, leading to oxidation or reduction of the analyte at the electrode. The rate of this step is dependent on the applied potential, this dependence being, at least to some extent, a function of the standard potential of the redox couple. In many amperometric sensors a potential can be chosen such that the rate of electron transfer is so high that the current is no longer limited by this process. If this condition is met, the rate-determining step may be diffusion or mass transfer or a (bio)chemical process.

If diffusion is the rate-determining step, in a quiescent solution the current gradually decays to zero according to the Cottrell equation:

$$y = nFAc(D/\pi t)^{1/2} \qquad \text{... (27.4)}$$

where, y is the current, n the number of electrons involved in the process, F the Faraday constant, A the area of the electrode, c the concentration of the analyte, D the diffusion constant, and t the time elapsed since application of the potential.

This decrease in current is due to the slow spread of the diffusion layer into the solution, which causes a decrease in the concentration gradient. But this process stops due to random convection in the solution, which results in a constant thickness of the diffusion layer and a constant current. In a quiescent solution this process is random, and thus most amperometric measurements are done by controlling the mass transfer, for example, by stirring the solution and/or applying an additional membrane. This membrane has two functions: First, it can cause the analyte to partition across the membrane/solution interface, leading to an enhancement or diminution of the analyte at the electrode surface. This is mainly related to special properties of polarity or charge of the membrane and the analyte. Second, if the membrane is relatively thick or has only small or few pores, it is a substantial diffusion barrier, reducing the effective value of the diffusion coefficient. This leads to an increase in the response time of the sensor, but at the same time the current is independent of the mass transfer in the solution (i.e. on the stirring speed), as the diffusion through this membrane is the rate-determining step. If this condition is given, the sensor response is said to be 'diffusion-controlled'.

These principles are valid if a substance that is already present in the solution is to be oxidised or reduced. In an enzyme electrode the situation is slightly different, because the electroactive compound is formed by the enzymatic reaction. Thus, electrochemical and diffusion kinetics are superposed with enzyme kinetics. If the activity of the enzyme layer is high or the diffusion membrane is thick, diffusion is still the rate-determining step. With low catalytic activity or a thin diffusion membrane, the sensor response is dependent on the enzyme activity, and the electrode operates in the kinetic mode. In this mode the sensor response can be enhanced by using more enzyme or an enzyme preparation with a higher specific activity. These sensors are generally less stable than sensors operating in the diffusion-controlled regime, because the sensor response strongly depends on the amount of active enzyme.

Kinetic parameters of the enzyme, its K_m value and V_{max}, influence the sensor response at high concentrations of the enzyme substrate because, due to enzyme saturation, the electrode response is not affected by the substrate concentration. These concentrations cannot be determined in routine testing, and sample dilution is required. If a sensor is to be designed for the analysis of high analyte concentrations, the analytical range can be increased by increasing diffusion limitations, but at the same time the sensitivity of the sensor decreases.

ELECTRODE MATERIALS AND FABRICATION

An essential step at the electrode is the heterogeneous electron transfer between the electrode and the analyte; electrode materials are chosen that allow a maximum rate of this step. The choice usually depends on the analyte that is to be oxidised or reduced. Enzyme electrodes used for food analysis are based mainly on the determination of oxygen, hydrogen peroxide, mediators, and NADH, in combination with oxidases and dehydrogenases (Equations 27.1 to 27.3). The most common electrode materials for these analytes are Pt (H_2O_2, O_2) and various forms of carbon (NADH, mediators), requiring oxidation potentials of 600–800 mV vs. an Ag/AgCl electrode for the H_2O_2 and NADH oxidation and a potential of −600 to −800 mV vs. an Ag/AgCl electrode in an oxygen electrode. The potentials required for mediator oxidation depend on the mediator used.

As these potentials are rather high, other electroactive compounds in food, such as ascorbic acid, may also cause an electrochemical signal. These interferences can be reduced by suitable sensor operation. First, the applied potential has to be as low as possible. Second, interfering substances can be kept away from the electrode by additional membranes. Third, the difference between the current obtained with the enzyme electrode and an electrode without active enzyme can be measured. In the case of the oxygen electrode, the specificity can be enhanced by applying a gas-permeable membrane on top of the electrodes.

Special properties were observed at electrodes made of platinised graphite; these electrodes show very high catalytic activity, allowing the oxidation of H_2O_2 and NADH at potentials as low as 300–400 mV vs. an Ag/AgCl electrode. But at the same time, other interfering substances are also readily oxidised; thus reducing the potential has no advantage. For the reference electrode any suitable material can be chosen (Ag, Pt, Au), but the Ag/AgCl electrode is the most common. The area of the counter electrode should be about 10 times that of the working electrode, the material being chemically inert, such as a noble metal or carbon modifications, such as graphite, glassy carbon or carbon paste.

The fabrication of amperometric setups is very easy. First experiments can be done with metal wires having a defined electrode area, in combination with a commercially available reference electrode. But for practical applications all electrodes should be combined within one device, such as a commercially available oxygen electrode or flow-through device. An alternative to the sensor fabrication based on conventional technologies is the use of production technologies of the microelectronics industry, such as thin- and thick-film technology. Figures 27.1 and 27.2 show examples of an oxygen electrode made with thin-film technology and an electrode suitable for hydrogen peroxide determination made using thick-film technology. These technologies allow the large-scale production of sensors and at the same time the miniaturisation of the devices. Thus it is possible to fabricate sensor arrays, for example, that combine 2000 oxygen electrodes in an area of 1.6 mm^2.

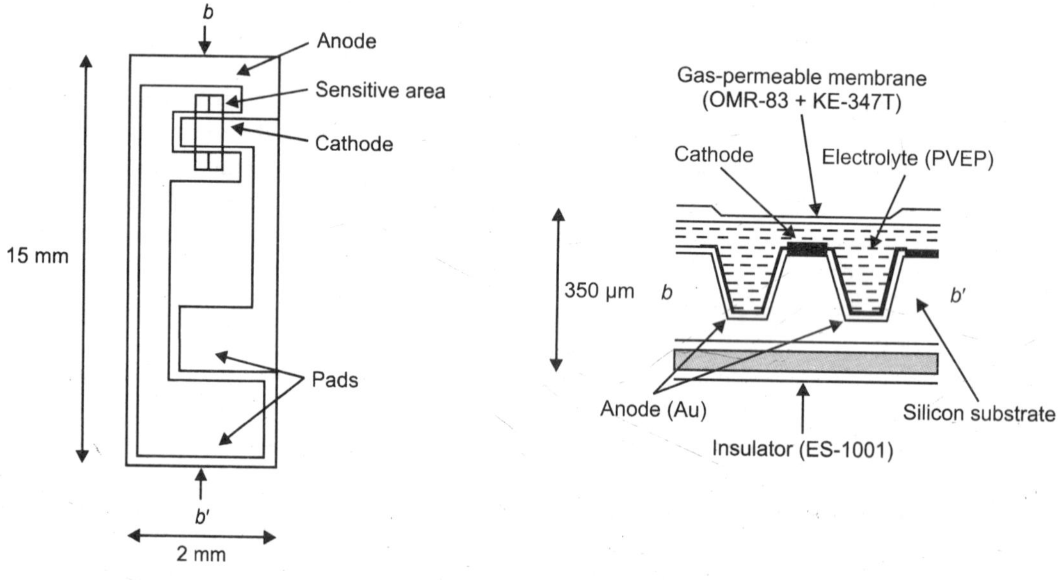

Fig. 27.1. Miniaturised Clark-type oxygen electrode.

DEVELOPMENT OF ANALYTICAL DEVICES

As explained earlier, the sensor response is dependent on several steps occurring in the analytical device. One of these steps is the biochemical reaction between the enzyme or micro-organism and the analyte, and thus one important parameter that has to be optimised is the method used to combine the enzyme or micro-organism with the amperometric device.

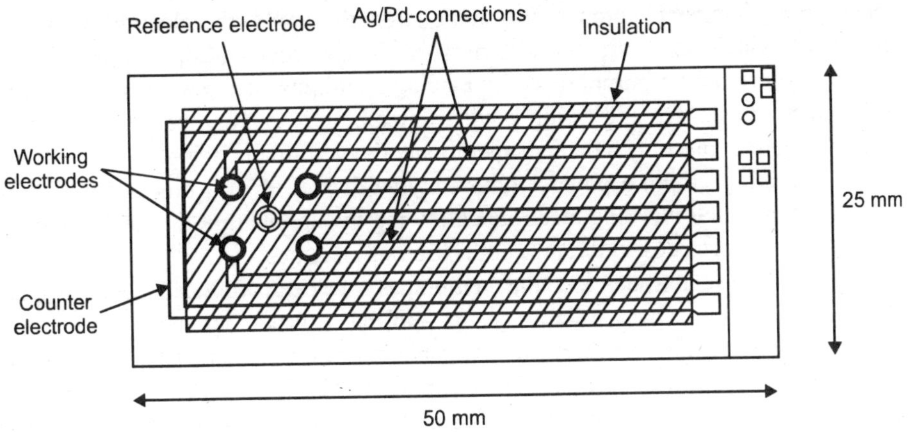

Fig. 27.2. Schematic diagram of a screen-printed electrode suitable for hydrogen peroxide oxidation.

A large variety of sensors have the enzyme or micro-organism immobilised directly on the (working) electrode. These sensors can be dipped in the sample and can form the basis of disposable sensors or portable devices. To achieve automated analysis, flow injection analysis (FIA) systems are developed using either enzyme or microbial electrodes as detectors or enzyme reactors in combination with amperometric flow-through cells. Both formats are based on the same biochemical and detection principles, and therefore both are described here.

Enzyme and Microbial Electrodes

Immobilisation methods used for enzyme electrodes that have been applied to food analysis can be classified into two major groups:

1. Methods that involve no covalent binding of the enzyme to the support. These immobilisation methods are based on adsorption or physical entrapment of the enzyme through van der Waals forces, ionic binding, or diffusion barriers.
2. Methods based on covalent binding of the enzyme to the support. This is achieved by the reaction between functional groups of the protein and the support material. Functional groups available in enzymes mainly originate from the amino acid's side chains. They include, for example, the ε-amino groups from lysine, carboxyl groups from aspartate and glutamate, sulphydryl groups from cysteine, and phenolic hydroxyl groups from tyrosine. In the case of glycoproteins such as glucose oxidase, aldehyde groups can be introduced by the oxidation of the sugar residue with $NaClO_4$. The support may be a suitable membrane or the electrode surface.

Compared to the immobilisation of enzymes, the immobilisation of living cells is much easier. Due to the size of the cells, a suspension can be filtered through a membrane with a suitable pore size (e.g. 0.45 μm). The filter is fixed to an oxygen electrode and covered with a dialysis membrane. Microbial membranes, which can be stored, are prepared by entrapping cells in a polymer such as polyvinyl alcohol or gelatine. The microbial membrane is fixed on the oxygen electrode and again covered with a dialysis membrane or a nylon net.

Immobilisation of enzymes through physical binding

Enzymes can be immobilised on a carrier without chemical binding if there are strong physical forces between the protein and the carrier, such as hydrophobic forces or ionic binding. As these forces are generally sensitive to changes in temperature, ionic strength, or pH, the electrodes are covered with dialysis membranes, which act as diffusion barriers for the enzymes and thus improve the stability of the sensors. The enzyme can be immobilised by placing an aliquot of a few microliters of the enzyme solution on the electrode surface and fixing it with a membrane, or it can be adsorbed at a membrane and this membrane held in place by a dialysis membrane fixed with an O-ring. Disposable sensors have been produced by mixing graphite powder with enzyme dissolved in a binder solution and screen-printing the paste on a planar working electrode, allowing the application of a technology suitable for the mass production of electrodes to the immobilisation of enzymes.

Alternatively to enzyme electrodes, enzyme membranes have been produced that have to be combined with the electrochemical setup by a suitable design and that can be changed when the enzyme loses activity. This principle is used in commercially available enzyme analysers, such as the Yellow Springs analysers, ESAT 6660 (Eppendorf-Netheler-Hinz GmbH, FRG), ECA 20 (Prüfgerätewerk Medingen, FRG), EKSAN analysers (USSR) and others. The enzyme is mixed with a gel or polymer, such as gelatine, spread on a (cellulose acetate) membrane, and covered with a second membrane. This principle can be modified by cross-linking the gel with glutaraldehyde, polyvinyl isocyanate, or gamma irradiation. Depending on the enzyme, some of these membranes proved to be very stable even in permanent use.

Immobilisation of enzymes through covalent binding

As described in the previous section for methods using physical binding, enzymes can be immobilised either directly on the working electrode or on a membrane fixed to the electrode assembly in a special way. To achieve covalent binding between the protein and the support, suitable functional groups are required.

Typical materials for working electrodes are noble metals and various forms of carbon. They can be activated for covalent binding of the protein by electrochemical oxidation at high potentials resulting in metal oxide groups (e.g. Pt-OH) and carboxylic groups, respectively. The metal surface has to be activated further by silanisation with aminopropyltriethoxysilane (ATS) or comparable compounds, modifying the metal surface with amino groups.

Glutaraldehyde, an aldehyde with several functional groups, is used as a cross-linking reagent between amino groups from the protein and the electrode surface. As the enzyme activity may be affected by this process, for example, if amino groups of the enzyme that are close to its active centre are involved, the enzyme is usually mixed with an inert protein such as bovine serum albumin to reduce the amount of amino groups from the enzyme involved in the reaction.

Carboxylic groups on the surface of carbon electrodes can be activated by dicyclohexyl carbodiimide according to the equation:

$$\text{Carrier-COOH} + C_6H_{11}\text{–N} = \text{C} = \text{N–}C_6H_{11} \longrightarrow \text{Carrier-COO–}\underset{\substack{|\\ \text{NH–}C_6H_{11}}}{\text{C}} = \text{N–}C_6H_{11} \quad \text{... (27.5a)}$$

$$\text{Carrier-COO–}\underset{\substack{|\\ \text{NH–}C_6H_{11}}}{\text{C}} = \text{N–}C_6H_{11} + H_2\text{N-enzyme} \longrightarrow \text{Carrier-CO–NH-enzyme} + C_6H_{11}\text{–NH–CO–NH–}C_6H_{11} \quad \text{... (27.5b)}$$

Electrochemical immobilisation methods have been developed specially for the preparation of microelectrodes. The best-known example is the electrochemical polymerisation of pyrrole, leading to a polypyrrole membrane, the thickness of which can be controlled by the electrolysis time. Enzymes can be entrapped in this membrane, or pyrrole can be modified so that enzymes can be bound covalently to the membrane. If it is possible to fix an enzyme membrane to the electrode, the enzyme can be immobilised on commercially available membranes that are already activated for enzyme immobilisation, such as Biodyne Immuno Affinity membranes, or on nylon membranes, which have to be activated prior to use with dimethyl sulphate at 75°C, lysine, and glutaraldehyde. In some cases it has proved to be advantageous to combine binding to the membrane with cross-linking of the enzyme molecule by glutaraldehyde.

Flow Injection Analysis (FIA)

Fundamentals

Flow injection analysis is a method of automated liquid handling. A defined sample volume is injected into a continuous carrier stream, which in most examples of biochemical analysis is a suitable buffer. If an enzyme reactor is inserted in the flow-through system, the enzyme substrate in the sample will be converted, and the reaction products can be measured with a suitable flow-through detector.

The enzyme reactor can be a small enzyme column containing small particles onto which the enzyme is immobilised or a nylon tube containing the enzyme on its inner surface. Typical materials for the enzyme column are glass, Kel-F, or any other inert polymer. The tubes generally have an inner diameter of 1–3 mm and are 1–10 cm long. The immobilised enzyme is held in this tube by nylon or stainless-steel netting. Alternatively, the enzyme can be immobilised in membranes that are placed directly on top of the electrode; thus an enzyme electrode is used as detector.

Typical sample volumes are 20–300 μl, and typical flow rates are 0.5–3.5 ml/min. During the transport through the flow-through system, the sample is mixed with the carrier stream through diffusion and convection processes, and the concentration profile of the analyte changes from a rectangle to a more or less tailored peak. The maximum concentration in this peak is lower than in the original sample. This effect is characterised by the dispersion D_{FIA}:

$$D_{FIA} = c_0/c_{max} \qquad \text{... (27.6)}$$

where c_0 is the original concentration in the sample and c_{max} the maximum concentration in the peak. The value of D_{FIA} depends on the setup and is typically greater than 2. The insertion of an enzyme column containing the enzyme immobilised on small particles will lead to greater dispersion than the use of a nylon tube with the enzyme immobilised on its inner surface. Additionally, the dispersion is decreased when the flow rate and/or sample volume are increased.

Flow injection analysis systems can be fully automated, allowing rapid analysis of a large number of samples. Depending on the dispersion of the system, flow rate, and enzyme activity, measuring frequencies of up to 60 samples/hr. can be achieved.

The typical form of the signals is a peak, with the peak height being related to the analyte concentration. The standard deviation of the data can be 2–5 per cent.

Immobilisation techniques used in FIA

As the purpose of the development of FIA systems is the automated analysis of a large number of samples, the device should exhibit considerable stability; this requires immobilisation of the enzyme through covalent binding to the support. Typical materials used in FIA systems as a support for the immobilisation of the enzymes are controlled pore glass (cpg), activated Sepharose, ion-exchange resins (e.g. Duolite), and oxiran-acrylic beads (e.g. Eupergit C). Of these, only the glass beads have to be activated prior to immobilisation of the enzyme. The other materials already contain suitable functional groups or the enzyme is bound through ionic binding.

In a first step the glass is activated through silanisation with aminopropyltriethoxysilane or other comparable compound which modifies the glass with amino groups. Particles that already contain amino groups are commercially available as aminocellulofine or aminopropyl-cpg. In the next steps glutaraldehyde and enzyme solution are added successively, leading to reactions of the aldehyde groups of glutaraldehyde (multifunctional aldehyde) with the amino groups on the glass and on the protein. These reactions are the same as those described for the binding of enzymes to electrode surfaces.

The amount of enzyme in an FIA system can be varied to a large extent by changing the size of the enzyme column. Thus, these systems usually work in the diffusion-controlled regime.

EXAMPLES OF ENZYME ELECTRODES APPLIED TO FOOD ANALYSIS

Evaluation of Fish (Meat) Freshness

Freshness is a very important aspect of fish quality, and rapid and simple methods for freshness estimation are urgently required in the food industries, especially in countries where raw fish is consumed. Fish quality is changed by processes that occur after the death of the fish. Immediately after the death of the fish, ATP (adenosine triphosphate) decomposition proceeds according to the sequence

$$\text{ATP} \rightarrow \text{ADP} \rightarrow \text{AMP} \rightarrow \text{IMP} \rightarrow \text{HxR} \rightarrow \text{Hx} \rightarrow \text{X} \rightarrow \text{U} \qquad \text{... (27.7)}$$

where, ADP is adenosine diphosphate, AMP is adenosine 5′-phosphate, IMP is inosine 5′-phosphate, HxR is inosine, Hx is hypoxanthine, X is xanthine, and U is uric acid. The relative concentrations of these compounds drastically change after the death of the fish. Hence, sensors for the specific determination of hypoxanthine and inosine were developed using the enzymes xanthine oxidase (XO) and nucleotide phosphorylase (NP) in combination with an oxygen electrode:

$$\text{Xanthine oxidase:} \quad \text{Hx} + 2\text{O}_2 \rightarrow \text{Uric acid} + 2\,\text{H}_2\text{O}_2 \qquad \text{... (27.8a)}$$

$$\text{Nucleoside phosphorylase:} \quad \text{HxR} + \text{P}_\text{i} \rightarrow \text{Hx} + \text{ribose} - \text{P}_\text{i} \qquad \text{... (27.8b)}$$

Both enzymes were co-immobilised, and the difference between the signal obtained with the addition of phosphate and that without phosphate was taken to determine HxR. Measurements were done in fish extracts, and a good correlation to the conventional enzymatic assay was observed.

More precise data are obtained if in addition to Hx and HxR the ratio of the various metabolites is determined:

$$K = (\text{HxR} + \text{Hx})/(\text{IMP} + \text{HxR} + \text{Hx}) \qquad \text{... (27.9)}$$

For several species, a substantial accumulation of inosine (HxR) has been observed and the K value approaches unity very quickly. For these fishes the hypoxanthine ratio (H value) was introduced as an indicator of fish freshness:

$$H = \text{Hx}/(\text{IMP} + \text{HxR} + \text{Hx}) \qquad \text{... (27.10)}$$

These ratios can be determined by a combination of three enzymes: 5′-nucleotidase (NT) converting inosine monophosphate to inosine, nucleoside phosphorylase, and xanthine oxidase (Equation 27.8). Several setups have been described and applied to the determination of K and H values of various fish species. The data were compared either to conventional enzymatic methods or capillary electrophoresis and showed a good correlation.

A second process related to the quality of fish is the degradation of proteins and amino acids by microbial, proteolytic, and decarboxylating activity to biogenic amines. Increasing levels of biogenic amines, of which putrescine and cadaverine are the most important, are regarded as chemical indicators of fish quality. Thus, enzyme electrodes based on immobilised putrescine oxidase (*Micrococcus rubens*) and thin-film hydrogen peroxide electrodes or oxygen electrodes were applied to amine determination during the storage of fish. The data showed the same trends as HPLC data.

Unspecific degradation of compounds present in fish samples by immobilised living spoilage-causing bacterium *Alteromonas putrefaciens* was observed with an oxygen electrode. The data obtained were compared to those obtained with the cultivation medium, and the ratio of the two correlated to the *K* value (Eq. 27.9). A linear correlation was observed during storage for two weeks.

Meat freshness could be related to amine formation due to microbial attack, comparable to fish quality evaluation. During the microbial process, glucose, which is normally present on the surface of the meat, is also consumed. Hence, a depletion of glucose at the surface relative to the bulk is suggested as an indicator of meat quality, and a biosensor array suitable for glucose determination at four different depths in the meat was presented. The data were compared to organoleptic properties.

Determination of Carbohydrates

Glucose sensors

Glucose sensors were the first biosensors described in the literature, and since then glucose has become the model analyte for new developments concerning enzyme sensors. Several thousand publications describe various formats, detection principles, production technologies, and applications of glucose sensors. Beyond that, glucose sensors are the basis of the determination of other carbohydrates, such as sucrose and lactose. In this chapter a few of the publications related to glucose determination in food are reviewed to demonstrate general principles.

All systems described here are based on the catalytic oxidation of glucose by oxygen using glucose oxidase and the electrochemical oxidation of the resulting hydrogen peroxide at platinum or platinised electrodes at potentials ranging from 350 to 700 mV vs. an Ag/AgCl reference electrode.

$$\text{Glucose} + O_2 \xrightarrow{\text{Glucose oxidase}} \text{Gluconolacton} + H_2O_2 \qquad \text{... (27.11a)}$$

Electrode reaction:

$$H_2O_2 \rightarrow 2H^+ + O_2 + 2e^- \qquad \text{... (27.11b)}$$

Flow injection analysis was used for the development of a three-parameter device, allowing the simultaneous determination of glucose, ethanol, and lactate. This system was based on the immobilisation of glucose oxidase on aminocellulofine, which was packed in a glass tube (2 mm *id* × 10 cm).

The other enzymes were immobilised in the same way, and three flow injection systems were combined to form one device by a 16-way switching valve. Oxidation of H_2O_2 was accomplished in a multichannel flow-through cell. A measuring frequency of 30 samples/hr. could be achieved. This system was applied to the analysis of beverages, which had to be diluted 125-fold (white wine) and 200-fold (sake). The correlation coefficient for data obtained with the photometric enzymatic test was 0.97.

As an alternative to automated devices, disposable sensors were developed. They were fabricated either as a needle-type sensor or as screen-printed electrodes suitable for the simultaneous determination of several compounds, because four working electrodes could be combined in one device. The enzyme glucose oxidase was immobilised by cross-linking with glutaraldehyde or by adsorption in an enzyme paste. The paste consisted of graphite modified with the mediator tetrathiafulvalene or finely divided platinum, polyvinylpyrrolidone or hydroxyethylcellulose as the binder, and the enzyme, and could be applied to the working electrodes by screen-printing. The sensors could be stored for up to 6 months and were applied to the analysis of fruit juices. The correlation coefficient for data obtained with HPLC was 0.989 for the needle-type sensors. The screen-printed electrodes showed standard deviations of 5–10 per cent.

A combination of thick-film technology and automated analysis is described by Scholze. A 12-channel enzymatic hybrid biosensor (Biobrid) is used as a detector in an automated flow-through system. The sensor response characteristic did not change during 2 months of permanent use. The system was again applied to juice analysis, and data were compared to the photometric enzymatic test.

Mason used the Model 27 industrial analyser for glucose determinations in cereals and compared data to those obtained with the AOAC HPLC method. Deviations of up to 30 per cent were found, with the glucose concentration determined by the enzyme electrode being generally higher than the values obtained with HPLC.

Sucrose sensors

Sucrose is the most important disaccharide consisting of glucose and fructose. It is not a reducing sugar, and thus it is hydrolysed even prior to chemical analysis. Hydrolysis can be done either chemically by acidic solutions or enzymatically by the enzyme invertase:

$$\text{Sucrose} + H_2O \xrightarrow{\text{Invertase}} \text{Fructose} + \alpha\text{-D-glucose} \qquad \text{... (27.12)}$$

Thus, a combination of invertase with a glucose sensor serves as a sucrose biosensor. To achieve quick conversion of α-D-glucose to β-D-glucose, the substrate of glucose oxidase, mutarotase is added as a third enzyme. The three enzymes were either coimmobilised on the electrode or invertase was added to the sample and a glucose determination was done after an incubation time of 30 minutes. The sensors were applied to sucrose determinations in various juices. As they measure the concentration of both glucose and sucrose, a glucose electrode must also be used to measure the glucose present in the sample. The difference between the two is the sucrose concentration according to the calibration curves obtained for sucrose and glucose.

Lactose sensors

Lactose is present in milk at a concentration of 4–7 per cent and is one of the reducing disaccharides. It is hydrolysed to D-galactose and α-D-glucose by the enzyme β-galactosidase:

$$\text{Lactose} + H_2O \xrightarrow{\beta\text{-Galactosidase}} \text{D-galactose} + \text{D-glucose} \qquad \text{... (27.13)}$$

Hence, lactose sensors can be made by a combination of β-galactosidase with a glucose sensor. Several β-galactosidases from different sources have been described, but usually the enzyme from *E. coli* is used. The enzymes have been immobilised in a gel, on a membrane, and on activated platinum or glassy carbon electrodes. Hamid studied the amplification of the signal by the combination of several enzymes.

The basic enzymatic reaction was the lactose hydrolysis described by Eq. 27.13. Glucose was oxidised using glucose oxidase (Eq. 27.11). In addition, galactose, which is also formed during the hydrolysis of lactose, was oxidised by the catalysis of galactose oxidase:

$$\text{D-Galactose} + O_2 \xrightarrow{\text{Galactose oxidase}} \text{galactohexodialdose} + H_2O_2 \qquad \text{... (27.14)}$$

Enhancement of the lactose signals was also attempted by adding mutarotase to accelerate the conversion of α-D-glucose to β-D-glucose. Sensors based on various combinations of these enzymes were characterised with respect to analytical range and sensitivity. The electrode containing β-galactosidase, mutarotase, and glucose oxidase proved to be the best one.

Galactose oxidase is not only suitable for the oxidation of galactose but also catalyses the oxidation of lactose:

$$\text{D-Lactose} + O_2 \xrightarrow{\text{Galactose oxidase}} H_2O_2 + \text{oxidation product} \qquad \text{... (27.15)}$$

Thus it is possible to develop lactose electrodes using only a single enzyme in combination with a hydrogen peroxide-sensing electrode. Characteristic features of the various lactose sensors are summarised in Table 27.1. All sensors were based on H_2O_2 oxidation and applied to the lactose determination in dairy products. Generally, no sample pre-treatment other than a dilution step was required. Only Hamid removed the proteins from the milk by a treatment with Carrez solution.

Table 27.1. Characteristic features of lactose electrodes applied to food analysis.

Device	*Linear range (mm)*	*Stability*	*Response time*	*Application*	*Relative standard deviations (%)*
EKSAN-Lac-I	5–220			Milk	2–5
ECA 20	0.02–3.0	> 3000 assays	100/hr.	Dairy products	2
Electrode		1 month	2 min.	Milk	
Thick-film electrode	<3			Dairy products	5
FIA	0.1–1.5	50% after 7 weeks	90 sec.	Milk	2
FIA	0.001–2		20 sec.	Milk	<1
Universal sensors (galactose oxidase)	0.01–100	1 month	<2 min.	Milk	<1
Yellow Springs (galactose oxidase)	–		2 min.	Cheese	2

Genetically manipulated bacteria *E. coli* K-12 recombinant PQ-37 and glucose oxidase were used by Svorc for the construction of a hybrid lactose sensor because of the hyperproduction of β-galactosidase by *E. coli* effected by a genotoxic agent. The major practical limitation of sensors that use whole cells is the presence of other enzymes in the cell. If these enzymes produce a common product, the selectivity of the sensor is adversely affected. The sensor proposed by Svorc could be used for the determination of lactose, glucose, mannose, sucrose, and maltose. But as these carbohydrates are not present in milk, it could be applied to lactose determinations in various dairy products. The correlation to data obtained with a photometric test was sufficient.

Fructose sensors

To date there is no fructose oxidase available; that is why the most commonly used enzymatic assay uses the enzyme sequence of hexokinase, phosphoglucose isomerase, and glucose-6-phosphate dehydrogenase:

$$\text{Fructose} + \text{ATP} \rightarrow \text{Fructose-6-phosphate} + \text{ADP} \qquad \text{... (27.16a)}$$

$$\text{Fructose-6-phosphate} \rightarrow \text{Glucose-6-phosphate} \qquad \text{... (27.16b)}$$

$$\text{Glucose-6-phosphate} + \text{NADP}^+ \rightarrow \text{NADPH} + \text{H}^+\text{-6-phospho-}\delta\text{-lactone} \qquad \text{... (27.16c)}$$

These reactions require ATP and NADP+ as cofactors. The measurement is a photometric, fluorimetric, or amperometric determination of NADPH.

Yamada reported on a fructose dehydrogenase that catalysed the oxidation of D-fructose to 5-keto-D-fructose in the presence of a mediator. It was found that this enzyme contains PQQ and heme *c* as redox active sites, PQQ being reduced by fructose. Re-oxidation can be achieved either electrochemically, which does not require mediator, or with artificial electron acceptors such as hexacyanoferrate(III). To achieve direct electron transfer from the enzyme to the electrode, a carbon paste electrode was prepared by mixing graphite powder and liquid paraffin and adsorbing the enzyme physically. The electrode surface was covered with a dialysis membrane. This sensor lost 70 per cent of its activity within two weeks. Measurements were done at 0.5 V against an Ag/AgCl electrode, so ascorbic acid was a strong interferent if it was not removed by ascorbate oxidase, which was co-immobilised with fructose dehydrogenase. The sensor was applied to fructose determination in fruit and showed good correlation with the conventional enzymatic assay. Using hexacyanoferrate(III) as electron acceptor, this mediator has to be added to each sample. Alternatively, an FIA system was developed in which the mediator was dissolved in the carrier stream. Interference due to ascorbic acid was avoided by using a precolumn of ascorbate oxidase. The relative standard deviations of the two systems were 0.68 and 1.8 per cent, respectively, and good correlations to the AOAC method and the enzymatic assay were observed when these systems were applied to the analysis of fruits and fruit juices.

Another approach to fructose determination is suggested by Park and Kim, who used living cells of *Zymomonas mobilis* immobilised on top of a pH electrode. The cells were made permeable with toluene to use the enzymes glucose-fructose oxidoreductase and gluconolactonase inside the cells. The sensor was specific to both glucose and fructose.

Sensors for the Determination of Acids in Food

Various organic and inorganic acids are present in food and influence quality, flavour, and the possibility of microbial growth. These compounds are determined by various chemical, chromatographic, or biochemical methods, including enzymatic assays based on soluble enzymes and enzyme electrodes or FIA systems based on immobilised enzymes. All enzyme electrodes described in this chapter are based on oxidases in combination with oxygen, hydrogen peroxide, or mediator determination.

Inorganic acids

The most important inorganic acid found in food and determined by biosensors is sulphite. Sulphite is used as a preservative and added to food products to inhibit bacterial growth and prevent oxidation. It can be found in beverages and dried products.

Enzymatic determination of sulphite is usually based on the oxidation of sulphite to sulphate, catalysed by sulphite oxidase:

$$SO_3^{2} + O_2 + H_2O \xrightarrow{\text{Sulphite oxidase}} SO_4^{2} + H_2O_2 \qquad \text{... (27.17)}$$

Sulphite oxidase has been immobilised on Sepharaose B, in a gel between membranes, and on a preactivated immunodyne nylon membrane and combined with oxygen or hydrogen peroxide determinations. Characteristic features of the different sensing systems are summarised in Table 27.2. The sensor proposed by Karube did not contain the purified enzyme but a hepatic microsome of rat liver S9 fraction immobilised by adsorption on an acetylcellulose membrane on top of an oxygen electrode.

Other anions of inorganic acids determined by enzyme electrodes are oxalate and the phosphate ion. Oxalate was oxidised to CO_2 by oxalate oxidase, which was immobilised on Biodyne immuno affinity membranes. These membranes were used instead of the glucose oxidase membranes in the Glucoproceseur. The sensor showed no decrease in sensitivity over a 4-month period. It was applied to oxalate determinations in juices, chocolate, spinach, and sorrel, and data were compared to those obtained with a spectrophotometric method. A correlation coefficient of 0.999 was given with the regression equation $y = 1.037x - 0.565$.

Table 27.2. Characteristic features of electrochemical biosensors for the determination of sulphite.

Detection principle	*Formate*	*Linear range*	*Stability*	*Application*	*Relative standard deviations*
O_2	Electrode	60–340 μM	6 months (stored at −20°C)	–	–
H_2O_2	FIA	1–10 ppm	20 days (stored at 5°C)	White wine	2.3% ($n = 10$)
H_2O_2	Electrode	5–550 μM	53 days (stored at 4°C)	Beverages and dried food	3.3% ($n = 8$)
O_2	Electrode of provesta bioanalyser	–	–	Beer	–
O_2	Electrode of provesta bioanalyser	Lower detection limit: 10 ppm	84 days (stored at 4°C)	Dried food, wine, juice	1%

The determination of phosphate is based on the same enzymes mentioned for fish freshness estimation. Inosine is converted in the presence of inorganic phosphate to hypoxanthine and ribose-1-phosphate by catalysis of nucleoside phosphorylase (Eq. 27.8); hypoxanthine again is oxidised by xanthine oxidase (Eq. 27.8). Both enzymes were co-immobilised on an immunodyne nylon membrane and attached to the tip of a hydrogen peroxide electrode. The assay was done in an FIA system with the carrier buffer containing 1 mm inosine. The system had a lower detection limit of 1.25 μM phosphate with a throughput of 10 samples/hr. Phosphate was determined in various foodstuffs requiring dilution façtors of 25–1000.

Organic acids

Among the organic acids present in food, glutamic acid, lactic acid, and ascorbic acid are of special importance.

L-Glutamate is a well-known taste component that is widespread in food and is also used as a seasoning or flavour enhancer in food products. All enzyme electrodes described here are based on glutamate oxidase from *Streptomyces* sp. catalysing the oxidation of L-glutamate to α-ketoglutarate:

$$\text{L-Glutamate} + O_2 + H_2O \xrightarrow{\text{Glutamate oxidase}} \alpha\text{-ketoglutarate} + NH_3 + H_2O_2$$

These electrodes were applied to glutamate determination in soya sauce, seasonings, and various other food products. The sensing systems were designed for automated analysis or as disposable sensors using mainly oxygen or hydrogen peroxide electrodes as detectors. But it is also possible to combine glutamate oxidase with a mediator-modified electrode, allowing glutamate determination even in a deoxygenated medium, which might be interesting for the application of glutamate sensors to *in vivo* monitoring in the brain. Very sensitive devices could be developed by co-immobilisation of glutamate oxidase and glutamate dehydrogenase:

$$\text{L-Glutamate} + NADP^+ + H_2O \xrightarrow{\text{Glutamate dehydrogenase}} \alpha\text{-ketoglutarate} + NH_4^+ + NADPH$$

The two-enzyme system causes cycling of L-glutamate, α-ketoglutarate, and NH_4^+ between the two enzymatic reactions in the presence of NADPH. But NADPH and α-ketoglutarate have to be added to each sample; therefore this method is recommended only for special applications. A summary of glutamate sensors is given in Table 27.3.

Table 27.3. Summary of sensing devices for glutamate determination.

Detection principle	*Linear range*	*Stability*	*Relative standard deviations (%)*	*Sampling rate (minute)*	*Application*
O_2, FIA	15–250 mg/l	–	<3	2	Consomme soup
H_2O_2, FIA	0.01–0.3 mm	15 days	1.7	1.5	Soup, soya sauce
H_2O_2, FIA	0.005–1 mm	6 months	<0.7	1	Seasoning
O_2	–	–	–	2	Soya sauce
O_2, H_2O_2	0.05-100 μM	50 days	–	1	Food products
O_2	0.1–1 mm	100 days	–	0.3	Soya sauce
H_2O_2, thick-film	29–290 mg/l	9 days	3	5–7	Soup, sauce

As the determination of lactic acid is important for medical applications, for example, cardiology, several enzyme electrodes designed for this purpose are commercially available, and most lactate electrodes are applied to serum and food analysis. All sensors are based on lactate oxidase from *Pediococcus* sp. catalysing the oxidation of lactate to pyruvate. Special attention to an optimised immobilisation procedure was paid by Hajizadeh, who immobilised the enzyme on a platinised graphite electrode in a matrix of poly(vinyl alcohol) cross-linked by gamma irradiation. They obtained a fast-response sensor (10–50 seconds) with a stable sensor response over a period of 80 days. Using an automated system, a measuring frequency of 40 samples/hr was obtained, and even in multianalyte FIA systems the time required for one sample was only 70 seconds in one report and 4 minutes in another. All systems showed good correlations to enzymatic spectrophotometric assays when applied to dairy products or white wine, with relative standard deviations of 1.1–3.2 per cent.

One of the most important interferents for amperometric determinations in food products is L-ascorbic acid because it can readily be oxidised at an H_2O_2 electrode to L-dehydroascorbic acid. Nevertheless, quantitative determination of L-ascorbic acid in food was achieved by using ascorbate oxidase in combination with an oxygen electrode. Another approach was suggested by Collin, who developed an FIA system with ascorbate oxidase immobilised in a packed-bed reactor. Collin took the difference of the signals obtained at a glassy carbon electrode at 850 mV vs. Ag/AgCl when the sample was passed first through a reactor with inactivated enzyme (interferents + ascorbic acid) and then through a reactor containing ascorbate oxidase, thus removing ascorbic acid from the sample (the signal is due only to interferents). A sample throughput of 15 samples/hr could be achieved.

ENZYME ELECTRODES FOR ALCOHOLIC COMPOUNDS

Ethanol Biosensors

Ethanol is one of the most important analytes in the food industry, and tax regulations require exact determination of ethanol content, particularly in spirits. Thus, a variety of biochemical ethanol-sensing systems have been proposed. They are based on the oxidation of ethanol either by oxygen using alcohol oxidase as a catalyst or by alcohol dehydrogenase with NAD required as cofactor.

Sensor systems using alcohol oxidase are based on the combination of the immobilised enzyme with either an oxygen electrode or hydrogen peroxide determination. The sensors were designed as disposable sensors based on screen-printing technology and immobilising the enzyme with glutaraldehyde. The standard deviation of sensors made by the same process was 5 per cent. Mason presented results of wine analysis using an enzyme membrane electrode in the Yellow Springs analyser. The enzyme was mixed with dithioerythritol and glutaraldehyde and sandwiched between a polycarbonate membrane and a cellulose acetate membrane. The overall coefficient of variation was 0.89 per cent when applied to various red and white wines. A fully automated device was introduced by Künnecke and Schmid. They achieved in-line dilution of the samples with a gas diffusion unit. By the choice of suitable membranes the linear range of the system could be adapted to the analytical problem, and samples of beer, wine, and spirits could be investigated without pre-treatment or dilution.

One drawback of sensors based on alcohol oxidase is the lack in specificity because methanol is the main substrate of the enzyme. Hence, alcohol sensors based on alcohol dehydrogenase were developed, this enzyme being mainly specific to ethanol. However, for these sensors the problem arises that the sensor has to be supplied with the cofactor NAD. This can be achieved by entrapment of NAD and the enzyme on top of the electrode behind a dialysis membrane. The major disadvantage of this sensor is its lack of stability, as it loses 30 per cent of its activity within 5 hours and has to be freshly prepared before use. This should be improved by an all-solid-state sensor, which was made by mixing NAD, alcohol dehydrogenase, and NMP^+TCNQ^- (used as mediator and electrode material) with a solution of PVC in tetrahydrofuran. This paste was pressed to form electrode pellets. These electrodes still lost 10 per cent of their activity within the first hour, but they could be stored for at least 2 weeks. Both sensors were applied to the analysis of wine. Biosensing systems based on alcohol dehydrogenase were more successful when based on the principle of flow-injection analysis with NAD added to the carrier solution. They also used soluble alcohol dehydrogenase stabilised with 2 per cent poly(ethylene glycol) as an alternative to an immobilised enzyme column.

More details concerning the ethanol sensors are given in Table 27.4.

Table 27.4. Summary of various enzyme electrodes for alcoholic compounds.

Analyte	*Enzyme*	*Detection principle*	*Stability*	*Application*	*Relative standard deviations (%)*
Ethanol	ADH	NADH at an NMPTCNQ electrode	10% drop per hr; 2 weeks storage	Wine	–
Ethanol	ADH	NADH at an NMPTCNQ electrode	30% loss in 5 hrs	Beverages	3.5
Ethanol	ADH	NADH, FIA	15% loss in 12 hrs	Wine	6
Ethanol	AOD	H_2O_2, FIA	8000 assays	Beverages	<0.5
Ethanol	AOD	H_2O_2, YSI	10 days	Wine	0.89
Ethanol	AOD	O_2	6.5 days	Beer	2.9
Ethanol	AOD	H_2O_2, thick-film	–	Beverages	–
Glycerol	GDH	NADH at Pt, FIA	> 60 days	White wine	0.5

(Contd...)

Analyte	*Enzyme*	*Detection principle*	*Stability*	*Application*	*Relative standard deviations (%)*
Tannin	Laccase	O_2, batch and FIA	500 assays	Tea	–
Cholesterol	Cholesterol oxidase	O_2, organic phase	50% after 4 or 5 assays	Butter, margarine	–

ADH = alcohol dehydrogenase; AOD = alcohol oxidase; and GDH = glycerol dehydrogenase.

Biosensors for other alcohols

There are other alcohols present in food besides ethanol, occurring as additives or as by-products during the process of food preparation. Glycerol is produced in wine as a by-product of alcoholic fermentation from sugar and is thought to be a quality factor of wine. Thus, an FIA system based on an enzyme reactor with glycerol dehydrogenase and amperometric NADH determination was proposed and applied to wine analysis. Wine samples had to be diluted 100-fold prior to analysis. Each sample could be analysed within 2 minutes. Deviations to the photometric enzymatic test of up to 6 per cent were observed.

Laccase catalyses the oxidation of polyphenols by oxygen with a broad substrate specificity. Thus, it was immobilised on DEAE-cellulose and combined with an oxygen electrode, either for operation in the batch mode or in flow injection analysis. These systems were applied to tannin determination in tea extracts. Biosensors for the determination of compounds that are not soluble in water are seldom described because of the additional problems of diffusion of the compound to the active centre of the enzymes. These problems are solved if the sensor is operated in an organic solvent. A suitable setup was proposed by Hall and Turner for cholesterol determination. Cholesterol oxidase was adsorbed on alumina on top of an oxygen electrode, and the assay was performed in chloroform/hexane. The sensor showed very poor stability; 50 per cent of the activity was lost after four or five assays. This may be due to the inactivation of the enzyme by H_2O_2, which is not soluble in the organic phase and will thus accumulate around the enzyme. The sensor was applied to the analysis of butters and margarines, giving results of the expected order of magnitude, but correlation to the photometric enzymatic test was very poor.

Characteristic features of these sensors are summarised in Table 27.4.

Various Biosensors

The artificial sweetener aspartame (*N*-L-α-aspartyl-L-phenylalanine methyl ester), known by the brand name NutraSweet, is most commonly determined by chromatographic procedures usually used for peptide analysis, such as HPLC and thin-layer chromatography. Two biochemical approaches are presented here that use different enzyme sequences.

The peptide bond in aspartame can be cleaved with a suitable enzyme purified from intestinal mucosa, yielding aspartate, the phenylalanine methyl ester, and phenylalanine. Aspartate can then be converted to glutamate, which was determined by a glutamate electrode based on glutamate oxidase:

$$\text{Aspartame} \xrightarrow{\text{APBC enzyme}} \text{Aspartate} + \text{phenylalanine/phenylalanine methyl ester}$$

$$\text{Aspartame} + \alpha\text{-ketoglutarate} \xrightarrow{\text{Aspartate aminotransferase}} \text{L-glutamate} + \text{oxaloacetate}$$

(APBC enzyme is an aspartate peptide bond-cleaving enzyme.)

Both enzymes were immobilised separately on aminopropyl glass beads packed into a 10-cm column of tygon tubing and integrated sequentially in an FIA system. Glutamate oxidase was immobilised on an immunodyne membrane and attached to the tip of a hydrogen peroxide electrode. Each assay could be performed in 7–8 minutes, the standard deviation being 2.2 per cent. The response of the sensor was stable for a period of 30 hours of permanent use. Data obtained in dietary products were in good agreement with those given by the manufacturer.

In the alternative method, the methyl ester group of aspartame is cleaved by α-chymotrypsin, and the resulting methanol is determined using an enzyme electrode. This method is based on the physical entrapment of alcohol oxidase on an oxygen electrode. Prior to hydrolysis the alcohol content of the sample must be determined, and the difference from the value obtained after hydrolysis is taken for aspartame determination. The relative standard deviation of this method is 0.83 per cent. This sensor has been applied to aspartame determination in various beverages.

Another food additive is lecithin, which is used as an emulsifying compound. Lecithin is a phospholipid that can be hydrolysed by phospholipase D, producing choline. Thus the proposed method is based on two enzymes, phospholipase D and choline oxidase. The former was used as soluble enzyme, and the latter was immobilised on nylon net and fixed on an oxygen electrode. The sensor was applied to the analysis of various drugs and foods, including chocolate, soya flour, and egg yolk, and data were compared to those obtained with an enzymatic spectrophotometric method. The correlation was acceptable, but the data differed from each other by up to 30 per cent.

The quality of milk is changed by the action of bacterial and milk enzymes, liberating, for example, short-chain fatty acids. These acids are responsible for the rancid flavour of stale milk. They can be determined by gas chromatography, or the sum of all acids can be measured by the acid degree value. A more specific determination is suggested by Ukeda using immobilised cells of *Arthrobacter nicotiana* on an oxygen electrode. With this microbial sensor inserted in an FIA system, a sample throughput of 20/hr could be achieved. The sensor was applied to fresh milk samples and milk samples stored at room temperature. A linear correlation was observed to the total concentration of fatty acids ($C_{4:0}$–$C_{12:0}$) determined by gas chromatography. Agitating of milk led to an exponential increase in the sensor response with storage time.

SECTION V

Special Topics

Chapter 28

Legal Considerations

INTRODUCTION

Enzymes are primarily ingested from food. All fresh fruits and vegetables contain active enzymes. Products of fermentation, including cheese, kefir, yogurt, raw sausage, sauerkraut, pickles, beer, and wine, contain both the native enzymes and the micro-organisms that produced the enzymes. These include a number of nonpathogenic microbes such as lactobacilli, micrococci, yeasts, streptomycetes, and moulds. Generally the fermentation products are not considered health hazards; many food chemistry studies have ascertained that such products contain no detectable amounts of toxic substances such as, for example, mycotoxins. A good example of the concept that enzymes used in food processing are nothing other than normal cell constituents was provided by Frank. Curing young Matjes herrings involves the native muscle enzymes and microbial exoenzymes. The latter are not native to the fish, but cannot be avoided because the fillets are not sterile. Thus, they become auxiliary agents in processing the Matjes herring or the current substitute, the Matjes-style herring fillet.

The active enzymes that are consumed are proteins and as such are equally harmless. Boiling food inactivates the enzymes, and the enzyme protein is denatured in the same way as other proteins. This may be the reason why some regulatory groups list the enzymes with food products. Enzyme technology utilises enzymes from animals, plants, and micro-organisms. In most of the processes discussed, the enzymes are heat-inactivated at the time the reaction is terminated. Traces of denatured protein that remain are catabolised during gastro-intestinal passage. On the other hand, enzyme processes have been described during which a carrier-bound enzyme comes in contact only with the substrate, and does not remain as part of the product. In these instances, the enzyme is considered a technical processing aid. However, other views hold that enzymes are technical aids even when they are denatured and remain in the food product in small quantities (~1 ppm).

Regardless of whether enzyme preparations are defined as food additives or technical aids, it must be established that they are not health hazards. Naturally, it could be possible to purify the enzymes to such an extent that only the 'pure enzyme protein' would be used. Such purified enzymes, used in medicine and laboratory analysis, are very expensive, and therefore the cost is prohibitive for processing food. Denner describes an example: 'It would be a terrible misuse of public funds if one would salt the streets in winter with reagent-grade table salt, and *vice versa* an extreme violation of public health to add rock salt that contains significant amounts of heavy metal to food.' Thus, specifications for food and food additives that consider the health factors first and the economics second will have to be developed.

REGULATORY REQUIREMENTS FOR ENZYME PREPARATIONS

The purpose of this brief review on the legal aspects of enzyme utilisation is to inform the reader about the measures that the scientific community and legislators have taken to provide maximum safety for both the user and the consumer of enzyme preparations. It should also be pointed out that the regulations for enzyme preparations and their use are by no means uniform throughout the world or even within the countries of the European Economic Community (EEC), therefore, much work is still needed to institute uniform regulations.

Employing recombinant DNA (rDNA) technology for the production of enzymes initiated discussion on the potential health hazards and the safety of such enzymes in the consumer marketplace. The EEC has published safety guidelines on the contained use of genetically modified micro-organisms, their deliberate release into the environment, and risk management of workers exposed to such organisms. Toet reported on the effects of rDNA technology relative to the safety requirements for enzyme preparations. According to AMFEP (Association of Microbial Food Enzyme Producers), it is necessary to regulate only those micro-organisms—genetically engineered or not—that serve as a source of enzymes used in food production and processing with no limitation other than good manufacturing practice (GMP).

WHO IS INVOLVED WITH ENZYMES?

Public health agencies in several countries have made an effort to establish regulatory guidelines for the use of enzymes in concert with legislation on food additives. The regulations vary greatly and often are not published in detail. Therefore, it was most commendable that the FAO/WHO Joint Expert Committee on Food Additives (JECFA) dealt with the problems associated with the safety of enzymes. A group of toxicologists discussed questions of toxicity and purity for enzymes used in food production. Reports from the Committee sessions held in 1971, 1981 became the guidelines for national laws or regulations. Session 21 generated the criteria for evaluating the health risks of enzymes according to their origin (FAO/WHO, 1978).

It was determined that enzymes from animal organs or plant parts require no toxicological studies for approval. This primarily concerns the industrial enzyme preparations from the pancreas, pepsin, rennet, papain, and bromelain. Some toxicological testing may be required for enzyme preparations originating from micro-organisms that occur as contaminants in food but generally are viewed as harmless. However, enzyme preparations from cultures of *Bacillus subtilis*, *Aspergillus niger*, and *A. oryzae* must be exceed for acute toxicity as well as in a short-term toxicological test. Long-term toxicological tests are required for enzymes from less well-known organisms. All pathogenic micro-organisms have been excluded from use in production systems. Furthermore, the Committee has established guidelines for purity of industrially used enzyme preparations.

The Code Committee on Food Additives (CCFA) is concerned itself with food additives. An open-ended list of additives and technical aids is continually examined by the Committee's study groups. If a new substance is added to the list, the CCFA can ask the Expert Committee (JECFA) for further testing.

The enzyme preparations are then listed with their general purity requirements, specifications, and analytic methods in the Food Chemical Codex (NAS, 1981).

The General Directorates (GD) in the EEC coordinate the various enzyme regulations in member countries. Responsibilities lie with GD III (internal market and industry), GD IV (agriculture), and GD XII (science, research, and technology). The latter oversees the developments in biotechnology, an important future field in the European Economic Community (EEC). Currently, no horizontal EEC directive regarding food enzymes exists; there are, however, a series of vertical directives that include requirements for the use of enzymes (e.g. Council Directives 79/337/EEC, 90/219/EEC, 90/220/EEC, 90/679/EEC).

One of these is the Fruit Juice Directive (EEC 726/75), which states in Article 4 that only pectinases, proteases, and amylases are allowed in juice production. This restricted enzyme use is based on an approved list that was adopted by legislation in Denmark and Germany. In contrast, the United Kingdom (UK) did not adopt Article 4, and enzymes can be used in juice production that are not contained in Article 4 of the EEC Directive. Cellulase, for example, can be employed in juice production in the United Kingdom.

The EEC Wine Regulation (EEC 79/337) permits treatment of grapes and grape must (unfermented or partially fermented) with pectolytic enzymes. An application for approval of β-glucanases in winemaking has been submitted.

The Association of Microbial Food Enzyme Producers (AMFEP), founded in 1977, represents an organisation of enzyme producers in western Europe. It has 12 members, including Germany (2), the United Kingdom (2), Finland (2), Belgium (1), Netherlands (1), France (1), Denmark (2), and Switzerland (1). Several study groups deal with questions of toxicity testing, microbial safety, and approval in general. Guidelines for the production and application of enzymes were published in 1983 in a brochure entitled Regulatory Aspects of Microbial Food Enzymes (AMFEP, 1988) and presented to the EEC and the individual governmental agencies. The Association of Manufacturers of Animal-derived Food Enzymes (AMAFE) is an organisation with members in Austria, Denmark, France, Germany, Holland, Italy, and Sweden. A proposal that will regulate enzymes from animal organs and edible plant parts for use in the food industry has been drafted.

TOXICITY AND ALLERGIES

Pure enzymes *per se* are not toxic, but as all proteins they can cause allergies. There is no reason why the consumption of enzyme proteins should result in more allergies than the consumption of any other protein. However, inhaled enzymes, especially protease concentrates, are highly effective allergens. Allergies from enzymes are in most cases inhalation allergies and rarely contact allergies. A sensitisation to enzymes that have been used as food additives is rare because:

1. The amounts added are small, in the range of a few parts per million of enzyme protein.
2. Only in a few cases are the enzymes consumed with food not thermally denatured. In breadmaking, fruit juice, and glucose production, all the enzymes are denatured on heating. It is a known fact, however, that consumption of cooked protein can trigger allergic reactants.

Enzyme producers themselves have taken the initiative to avoid or minimise these problems. Enzyme dust can be avoided by having the enzymes granulated, enclosed in microcapsules, and used as liquid preparations.

TOXINS

Toxic Activities

It has been known for some time that micro-organisms produce toxic metabolites. One group of mycotoxins produced by fungi, the aflatoxins, have drawn attention recently because of severe cases of poisoning. Worldwide, many people die annually from poisoning by the botulism toxin in spoiled food. It is also known that these toxins are carcinogenic.

A large number of such toxins have been detected over the past few years by highly refined analytic methods. Their chemical structures have been determined and sensitive analytic methods for their detection have been developed. The level of detection has been lowered to the part-per-billion range, that is, to the level of 1 mg per metric tonne of foodstuff.

What must enzyme producers do to ensure that their products are free of such toxins? Initially, the producer will exclude all strains of micro-organisms known to generate toxins from the production process, even though such toxins are not formed during exponential growth of the micro-organisms. However, this is the phase during which the enzymes are generated. It is standard procedure to test for toxins in exploratory laboratory experiments when screening micro-organisms.

Currently, tests for mycotoxins are conducted during product quality control.

Antibiotic Activities

It is well known that micro-organisms synthesise antibiotics. To assure the absence of antibiotic activity in an enzyme preparation, JECFA has recommended series of tests employing specific micro-organisms. These organisms and the associated test methods are listed in the AMFEP brochure.

Documentation and Classification

The standards for the classification of commercial enzyme products were, as mentioned earlier, developed by FAO/WHO committee experts. The various enzyme producers did, in turn, draft procedures for documentation and classification of enzymes that will be used in the approval process by the national regulatory agencies. The AMFEP has derived its General standards for enzyme regulations from this information. These standards serve as guidelines for licensing enzymes in the EC member countries.

The standard describes industrial enzymes and their nature and properties. It further contains suggestions for the requirements of free and carrier-bound enzymes and includes a list of micro-organisms that produce the enzymes in question. This list is based on JECFA recommendations. It groups the organisms into three categories which are discussed below:

1. Micro-organisms traditionally used in food production:
   (a) *Bacillus subtilis*, including strains such as *Bacillus mesentericus*, *B. natto*, *B. amyloliquefaciens* (amylases, proteases, β-glucanase, hemicellulases).
   (b) *Aspergillus niger*, including strains of *A. aculeatus*, *A. awamori*, *A. ficuum*, *A. foetidus*, *A. japonicus*, *A. phoenicis*, *A. saitoi*, and *A. usamii* (α-amylases, amyloglucosidase, cellulase, catalase, β-glucanase, lipase, pectinases, proteases, pentosanases, lactase, α-glucosidases, β-glucosidase, glucose oxidase, etc.).
   (c) *Aspergillus oryzae*, including *A. effusus*, and *A. sojae* (α-amylases, amyloglucosidase, glucanases, cellulase, pectinases, proteases, lactase, etc.).
   (d) Various other micro-organisms: (i) *Mucor javanicus* (lipase), (ii) *Rhizopus arrhizus* (lipase, amyloglucosidase), (iii) *Rhizopus oligoporus*, (iv) *Saccharomyces cerevisiae* (invertase), (v) *Kluyveromyces fragilis* (lactase), (vi) *Kluyveromyces lactis* (lactase), and (vii) *Leuconostoc oenus* (maleic acid decarboxylase).

   This group also includes the inoculation cultures used in the production of fermentation.
2. Micro-organisms recognised as safe based on scientific evidence from toxicological testing:
   (a) *Bacillus stearothermophilus* (proteases).
   (b) *Bacillus licheniformis* (proteases, amylases).
   (c) *Bacillus coagulans* (glucose isomerase).
   (d) *Bacillus megaterium* (β-amylase).

(e) *Bacillus circulans* (β-glucanase).
(f) *Klebsiella aerogenes* (pullulanase).

3. Micro-organisms not listed in sections 1 and 2 above: The recommendations of the Deutsche Forschungsgemeinschaft (DFG) refer to 'enzyme preparations from lesser known organisms which commonly are not known to be associated with food'. Regarding this statement, that this classification is unclear from a food microbiologist's point of view and, therefore, unlikely to be used:
   (a) *Bacillus acidopullulyticus*.
   (b) *Bacillus cereus* (protease, β-amylase).
   (c) *Mucor miehei* (rennet, lipases).
   (d) *Mucor pusillus* (rennet).
   (e) *Endothia parasitica* (rennet).
   (f) *Actinoplanes missouriensis* (glucose isomerase).
   (g) *Streptomyces albus* (glucose isomerase).
   (h) *Trichoderma reesei* (*T. viride*) (cellulase, β-glucanase).
   (i) *Trichoderma harzianum* (β-glucanases).
   (j) *Penicillium lilacinum* (dextranase).
   (k) *Penicillium emersonii* (β-glucanase).
   (l) *Penicillium funicolosum* (dextranase).
   (m) *Penicillium simplicissimum* (pectinase).
   (n) *Sporotrichum diamorphosporum* (cellulase, hemicellulase).
   (o) *Streptomyces olivaceus* (glucose isomerase).
   (p) *Streptomyces raserubiginosus* (glucose isomerase).

On the basis of this list, recommendations were made for the toxicological examination of enzyme preparations from micro-organisms in groups (1) and (3); in essence, these represent a more precise definition of the JECFA recommendations. The study of the DFG has incorporated these recommendations.

DOCUMENTATION

The documentation prepared in support of an enzyme approval application should include the following:

1. Information on the enzyme source (referring to the preceding list of micro-organisms).
2. Major enzyme activities (EC numbers); the most common coenzymes should also be listed.
3. Data obtained from toxicological tests.
4. Methods used for production and product quality control.
5. Information about the area of application and quantities required.
6. Information about the activity and stability of the product.
7. Information about contaminants, based on the JECFA recommendations:
   (a) Arsenic, not exceeding 3 mg/kg.
   (b) Lead, not exceeding 10 mg/kg.
   (c) Heavy metals, not exceeding 40 mg/kg, calculated on the basis of lead.
   (d) Hygiene requirements; microbes:
      (i) *Salmonella*, not present in a 25-gram sample.
      (ii) *E. coli*, not present in a 25-gram sample.
      (iii) Coliforms, not exceeding 30/g.
      (iv) Total live bacteria, not exceeding 5×10^4 g^{-1}.

The Society of German Chemists (GDCH) standards provide a practical supplement; specifically, enzyme preparations must be devoid of viable pathogens that could damage human health. Because the total permissible bacterial counts in food are defined, only those enzyme preparations that do not exceed the maximum bacterial count permitted are employed. Determining this is important because enzymes can be blended into food for which bacterial counts have not been established.

(e) Mycotoxins: The German standard permits not more than 10 μg of total aflatoxins per kilogram and not more than 5 μg of aflatoxin B_1 per kilogram.

(f) Antibiotic activity: Test negative.

8. Data on carrier material and blending agents, as well as antioxidants, stabilisers, and preservatives.

In addition to information about the enzyme manufacturer, items 1 through 8 should also be part of the enzyme specification for enzyme users.

SOME NATIONAL REGULATIONS

Argentina: There is no specific legislated regulation for enzymes. Pectinases are permitted in winemaking. For licensing, data on activity and purity must be submitted.

Australia: Only papain is included in the list of food additives. Enzymes are authorised as technical aids in winemaking. Use of amylases in the starch industry is permitted as well.

Belgium: No special legislation exists for food enzymes; yet there is a general regulation for food additives that includes enzymes.

Denmark: An approved list exists for food additives. The registry format includes data on enzyme characteristics, source, and production methods and requires submission of toxicological data. Enzymes from animal organs and edible plant parts, as well as immobilised enzymes, require no toxicological data. Tests of enzymes from other sources follow JECFA recommendations.

Germany: Enzymes are additives according to 'food and consumer goods'. Therefore, enzymes can be used in food without additional approval. Some rules and regulations provide exceptions and permit the use of only very specific enzymes.

Restricted applications are specified in the Dairy Act of December 16, 1986, and in the cheese regulation of April 14, 1986. For cheese production, only rennet from animal and microbial origin is permitted; and for lactose hydrolysis, only lactase.

For fruit juice production, the fruit juice regulation permits the use of pectinases put not cellulases and hemicellulases, which, however, can be used in vegetable processing. The enzymes employed in winemaking are set forth in an EC regulation and at present include pectinases and β-glucanase preparations.

Netherlands: Enzymes are not specifically regulated, but their purity requirements correspond to JECFA recommendations. Their application must agree with the individual food standard.

France: Enzymes are considered as additives. Applications of certain enzyme preparations are regulated by decrees. All enzyme preparations must be registered with the 'Repression des Fraudes' agency. Testing of new enzyme preparations from organismal sources that have not yet been registered will be conducted by the expert committee 'Conseilles Supérieur de la Hygiène'. There is an approved list giving the source and application of microbial enzymes.

Preparations currently registered are presented in Table 28.1.

Table 28.1. Enzymes approved in France.

Enzyme	Source	Application: Food or beverage
	1. Function: Glycosidic bond hydrolysis in polysaccharides	
α-Amylase	*Bacillus subtilis*	Starch products
	Bacillus licheniformis	Cookies, pastries, viennese pastries, rusks
	Aspergillus niger	Bread production
	Aspergillus oryzae	Bread production, beer, fruit juices and concentrates, vegetable juices, fruit nectars, syrups
Amyloglucosidase	*A. niger*	Starch products, cookies, pastries, viennese pastries, beer
	2. Function: Sucrose hydrolysis	
Invertase	*Saccharomyces*	Invert sugar, sweets
	3. Function: Peptide bond hydrolysis in proteins	
Papain	*Carica papaya*	Beer
Proteases	*B. subtilis*	Cookies, pastries, viennese pastries
		Beer
	A. oryzae	Fruit juices and concentrates, nectars
	A. wentii	
Acid proteases	*Mucor pusillus*	Cheese production from cow milk only
	Mucor miehei	
	Endothia parasitica	
	4. Function: α-1,4-Galacturonic acid bond hydrolysis in pectins	
Pectinases	*A. niger*	Fruit juices and concentrates
	A. wentii	Fruit pulps, apple and pear cider
	5. Function: Glucose isomerisation	
Glucose isomerase (free, or immobilisation by glutaraldehyde)	*Streptomyces violaceoniger* *Streptomyces olivochromogenes* *B. coagulans*	Glucose syrup with increased fructose content

Italy: Enzymes are not specifically regulated. Enzymes are addressed in some governmental ordinances. In the regulation on fruit juice production, pectinases, proteases, and amylases are mentioned. Rennet, microbial rennet, and lipase are permitted in cheese production; pectinases, in winemaking. Malt can be treated with proteases. Papain and microbial proteases are permitted in brewing, and amylases, in breadmaking.

United Kingdom: Enzymes are not specifically regulated. An independent group of experts, the 'Committee on Food Additives and Contaminants' (FACC) issued a report dealing with enzyme preparations in 1982. They recommended that an approved list be developed to regulate enzymes. Special identifying test data must be presented for each enzyme. Future regulations will include not only the JECFA recommendations but also the suggestions of the FACC. The use of enzymes in fruit production is permitted and not limited to the enzymes listed in the EEC directive.

The regulation for cheese production permits the use of milk-clotting enzymes as well as the enzymes used for ripening cheese and making processed cheese.

Canada: A long list of authorised enzymes and their applications is available. Because this list provides a compendium of enzyme applications, it is excerpted. It contains the name and source of the enzyme, and the authorised use or area of application, and the maximum dose recommended for the given application. The dose was excluded from this list (Table 28.2) because good manufacturing practice will not administer a given enzyme in larger quantities than absolutely necessary.

Table 28.2. Enzymes approved in Canada.

Enzymes	*Source*	*Application*
Amyloglucosidase	*Aspergillus niger* var.	Ale, beer, light beer, malt beer, porter, malt liquor
	A. oryzae var.	Bread, flour
	Rhizopus oryzae var.	Chocolate syrup
		Distillery mash
		Precooked instant cereals
		Starch for producing dextrins, glucose, glucose syrup, powdered glucose
		Nonstandardised baked goods
Amyloglucosidase	*Rhizopus niveus* var.	Distillery mash
	R. delemar var.	Vinegar mash
Bromelain	*Ananas comosus*	Beer, light beer, malt extract, porter, malt liquor
	Ananas bracteatus	Bread, flour, whole wheat flour
		Edible collagen sausage casings
		Hydrolysing meat, milk, plant proteins
		Tenderising cuts of meat
		Meat tenderising and processing
		Beef marinades
Catalase	*Aspergillus niger*	Lemonades
	Calf liver	Egg whites
Cellulase	*Aspergillus niger* var.	Distillery mash
		Liquid coffee concentrates
		Spice extracts
Ficin	Latex of *Ficus* sp.	See 'Bromelain'
Glucanase	*Aspergillus niger* var.	Ale, beer, light beer, malt beer, porter, malt liquor
	Bacillus subtilis var.	Distillery mash
		Vinegar mash
		Degerming
		Nonstandardised baked goods
Glucose isomerase	*Bacillus coagulans* var.	Glucose isomerisation
	S. olivochromogenes	
	S. olivaceus	
	Actinoplanes missouriensis var.	
Glucose oxidase	*Aspergillus niger* var.	Lemonades
		Liquid whole egg, egg white

(Contd ...)

Enzymes	*Source*	*Application*
Hemicellulase	*Bacillus subtilis* var.	See 'Cellulase'
Invertase	*Saccharomyces* sp.	Confections with soft or liquid centres
		Nonstandardised baked goods
Lactase	*Aspergillus niger* var.	Lactose reducing enzymes
	Saccharomyces sp.	Milk used in ice cream
Lipase	*Aspergillus niger*	Flavours from milk products
	Pregastric	Liquid and powdered egg, or egg white
	Pancreatic	Romano cheese
Milk-coagulating enzymes	*Mucor miehei*	See 'Rennet'
	M. pusillus	Swiss cheese
	Enthia parasitica	
Papain	*Carica papaya* (fruit)	See 'Bromelain'
Pectinase	*Aspergillus niger* var.	Cider, wine
	Rhizopus oryzae var.	Distillery mash
		Fruit juices
		Extracting pigments and flavours
		Treating citrus peels for jams, marmalades, and candied fruits
		Utilising vegetables for soups
Pentosanase	*Aspergillus niger* var.	Ale, beer, light beer, malt extract, porter, malt liquor
	Bacillus subtilis var.	Degerming
		Distillery mash
		Vinegar mash
		Nonstandardised baked goods
Pepsin	Pig gasters	See 'Protease': 'Beer', etc.
		See 'Rennet': 'Cheese', etc.
		Fat-free soya flour
		Precooked instant cereals
Protease	*Aspergillus oryzae* var.	Beer, light beer, malt extract, porter, malt liquor
	Aspergillus niger var.	Bread and flour
	Bacillus subtilis var.	Flavours from milk products
		Distillery mash
		Edible collagen sausage casings
		Hydrolysing proteins
		Spray-dried powdered cheese
		Meat tenderiser
		Pre-cooked instant cereals
		Nonstandardised baked goods
Rennet	Rennet, stomach of calves, sheep, goats (aqueous extracts)	Cheese, cottage cheese, cheese spread, cream cheese

United States of America: Certain substances in the United States are classified as 'generally recognised as safe' (GRAS). Such substances can be used in the production of food, provided that 'good manufacturing practice' is applied. The GRAS status for an enzyme preparation can be granted by the Food and Drug Administration (FDA) on the basis of the documentation presented. This status, however, can also be issued by an expert panel of scientists that is qualified to judge the safety of food additives. For example, the use of papain and several fungal proteases as meat tenderisers was approved by the US Department of Agriculture. GRAS status pertains to enzymes that are obtained from plants and animals and also those from micro-organisms that have been used as enzyme sources for a long time. This includes carbohydrases, lipases, catalase, and glucose oxidase from *Aspergillus niger* and the carbohydrases, lipases, and proteases from *A. oryzae*. Enzymes without GRAS status can be approved for a specific application by means of a special food additive petition (GRASP).

'Standards for Identity' have been developed in the United States for many foodstuffs in which the number and the type of additives have been specified for the production of a given food. Such food products are flour, bread, cheese, soft drinks, liquid egg, and egg white.

In breadmaking, malted wheat, malted barley, malt extract, *A. oryzae* amylase, or protease, papain, and bromelain can be utilised. It is interesting to note that while the protease from *A. niger* has GRAS status, it is not included in the 'Standard for Identity' discussed earlier.

Table 28.3 lists enzymes that have qualified for and obtained GRAS status. Also included are enzymes for which a GRAS petition (GRASP) has been filed and others, which, because of their acceptance in the Code of Federal Regulations (CFR, June 1983), have been approved.

Table 28.3. Enzymes with-GRAS, GRASP, and CFR status in the United States.

Enzyme	*Source*	*Status*[a]
α-Amylase	*Aspergillus niger*	GRAS
	Aspergillus oryzae	GRAS
	Bacillus licheniformis (mixed amylase-protease)	21 CFR 184.1027
	B. subtilis	GRAS
	Barley malt	GRAS
β-Amylase	Barley malt	GRAS
Bromelain	*Ananas comosus*	GRAS
	Ananas bracteatus	GRAS
Catalase	*A. niger*	GRAS
	Beef liver	GRAS
	Micrococcus lysodeikticus	21 CFR 173.135
Cellulase	*A. niger*	GRAS
	Trichoderma reesei	GRASP 9G0260
Ficin	*Ficus* sp.	GRAS
α-Galactosidase	*Morteirella vinacea*	21 CFR 173.145
β-Galactosidase	*A. niger*	GRAS
	Candida pseudotropicalis	GRASP 2GO282
	Kluyveromyces fragilis	GRAS
	K. lactis	GRASP 6GOO88, 7GOO88

(Contd ...)

Enzyme	*Source*	*Status*[a]
Glucoamylase	*A. niger*	GRAS
	A. oryzae	GRAS
	Rhizopus oryzae	21 CFR 173.145
	Rhizopus niveus	21 CFR 173.110
Glucose isomerase	*Actinoplanes missouriensis*	21 CFR 184.1372
	Arthrobacter globiformis	GRASP 7GO087
	Streptomyces olivaceus	21 CFR 184.1372
	S. olivochromogenes	21 CFR 184.1372
		GRASP 1GO277
Glucose oxidase	*A. niger*	GRAS
Invertase	*S. cerevisiae*	GRAS
Lipase	Gaster of calf, sheep, and goat	GRAS
	Pancreas	GRAS
	A. niger	GRAS
	A. oryzae	GRAS
Lipase esterase	*Mucor miehei*	21 CFR 173.140
Papain	Papaya	GRAS
Pectinase	*A. niger*	GRAS
Pepsin	Gaster of pig and cattle	GRAS
Protease, generic	*A. niger*	GRAS
	B. licheniformis (mixed amylase-protease)	21 CFR 184.1027
Rennet	Gaster of ruminants	GRAS
	Endothia parasitica	21 CFR 173,150
	B. cereus	21 CFR 173.150
	Mucor miehei	21 CFR 173.150
	Mucor pusillus	21 CFR 173.150
Trypsin	Pancreas	GRAS

[a] GRAS, generally recognised as safe; GRASP, GRAS petition; CFR, Code of Federal Regulation.

Japan: Enzymes are considered natural products and are subject to existing regulations for food additives. The legislative and regulatory bodies strongly promote innovation. The Japanese enzyme producers are presented in the 'Japanese Enzyme Organisation'.

Chapter 29

Economic Considerations for the Use of Technical Enzymes

INTRODUCTION

The economic benefit of using technical enzyme preparations lies in lowered process costs, often increasing product quality, reducing environmental impact, and making use of renewable resources in technology and food possible (Table 29.1). The significance of these considerations can be demonstrated with a few examples. There will, of course, be some overlapping between those situations in which utilisation of natural products go along with cost reduction.

Table 29.1. Economic considerations for the use of technical enzymes.

Objectives	*Measures*	*Products/Processes*
Cost reduction	Increase yields	Glucose, glucose-fructose syrup, fruit processing
	Upgrade raw-materials utilisation	Fruit and vegetable juices, meat processing
	Process costs	Glucose and fructose syrups, removing sizing from textiles
	Filtration costs	Beer, wine, fruit juices
Quality improvement	Preservation	Fruit juice concentrates, lemonade stabilisation, silage
	Alter technical properties	Protein modification flour, baked goods, trans-esterification of lipids
	Flavour	Milk products, cheese, flavour concentrates
	Cleansing power	Laundry and cleansing detergents
	Process control	Leather
Utilisation of raw materials	Whey, hydrolysis of starch	Beverages, baking aids
	and cellulose	Ethanol
		Animal feed
Lowering environmental impact		Leather production, whey utilisation

INCREASING YIELDS AND IMPROVING RAW-MATERIALS UTILISATION

Improving Product Yields

In food and beverage manufacturing many examples of enhanced product yields can be cited such as processing fruits and vegetables into juices and canned foods, or producing glucose syrup from starch.

Glucose yields could be increased from about 75 to 90 per cent upon substituting enzymatic starch hydrolysis for acid hydrolysis and, in addition, process costs could be lowered.

Raw-Materials Utilisation

Various fruit and vegetable varieties can, except for small residual amounts, be completely hydrolysed with the combined use of pectinases and cellulases. This represents optimal utilisation of a staple commodity. It has been shown that the action of enzymes can generate useful food industry products from whey, which previously was considered only a waste product. Because some countries still emit large quantities of whey into waste-water systems, processing whey into food or fodder will reduce the environmental burden. Raw materials for fodder can be digested by enzymes so that the nutrient value of food is improved. Ethanol for fuel can be derived from starchy raw materials. An objective of modern enzyme technology is to replace starch with cellulose for ethanol fuel.

LOWERING COSTS

Process Costs

Starch hydrolysis with amylases can again be mentioned. Even greater advances can be possible when raw starch can be hydrolysed without prior temperature treatment. In fact, all processes in which energy can be conserved through the use of alternative methods that employ enzymes can be included, such as lipid hydrolysis with lipases rather than with high temperatures and pressures.

Filtration Costs

These process costs are listed separately because they are essential for certain procedures. Filter clarification and sterilisation of beverages or fermentation broths are processes that entail considerable costs. These include:

1. Expenditures for filter material, filter plates, membranes, or other filtering devices.
2. Expenditures for labour in servicing and operating the facilities.
3. Loss of liquids in the filters.

Highly hydrophilic substances such as pectins, β-glucans, pentosans, or glycoproteins form very viscous solutions in water and make filtration difficult. Applied technology has demonstrated that enzymes can be employed here to great advantage. Often, filtration efficacy can be doubled, if not increased ten-fold.

ALTERING TECHNICAL PROPERTIES

This section includes methods in which the characteristics of natural products are altered to fit nutritional or technological needs, for example, changing the properties of wheat gluten to produce various kinds of baked goods or improving a protein's ability to be whipped and emulsified on enzymatic partial hydrolysis. Similarly, altering the properties of lipids discussed earlier should also be mentioned in this context.

IMPROVING PRESERVATION, FLAVOUR AND CLEANSING ACTION

Preservation

Preservation makes a significant impact on quality, especially in fruit juice production. As shown earlier, modern processes convert juices into concentrates that, except for aroma, can be stored for a long time without loss in quality. Stabilising flavour and colour in beverages with glucose oxidase is also an example of improved preservation.

Flavour

It is well known that aroma formation is aided by lipases and peptidases. However, flavour improvements can also be obtained on partial hydrolysis of proteins or digestion of cell material.

Cleansing Action

This topic, which includes enzymes in laundry detergents, has become especially relevant today because new efforts are required to reduce the environmental burden. Enzyme-powered detergents are especially significant for removing stubborn, ground-in dirt, which in the past could be removed only with strong alkali, surfactants, or solvents.

Enzymes should be considered for such future applications, although current detergent manufacturing processes utilising enzymes are still expensive in Europe. However, laundry detergents with enzyme power are successfully marketed in the United States and Canada.

Numerous examples of this topic can be cited from the previous chapters. Leather manufacturing, as an example, is one area in which process control can be accomplished by introducing enzymes. The advantages of enzyme use are reducing losses, significant quality improvement, and last, but not least, a lowered environmental impact by limiting chemical effluents into waste-water.

Costs of Technical Enzymes

The costs of technical enzymes are determined primarily by micro-organismal productivity, energy, and investment requirements. Because enzymes are employed in relation to their activity, it is not the price per kilogram but rather the price per unit activity that determines the cost. Unfortunately, the activity data from the various enzyme producers are, with some exceptions, not comparable. If users wish to compare enzyme products from various suppliers, they must conduct their own comparison test for their specific applications.

Users require information on how enzyme costs will affect their process costs as compared to the costs presently incurred with a conventional enzyme-free process. Enzyme costs related to value of the product generated are shown in Table 29.2.

Table 29.2. Quantities and cost of enzymes in relation to product value.

Product	*Amount of enzyme*[a] *as ppm of product*	*Enzyme cost as percentage of product value*
Laundry detergent	150–200	1
Glucose	150–200	1
High-fructose corn syrup	150–200	3
Ethanol	300–400	3
Cheese	3–6	0.1–0.3

[a] Pure enzyme.

Table 29.3 listed the percentage of enzyme product present in technical product.

Table 29.3. Percentage of enzyme product present in technical product.

Enzyme	*Per cent enzyme in technical preparation*
Amyloglucosidase	5–30
Bacterial α-amylase	5–60
Laundry detergent protease	3–6
Glucose isomerase	5–10
Fungal α-amylase	1–25
Chymosin	0.1–0.2
Microbial rennet	0.2–0.4

Amyloglucosidase has become a surprisingly inexpensive enzyme after a long developmental phase involving expensive strain, culturing, and process optimisations. Enriched technical-grade bacterial α-amylase contains approximately 50 per cent pure enzyme. Because microbial rennet is based on technical enzyme protein, it costs only one-tenth as much as animal-based rennet.

MARKET OF TECHNICAL ENZYME PREPARATIONS

This market emcompasses the so-called food enzymes and enzyme preparations for industrial processes that include detergent enzymes. In both areas, the market can be divided into two sectors: (i) enzyme concentrates sold by the enzyme producer to manufacturers who produce custom formulations, and (ii) preparations for specific consumers. But enzyme producers now increasingly formulate such products themselves.

Well-known marketing studies were conducted by Hepner or by Frost and Sullivan who presumably evaluated data from genuine enzyme producers.

Formulations are prepared for specific applications in which more than one enzymatic activity is required for the desired effect. Furthermore, use of enzyme concentrates is often difficult in practical applications, such as when introducing small enzyme quantities into large substrate volumes. This is the case in leather manufacturing, where excess dosage can result in great damage. Also, when treating flour with enzymes at the mill, it is important to work with premixes to ascertain homogeneous distribution. Companies that produce wine- and juice making aids offer their customers diluted enzyme combinations tailored to specific processes.

Chapter 30

Patenting Biotechnology Inventions

INTRODUCTION

The principle objective of biotechnology is to produce commercial products for economic gain. However, any industry will not initiate high-risk, long-term projects without knowing that the results of its research efforts can be legally protected from competitors. At the same time, society at large has a stake in encouraging industrial innovation. A strategy that meets both of these objectives is for the government to grant inventors exclusive rights to the novel products or processes that they develop. Collectively, these sanctioned privileges are called intellectual property rights and include trade secrets, copyrights, trademarks, and patents.

Trade secrets comprise private information about specific technical procedures and formulations that a company wishes to protect from others. Copyrights protect the authorship of published works from unauthorised use. Trademarks can be either words or symbols that identify a particular product or process of one company. For example, the name Gigapack™ has been legally designated to describe a bacteriophage λ *in vitro* packaging extract that is sold by the biotechnology company Stratagene. Other companies sell similar kits that have their own protected names.

For biotechnology, patents are the most important form of intellectual property. A patent is a legal document that gives the patent holder exclusive rights to implement the described invention commercially. Moreover, on the basis of the extent of the claims of the patent, the patent holder can develop other products that are directly derived from the original invention, while competitors would have to purchase the right to use the invention in order to develop a product based on it.

On the other hand a patent is a public document that must contain a detailed description of the invention, so it informs others about the nature and limits of the invention, allowing them to decide whether they should continue working in a particular direction or try to use the patented invention as a springboard to other possible innovations.

Patent laws vary considerably from country to country, although there are ongoing attempts to develop international standards. In the United States the duration of the exclusive rights of a patent is 17 years after it is officially awarded; in Canada it is 20 years; and in many European countries it is 20 years from the date of the filing of the patent application.

Usually it takes 2 to 5 years following the filing of the initial patent application before a patent is granted. In every jurisdiction, the holding of a patent can be of considerable economic value, and it is not a trivial matter to receive one. For this reason, both the patent application and the invention must meet a very strict set of criteria.

WHAT CONSTITUTES A PATENT?

Generally, for either a product or a process to be patentable, it must satisfy three fundamental requirements:

1. The invention, after having been shown to work ('reduced to practice') must be 'novel', meaning that the invention does not exist as another patent that is held by someone else in another country; is not an existing product or process; or (outside of the United States) has not appeared in some published form prior to the submission of the patent application. In the United States an inventor has 1 year following publication in which to apply for a patent.
2. A patent cannot be granted for something that was merely previously unknown; but rather, the invention must contain, as judged by the patent office, an inventive step that was 'not obvious' to other workers in the field.
3. The invention must be 'useful' in some way, whether it is a process, an instrument, a compound, a micro-organism, or a multicellular organism.

In addition, every patent application must contain a description of the invention that is sufficiently thorough so that a person knowledgeable in the same field can implement it.

Inventions and discoveries that are not patentable include scientific theories, mathematical methods, aesthetic creations, and therapeutic treatments of humans or animals. In addition, a consistent principle of patent law is that a patent cannot be granted for anything that is a 'product of nature'. The notion here is that it is not appropriate for society to give a monopoly to someone for something that occurs naturally and therefore belongs to the public. Often companies and individuals skirt around this constraint by applying for a patent that covers the process of purification of a product, thereby avoiding the direct question of ownership of either a natural substance or an organism that produces the product.

There is no simple, immediate system for the granting of a patent. The application must be prepared by an expert—normally a patent lawyer—and is organised following a defined pattern. In the United States the application must have a title; an abstract describing succinctly the nature of the application; a section on the background of the invention that includes a full and open description of the current 'state-of-the-art' in the field of the invention; a comprehensive summary of the invention with, if considered helpful, figures and schematic representations; sections that explain the nature of the invention and describe how the invention works; and, finally, a list of claims about the invention and how the invention may be used. The application is sent to the US Patent and Trademark Office, where it is reviewed by an examiner for novelty, non-obviousness, utility, feasibility, and general acceptability as a patentable invention.

If an examiner agrees that the invention meets all the criteria for patentability, then a patent is awarded. However, the receipt of a patent is not a license to produce and sell the invention. All statutory regulations must be met before any product can be marketed. For example, if a patent is granted for a genetically engineered micro-organism, the manufacturer must satisfy the recombinant DNA regulations for its production, distribution, and release. Protection of patent rights is the responsibility of the patent holder, and generally that means bringing a lawsuit(s) against those who are presumed to be infringing on the patent. These disputes are decided by the courts and not the patent office. Similarly, if a person or company feels that an awarded patent is inappropriate, the legitimacy of the patent can be challenged by a lawsuit.

If a patent application is rejected by an examiner, then the applicant can appeal the decision to a Patent Appeals board. If this appeal is turned down, then the decision can be challenged legally. For some applicants, patenting can be a frustrating experience. There are a number of cases where the stakes were considered to be so high that costly court battles have gone on for years. Thomas Edison, who held more than a thousand patents in his lifetime, once described a patent as 'an invitation to a lawsuit'.

Product patents and process patents make up the two major categories of patents. Products include homogeneous substances, complex mixtures, and various devices; processes include preparative procedures, methodologies, or actual uses (Table 30.1). The patenting of biotechnology innovations has been based on the historical experience of patenting inventions by the agricultural, fermentation, pharmaceutical, and medical industries. For example, Louis Pasteur received a patent for a process for fermenting beer. Today, most biotechnology patent applications are straightforward, and patents are granted without any significant problems. However, the first time that a scientist attempted to patent a genetically modified micro-organism (one engineered by the introduction of different plasmids, each of which provided a separate hydrocarbon degradative pathway), the case was highly controversial.

Table 30.1. Common types of patent categories, with examples from recombinant DNA technology.

Categories	*Examples*
Product patents	
Substance	Cloned genes, recombinant proteins, monoclonal antibodies, plasmids, promoters, vectors, cDNA sequences, and monovalent vaccines
Compositions of matter	Multivalent vaccines, biofertilisers, bioinsecticides, pharmaceutical mixtures, micro-organisms, and transgenic organisms
Devices	Pulse field gel electrophoresis apparatus, DNA sequencing apparatus, and micro-projectile gene transfer machine
Process patents	
Process of preparation	DNA isolation, synthesising double-stranded DNA, vector-insert construction, PCR applications, and purification of recombinant protein
Method of working	Nucleic acid hybridisation assays, diagnostic procedures detection systems using PCR, and mutant assays
Use	Applying biofertilisers and bioinsecticides, fermentation of genetically modified organisms, and non-therapeutic animal treatment systems

The genetically modified bacterium, which was capable of breaking down many of the components of crude oil, was developed by A. Chakrabarty, who was at the time working for the General Electric Corporation. However, its potential usefulness in cleaning up oil spills notwithstanding, an attempt by Chakrabarty to patent this bacterium was rejected by the US Patent and Trademark Office on the grounds that micro-organisms are products of nature and, as living things, are not patentable subject matter. In 1980, in a landmark decision, the US Supreme Court decided that this organism was patentable according to the US Patent Statute, arguing that 'a live, human-made micro-organism is patentable subject matter . . . as a manufacture or composition of matter'.

The argument against patenting this genetically engineered micro-organism tended to center on how the organism was developed. In the past, induced mutation and subsequent selection for novel properties was an acceptable way to create a patentable living organism. However, genetic engineering was considered by some to be 'tampering with nature'. Consequently, it was argued that no inventor should benefit from manipulating 'products of nature'. This position was not upheld. Thus, in the United States from 1980 onward—later in other countries—organisms, regardless of the means that were used to develop them, must be judged by the standard criteria of novelty, non-obviousness, and utility to determine if they are patentable.

PATENTING IN DIFFERENT COUNTRIES

The patenting process is inconsistent, varying from country to country. For example, in 1989 the biotechnology company Genentech applied for a patent in the United Kingdom (UK) for, among other things, the production of human tissue plasminogen activator (tPA) by recombinant DNA processes. This protein exists in small amounts in the human body and converts plasminogen to plasmin. Plasmin is an active enzyme that degrades the fibrin of a blood clot. Consequently, human tPA has been considered as a possible therapeutic agent for the prevention and treatment of coronary thrombosis. Genentech, after considerable effort, assembled a complete version of a human tPA cDNA and cloned this cDNA into *E. coli* for the production of large amounts of pure tPA. In its patent application, Genentech claimed-rights to human tPA as a product based on certain procedures of recombinant DNA technology that they developed, the cloning vector system, and the transformed micro-organism. As a part of the 'process' category, protection for the use of human tPA as a pharmaceutical agent was also sought by Genentech. A total of 20 claims were presented in the original patent application. Some of these were broad and others were narrow in scope. The patent was rejected by the UK patent office. Genentech then appealed to the UK Court of Appeals, which, after considerable deliberation invalidated all of the claims for a variety of reasons. The judgment concluded that the patent was novel, but some of the judges argued that the submission was obvious; therefore, it could not be patented.

In contrast, Genentech was readily awarded a patent for human tPA in the United States. The US patent not only protects the form of human tPA that was to be marketed by Genentech but also gives them exclusive rights to all similar, but not identical, active forms of human tPA. Genentech won a lawsuit against two other biotechnology companies who were found to be infringing on its tPA patent, although they were selling non-identical forms of tPA.

The Japanese version of Genentech's tPA patent is limited to the amino acid sequence of the human tPA that was cloned and patented by Genentech. In Japan, other companies can sell variant forms of human tPA. Thus basically the same patent application was rejected; approved and given a broad interpretation; and approved and given a narrow interpretation by three different patent offices. For the present, at least, there are divergent views about what is or is not a patentable invention.

PATENTING MULTICELLULAR ORGANISMS

The patenting of multicellular organisms continues to raise ethical and social concerns. However, there is nothing intrinsically new about the exclusive ownership of living material. In the past, micro-organisms were routinely patented, and specific laws were promulgated to give plant breeders the right to own various plant varieties. To date, a transgenic mouse ('oncomouse') carrying activatable genes that make it susceptible to tumour formation has been patented in both the United States and Europe. In addition, genetically manipulated plants have also been accepted as patentable inventions.

It is likely that the most vigorous challenges to patenting transgenic animals will be put forward on moral grounds. In other words, the issue is whether society considers this form of patenting unacceptable. From a historical perspective, it is unlikely that a position based on ethical considerations will be completely successful in preventing the patenting of all transgenic animals. For example, if an invention purports to facilitate a new treatment for human disease, the prevalent current view is that human rights and needs supersede those of animals. However, patenting is not an absolute right; and governments, by passing specific laws, can determine what can or cannot be patented. If an invention is considered by various special interest groups to have a potential negative economic impact on an existing agricultural practice, then it is quite possible that a law could be passed that would prevent the implementation of the new technology. For example, the Danish parliament recently created a statute that exempted animals from becoming intellectual property. Therefore, they will not be subject to patent protection.

PATENTING AND FUNDAMENTAL RESEARCH

Not everyone believes that patenting is worthwhile. Some opponents argue that awarding a monopoly restricts competition, leads to higher prices, curtails new inventions, and favours large corporations at the expense of the individual inventor or small companies. Notwithstanding these concerns, the patent system is well established and is here to stay. Moreover, patent ownership does not appear to prevent significant research development by other researchers and companies. Indeed, it might be argued that if patents were serious impediments to innovation, then US patent 4,237,224, which was granted to Stanley Cohen and Herbert Boyer in 1980 for recombinant DNA technology for both the use of viral and plasmid vectors and the cloning of foreign genes should have seriously constrained the development of recombinant DNA technology. Obviously, no such hindrance has occurred.

The first claim of US patent 4,237,224 granted to S. Cohen and H. Boyer on December 2, 1980, entitled, 'Process for producing biologically functional molecular chimeras' is reproduced below:

> A method for replicating a biologically functional DNA, which compromises: transforming under transforming conditions compatible unicellular organisms with biologically functional DNA to form transformants; said biologically functional DNA prepared *in vitro* by the method of: (i) cleaving a viral or circular plasmid DNA compatible with said unicellular organism to provide a first linear segment having an intact replicon and termini of a predetermined character; and (ii) combining said first linear segment with a second linear DNA segment, having at least one intact gene and foreign to said unicellular organism and having termini ligatable to said termini of said first linear segment, wherein at least one of said first and second linear DNA segments has a gene for a phenotypical trait, under joining conditions where the termini of said first and second segments join to provide a functional DNA capable of replication and transcription in said unicellular organism; growing said unicellular organisms under appropriate nutrient conditions; and isolating by means of said phenotypical trait imparted by said biologically functional DNA.

In the past, patenting and patent enforcement were rarely of interest to academic researchers working in the biological sciences. Now, however, there is a view within the academic scientific community that patents and the consequences of patenting may be detrimental to established scientific values. Traditionally, science, especially university-based research, has been an open system with a free exchange of ideas and materials through publications and personal communications. The ideas of others have been respected, and contributions to the technical development of an area of study has, in many instances, been a shared enterprise. However, more recently, some scientists have begun to feel that the integrity of traditional scientific inquiry has become secondary to self-interest, where public recognition and financial gain from innovations are the prime motivations for conducting scientific research. It is argued that today research is often carried out secretly and has created elite, non-cooperating research groups.

In the past, there was a tendency to avoid secrecy in basic research. The belief was that scientific knowledge would grow if research results were published as articles in journals that could be read by anyone, thereby enabling researchers to direct their studies in appropriate directions and to benefit from the discoveries of others. With secrecy, time and effort may be wasted on repeating experiments that, unbeknownst to the researcher, have already been done. Now, scientists are advised by patent lawyers to keep their work secret until a patent is filed. Consequently, the lure of patenting has made a large number of scientists reluctant to talk about their work, at least until after the patent application has been filed.

The enthusiasm for patenting and patent protection has elicited the perception that traditional science may become hostage to patent holders and that research will become less fruitful. Others feel that the traditional way of doing science is an outmoded, inefficient, and indulgent exercise and that patent ownership and the drive for ownership will spur new discoveries. This controversy will not be readily resolved. It is clear that the emergence of molecular biotechnology has raised far-reaching considerations, even including how scientific inquiries ought to be conducted.

To sum up, there are a number of reasons for issuing patents. For example, the patent holder obtains exclusive rights for the invention, rights that are a reward for developing an innovative procedure, compound, or apparatus; the public learns about the details of an invention, knowledge that will both prevent loss of time and energy in the pursuit of something that has already been invented and stimulate further research; and the opportunity for financial gain drives companies and researchers to create and develop new ideas. Patents are intended to spur innovation and to recognise the rights of the innovators.

For a patent to be granted, an invention must be novel, not obvious, and useful. In addition, the invention should not be a 'product of nature'. The key case that established that genetically engineered micro-organisms were patentable was brought forward by A. Chakrabarty. In 1980 the US Supreme Court ruled that a bacterium that had been created by a form of genetic manipulation could be patented. More recently, a US patent was granted for a cancer-prone transgenic mouse. However, in some other countries there is a reluctance to patent genetically engineered animals.

With the development of molecular biotechnology, questions about whether private industries should be allowed to own or patent genetically engineered organisms have been raised. On the one hand, it has been argued that without such proprietary rights, biotechnology companies would not have the incentive to develop and market novel products. On the other hand, some critics find this type of privilege to be morally unacceptable, and they contend that patents of this sort inhibit research and constrain innovation.

References

Arbige, M.V., *Industrial Enzymology*, Applied Science Publishers, London.

Bailey, J.E., *Biochemical Engineering Fundamentals*, McGraw-Hills, New York.

Bilitewski, V., *Biosensors and Their Application in Enzymes*, Marcel Dekker, New York.

Burtis, S.M., *Enzymes and Their Role in Cereal Technology*, Prentice-Hall Inc., Englewood Cliffs, New Jersey, USA.

Cornish-Bowden, A., *Enzyme-catalysed Modification of Oils and Fats*, Cambridge University Press, Cambridge.

Drauz, K., *Enzyme Catalysis in Organic Synthesis*, Peragam, Oxford.

Edelson, S.S., *Commercialisation of Engineered Industrial Enzyme*, Progress Publishers, Moscow.

Eisenthal, R., *Safety Evaluation of Genetically Engineered Enzymes*, Academic Press, London.

Fersht, A., *Enzyme Structure and Mechanism*, John Wiley & Sons, New York.

Fernie, T., *Industrial Enzymes*, Marcel Dekker Inc., New York.

Guilbault, G.G., *Analytical Uses of Immobilised Enzymes*, University of Wisconsin Press, Madison.

Holton, J.B., *Application of Enzyme Preparation in Engineering*, Humana Press Inc., New Jersey.

Jackwerth, N.P., *Principles of Enzyme and Its Activity*, McGraw Hill, Columbus, Ohio, USA.

Koch, T.R., *Principles of Enzymatic Analysis*, Pergamon Press, Oxford, New York.

Kraut, J., *Analytical Methods for Measuring Enzymes*, Cold Spring Harbour Press, UK.

Lewis, R., *Enzymes in Wheat, Floor and Bread*, Science Publishers, New York.

Linderstrom, K., *Enzymic Hydrolysis of Food Proteins*, Tata McGraw Hill, New York.

Mann, J., *Chemical Aspects of Biosynthesis*, Oxford University Press, Oxford.

Nelson, D.L., *Lehninger Principles of Biochemistry*, Reston Publishing Co., Reston, Virginia.

Ryan, O.M., *Carbohydrate-degrading Enzymes in Food Industries*, Prentice-Hall of India Pvt. Ltd., New Delhi.

Stryer, L., *Enzyme Activity and Effectiveness*, Pearson Education, Singapore.

Trevor, P., *Enzymes: Biochemistry, Biotechnology and Chemical Chemistry*, Affiliated East-West Press Pvt. Ltd.

Umetsu, H., *Industrial Enzymes and Their Characteristics*, Plenum Publishing Corporation, London.

Watanabe, M., *Legislation and Regulation in Industrial Enzymology*, John Wiley & Sons, New York.

Watson, J.D., *Molecular Biology*, Routledge, New York.

Wiseman, A., *Handbook of Enzyme Biotechnology*, Ellis Harwood, New York.

Index

C

D

F

G

H

I

J

K

L

M

N

V

W

X

Y

Reader's Notes

Reader's Notes